Operator Theory: Advances and Applications
Vol. 147

Operator Theory: Advances and Applications
Vol. 147

Operator Theoretical Methods and Applications to Mathematical Physics

The Erhard Meister Memorial Volume

Israel Gohberg
António Ferreira dos Santos
Frank-Olme Speck
Francisco Sepúlveda Teixeira
Wolfgang Wendland
Editors

Springer Basel AG

Editors:

Israel Gohberg
School of Mathematical Sciences
Raymond and Beverly Sackler
Faculty of Exact Sciences
Tel Aviv University
IL – Ramat Aviv 69978
Israel
e-mail: gohberg@post.tau.ac.il

Wolfgang Wendland
Mathematisches Institut A
Universität Stuttgart
Pfaffenwaldring 57
70569 Stuttgart
Germany
e-mail: wendland@mathematik.uni-stuttgart.de

António Ferreira dos Santos
Frank-Olme Speck
Francisco Sepúlveda Teixeira
Departamento de Matemática
Instituto Superior Técnico, U.T.L.
Avenida Rovisco Pais
1049-001 Lisboa
Portugal
e-mail: afsantos@math.ist.utl.pt
fspeck@math.ist.utl.pt
fteix@math.ist.utl.pt

2000 Mathematics Subject Classification 01A70, 35-xx, 45-xx, 46-xx, 47-xx, 74-xx, 76-xx, 78-xx

A CIP catalogue record for this book is available from the
Library of Congress, Washington D.C., USA

Bibliographic information published by Die Deutsche Bibliothek
Die Deutsche Bibliothek lists this publication in the Deutsche Nationalbibliografie; detailed bibliographic data is availab
in the Internet at <http://dnb.ddb.de>.

ISBN 978-3-0348-9623-8 ISBN 978-3-0348-7926-2 (eBook)
DOI 10.1007/978-3-0348-7926-2

Originally published by Birkhäuser Verlag, Basel, Switzerland in 2004
Softcover reprint of the hardcover 1st edition 2004
Printed on acid-free paper produced from chlorine-free pulp. TCF ∞
Cover design: Heinz Hiltbrunner, Basel

ISBN 978-3-0348-9623-8

9 8 7 6 5 4 3 2 1

www.birkhäuser-science.c

Erhard Meister
1930–2001

Table of Contents

List of photographs and sources

(1) provided by the Meister family,
(2) provided by F.-O. Speck

Preface

This volume is devoted to the life and work of the applied mathematician Professor Erhard Meister (1930–2001).

He was a member of the editorial boards of this book series Operator Theory: Advances and Applications as well as of the journal Integral Equations and Operator Theory, both published by Birkhäuser (now part of Springer-Verlag). Moreover he played a decisive role in the foundation of these two series by helping to establish contacts between Birkhäuser and the founder and present chief editor of this book series after his emigration from Moldavia in 1974.

The volume is divided into two parts. Part A contains reminiscences about the life of E. Meister including a short biography and an exposition of his professional work. Part B displays the wide range of his scientific interests through eighteen original papers contributed by authors with close scientific and personal relations to E. Meister.

We hope that a great part of the numerous features of his life and work can be re-discovered from this book.

July 2003 The editors

Introduction

On 5 July 2001 Professor Erhard Meister passed away at the age of 71. We all remember him as a distinguished mathematician with a strong and impressive nature. He was interested in practically all fields of mathematical physics and its theoretical background like partial differential equations, singular integral and pseudodifferential equations as well as in the interplay with operator theory and complex analysis. We knew him as a favorite teacher for several generations of students, as a mentor of foreign colleagues and young people, within and beyond his group, as an unselfish and indefatigable worker, as a referee, editor and organizer of conferences, workshops and research projects – in short, someone who deserved to enjoy the fruits of his life's work much longer.

Erhard Meister was 36 when he received his first offer of a Chair in the Mathematics Department of the Technical University of Berlin as a consequence of his pioneering work in mathematical airfoil theory. During the period 1966–70 he started to found a school with the first group of his PhD students, working on singular integral equations, Riemann boundary value problems and related topics. Only four years later, he became the main representative of applied mathematics at the University of Tübingen in 1970–74, attracting plenty of students and extending his fields of interest in various directions such as scattering theory, nonlinear wave propagation, distribution theory etc. Finally he occupied the Chair of Mathematical Physics in Darmstadt during the years 1974–98 until his retirement. Continuing his investigations in the previous areas, his main research was then centered on diffraction problems in electromagnetic and elastic wave propagation, factorization of matrix functions and linear operators, spectral analysis in aperiodic scattering theory and quaternion analysis.

His greatest efforts were always devoted to bringing together the expertise of mathematicians, physicists and engineers for the progress of science. This is very evident from the contributions of the present volume; see, for instance, his own report from 1993 on the international relations of his research group, the titles in the list of his PhD students and the reminiscences of his friends and colleagues. The outstanding kindness of his personality impressed almost everybody around him – as much as his unlimited ambition.

The idea of honoring Professor Meister by the publication of this book came up during a visit by Israel Gohberg to the Mathematics Department of the Instituto Superior Técnico, Universidade Técnica de Lisboa, in March 2002. The close relations of the editors to Erhard Meister and his correspondents caused much interest in the project, with the result that the page limit was rapidly achieved with a nice representative collection of scientific papers and reminiscences. Shortly afterwards, Hans-Dieter Alber and Reinhard Farwig from the Darmstadt University of Technology started together with one of the editors to organize a Workshop in honor of Erhard Meister, and this took place in Darmstadt in June 2003. Owing to the great interest shown by the colleagues and friends of E. Meister, a number of additional dedicated articles will appear in two special issues of the journal Mathematische Nachrichten. At this workshop the collected works of E. Meister were able to be displayed for the first time. They consist of four volumes assembled by Steffen Roch and can be viewed in the library of the Mathematics Department of the Darmstadt University of Technology. In this context it might be mentioned that two short obituaries have already appeared in Mathematical Methods in the Applied Sciences 25/1 (2002) and in the conference proceedings of the IWOTA-Portugal 2000 which was published as part of this book series, OT 142 (2003). There have also been local publications, like the article entitled "In memoriam Erhard Meister" by Hans-Dieter Alber in *intern* – Informationen und Meinungen der Technischen Universität Darmstadt 22/6 (2001).

Let us give some additional information about the origins of the contributions to this volume. Part A begins with an obituary by Wolfgang Wendland, E. Meister's closest colleague and friend who worked with him in the same department for many years in Berlin and Darmstadt. It is based upon his German funeral oration and the opening speech of the Darmstadt workshop. The report which follows on scientific international relations, was written by Erhard Meister himself in 1993 as a project application annex and is published here for the first time; the editors added just one page so as to include some information about the more recent years. That report is followed by Meister's own brief curriculum vitae. The editors could not find any extended version; obviously E. Meister preferred to reveal findings of the full group in survey articles rather than in single results by himself. As a result, it took almost one year to complete the list of publications. It is based on research by Steffen Roch with the help of Frank-Olme Speck and colleagues from Darmstadt including the secretary Mrs. Ursula Röder. The same observation holds for the list of PhD dissertations supervised by E. Meister.

The short biography entitled "In Memory of Erhard Meister" is based on some correspondence between the author and family members, colleagues and friends of Erhard Meister as well as on various sources from his estate provided by Ljudmila Meister. The above-mentioned collected works were very helpful in describing the thread running through E. Meister's scientific life. The text is partly based on the author's research undertaken for his colloquium speech on the occasion of Meister's retirement in February 1998. However, the second ceremonial address at that colloquium, a humorous and exciting lecture by Douglas Jones, on the theme "Is Electromagnetism any Use?" is reproduced almost entirely in its original form.

Reminiscences in the true sense of the word start with the memories of Erhard Meister's second wife Ljudmila from Moscow and the charming story of how they actually found each other. It is pointless to comment on all of the other reminiscences full of honest feelings – it is enough to read them through for they give a real picture of our friend Erhard.

The editors would like to acknowledge the pleasant collaboration with all the book's contributors, particularly with the members of the Meister family: Ljudmila, Olaf and his wife Claudia. We are grateful for the help of the Darmstadt colleagues Hans-Dieter Alber, Reinhard Farwig, Steffen Roch, their secretaries Ursula Röder and Magdalena Tabbert, and to Professor Karl Heinrich Hofmann who gave permission for the inclusion of his poster painted for Meister's colloquium talk of 1998 in the biography. We very much appreciated the help of João Palhoto de Matos in preparing the TEX files. Last but not least we would like to thank Thomas Hempfling and Sylvia Lotrovsky from Birkhäuser for their prompt and efficient help with various technical questions concerning the inclusion of photos and the general composition of the book.

July 2003 The editors

PART A

Reminiscences on the Life and Work of *Erhard Meister*

The following contributions were written by close friends and colleagues of E. Meister to honour him as a scientist with a likable character and unlimited enthusiasm for applied mathematics.

Operator Theory
Advances and Applications, Vol. 147, 3–6

Obituary

Wolfgang Wendland

In Erhard Meister we have lost a faithful, sincere, reliable, thoughtful and fatherly friend.

I first met Erhard Meister in 1965 at the Technical University in Berlin–Charlottenburg after his habilitation in Saarbrücken, when he was temporarily holding E. Kähler's Chair. He then spent half a year as a guest professor at the University of Strathclyde in Glasgow. In October 1966 Erhard Meister returned to become a Full Professor in succession to E. Kähler. With his doctoral students G. Hery, N. Latz, M. Kremer, E. Gerlach, G. Eisenhardt, J. Donig and F.-O. Speck, he soon started to organize excellent seminars on singular integral equations. In 1967 I was very lucky to be among the participants of the unforgettable seminar on Wiener–Hopf operators where we studied the AMS-translations of Kreĭn's and Gohberg's–Kreĭn's famous articles and the corresponding mathematical background material. Perhaps also due to similar former experiences in East Germany we became close friends — Erhard was born in 1930 in the East German town of Bernburg, and the communists did not allow him to study at any of the East German universities. He began an apprenticeship as an electrician which he completed in Mannheim, West Germany, in 1951, after he had escaped to the West. I am sure that his first profession brought him happiness — he began to study the secrets of electromagnetic waves — and also calamity — the intensive use of asbestos in those times germinated the deadly cancer that ended his life.

His seminars on brand new publications attracted all of us; many participants became Erhard Meister's scholars. He acted as the father of this pretty large family and he cared for all its members. Already in those times he had started to weave a remarkably dense net of personal and professional relations which he always used to the benefit of science. He used to offer encouragement and support whenever necessary. A few years later when I was told that my position at the Hahn–Meitner–Institute was about to be abolished, he offered me an associate professorship at his Chair in 1969. For my courses he even delegated one of his pupils to assist and to help me. As in any healthy family, every happy event like an exam or a private affair was heartily celebrated. I also remember the huge amounts of apples and pears from Meister's garden, and the excellent juices we enjoyed year after year.

Erhard Meister had visions and principles which he always stood up for: the quality of mathematical work as well as service to the community. He always encouraged teaching and research, had a fabulous memory and was a never-ending source of references and research suggestions. You could also be sure that he noticed

The couples Meister and Wendland with
Mrs. U. Röder in 1998

flaws and failures, but he was a very sensitive gentleman — a single kind question of his made one aware that he knew exactly what was going on and one was ashamed; but Erhard never went on about it!

In the late 60s — quite similar to our current times — the university professors in West Berlin were blamed for almost every problem of society, and the politicians imposed rather harsh so-called reforms on the universities. Erhard Meister protested and spoke up, and in 1970 he accepted an offer from the famous and traditional university in Tübingen and left Berlin. There he became the leading applied mathematician which made him rather lonesome among all his pure colleagues but rather popular among the students, and there, with his assistants, he again created a splendid group of new scholars including H.-D. Alber, H.-J. Sommer, M. Costabel, G. Buggle, G. Eberlein, G. Thelen, W. Peterhänsel and the lecturer H. Sohr. No wonder, since Erhard Meister had an amazingly abundant range of interests and expertise — in mathematics, physics, astronomy, sociology and politics — and he became one of the internationally leading specialists in the theory of diffraction and scattering of electromagnetic waves and the Sommerfeld problem. During the 70s Erhard Meister was one of the spokesmen of the angry young applied mathematicians like B. Brosowski, K. Kirchgässner, R. Leis, E. Martensen, J. Polasek, W. Törnig, W. Velte, P. Werner. In Oberwolfach, as a youngster, I was lucky enough to sit next to that group when they discussed the founding of the GAMM committee "Mathematical Physics" in order to make modern applied mathematics better heard in society. Erhard Meister's subsequent negotiations with Dr. Spuhler from the Teubner publishing house in Stuttgart resulted in the birth of the esteemed journal "Mathematical Methods in the Applied Sciences".

In 1974 we were lucky to win Erhard Meister for the Technische Hochschule Darmstadt where he significantly shaped the Mathematics Department for almost thirty years. Due to him, international cooperation strongly intensified. His friends in Poland, the Czech and Slovak Republics, in Russia, in the Ukraine, in Georgia, the United Kingdom, U.S.A., Portugal, France, Israel, Canada, Sweden, Turkey, Greece, China and his homeland miss him — mathematicians have lost a great colleague!

In Darmstadt, he again attracted an excellent group of new students such as S. Minutsher, B. Schüppel, E. Penzel, R. Schneider, T.v. Petersdorff, H. Marschall, C. Erbe, R. Hinder, K. Rottbrand, K. Mihalincic, J. Mark, A. Passow, W. Eckert and the lecturers W. Varnhorn, P. Deuring and F. Ali Mehmeti. During our common times in Darmstadt we became very good friends, and my own work was strongly influenced by him and some of his former students who joined my research group. Moreover, he helped our young scholars to find positions by using his network of relations with colleagues all over the world. Eleven of Erhard Meister's students are now professors at universities.

Erhard Meister was an admirably hard worker. I possess three books written by him and also gratefully remember various conferences organized by him where he succeeded in attracting and bringing together colleagues from the East and the West. In his publication list one finds more than 100 titles, about 80 of which have appeared in international, refereed journals. The last major topic he published was on quarternionic analysis. This was in collaboration with Ljudmila Meister who had become his second wife. Erhard Meister also helped to found and participated in the priority research program "Boundary Element Methods" of the German Research Foundation DFG 1990–1995 where engineers and mathematicians were working together. He also initiated the collaborative research centre on "Deformation and Failure of Metallic and Granular Structures", the "Sonderforschungsbereich 298" between mathematics and engineering in Darmstadt, was an elected referee for Applied Mathematics and a member of the committee for "Sonderforschungsbereiche" in the German Research Foundation DFG.

As a faculty member of Mathematics Departments he was a sensible and honest, but not always easy partner: outspoken and having strong, sometimes inconvenient principles, since dedication to mathematics and a high quality of teaching and research were his firm criteria.

After the German re-unification he was immediately ready to help with advice and action and assisted in the restoration of the Technical University of Chemnitz as well as in search committees at the Martin-Luther University in Halle. There he won deep respect for his clear arguments and principles.

I remember with all my heart and in grateful admiration the time I shared with Erhard Meister for his unremitting striving on behalf of mathematicians, his students, his friends and his family; deep gratitude is my strongest feeling.

We have lost a great man.

Institut für Angewandte Analysis und Numerische Simulation
Universität Stuttgart
Pfaffenwaldring 57
D-70550 Stuttgart
Germany
e-mail: wendland@mathematik.uni-stuttgart.de

Operator Theory
Advances and Applications, Vol. 147, 7–12

Report on the Scientific International Relations of the Research Group of E. Meister

Erhard Meister[1]

The first international contacts developed through common scientific interests in mathematical problems of flows through staggered cascades which form part of the subject area of E. Meister's PhD dissertation (1958) and habilitation (1963) in Saarbrücken. The methods involved, like conformal mapping, Riemann boundary value problems, singular integral equations and the Wiener-Hopf technique, were intensively used at that time at the Technical University of Prague, the University of Strathclyde in Glasgow and at the Institute of the Georgian Academy of Sciences in Tbilisi. The first contacts were established in 1963 with Professor D.C. Pack, Glasgow, and in 1964 with Professor J. Polacek, Prague, during their visits to the conferences of the Gesellschaft für Angewandte Mathematik und Mechanik (GAMM).

In Germany only a few specialists existed at that time in the above-mentioned areas, such as E. Martensen, K. Nickel, H. Schubert (Halle) and E. Meister's PhD supervisor H. Söhngen (Saarbrücken). The identification of common research interests in the context of flows through turbines and compressors led to frequent meetings at subsequent GAMM and Oberwolfach conferences on "Methoden und Verfahren der Mathematischen Physik" which started to be organized annually from 1969 onwards by B. Brosowski and E. Martensen and later by E. Meister, J. Polacek and others.

During the 1968 GAMM conference in Prague, the first visiting program was prepared with members of the State Research Institute of Mechanical Engineering in Prague. The "Prague Spring" had raised people's hopes.

After the tragic events of 1968, new possibilities were explored of continuing to meet in Prague whilst reducing the meetings in Germany. The next visit of E. Meister to Prague took place in 1975 for the examination of a Humboldt fellowship for J. Neustupa to Würzburg (Navier-Stokes equations and bifurcation theory with W. Velte). In 1979, a written agreement was drawn up for the cooperation of the

[1] The text is taken from a German typescript of 1993 that was appended to a project application of E. Meister and translated into English by the editors. Its purpose was apparently to outline the existing international relations with institutions and colleagues, without describing the mathematical side. The part until 1994 has been kept in its original form, so that it has a certain historical interest. Only the last page has been added with some information about the period of 1994–2001 in order to round out the text a little, albeit lacking in the detail that E. Meister could have provided.

Faculty of Mechanical Engineering with the chair of J. Polacek at the Technical University (TU) of Prague, and the Department of Mathematics of the Technische Hochschule (TH) Darmstadt. The official partnership of the TU Prague and the TH Darmstadt started in October 1987. Despite the increasingly frosty East-West political relations, an intensive exchange developed with the special agreements of GAMM for Eastern scientists. Now Prof. J. Polacek and his collaborators, as well as the lecturers J. Neustupa, K. Kozel and J. Benda, along with young collaborators, participated regularly whenever they received travel permission. From then onwards, close cross-connections also grew with the analysts M. Feistauer and J. Necas of the Charles University of Prague, who worked in fluid mechanics, and with the Mathematical Institute of the Academy. In the framework of this partnership and also of the international "Equadiff Conferences", the colleagues of the Darmstadt group and E. Meister delivered various lectures (sometimes the main lectures) during their various stays: 1980 in Prague, 1981 and 1984 in Bratislava, 1985 in Brno, 1990, 1991 and 1993 in Prague. Colleagues from the TU Prague came to Darmstadt periodically after the borders opened, in order to exchange ideas on analytical and numerical problems for Navier-Stokes equations, bipolar fluids and turbine flows. J. Neustupa visited Darmstadt from July to December 1991 during his second Alexander-von-Humboldt scholarship. In the meantime personal contacts amongst the young generation of assistants developed in both places (on the one hand W. Varnhorn, P. Deuring, numerical analysts of the group of W. Törnig and co-workers of J.H. Spurk, on the other hand J. Benda, S. Kraczmar, S. Necasova-Matusu and J. Fort). Now various extended visits of younger colleagues were planned on both sides.

In the year 1978, the second Congress on Engineering Education took place at the TH Darmstadt where J. Polacek was one of the main speakers.

In 1991/92 a TEMPUS project for "Mathematical Models in Fluid Mechanics and for Turbines" was drawn up with the partners being the TH Darmstadt, Trinity College of Dublin, the University of Oxford, the TU Prague and Charles University, but unfortunately could not be carried out by European community funds due to budget limitations.

Another centre of cooperation was formed with three institutions in Warsaw: the Academy of Sciences with the Institute of Fundamental Technological Research (H. Zorski, S. Przezdziecki), the Banach Center (B.W. Bojarski, C. Olech, S. Rolewicz), and Politechnika Warszawska (from 1980 onwards within the framework of a partnership with the TH Darmstadt: Profs. J. Wolska-Bochenek, A. Borzymowski and their co-workers). E. Meister participated in the tri-annual December conferences honoring W. Pogorzelski – a leading specialist in the field of singular integral equations. Besides giving numerous opinions for journals edited by colleagues from Warsaw, E. Meister was involved in a habilitation process at Politechnika. Another TEMPUS project was formulated with groups at the three Warsaw institutions working on analytic-mechanical problems and another one involving the University of Strathclyde in Glasgow but they could not be implemented for the reasons mentioned before.

The third Eastern European partner nation in the area of continuum mechanical problems and their mathematical methods was the Soviet Union. Ever since the International Symposium on Function Theoretic Methods for Partial Differential Equations held at Darmstadt in 1976 (see the Springer Lecture Notes in Math., Vol. 561), there had been close connections with outstanding experts in Tbilisi (A.V. Bitsadze, V.D. Kupradze, I.N. Vekua etc.), in Leningrad (V.G. Maz'ya, S.G. Mikhlin, V.A. Solonnikov) and in Moscow (R.V. Goldstein, O.A. Oleinik, M.I. Vishik etc.). In 1976 the organizers managed to bring some of the most prominent representatives to Darmstadt. In spring 1979, E. Meister was able to travel to Tbilisi at the invitation of the Georgian Academy and he gained a young colleague, R. Duduchava, for an extended stay in Darmstadt as an Alexander-von-Humboldt fellow in 1980/81 and 1985/86. In 1991, E. Meister traveled to Tbilisi to give a main lecture in a conference to honor N.I. Mushkelishvili on the occasion of the anniversary of his hundredth birthday, coming directly from the Petrovski conference in Moscow to which he had been invited by Mrs. O.A. Oleinik, who had presented already several lectures at Darmstadt.

In July 1993, E. Meister was a guest of the Departments of Mathematics and Physics of the University of St. Petersburg in order, among other activities, to prepare a European Community exchange program in the area of crack mechanics with institutions there (N.F. Morozov, S.A. Nazarov and colleagues) together with others in Moscow (R.V. Goldstein), the TH Darmstadt and the University of Strathclyde in Glasgow. In the meantime, the author had also written various opinions on behalf of Soviet mathematicians applying for a research project of the Soros Foundation (USA).

These activities directed towards the East made it possible, during the visits to the Eastern European countries, to develop valuable contacts with former GDR representatives of the field of singular integral equations and their application in continuum mechanics (in Berlin, Chemnitz, Freiberg, Halle and other places). Furthermore, active exchange relations developed from 1966 onwards with Glasgow, where E. Meister was invited as a visiting lecturer in the summer semester of 1966, as were two assistants from Darmstadt later on as research fellows (J. Donig 1978, F.-O. Speck 1980). The head of the Mathematics Department at Glasgow, Professor D.C. Pack, became a DFG-funded visiting professor at the TH Darmstadt in 1981.

From 1988 onwards contacts were developed with the Technical University of Lisbon (A.F. dos Santos and colleagues) with rapidly increasing exchanges, resulting in numerous joint publications about Wiener-Hopf problems in diffraction theory, three visits of E. Meister to Lisbon and short lecture courses by F. Penzel, F.-O. Speck and W. Varnhorn. F.-O. Speck started to work there as an invited full professor from 1990, W. Varnhorn was invited to take up an associate professor position in 1992, and various visits of PhD students followed (C. Erbe, K. Rottbrand and others). Conversely, F.S. Teixeira spent a one-year sabbatical in the Darmstadt research group sponsored by the DFG program, and several short stays by A.B. Lebre and P.A. Santos followed. In addition, E. Meister and F.-O.

Speck were referees for the doctoral dissertations of F.S. Teixeira (1989) and A.B. Lebre (1990) respectively. In the years 1990/91 a common DAAD/IMC research program was financed and another one was projected for 1994/95.

Further connections and joint research activities in the field of distributional integral transformations and diffraction problems have existed since 1975 with the University of Saskatchewan in Regina (E.L. Koh, who was guest professor at the THD in the academic year 1975/76); with Edmonton (R.P. Millar), where E. Meister was a guest in 1978, and with Ottawa, from where A. Hurd came to spend the year 1985/86 as a DFG guest professor at Darmstadt and at the Deutsche Versuchsanstalt für Luft- und Raumfahrt in Oberpfaffenhofen.

Relations with Israel began with Professor I. Gohberg from Tel Aviv, in 1975, when his book, written with N. Krupnik in Kishinev (Moldavia, former USSR) and published in 1973 in Russian, was translated in Darmstadt and published in German by Birkhäuser in 1979. His big research group in Kishinev, which at one time also included East-German PhD students, broke up and became dispersed all over Israel and the USA. The journal and book publications edited by I. Gohberg for the Birkhäuser publishing house arose from E. Meister's negotiations during the above-mentioned book translation. E. Meister and M. Costabel were invited speakers at the Toeplitz memorial conference in Tel Aviv in 1981, and in 1993/94, I. Gohberg spent a sabbatical year in Germany funded by a Humboldt award.

There are plenty of other collaborations, particularly on Wiener-Hopf problems and singular integral equations, with Delft, Nantes, Paris and Rennes, Ann Arbor, Newark (Delaware), Ankara and Adana (Turkey).

E. Meister also coordinated the Erasmus program in mathematics between Darmstadt and Dublin.

From 1994[2] onwards contacts with colleagues from the former Soviet Union developed rapidly and more intensively, in particular with O.A. Ladyzenskaja, O.A. Oleinik, M.I. Visik, A.I. Snirelman (Moscow) and D.P. Kouzov, S.A. Nazarov, N.F. Morozov, B.A. Plamenevsky and A.I. Koshelev (St. Petersburg) who visited Darmstadt and other places in Germany. I. Surovtseva came as a PhD student from St. Petersburg to Darmstadt in 1997–98. A number of visits by R. Duduchava and D. Natroshvili (Tbilisi) took place at the instigation of E. Meister and yielded a fruitful collaboration with colleagues from Darmstadt and Stuttgart.

Relations with Mexico started and were based upon contacts with V.S. Rabinovich and N.L. Vasilevskij. The entire collaboration was supported by research projects coordinated by E. Meister.

A highlight of the cooperation with I. Gohberg (Tel Aviv) was an honorary PhD degree awarded to him by the TH Darmstadt in June 1997. From the author's laudatio: *The relations between Gohberg and the Darmstadt mathematicians became extremely fruitful through various conferences held at the Mathematisches Forschungsinstitut in Oberwolfach, which were organized by Gohberg, Kaashoek*

[2]This final part was added by the editors and is mainly based upon information supplied by L. Meister.

Participants of the Tel Aviv conference in 1995
including ten authors of this book

and myself. We all belonged to the editorial committee of the international mathematical journal "Integral Equations and Operator Theory" and the book series "Operator Theory: Advances and Applications" which are published by Birkhäuser Verlag. Subsequently, these close relations were strengthened by various visits of I. Gohberg to conferences and workshops in Darmstadt and by return visits of my collaborators and myself to the "Toeplitz conferences" in Tel Aviv in 1981 and 1995... In March 1995 and 1997 we all met in Berlin, with Gohberg, his wife Bella, my wife Ljudmila, the Vishiks and Prössdorf. So a large circle was completed – also for us.

The cooperation with Portugal continued through two joint projects with A.F. dos Santos, F.-O. Speck and F.S. Teixeira, through frequent mutual visits, particularly by young colleagues, and through a remarkable scientific output. This cooperation helped to establish the area of Operator Theory and Applications in the scientific landscape of Portugal during the 90s.

Frequent meetings with V. Maz'ya resulted in an Alexander von Humboldt award for him in 1999 which supported a nine-month stay in Germany that helped also to strengthen relations with the Weierstrass Institute for Applied Analysis and Stochastics in Berlin.

From 1994 onwards the presence of Ljudmila Meister (born Aramanovitch) led to contacts with colleagues working in the field of Quaternionic Analysis from various countries as well as the growing cooperation with Freiberg and Chemnitz: some of the people involved are J. Ryan, J. Snygg (USA), H.R. Malonek (Portugal), M. Mitrea (Romania), M.V. Shapiro and N.L. Vasilevski (Mexico). Results of E. Meister's group in this research area were presented at various international conferences and finally at the International Conference on Clifford Analysis, Its Applications and Related Topics in Beijing during the World Mathematical Year 2000.

Operator Theory
Advances and Applications, Vol. 147, 13–14

Curriculum Vitae

Volkmar Erhard Meister

Personal Information

Marital Status:	married with Dr. Ljudmila Meister, née Aramanovitch, three own children: Frauke (born 1957), Frithjof (born 1963), Olaf (born 1964) and two stepsons: Andrew Comech (born 1970) and Sergej Komech (born 1978).
Nationality:	German
Date of Birth:	12 February 1930
Place of Birth:	Bernburg/Saale

Education

School training:	1936–1948 Bernburg with final exam
Practical training:	Großkraftwerk Mannheim: electrical engineering apprenticeship (9/1948–4/1951)
University training:	University of Heidelberg (4/1951–6/1956) Subjects: Mathematics and Physics
Diploma:	University of Heidelberg (1956)
Ph.D.:	University of Saarbrücken (1958)
Habilitation:	University of Saarbrücken (1963)

Work Experience

DVL (German Research Establishment of Aeronautics) at Mülheim/Ruhr:
Scientific employee (7/1956–4/1958)

University of Saarbrücken:

Assistant at Institute of Mathematics (5/1958–4/1964)
Dozent (Lecturer) at Institute of Mathematics (4/1964–9/1966)
with leave of absence:
To Technical University of Berlin (10/1965–3/1966)
To University of Strathclyde at Glasgow (4/1966–9/1966)

Technical University of Berlin:

Full Professor at the Faculty of Mathematical Sciences (10/1966–8/1970)

This short CV was used by E. Meister for project applications

University Tübingen:
Full Professor in the Department of Mathematics (9/1970–6/1974)

Technical University of Darmstadt:
Full Professor in the Department of Mathematics since 6/1974
with leave of absence:
to University of Regina, Sask./Canada (2–3/1978)
to University of Alberta at Edmonton/Canada (3/1978)

Extracurricular Activities

Chairman of the GAMM committee on Applied Analysis and Mathematical Physics (since 1971)
Member of the Scientific Board of GAMM: 1974–1977
Elected vice-referee for applied Mathematics for DFG, acting as main referee (8/1980–3/1981)
Member of the DFG-committee "Sonderforschungsbereiche" 1982–1987
Organizer of several conferences in Oberwolfach
Organizer of the International Symposium on Function Theoretic Methods for Partial Differential Equations, 12–15 April in Darmstadt, 1976 (together with N. Weck and W.L. Wendland)
Organizer of the Sommerfeld-Workshop on Modern Mathematical Methods in Diffraction Theory and its Applications in Engineering, Freudenstadt, 30.9.–4.10.1996

Member of Editorial Boards of Journals

Applicable Analysis, ed. by R.P. Gilbert, Newark/Del., USA, since 1985
LAMM (Leitfäden der Angewandten Mathematik und Mechanik) since 1973
Mathematical Methods in the Applied Sciences, ed. by B. Brosowski, Frankfurt, since 1979
Proceedings of the Royal Society, Edinburgh, 1974–1978
Integral Equations and Operator Theory, ed. by I. Gohberg, Tel Aviv, since 1977
Asymptotic Analysis, ed. by L.S. Frank, Reims, since 1989
Mathematische Nachrichten, ed. by R. Mennicken, Regensburg, since 1992
Demonstratio Mathematica, ed. by M. Maczynski, Warschau, since 1995

Operator Theory
Advances and Applications, Vol. 147, 15–24

Publications of Erhard Meister

Collected by S. Roch[1]

Books

1. Randwertaufgaben der Funktionentheorie. Mit Anwendungen auf singuläre Integralgleichungen und Schwingungsprobleme der mathematischen Physik.– Leitfäden der Angewandten Mathematik und Mechanik, Vol. 59, B.G. Teubner, Stuttgart 1983, 320 pp.
2. Integraltransformationen mit Anwendungen auf Probleme der mathematischen Physik. – Methoden und Verfahren der Mathematischen Physik, Vol. 28, Verlag Peter Lang, Frankfurt am Main 1983, 288 pp.
3. Partielle Differentialgleichungen. Eine Einführung für Physiker und Ingenieure in die klassische Theorie. – Akademie-Verlag, Berlin 1996, 293 pp.

Edited Books and Proceedings

1. Function Theoretic Methods for Partial Differential Equations. Proceedings of the International Symposium, held at Darmstadt, Germany, in April 1976. Lecture Notes in Mathematics, Vol. 561, Springer-Verlag, Berlin 1976, 520 pp. (with N. Weck and W.L. Wendland).
2. Einführung in die Theorie der eindimensionalen singulären Integraloperatoren (by I. Gohberg and N. Krupnik). Birkhäuser Verlag, Basel 1979, 379 pp. (German edition by E. Meister, translation from the Russian original of 1973 by B. Schüppel.)
3. Mathematical Methods in Fluid Mechanics. Proceedings of a Conference at Oberwolfach, Germany, in November 1981. Methoden und Verfahren der Mathematischen Physik, Vol. 24, Verlag Peter Lang, Frankfurt am Main 1982, 266 p. (with K. Nickel and J. Polasek).
4. Modern Mathematical Methods in Diffraction Theory and its Applications in Engineering. Proceedings of the Sommerfeld'96 Workshop. Freudenstadt, Germany, in September–October 1996. Methoden und Verfahren der Mathematischen Physik, Vol. 42, Verlag Peter Lang, Frankfurt am Main 1997, 308 pp.

[1]The collected works of E. Meister can be found in the library of the Mathematics Department of Darmstadt University of Technology

Articles in Journals and Proceedings

1. Beitrag zur Aerodynamik eines schwingenden Gitters I. – Z. Angew. Math. Mech. **38**(1958), 442–465 (with H. SÖHNGEN).
2. Über ein Randwertproblem aus der Aerodynamik eines Gitters. – Z. Angew. Math. Mech. **39**(1959), 433–435.
3. Flow of an incompressible fluid through an oscillating staggered cascade. – Arch. Ration. Mech. Anal. **6**(1960), 198–230 (Summary of the PhD Dissertation).
4. Ein Eindeutigkeitsbeweis für ein gemischtes Randwertproblem der Schwingungsgleichung. – Math. Z. **77**(1961), 38–44.
5. Beitrag zur Aerodynamik eines schwingenden Gitters II. (Unterschallströmung.) – Z. Angew. Math. Mech. **42**(1962), 9–31.
6. Beitrag zur Aerodynamik eines schwingenden Gitters III. (Unterschallströmung.) – Z. Angew. Math. Mech. **42**(1962), 245–254.
7. Zum Dirichlet-Problem der Helmholtzschen Schwingungsgleichung für ein gestaffeltes Streckengitter. – Arch. Ration. Mech. Anal. **10**(1962), 67–100.
8. Zum Neumann-Problem der Helmholtzschen Schwingungsgleichung für ein gestaffeltes Streckengitter. – Arch. Ration. Mech. Anal. **10**(1962), 127–148.
9. Beitrag zur Theorie des Strahlungsfeldes dielektrischer Antennen. – Forschungsbericht 1175 des Landes Nordrhein-Westfalen, Westdeutscher Verlag Köln und Opladen 1963, 43 pp. (with K.-D. Becker).
10. Die instationäre Unterschallströmung durch ein schwingendes, gestaffeltes Gitter mit halbunendlich tiefen Profilen, Teil I. – Deutsche Versuchsanstalt für Luft- und Raumfahrt e.V., Bericht 246, June 1963, 79 pp.
11. Die instationäre Unterschallströmung durch ein schwingendes, gestaffeltes Gitter mit halbunendlich tiefen Profilen, Teil II. – Deutsche Versuchsanstalt für Luft- und Raumfahrt e.V., Bericht 247, March 1964, 61 pp.
12. Zur Theorie instationärer Gitterströmungen. – Z. Angew. Math. Mech. **43**(1963), Proceedings of the GAMM Conference at Karlsruhe, T150–T152.
13. Ein System singulärer Integralgleichungen aus der Theorie der Beugung elektromagnetischer Wellen an dielektrischen Keilen. – Z. Angew. Math. Mech. **44**(1964), Proceedings of the GAMM Conference at Giessen, T47–T49 (with N. LATZ).
14. Die Beugung ebener elektromagnetischer Wellen an einem Parallelplattengitter. – Z. Angew. Math. Mech. **45**(1965), Proceedings of the GAMM Conference at Vienna, T57–T59.
15. Zur Theorie der ebenen, instationären Unterschallströmung um ein schwingendes Profil im Kanal. – Z. Angew. Math. Physik **16**(1965), 770–780.

16. Zur Theorie der ebenen Unterschallströmung durch ein schwingendes, gestaffeltes Gitter. Teil I: Lösung des Problems in der Bildebene. – Deutsche Luft- und Raumfahrt, Forschungsbericht 66-12, March 1966, 77 pp.

17. Zur Theorie der ebenen Unterschallströmung durch ein schwingendes, gestaffeltes Gitter. Teil II: Rücktransformation der Bildfunktion und Eigenschaften ihrer Urbilder. – Deutsche Luft- und Raumfahrt, Forschungsbericht 66-13, March 1966, 85 pp.

18. Zur Theorie der ebenen Unterschallströmung durch ein schwingendes, gestaffeltes Gitter. Teil III: Eigenschaften der formalen Lösung. Berechnung der Luftkräfte. – Deutsche Luft- und Raumfahrt, Forschungsbericht 66-14, March 1966, 65 pp. (Habilitation thesis).

19. Neuere Ergebnisse der mathematischen Theorie instationärer Gitterströmungen. – Acta Mech. **3**(1967), 325–341.

20. Unterschallströmung durch ein schwingendes Gitter in einem ebenen Kanal. – Z. Angew. Math. Mech. **48**(1968), Proceedings of the GAMM Conference at Praha, T213–T216.

21. Theorie instationärer Unterschallströmungen durch ein schwingendes Gitter im Windkanal. – Z. Angew. Math. Mech. **49**(1969), 481–494.

22. Ein System von Dreiteil-Wiener-Hopf-Funktionalgleichungen aus der Aerodynamik schwingender Gitter. – Methoden Verfahren Math. Phys. **1** (1969), 85–99.

23. Über eine Strömung durch ein ebenes Streckengitter mit Kavitation. – Methoden Verfahren Math. Phys. **3**(1970), BI-Hochschulskripten 722/722 a, 33–71 (with G. Hery and J. Wurms).

24. Randwertprobleme aus der Beugungstheorie ebener Wellen an Parallelplattengittern. – Methoden Verfahren Math. Phys. **3**(1970), BI-Hochschulskripten 722/722 a, 85–116.

25. Das Riemannsche Randwertproblem. Ergebnisse und Anwendungen.– Überblicke der Mathematik **6** (Ed. D. Laugwitz), Bibliographisches Institut AG Mannheim, Wien, Zürich 1973, 113–178.

26. Ein Überblick über analytische Methoden zur Lösung singulärer Integralgleichungen. – Z. Angew. Math. Mech. **57**(1977), Proceedings of the GAMM Conference at Graz in 1976, T23–T35.

27. Einige gemischte Randwertprobleme für Unterschallströmungen um schwingende Profile. – Z. Angew. Math. Mech. **58**(1978), Proceedings of the GAMM Conference at Lyngby, T383–T386.

28. Some mixed boundary value problems in the theory of subsonic flow past oscillating profiles. – Complex Analysis and its Applications, Coll. Articles in Honour of Academician I.N. Vekua's 70th Anniversary, Akad. Nauk SSR,

Steklow Math. Inst., Nauka, Moscow 1978, 346–362 (extended version of no. 27).

29. Some multidimensional Wiener-Hopf equations with applications. – Trends in Applications of Pure Mathematics to Mechanics, Vol. II, Proc. Symp. Kozubnik/Poland, September 1977 (Ed. H. Zorski), Pitman, London 1979, 217–262 (with F.-O. Speck).

30. Ein algebraisches Verfahren zur Lösung von gemischten Anfangsrandwertproblemen aus der Theorie instationärer Gitterströmungen. – Math. Methods Appl. Sci. **1**(1979), 158–186 (with H.-J. Sommer).

31. Einige neuere Ergebnisse über Integralgleichungen vom Faltungstyp. – Helmut Pachale in memoriam: Ansprachen und Vorträge des Gedenkkolloquiums vom 14. Juli 1979 (Hrsg. H. Begehr), Freie Universität Berlin 1980, 20–22.

32. Gemischte Randwertprobleme in der Theorie instationärer Triebwerksströmungen. – Preprint 484 TH Darmstadt, July 1979, 94 pp. (with G. Buggle, M. Jirman, H.-J. Sommer).

33. Wiener-Hopf operators on three-dimensional wedge-shaped regions. – Appl. Anal. **10**(1980), 31–45 (with F.-O. Speck).

34. Some classes of integral and integro-differential equations of convolutional type. – Ordinary and Partial Differential Equations, Proc. 5th Confer. Differ. Equations, Dundee, Scotland 1978 (Ed. W.N. Everitt), Springer Lecture Notes Math. **827**(1980), 182–228.

35. Multiple-part Wiener-Hopf operators with some applications in mathematical physics. – Toeplitz Centennial, Toeplitz Mem. Conf., Tel Aviv 1981, Operator Theory: Adv. and Appl. **4**, Birkhäuser Verlag, Stuttgart, Basel 1982, 427–447.

36. Generalized Sommerfeld half-plane problems. – Differential equations and their applications, Equadiff 5, Proc. 5th Czech. Conf., Bratislava 1981, Teubner-Texte Math. **47**, 1982, 264–267.

37. Zur Theorie rotierender und schwingender Schaufelkränze in einer Unterschallströmung durch einen Ringkanal. – Math. Methods Appl. Sci. **4** (1982), 454–479 (with G. Buggle).

38. Zur Theorie rotierender und schwingender Schaufelkränze in einer Unterschallströmung durch einen Ringkanal. – Z. Angew. Math. Mech. **63** (1983), Proceedings of the GAMM Conference at Hamburg, T233–T235 (with G. Buggle).

39. Multiple-part Wiener-Hopf operators. – Trends in Applications of Pure Mathematics to Mechanics IV, Proc. 4th Symp., Bratislava 1981, Monogr. Stud. Math. **20**(1983), 152–175.

40. On the transmission problem of the Helmholtz equation for quadrants. – Math. Methods Appl. Sci. **6**(1984), 129–157 (with N. Latz).

41. Some multiple-part Wiener-Hopf problems in mathematical physics. – Mathematical Models and Methods in Mechanics, Banach Cent. Publ. **15**, Warsaw 1985, 359–407.

42. Über ein System von Integralgleichungen zum Transmissionsproblem der Helmholtzgleichung für Quadranten. – Z. Angew. Math. Mech. **65**(1985), 581–583 (with N. Latz).

Also in MATHEMATICA, Festschrift zum 75. Geburtstag von E. Mohr, TU Berlin 1985, 191–198.

43. The Moore-Penrose inverse of Wiener-Hopf operators on the half axis and the quarter plane. – J. Integral Equations **9**(1985), 45–61 (with F.-O. Speck).

44. Integral equations for the Fourier-transformed boundary values for the transmission problems for right-angled wedges and octants. – Math. Methods Appl. Sci. **8**(1986), 182–205.

45. Diffraction problems with impedance conditions. – Appl. Anal. **22**(1986), 193–211 (with F.-O. Speck).

46. Some solved and unsolved canonical problems of diffraction theory. – Differential Equations and Their Applications, Equadiff 6, Proc. 6th Int. Conf. Brno 1985 (Eds. J. Vosmansky, M. Zlamal), Lect. Notes Math. **1192**(1986), 393–398.

47. Some solved and unsolved canonical problems of diffraction theory. – Differential Equations and Mathematical Physics, Proc. Int. Conf. Birmingham/Alabama 1986 (Eds. I.W. Knowles, Y. Saito), Springer Lect. Notes Math. **1285**(1987), 320–336.

48. Einige gelöste und ungelöste kanonische Probleme der mathematischen Beugungstheorie. – Expo. Math. **5**(1987), 193–237.

49. Einige Randwert-Transmissionsprobleme der Beugungstheorie für Keile bei mehreren Medien. – Z. Angew. Math. Mech. **67**(1987), T432–T434 (with R. Penzel).

50. Scalar diffraction problems for Lipschitz and polygonal screens. – Z. Angew. Math. Mech. **67**(1987), T434–T435 (with F.-O. Speck).

51. Einige Randwert-Transmissionsprobleme der Beugungstheorie für Keile bei mehreren Medien. – Kleinheubacher Berichte **30**(1987), Fernmeldetechn. Zentralamt Darmstadt, 429–438 (with R. Penzel).

52. Boundary integral equation methods for canonical problems in diffraction theory. – In: Boundary Elements IX, Vol. 1 (Eds. C.A. Brebbia, W.L. Wendland, G. Kuhn), Springer, Berlin 1987, 59–77 (with F.-O. Speck).

53. Generalized lens problems. – Preprint 1037 TH Darmstadt, März 1987, 12 pp. (with R.A. Hurd).

54. Generalized waveguide bifurcation problems. – Q. J. Mech. Appl. Math. **41**(1988), 127–139 (with R.A. Hurd).

55. A contribution to the quarter-plane problem in diffraction theory. – J. Math. Anal. Appl. **130**(1988), 223–236 (with F.-O. Speck).

56. On some generalized Sommerfeld half-plane problems. – Z. Angew. Math. Mech. **68**(1988), T468–T470 (with F.-O. Speck).

57. Sommerfeld diffraction problems with third kind boundary conditions II. – SIAM J. Math. Anal. **20**(1989), 589–607 (with F.-O. Speck, R.A. Hurd).

58. Diffraction by a partially-coated dielectric slab. – Preprint 1114 TH Darmstadt, February 1988, 34 pp. (with E. Lüneburg, R.A. Hurd).

59. Elastodynamic scattering and matrix factorization. – Elastic Wave Propagation. Proc. IUTAM Conf. Galway/Ireland 1988 (Eds. M.F. McCarthy, M.A. Hayes), Elsevier Sci. Publ. B.V. (North Holland), 1989, 373–380 (with F.-O. Speck).

60. The explicit solution of elastodynamical diffraction problems by symbol factorization. – Z. Anal. Anwend. **8**(1989), 307–328 (with F.-O. Speck).

61. Wiener-Hopf factorization of certain non-rational matrix functions in mathematical physics. – The Gohberg Anniversary Collection, Volume II: Topics in Analysis and Operator Theory, Operator Theory: Advances and Applications **41** (Eds. H. Dym, S. Goldberg, M.A. Kaashoek, P. Lancaster), Birkhäuser 1989, 385–394 (with F.-O. Speck).

62. Matrix factorization for canonical Wiener-Hopf problems in elastodynamic scattering theory. – Problems and methods in mathematical physics, Proc. 9th Conf. TMP, Karl-Marx-Stadt/GDR 1988 (Eds. F. Kuhnert, B. Silbermann), Teubner-Texte Math. **111**, Teubner, Leipzig 1989, 175–184 (with F.-O. Speck).

63. Modern Wiener-Hopf methods in diffraction theory. – Ordinary and partial differential equations, Vol. II, Proc. 10th Dundee Conf., Dundee/UK 1988 (Eds. B.D. Sleeman, R.J. Jarvis), Pitman Res. Notes Math. Ser. **216**(1989), 130–171 (with F.-O. Speck).

64. Matrix factorization for canonical Wiener-Hopf problems in elastodynamic scattering theory. – Proc. Workshop on Integral and Field Equation Methods in Fluid-Structure Interactions (Eds. G.C. Hsaio, R.E. Kleinman), University of Delaware 1989, 79–85 (short version of no. 62 with F.-O. Speck).

65. Double-knife edge problems with elastic/plastic type screens. – Integral Equations Oper. Theory **14**(1991), 373–389 (with A.D. Rawlins, F.-O. Speck).

66. Diffraction by an acoustically transmissive or an electromagnetically dielectric half-plane. – Math. Methods Appl. Sci. **14**(1991), 387–402 (with A.D. Rawlins, F.-O. Speck).

67. Wiener-Hopf equations for waves scattered by a system of parallel Sommerfeld half-planes. – Math. Methods Appl. Sci. **14**(1991), 525–552 (with K. Rottbrand, F.-O. Speck).

68. On the reduction of the factorization of matrix functions of Daniele-Khrapkov type to a scalar boundary value problem on a Riemann surface. – Complex Variables, Theory Appl. **18**(1992), 63–71 (with F. Penzel).

69. Wiener-Hopf-Hankel operators for some wedge diffraction problems with mixed boundary conditions. – J. Integral Equations Appl. **4**(1992), 229–255 (with F.-O. Speck, F.S. Teixeira).

70. Two-media scattering problems in a half-space. – Partial Differential Equations with Real Analysis. Dedicated to Robert Pertsch Gilbert on the Occasion of his 60th Birthday (Eds. H. Begehr, A. Jeffrey), Pitman Res. Notes Math. Ser. **263**, Harlow: Longman Scientific & Technical, 1992, 122–146 (with F. Penzel, F.-O. Speck, F.S. Teixeira).

71. A Sommerfeld-type diffraction problem with second-order boundary conditions. – Z. Angew. Math. Mech. **72**(1992), 621–630 (with P.A. Santos, F.S. Teixeira).

72. Some results on the invertibility of Wiener-Hopf-Hankel operators. – Z. Anal. Anwend. **11**(1992), 57–76 (with A.B. Lebre, F.S. Teixeira).

73. Operatorfaktorisierung in der Beugungstheorie für kanonische halbunendliche Gebiete. – Mitt. Ges. Angew. Math. Mech. **16**(1993), 111–132.

74. Two-media scattering problems in a half-space. – Z. Angew. Math. Mech. **73**(1993), T793–T795 (with F. Penzel, F.-O. Speck, F.S. Teixeira).

75. Some interior and exterior boundary-value problems for the Helmholtz equation in a quadrant. – Proc. R. Soc. Edinb., Sect. A **123**(1993), 275–294 (with F. Penzel, F.-O. Speck, F.S. Teixeira).

76. Matrix factorization for canonical Wiener-Hopf problems in elastodynamic scattering theory. – Continuum Mechanics and Related Problems of Analysis. Proc. Int. Symp. dedicated to the centenary of Academician N. Muskhelishvili, Tbilisi/Georgia 1991 (Ed. M. Balavadze), Metsniereba, Tbilisi 1993, 280–289 (reprint of no. 62 with F.-O. Speck).

77. Spectral analysis of a transmission problem for the Helmholtz equation on the half-space. – Rend. Mat., Ser. VII, Vol. 13(1993), 751–772 (with N. Latz, J. Scheurer).

78. Some solved and unsolved canonical transmission problems of diffraction theory. – Problems and Methods in Mathematical Physics, Proc. 10th Conf. TMP, Chemnitz/FRG 1993 (Eds. L. Jentzsch, F. Tröltzsch), Teubner-Texte Math. **134**, Teubner, Leipzig 1994, 89–99.

79. Two canonical wedge problems for the Helmholtz equation. – Math. Methods Appl. Sci. **17**(1994), 877–899 (with F. Penzel, F.-O. Speck, F.S. Teixeira).

80. Modern Wiener-Hopf methods in diffraction theory. – Proc. 5th Int. Colloq. on Differential Equations, Plovdiv/Bulgaria 1994 (Ed. D. Bainov), VSP, Utrecht 1995, 245–256.

81. Initial-boundary value problems in linear viscoelasticity on the half-space. – Math. Methods Appl. Sci. **18**(1995), 1181–1214 (with J. Mark).

82. Stand und Trends der Angewandten Analysis. – Preprint 1711 TH Darmstadt, Januar 1995, 10 pp.

83. Elastodynamical scattering by N parallel half-planes in $\mathbb{R}^3$: I. Equivalent matrix Wiener-Hopf factorization problems. – Math. Nachr. **177**(1996), 189–232 (with K. Rottbrand).

84. Applied analysis: Some trends connected with factorization problems in diffraction theory. – Proc. Symp. Analytical and Numerical Methods in Quaternionic and Clifford Analysis, Seiffen/FRG 1996 (Eds. W. Sprössig et al.), TU Bergakademie Freiberg, Freiberg 1996, 121–127.

85. Factorization problems in diffraction theory for canonical domains. – Direct and Inverse Electromagnetic Scattering. Proc. Workshop Gebze/ Turkey 1995 (Eds. A.H. Serbest et al.), Pitman Research Notes Math. Ser. **361**, Longman, Harlow 1996, 69–81.

86. Integral equations arising from instationary flows around thin wings. – Boundary Element Topics. Proc. Final Conference of the Priority Research Programme Boundary Element Methods 1989 – 1995 of the German Research Foundation, Stuttgart/FRG 1995 (Ed. W. Wendland), Springer, Berlin 1997, 167–188 (with R. Hinder).

87. Einige Klassen singulärer Integral- und Integro-Differentialgleichungen auf der Halbachse. – Mathematik aus Berlin. Sammelband von Vortragsausarbeitungen zu "Wissenschaft und Stadt. Publikationen der Freien Universität Berlin aus Anlass der 750-Jahr-Feier Berlins 1987" (Ed. H. Begehr), Weidler Buchverlag, Berlin 1997, 265–290.

88. Neuere Entwicklungen in der Angewandten Analysis – Am Beispiel Kanonischer Beugungsprobleme. – Preprint 1839 TH Darmstadt, Mai 1996, 12 pp. (Lecture of the memorial colloquium for W. Haack).

89. Spectral theory for the wave equation in two adjacent wedges. – Math. Methods Appl. Sci. **20**(1997), 1015–1044 (with F. Ali Mehmeti, K. Mihalincic).

90. Generalized eigenfunctions for canonical boundary-transmission problems in diffraction theory. – Differential Equations, Asymptotic Analysis, and Mathematical Physics. Papers associated with the International Conference on Partial Differential Equations, Potsdam/FRG 1996 (Eds. M. Demuth et al.), Akademie-Verlag, Berlin 1997, 242–251.

91. Regarding some problems of the Kutta-Joukovskii condition in lifting surface theory. – Math. Nachr. **184**(1997), 191–228 (with R. Hinder).

92. Elastodynamical scattering by N parallel half-planes in $\mathbb{R}^3$: II. Explicit solutions for $N = 2$ by explicit symbol factorization. – Integral Equations Oper. Theory **29**(1997), 70–109 (with K. Rottbrand).

93. Stable desmearing of slit-collimated SAXS patterns by adequate numerical conditioning. – Acta Cryst. **A54**(1998), 675–681 (with M. Soliman, B.-J. Jungnickel).

94. Scattering of Acoustical and Electromagnetic Waves by some Canonical Obstacles. – Partial Differential Equations: Theory and Numerical Solution, Proc. ICM'98 Satellite Conference at Prague 1998 (Eds. W. Jäger et al.), Boca Raton, Chapman & Hall/CRC, Res. Notes Math. **406** (2000), 254–272 (with A. Passow).

95. Initial Dirichlet problem for half-plane diffraction: Global formulae for its generalized eigenfunctions, explicit solution by the Cagniard-de Hoop method. – Z. Anal. Anwend. **18**(1999), 307–330 (with K. Rottbrand).

96. Funktionentheorie und Funktionalanalysis. – Preprint 1885 TH Darmstadt, Januar 1997, 19 S. (Lecture on the occasion of the 60th birthday of W.L. Wendland).

97. Some interior and exterior boundary-value problems for the Helmholtz equation in a quadrant. – Operator Theory: Advances and Applications **102**, Birkhäuser Verlag, Basel 1998, 169–178 (short version of no. 75 with F. Penzel, F.-O. Speck, F.S. Teixeira).

98. New results on wave diffraction by canonical obstacles. – The Maz'ya Anniversary Collection. Vol. 2: Rostock Conference on Functional Analysis, Partial Differential Equations and Applications, Rostock/Germany 1998 (Eds. J. Rossmann et al.), Operator Theory: Adv. Appl. **110**, Birkhäuser Verlag, Basel 1999, 235–256 (with A. Passow, K. Rottbrand).

99. Asymptotics of solutions to Joukovskii-Kutta-type problems at infinity. – Applicable Analysis **76**(2000), 153–166 (with R. Hinder, S. Nazarov).

100. Quaternion boundary value problems for canonical objects in continuum physics. – Proc. Conf. Clifford Analysis and Appl. to Math. Physics and Related Topics, Beijing, August 2000 (with L. Meister).

101. Some initial boundary problems in elastodynamics for canonical domains in quaternions. – Math. Bohemica **126**(2001), 429–442 (with L. Meister).

102. Aperiodic wave diffraction by semi-infinite obstacles. – Singular Integral Operators, Factorization and Applications, Proc. Int. Workshop Oper. Theory Appl. at Faro 2000, Operator Theory: Advances and Applications **142** (Eds. A. Böttcher, M.A. Kaashoek, A. Lebre, A.F. dos Santos, F.-O. Speck), Birkhäuser Verlag, Basel 2003, 253–261.

Further Publications

1. Angewandte Mathematik: Ihre Rolle am Beispiel der Aerodynamik. – Antrittsvorlesung, held on 14.01.1971 at the University of Tübingen.
2. Minisymposium 1: Angewandte Analysis. – Z. Angew. Math. Mech. **72** (1992), T 427 (Conference Report).
3. Krupnik, N.Y., Banach Algebras with a Symbol and Singular Integral Operators. – Jahresber. DMV **93**(1991), 8–11 (Book Review).
4. Nachruf Siegfried Prössdorf (2.1.1939–19.7.1998). – Math. Nachr. **201**(1999), i–ii.
5. Die Wissenschaft am Ende des 20. Jahrhunderts: Positionen, Probleme, Perspektiven – Die Angewandte Mathematik. Typescript.
6. Abschlussbericht zum Forschungsprojekt ME 261/13-1, 2: Analytische und spektraltheoretische Untersuchung zeitabhängiger Streuprobleme an kanonischen halbunendlichen Objekten. – GAMM-Mitteilungen **23**(2000), 93 – 117.

Operator Theory
Advances and Applications, Vol. 147, 25–26

Supervision of PhD Dissertations

Collected by S. Roch

1. G. Hery, Über eine Klasse von Integralgleichungen mit Cauchyschem Hauptwert (University of Saarbrücken, 1965)
2. D. Schulze, Konstruktive Verfahren zur Lösung linearer Gleichungen im Hilbert-Raum (Technical University of Berlin, 1967)
3. N. Latz, Untersuchungen über ein skalares Übergangswertproblem aus der Theorie der Beugung elektromagnetischer Wellen an dielektrischen Keilen (University of Saarbrücken, 1968)
4. M. Kremer, Über eine Klasse singulärer Integralgleichungen vom Faltungstyp (Technical University of Berlin, 1969)
5. E. Gerlach, Zur Theorie einer Klasse von Integrodifferentialgleichungen (Technical University of Berlin, 1969)
6. G. Eisenhardt, Das schwingende Profil unter der freien Oberfläche eines endlich tiefen Stromes (Technical University of Berlin, 1973)
7. J. Donig, Zur Theorie einer Klasse elliptischer singulärer Integro-Differentialoperatoren in Grund-und Distributionenräumen (University of Tübingen, 1973)
8. F.-O. Speck, Über verallgemeinerte Faltungsoperatoren und eine Klasse von Integrodifferentialgleichungen (Technical University of Darmstadt, 1974)
9. H.-D. Alber, Ein quasiperiodisches Randwertproblem für die Helmholtzsche Schwingungsgleichung bei verschiedenen Ausstrahlungsbedingungen (Technical University of Darmstadt, 1976)
10. H.-J. Sommer, Die Nonstandardtheorie als Alternative zur Fouriertransformation bei der Lösung gemischter Randwertprobleme (Technical University of Darmstadt, 1977)
11. G. Buggle, Die Kontorovich-Lebedev-Transformation und die Mehler-Fok-Transformation für Klassen verallgemeinerter Funktionen mit Anwendungen auf Probleme der mathematischen Physik (Technical University of Darmstadt, 1977)
12. M. Costabel, Kausale Distributionen und ihre Anwendung auf eine Klein-Gordon-Gleichung mit Potentialterm (Technical University of Darmstadt, 1977)

13. W. Peterhänsel, Zur Theorie einer Klasse singulärer Integralgleichungen auf kompakten Mannigfaltigkeiten ohne und mit Rand (Technical University of Darmstadt, 1980)
14. G. Thelen-Rosemann-Niedrig, Zur Fredholmtheorie singulärer Integro-Differentialoperatoren auf der Halbachse (Technical University of Darmstadt, 1984)
15. S. Minutschehr, Entwicklungen nach Streueigenelementen bei Schrödinger-Operatoren für Teilchen mit Spin (Technical University of Darmstadt, 1984)
16. F. Penzel, Zur Theorie verallgemeinerter Abelscher Integralgleichungen (Technical University of Darmstadt, 1986)
17. R. Schneider, Contributions to the Theory of Elliptic Pseudodifferential Operators on Manifolds with Nonsmooth Boundaries (Technical University of Darmstadt, 1989)
18. T. v. Petersdorff, Randwertprobleme für Polyeder-Singularitäten und Approximation mit Randelementmethoden (Technical University of Darmstadt, 1989)
19. H. Marschall, Über ein gemischtes Randwertproblem beim Zusammenfluss zweier Poiseulle-Ströme einer zähen, inkompressiblen Flüssigkeit in einem geteilten Kanal (Technical University of Darmstadt, 1991)
20. C. Erbe, Zum Sommerfeldschen Schirmproblem in der Thermoelastizität (Technical University of Darmstadt, 1994)
21. R. Hinder, Zur Problematik der Kutta-Bedingung (Technical University of Darmstadt, 1994)
22. K. Rottbrand, Matrix-Wiener-Hopf-Faktorisierungsprobleme in der Elastodynamik bei mehreren parallelen Rissen (Technical University of Darmstadt 1994)
23. K. Mihalincic, Time Decay Estimates for the Wave Equation with Transmission and Boundary Conditions (Darmstadt University of Technology, 1998)
24. J. Mark, Initial-Boundary Value Problems in Linear Viscoelasticity Using Wiener-Hopf Methods (Darmstadt University of Technology, 1998)
25. A. Passow, Sommerfeld-Halbebenenproblem mit elektromagnetisch anisotropen Leontovich-Randbedingungen (Darmstadt University of Technology, 1999)
26. W. Eckart, Theoretische Untersuchungen von elektrorheologischen Flüssigkeiten bei homogenen und inhomogenen elektrischen Feldern (Darmstadt University of Technology, 1999)

Operator Theory
Advances and Applications, Vol. 147, 27–46

In Memory of Erhard Meister

Frank-Olme Speck

This short biography is mainly about aspects of the scientific part of Erhard Meister's life, its structure and context, results and impacts. The reader is referred to the list of publications included in this volume and to complementary sources for more extended studies of the mathematical details. The areas of research and some basic ideas are outlined in a form that makes the text accessible to a wide group of readers.

1. Life phases

Erhard Meister was born in Bernburg-on-Saale on 12 February 1930 and attended school in Bernburg from 1936 to 1948. His parents were high school teachers and encouraged his inborn interest in science. At the time of his "Abitur" (final high school exam) in the Soviet-occupied zone (transformed subsequently into the GDR), he was convinced that he should study electrical engineering.

E. Meister in 1948

Because of the limited possibilities at home, he escaped to the West and first completed a three-year apprenticeship at the main power station (Grosskraftwerk) in Mannheim. His parents followed some ten years later and moved to Hameln in Northern Germany. Changing his mind, he began university studies in mathematics and physics at the University of Heidelberg in the spring of 1951. In 1954 he married Christa Rienecker. He obtained the diploma of mathematics in 1956 in Heidelberg. His supervisor was Heinz Söhngen – a lecturer from Darmstadt who represented the Chair of Applied Mathematics at Heidelberg in 1955–57. Söhngen was a pupil of Artin, Blaschke, Hausdorff and Toeplitz, and this fact undoubtedly also influenced Meister's thoughts on and interests in complex and functional analysis.

For the following two years Erhard Meister worked as a scientist in the German Research Institute of Aeronautics in Mühlheim-on-Ruhr. The work that he did there later on became the basis for his doctoral thesis. In 1957 his daughter Frauke was born.

In 1958 he started working as an assistant at the University of Saarbrücken and in the same year he finished his PhD under the supervision of Professor Heinz Söhngen who then held a Chair in the Mathematics Department. Five years later Erhard Meister completed his habilitation and became a lecturer at the Institute of Mathematics in 1964. During the same period of his life, his first and second sons, Fritjof and Olaf, were born, in 1963 and 1964 respectively.

During the winter of 1965/66, Erhard Meister was a visiting professor at the Technical University of Berlin and, in spring 1966, at the University of Strathclyde in Glasgow. During his stay in Scotland he was offered a full professorship from the Technical University of Berlin because of his pioneering work in mathematical airfoil theory (see 3.1). He accepted the position and the family moved to West Berlin. At the Technical University he delivered many lectures on different subjects, on which he later based his books. During this period he also acquired his first generation of PhD students. However the political situation in West Berlin made scientific work rather difficult.

In 1970 Professor Meister accepted an invitation from the University of Tübingen to a Chair of Mathematical Physics in the Department of Mathematics. Mathematical Physics became a new research area at this institute. During the next few years Erhard Meister was the principal representative of applied mathematics in Tübingen. He ran very successful seminars; the number of Master and PhD students increased rapidly as did his teaching load. He was willing to assume administrative duties; he served as head of the mathematics department in 1972/73 and he was elected to several university committees. As head of department, well beyond his own domain of research, he established close ties with the science departments.

From 1974 until 1998, the longest period of his life, he was a professor in the Department of Mathematics at the Technical University of Darmstadt (known until 1997 as the Technische Hochschule or TH Darmstadt). There he found excellent possibilities of working together with colleagues from mathematics, engineering and the applied sciences. As he said himself, during this period a wide network of international relations was built up (see his 1993 report in this volume); various research projects were carried out, and helped many young people to complete their doctoral degrees (see the impressive list of his PhD students). Besides teaching basic and advanced courses eight hours per week, Erhard Meister worked successfully as an author, editor, referee, dean, initiator of international exchanges and organizer of scientific events, to name but a few of his roles. He participated in many conferences and reported on the research results of his school.

In May 1993 his wife Christa passed away suddenly due to a heart failure during the night. At that time, the children already lived independently in different places with their families and the first two grandchildren were born a year later, in December 1994.

In August 1994 Erhard Meister married Ljudmila Aramanovitch, a mathematician from Moscow working on quaternion analysis. The story of how they met is reported in a lively fashion in her reminiscences. Their union through marriage also established a new and unexpected scientific cooperation.

Erhard Meister retired in 1998 at the age of 68 in accordance with the prevailing legislation. He managed to keep his office in the Mathematics Department in Darmstadt with access to the library and secretarial services. Until the very last days of his life he remained busy with scientific work and in particular enjoyed his participation in conferences together with his wife Ljudmila.

2. Teaching

Erhard Meister was a fascinating teacher who offered an outstanding choice of lectures in the West Germany of the 60's. For instance, he offered the course on "Höhere Funktionentheorie" (Advanced Complex Analysis) in 1967 at the TU Berlin; it contained a crash course of complex analysis, immediately followed by applications to continuum mechanics and, in the third chapter, he presented problems for analytical functions: the Riemann boundary value problem, singular integral equations of Cauchy type and Wiener-Hopf equations. Many students were attracted by the content and style of the course, which was extremely clear, stimulating and led to interesting topics for theses – as should be the case.

One of the reasons for the success was that he prepared all his lectures from scratch and almost completely *before* each term, including the most recent literature; for repeated courses he revised his lecture notes by updating the literature part. He developed an enormous capability for transforming mountains of new material into interesting and up-to-date lectures. The students faced some 20-30 large blackboards perfectly filled with formulas in 75 minutes' time. It was customary for an assistant to clean the blackboard (a time optimization problem). If the assistant was not at hand, some students did the service *voluntarily.*

All teaching material was chosen from the viewpoint of applications. The "Antrittsvorlesung" (inauguration lecture) at Tübingen in January 1971 demonstrated his philosophy and the wide spectrum of his interests and knowledge. The title was "Applied Mathematics: its role illustrated by aerodynamics" and it covered the history from Newton's "Principia" to the Leray-Schauder theory, as well as stories about flying rockets, well-posed problems and Fredholm operators.

His favorite topics for undergraduate lectures included partial differential equations, potential theory, linear and non-linear wave propagation, mathematical physics, integral equations, functional analysis, special functions, integral transformations, conformal mappings and, moreover, the theory of distributions, probability and statistics, measure and integration (real analysis) for instance. His constant teaching resulted in three textbooks and attracted a large number of diploma and PhD students listed before in this book.

3. Scientific work: publications

Professor Meister's research covered a wide range of topics and methods in mathematical physics. He used to ask himself and others: "What kind of engineer's questions can mathematicians answer?"

We will comment on this question by examining the main lines of research of Erhard Meister and his group. The following references correspond to the preceding list of publications: [1, 2, 3] for articles, [B1, B2, B3] for books, [D1, D2, D3] for dissertations, etc.

3.1. Aerodynamics of wings and cascades

During the period 1958–70 (Saarbrücken and Berlin) most publications were devoted to the area of aero- and fluid dynamics, subsonic flows around profiles (2D), oscillating wings and staggered cascades. The work started with a careful modeling of physically consistent boundary value problems by complex analysis [1–3] to determine the velocity fields of incompressible and non-viscous flows. "The solution of this problem requires a series of partial investigations", he wrote in [1], p. 445, to announce a paper on the detailed study of Hilbert problems, singular integral equations based on the work of N.I. Muskhelishvili, conformal mappings, harmonic analysis and asymptotic methods. In this paper, highly interesting physical problems were investigated systematically including the determination of the lift and moments of profiles, and the description of the trailing vortices [3], to name but a few.

A more intensive study of real analysis formulations in connection with "die Schwingungsgleichung" (the Helmholtz equation), led to his first contact with radiation conditions and the papers of A. Sommerfeld, F. Rellich, D.S. Jones and R. Leis [4]. The latter were important for uniqueness proofs and for the correct form of asymptotic expansions.

These fundamental investigations were successfully extended to various other problems such as profiles and various types of profile systems in non-stationary and compressible flows, in channels and ring channels, with cavities, etc. The Laplace transformation was employed [5–12] and the Wiener-Hopf technique for semi-infinite configurations [10–12] was developed from the classical books of D.S. Jones and B. Noble.

For all this work see [1–8, 10–12, 15–23] and [D6]. The resulting habilitation thesis was published in full length as a DLR research report [16–18]. This important work led to the offer from Berlin when Meister was only 36 years old. The first survey articles about aerodynamics appeared in 1967 [19] and later in 1978 [27, 28] when he was already in Darmstadt. Some of the material can be also found in [B1], Chap. 5.

Erhard Meister returned to this field at various times. In 1979–83 he did research with his PhD students G. Buggle, M. Jirmann and H.-J. Sommer and published a series of papers [30, 32, 37, 38] which included the study of time-dependent, so-called *aperiodic problems*. These investigations were based on a wide spectrum of methods: the Mikusinski calculus [D10], the Kontorovich-Lebedev and Mehler-Fock transformations [D11], numerical schemes, and distributional and non-standard analysis.

Many years later, in 1997–99, he studied with R. Hinder and S. Nazarov three-dimensional non-stationary flows and, in particular, the Kutta-Joukovski condition arising in the linearized lifting surface theory in order to find the physically meaningful spaces of solutions [D21, 86, 91, 99].

3.2. Diffraction theory – analytical methods, electrodynamics

The study of the Helmholtz equation led to an early contact with linear diffraction and scattering theory: "In aerodynamics or diffraction theory one comes across boundary value problems for $\Delta f + k^2 f = 0\ldots$" [4], p. 38. The fundamental boundary value problems of Dirichlet and of Neumann type were studied for special geometries (see above) now in the context of propagation of electromagnetic waves which was one of the most important applications in those times (radar technique, etc., see D.S. Jones' paper).

In the articles [7, 8] he presented formulations of the precise edge and radiation conditions, proofs of uniqueness, constructive solution methods via Laplace transformation, and the asymptotic (rather hard) analysis to verify formal solutions.

Some joint work with K.-D. Becker and N. Latz was started on related problems of diffraction by antennas, wedges and systems of parallel plates [9, 13–14, 24]. A spectral theoretic approach to periodic structures was considered in the thesis of H.-D. Alber [D9].

3.3. Riemann problems, singular integral equations, Wiener-Hopf operators

The move to Berlin in 1967 changed Erhard Meister's lifestyle and his possibilities of writing research papers as well. Busy with big courses, administration, and heavy discussions resulting from the students' revolt in Berlin (1967–70), Erhard Meister concentrated on a different form of publication, namely writing survey articles in a field where he was working with his group or where he wanted to start working. These also served as a direct source for his classes, which made them well up-to-date. Moving to Tübingen, as early as 1970, brought him a large number of diploma and PhD students, who were attracted by his lectures and seminars in mathematical physics and related areas.

This development is reflected in his survey articles on Riemann problems, singular integral equations, and Wiener-Hopf operators [25, 26, 29] published in 1973–79 as well as in the research articles [31, 33, 35, 39]. His great interest in these fields was partly inherited from H. Söhngen and highly stimulated by their usefulness in applications; also, perhaps, by the proximity of West Berlin to the Eastern bloc: he was impressed by the work of N.I. Muskhelishvili, M.G. Kreĭn, M.S. Agranovich, M.I. Vishik, G.I. Eskin and many other colleagues from the former Soviet Union. A number of (more theoretical) PhD dissertations demonstrate this permanent interest in the field of singular integral equations and operator theory [D1, D2, D4, D5, D7, D8, D11, D13, D14, D16]. Much of that work was presented in the extended surveys [34] and [87].

Furthermore, the preparation of his advanced lectures (see Sec. 2) provided the basis for Erhard Meister's first two books on "Boundary Value Problem in Complex Analysis" and "Integral Transformations", which appeared in 1983 [B1, B2].

3.4. Sommerfeld half-plane problems

As Erhard Meister pointed out many times, he was convinced that the Arnold Sommerfeld paper "Mathematische Theorie der Diffraction" in Mathematische Annalen 47 (1896), 317–374, and its generalizations in connection with modern functional analytical methods were of extreme importance, see e.g. [B1], p. 258, or the introduction of [E4] and some of the survey papers as well, for the following two good reasons: (1) the practical use of the knowledge of the *near and far field pattern* of the solutions, particularly of the *order of edge singularities* and sharp *radiation conditions*, (2) the joy of seeing that the model problems were explicitly solvable by analytical methods. D.S. Jones pointed this out in the introduction of a recent paper: *Analytical solutions to the scattering of electromagnetic waves by dielectric obstacles are few in number. Yet they are of considerable importance because of practical applications and because they serve as benchmarks against which to check numerical procedures. Therefore any increase in the scope of analytical solutions is a useful addition to knowledge* (see D.S. Jones: "Rawlins' method and the diaphanous cone". Q. J. Mech. Appl. Math. 53 (2000), 91–109).

The basic Sommerfeld half-plane problem consists of the solution of the Helmholtz equation in a domain that is the exterior to a half-plane in three-dimensional space together with the traces of the field given on the two banks of the half-plane (Dirichlet conditions).

Mathematically, similar basic problems appear in the diffraction and scattering theory for acoustic, electromagnetic and elastodynamic wave propagation. They can be formulated as mixed boundary and transmission problems or as initial-boundary value problems respectively, on so-called *canonical* domains, with particular configurations (axial-parallel, rectangular, circular etc., see the paper by L. Castro et al.). The studies on the asymptotic behavior of the solutions in the neighborhood of geometric singularities on the one hand, and for low and high frequencies on the other hand, are used in several applied disciplines like radar techniques, computer tomography, material science, geophysics and so on.

A great impact was made by a short paper by A.D. Rawlins entitled "The explicit Wiener-Hopf factorisation of a special matrix" in ZAMM 61 (1981), 527–528, which solved a key problem, namely the Sommerfeld half-plane problem with the Dirichlet condition on one side and the Neumann condition on the other side of the screen, and which had been proposed three years earlier by E. Meister in [28]. Independently E. Meister [36, 41] and also A. Heins in Math. Meth. Appl. Sci. 5 (1983), 14–21, solved the same problem in a similar way (unfortunately, due to the prevailing martial law in Poland, the Banach Center took almost five years to publish Meister's paper [41]). Now the problem is known as *Rawlins' problem* because he had tackled and solved it back in 1975 by an *ad hoc integral representation approach*, cf. [41], Sec. 1.3, and [63], Sec. 3 – without any matrix factorization. See also the introduction of E. Lüneburg's paper in this volume.

A renaissance of diffraction theory was initiated by the new matrix factorization approach in which a number of scientists were involved. Erhard Meister at the TH Darmstadt was at the centre of this group because he knew *both* techniques very well, namely how to work classically (in the sense of the British school, i.e., to understand asymptotic results, e.g., in terms of Abel and Tauber type theorems for Laplace integrals and limiting absorption principles, etc.) and also how to deal with weak formulations in Sobolev spaces (for an adequate solvability theory, well-posedness, etc.); see the survey [63]. In particular, he contributed to the explicit analytic solution of half-plane problems with impedance conditions (jointly with the author [45]), general third kind conditions (with A. Hurd and the author [57]), considering dielectric half-planes (with A.D. Rawlins and the author [66]), higher order boundary conditions (with P.A. Santos and F.S. Teixeira [71]), and their spectral analysis (with N. Latz and J. Scheurer [77]).

3.5. Canonical screen problems in diffraction theory

For the diffraction of linear waves by plane screens and other canonical configurations, boundary value and transmission problems can be reduced to systems of pseudodifferential equations and can often be solved by analytical formulas. One of the most challenging problems was the famous *quarter-plane problem in diffraction theory*, which was explicitly solved (for the first time) in a rather subtle form by infinite products of operators by Erhard Meister and the author in [55], after some irritation in the literature (papers of J. Radlow from the 60's) and also with the help of Moore-Penrose inverses [43, 50]. These contributions found a strong echo in the book by V.B. Vasil'ev on "Wave Factorization of Elliptic Symbols: Theory and Applications", Kluwer, Dordrecht 2000. They also initiated the development of the theory of Bessel potential and pseudodifferential operators on the quarter-plane by R. Schneider [D17], and by R. Duduchava and the author in the early 80's.

Another class of problems in connection with parallel plates was treated extensively (see papers with R.A. Hurd, K. Rottbrand and F.-O. Speck [52, 53, 54, 56, 58, 65, 67, 83, 92] and [D22]). Overview articles for this class of problems appeared in [46, 47, 48, 49, 78], together with further results in diffraction theory. In short, these problems reduce to Wiener-Hopf systems with Fourier symbol matrices of high complexity, containing exponentially decreasing and oscillating functions. Only a few of the matrices can be factorized. However a sophisticated matrix symbol calculus of the general cases led us to results about well-posedness via fixed-point arguments and to asymptotic results as well. Waveguides represent a physically important subclass of these problems which involve surface waves, see [54].

3.6. Elastodynamics

Besides the linear wave phenomena in acoustics, optics and (polarized) electrodynamics, which are governed by the Helmholtz equation, there was much interest in applications to elasticity such as cracks, earthquakes and detection problems. The basic (model) half-plane problems for the Lamé equations as formulated in the well-known book by V.D. Kupradze were considered in [59, 60, 62, 76] where much deeper constructive methods of matrix factorization had to be developed. In this context, the decomposition into shear and pressure waves corresponds to unbounded operators given by projection matrices containing certain poles, which are important from both the mathematical and the physical point of view.

This part of the research was extended by C. Erbe to thermo-elastic phenomena in her PhD dissertation [D20].

3.7. Factorization theory

As mentioned earlier, the papers by A.D. Rawlins, A. Heins and E. Meister, sparked off a 20-year research interest in constructive matrix factorization related to diffraction theory (see [61, 62, 64, 76], the surveys [63, 80] and a remarkable paper with F. Penzel [68] on the reduction of the factorization problem for the difficult part of the class of Daniele-Khrapkov matrices to Riemann problems on Riemann surfaces) – still a challenging topic. Various research groups in the world have been and are still busy with hard open problems from this field. Some general aspects and literature are set out in the recent survey by I. Gohberg, M.A. Kaashoek and I.M. Spitkovsky on "An overview of matrix factorization theory and operator applications" in Factorization and Integrable Systems (Eds. I. Gohberg et al.), Birkhäuser, Basel 2000, p. 1–102.

There is a strong motivation for generalized factorization, since it allows a rigorous operator theoretical approach: a solvability theory in Sobolev spaces and the construction of generalized inverses, which are not so evident from the classical Wiener-Hopf technique. The point is that explicit factorization of non-rational matrix functions is a very difficult subject. E. Meister and the author have contributed to the factorization of the so-called Daniele-Khrapkov matrix functions and those which appear in the elastodynamic theory, mentioned above.

This research also attracted S. Prössdorf from the Weierstrass Institute in Berlin, resulting in a joint paper with the author in 1989 (before the fall of the Wall) and it was very successfully continued by the group of A.F. dos Santos in Portugal with several publications.

One of the most important results is a complete discussion of the *orders of singularities* for Sommerfeld problems, which depend directly on the increase of the so-called *plus factor* in the generalized factorization; see [63], which outlines the results for scalar symbols and for some examples in the matrix case. For a long time it was not clear whether the *edge condition* in general should be formulated in the form $grad\, u \sim r^{-1/2}$ (which is convenient in the study of the Dirichlet and the Neumann problem) or in a different way. The solution of the Rawlins problem shows the surprising asymptotic behavior $grad\, u \sim r^{-3/4}$ which forced the engineers to discuss the conventional square root singularity in the edge condition. Now we know that there are examples of Sommerfeld problems with two different boundary conditions on the two sides of the screen which exhibit a singularity of any order $\delta \in [1/2, 1[$. This can be considered as a consequence of the *finite energy condition* $u \in H^1(\Omega)$, as already suggested by E. Meister in his earlier work.

3.8. Wedge problems and multimedia problems

The *dielectric wedge problem* attracted researchers in diffraction theory as early as the 1930's. Around 1980 Erhard Meister with N. Latz began research on this kind of problem (see [D3]), and also considered further transmission problems for wedges [40, 42]. Multimedia problems, where the space is divided into four rectangular wedges, have a similar structure, which he studied with R. Penzel in [44, 51]. During the visit of Francisco Teixeira to Darmstadt in 1989, a rigorous approach in Sobolev spaces was started and extended in joint work with the *three Franks*, namely F. Penzel, F.-O. Speck and F.S. Teixeira [69, 70, 74, 75, 79] and with A.B. Lebre and F.S. Teixeira [72]. Meister reported on these topics in [73] and, more recently, in [97]. He investigated this class of problems in a wider context in [82, 84, 85, 88] and partly in [86, 87, 96]. It became one of his very favourite topics. His *retirement lecture* in the Darmstadt Mathematical Colloquium on 11 February 1998 was entitled "Ecken und Kanten nah und fern – mathematische Probleme der Streutheorie für kanonische Gebiete" (Edges and wedges near and far – mathematical problems of diffraction theory for canonical domains).

After a formulation of the problems in a Sobolev space setting, they could be reduced to systems of Wiener-Hopf-Hankel equations where certain compatibility conditions for the given data were crucial for the construction of *analytically strong relations* between the associated operators. In several physically relevant cases the authors succeeded in constructing analytic solutions by discovering new composition formulas for Wiener-Hopf-Hankel operators and using recent matrix factorization techniques. Part B of this volume contains an article by L.P. Castro, F.S. Teixeira and the author that extends corresponding joint work with E. Meister and also reports on the present state-of-the-art in this field. However, the complete analytical solution of the dielectric wedge problem is still unknown.

3.9. Spectral theory and time-dependent problems, scattering theory

Many times E. Meister claimed that it is necessary also to solve the spectral problems or *time-dependent* initial-boundary value problems (rather than the *time-harmonic*) in order to understand scattering completely. So he tried to find a way to generalize the results on time-harmonic problems to general time-dependent ones, the so-called *aperiodic problems.* One can say that the latter problems were successfully investigated by the last generation of young people in his group, after quite a long period of research (see the papers on spectral theory [77, 89, 90] and the aperiodic case [81, 94, 95] as well as the surveys [98, 102] and, last but not least, the four PhD dissertations [D23-D26]). All of this work is carefully presented in the report on "Analytical and spectral theoretical investigation of time-dependent scattering problems with canonical semi-infinite obstacles" published in GAMM-Mitteilungen 23 (2000) 93–117 [FP6] – up to including his final work [102] that he presented at the IWOTA in Faro in September 2000 (see the photograph towards the end of part A).

3.10. Quaternion boundary value problems

In his last years, Erhard Meister also moved into the field of quaternionic analysis. In this area, in collaboration with his wife Ljudmila, he made interesting observations particularly in electromagnetic theory, and also for canonical boundary-transmission problems in elastodynamics and visco-elastodynamics [100, 101]. These new formulations led to new results on solvability and contain some key ideas for further analytical treatment.

4. Research projects

The research projects undertaken by E. Meister strongly stimulated the progress of the group. They helped to sponsor foreign scientists and to support young researchers. Let us look at some of the most important enterprises in the different areas of research.

The Deutsche Versuchsanstalt für Luft- und Raumfahrt (DVL) provided the first and most important financial support for E. Meister personally during the period 1956–58 whilst he prepared his PhD dissertation, and until the mid-60's he gratefully acknowledged this support in various papers up to the publication of his habilitation thesis in the series of DVL reports [16–18]. Besides this aid, he received a grant from the European Office of the US Army [3, 6, 7, 8, 10, 11] due to the practical relevance of his research. It seems that this experience convinced him that he had a duty in his turn to provide substantial support for his young collaborators later in his life (and to ask them to treat real world problems).

Some of the most active DFG projects were:

- Instationäre Triebwerkströmungen. DFG project Me 261/3, 1975–1979. See [30, 32, 37] including H.-J. Sommer, G. Buggle and M. Jirmann.

- Nichtklassische Wiener-Hopf-Probleme mit Anwendungen in der Beugungstheorie. DFG project ME 261/4. See various publications in 1985–89 [48, ..., 60] with partial support for M. Costabel, F. Penzel, R. Penzel, R. Schneider, full support for the author and collaboration with N. Latz and E. Lüneburg.
- Non-classical Wiener-Hopf problems with applications in fracture mechanics. Subproject in the Forschergruppe Ingenieurwissenschaftliche und mathematische Analyse bruchmechanischer und inelastischer Probleme. DFG project KO 634/32, 1988–96 with support for young members of the group like K. Rottbrand [67, 83, 92], F. Penzel [68], J. Mark [81] and two visits by F.S. Teixeira in 1989/90, see [69–72, 74, 75, 79, 97] partly published with A.B. Lebre, P.A. Santos, F. Penzel and the author.

The cooperation with the Portuguese colleagues was continued through two projects jointly undertaken with the author and directly supported by the German Academic Exchange Service (DAAD), the Federal Ministry for Research and Technology (BMFT) and the Portuguese National Science Foundation (INIC and JNICT, respectively):

- Wave diffraction phenomena, Sobolev space methods, and approximation theory. INIDA project 311 sponsored by DAAD and INIC (Portugal) in 1991–92 with exchange visits by E. Meister, B. Silbermann, F.-O. Speck and members of their groups like C. Erbe, T. Ehrhardt and others already mentioned before.
- Singular operators – new features and applications. JNICT/BMFT Project no. 423/1 in 1995–98 with exchange visits by L. Castro, J. Mark, A. Passow, A. Moura Santos (various publications resulting in four PhD dissertations) and F.-O. Speck.

Another DFG project should be mentioned in connection with the research described in Sec. 3.9:

- Analytische und spektraltheoretische Untersuchungen zeitabhängiger Streuprobleme an kanonischen halbunendlichen Objekten. DFG project ME 261/13-1, 2, duration 1997–2000. See the detailed report [F6] and papers, e.g., with A. Passow and K. Rottbrand [94, 95, 98] and Meister's last publication [102].

Some members of the group were supported by single applications like R. Hinder, sponsored by the VW foundation and the DFG [91, 99] and M. Soliman and B.-J. Jungnickel [93], who were sponsored by the German Federal Ministry of Trade and Commerce.

5. Guests, international contacts

Convinced of the importance of international contacts, E. Meister promoted many visits by foreign colleagues to Germany, especially from the Eastern bloc. He and his guests enabled many young people to widen their scientific horizons by mutual exchanges.

Some of the most important contacts were connected with three Alexander von Humboldt awardholders proposed by Erhard Meister: R. Duduchava, I. Gohberg and V. Maz'ya. We refer the reader to the corresponding reminiscences that are included in the next part of the book, which contains a variety of contributions with quite different points of view. Naturally, they cannot reflect all the relations that existed, so that some complementary information can be given here from the author's memory (in addition to Meister's report of 1993).

There was for instance E.L. Koh from the University of Regina, Saskatchewan, Canada, who spent the year 1975–76 in Darmstadt on a National Research Council Travel Fellowship. As a specialist in integral transformations, he influenced the work of E. Meister with G. Buggle [32, 37, 38] and stimulated contacts with Canada, see Sec. 7.

M. Shinbrot from Montreal (later he moved to Victoria) stopped off in Darmstadt in May 1979 and talked about "Difference kernels". The discussion was very lively and had some impact on our work on "General Wiener-Hopf Operators" during the following years; see [43], for example.

A. Heins from Ann Arbor, Michigan, was invited to Darmstadt in the summer of 1980 and gave a lecture course on "Some representation theorems for diffraction problems". He was one of the first who thought about matrix factorization taking the logarithm of a matrix by series expansion (there is a remark in a 1950 paper by him on "Systems of Wiener-Hopf equations…"). The idea gained some importance later in the context of Daniele-Khrapkov and Jones form matrix functions.

Three DFG fellowship visits of one year each took place in the 80's (see also the reminiscences section): D.C. Pack came in 1980–81 to continue work on the Boltzmann equation, and also with his contacts in the Mechanics Department of the TH Darmstadt. R.A. Hurd spent the academic year 1986–87 in Meister's group with nice results on diffraction problems [53, 54, 57, 58] and an interesting review on "The explicit factorization of 2×2 Wiener-Hopf matrices" which appeared only as a departmental preprint no. 1040 in March 1987, but attracted a remarkable amount of attention in the Darmstadt group and its scientific vicinity.

The stay of A.D. Rawlins in 1988–89 occasioned a useful comparison of the methods of the British school and the operator theoretical approach resulting in two joint papers [65, 66].

S. Prössdorf from the Karl Weierstrass Institute of the Academy of Sciences in (East) Berlin was invited for *half a year* in 1988 and, after a lengthy correspondence with the GDR Foreign Ministry, he succeeded in carrying out his first visit to West Germany in March 1988 – for *one month.* He gave various talks in Darmstadt, Bonn, Stuttgart and other places recommended by E. Meister. The visit was the basis for an intensive *international relation* that unexpectedly turned into a *domestic* one two years later. Despite having no time, he succeeded in publishing a well-received paper about matrix factorization together with the author in the summer of 1989, a few months before the fall of the Berlin Wall.

Meister with Prössdorf and Speck in Feb. 1998

After the political liberalization at the beginning of the 90's, many contacts grew rapidly with colleagues from the Eastern bloc countries as reported in various reminiscences. Relatively few of these resulted in joint publications, an exception being the collaboration with S. Nazarov [99] who spent one month at Darmstadt in 1997. On the other hand, the author's departure for Portugal yielded a high number of projects and publications as a natural continuation of existing activities reported above; see also the reminiscences of A.F. dos Santos and F.S. Teixeira. Various existing contacts of E. Meister were extended and continued to Portugal and France, e.g., with R. Duduchava; see the list of his publications (now easily accessible via the internet) with M.A. Bastos, A.F. dos Santos, L. Castro and the author, and with M. Costabel and M. Dauge and others.

It is impossible to mention here all the guests of E. Meister – just a brief glance into the files of the 80's gives some impression of their variety: we can find names like B. Bojarski, L. Boutet de Monvel, L. Coburn, T. Ha Duong, A. Dzuraev, W.N. Everitt, J. Giroire, A.T. de Hoop, M.A. Kaashoek, H. Langer, R.C. MacCamy, G. Mendoza, M.W. Pautschko, A. Piscorek, D. Przeworska-Rolewicz, B.-W. Schulze, M.A. Shubin, S.R. Simanca, I. Spitkovsky, J. Wolska-Bochenek and many, many others.

6. Supervision, PhD students and habilitations

The list of Meister's PhD students can be divided into five generations originating from Saarbrücken (no. 1–5), Berlin (6–8 and 13), Tübingen (9–12 and 14) and Darmstadt – the latter split into three groups (15–19, 20–22, 23–26). It is interesting to compare the topics of the seminars, theses and publications, see, e.g., the reminiscences of N. Latz, and to observe the steady progress within the group, guided in a fatherly way by Erhard Meister.

Being a humorous man, he started during his time in Tübingen to talk about filling the alphabet with his students: Alber, Buggle, Costabel, Donig, Eisenhardt, Frank (Speck), after having already started further along the alphabet with Gerlach, Hery, (I, J missing), Kremer, Latz. He did not succeed in getting all the letters, but he did in fact get 26 PhD students – the number of letters in the German alphabet.

There are also a considerable number of habilitations among group members, which might be mentioned here since E. Meister helped their progress:

N. Latz, TU Berlin 1975, now prof. at the TU Berlin
H. Sohr, Tübingen 1975, prof. in Paderborn
H.-D. Alber, Bonn 1982, prof. in Darmstadt
J. Donig, Darmstadt 1982, prof. in Duisburg
F.-O. Speck, Darmstadt 1984, prof. at the TU Lisbon
M. Costabel, Darmstadt 1984, prof. in Rennes
J. Gwinner, Darmstadt 1989, prof. in Munich
W. Varnhorn, Darmstadt 1992, prof. in Kassel
P. Deuring, Darmstadt 1993, prof. in Calais
F. Penzel, Darmstadt 1993, PD in Darmstadt
R. Schneider, Darmstadt 1995, prof. in Chemnitz
F. Ali Mehmeti, Darmstadt 1995, prof. in Valenciennes
K. Rottbrand, Darmstadt 2000, consultant in Cologne

Most of them are working as professors in various universities, some in polytechnic schools like M. Kremer, who became rector of the Fachhochschule Darmstadt as long as twenty years ago. T. von Petersdorff went to the United States directly after his PhD exam and is now professor at the University of Maryland at College Park.

7. Sabbaticals, journeys, participation in conferences

There is no systematic report on traveling activities in E. Meister's CV although he enjoyed meeting his colleagues all around the world. So this part of these notes highlights only a few of his trips (referring to various reminiscences again).

The first conferences he visited were the GAMM meetings from 1958 onwards (see R. Leis, D.C. Pack), and a number of E. Meister's publications appeared in the Zeitschrift für Angewandte Mathematik und Mechanik (ZAMM), being one of the principal journals of applied and engineering mathematics. Later he was invited repeatedly to deliver a main lecture at the annual meetings of the Society. Thus he spoke about "Ein Überblick über analytische Methoden zur Lösung singulärer Integralgleichungen" in Graz in 1976 and about "Operatorfaktorisierung in der Beugungstheorie für kanonische halbunendliche Gebiete" in Dresden in 1993.

The invitation by D.C. Pack to spend a sabbatical year as a guest professor at Strathclyde University in Glasgow represented his first opportunity to work outside Germany. The stay took place in the summer of 1966, but only for one semester, because of the professor position at the TU Berlin that he was offered at the same time (see N. Latz, D.C. Pack). A two-hour lecture course on "Hilbert's boundary value problem with applications" was delivered for the first time, forming the basis of one of his favorite future lectures.

In the late 60's the famous Mathematisches Forschungszentrum at Oberwolfach in the Black Forest started to organize conferences on "Methoden und Verfahren der mathematischen Physik" initiated by B. Brosowski and E. Martensen who published the proceedings under the same title in a new book series of the Bibliographisches Institut Mannheim. E. Meister immediately belonged to the inner circle and profited a great deal from the corresponding contacts (see his report and R. Leis).

He participated in some thirty conferences at Oberwolfach and in at least the same number of national and international meetings, frequently as invited main speaker. Among them one can find various important conferences in Eastern Europe (see his report), e.g., in Kozubnik, Poland, in 1977 [29] and in Tbilisi, Georgia, in 1979. He attended the Toeplitz Memorial Conference in Tel Aviv in 1981 [35], the IWOTA conferences in Amsterdam '85, Rotterdam '89, Regensburg '95 and Faro in 2000, the tenth Dundee conference [63], and the Calgary conference on the occasion of the 70th birthday of I. Gohberg in 1988 [61], the International Workshop on Direct and Inverse Electromagnetic Scattering at Gebze/Istanbul in 1995 [85], and the ICM in Peking in 2000 [100] – just to mention a few milestones.

The second sabbatical term in February–March 1978 served for a quite busy journey to Canada. It included visits to E.L. Koh and W.D. Montgomery in Regina where he gave a two-hour lecture course on "Singular integral equations of convolutional and Cauchy type", another visit to R.F. Millar in Edmonton, Alberta, and further talks in Saskatoon and Vancouver. On the way back he participated in the Fifth Conference on Differential Equations in Dundee, Scotland [34].

In the spring of 1981, E. Meister was an invited speaker at the XVII Banach Center scientific meeting on Mathematical Methods and Models in Contemporary Mechanics [41], see Sec. 3.3. Another sabbatical term was spent at the University of Delaware from September to December 1981 where he prepared a paper with N. Latz [40].

Finally, it should be mentioned that E. Meister spent several one-month stays in Oberwolfach being a good and quiet place for research and project preparations, within the framework of Research in Pairs (RiP), together with the author in 1985, 1987 and 1998.

8. Organization of conferences, workshops and colloquia

"Function Theoretic Methods for Partial Differential Equations" was the title of the first international conference organized by E. Meister jointly with N. Weck and W. Wendland in Darmstadt in 1976. It brought together some of the great pioneers of the area such as S. Bergman, W. Haack and I.N. Vekua, who met for the first time in their lives, and made possible exciting discussions between A.W. Bidzadse and R.P. Gilbert, G. Fichera, V.D. Kupradze, B. Bojarski and many others. The symposium represented one of the most interesting East-West meetings in the area at that time. The proceedings were published in the Springer series Lecture Notes in Mathematics [E1].

Oberwolfach saw various conferences organized by E. Meister: "Integral Equation Methods in Engineering" jointly with W. Wendland in 1980, "Mathematical Methods in Fluid Mechanics" with K. Nickel and J. Polasek in 1981 [E3], "Wiener-Hopf Problems with Applications" with I. Gohberg and M.A. Kaashoek in 1986, "Practical Treatment of Integral Equations, Boundary Element Methods and Singular Integral Equations" with G. Hämmerlin and W. Wendland in 1988, "Wiener-Hopf Problems, Toeplitz Operators and Applications" with I. Gohberg and M.A. Kaashoek in 1989, and "Singular Integral and Pseudo-Differential Operators and its Applications" with S. Prössdorf and B. Silbermann in 1994. As one can see, integral equations and operator theory were the main topics of these meetings, alternating between pure theoretical and very practical themes. Also the East-West exchange was very strong as highlighted by the reminiscences of A. Böttcher, B. Silbermann, L. von Wolfersdorf, and the colleagues from Prague, for instance.

As early as the 80's a very clear plan had been conceived by E. Meister: to celebrate the 100th anniversary of the publication of Arnold Sommerfeld's famous paper on diffraction theory mentioned before in Sec. 3.4. His plan was carried out at the conference on "Modern Mathematical Methods in Diffraction Theory and its Applications in Engineering" in Freudenstadt/Black Forest in 1996 with eighty participants. In the introduction of the proceedings [E4] we find a passage which sounds like a confession: *The aim was to bring together experts from all over the world with young scientists so as to encourage them to cooperate in the fields of microwave-propagation and scattering as mathematicians, physicists, and engineers following the lines of Sommerfeld's ideas. He was one of the first people to formulate diffraction problems as boundary value problems for canonical domains, like half-planes and wedges. Many analytical and numerical methods sprang up from his ideas, but a lot of unsolved old and new problems still exist.*

E. Meister also organized plenty of small workshops in Darmstadt and other places, sometimes informal and ad hoc, if there was a group of people around with similar interests, see, e.g., [F2]. He had a strong desire to stimulate the exchange of ideas for the progress of science without thoughts of personal gain. In the 90's some of the meetings were jointly organized with colleagues from the DFG Forschergruppe mentioned in Sec. 4. Also the success of his last big project was celebrated by three workshops in 1997–2000 as reported in Sec. C.1 of [F6].

At least two very special meetings should be mentioned that were organized *to honour* E. Meister. The 60th birthday colloquium at the TH Darmstadt in February 1990, jointly dedicated to his departmental colleague Willi Törnig who was born one day earlier, gathered together two hundred guests from all over Germany and surrounding countries, including the first president of the European Mathematical Society, Friedrich Hirzebruch.

Second, the workshop on "Analytical Problems of Continuum Mechanics" on the occasion of E. Meister's "Emeritierung" (retirement) in February 1998 was carefully prepared by H.-D. Alber and his colleagues and brought together the majority of his pupils, friends and colleagues in the so-familiar lecture rooms of the (re-named) Darmstadt University of Technology. E. Meister gave a colloquium talk that was announced by the poster of Prof. K.-H. Hofmann shown in the next two pages. Some photos of that meeting are included in this volume as well as the exciting ceremonial address by Douglas Jones (E. Meister liked it so much) and parts of the author's lecture on "Some operator theoretical aspects in diffraction problems" within this short biography.

In June 2003 there was another workshop to honour E. Meister entitled "Operator Theory in Analysis and Engineering" which took place at the Darmstadt University of Technonlogy with the proceedings to be published in Mathematische Nachrichten.

9. Editorial and referee work

E. Meister was a member of the editorial boards of eight scientific journals and periodicals; see his CV. He was founder member of Mathematical Methods in the Applied Sciences in 1973, Integral Equations and Operator Theory in 1977 and of the Birkhäuser book series Operator Theory: Advances and Applications in 1979, and belonged to the editorial boards until his death. After the unification of Germany, he played a substantial role saving Mathematische Nachrichten as one of the leading journals of mathematical analysis in Germany.

Moreover, he was involved in a lot of other extracurricular activities. Thus he was chairman of the GAMM committee on Applied Analysis and Mathematical Physics, worked as elected vice- and main referee for applied mathematics for the DFG and for several years was a member of the DFG committee "Sonderforschungsbereiche".

In addition he acted many times as second referee in doctoral exams in mathematics and engineering, and wrote countless reviews for Zentralblatt and Mathematical Reviews, letters of recommendation and opinions for all possible purposes.

10. Seminars and life in the group

I am not a collector Erhard Meister used to claim meaning stamps and things like that. But mathematics was a different story; witness his interest in collecting PhD

K.H. Hofmann poster

students. In his house, you can still find tens of thousands of pages of notes written down in seminars and conferences. He always took as many notes as possible in lectures – complaining about the use of overhead projectors when they appeared. On being asked why he did that, even when there existed manuscripts of the talk, he answered with a smile "I am afraid of falling asleep", but the truth was that he always liked to take with him as much as he could, extending his memory (which was excellent anyway). Sometimes he would say in a discussion: Yes, this question we considered some 15 years ago after a talk of Prof. X ... let us have a look at my notes.

All his life he ran a busy research seminar for collaborators, postgraduates, occasional guests etc. in addition to the ordinary seminars for students. Moreover he initiated and supported interdisciplinary joint seminars with colleagues from other departments, e.g., with K.P. Hadeler (Biomathematics) at Tübingen in the early 70's. There were mutual seminars with P. Werner from Stuttgart, e.g., with lecture courses by C.M. Wilcox who was there on sabbatical leave in 1974. Prof. Wilcox was a specialist in scattering theory, spectral theory and waveguides, topics that were later treated intensively in Meister's group. He used to appear with a sporting outfit as E. Meister on walking tours.

K.H. Hofmann poster (2nd part)

In Darmstadt Erhard Meister started having contacts and joint seminars with G. Piefke (Electrotechnics) in the mid 70's. Later he founded the Colloquium on Fluid Mechanics, jointly organized by E. Becker, E. Meister, K.G. Roesner and J.H. Spurk in the Department of Mechanical Engineering at the second campus of the THD called "Lichtwiese", which existed for some twenty years. Prof. Spurk became the supervisor of E. Meister's son Olaf. In the 90's, the DFG "Forschergruppe" created the "Kolloquium Festkörpermechanik" (Colloquium on rigid body mechanics) organized by Meister, Spellucci and Törnig (Mathematics), Gross and Haupt (Mechanics) and Kollmann, Schmöckel and Wissmann (Mechanical Engineering), constituting the largest interdisciplinary seminar, bringing together seven research groups.

E. Meister's universal interest was reflected in two habits: always taking notes and asking questions after every talk; see some quite original observations in the reminiscences of L. Meister, I. Gohberg and M.A. Kaashoek. His liking for discussions also resulted in the daily lunches with any colleagues and assistants at hand at 12h30 sharp. In Berlin's restaurants near "Savigny Platz" one could hear the melody of the Saarland language, in Tübingen, the inn "Die Rose" at Lustnau benefited from a little group discussing in Berlin slang and in "Bayerischer Hof" in Darmstadt, Swabian and Hessian sounds mingled with remnants of Saarland and Berlin tones.

One nice tradition was to have supper with the guests after their seminar or colloquium talks in an Italian restaurant, which suited every taste, and often the groups went on talking till midnight. The younger colleagues created a Friday afternoon beer circle in "Hotzenplotz" or some "Biergarten", and many guests also frequented and enjoyed the private weekend parties where everybody was welcome and university life mingled with real life. Jokes from Tbilisi, Warsaw, and Israel supplemented the pleasure of German specialities like sausages, garlic butter, local beer and Alsace wine, and sometimes the other way around, stories from the *place we met before* were conjured up by the taste of Pilmeini, Adjika and Polish Wodka.

Once or twice a year E. Meister asked his assistants to organize a work outing for his group called *AG 12* (Arbeitsgruppe 12) at Darmstadt. It was usually a hiking tour – one of his favorite sporting activities as everybody realized from his clothing, cf. the memories of R.A. Hurd, Meister's son Olaf and many others. Sometimes friendly groups like that of W. Jaeger from Heidelberg (see the photo) and, of course, the *AG 6* (of Prof. Wendland) were invited to take part in the tour. Some people therefore spoke of *AG 18*. Typically this ended after a four-hour walk in a nice inn where the discussions continued in more comfortable surroundings. The Oberwolfach *walking and talking tours* on Wednesday afternoons were most pleasant and successful when Erhard Meister was the guide – a master guide (nomen est omen).

Meister's marriage with Ljudmila brought university life to his home more and more. Various colleagues were invited to stay there during their visits; see the comments by V. Maz'ya, S. Nazarov and D.C. Pack for instance. And many times we met there with his guests, colleagues and friends enjoying a little concert in the large living room or sitting in the garden and talking about the sense and pleasure of doing mathematics.

Acknowledgement: The author would like to thank Ljudmila Meister and Olaf Meister and his family for their friendly support during the preparation of this short biography. It was a great help to have the list of publications and PhD dissertations and to see the collected works that was provided by Steffen Roch. The suggestions of the editors are gratefully acknowledged as well as the proof-reading by Mrs. Susan Picken and comments from various colleagues and my wife Gaby.

Departamento de Matemática
Instituto Superior Técnico
Av. Rovisco Pais
1049-001 Lisboa
Portugal
e-mail: fspeck@math.ist.utl.pt

Operator Theory
Advances and Applications, Vol. 147, 47–61

Is Electromagnetism any Use?

Douglas S. Jones

It is a great pleasure for me to have this opportunity of honoring Professor Meister on this special occasion. He has made many important and valuable contributions to the theory of electromagnetism as you will have gathered from the fascinating lecture by Professor Speck and from yesterday's and today's workshop. However, it is not my task to describe again the power of Erhard's mathematics but to try to explain the significance (if any) of electromagnetism for life today. I shall attempt to do this by tracing how it has developed over the past century or so and the impact of those developments on society – a sort of potted history in non-technical terms. This will mean that I shall be omitting the technical details of the contributions of many distinguished theorists and, regrettably, that includes Professor Meister. Anyhow he is far too active to be consigned to the history books yet.

Tracing the origins of electromagnetism is like finding the source of a river with many tributaries because numerous international scientists are involved. Some would say that it started about 130 years ago with that well-known Scotsman, Maxwell, but I think that the beginning is much earlier. There is one form of electromagnetism with which we are all familiar, namely, light. And some 300 years ago a Dutchman Huygens explained how waves of light behaved. His house is still preserved in the neighborhood of Den Haag. Mind you, a Dutchman once told me that everywhere in the Netherlands is in a neighborhood of Den Haag. I guess he did not come from Friesland.

I shall now describe Huygens' theory so that you can discover all you need to know about waves and perhaps a good deal more than you ever wanted to know. What Huygens said was that, if a wave is started from a source, it will begin to spread at a certain speed and after a time will reach position 1 (Fig. 1); it will move on coming to position 2 eventually and keep going. The sound from my mouth is doing exactly that, reaching you row by row, and is covering a kilometre roughly every three seconds.

Some people say they can detect that the sound from the brass and strings of a large orchestra arrive at slightly different times though I cannot. Another illustration that waves travel at a definite speed is the time you have to wait for an echo in a valley with steep sides. Bats, by sending out pings at regular intervals, use this property to avoid obstacles; they can distinguish between two objects separated by the width of a pencil line.

Lecture on the occasion of the retirement of E. Meister on 13 February 1998 at Darmstadt.

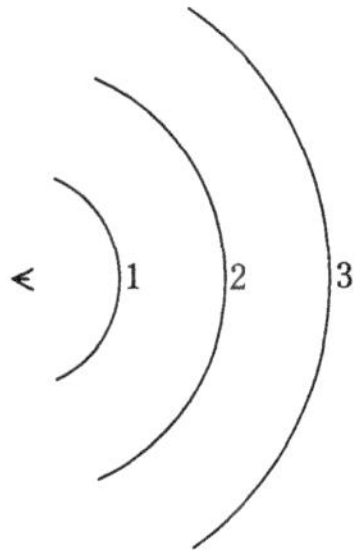

FIGURE 1

Not all waves travel at the same speed; electromagnetic waves travel very fast. When you look in a mirror you see yourself at once; yet the light has gone from your body to the mirror and back again to your eyes. Thunderstorms are a good example, when a discharge takes place you see the lightning immediately but the thunder comes later; every 3 seconds delay adds a kilometer to your distance from the flash. So if you want to know how far away the storm is you count the number of seconds between the flash and first rumble, then divide by 3. Gives you an idea of when to take cover or start running.

There is a useful practical application. If you measure how long it takes for a radio signal to go to A and back again you know how far you are from A. (Fig. 2(a)). If you do the same with B you can pinpoint your position if you know whether you are north or south of AB (Fig. 2(b)). Actually it is more reliable to have a third station so that you get a precise fix without worrying whether you are north or south. Applying a bit of mathematics you get a method that was used to navigate aircraft in the Second World War. Then if you are living after 1957 the idea occurs that wider coverage would be achieved if you put A in the sky on a satellite and eventually you come up with a system which, today, enables you to locate your position on the earth to within a meter or two by electromagnetic

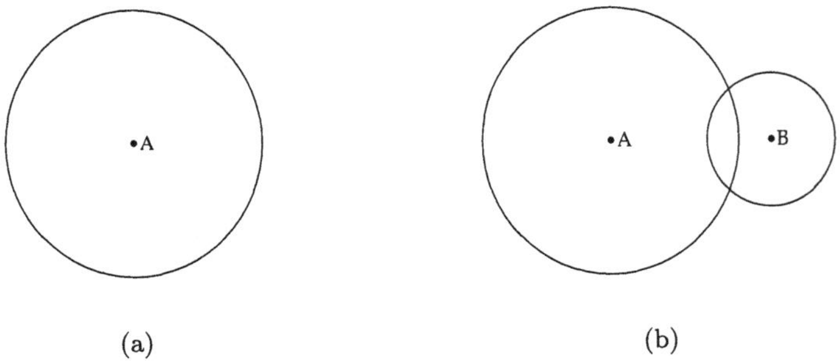

FIGURE 2

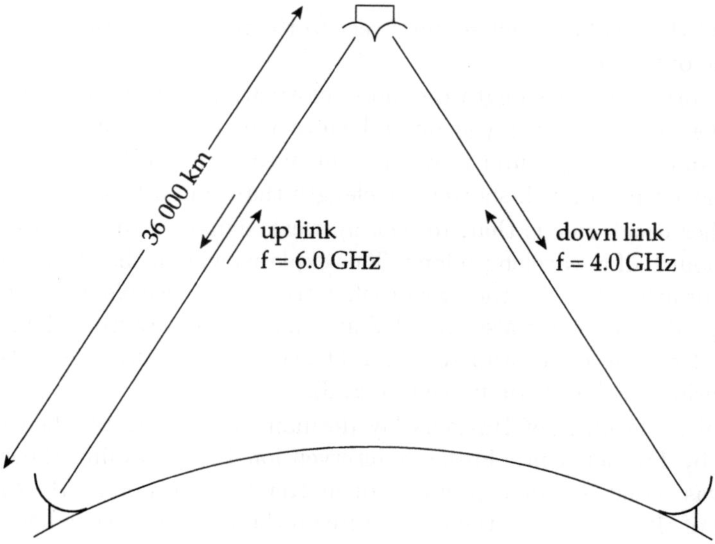

FIGURE 3

waves (Fig. 3). Once such links are established why not use them for the telephone – the Inmarsat Mini-M service does that though the phones cost some 4000 marks at the moment and then you can call from anywhere in the world to anywhere in the world for 4 marks a minute. An offshoot is a gadget called MoBIC produced by British Telecom and the German company FHP which tells a blind man his position and route to his destination, and warns him if he deviates from the route. Perhaps there'll be a button to let others know where he is.

Now, if you could see the sound wave as it went by what would you observe? Roughly speaking it would go up and down; the distance between the peaks is called the wavelength (Fig. 4).

Middle C on a piano has a wavelength of about $1\frac{1}{3}$ metres, top C is about 8 cm. The wavelength of an average voice is about 2 m varying from 4 m for low voices to 20 cm for high-pitched speech. Some ears can hear wavelengths as short as 2 cm and as long as 20 m. Compare the size of the ear, but remember that

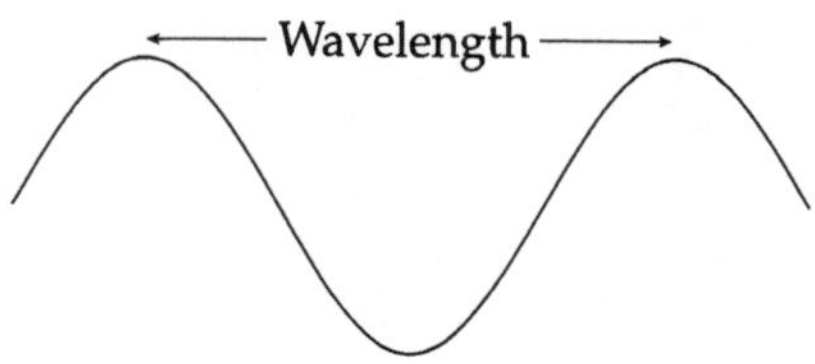

FIGURE 4

the bones of the skull conduct sound direct to the inner ear without having to go through the outer ear.

The shorter the wavelength the more information a wave can carry. Pictures speak louder than words; if you don't believe me try describing a painting by Picasso or an etching by Dürer in words. This means that TV waves, which carry pictures, must have a much shorter wavelength than sound broadcasts.

Another rough rule is that, to pick up weak signals, a rod or wire is needed which is about half a wavelength long. This explains why in the 1920's houses had long wires draped around – the broadcasts were on long wavelength and not very strong (Fig. 5). It explains also why TV antennas some way from the station are composed of rods a few centimeters long at least in countries like Britain where cable television has had little impact (Fig. 6).

Actually the ideas of Huygens lay dormant for 150 years when they were picked up by Fresnel. Until Fresnel's intervention the prevailing theory, due to Newton, was based on small particles of matter floating about. By the time of the revival of Huygen's ideas the next figure on the historical scene had appeared.

FIGURE 5

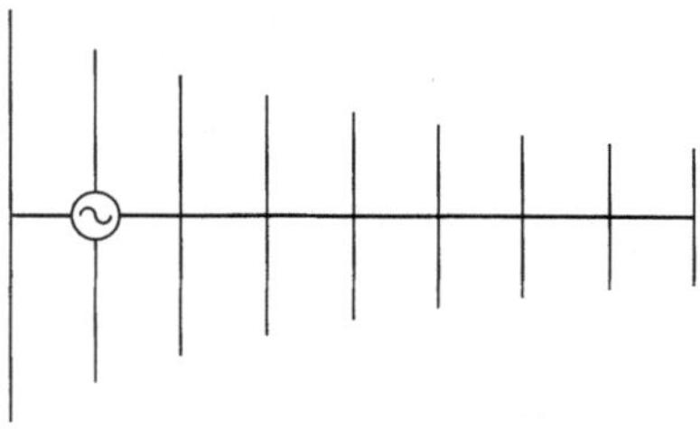

FIGURE 6

That was Faraday at the University of London. He has been called the father of electricity and the bicentenary of his birth occurred about 6 years ago. When the Finance Minister asked him what the value of science was he replied "Whatever it is I am sure you will find a way to tax it". By the middle of the last century Faraday's researches had led to the creation of the telegraph which sent messages by Morse code. The Scotsman Bowman Lindsay of Dundee, called by Marconi the father of wireless telegraphy, had made transmissions in 1853. It would be another twenty years or so before the microphone and telephone were invented by Bell.

This is where Maxwell comes on the scene, though his entry was by accident. By the way, in addition to his scientific achievements, Maxwell also wrote humorous verse and a collection of his poems has been published by Markings Heritage. Maxwell had been educated at the Universities of Edinburgh and Cambridge where he was taught by the same tutor as taught Lord Kelvin, Stokes, Cayley, Tait to mention a few eminent scientists and rated best of the lot (in those days the Cambridge examination was 16 papers each of 3 hours taken on 8 consecutive days – I wonder what today's students would think of such an exam). At the age of 25 he was appointed Professor of Natural Philosophy at Marischal College in Aberdeen, Scotland. Here he made an outstanding reputation with contributions in astronomy, color vision and the theory of gases.

Aberdeen also housed King's College and in 1860 the government decided on an economy measure – the British government never seems to have lost the habit of economising on higher education; perhaps the German government is wiser. The economy measure was to combine the two colleges into a single institution, the University of Aberdeen, and halve the number of academic posts. This was fierce even for a city like Aberdeen, reputed to be very careful with money – inhabitants don't like using blinkers on their cars in case they have to replace a bulb. There were two Professors of Natural Philosophy; one had to go and, politics being what it is, it was Maxwell who went. He tried for Edinburgh but Tait was appointed. Eventually, he found a position as Professor at King's College, London.

In London he had Faraday as a colleague and the regular contact caused him to take a serious interest in electromagnetism. At the time the governing equations were in skeleton form (Fig. 7).

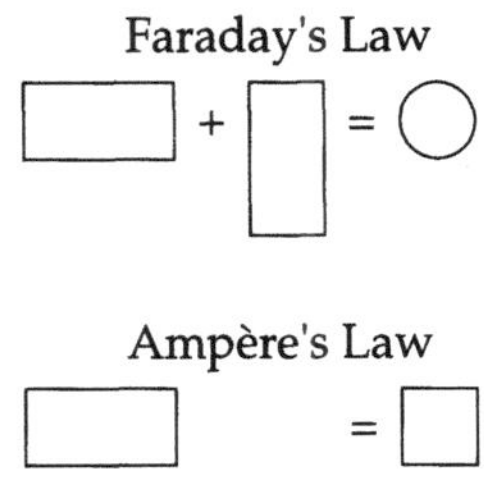

FIGURE 7

Faraday's Law explains how dynamos, electric motors, hairdryers and electric razors work. Talking of hairdryers, when I checked in at the hotel where I am staying the receptionist said that a special room had been reserved for me. So I asked her "What's special about this room?" She replied "It has a hairdryer."

Ampère's Law indicates that current produces magnetic field.

What Maxwell said, in a highly simplified argument, was: according to Faraday electricity and magnetism occur together but there is no tall box in the second equation so we must add one to get Fig. 8

For those who are worried that there is no square in the first equation, you can add one but it is an optional extra. For the record, Maxwell's equations are

$$\text{curl}E + \frac{\partial B}{\partial t} = 0$$

$$\text{curl}H - \frac{\partial D}{\partial t} = J$$

Undergraduates are offered a simple explanation (somewhat deeper than the one just given) for the addition, but is well to recall that, in his original paper, Maxwell listed some 20 reasons for adding the box. Maxwell's arguments were purely theoretical; no practical method of generating electromagnetic waves existed then and, indeed, 20 years were to elapse before their production.

For the next step we have to move to Berlin and 1879. In Berlin the big chief was Helmholtz (the same Helmholtz as is associated with one of the notations for musical scales). Helmholtz did not hold with Faraday's theories and therefore distrusted Maxwell's equations. He had his own rival theory. Under his supervision was a doctoral student named Hertz.

Helmholtz gave Hertz a problem which, in essence, would decide between the two opposing theories. Funny sort of supervisor. I mean if a supervisor asks a graduate student to decide between the supervisor's theory and a rival's you know what the answer must be. Hertz worked on it for a couple of months, concluded it was too hard and returned to completing his doctoral thesis. Shortly afterwards, he moved to Kiel as a Privatdozent but he found the laboratory facilities so inadequate that he could not carry out his research properly. Fortunately, in 1885, he was appointed Professor of Physics at Karlsruhe and here he flourished. Using a wavelength of about 24 cm he established the speed of electromagnetic waves,

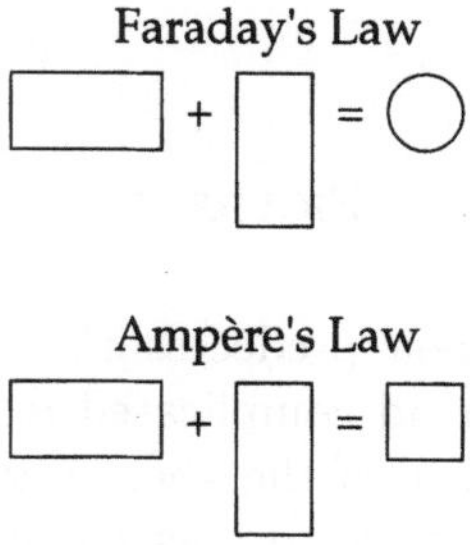

FIGURE 8

constructed dishes like those on today's satellite receivers and made a cavity resonator. He described rays, diffraction, polarisation and refraction; solved the dipole problem and found its far field. All this in the space of some 3 years – a truly staggering achievement. Do you think he ever found time to sleep? He must have had as much energy as Professor Meister. His conclusion was and I quote "The results of the experiments on rapid electric oscillations which I have carried out appear to me to confer upon Maxwell's theory a position of superiority to all others". While there is controversy over who first produced electromagnetic waves there can be no doubt that the remarkable Hertz placed everything on a firm footing both theoretically and experimentally. His name is in constant use today when we say that the electricity supply is 50 Hz or a computer operates at 200 MHz.

The translation of Hertz's laboratory experiments into a practical operating system was accomplished by Marconi. If this meeting had been held 2 months earlier we would have been celebrating the centenary of the first wireless message sent outside a laboratory. Marconi had moved to England because the Italian government did not think it was worth financing wireless transmission. In December 1897 he succeeded in sending a message in Morse by wireless from the English coast to a ship some 5 km away.

In the meantime the theorists had not been idle.

Following Hertz they had set to work to find out that Maxwell's equations predicted about electromagnetic waves in a variety of circumstances. I shall mention only some of the relevant names: Poynting, Kirchhoff, Helmholtz, Heaviside, Rayleigh, Sommerfeld, Poincaré and MacDonald – many of the papers are still worth reading if you can get hold of a copy.

I should mention a couple of things about that great man Sommerfeld. First, I have been in this business so long that some people think I was a contemporary of Sommerfeld at university. I was not; he attended a few years earlier. The second item is his legacy of what I call Sommerfeld's ghost wave. In his famous paper on propagation over the earth there is a misprint in the final formula so that a sign is printed incorrectly. With the incorrect sign two terms combine to produce a wave; with the right sign the terms cancel and there is no wave. That is why I call it a ghost wave. Subsequent authors did not bother to read the paper –

FIGURE 9

do today's writers read what their predecessor's have written I wonder (it saves a lot of trouble not having to read complicated mathematics) – quoted the final result and predicted the existence of the ghost wave. At the time apparatus was too primitive to decide the issue and it was not until 1935 that Norton pointed out the misprint and gave the proper interpretation. Norton's predictions were confirmed experimentally by Burrows in 1937. You would think that settled the matter but the ghost wave continues to haunt at regular intervals – only last year I was involved in an argument with some authors who believe in its existence – plus ça change, plus c'est la même chose.

As I have remarked early apparatus was primitive and some of the massive set of theoretical results had to wait 30 years before verification, for example, by Savornin's experiments on razor blades and Burrows' experiments on earth effects. In 1908 came the solution by Mie of what happens when an electromagnetic wave hits a sphere or ball. This was important because it explains the color of a sunset, the formation of a rainbow and how radar is scattered by rain thus enabling modern aircraft to carry a radar which detects the presence of storm clouds.

Another thing which Mie's solution tells you is how a wave propagates from one point of the earth to another since the earth is a sphere. The theorists concluded that the wave did not penetrate beyond the visible horizon (Fig. 9). This was a disaster because Marconi had managed already to send signals between Europe and Newfoundland. Could Maxwell's equations be wrong? The resolution took 10 years. It was nearly 1920 before Watson and White showed that the theorists had made a mistake in their sums and that waves would pass the horizon travelling further the longer the wavelength (Fig. 10). This explains why to cover a country many ground stations are required for TV or FM radio, not so many for

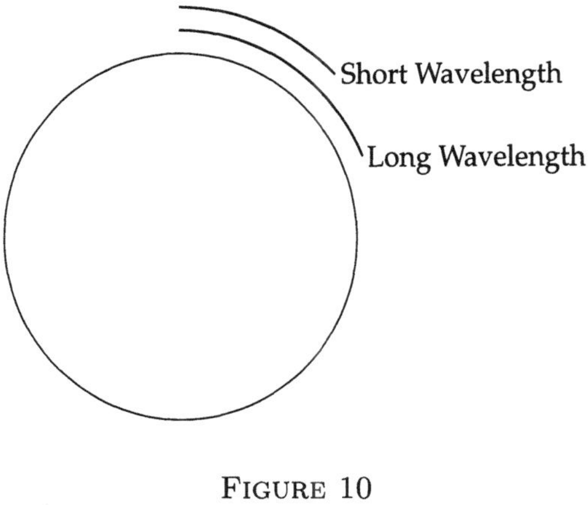

FIGURE 10

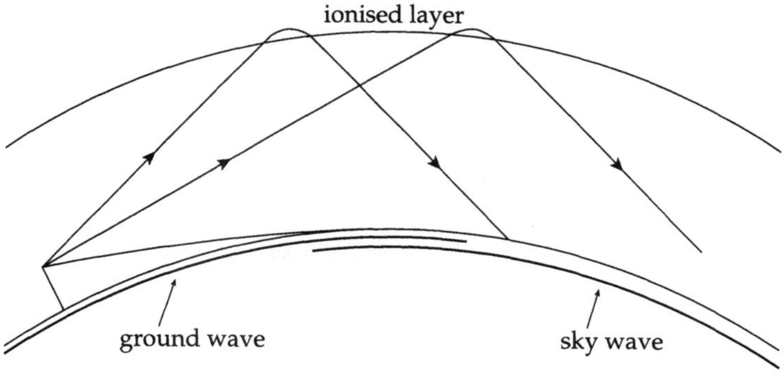

FIGURE 11

the radio medium waveband and ever fewer for the long waveband. The episode reminds us that the motto of the Royal Society of London is NULLIUS IN VERBA which means "Don't believe a word until you see the evidence". Theoretical predictions not backed by experimental confirmation should be treated with the same deference as the statement that a political party will halve taxes or that all Roman Catholic priests are celibate.

The 1920's brought further advances. Appleton discovered a layer in the sky which would bounce back radio waves so that, like Heineken beer, signals could reach parts that other waves could not reach (Fig. 11). The first public broadcast of sound took place in Britain. It was on a wavelength of 1500 metres so that it would go long distances and was received by a long wire attached to a crystal set connected to a pair of earphones. A thin wire, called a cat's whisker, joined the crystal to the earphones (Fig. 12). You wiggled the cat's whisker about on the surface of the crystal until you got acceptable sound. Although the device would be superseded by another invention the crystal was to play an important part in the Second World War and a development 20 years later.

Another arrival in the 1920's was the valve (British) or tube (US) also known as a tube (Röhre) in Germany, I believe. It consisted of a tungsten wire which was heated by an electric current so as to emit electrons. The electrons were pulled past a grid of wires by a high voltage (Fig. 13). By this means a small wave on the grid was turned into a much larger one at the top (anode, plate). Thus amplification became available and it was possible to design hi-fi sets with loudspeakers instead of earphones. It took time for adoption because in those days not all households were connected to electrical power lines and a good supply of electricity was necessary to run the valves.

With valves much stronger signals became available. They enabled the first transmission of television pictures by Baird, another Scotsman, followed by the BBC's television service in the 1930's. They also made possible the first demonstration of radar detecting flying aircraft. Another valuable consequence was communication between ground and aircraft by voice instead of by Morse code via a

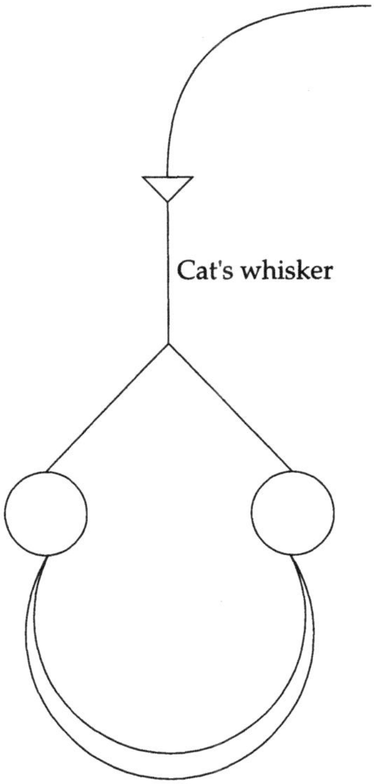

FIGURE 12

long trailing wire. The illustration in Fig. 14 from a book of 1935 on aircraft communication shows how, up till then, the crew had to winch out a wire to receive messages from the ground. Both of these developments were to play a crucial role in World War II.

World War II brought an enormous amount of research into electromagnetism, but I shall mention only a couple of things. There were infrared goggles to assist nightfighters in spotting enemy aircraft. They are used today by those who ski at night. Are they called nightflighters, I wonder?

A very significant invention was the magnetron at the University of Birmingham in England. I expect that many of you will have found a magnetron very useful at home. In a microwave oven the food is heated by electromagnetic waves and the waves are produced by a magnetron.

Actually, the magnetron had been known about for 20 years, but it was unstable. What the University of Birmingham did was to discover how to make it controllable. The invention was given away free of charge to the US. Imagine what royalties would have accrued if it had been patented by the University.

The magnetron generates high power at short wavelengths, e.g., 10 or 3 cm. It allows the concentration of waves into beams like searchlights by means of a

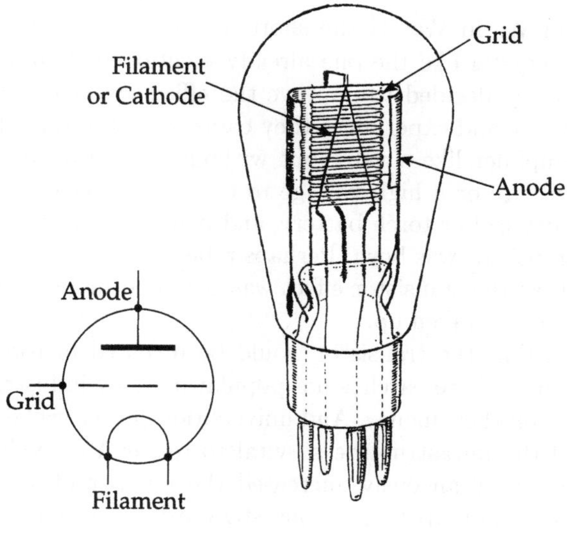

FIGURE 13

dish such as is used in satellite reception today. The short wavelength also permits the picking out of detail in anything which reflects the waves. Today this property is employed by satellites to determine changes in vegetation, movement of ground objects and sometimes prospecting for mineral deposits. Some industries use it to check for flaws in a manufactured item without destroying the product – non-destructive testing.

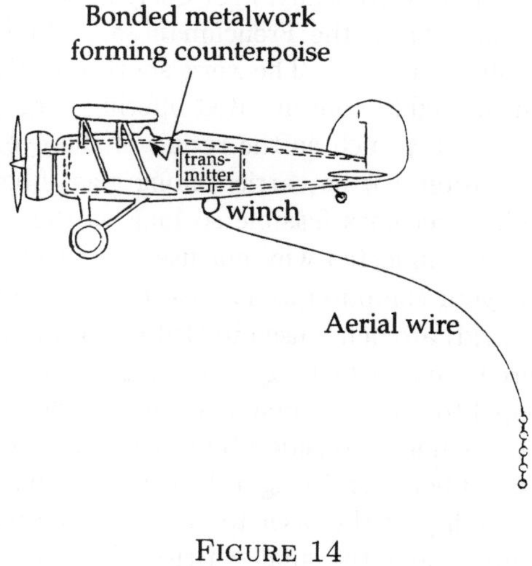

FIGURE 14

At the end of World War II the short wavelength receivers contained, for various reasons, a crystal like the one already mentioned. Bell Telephone Laboratories in New Jersey decided to improve the efficiency of the crystal. As they proceeded with theory and experiment they came to realise that they could make it behave as an amplifier like a valve but without any need for a heavy current to heat a tungsten wire or a high voltage to draw the electrons. It could be run from an ordinary dry cell or torch battery, and it was much smaller than a valve. So, in 1948, the transistor was born. Perhaps rebirth would be more correct since I have been told that the transistor effect was patented in 1925 by the Canadian Lilienfield, though never taken up.

Do you think that the transistor would be invented in today's climate for research? Long-term basic research is not popular in many industries; accountants like quicker returns on their money. And universities, pretty well everywhere, have suffered erosion of the infrastructure so vital to research as politicians have cut basic funding and, simultaneously, increased the number of students. Teaching extra students, many not up to previous standards, consumes valuable faculty effort though the popularity of science in our schools is declining.

Once cheap reliable manufacture was established the transistor became very popular and had replaced the valve practically everywhere by the 1960's. Walkie-talkies appeared, hand-held radio receivers, bleepers and mobile phones emerged, though not everyone in the proximity of people with mobile phones is convinced that they represent progress especially when they are in one of your lectures. Color television and video recorders were affordable by the masses. Combined with the LCD (Hull and Strathclyde) the cheap hand calculator could be manufactured. And 25 years ago saw the advent of the first PC computer, the forerunner to today's PC with its millions of transistors.

Another important electromagnetic invention is the laser which originated from the Hughes Research Laboratory in California in 1960. The advantage of the laser is that it controls light electronically. Previously light could only be controlled by switching it on or off – or as the Frenchman said "Light the candle in the window when your husband is out". The very short wavelength of light means that a laser beam can be highly concentrated offering surgeons the opportunity of delicate repairs of the body such as the retina of the eye. It means also that large amounts of information can be carried along glass fibers not much thicker than a hair. So cable TV becomes feasible so long as the cable is connected to your house. But once it is connected why not use it for your telephone as well or for communications via your computer and access to the Internet. Already electric power lines (called the grid) are being used in Britain for high-speed connection to the Internet. The alternative of attaching something called ADSL (Asynchronous Digital Subscriber Loops) to enable ordinary telephone lines to transmit television or send high-speed data is under consideration in Britain. Another application is the endoscope which is a fiber permitting a doctor to see inside your body.

There is also the ability of the laser to create three-dimensional pictures in the air, called holograms. One of the smaller British horses is illustrated (Fig. 15).

FIGURE 15

In fact, the horse is a hologram produced in Trafalgar Square in the center of London, with the National Gallery visible through the horse. You might think that this is of interest only to those of an artistic nature but surgeons are finding them helpful in treating brain tumours, skull fractures and vascular diseases, while dentists use them in dealing with awkward teeth.

The full benefits of the laser are probably not yet explored. The Dutch invention of the CD demonstrates its efficiency at storing information. If optical or quantum devices can be found equivalent to the transistor in convenience and price maybe the transistor will be as unknown in the future as the valve is today.

Applications such as I have been describing often require a knowledge of how an object affects an electromagnetic wave. Much time has been spent by theorists and experimentalists in trying to acquire this knowledge, yet plenty of problems remain. Nature is not always amenable to analysis and sometimes we want to modify its behavior.

For example, efforts have been made to alter the response of an object to an interrogating signal. In some cases the aim is an enhanced response. This might be to assist navigation by indicating the position of a landmark or beacon, or perhaps to identify a friendly aircraft from enemies so long as the enemies do not take advantage of this means of identification.

The other extreme is to try to conceal a target as in the U.S. Stealth aircraft. When asked if a target can be made invisible I have always replied with a firm No because all known methods of reducing the response are frequency dependent. Therefore transmission of a δ-function of sufficient power will render any target visible. I am sure that this is the correct engineering answer but I am not sure that a mathematician would agree. After all it is known that there can be zeros in radiation patterns and that, in water waves and acoustics, there are certain situations where no reflected waves are produced. Perhaps the question is a suitable piece of homework for Professor Meister when he retires.

You might argue that you cannot generate a δ-function in real life. But, and I know this will please Mrs. Meister, recently Russian and British scientists have collaborated to make a backward oscillator using a superconducting magnet which can operate at 2 gigawatts of pulse power. To put 2 gigawatts in context note that the maximum consumption of electricity on a winter's day in Britain is about 40 gigawatts. The device is capable of producing 5 nanosecond pulses of 500 megawatts at a wavelength of 3 cm. Its resolution is so sharp that you can distinguish the blades of a helicopter and its rear rotor at 35 km. In fact, it can separate the responses of two scatterers which are as little as a 1/2 metre apart. So it is a pretty fair approximation to a δ-function. The existence of such a powerful transmitter does prompt a thought. What would happen if the beam were directed at a nearby network of computers? Would it not put the network out of action?

Another area where electromagnetism will have an impact on society is the introduction this year of digital video. Digital radio started about a year ago in various parts of Europe, including Germany and Britain. For those who can afford the equipment and live in the right place, digital radio offers sound of the quality of a compact disc and highly robust reception even on portable and mobile receivers such as a car radio. Moreover, a top quality stereo transmission in digital radio required only about one tenth of the spectrum of an FM stereo service so that there is the prospect of more choice of programme and additional facilities, assuming that the broadcasters are up to providing the requisite variety. Will it not be marvellous – ten more channels devoted to tennis, golf and football?

Of course, not all developments are beneficial because electromagnetic waves can be interfered with. When transmitting a TV signal freckles can be added to the face of a TV presenter or removed. Signals can be added to modify or destroy information. A strong electromagnetic disturbance can paralyse a computer or network – make sure your machines are properly protected.

Again, once an electromagnetic wave is in the air it can be picked up by anyone with the right apparatus in the right place. Remember that during your next confidential conversation on a mobile or cordless phone. Some shops when they check your credit card by swiping rely on electromagnetic waves and the information can be intercepted by someone wishing to tap into your account. Likewise codes on remote car lockers can be picked up and used by thieves unless they are constantly changed. Indeed, it is said that intelligence agencies of several countries monitor all electromagnetic transmissions as well as all electronic mail.

Electromagnetic waves may damage your pocket.

It is difficult to imagine what society would be like today without electromagnetism. It pervades almost every aspect of our lives. I hope that what I have said has shown you how all the facilities we enjoy have been made possible by a conjecture of 150 years ago and the efforts of many scientists and engineers.

You may wonder what is the role of an applied mathematician like Erhard in all these developments. He starts with a practical problem such as "What is the size of the signal from a mobile phone when it reaches another mobile phone in a town?" In this generality the problem is extremely difficult because the signal encounters buildings of various shapes and sizes, moving vehicles and other obstacles. So the mathematician breaks it down into simpler problems from which general principles can be inferred and applied to the original problem. For instance, he might examine the effect of a wall or how the signal goes over a roof. Even so the resulting mathematical problem is at best formidable, as you can see from Maxwell's equations, and at worst insoluble if the approximations are unsuitable.

Professor Meister was a brilliant expert in choosing conditions for entirely new situations and showing that his formulation led to soluble problems by intricate and deep methods. Moreover, by extremely powerful and original techniques, he went on then to find the solution. The methods he devised were so versatile and effective that they allowed significant progress to be achieved in other fields such as the vibrations of solid bodies and aerodynamics.

Professor Meister has been an important scientist with a considerable influence throughout applied mathematics. His name will be cited many times in the future. It is a privilege and honor to dedicate this lecture to him.

Department of Mathematics
The University
Dundee DD1 4HN
Scotland
United Kingdom
e-mail: `dross@maths.dundee.ac.uk`

Operator Theory
Advances and Applications, Vol. 147, 63–65

How Could it Happen?

Ljudmila Meister

We were asked many times: "How did you meet?" Indeed, it is not quite obvious how it could happen...

In 1993 I got a DAAD scholarship and an invitation to spend two months in the Institute of Physical Geodesy at the Technical University of Darmstadt in Germany. At that time I was an Assistant Professor at the Moscow University of Geodesy and Cartography and this scholarship gave me the possibility of travelling abroad which was still not very common for Russian scientists at that time. I could tell a whole separate story about the adventures I had to get a passport, visa, etc. Finally, the day came when all the difficulties were behind me and on the first of October 1993 I arrived in Germany. It was my first visit to a "capitalist country". All this was only possible because of a recent Russian innovation called "Perestrojka".

I worked at the Institute of Physical Geodesy all day long and spent my spare time looking around and trying to understand this absolutely new life. It was like being in a pink fog. Once I got an e-mail from my elder son Andrew, who was a PhD student in mathematics at Columbia University in New York at that time. He wrote to me: "Mother, there is a famous Professor named Erhard Meister in Darmstadt. Please, meet him and ask about the possibility of my having scientific contacts with him". Well, I found in the university address book that Prof. Dr. Erhard Meister was the head of the section "Mathematical Methods in Physics" and I went directly to his department, since I was not confident about speaking English on the phone. It seems me like only yesterday...

I remember very well that rainy November day when I asked a man with very kind eyes, "Where can I find Professor Meister?" and he answered: "It's me. Come in." Erhard liked this story very much. He asked me to repeat it again and again with all the details. He did not belong to any faith, but he was sure (we both were sure) that our meeting was predestined. We married on the 31st of August 1994 and did not part until his last minute.

Our life was centred around his job, that is, around mathematics. He was interested in everything that was going on in the mathematical world, particularly in any progress in mathematics. New solutions, new problems, new names attracted and delighted him. Of course, all aspects of mathematics were the main subject of our talks at home. However we discussed not only mathematical problems. He was interested in practically everything: philosophy and biology, history and modern politics, etc. Days were too short for us. Fortunately, we had no language problem, we spoke English and understood each other even without any language. One of the results of our long discussions was my talk in the Colloquium of the Mathematics Department in 1997 on "Mathematical education in Russia before and after Perestrojka".

Erhard was very glad when guests came to our house. There were well-known mathematicians like Gohberg, Vishik and others among our guests and young people who had just started making their way in science. We used to sit on our terrace, drinking good wine and talking about interesting things and Erhard was happy.

Quaternions, quaternionic analysis and all kind of applications were the main theme of our joint mathematical interests. In the summer and winter terms of 1996–1997 Erhard organized seminars for students of the Technical University of Darmstadt on these subjects. He gave the main introductory talks and I was responsible for advising the students on their lecture preparation. Erhard had the idea of applying relatively new methods of quaternion analysis (or Clifford analysis in a wider sense) to solve well-known problems of visco-elasticity and plasticity on the one hand and problems of electrodynamics on the other. He tried to convince me to concentrate on one of these directions, but since as my previous work was connected with applications of quaternions in geodesy and navigation problems, I wanted to continue research also in that field. The results of our long and fruitful discussions were my Habilitation with a thesis on "Quaternions and their applications in photogrammetry and navigation" presented at the Technical University "Bergakademie" Freiberg in 1998 (Erhard was very proud of it) and, moreover, our joint papers entitled "Some initial boundary problems in electrodynamics for canonical domains in quaternions" and "Quaternion boundary value problems for canonical objects in continuum physics". They were presented at conferences in Olomouz in 1999 and Beijing in 2000, respectively.

The best place for our joint work was the spa Bad Füssing in Germany. We went there many times. After finishing breakfast and all the morning spa duties, we began to talk and write down new ideas. We were both free from everyday duties and nothing disturbed us except the bells from the church nearby. I was angry about this permanent noise but Erhard just laughed about it and so I started laughing, too.

In February 2000 (Erhard's 70th birthday) Professor Wendland organized a celebration with a so-called "Meister's Colloquium" which included two talks. First came my talk about general properties of quaternions and then after a short break came the main talk, by Erhard, about quaternionic analysis in mathematical physics.

We took part together in several Conferences on Clifford Algebra and Analysis (in Seiffen, Germany 1996; Delaware, USA 1997; Calabria, Italy 1998; Beijing, China 2000). His wide view and deep understanding of the problems immediately made Erhard a central figure in discussions in this relatively new field for him. When the next scheduled talk was finished and the words "Does anyone have any questions?" were pronounced, all heads turned to Professor Meister. Of course, he had questions and the next interesting discussion started.

Erhard and Ljudmila Meister at home in 1998

Erhard considered it his duty to participate in every conference until the last talk was given. Sometimes I asked him to go for a walk with me instead of listening to all the lectures, but every time he answered that it was his duty towards young mathematicians. He told me that his own first talk at a conference in Oberwolfach (around 1958) had also been the last one on the schedule, and he had been so pleased that Professor R. Courant had heard and encouraged him. Erhard tried to help young mathematicians as much as he could. The last three-day workshop he organized in TU Darmstadt was solely for young mathematicians from different countries: Germany, the Netherlands, Russia. His students J. Mark, K. Rottbrand, K. Michalincic and I. Sourovtzeva also gave talks there.

Once, in May 2001 when he was already ill, we were watching TV. There was a story about a young mathematician who got the first prize in some competition. Suddenly, Erhard told to me: "Our life was not in vain, when such young people follow us."

We both knew that we were destined for each other and I am thankful for every day I had with him.

Ljudmila Meister
Tieckstrasse 6
D-64291 Darmstadt, Germany
e-mail: meister.mat@t-online.de

Operator Theory
Advances and Applications, Vol. 147, 67–67

Erhard Meister – Friend and Colleague

Albrecht Böttcher and Bernd Silbermann

Although we had known the work of Erhard Meister since the early 70s, we only met him in person in 1984 at the Conference on Problems and Methods of Mathematical Physics in Chemnitz. This meeting left an indelible imprint on our memory. We remember perfectly that we immediately had the feeling of meeting a good friend whose views on mathematics and opinions on academic life were in astonishing agreement with our own. His kindness and frankness were truly winning characteristics. Later on we discovered that the congruence of our minds extended to nearly all vital questions. As we were living in different political systems, this compatibility was by no means self-evident.

We feel deeply grateful for the great effort Erhard Meister put into helping our personal careers and making the Faculty of Mathematics of the Chemnitz University of Technology flourish. Thanks to his inexhaustible enthusiasm, our personal and scientific contacts with him continued right up to his death. For example, he took the initiative to meet us at a conference in Warsaw, and he was one of the organizers of the memorable Oberwolfach conference on Wiener-Hopf problems in December 1989. The latter conference was our first opportunity to go to the Western part of Germany. At this conference, we met numerous colleagues whose names had been well known to us from their work for a long time but whom we had never seen before and, moreover, we came together again with good friends, such as Israel Gohberg, who had left the Eastern hemisphere many years before and who, at the time, seemed to be out of our reach forever.

In the subsequent years, we exchanged many visits with Erhard Meister. We remember with great pleasure the variety of discussions on mathematics and all topics under the sun. Erhard Meister's most striking traits were his genuine integrity and altruism. He helped us wherever possible, he established contacts between us and other colleagues, and he worked actively in the commission that directed the reorganization of our faculty after the political changes in Germany. He was a true stroke of luck for our faculty.

His death is a harsh loss for us. All his friends greatly valued his extraordinarily broad spectrum of mathematical interests and experience, his enthusiasm for mathematics and his zest for discourse. It is inconceivable that we can no longer relish all that.

Fakultät für Mathematik, TU Chemnitz, D-09107 Chemnitz, Germany
e-mail: aboettch@mathematik.tu-chemnitz.de
e-mail: Bernd.Silbermann@Mathematik.TU-Chemnitz.de

Operator Theory
Advances and Applications, Vol. 147, 69–70

Thank You – Erhard

Roland Duduchava

My first meeting with Erhard was in the autumn of 1979, when he arrived in Tbilisi as a guest of the Academy of Sciences of Georgia. The reason for the invitation was the intersection of interests in mathematics we had found in the preceding years. V. Kupradze, the chairman of the Department of Mathematics and Physics, supported this visit (they had met before during a GAMM conference and also at the International Conference on Function Theoretic Methods for Partial Differential Equations organized by Erhard Meister, Norbert Weck and Wolfgang Wendland at Darmstadt in 1976). It was the starting point of a long friendship and collaboration. During the stay in Tbilisi, Erhard showed himself to be not only a strong mathematician (he gave several excellent talks there), but also a strong adept of Georgian traditional meals: he was definitely ranked a Long Distance Runner in the competition of proposing traditional Georgian toasts inspired by the taste of Georgian wines.

During his visit, Erhard advised me to apply for a Humboldt scholarship, an idea which I gladly followed. In 1980 I arrived in Darmstadt for one year. There I made some of the closest friends I have to this day. My first visit to Darmstadt (followed by many others, of long or short duration) coincided with a very fruitful period of collaboration with the two groups of E. Meister and W. Wendland. Sometimes it was difficult to understand who belonged to which group. Excellent lectures and seminars as well as many Friday meetings ending in the "Kneipe" (pub) nearby and nice week-end parties belong to the best memories in my social life that I can recall. The atmosphere of those parties was so inspiring that some of our older colleagues, who arrived with the intention of leaving after an hour or two, were still hanging around in the early morning, when most of the younger participants had already gone. The group is now spread over the whole of Europe: M. Costabel in France, F.-O. Speck in Portugal, O. Liess in Italy, W. Wendland, R. Schneider and E. Stephan in different places in Germany etc. and when we meet we always recall that time as unforgettable.

These connections, inspired by Erhard, helped me and many of my colleagues in Tbilisi to survive as professional mathematicians during the turbulent years of the disintegration of the Soviet Union. It is essential to note that Erhard always had a great heart for friends and colleagues and, especially, for those who were victims of circumstances. This emerged from his individual experience, because the start of his career after school was not a straightforward and easy one.

What were the most attractive features of Erhard's character? Besides being a highly educated man with an encyclopedic knowledge, he had an infectious

E. Meister with R. Kleinman, R. Duduchava, F.-O. Speck and V. Maz'ya at Oberwolfach in 1986

enthusiasm and energy. He was the organizer of many international meetings, both formal and informal. His friends are to be found all around the globe and he took pleasure in stimulating contacts and exchanges of people.

I met him for the last time during the IWOTA conference in Faro, when he was already suffering from his deadly disease. But even there I was sure we would meet many times in the future and that his health problem would vanish like smoke in a strong wind. I would like to believe still that at the next conference I attend, Erhard will show up and we will have a nice meal with long conversations about life and mathematics, with some glasses of nice wine. I very much miss these meetings and conversations with Erhard.

Roland Duduchava
A. Razmadze Mathematical Institute
Academy of Sciences of Georgia
1, M. Alexidze St.
Tbilisi 380093
Georgia
e-mail: duduch@rmi.acnet.ge

Operator Theory
Advances and Applications, Vol. 147, 71–72

Our First Visit to Oberwolfach

Miloslav Feistauer, Karel Kozel, and Jiří Neustupa

It was in December 1977 that we had our first opportunity to visit the Federal Republic of Germany. After we finished our Ph.D. studies, our teacher and the chairman of the Department of Applied Mathematics and Computer Science at the Faculty of Mechanical Engineering, Prof. Dr. Jan Polášek, in collaboration with his German colleagues, had managed to get an invitation for us to the conference "Methoden und Verfahren der mathematischen Physik" in Oberwolfach. We will never forget the moment when we arrived in Oberwolfach. Sitting in the conference house near the big chess board painted on the floor, we were observing everything with curiosity. Suddenly Prof. Polášek jumped up and started to greet joyfully and loudly a tall, smiling man. It was Prof. Dr. Erhard Meister, who was one of the organizers of the conference. Prof. Meister's expressive personal charm, cordiality and optimism immediately made a strong impression on us. Moreover, in the days that followed, we found Prof. Meister to be an excellent mathematician. There did not appear to be many people in the mathematical community whose wide knowledge, interest and brilliant thinking gave them the ability to understand immediately the merit of a scientific problem, or to formulate an opinion and discuss it. Prof. Meister was one of these outstanding people.

We were really grateful to Prof. Meister for his help in many situations, for his friendship and for his great help in expanding the relations and cooperation of scientists from our two (and also other) countries. Prof. Meister was, together with Prof. Polášek, the initiator of a long-term collaboration between the Technical University of Darmstadt and the Czech Technical University and the Charles University in Prague. Many of our colleagues were able to visit the TU Darmstadt and other German universities and take part in various conferences in Germany owing to Prof. Meister's support. Thanks to Prof. Meister's influence, we have also had the pleasure of welcoming many German mathematicians and engineers to Prague. We are glad that the cooperation and our contacts live on and continue on many levels – official, scientific and personal.

Although we had many further opportunities to meet with Prof. Meister after the year 1977, our first positive impressions never changed.

Miloslav Feistauer
Department of Numerical Mathematics
Faculty of Mathematics and Physics
Charles University in Prague
Ke Karlovu 3
121 16 Praha 2
Czech Republic
e-mail: feist@karlin.mff.cuni.cz

K. Kozel
Department of Technical Mathematics
Faculty of Mechanical Engineering
Technical University Prague
Karlovo nam. 13
121 35 Praha 2
Czech Republic
e-mail: kozelk@fsik.cvut.cz

Jiří Neustupa
Department of Technical Mathematics
Faculty of Mechanical Engineering
Technical University Prague
Karlovo nam. 13
121 35 Praha 2
Czech Republic
e-mail: neustupa@math.niu.edu

Operator Theory
Advances and Applications, Vol. 147, 73–75

Our Meetings with Erhard Meister

Israel Gohberg and Marinus A. Kaashoek

Professor Erhard Meister was an outstanding expert in many areas of mathematics and a wonderful person. He was a pleasant colleague and a good friend of both of us. We shall miss him.

We met Erhard Meister regularly at colloquia, conferences and seminars, in Germany, in Israel, in the Netherlands, and in other countries. Our meetings with him were important for us. During these meetings we discussed progress and problems in the area of singular integral equations and Wiener-Hopf operators, with an emphasis on recent developments in factorization. But not only that, we also discussed organizational matters, including conferences, journals, books, and invitations.

The first author recalls with gratitude his meeting with Meister in the Spring of 1975 when he visited West Germany for the first time after leaving the former Soviet Union. During this first meeting, which took place in Darmstadt, Meister put the first author in touch with C. Einsele, a member of the Birkhäuser family, who was then the head of the editorial branch of the publishing company Birkhäuser Verlag Basel. At that time Birkhäuser Verlag was preparing the translation of the Gohberg-Krupnik book[1] into German, under Meister's supervision and editorship. The connection with Einsele was very important and led to the foundation of the journal *Integral Equations and Operator Theory* in 1978. From the very beginning, Meister was a member of the editorial board of the journal, as well as of the later book series *Operator Theory: Advances and Applications.*

Meister was knowledgeable in pure and applied mathematics. His scientific achievements in both areas are many and essential. He attracted students and co-authors, and was influential both in the German mathematical community and abroad. His wide knowledge made him a valuable contributor to discussions in conferences and seminars where he often raised unexpected questions and made interesting remarks.

Together with Meister, the two of us organized two one-week Oberwolfach meetings on "Wiener-Hopf Problems", one in 1986 and the other in 1989. These were very special meetings which attracted participants from the West (including the USA) and from Eastern Europe. The second meeting was one of the first where experts from East Germany met their colleagues from the West in person. These

[1] I. Gohberg, N. Krupnik. *Einführung in die Theorie der Eindimensionalen Singulären Integraloperatoren.* Birkhäuser Verlag, Basel, (translated from Russian E. Meister (Ed.)) 1979.

E. Meister with I. Gohberg and M.A. Kaashoek in 1997

were exciting events which had a considerable influence on the further development of Wiener-Hopf theory and its applications.

Meister was also an active participant in other Oberwolfach meetings on operator theory. At all these meetings his positive attitude and cheerful personality helped to create a stimulating atmosphere and an environment of friendship.

At the Oberwolfach meetings he had a double role: first as a scientific leader, and secondly as the chief organizer and guide of the hikes on Wednesday afternoons. In his second role he usually appeared in knickerbockers and colorful long socks, the traditional German outfit for hikers, and an impressive hat. He knew the geography of the Black Forest well, including the right cafes with good coffee and delicious tortes. His guiding was precise, and we always were back on time at the institute.

An unexpected meeting, which Meister organized, was a small one day conference in Darmstadt in 1997 when the first author received a honorary doctoral degree from the Technische Hochschule at Darmstadt, and the second author presented the laudatio lecture. Later we learned that Erhard had been preparing this event for many years, skilfully guiding his proposal through the university channels.

We remember gratefully his support for and participation in the International Workshops on Operator Theory and Applications (IWOTA). At the IWOTA meeting at Faro (Portugal) he could not give his plenary talk, and his lecture was presented by his wife Ludmilla. But even then no one expected that this IWOTA meeting would be his last public appearance at an international conference. His death is a great loss to the mathematical community and to his many friends.

Israel Gohberg
School of Mathematical Sciences
Ryamond and Beverly Sackler Faculty of Exact Sciences
Tel-Aviv University
69978 Tel Aviv, Israel
e-mail: gohberg@post.tau.ac.il

Marinus A. Kaashoek
Afdeling Wiskunde
Faculteit der Exacte Wetenschappen
Vrije Universiteit
De Boelelaan 1081a
NL-1081 HV Amsterdam
The Netherlands
e-mail: kaash@few.vu.nl

Operator Theory
Advances and Applications, Vol. 147, 77–78

Some Canadian Collaboration

Robert Allan Hurd

I first met Erhard Meister in 1982 when at his invitation I paid a visit to the Mathematics Department of the Darmstadt Technical University. But I had known him long before as a reviewer of Wiener-Hopf subjects for the Mathematical Reviews. His comments there were always perceptive and charitable.

Our common interest in Wiener-Hopf problems led to an invitation to spend a year at the University in 1986. For this he obtained for me a Deutsche Forschungsgemeinschaft fellowship. It was during this year that our friendship was established. It lasted till his unfortunate death. I came to admire him in many ways – not the least of which was his ability to make telling comments and ask searching questions in many mathematical fields. During this year we collaborated on

Hiking in the Odenwald with the groups of E. Meister and W. Jäger in 1986

the application of Wiener-Hopf techniques to diffraction problems, resulting in 4 papers and published reports.

Looking back on that year, it is amusing to remember that our initial collaboration was fraught with misunderstandings. I was surprised to find that to a large degree we spoke different mathematical languages when we spoke of Wiener-Hopf. For my part, I encountered a completely different notation which took me quite a long time to feel comfortable with. He, I think, was encountering for the first time some new matrix Wiener-Hopf factorization methods. Each of us assumed the other knew all these things. Happily it was all cleared up and 4 papers and our long friendship resulted.

One highlight of the year in Darmstadt was a conference at Oberwolfach where Erhard revealed himself as an athlete. Hitherto, I had not thought of him as an outdoorsman. But suddenly he appeared in hiking costume and proceeded to lead us over beautiful hill and dale for many kilometers before ending up at a pub for restoration.

With my departure from Darmstadt our collaboration ended; but not our association. We met at conferences at Gebze in Turkey and at the Sommerfeld Centenary in Freudenstadt. But he also twice visited my wife and me at our seashore house in Nova Scotia. The second time was with his wife Lyudmila. I still can see them, arm-in-arm, admiring the sunset – just as I admired him.

Robert Allan Hurd
RR 1 Cardinal, Ontario
K0E 1E0, Canada
e-mail: hurd@recorder.ca

Operator Theory
Advances and Applications, Vol. 147, 79–80

Reminiscences of Erhard Meister at Saarbrücken and Berlin

Norbert Latz

Erhard Meister came to Saarbrücken from the German Research Establishment of Aeronautics at Mühlheim/Ruhr in April 1958 to work as an assistant of Heinz Söhngen, who had accepted a chair of mathematics at the Faculty of Mathematics and Natural Science of the University of Saarland.

During the summer term of 1958, Erhard Meister was responsible for exercise classes corresponding to the analysis lectures of Professor Söhngen given to students of mathematics, physics and chemistry. Erhard Meister was a very severe supervisor, especially when trying to find out which students had solved the weekly exercises and which were only copying. But he was not very successful, because most of us knew each other pretty well, being from the Saar region.

In July 1958 Erhard Meister received his doctorate from the University of Saarland presenting a thesis entitled "Berechnung des Geschwindigkeitsfeldes und der Luftkräfte an einem schwingenden, gestaffelten Gitter" (Calculation of the velocity field and the forces on an oscillating staggered cascade).

At the beginning of the summer term of 1960 and during three successive terms, Erhard Meister was the assistant responsible for the Seminar on Wiener-Hopf Methods with Applications in Mathematical Physics. This series of seminars was initiated by Heinz Söhngen and Gottfried Eckart, Director of the Institute of Applied Physics and Electrotechnics. The talks were proposed and supervised by Erhard Meister.

Besides Heinz Söhngen and Gottfried Eckart, another scientist who stimulated Erhard Meister to take his habilitation was Udo Wegner, Director of the Institute of Mechanics. In July 1963 Erhard Meister presented a habilitation thesis entitled "Beitrag zur Theorie der ebenen instationären Unterschallströmung um ein Profil im Kanal oder im Gitterverband" (Contribution to the theory of planar non-stationary subsonic flow around a profile in a channel or in a staggered system). At the end of the summer term of 1963, Erhard Meister became a "Privatdozent". The students and assistants of the Wiener-Hopf seminar were proud to be invited by him to wine and dine at the "Schlossrestaurant Saarbrücken".

Erhard Meister and his young family moved from Riegelsberg to Saarbrücken. A little later, he was invited to give a course at the Technical University of Berlin during the winter term of 1965/66. I myself joined him as an assistant there. Erhard Meister lectured on Advanced Mathematics for Engineers. He developed good contacts with his colleagues, especially with Ernst Wienholtz and Klaus Habetha.

During the summer term of 1966, Erhard Meister was invited by Professor D.C. Pack to lecture at the University of Strathclyde in Glasgow. I myself went there as a DAAD scholar. It was a most exciting time. Besides research and teaching there was also time to discover the beauty of Scotland.

Then Erhard Meister was offered the Chair for Pure and Applied Mathematics at the Technical University of Berlin previously occupied by the famous mathematicians Werner Schmeidler and Erich Kähler. It was a great day in the life of Erhard Meister.

He accepted for the winter term of 1966/67 and was received with great cordiality, especially by the "big shots" of that time: Alexander Dinghas from the Free University, Wolfgang Haack from the Hahn-Meitner Institute and Ernst Mohr from the Technical University.

During the years that followed, Erhard Meister was engaged in teaching, research and administration. The following collaborators passed their doctorate under his supervision: Gerhard Hery, Dieter Schulze, Norbert Latz, Manfred Kremer, Echter Gerlach, Gottfried Eisenhardt.

Wolfgang Haack and Erhard Meister motivated Wolfgang Wendland to take his habilitation in 1969.

Political restrictions at the end of the sixties induced perturbations into the universities of West Berlin. Erhard Meister was offered a chair of mathematics at the venerable University of Tübingen. He accepted in 1970. A new stage in the university life of Erhard Meister began.

Norbert Latz
Institute of Mathematics
Technical University of Berlin
Strasse des 17.Juni 135
D-10623 Berlin
Germany
e-mail: latz@math.tu-berlin.de

Operator Theory
Advances and Applications, Vol. 147, 81–82

More than Forty Years with Erhard Meister

Rolf Leis

Erhard Meister and I we first met in 1958 at a GAMM meeting in Saarbrücken. For both of us it was the first attendance at such a conference, as far as I remember. The Saar territory had only shortly before been integrated into the Federal Republic a milestone in the process of Franco-German reconciliation. Erhard Meister was moving to Saarbrücken at that time to be an assistant of H. Söhngen, who had just received a Chair there. As for me, my first visit to the USA was coming up.

As a result we were both looking ahead full of expectation and immediately got along with each other very well. Since that time – and until his death – we met each other on numerous occasions and visited each other regularly, having the opportunity for long interesting talks and discussions, and stimulating exchanges. I shall also never forget the many excursions and short walking tours during or after meetings which we both attended. Erhard Meister came from Sachsen-Anhalt which at that time belonged to the GDR. If only for that reason, he had a great interest in what was going on in the East and he had a lot of contacts there. Thus he helped me, coming as I did from the Rheinland, making the East more accessible to me. I will never forget my first visit to Poland in 1977 attending a conference in Kozubnik near Porabka in the Beskiden. Erhard Meister was also there and afterwards we decided to drive through Poland in my beetle together, going to Warsaw via Auschwitz and Tschenstochau. The journey made a big impression on both of us.

Erhard Meister was extremely knowledgeable and had broad mathematical interests. He also knew how to manage organizing details very well. He liked to give of his best and he was always ready to pass on his knowledge and experience to others. That made him become a much sought-after adviser. I might mention, for example, his longstanding chairmanship of the GAMM Fachausschuss for Applied Analysis, his activities in the group of founders of the "Mathematical Methods in the Applied Sciences" and his activities for the DFG. He served as a member of the "Senatsausschuss für die Angelegenheiten der Sonderforschungsbereiche" for six years (1981–1986) and he repeatedly acted as a referee. From 1983 till 1992 he acted several times as an expert or referee for our Sonderforschungsbereiche in Bonn.

As I have already said, Erhard Meister always did his very best – for his family, for his students and of course for his own field – and he went on doing this to the very end. I do not want to reflect on his mathematical activities in detail now. That will be done elsewhere. But I want once more to stress his enthusiasm for mathematics and his vigorous attempts to promote the usage of mathematical methods in other fields of the applied sciences, especially in the engineering sciences. His death – much too early, we feel – hurts a lot.

Rolf Leis
Institut für Angewandte Mathematik der Universität Bonn
Wegelerstr. 10
D-53115 Bonn
Germany
e-mail: leis@uni-bonn.de

Operator Theory
Advances and Applications, Vol. 147, 83–85

Remembering Erhard Meister

Vladimir Maz'ya

At the end of the 1970s, big envelopes from Darmstadt filled with red preprints started to come to my address in Leningrad. Like many others in the USSR at that time, I was isolated from Western mathematicians and the arrival of every parcel symbolized a kind of personal contact for me. The envelopes were charged with sympathy. That was the beginning of my acquaintance with Erhard Meister, the head of the young Darmstadt school of applied mathematics. He and his colleagues Wendland, Speck, Costabel, Stephan, Varnhorn and others worked on the problems I was also interested in. Their papers dealt with numerical methods of solving partial differential equations, scattering theory, boundary integral equations, elliptic boundary value problems in non-smooth domains etc.

It became possible for me to accept Meister's invitation much later during *perestroika.* Coming to Darmstadt for the first time in 1990, the year of his 60th birthday, I met a big, strong, energetic man, speaking enthusiastically and with a broad smile. There were no spare offices at the Department of Mathematics and he offered me a part of his own. I was seated on a sofa, with my papers on a long, low table perpendicular to Meister's. From time to time I had to leave for the library, since he had meetings.

As a young man, Meister was very interested in the engineering sciences, but he changed his mind and studied mathematics and physics at Heidelberg. He had a unique ability to speak expertly both on topics in engineering and pure mathematics. Following the course of modern technologies and materials, he liked to discuss issues of their mathematical modeling. Some of his work dealt with rotating and vibrating blade rows, subsonic flow, dielectrical antennas, scattering of acoustic and electromagnetic waves. At that time, I was running an engineering laboratory in Leningrad and was involved in some of these topics. I spent hours listening to him and was impressed by his knowledge of both classical structural mechanics and the most recent theories of coupled fields.

Unselfishness characterised Meister's devotion to science. He loved the science itself and not himself in the science. Always surrounded by students, he had great influence in shaping the careers of many applied mathematicians.

I saw clearly that Meister was greatly respected by his colleagues at the Department, at the Darmstadt Technical School, in Germany and beyond. That was not only because of his enormous erudition but also owing to his openness, optimism, and readiness to help. My late friend Siegfried Prössdorf told me that Meister's role in saving the Weierstrass Institute in Berlin after the unification of

E. Meister with V. Maz'ya and T. Shaposhnikova in Feb. 1998

Germany was very important. My personal profound indebtedness goes to Meister for his support when I told him about my intention to leave the USSR. I keep with gratitude his strong recommendation letter for my Linköping position.

In 1991 and 1994, I invited Meister and Wendland to be plenary speakers at the Linköping Conference on Applied and Industrial Mathematics. My impressions of their visits are unforgettable. Neither missed a single talk, always making notes and bombarding every speaker with questions. The range of their knowledge seemed to be boundless and both conferences owed most of their success to their participation.

Meister was an inexhaustible source of stories, personal, professional, and historical. Some of his reminiscences were exceptionally interesting and it is a pity that they are not preserved. For example, he was an expert on R. Courant's relations with German colleagues after World War II, the history of the Oberwolfach Institute, and the lives and works of many in the generation of his teachers.

Years passed, our relations became warmer and meetings more frequent. My first Humboldt visit to Darmstadt in 1999 was especially memorable. We made plans for joint research in crack propagation. He and I, and our wives Mila and Tanya, spent much time together. The Meisters showed us picturesque neighboring towns and invited us to their hospitable home with a collection of teddy bears which he was fond of. They were a happy couple, caring for each other. Their house, open to numerous friends, was a place where one could enjoy both vivid discussions and tasty food. Erhard once said to me that he had been lucky to meet Mila, a woman of extraordinary kindness.

In the summer of 2000, the four of us planned to spend some time together on the sea shore in Portugal after the IWOTA workshop in Faro but this plan could not be fulfilled. The workshop was overshadowed by his illness and became his last mathematical conference.

Together with many others, I will greatly miss Erhard Meister's friendship, humor, and inspiration.

Vladimir Maz'ya
Department of Mathematics
University of Linköping
SE-581 83 Linköping
Sweden
e-mail: vlmaz@mai.liu.se

Operator Theory
Advances and Applications, Vol. 147, 87–89

The Family Man

Claudia Meister and Olaf Meister

When our father died after a short but intense illness, many colleagues and friends spontaneously decided to honor him by giving lectures, organising workshops and writing articles. All the members of our family are extremely thankful for these signs of friendship and gratitude.

We as his family were asked to contribute by describing the family man Erhard Meister. With our words we want to paint a picture but do not wish to present an exact curriculum vitae. When we think of our father, many short stories and memories come to our minds.

Here are some of them:

Everybody in our family called him "Ata". Why? When his daughter Frauke was a little child, he very often took her for a short walk outside. In Germany this kind of activity is usually called in children's language "Ata-ata-gehen". So, the nickname of our father was born: "Ata". Many years later, he was proud of this name because of its relationship to "Atatürk"; like the father of the Turkish people, he was the father of the "Meister-family".

Ata was always very interested in what was going on in the world. This was why he never wanted to miss watching the 7 p.m. news on TV. This was like a ritual. Ata came home approximately at 5 to 7. Then, he went straight to the TV, watched the news and meanwhile ate his dinner and drank his beer. It was really the same procedure every day and nobody could prevent him from doing this.

His interest in political and historical affairs was so strong that it was nearly impossible to play games with him like "Trivial Pursuit". Why? After a short time, he usually found a question which interested him so much that we forgot to play on because we researched this question very intensively by looking in different encyclopaedia and studying maps. After a while we realised that we had forgotten to continue our game.

On the other hand, most sport themes bored him. It never was possible to talk to him about soccer or skiing news.

Ata was not really a sportsman, but he liked riding a bike and hiking through mountains. We recall many very nice bicycle tours together with him in the surroundings of Tübingen and Darmstadt. The mountain tours were particularly memorable, for example the very hard and impressive "Alm-Marsch" of about 25 km around Sölden, or several mountain tours in the Stubai valley.

Although Ata himself did not practice sports, apart from the exceptions described above, he gave us every opportunity to find out what we – his children

The Meister family in 1967 with E. Meister's mother

– wanted to do. Not only in sports, but also for every other situation in life. Most important of all is that our father encouraged us children to choose our own profession according to our own interests and knowledge.

Every year in early spring when the first flowers were coming out, Ata started his hard and dedicated work in the garden. His first important objective was to kill and destroy all the weeds in the garden. But this passion did not end at the border of our garden, no: he hated weeds so much that he continued this work everywhere he happened to walk. His killing of weeds was so radical that very often flowers became his victims too. This resulted in some long discussions with our mother and in some "reconstruction" of the garden. Furthermore, gardening was a dangerous adventure for him as he sometimes got injured, so that he ended up being a victim of himself.

Many people don't know that Ata was not only a head worker but also had a talent for practical work. When he was young, for instance, he carried out all the electrical installation in the house and he often joined us, the children, when we played with our model railway. But we didn't just play; we also built many of the little houses, the railway-stations, the little mountains etc. ourselves.

Because of his great commitment to his professional work (including his honorary work), he had very little time for his family. But his family was very important to him, providing him with great support. The most important people in his family life were our mother and our grandmother. They took care of us children and together they managed the household.

After the unexpected death of our mother, Ata had the great fortune to meet Ljudmila. With his second love he was given again several years full of joy and happiness.

Claudia Meister and Olaf Meister
Weinbergstrae 42
D-63925 Laudenbach
Germany
e-mail: olaf.meister@t-online.de

Operator Theory
Advances and Applications, Vol. 147, 91–92

Recalling Meetings with Erhard Meister

Sergueï A. Nazarov

It was in Prague during one of the EQUADIFF conferences that Professor Erhard Meister agreed to discuss a mathematical question with me while we were walking from our hotel to the lecture rooms. Although he was really very friendly and well disposed, it was definitely difficult to follow him in two senses; the stride of the tall and strong man was too long and the perfect English of the bilingual mathematician was too quick for me. Being an open person with a likeable smile he attracted everybody, but at that time, I could hardly understand his rapid torrent of speech and, being deeply daunted by our first dialogue, I tried to avoid mathematical topics when meeting him again at two conferences in Germany.

Fortunately for me, our relationship underwent a crucial change during the Petrovskii conference in Moscow. It was a bit strange that, having a lot of foreign guests, lectures were mainly presented in Russian. Occasionally we were sitting alongside each other and once Professor Meister asked me to translate a fragment of the talk by V.A. Kondratiev. What was surprising to me was his understanding of the principal ideas even without translation. Afterwards, when having lunch together, we continued our discussion and I was greatly impressed by his knowledge of both the engineering sciences and related topics in pure mathematics. His deep comments on my questions concerning hydrodynamics and the diffraction theory were indeed very helpful for my future research.

Later, while visiting Professor D.P. Kouzov in Saint-Petersburg, Erhard Meister kindly accepted an invitation to my house. I remember that he liked cold vodka and hot *pelmeni* (Russian meat-dumplings cooked by my wife). This time we did not argue about mathematics at all...

During a two month stay in Paderborn at the invitation of Professor Hermann Sohr, a former student of Erhard Meister, I visited the Darmstadt Technical University with my wife for a few days. He showed us Darmstadt, a nice small town with a tiny Orthodox church, and we walked along colorful streets and through a cheerful Christmas fair, and his stride was not as long and impetuous as usual, since it had to be adapted to a woman's stroll.

We also had many discussions, especially on a rigorous formulation of the Kutta-Joukovskii conditions at the trailing edge of a three-dimensional airfoil. This resulted in a joint paper, by the two of us and one of his pupils, Rainer Hinder, as well as several separate papers by him, by me, and by him in collaboration with Hinder.

It was virtually a tradition for me to go at least for a day to Darmstadt whenever I was in Germany. Once during my talk at the Darmstadt Technical

University, I found a graceful woman sitting next to him, who also attended the "after-lecture" dinner in an Indonesian restaurant. My wife, Nadia, was convinced of the existence of a romantic love-affair and, indeed, within a year we were informed that Luydmila and Erhard had been married.

Our joint work on the Kutta-Joukovskii conditions was especially intensive in 1996/97. I had a half-year guest professorship at Stuttgart University in the group of Professor Wolfgang Wendland. I often used to make one-day trips to Darmstadt in order to clarify mathematical details of our work. Eventually, I was invited to the Darmstadt Technical University to lecture a short course on the propagation of cracks in three-dimensional solids. Erhard, as he asked me to call him, kindly proposed that Nadia and I should live in his house during this month. It was a wonderful time, especially in the evenings... We, together with Ljudmila and Nadia, would discuss the most varied topics – books, movies, art, politics. A comet could be seen in the night sky above the roof of Meister's house.

I met Erhard for the last time at the Instituto Superior Técnico in Lisbon. He had a very short but busy visit to Portugal and we could only have a very brief conversation. A year later, Roland Duduchava came into my office at the same institution and said that he had some very sad news...

Erhard Meister was an outstanding scientist, he enjoyed scientific work and his energy and enthusiasm was infectious, and inspired many mathematicians and mechanicians. At the same time he had a great personality, forceful and good-hearted. Even his criticism was amiable and thus easily acceptable. His sudden death was a heartbreaking blow to all his friends.

Sergueï A. Nazarov
Laboratory of Mathematical Methods in Mechanics of Materials
Institute of Problems of Mechanical Engineering
Russian Academy of Sciences
V.O., Bol'shoy pr., 61
St. Petersburg, 199178
Russia
e-mail: serna@snark.ipme.ru

Operator Theory
Advances and Applications, Vol. 147, 93–94

A View from Scotland

Donald C. Pack

My first meeting with Erhard was at the GAMM conference in Karlsruhe in 1964, in which I took part in acknowledgment of having been made a member not long before. After my talk I was invited by Erhard to have dinner with him and one of his students from the University of Saarbrücken, Benno Schorr, whose research related to mine at that time. We got on very well together and I invited him to spend the 1965–66 session with us at the University of Strathclyde as a Guest Lecturer. Before he could come he received a call to a Chair at the Technical University of Berlin and, since he was not already a Professor, he felt that he had to accept the offer and move to Berlin. In consequence he came to Glasgow for the first half of 1966. Before he left us he travelled widely in Scotland by public transport, and when I was his guest in Berlin in the Spring of 1967 he told me of

Two friends from Scotland: D.C. Pack and D.S. Jones
at the retirement meeting in Feb. 1998

several beautiful places in Scotland that he had visited and we to that date had not! These exchanges were the beginning of a friendship that lasted for the rest of Erhard's lifetime, a period of more than 35 years.

Erhard's first wife, Christa (who died in 1993), having fled from the DDR, was anxious about living in West Berlin, and was relieved when Erhard moved to Tübingen. I stayed at their home there; before bedtime Erhard and I went walking with his large wolf-hound, Growly, and it was amusing to see that this big man was walked by his dog, rather than the other way round!

The connection with Erhard opened up the possibility of other links between colleagues and these developed greatly over the succeeding years, with lecture visits, longer stays and workshops.

When Erhard came to Glasgow he usually stayed with us; he would greet my wife, Connie, with a big "bear-hug", as she called it. In Germany I was "Onkel Pack" to his children. I was his guest for 5 months in 1981 in Darmstadt staying with Connie in the University's house in the Dieburgerstrasse. Erhard and I would return to his office for a cup of coffee after lunch in the Mensa, and talk over a wide range of subjects from mathematics to the problems of Departments and the difficulties facing universities in our two countries. On one occasion he allowed me to sit in on the final oral examinations of some of his students (always with their individual permission), which was an interesting experience since at that time our passes were almost entirely dependent on written papers.

The last time I saw Erhard was at his retirement celebrations in Darmstadt. I stayed with him and his second wife, Ljudmila, whom I had not previously met. After the ceremonies she prepared a lovely meal at their house for some of his colleagues and friends, and a lively evening followed. On the Sunday morning Erhard and I went for a long walk in the neighborhood and talked of many things. In the afternoon I flew home.

Erhard was not only a fine mathematician but also a very kind and genial man, whose letters, phone-calls and friendship I do indeed miss.

Donald C. Pack
18 Buchanan Drive
Bearsden
Glasgow G61 2EW
Scotland
United Kingdom

Operator Theory
Advances and Applications, Vol. 147, 95–97

The Lisbon Connection

António F. dos Santos and Francisco S. Teixeira

We recall, with emotion, our first meeting with Erhard Meister, in August 1988, in Calgary, during the IWOTA conference to commemorate the 60th anniversary of Israel Gohberg. Erhard Meister had a charming personality, something which made a strong impact on our group. His invitation for further contacts was promptly accepted by us and led to a long-lasting cooperation between the Darmstadt and the Lisbon groups.

The visit to Lisbon in September of that year of a member of his group at the time, Frank-Olme Speck, was the first step in a close collaboration that continued until the death of Erhard Meister.

Motivated by common scientific interests centered around the theory of Wiener-Hopf operators, factorization and applications to wave diffraction, an exchange (by post) of views and plans for future cooperation was started, culminating in a visit by E. Meister to the Department of Mathematics at the Instituto Superior Técnico (IST) in April 1989 during which he delivered two lectures entitled "Modern Wiener-Hopf Methods in Diffraction Theory" and "On Some Generalized Sommerfeld Half-Plane Problems". This visit was the starting point for several joint projects.

At the time of this visit the second author was concluding his Ph.D. thesis under the supervision of the first author. The interest shown by Erhard Meister in the work that was being done led to Prof. Meister's being invited to be the external referee in the Ph.D. examination that took place in October 1989. His participation in the examination jury led in turn to his inviting the new Ph.D. member of the Lisbon group to visit the Technische Hochschule Darmstadt (THD) for a two-month period following the examination.

This took place in November/December 1989. This visit allowed the second author to establish work and friendship relations with members of Meister's research group (M. Costabel, W. Varnhorn, F. Penzel and R. Schneider, as well as F. Speck). In that period we participated together in a workshop in Oberwolfach organized jointly by E. Meister, I. Gohberg and M. Kaashoek on "Wiener-Hopf Problems, Toeplitz Operators and Applications". The great mathematical background of Erhard Meister and his enthusiasm were evident from his participation in and organization of the meeting. This meeting was crucial for us in establishing personal contacts with several scientists working in the area of Operator Theory and Mathematical Physics. We would mention, in particular, contacts with S. Prössdorf, B. Silbermann, G. Litvinchuk and W. Wendland.

E. Meister's last lecture at the Faro conference
in September 2000

During 1990 several members of Meister's group visited Lisbon: M. Costabel in May, giving two lectures on "Boundary Integral Equation Methods", F. Penzel in July, giving a short course on "Projection Methods and Numerical Solution of Integral Equations" as part of the program of our Master's Course on Applied Mathematics for that year and F. Speck who gave a series of four lectures on "Multidimensional Singular Integral Operators".

Starting in September 1990, F. Speck became Invited Full Professor in our Department, a position that he still retains. At the same time the second author accepted a position of Invited Researcher, financed by the DFG, in the Department of Mathematics of the THD, as part of E. Meister's group. During this period a number of papers were published or submitted for publication together with E. Meister, F. Speck, A. Lebre (who visited Darmstadt in April/May 1991), F. Penzel and P. A. Santos, then a student in the course of Applied Mathematics at the IST (these publications are included in "Publications by Erhard Meister", refs. [69], [70], [71], [72], [74], [75], [79]).

In 1991 the Darmstadt and Lisbon groups succeeded in getting approval for a two-year project (1991–92), "Wave diffraction phenomena – Sobolev space methods and approximation theory" financed jointly by INIC (Portugal) and DAAD

(Germany). This facilitated the research exchange between the two groups and made possible another visit by Meister to Lisbon. During this period the authors of this note participated in workshops in Darmstadt and another in Oberwolfach organized or co-organized by E. Meister. From 1991 up to the IWOTA conference in Faro (September 2000) several meetings took place between members of the Lisbon group and E. Meister. We recall his pleasure in talking about the careers of his ex-students and the enjoyable walks that he liked to organize, specially at the time of workshops. On all non-formal occasions he was always a great friend, and a master in creating a pleasant atmosphere around him.

Between 1994 and 1998 another exchange project strengthening the ties Darmstadt–Lisbon was approved jointly by JNICT (Portugal) and BMFT (Germany) entitled "Singular Operators-New Features and Applications". Within the framework of this project younger researchers from both groups were able to visit the other group. L. Castro, A. M. Santos, K. Rottbrand, C. Erbe, A. Passow and J. Mark participated in this exchange programme and concluded their Ph.Ds. within the period 1994–1998.

In 1996 E. Meister agreed to become a member of the Advisory Committee of the Centro de Matemática Aplicada of the Instituto Superior Técnico, a position that he occupied until his death. A meeting of this committee motivated his last visit to Lisbon in May 1999. On this visit he was accompanied by his wife Ljudmila. They both gave seminars in the Mathematics Department of the IST.

We recall, with emotion, our last meeting with Erhard in September 2000, in Faro, at the IWOTA conference. He was already debilitated by the illness that took his life but, nevertheless, he made an effort and succeeded in giving his lecture, helped by his wife. This was the closing lecture of the conference. At the end of this lecture he made a brilliant summary of his area of research in reply to questions put by several participants in the conference.

The image of a great friend and Mathematician, to whom we feel grateful, remains in our memory.

Francisco S. Teixeira
Departamento de Matemática
Instituto Superior Técnico
Av. Rovisco Pais
1049-001 Lisboa
Portugal
e-mail: `Antonio.Ferreira.Santos@math.ist.utl.pt`
e-mail: `fteix@math.ist.utl.pt`

Operator Theory
Advances and Applications, Vol. 147, 99–100

Prof. E. Meister's Relations with Freiberg

Lothar von Wolfersdorf

It is with great pleasure that I have accepted the invitation of the editors to write a few words about the relations of Prof. Meister with Freiberg and the mathematicians at the Mining Academy. The relations between Prof. Meister and myself had their origin in the tradition of our teachers. Prof. Meister was an assistant of Heinz Söhngen in Saarbrücken and I am a pupil of Hans Schubert in Halle. Both our teachers together with Johannes Weissinger, later in Karlsruhe, Kurt Schröder and also Wolfgang Haack in Berlin, represented a group of applied mathematicians who worked during the second world war at the Deutsche Versuchsanstalt für Luftfahrt (DVL) in Adlershof (Berlin) on mathematical problems in aerodynamics, in particular on singular integral equations in airfoil and propeller theory. After the war Prof. Söhngen and Prof. Schubert continued in scientific contact, so that even as a young assistant in Halle, I was aware of the interesting results by Prof. Meister and Prof. Söhngen on singular integral equations from the theory of potential flows in lattices of profiles.

I met Erhard Meister in person at the interesting conferences on integral equations organized by Prof. Janina Wolska-Bochenek and her colleagues and coworkers from the Technical University of Warsaw together with Prof. Danuta Przeworska-Rolewicz from the Polish Academy of Sciences. At these conferences Erhard Meister, Wolfgang Wendland and Heinrich Begehr made a point of establishing good personal contacts with their colleagues from East Germany which was not always easy in Germany in the times of the GDR. Later on, the relationship between Prof. Meister and myself and my wife was strengthened through visits by Prof. Meister to Freiberg with lectures at the department and trips to Dresden and "Saxonian Switzerland". Prof. Meister helped my coworkers and me with his great knowledge of the literature and with publishing our papers in Mathematical Methods in the Applied Sciences, something that was important for our reputation in the mathematical community.

After the political changes in East Germany, Prof. Meister was one of the first to support us in Freiberg in word and deed in the reorganization of mathematical life at our university. In particular he gave advice on the structure of our department and helped us with scientific opinions concerning the new appointments to professorships in our university, among them an important one for the author. He was also deeply interested in the new official twinning of our towns Freiberg and Darmstadt. In those enthusiastic days in Germany, Prof. Meister's invitations to several conferences on integral equations and function theoretic methods in partial differential equations at the Mathematisches Forschungsinstitut in Oberwolfach and to the colloquium on the occasion of his 60th birthday in Darmstadt, as well as the joint visits to the GAMM conference in Krakow (Poland) and the workshop on singular integral equations for German and Israeli mathematicians in Tel-Aviv were highlights of our mathematical life.

Prof. Meister's relationship with Freiberg adquired a new dimension due to the scientific interaction between my colleague Prof. W. Sprößig and Prof. Meister's wife Ljudmila in the field of quaternionic analysis. These contacts started with a lecture by Mrs. Meister in Freiberg and culminated in her habilitation at our university. On these occasions we had a good feeling that Prof. Meister enjoyed being in Freiberg. At the very pleasant evening get-together with colleagues following his wife's habilitation he remarked that he could imagine living in Freiberg. This we took as a great commendation of our town and university. We also greatly appreciated the invitation to Prof. Sprößig to join the editorial board of Mathematical Methods in the Applied Sciences, and later to become a managing editor.

Further remarkable events in our relationship with Prof. Meister were our visits to the very successful Sommerfeld workshop at Freudenstadt (Black Forest) in 1996 and to the colloquium on the occasion of his retirement in Darmstadt including a nice birthday party at his house. My colleagues Prof. W. Sprößig, Prof. E. Wegert and myself are proud and thankful that we were lucky enough to enjoy very good contacts with Prof. Meister for a very long time. And we are glad to see that some of his pupils are continuing this positive relationship with Freiberg.

Lothar von Wolfersdorf
Institute for Applied Analysis
Faculty for Mathematics and Computer Science
Technical University Bergakademie Freiberg
D-09596 Freiberg
Germany
e-mail: wolfersdorf@mathe.tu-freiberg.de

Taking notes at the Darmstadt workshop in Feb. 1998

PART B

Invited Papers from the Areas of Interest of
Erhard Meister

The second part of the book contains scientific papers from various fields that belonged to the research interests of E. Meister such as mathematical physics, integral equations, operator theory, matrix analysis, the theory of function spaces and related topics.

Operator Theory
Advances and Applications, Vol. 147, 105–129

Quasistatic Problems in Viscoplasticity Theory I: Models with Linear Hardening

Hans-Dieter Alber and Krzysztof Chełmiński

In memory of the dedicated mathematician Erhard Meister

Abstract. We study the existence theory to quasistatic initial-boundary value problems with internal variables, which model the viscoelastic or viscoplastic behavior of solids at small strain. In these problems a system of linear partial differential equations coupled with a nonlinear system of differential equations or differential inclusions must be solved. The solution theory is based on monotonicity properties of the differential equations or differential inclusions. The differential inclusions considered in this article belong to the class of constitutive relations of monotone type with positive definite free energy and typically model solids with linear hardening behavior. Models for solids without hardening or with nonlinear hardening are considered in a companion article.

1. Introduction and statement of results

In this article we study existence and uniqueness of solutions to initial-boundary value problems, which model the viscoelastic or viscoplastic deformation behavior of solids at small strain. The initial-boundary value problems we study use constitutive relations modelling the dependence of the stress on the strain history, which consist of differential equations or differential inclusions. We restrict our attention to the class of constitutive relations of monotone type with positive definite free energy introduced in [3]. Typically, these constitutive relations can model solids with linear hardening behavior. Existence of solutions to constitutive relations of monotone type with positive semi-definite free energy is studied in the companion article [1].

To formulate the initial-boundary value problem let $\Omega \subseteq \mathbb{R}^3$ be an open bounded set, the set of material points of a solid body. $\mathcal{S}^3$ denotes the set of

---

2000 *Mathematics Subject Classification.* Primary 74H20, 74H25, 74D10; Secondary 74C05, 74C10, 35Q72.

Key words and phrases. Quasistatic problems, viscoplasticity, internal variables, linear hardening, evolution equations to monotone operators.

Supported by the DFG-Sonderforschungsbereich 298 *Deformation und Versagen bei metallischen und granularen Strukturen.*

symmetric 3×3-matrices. Unknown are the displacement $u(x,t) \in \mathbb{R}^3$ of the material point labeled x at time t, the Cauchy stress $T(x,t) \in \mathcal{S}^3$ and the vector of internal variables $z(x,t) \in \mathbb{R}^N$. The model equations are

$$-\operatorname{div}_x T(x,t) = b(x,t), \tag{1.1}$$

$$T(x,t) = \mathcal{D}(\varepsilon(\nabla_x u(x,t)) - Bz(x,t)), \tag{1.2}$$

$$\frac{\partial}{\partial t} z(x,t) \in f(\varepsilon(\nabla_x u(x,t)), z(x,t)), \tag{1.3}$$

which must hold for $(x,t) \in \Omega \times [0,\infty)$. The unknowns must satisfy the initial condition

$$z(x,0) = z^{(0)}(x), \quad x \in \Omega, \tag{1.4}$$

and either the Dirichlet boundary condition

$$u(x,t) = \gamma_D(x,t), \quad (x,t) \in \partial\Omega \times [0,\infty), \tag{1.5}$$

or the Neumann boundary condition

$$T(x,t)n(x) = \gamma_N(x,t), \quad (x,t) \in \partial\Omega \times [0,\infty). \tag{1.6}$$

Here $\nabla_x u(x,t)$ denotes the 3×3-matrix of first order derivatives of u, the deformation gradient, $(\nabla_x u(x,t))^T$ denotes the transposed matrix, and

$$\varepsilon(\nabla_x u(x,t)) = \frac{1}{2}\Big(\nabla_x u(x,t) + (\nabla_x u(x,t))^T\Big) \in \mathcal{S}^3,$$

is the strain tensor. $B : \mathbb{R}^N \to \mathcal{S}^3$ is a linear mapping, which assigns to the vector $z(x,t)$ the plastic strain tensor $\varepsilon_p(x,t) = Bz(x,t)$. One often assumes that all six independent components of the symmetric matrix ε_p belong to the components of z, in which case B would be the projection to these six components. Moreover, $\mathcal{D} : \mathcal{S}^3 \to \mathcal{S}^3$ is a linear, symmetric, positive-definite mapping, the elasticity tensor.

The given data of the problem are $b : \Omega \times [0,\infty) \to \mathbb{R}^3$, the volume force, $\gamma_D : \partial\Omega \times [0,\infty) \to \mathbb{R}^3$, the boundary displacement, $\gamma_N : \partial\Omega \times [0,\infty) \to \mathbb{R}^3$, the traction at the boundary, and $z^{(0)} : \Omega \to \mathbb{R}^N$, the initial data. $n(x)$ in the Neumann boundary condition denotes the exterior unit normal to $\partial\Omega$ at x.

The equation (1.2) and the differential inclusion (1.3) with a given function $f : \mathcal{S}^3 \times \mathbb{R}^N \to 2^{\mathbb{R}^N}$ together determine the dependence of the stress $T(x,t)$ on the strain history $s \to \varepsilon(\nabla_x u(x,s))$. They are the constitutive relations which model the inelastic behavior of the body. Clearly, these constitutive relations cannot be chosen arbitrarily. Instead, thermodynamical and mathematical requirements restrict the choice of f. In this article we assume that (1.3) belongs to the class of constitutive relations of monotone type with positive-definite free energy, a class for which these requirements are satisfied. For this class there is a quadratic positive-definite free energy

$$\rho\psi(\varepsilon, z) = \frac{1}{2}[\mathcal{D}(\varepsilon - Bz)] \cdot (\varepsilon - Bz) + \frac{1}{2}(Lz) \cdot z \tag{1.7}$$

and a monotone function $g : \mathbb{R}^N \to 2^{\mathbb{R}^N}$ with $0 \in g(0)$, such that f in (1.3) is of the form

$$f(\varepsilon, z) = g\Big(-\rho\nabla_z \psi(\varepsilon, z)\Big), \quad (\varepsilon, z) \in \mathcal{S}^3 \times \mathbb{R}^N. \tag{1.8}$$

The scalar product of two 3×3-matrices in (1.7) is defined by

$$\sigma \cdot \tau = \sum_{i,j=1}^{3} \sigma_{ij}\tau_{ij}\,, \quad |\sigma| = \sqrt{\sigma \cdot \sigma}, \quad \sigma, \tau \in \mathcal{S}^3,$$

and L denotes a symmetric positive definite $N \times N$-matrix.

The aim of this article thus is to study existence and uniqueness of solutions to the quasistatic initial-boundary value problems (1.1)–(1.6) with constitutive relations of monotone type to positive-definite free energy. In addition to the Dirichlet condition (1.5) and the Neumann condition (1.6) we also consider a mixed boundary condition stated later.

Using that (1.2) and (1.7) together yield $-\rho\nabla_z \psi(\varepsilon, z) = B^T T - Lz$, where $B^T : \mathcal{S}^3 \to \mathbb{R}^N$ is the mapping adjoint to B, these initial-boundary value problems can be written in the transparent form

$$\begin{aligned} -\mathrm{div}_x\, T(x,t) &= b(x,t), && (1.9)\\ T(x,t) &= \mathcal{D}(\varepsilon\big(\nabla_x u(x,t)\big) - Bz(x,t)), && (1.10)\\ z_t(x,t) &\in g(B^T T(x,t) - Lz(x,t)), && (1.11)\\ z(x,0) &= z^{(0)}(x), && (1.12) \end{aligned}$$

either with the Dirichlet boundary condition

$$u(x,t) = \gamma_D(x,t), \quad (x,t) \in \partial\Omega \times [0,\infty), \tag{1.13}$$

or with the Neumann boundary condition

$$T(x,t)n(x) = \gamma_N(x,t), \quad (x,t) \in \partial\Omega \times [0,\infty), \tag{1.14}$$

or with mixed boundary conditions.

The class of constitutive relations of monotone type was introduced in [3]. It extends the class of constitutive relations of generalized standard materials defined by B. Halphen and Nguyen Quoc Son in [16]. For a generalized standard material the function g is the gradient or subdifferential of a convex function. Extension of the class of constitutive relations of generalized standard materials is necessary, since, as shown in [3] by studying examples, most constitutive equations from engineering do not belong to this class. The definition of the class of constitutive relations of monotone type is motivated by the "good" mathematical properties it has, and it includes the classical constitutive models like the Norton-Hoff and the Prandtl-Reuss laws. However, in [3] it is also shown that even this class is too small to include most models from engineering for the inelastic behavior of metals. It would thus be necessary to define and study still larger classes.

Nevertheless, it is an important mathematical goal to understand the existence theory of initial-boundary value problems to constitutive equations of monotone type, and in this article and in the companion article [1] we develop this existence theory. For mathematical reasons the class of constitutive relations of monotone type is separated in the subclass of constitutive relations with positive definite free energy (1.7), and in the subclass where the free energy is only positive semi-definite. The existence theory to the former subclass, which is rather satisfactory and complete, is studied here, the existence theory to the latter subclass, which presents greater mathematical difficulties and is less complete, is studied in [1].

Typically, the free energy is positive definite if the constitutive relations include terms modelling linear hardening, as in the model of Melan-Prager for example, whereas the free energy is only positive semi-definite if terms modelling nonlinear hardening are included, or if no hardening at all is modeled. The Prandtl-Reuss and Norton-Hoff laws, for example, do not model hardening, and thus belong to the class of constitutive relations of monotone type with positive semi-definite free energy. The results of the present article therefore can be applied to the Melan-Prager model, but not to the Prandtl-Reuss and Norton-Hoff models.

Since the article [1] contains an extended discussion of the literature to the existence theory for problems from plasticity and viscoplasticity, we mention here only a few references specially related to the topic of the present article:

The existence theory for quasistatic problems to classical constitutive models started with [21] and [12], and was continued in [17, 14], in which elasto-plastic constitutive equations were studied, and in [18, 15], where the existence theory for constitutive equations with hardening was investigated. In particular, in [15] it was noticed that the presence of linear hardening simplifies the existence proofs.

The existence theory to dynamic initial-boundary value problems, where the equation (1.1) is replaced by the equation $\rho u_{tt} - \operatorname{div}_x T(x,t) = b(x,t)$, is considered in [20], where special models from viscoplasticity are studied, and in [3], where the existence theory for dynamic problems to constitutive equations of monotone type with positive definite free energy was started. Both of these references use methods similar to the ones employed here for the existence proof. The investigation of dynamic problems to models of monotone type is continued in [10, 11, 8] and extended to special classes of models with positive semi-definite free energy.

The existence theory for quasistatic problems to models of monotone type with positive definite free energy was started in [4]. The results of the present article complete these investigations and they extend the results for quasistatic problems in [9, 13]. Moreover, our results form the basis of the investigations in [5], where the homogenisation of initial-boundary value problems to constitutive equations of monotone type is studied and justified.

The main results are precisely stated in the remainder of this introduction.

Statement of the main results. We need some notations and definitions: We always assume that $\Omega \subseteq \mathbb{R}^3$ is a bounded open set with C^1-boundary $\partial\Omega$. T_e denotes a positive number (time of existence), and for $0 \le t \le T_e$ we set

$$Z_t = \Omega \times [0,t], \quad Z = Z_{T_e}.$$

If w is a function defined on Z_t and if $0 \le s \le t$, we denote the function $x \mapsto w(x,s)$ by $w(s)$. The norms on $L^p(\Omega, \mathcal{S}^3)$ and $L^p(\Omega, \mathbb{R}^N)$ are denoted by $\|u\|_{p,\Omega}$, and for the scalar products on the spaces $L^2(\Omega, \mathcal{S}^3)$, $L^2(\Omega, \mathbb{R}^N)$and $L^2(Z, \mathcal{S}^3)$, $L^2(Z, \mathcal{S}^3)$ we use the symbols

$$(\sigma, \tau)_\Omega = \int_\Omega \sigma(x) \cdot \tau(x)\, dx\,, \quad (\sigma, \tau)_Z = \int_Z \sigma(x,t) \cdot \tau(x,t)\, d(x,t)\,.$$

Since $\mathcal{D} : \mathcal{S}^3 \to \mathcal{S}^3$ is symmetric and positive definite, other scalar products on these spaces are

$$[\sigma, \tau]_\Omega = (\mathcal{D}\sigma, \tau)_\Omega\,, \quad [\sigma, \tau]_Z = (\mathcal{D}\sigma, \tau)_Z\,.$$

$H_1^p(\Omega, \mathbb{R}^n)$ is the Banach space of functions in $L^p(\Omega, \mathbb{R}^n)$, for which the components have weak derivatives in $L^p(\Omega, \mathbb{R}^n)$, and $\overset{\circ}{H}{}_1^p(\Omega, \mathbb{R}^n)$ denotes the closure in $H_1^p(\Omega, \mathbb{R}^n)$ of the space $C_0^\infty(\Omega, \mathbb{R}^n)$ of all infinitely differentiable functions with compact support contained in Ω. The norm of $H_1^p(\Omega, \mathbb{R}^n)$ is $\|u\|_{1,p,\Omega}$. Clearly, for Z instead of Ω we use the same notations.

We need solutions of the Dirichlet and Neumann boundary value problems of linear elasticity theory. With $\hat{b}, \hat{\varepsilon}_p \in L^2(\Omega, \mathbb{R}^3)$, $\gamma_D \in H_1^2(\Omega, \mathbb{R}^3)$ and $\gamma_N \in L^2(\partial\Omega, \mathbb{R}^3)$ these problems are given by the equations

$$-\operatorname{div} T(x) = \hat{b}(x)\,, \quad x \in \Omega, \tag{1.15}$$

$$T(x) = \mathcal{D}\Big(\varepsilon\big(\nabla u(x)\big) - \hat{\varepsilon}_p(x)\Big), \quad x \in \Omega, \tag{1.16}$$

with the Dirichlet condition

$$u(x) = \hat{\gamma}_D(x), \quad x \in \partial\Omega, \tag{1.17}$$

or the Neumann condition

$$T(x)\, n(x) = \hat{\gamma}_N(x), \quad x \in \partial\Omega\,. \tag{1.18}$$

To define a weak solution of the Dirichlet problem, we insert (1.16) into (1.15), multiply the resulting equation with $v \in \overset{\circ}{H}{}_1^2(\Omega, \mathbb{R}^3)$ and integrate by parts. A weak solution is sought in the form $u = w + \hat{\gamma}_D$ with $\hat{\gamma}_D \in H_1^2(\Omega, \mathbb{R}^3)$ and $w \in \overset{\circ}{H}{}_1^2(\Omega, \mathbb{R}^3)$. Insertion of this representation in the resulting integral identity yields

$$\Big(\mathcal{D}\big(\varepsilon(\nabla w) - \hat{\varepsilon}_p\big), \varepsilon(\nabla v)\Big)_\Omega = (\hat{b}, v)_\Omega - \Big(\mathcal{D}\big(\varepsilon(\nabla \hat{\gamma}_D)\big), \varepsilon(\nabla v)\Big)_\Omega, \tag{1.19}$$

where we used that $\mathcal{D}((\varepsilon \nabla u(x)) - \hat{\varepsilon}_p(x))$ is a symmetric matrix. Accordingly, a function $(u, T) \in H_1^2(\Omega, \mathbb{R}^3) \times L^2(\Omega, \mathcal{S}^3)$ is called a weak solution of the Dirichlet problem (1.15)–(1.17), if (1.16) holds and if $w \in \overset{\circ}{H}{}_1^2(\Omega, \mathbb{R}^3)$ exists with $u = w + \hat{\gamma}_D$, such that (1.19) is satisfied for all $v \in \overset{\circ}{H}{}_1^2(\Omega, \mathbb{R}^3)$.

Analogously, to define weak solutions of the Neumann problem we use (1.18) in the process of partial integration and obtain for $v \in H_1^2(\Omega, \mathbb{R}^3)$

$$\Big(\mathcal{D}\big(\varepsilon(\nabla u) - \hat{\varepsilon}_p\big), \varepsilon(\nabla v)\Big)_\Omega = (\hat{b}, v)_\Omega - \int_{\partial\Omega} \hat{\gamma}_N \cdot v\, dS. \tag{1.20}$$

$(u,T) \in H_1^2(\Omega,\mathbb{R}^3) \times L^2(\Omega,\mathcal{S}^3)$ is called a weak solution of the Neumann problem (1.15), (1.16), (1.18), if (1.16) and the equation (1.20) hold for all $v \in H_1^2(\Omega,\mathbb{R}^3)$.

For reference we state the following classical result:

Lemma 1.1. *There is a constant $C > 0$ with the following properties: To all $\hat{b} \in L^2(\Omega,\mathbb{R}^3)$, $\hat{\varepsilon}_p \in L^2(\Omega,\mathcal{S}^3)$ and $\hat{\gamma}_D \in H_1^2(\Omega,\mathbb{R}^3)$ there is a unique weak solution (u,T) of the Dirichlet problem (1.15)–(1.17). This solution satisfies*

$$\|u\|_{1,2,\Omega} \le C(\|\hat{b}\|_{2,\Omega} + \|\hat{\varepsilon}_p\|_{2,\Omega} + \|\hat{\gamma}_D\|_{1,2,\Omega})\,. \tag{1.21}$$

Moreover, there is a weak solution (u_0,T) of the Neumann problem (1.15), (1.16), (1.18) if and only if $\hat{b} \in L^2(\Omega,\mathbb{R}^3)$, $\hat{\varepsilon}_p \in L^2(\Omega,\mathcal{S}^3)$ and $\hat{\gamma}_N \in L^2(\partial\Omega,\mathbb{R}^3)$ satisfy the equation

$$\int_\Omega \hat{b}(x)\cdot(a+\omega\times x)dx = \int_{\partial\Omega} \hat{\gamma}_N(x)\cdot(a+\omega\times x)dS$$

for all $a,\omega \in \mathbb{R}^3$. This solution can be chosen such that

$$\|u_0\|_{1,2,\Omega} \le C(\|\hat{b}\|_{2,\Omega} + \|\hat{\varepsilon}_p\|_{2,\Omega} + \|\hat{\gamma}_N\|_{2,\partial\Omega}) \tag{1.22}$$

holds. All solutions of the Neumann problem are obtained in the form $(u_0 + a + \omega\times x, T)$ with $a,\omega \in \mathbb{R}^3$.

The proof of these assertions is based on Korn's inequality, cf. [24]. We note that for the Neumann problem the symmetric gradient $\varepsilon(\nabla u)$ is uniquely determined even if u is not, since $\varepsilon(\nabla_x(a+\omega\times x)) = 0$. The function $a+\omega\times x$ represents an infinitesimal rigid motion.

Next we give the definition of weak and strong L^2-solutions of the initial-boundary value problems (1.9)–(1.14).

Definition 1.2. *Assume that the volume force $b \in L^1\big(0,T_e;L^2(\Omega,\mathbb{R}^3)\big)$, the initial data $z^{(0)} \in L^2(\Omega,\mathbb{R}^N)$ and the boundary data $\gamma_D \in L^1\big(0,T_e;H_1^2(\Omega,\mathbb{R}^3)\big)$, $\gamma_N \in L^1\big(0,T_e;L^2(\partial\Omega,\mathbb{R}^3)\big)$ are given.*

A function $(u,T,z) \in L^2(0,T_e;H_1^2(\Omega,\mathbb{R}^3))\times L^2(Z,\mathcal{S}^3)\times C\big([0,T_e],\,L^2(\Omega,\mathbb{R}^N)\big)$ is a strong L^2-solution of the Dirichlet problem (1.9)–(1.13) or the Neumann problem (1.9)–(1.12), (1.14), respectively, if

(i) *for almost every $t \in [0,T_e]$ the function $\big(u(t),T(t)\big)$ is a weak solution of the Dirichlet problem (1.15)–(1.17) or of the Neumann problem (1.15), (1.16), (1.18), respectively, with $\hat{\varepsilon}_p = Bz(t)$, $\hat{b} = b(t)$, $\hat{\gamma}_D = \gamma_D(t)$ and $\hat{\gamma}_N = \gamma_N(t)$.*

(ii) *$t \mapsto z(t) : [0,T_e] \to L^2(\Omega,\mathbb{R}^N)$ is absolutely continuous on every compact subset of $(0,T_e)$.*

(iii) *$g(B^T T(t) - Lz(t)) \in L^2(\Omega,\mathbb{R}^N)$, for almost every $t \in [0,T_e]$.*

(iv) *(1.11) is satisfied almost everywhere in Z.*

(v) *$z(0) = z^{(0)}$.*

The function (u, T, z) is a weak L^2-solution of the Dirichlet initial-boundary value problem (1.9)–(1.13), *or of the Neumann initial-boundary value problem* (1.9)–(1.12), (1.14), *respectively, if there exists a sequence $\{k_n\}_{n=1}^{\infty}$ of functions $k_n \in L^1(0, T_e; L^2(\Omega, \mathbb{R}^N))$, which converges in $L^1(0, T_e; L^2(\Omega, \mathbb{R}^N))$ to 0, and if there exists a sequence $\{(u_n, T_n, z_n)\}_{n=1}^{\infty}$ of strong solutions of the Dirichlet or Neumann problem with* (1.11) *replaced by*

$$\frac{\partial}{\partial t} z_n(x,t) \in g\Big(B^T T_n(x,t) - L z_n(x,t)\Big) + k_n(x,t)\,, \tag{1.23}$$

which converges in $L^2(0, T_e; H_1^2(\Omega, \mathbb{R}^3)) \times L^2(Z, \mathbb{R}^3) \times C([0, T_e], L^2(\Omega, \mathbb{R}^N))$ to (u, T, z).

We remark that if X is a Hilbert space (more generally, if it is a reflexive Banach space), then every absolutely continuous function $t \mapsto h(t) : (0, T_e) \to X$ is almost everywhere differentiable, cf. [7, p. 145].

Next we state our two main theorems. We specially remark here that the free energy ψ defined in (1.7) is positive definite if the symmetric $N \times N$-matrix L is positive definite; ψ is positive semi-definite if L is positive semi-definite. We leave the obvious verification to the reader.

The first theorem concerns the initial-boundary value problems to the Dirichlet and Neumann boundary conditions:

Theorem 1.3. *Assume that the matrix L in* (1.7) *is positive definite and that $g : \mathbb{R}^N \to 2^{\mathbb{R}^N}$ is a maximal monotone function satisfying $0 \in g(0)$.*

For $i = 1$ or $i = 2$ suppose that $b \in H_i^1(0, T_e; L^2(\Omega, \mathbb{R}^3))$, $\gamma_D \in H_i^1(0, T_e; H_1^2(\Omega, \mathbb{R}^3))$ and $\gamma_N \in H_i^1(0, T_e; L^2(\partial\Omega, \mathbb{R}^3))$. Moreover, for the Neumann problem assume that b and γ_N satisfy

$$\int_\Omega b(x,t) \cdot (a + \omega \times x) dx = \int_{\partial\Omega} \gamma_N(x,t) \cdot (a + \omega \times x) dS_x \tag{1.24}$$

to all vectors $a, \omega \in \mathbb{R}^3$ and almost all t in $[0, T_e]$. Finally, assume that $z^{(0)} \in L^2(\Omega, \mathbb{R}^n)$ and that there is $\zeta \in L^2(\Omega, \mathbb{R}^N)$ such that

$$\zeta(x) \in g(B^T T^{(0)}(x) - L z^{(0)}(x)) \quad \text{a.e. in } \Omega\,, \tag{1.25}$$

where $(u^{(0)}, T^{(0)})$ is a weak solution of the Dirichlet or Neumann problem (1.15)–(1.18), *respectively, to the data $\hat{b} = b(0)$, $\hat{\varepsilon}_p = B z^{(0)}$, $\hat{\gamma}_D = \gamma_D(0)$, $\hat{\gamma}_N = \gamma_N(0)$.*

Then, for $i = 1$ there is a unique weak L^2-solution and for $i = 2$ a unique strong L^2-solution of the Dirichlet initial-boundary value problem (1.9)–(1.13). *Furthermore, for $i = 1$ there is a weak L^2-solution and for $i = 2$ a strong L^2-solution of the Neumann initial-boundary value problem* (1.9)–(1.12), (1.14). *If (u_0, T, z) is a weak or a strong solution of the Neumann problem, then all solutions are obtained in the form $(u_0 + a + \omega \times x, T, z)$ with $a, \omega \in L^2([0, T_e], \mathbb{R}^3)$.*

This theorem is proved in Sections 2. For the proof we reduce the initial-boundary value problem to an evolution equation in the Hilbert space L^2 with a maximal monotone evolution operator. Theorem 1.3 then follows from well-known results for such abstract evolution equations. In the reduction we use projection operators defined by the boundary value problems of linear elasticity theory.

We next study existence and uniqueness of global in time solutions to the problem (1.9)–(1.12) with a positive definite free energy function ψ to the mixed boundary condition

$$u(x,t) = \gamma_D(x,t) \text{ for } x \in \Gamma_1 , \quad T(x,t)n(x) = \gamma_N(x,t) \text{ for } x \in \Gamma_2 , \ t > 0, \quad (1.26)$$

where $\Gamma_1, \Gamma_2 \subseteq \partial\Omega$ are relatively open subsets of $\partial\Omega$ satisfying $\Gamma_1 \cap \Gamma_2 = \emptyset$, $\partial\Omega = \overline{\Gamma}_1 \cup \overline{\Gamma}_2$ and $|\Gamma_1| > 0$. Here $|\Gamma_1|$ denotes the 2-dimensional boundary measure of Γ_1. The proof of the following theorem given in Section 3 is based on a different method, which uses the partial Yosida approximation:

Theorem 1.4. *Let us suppose that the boundary data γ_D, γ_N and the external force b possess the regularity*

$$b \in H_2^\infty(0, T_e; L^2(\Omega; \mathbb{R}^3))$$

$$\gamma_D \in H_3^\infty(0, T_e; H^2_{1/2}(\Gamma_1; \mathbb{R}^3)) , \quad \gamma_N \in H_2^\infty(0, T_e; H^2_{-1/2}(\Gamma_2; \mathbb{R}^3)) ,$$

for all $T_e > 0$, and that to the initial data $z^0 \in L^2(\Omega; \mathbb{R}^N)$ there is $\zeta \in L^2(\Omega, \mathbb{R}^N)$ such that

$$\zeta(x) \in g(B^T T^{(0)}(x) - Lz^{(0)}(x)) \quad \text{a.e. in } \Omega ,$$

where $(u^{(0)}, T^{(0)})$ is a weak solution of the problem formed by (1.15), (1.16), (1.26) with $t = 0$, $\hat{b} = b(0)$ and $\hat{\varepsilon}_p = Bz^{(0)}$. If the considered model is of monotone type with the maximal monotone constitutive function $g : \Delta(g) \subset \mathbb{R}^N \to 2^{\mathbb{R}^N}$, which satisfies $0 \in g(0)$, and with a positive definite matrix L, then the system (1.9)–(1.12) with the boundary condition (1.26) possesses a global in time, unique solution

$$(u, T, z) \in H_1^\infty(0, T_e; H_1(\Omega; \mathbb{R}^3) \times L^2(\Omega; \mathcal{S}^3 \times \mathbb{R}^N)) \text{ for all } T_e > 0 .$$

2. Proof of Theorem 1.3

In this section we prove Theorem 1.3. First we reduce the initial-boundary value problems (1.9)–(1.14) to evolution equations. To this end we need projection operators whose definition is based on Lemma 1.1:

Definition 2.1. *Let the operators $P_D : L^2(\Omega, \mathcal{S}^3) \to L^2(\Omega, \mathcal{S}^3)$, $P_N : L^2(\Omega, \mathcal{S}^3) \to L^2(\Omega, \mathcal{S}^3)$ be defined by*

$$P_D\varepsilon_p = \varepsilon(\nabla u_D) , \qquad P_N\varepsilon_p = \varepsilon(\nabla u_N) ,$$

for every $\varepsilon_p \in L^2(\Omega, \mathcal{S}^3)$, where $u_D \in \overset{\circ}{H}{}_1^2(\Omega, \mathbb{R}^3)$ is the first component of the weak solution of the Dirichlet boundary value problem (1.15)–(1.17) to the given functions $\hat{b} = 0$, $\hat{\varepsilon}_p = \varepsilon_p$, $\hat{\gamma}_D = 0$, and $u_N \in H_1^2(\Omega, \mathbb{R}^3)$ is the first component of a weak solution of the Neumann boundary value problem (1.15), (1.16), (1.18) to the given functions $\hat{b} = 0$, $\hat{\varepsilon}_p = \varepsilon_p$, $\hat{\gamma}_N = 0$.

By the remark after Lemma 1.1 P_N is well defined, since $\varepsilon(\nabla u_N)$ is unique even if the solution of the Neumann problem is not.

Lemma 2.2.

(i) *The operators* P_D *and* P_N *are bounded projection operators orthogonal with respect to the scalar product* $[\sigma, \tau]_\Omega$ *on* $L^2(\Omega, \mathcal{S}^3)$.

(ii) *The operators* $B^T \mathcal{D} P_D B$, $B^T \mathcal{D} P_N B : L^2(\Omega, \mathbb{R}^N) \to L^2(\Omega, \mathbb{R}^N)$ *are self-adjoint and non-negative with respect to the scalar product* $(z, \hat{z})_\Omega$.

Remark. In the following we drop the indices D and N if a statement is valid for P_D and P_N and write P instead.

Proof: (i) The boundedness of P follows from (1.21) and (1.22). To prove that P_D is a projection assume that $\hat{\varepsilon}_p$ belongs to the range of P_D, hence $\hat{\varepsilon}_p = \varepsilon(\nabla u)$ for a suitable $u \in \overset{\circ}{H}{}^2_1(\Omega, \mathbb{R}^3)$. Thus, by definition of P_D we have $P_D \varepsilon(\nabla u) = P_D \hat{\varepsilon}_p = \varepsilon(\nabla u_D)$, where u_D is the unique function in $\overset{\circ}{H}{}^2_1(\Omega, \mathbb{R}^3)$ which satisfies (1.19) to $\hat{b} = \hat{\gamma}_D = 0$ for all $v \in \overset{\circ}{H}{}^2_1(\Omega, \mathbb{R}^3)$. Clearly u satisfies (1.19), hence $u_D = u$ and $P_D \hat{\varepsilon}_p = \varepsilon(\nabla u) = \hat{\varepsilon}_p$, which shows that P_D is a projection. In the same way it follows that P_N is a projection.

To prove that P is orthogonal, note that for all $\sigma \in L^2(\Omega, \mathcal{S}^3)$ the function $P\sigma$ is of the form $\varepsilon(\nabla u)$. Therefore we can plug $P\sigma$ in the second argument of the scalar products in (1.19) and (1.20) and obtain by definition of P for all $\tau \in L^2(\Omega, \mathcal{S}^3)$

$$[P\tau - \tau, P\sigma]_\Omega = 0.$$

Interchanging the roles of τ and σ yields

$$[P\sigma - \sigma, P\tau]_\Omega = 0.$$

From these equations we conclude for all $\sigma, \tau \in L^2(\Omega, \mathcal{S}^3)$

$$[\sigma, P\tau]_\Omega = [P\sigma, P\tau]_\Omega = [P\tau, P\sigma]_\Omega = [\tau, P\sigma]_\Omega = [P\sigma, \tau]_\Omega .$$

This yields $P^* = P$, whence P is self-adjoint. Therefore P is an orthogonal projection.

(ii) For $z, \hat{z} \in L^2(\Omega, \mathbb{R}^N)$ we have

$$\begin{aligned} (B^T \mathcal{D} P B z, \hat{z})_\Omega &= (\mathcal{D} P B z, B\hat{z})_\Omega = [PBz, B\hat{z}]_\Omega \\ &= [PBz, PB\hat{z}]_\Omega = [Bz, PB\hat{z}]_\Omega = (\mathcal{D} Bz, PB\hat{z})_\Omega \\ &= (Bz, \mathcal{D} PB\hat{z})_\Omega = (z, B^T \mathcal{D} PB\hat{z})_\Omega , \end{aligned}$$

which implies that $B^T \mathcal{D} P B$ is self-adjoint and non-negative. This completes the proof.

After these preparations we can reduce the initial-boundary value problems (1.9)–(1.14) to an evolution equation. Note first that (1.10) yields

$$B^T T - Lz = B^T \mathcal{D}(\varepsilon - Bz) - Lz = B^T \mathcal{D}\varepsilon - Mz ,$$

with the symmetric, positive semi-definite $N \times N$-matrix $M = B^T \mathcal{D} B + L$. Therefore (1.11) can be written as

$$z_t \in g\big(B^T \mathcal{D}\varepsilon(\nabla_x u) - Mz\big). \tag{2.1}$$

Next, let (u, T, z) be a solution of the Dirichlet or Neumann initial-boundary value problem (1.9)–(1.14). If $z(t)$ is known, then (1.9), (1.10), (1.13) or (1.9), (1.10), (1.14) form boundary value problems for the components $u(t)$ and $T(t)$ of the solution. These functions are obtained in the form

$$(u(t), T(t)) = (\tilde{u}(t), \tilde{T}(t)) + (\hat{u}(t), \hat{\sigma}(t)),$$

where $(\hat{u}(t), \hat{\sigma}(t))$ is a solution of the Dirichlet or Neumann boundary value problem (1.15)–(1.18) to the data $\hat{b} = b(t)$, $\hat{\varepsilon}_p = 0$, $\hat{\gamma}_D = \gamma_D(t)$, $\hat{\gamma}_N = \gamma_N(t)$, and $(\tilde{u}(t), \tilde{T}(t))$ is a solution of the same boundary value problem to the data $\hat{b} = 0$, $\hat{\varepsilon}_p = Bz(t)$, $\hat{\gamma}_D = \hat{\gamma}_N = 0$. The Definition 2.1 of the projector P thus yields

$$\varepsilon\big(\nabla_x u(t)\big) = \varepsilon\big(\nabla_x \tilde{u}(t)\big) + \varepsilon\big(\nabla_x \hat{u}(t)\big) = PBz(t) + \varepsilon\big(\nabla_x \hat{u}(t)\big). \tag{2.2}$$

Using that $\hat{\sigma}(t) = \mathcal{D}\varepsilon\big(\nabla_x \hat{u}(t)\big)$, we obtain by insertion of this equation into (2.1) that

$$z_t(t) \in g\Big((B^T \mathcal{D} P B - M)z(t) + B^T \hat{\sigma}(t)\Big). \tag{2.3}$$

Since $\hat{\sigma}$ is computed from the data b, γ_D, γ_N, and thus is known, (2.3) is an evolution equation for z. With the evolution operator $A(t)$ defined by

$$A(t)z = -g\Big((B^T \mathcal{D} P B - M)z + B^T \hat{\sigma}(t)\Big),$$

this evolution equation can be written as

$$z_t(t) + A(t)z(t) \ni 0.$$

This is a non-autonomous evolution equation. In the following we show that in the case of positive definite free energy it is possible to transform this equation into an autonomous evolution equation. Moreover, we prove that the resulting time independent evolution operator is maximal monotone in the Hilbert space $L^2(\Omega, \mathbb{R}^N)$. This permits to apply the strong existence results known for such evolution equations. This good behavior of the initial-boundary value problem with positive definite free energy is a consequence of the following property of the linear mapping $B^T \mathcal{D} P B - M$ in the evolution equation (2.3):

Lemma 2.3. *Assume that the symmetric matrix L from (1.7) is positive definite, let $M = B^T \mathcal{D} B + L$, and let $P = P_D$ or $P = P_N$ be the projection in $L^2(\Omega, \mathcal{S}^3)$ corresponding to the Dirichlet or Neumann boundary condition. Then the linear operator $M - B^T \mathcal{D} P B : L^2(\Omega, \mathbb{R}^N) \to L^2(\Omega, \mathbb{R}^N)$ is bounded, self-adjoint and positive definite with respect to the scalar product $(z, \hat{z})_\Omega$.*

Proof: $M - B^T \mathcal{D} P B$ is obviously bounded. By Lemma 2.2 the operator $B^T \mathcal{D} P B$ is self-adjoint. Since M is symmetric, we conclude that also $M - B^T \mathcal{D} P B$ is self-adjoint. To see that this mapping is positive definite, note that for $z \in L^2(\Omega, \mathbb{R}^N)$

$$\begin{aligned} &((M - B^T \mathcal{D} P B)z, z)_\Omega = (Lz, z)_\Omega + (\mathcal{D}(I-P)Bz, Bz)_\Omega \\ = \ & (Lz, z)_\Omega + \Big[(I-P)Bz, Bz\Big]_\Omega = (Lz, z)_\Omega + \Big[(I-P)Bz, (I-P)Bz\Big]_\Omega \\ \geq \ & (Lz, z)_\Omega \geq \mu \|z\|_\Omega^2, \end{aligned}$$

with a positive constant μ. Here we used that P is orthogonal with respect to the scalar product $[\sigma, \tau]_\Omega$ on $L^2(\Omega, \mathcal{S}^3)$ and that L is positive definite. This proves the lemma.

In particular, this lemma implies that $M - B^T \mathcal{D} P B$ has a bounded, self-adjoint, positive definite inverse.

To transform (2.3) to an autonomous equation we insert

$$h = (B^T \mathcal{D} P B - M)z + B^T \hat{\sigma}$$

into (2.3) and obtain

$$h_t \in (B^T \mathcal{D} P B - M) g(h) + B^T \hat{\sigma}_t \,. \tag{2.4}$$

Definition 2.4. *Let the evolution operator $\mathcal{C} : L^2(\Omega, \mathbb{R}^n) \to 2^{L^2(\Omega, \mathbb{R}^N)}$ be defined as follows: For $h \in L^2(\Omega, \mathbb{R}^N)$ set*

$$\mathcal{C}h = \Big\{ (M - B^T \mathcal{D} P B)\zeta \ \Big| \ \zeta \in L^2(\Omega, \mathbb{R}^N), \ \zeta(x) \in g(h(x)) \ \text{a.e. in } \Omega \Big\}.$$

The domain of $\mathcal{C}$ is

$$\Delta(\mathcal{C}) = \{ h \in L^2(\Omega, \mathbb{R}^N) \,|\, \mathcal{C}h \neq \emptyset \}.$$

With this operator $\mathcal{C}$ the evolution equation (2.4) on $L^2(\Omega, \mathbb{R}^N)$ can be written as

$$h_t(t) + \mathcal{C}h(t) \ni B^T \hat{\sigma}_t(t). \tag{2.5}$$

We prove now that the operator $\mathcal{C}$ is maximal monotone if the vector field g is maximal monotone. The standard theory for evolution equations to such operators implies then that (2.5) has a unique solution for suitable functions $\hat{\sigma}_t$. To prove this, we need that by Lemma 2.3

$$\langle z, \hat{z} \rangle_\Omega = \Big((M - B^T \mathcal{D} P B)^{-1} z, \hat{z} \Big)_\Omega$$

is a scalar product on $L^2(\Omega, \mathbb{R}^N)$. Well-known considerations show hat

$$\|(M - B^T \mathcal{D} P B)^{-1}\|^{-1} \langle z, z \rangle_\Omega \leq \|z\|_{2,\Omega}^2 \leq \|M - B^T \mathcal{D} P B\| \langle z, z \rangle_\Omega.$$

Whence the associated norm

$$|z|_\Omega = \langle z, z \rangle_\Omega^{1/2}$$

is equivalent to $\|z\|_{2,\Omega}$.

Theorem 2.5.
(i) *Let the mapping* $g : \mathbb{R}^N \to 2^{\mathbb{R}^N}$ *be monotone. Then the operator* $\mathcal{C}$ *is monotone with respect the scalar product* $\langle z, \hat{z}\rangle_\Omega$.
(ii) *If* g *is maximal monotone and satisfies* $0 \in g(0)$, *then* $\mathcal{C}$ *is maximal monotone with respect to this scalar product.*

Proof: (i) Let $z_1, z_2 \in \Delta(\mathcal{C})$. Then for $\zeta_1 \in \mathcal{C}z_1$, $\zeta_2 \in \mathcal{C}z_2$ the functions $\xi_i = (M - B^T\mathcal{D}PB)^{-1}\zeta_i$ satisfy $\xi_i(x) \in g(z_i(x))$ a.e. in Ω, for $i = 1, 2$. Thus,

$$\begin{aligned} \langle \zeta_1 - \zeta_2, z_1 - z_2\rangle_\Omega &= \Big((M - B^T\mathcal{D}PB)^{-1}(\zeta_1 - \zeta_2), z_1 - z_2\Big)_\Omega \\ = (\xi_1 - \xi_2, z_1 - z_2)_\Omega &= \int_\Omega \Big(\xi_1(x) - \xi_2(x)\Big)\cdot\Big(z_1(x) - z_2(x)\Big)dx \geq 0\,, \end{aligned}$$

since the monotonicity of g implies $(\xi_1(x) - \xi_2(x))\cdot(z_1(x) - z_2(x)) \geq 0$. Therefore $\mathcal{C}$ is monotone.
(ii) We first show that the maximal monotone mapping $g : \mathbb{R}^N \to 2^{\mathbb{R}^N}$ with $g(0) = 0$ defines a mapping $G : L^2(\Omega, \mathbb{R}^N) \to 2^{L^2(\Omega,\mathbb{R}^N)}$, which is maximal monotone with respect to the scalar product $(z, \hat{z})_\Omega$. For $z \in L^2(\Omega, \mathbb{R}^N)$ we set $Gz = \{\zeta \in L^2(\Omega, \mathbb{R}^N) \mid \zeta(x) \in g(z(x)) \text{ a.e. in } \Omega\}$. The domain is $\Delta(G) = \{z \in L^2(\Omega, \mathbb{R}^N) \mid Gz \neq \emptyset\}$. Monotonicity of G is seen similarly as in (i). The mapping G is maximal monotone if and only if $I + G$ is surjective. To prove surjectivity, it must be shown that to every $h \in L^2(\Omega, \mathbb{R}^N)$ the equation

$$h \in z + Gz \tag{2.6}$$

has a solution $z \in L^2(\Omega, \mathbb{R}^N)$. Since g is maximal monotone, $(I + g) : \mathbb{R}^N \to 2^{\mathbb{R}^N}$ has an inverse $j : \mathbb{R}^N \to \mathbb{R}^N$, which satisfies $|j(\zeta) - j(\zeta')| \leq |\zeta - \zeta'|$ for all $\zeta, \zeta' \in \mathbb{R}^N$. Thus j is Lipschitz continuous. From $0 \in g(0)$ it follows that $j(0) = 0$, whence

$$|j(\zeta)| = |j(\zeta) - j(0)| \leq |\zeta - 0| = |\zeta|. \tag{2.7}$$

For $h \in L^2(\Omega, \mathbb{R}^N)$ we now define $z(x) = j(h(x))$ for all $x \in \Omega$. Clearly, z solves (2.6) if $z \in L^2(\Omega, \mathbb{R}^N)$. To see that z belongs to this space, note that z is measurable, since h is measurable and j is Lipschitz continuous. Moreover, (2.7) implies $|z(x)| \leq |h(x)|$, hence $z \in L^2(\Omega, \mathbb{R}^N)$. Consequently, z solves (2.6) and we conclude that $I + G$ is surjective. Hence G is maximal monotone.

With this result we can show that $I + \mathcal{C}$ is surjective. Since $\mathcal{C} = (M - B^T\mathcal{D}PB)G$, it must be shown that to every $h \in L^2(\Omega, \mathbb{R}^N)$ there is a solution $z \in L^2(\Omega, \mathbb{R}^N)$ of

$$h \in z + (M - B^T\mathcal{D}PB)Gz. \tag{2.8}$$

We apply $(M - B^T\mathcal{D}PB)^{-1}$ to (2.8) and obtain the equivalent equation

$$(M - B^T\mathcal{D}PB)^{-1}h \in (M - B^T\mathcal{D}PB)^{-1}z + Gz. \tag{2.9}$$

To prove that (2.9) has a solution z we note that since $(M - B^T\mathcal{D}PB)^{-1}$ is positive definite, there is $\mu > 0$ such that

$$(M - B^T\mathcal{D}PB)^{-1} - \mu I$$

is positive definite on $L^2(\Omega, \mathbb{R}^N)$. As a bounded mapping, it is Lipschitz continuous. Thus, $G + ((M - B^T\mathcal{D}PB)^{-1} - \mu I)$ is maximal monotone as sum of the maximal monotone mapping G and a Lipschitz continuous monotone mapping, cf. [7, p. 34]. This implies that

$$G + ((M - B^T\mathcal{D}PB)^{-1} - \mu I) + \mu I = G + (M - B^T\mathcal{D}PB)^{-1}$$

is surjective. Therefore (2.9) and as a consequence also (2.8) has a solution $z \in L^2(\Omega, \mathbb{R}^N)$. This implies that $I + (M - B^T\mathcal{D}PB)G = I + \mathcal{C}$ is surjective, whence $\mathcal{C}$ is maximal monotone. The proof of Theorem 2.5 is complete.

With this result we can apply standard existence results to the evolution equation (2.5). To state two of these results we need the following definitions (cf. [7, p. 64], [6, p. 123, 134]):

Definition 2.6. *Let $k \in L^1(0, T_e; L^2(\Omega, \mathbb{R}^N))$. A function $h \in C([0, T_e], L^2(\Omega, \mathbb{R}^N))$ is called a strong solution of the equation*

$$\frac{dh}{dt} + \mathcal{C}h \ni k, \tag{2.10}$$

on the interval $(0, T_e)$, if it satisfies

(i) *$t \mapsto h(t)$ is absolutely continuous on every compact subset of $(0, T_e)$,*

(ii) *$h(t) \in \Delta(\mathcal{C})$, a.e. on $(0, T_e)$,*

(iii) *$\frac{dh}{dt}(t) + \mathcal{C}h(t) \ni k(t)$, a.e. on $(0, T_e)$.*

As we already remarked after Definition 1.2, (i) implies that $t \mapsto h(t)$ is almost everywhere differentiable.

Definition 2.7. *Let $k \in L^1(0, T_e; L^2(\Omega, \mathbb{R}^N))$. A function $h \in C([0, T_e], L^2(\Omega, \mathbb{R}^N))$ is called a weak solution of the equation (2.10), if there exists a sequence $\{k_n\}_{n=1}^{\infty}$ of functions $k_n \in L^1(0, T_e; L^2(\Omega, \mathbb{R}^N))$, which converges in $L^1(0, T_e; L^2(\Omega, \mathbb{R}^N))$ to k, and if there exists a sequence $\{h_n\}_{n=1}^{\infty}$ of strong solutions $h_n \in C([0, T_e], L^2(\Omega, \mathbb{R}^n))$ of $\frac{dh_n}{dt} + \mathcal{C}h_n \ni k_n$, which converges to h in $C([0, T_e], L^2(\Omega, \mathbb{R}^N))$.*

Theorem 2.8. *If $g : \mathbb{R}^N \to 2^{\mathbb{R}^N}$ is a maximal monotone function satisfying $0 \in g(0)$, then the following assertions hold:*

(i) *Let $\hat{\sigma} \in H_1^1(0, T_e; L^2(\Omega, \mathcal{S}^3))$ and let $h^{(0)} \in \Delta(\mathcal{C}) \subseteq L^2(\Omega, \mathbb{R}^N)$. Then the evolution equation*

$$h_t + \mathcal{C}h \ni B^T\hat{\sigma}_t$$

has a unique weak solution on $[0, T_e]$ satisfying

$$h(0) = h^{(0)}.$$

(ii) *Let $\hat{\sigma}_1, \hat{\sigma}_2 \in H_1^1(0, T_e; L^2(\Omega, \mathcal{S}^3))$, and let h_1, h_2 be weak solutions of*

$$h_{it} + \mathcal{C}h_i \ni B^T\hat{\sigma}_{it}, \quad i = 1, 2.$$

Then

$$|h_1(t) - h_2(t)|_\Omega \le |h_1(s) - h_2(s)|_\Omega + \int_s^t |B^T(\hat{\sigma}_{1t}(\eta) - \hat{\sigma}_{2t}(\eta))|_\Omega \, d\eta\,, \tag{2.11}$$

for all $0 \le s \le t \le T_e$.

Proof: Theorem 2.5 shows that $\mathcal{C}$ is maximal monotone with respect to the scalar product $\langle z, \hat{z}\rangle$ under the assumptions of this theorem. Since for $\hat{\sigma} \in H_1^1(0, T_e; L^2(\Omega, \mathcal{S}^3))$ the function $B^T\hat{\sigma}_t$ belongs to $L^1(0, T_e; L^2(\Omega, \mathbb{R}^N))$, this theorem is therefore an immediate consequence of [7, Lemma 3.1 and Theorem 3.4, p. 64, 65]. The norm in (2.11) is associated to the scalar product $\langle z, \hat{z}\rangle$.

Theorem 2.9. *Suppose that* $g : \mathbb{R}^N \to 2^{\mathbb{R}^N}$ *is maximal monotone and satisfies* $0 \in g(0)$. *Also, let* $\hat{\sigma} \in H_2^1(0, T_e; L^2(\Omega, \mathcal{S}^3))$ *and let* $h^{(0)} \in \Delta(\mathcal{C})$. *Then the evolution equation*

$$h_t + \mathcal{C}h \ni B^T\hat{\sigma}_t \tag{2.12}$$

has a unique strong solution on $[0, T_e]$ *with*

$$h(0) = h^{(0)}. \tag{2.13}$$

Moreover, h *belongs to the space* $H_1^\infty(0, T_e; L^2(\Omega, \mathbb{R}^N))$ *and satisfies*

$$|h_t(t)|_\Omega \le |||\mathcal{C}h^{(0)} + B^T\hat{\sigma}_t(0)|||_\Omega + \int_0^t |B^T\hat{\sigma}_{tt}(s)|_\Omega \, ds \quad a.e.,$$

where $|||\mathcal{C}h^{(0)} + B^T\hat{\sigma}_t(0)|||_\Omega = \inf\{|\zeta|_\Omega \mid \zeta \in \mathcal{C}h^{(0)} + B^T\hat{\sigma}_t(0)\}$.

We remark that for a Hilbert space X (more generally, for a reflexive Banach space X), the space $H_1^\infty(0, T_e; X)$ coincides with the space of Lipschitz continuous functions on $[0, T_e]$ with values in X, cf. [7, p. 143–145]. Therefore the solution in this theorem is Lipschitz continuous.

Proof: Since $\mathcal{C}$ is maximal monotone and since for $\hat{\sigma} \in H_2^1(0, T_e; L^2(\Omega, \mathcal{S}^3))$ the function $B^T\hat{\sigma}_t$ belongs to $H_1^1(0, T_e; L^2(\Omega, \mathbb{R}^n))$, this theorem is an immediate consequence of [6, Theorem 2.2, p. 131].

Proof of Theorem 1.3: We first prove that strong L^2-solutions of the Dirichlet problem are unique and strong L^2-solutions of the Neumann problem are of the multiplicity stated in Theorem 1.3. Thus, assume that (u_i, T_i, z_i) for $i = 1, 2$ are two strong solutions of the Dirichlet problem or of the Neumann problem to the data b, γ_D, γ_N and $z^{(0)}$. We set $(\overline{u}, \overline{T}, \overline{z}) = (u_1 - u_2, T_1 - T_2, z_1 - z_2)$.

By Definition 1.2 of strong L^2-solutions, $(\overline{u}(t), \overline{T}(t))$ is a weak solution of the Dirichlet or Neumann problem (1.15)–(1.18) to the data $\hat{b} = 0$, $\hat{\varepsilon}_p = B\overline{z}(t)$ and $\hat{\gamma}_D = \hat{\gamma}_N = 0$. We have $\overline{u} \in L^2(0, T_e; \mathring{H}_1^2(\Omega, \mathbb{R}^3))$ for the the Dirichlet problem and $\overline{u} \in L^2(0, T_e; H_1^2(\Omega, \mathbb{R}^3))$ for the Neumann problem. Since, again by Definition 1.2, the function $t \mapsto z(t) : [0, T_e] \to L^2(\Omega, \mathbb{R}^N)$ is absolutely continuous on every compact subset of $(0, T_e)$ and $\overline{z}_t(t)$ exists in $L^2(\Omega, \mathbb{R}^N)$ for almost all t, we conclude from the regularity properties of the linear elliptic boundary value problems (1.15)–(1.18) that also $t \mapsto \overline{u}(t) : [0, T_e] \to H_1^2(\Omega, \mathbb{R}^3)$ is absolutely continuous on every compact subset of $(0, T_e)$, and that for almost all $t \in (0, T_e)$ the derivative $\overline{u}_t(t)$ exists with $\overline{u}_t(t) \in \overset{\circ}{H}{}_1^2(\Omega, \mathbb{R}^3)$ for the Dirichlet problem and $\overline{u}_t(t) \in H_1^2(\Omega, \mathbb{R}^3)$ for the Neumann problem. In the case of the Neumann problem we possibly have to add a function $a(t) + \omega(t) \times x$ with suitable coefficients a and ω to obtain a solution with these regularity properties. Using that weak solutions satisfy (1.19) or (1.20) and that $\overline{u}(0) = \overline{z}(0) = 0$, we thus obtain

$$
\begin{aligned}
&\int_\Omega \rho\psi(\varepsilon(\nabla_x \overline{u}(t)), \overline{z}(t))\, dx \\
&= \int_0^\infty \int_\Omega (\mathcal{D}(\varepsilon(\nabla_x \overline{u}(s)) - B\overline{z}(s))) \cdot \varepsilon(\nabla_x \overline{u}_t(s)) \\
&\quad + \nabla_z \psi(\varepsilon(\nabla_x \overline{u}(s)), \overline{z}(s)) \cdot \overline{z}_t(s)\, dx ds \\
&= -\int_0^\infty \int_\Omega \Big[B^T \mathcal{D}(\varepsilon(\nabla_x \overline{u}(s)) - B\overline{z}(s)) - L\overline{z}(s)\Big] \cdot \overline{z}_t(s)\, dx ds \qquad (2.14) \\
&= -\int_0^\infty \int_\Omega \overline{(B^T T(s) - Lz(s))} \cdot \overline{z}_t(s)\, dx ds \le 0.
\end{aligned}
$$

Here we used that the relation $z_{it}(x,t) \in g(B^T T_i(x,t) - Lz_i(x,t))$ a.e. and the monotonicity of g together imply $\overline{(B^T T(x,t) - Lz(x,t))} \cdot \overline{z_t(x,t)} \ge 0$ a.e. (2.14) implies $\int_\Omega \rho\psi(\varepsilon(\nabla_x \overline{u}(t)), \overline{z}(t))\, dx = 0$, whence $\psi(\varepsilon(\nabla_x \overline{u}(x,t)), \overline{z}(x,t)) = 0$ almost everywhere. This yields $\varepsilon(\nabla_x \overline{u}) = \overline{z} = 0$, since ψ is positive definite. In the case of the Dirichlet problem we have $\overline{u(t)} \in \overset{\circ}{H}{}_1^2(\Omega, \mathbb{R}^3)$, hence we can use Korn's first inequality to conclude from $\varepsilon(\nabla_x \overline{u}) = 0$ that $\overline{u} = 0$. In the case of the Neumann problem we conclude from Korn's second inequality and from $\varepsilon(\nabla_x \overline{u}) = 0$ that $\overline{u} = a(t) + \omega(t) \times x$ with $a, \omega \in L^2([0, T_e], \mathbb{R}^3)$. This proves the uniqueness statements for strong L^2-solutions in Theorem 1.3. For the inequalities of Korn used here we refer to [19], [22] or [24, pp. 278].

To prove existence of strong solutions, let b, γ_D, γ_N and $z^{(0)}$ satisfy the assumptions of Theorem 1.3 with $i = 2$, and let $(\hat{u}(t), \hat{\sigma}(t)) \in \overset{\circ}{H}{}_1^2(\Omega, \mathbb{R}^3) \times L^2(\Omega, \mathcal{S}^3)$ or $(\hat{u}(t), \hat{\sigma}(t)) \in H_1^2(\Omega, \mathbb{R}^3) \times L^2(\Omega, \mathcal{S}^3)$, respectively, be a solution of the Dirichlet problem (1.15)–(1.17) or of the Neumann problem (1.15), (1.16), (1.18), respectively, to the data $\hat{b} = b(t), \hat{\gamma}_D = \gamma_D(t), \hat{\gamma}_N = \gamma_N(t)$ and $\hat{\varepsilon}_p = 0$, for almost every t. From the regularity properties of these elliptic boundary value problems we conclude under the assumptions for b, γ_D and γ_N of Theorem 1.3 that $\hat{\sigma} \in H_2^1(0, T_e; L^2(\Omega, \mathcal{S}^3))$. With a solution $(\tilde{u}(0), \tilde{T}(0))$ of the Dirichlet or Neumann problem (1.15)–(1.18) to the data $\hat{b} = 0$, $\hat{\varepsilon}_p = Bz^{(0)}$, $\hat{\gamma}_D = \hat{\gamma}_N = 0$ we obtain a solution $(u^{(0)}, T^{(0)}) = (\tilde{u}(0) + \hat{u}(0), \tilde{T}(0) + \hat{\sigma}(0))$ of the equations (1.9), (1.10), (1.13) or (1.9), (1.10), (1.14) for $t = 0$, and the Definition 2.1 of the projector P

yields

$$T^{(0)} = \mathcal{D}(\varepsilon(\nabla_x \tilde{u}(0)) - Bz^{(0)}) + \mathcal{D}\varepsilon(\nabla_x \hat{u}(0)) = \mathcal{D}(P-I)Bz^{(0)} + \hat{\sigma}(0).$$

With $M = B^T\mathcal{D}B + L$ we thus conclude that $h^{(0)}$ defined by

$$h^{(0)} = (B^T\mathcal{D}PB - M)z^{(0)} + B^T\hat{\sigma}(0)$$

satisfies $h^{(0)} = B^T\mathcal{D}(P-I)Bz^{(0)} - Lz^{(0)} + B^T\hat{\sigma}(0) = B^TT^{(0)} - Lz^{(0)}$, and thus belongs to $\Delta(\mathcal{C})$, by (1.25). Whence the assumptions of Theorem 2.9 are satisfied, and consequently there is a unique strong solution h of (2.12), (2.13). Set $z = (B^T\mathcal{D}PB - M)^{-1}(h - B^T\hat{\sigma})$, and for almost every t let $(\tilde{u}(t), \tilde{T}(t))$ be a weak solution of the Dirichlet or Neumann boundary value problems (1.15)–(1.18) to the data $\hat{b} = 0$, $\hat{\varepsilon}_p = Bz(t)$, $\hat{\gamma}_D = \hat{\gamma}_N = 0$. We leave it to the reader to verify that $(u, T, z) = (\tilde{u} + \hat{u}, \tilde{T} + \hat{\sigma}, z)$ is a strong solution of the Dirichlet or Neumann problems (1.9)–(1.14). This concludes the proof of the statements of Theorem 1.3 for strong solutions.

To prove the statements for weak solutions, we note that from the preceding uniqueness and existence proof it follows in particular that (u, T, z) is a strong solution of the initial-boundary value problems (1.9)–(1.14) if and only if $h = (B^T\mathcal{D}PB - M)z + B^T\hat{\sigma}$ is a strong solution of (2.12), (2.13). For $k_n \in L^1(0, T_e; L^2(\Omega, \mathbb{R}^N))$ it can be shown in exactly the same way that (u_n, T_n, z_n) is a strong solution of the problems (1.9)–(1.14) with (1.11) replaced by (1.23) if and only if $h_n = (B^T\mathcal{D}PB - M)z_n + B^T\hat{\sigma}$ is a strong solution of $\frac{d}{dt}h_n + \mathcal{C}h_n \ni B^T\hat{\sigma}_t + (B^T\mathcal{D}PB - M)k_n$. Using this result it can be verified in a straightforward way that (u, T, z) is a weak solution of the problems (1.9)–(1.14) if and only if $h = (B^T\mathcal{D}PB - M)z + B^T\hat{\sigma}$ is a weak solution of $h_t + \mathcal{C}h \ni B^T\hat{\sigma}_t$, $h(0) = h^{(0)}$. The statements about weak solutions in Theorem 1.3 now follow from Theorem 2.8.

3. Proof of Theorem 1.4

The method used in this section to prove Theorem 1.4 is based on the use of the Yosida approximation G_λ of the maximal monotone constitutive multifunction g, and thus differs from the method of the preceding section. We proceed similarly as in [8, 11], where the dynamical problem was investigated. In the first step of the proof we replace $g(B^T - Lz)$ by a global Lipschitz function f and show that the theory of differential equations in Banach spaces yields global in time solutions of the resulting initial-boundary value problem. This result is formulated in Theorem 3.1. In the second step we use this result to construct a sequence of approximate solutions $\{(u^\lambda, T^\lambda, z^\lambda)\}_{\lambda>0}$ and prove a priori estimates, which allow to pass to the limit $\lambda \to 0^+$. These estimates are formulated in Theorem 3.2 and Theorem 3.3.

For brevity we use in this section the notation

$$\varepsilon(u(x,t)) = \frac{1}{2}(\nabla_x u(x,t) + \nabla_x^T u(x,t))\,.$$

Since we study positive definite free energy ψ in this section, the matrix L is assumed to be positive definite.

Let us start our investigation with the following problem of the type (1.9)–(1.12) containing only global Lipschitz nonlinearities

$$\begin{aligned}\operatorname{div}_x T(x,t) &= -b(x,t)\,, &(3.1)\\ T(x,t) &= \mathcal{D}(\varepsilon(x,t) - Bz(x,t))\,, &(3.2)\\ \varepsilon(x,t) &= \frac{1}{2}(\nabla_x u(x,t) + \nabla_x^T u(x,t))\,, &(3.3)\\ z_t(x,t) &= f(\varepsilon(x,t), z(x,t))\,, &(3.4)\\ z(x,0) &= z^{(0)}(x)\,, &(3.5)\end{aligned}$$

where the nonlinear constitutive function f is everywhere defined and global Lipschitz, which means that

$$\exists\, L > 0 \quad \forall\, (\varepsilon^1, z^1),\ (\varepsilon^2, z^2) \in \mathcal{S}^3 \times \mathbb{R}^N$$
$$|f(\varepsilon^1, z^1) - f(\varepsilon^2, z^2)| \leq L(|\varepsilon^1 - \varepsilon^2| + |z^1 - z^2|)\,. \quad (3.6)$$

We are going to prove existence of global in time solutions for the system (3.1)–(3.5) with the boundary conditions (1.26). Let us denote by $B^{\perp}$ the orthogonal projector of $\mathbb{R}^N$ onto the subspace $\{(\varepsilon_p, y) \in \mathbb{R}^N : \varepsilon_p = 0\}$. Then equation (3.4) can be written in the components (ε_p, y)

$$\begin{aligned}-\mathcal{D}^{-1}T_t(x,t) + \varepsilon(u_t(x,t)) &= B\, f(\varepsilon(u(x,t)), z(x,t))\,, &(3.7)\\ y_t(x,t) &= B^{\perp} f(\varepsilon(u(x,t)), z(x,t))\,, &(3.8)\end{aligned}$$

where $z = (\varepsilon(u) - \mathcal{D}^{-1}T, y)$. Hence, this system describes the evolution of the triple (u, T, y). Obviously, in general we cannot expect that the system (3.7)–(3.8) possesses unique solutions. However, if we add to this system the equation of motion (3.1) then we can use the "orthogonality" of the functions T and $\varepsilon(u)$. Let us define some Hilbert spaces in which we will solve our problem:

$$\begin{aligned}H_{1,\Gamma_1}(\Omega;\mathbb{R}^3) &= \{u \in H_1(\Omega;\mathbb{R}^3) : u_{|\Gamma_1} = 0\},\\ L^2_{\Gamma_2,\mathrm{sol}}(\Omega;\mathcal{S}^3) &= \{T \in L^2(\Omega;\mathcal{S}^3) : \operatorname{div} T = 0,\ T n_{|\Gamma_2} = 0\}.\end{aligned}$$

$H_{1,\Gamma_1}(\Omega;\mathbb{R}^3)$ is the space of vectorial Sobolev functions with traces vanishing on Γ_1, and $L^2_{\Gamma_2,\mathrm{sol}}(\Omega;\mathcal{S}^3)$ is the space of L^2-functions with vanishing divergence and with trace in the normal direction to the boundary vanishing on Γ_2. Note that for functions $T \in L^2(\Omega;\mathcal{S}^3)$ with $\operatorname{div} T \in L^2(\Omega;\mathbb{R}^3)$ the trace in the normal direction to the boundary is well defined and belongs to the space $H^2_{-1/2}(\partial\Omega;\mathbb{R}^3)$ (see for example [23] p. 14). Moreover, let us denote by

$$\mathbb{H}(\Omega) = H_{1,\Gamma_1}(\Omega;\mathbb{R}^3) \times L^2_{\Gamma_2,\mathrm{sol}}(\Omega;\mathcal{S}^3) \times L^2(\Omega;\mathbb{R}^{N-6})$$

the Hilbert space in which we solve the system (3.7)–(3.8). In this space we redefine the scalar product as follows

$$\langle (u^1,T^1,y^1),\ (u^2,T^2,y^2)\rangle_{\mathbb{H}} = (\mathcal{D}\varepsilon(u^1),\varepsilon(u^2))_\Omega+(\mathcal{D}^{-1}T^1,T^2)_\Omega+(y^1,y^2)_\Omega\,. \quad (3.9)$$

By the assumption $|\Gamma_1|>0$ the scalar product $\langle\cdot\,,\,\cdot\rangle_{\mathbb{H}}$ is equivalent to the standard scalar product.

The next step in our investigation is the cancellation of the given data (the external force and the boundary data). Hence, let us assume that our data b,γ_D,γ_N have the following regularity

$$b\in C^1(\mathbb{R}_+;L^2(\Omega;\mathbb{R}^3))\,, \quad (3.10)$$

$$\gamma_D\in C^1(\mathbb{R}_+;H^2_{1/2}(\Gamma_1;\mathbb{R}^3))\,,\quad \gamma_N\in C^1(\mathbb{R}_+;H^2_{-1/2}(\Gamma_2;\mathbb{R}^3)) \quad (3.11)$$

and denote by (u^*,T^*) the global in time solution of the linear problem

$$\operatorname{div}_x T^*(x,t) = -b(x,t)\,, \quad (3.12)$$

$$T^*(x,t) = \mathcal{D}\varepsilon(u^*(x,t))\,, \quad (3.13)$$

$$u^*(x,t)_{|\Gamma_1} = \gamma_D(x,t)\,,\quad T^*(x,t)\,n(x)_{|\Gamma_2}=\gamma_N(x,t)\,. \quad (3.14)$$

The solution of this problem has the regularity

$$u^*\in C^1(\mathbb{R}_+;H_1(\Omega;\mathbb{R}^3))\,,\ T^*\in C^1(\mathbb{R}_+;L^2(\Omega;\mathcal{S}^3))\,. \quad (3.15)$$

Thus for the differences $\tilde u=u-u^*$ and $\tilde T=T-T^*$ we obtain the following evolution problem in the space $\mathbb{H}(\Omega)$

$$(\tilde u_t,\tilde T_t,y_t)=\mathcal{A}(t)(\tilde u,\tilde T,y)\,, \quad (3.16)$$

where the operator $\mathcal{A}(t)$ is defined by:

$$\forall\ (v,S,w)\in\mathbb{H}(\Omega) \quad (3.17)$$

$$\langle\,\mathcal{A}(t)(\tilde u,\tilde T,y),(v,S,w)\,\rangle_{\mathbb{H}} = (\mathcal{D}B\,f(\varepsilon(\tilde u)+\varepsilon(u^*(t)),\tilde z+z^*(t))\,,\,\varepsilon(v))$$
$$-(B\,f(\varepsilon(\tilde u)+\varepsilon(u^*(t)),\tilde z+z^*(t))\,,\,S)+(B^\perp\,f(\varepsilon(\tilde u)+\varepsilon(u^*(t)),\tilde z+z^*(t))\,,\,w)\,,$$

where $\tilde z=(\varepsilon(\tilde u)-\mathcal{D}^{-1}\tilde T,y)$ and $z^*(t)=(\varepsilon(u^*(t))-\mathcal{D}^{-1}T^*(t),0)$. The initial value for the function y is known and equal to $B^\perp z^{(0)}$, but for the functions $\tilde u$ and $\tilde T$ we have to compute the initial values from the initial value for the plastic strain $\varepsilon_p(0)=B\,z^{(0)}$. Hence, $\tilde u(0)$ is the unique solution of the problem

$$\operatorname{div}_x\mathcal{D}(\varepsilon(\tilde u(0))-\varepsilon_p(0))=0\,, \quad (3.18)$$

$$\tilde u(0)_{|\Gamma_1}=0\,,\quad \mathcal{D}(\varepsilon(\tilde u(0))-\varepsilon_p(0))\,n_{|\Gamma_2}=0, \quad (3.19)$$

and we define $\tilde T(0)=\mathcal{D}(\varepsilon(\tilde u(0))-\varepsilon_p(0))$.

Theorem 3.1. *For initial data* $(\varepsilon_p(0), y(0)) \in L^2(\Omega; \mathbb{R}^N)$ *the problem (3.16) possesses a global, unique solution*

$$(\tilde{u}, \tilde{T}, y) \in C^1(\mathbb{R}_+; \mathbb{H}(\Omega)). \tag{3.20}$$

Proof: Let us choose $t_1, t_2 \in \mathbb{R}_+$ and (u^1, T^1, y^1), (u^2, T^2, y^2), $(v, S, w) \in \mathbb{H}(\Omega)$. The Lipschitz continuity of the function f yields

$$\begin{aligned}
&\langle \mathcal{A}(t_1)(u^1, T^1, y^1) - \mathcal{A}(t_2)(u^2, T^2, y^2), (v, S, w) \rangle_{\mathbb{H}} \\
&\quad \le C \int_\Omega \Big\{ |\varepsilon(u^1) - \varepsilon(u^2)| + |T^1 - T^2| + |y^1 - y^2| + |\varepsilon(u^*(t_1)) - \varepsilon(u^*(t_2))| \\
&\qquad\qquad + |T^*(t_1) - T^*(t_2)| \Big\} \cdot (|\varepsilon(v)| + |S| + |w|)\, dx \\
&\quad \le L \Big\{ \|(u^1, T^1, y^1) - (u^2, T^2, y^2)\|_{\mathbb{H}} \\
&\quad + \sup_{\tau \in (t_1, t_2)} (\|\varepsilon(u_t^*(\tau))\|_{2,\Omega} + \|T_t^*(\tau)\|_{2,\Omega}) |t_1 - t_2| \Big\} \|(v, S, w)\|_{\mathbb{H}},
\end{aligned}$$

where the constants C and L do not depend on t. Consequently the family of operators $\mathcal{A}(t)$ is continuous with respect to t and global Lipschitz in the space $\mathbb{H}(\Omega)$ with the Lipschitz constant L independent of t. Thus from the theory of differential equations in Banach spaces it follows that the evolution problem (3.16) possesses global in time, unique solutions having the regularity (3.20).

We use Theorem 3.1 to construct a sequence of approximate solutions to the problem (1.9)–(1.12). To this end we replace g by the Yosida approximation $G_\lambda = \lambda^{-1}(I - J_\lambda)$, where $J_\lambda = (I + \lambda g)^{-1}$ and $\lambda > 0$. This yields the following sequence of approximate problems

$$\operatorname{div}_x T^\lambda(x,t) = -b(x,t), \tag{3.21}$$

$$T^\lambda(x,t) = \mathcal{D}(\varepsilon^\lambda(x,t) - Bz^\lambda(x,t)), \tag{3.22}$$

$$\varepsilon^\lambda(x,t) = \frac{1}{2}(\nabla_x u^\lambda(x,t) + \nabla_x^T u^\lambda(x,t)), \tag{3.23}$$

$$z_t^\lambda(x,t) = G_\lambda\Big(B^T T^\lambda(x,t) - Lz^\lambda(x,t)\Big), \tag{3.24}$$

$$z^\lambda(x,0) = z^{(0)}(x) \tag{3.25}$$

with the boundary condition (1.26). Assuming that the external force b and the boundary data γ_D, γ_N have the regularity required in (3.10), (3.11) and noting that G_λ is globally Lipschitz continuous we conclude from Theorem 3.1 that for all initial data $z^{(0)} \in L^2(\Omega; \mathbb{R}^N)$ the system (3.21)–(3.25) possesses global in time solution

$$(u^\lambda, T^\lambda, z^\lambda) = (\tilde{u}^\lambda, \tilde{T}^\lambda, z^\lambda) + (u^*, T^*, 0) \ \text{ with } (\tilde{u}^\lambda, \tilde{T}^\lambda, B^\perp z^\lambda) \in C^1(\mathbb{R}_+; \mathbb{H}(\Omega)).$$

Let us assume that $(\varepsilon, z) \in L^2(\Omega; \mathcal{S}^3 \times \mathbb{R}^N)$ and define the total energy associated with the problem (1.9)–(1.12) by

$$\begin{aligned} \mathcal{E}(\varepsilon, z) &= \int_\Omega \rho\psi(\varepsilon, z)\, dx \\ &= \frac{1}{2}\int_\Omega \mathcal{D}(\varepsilon - Bz)\cdot(\varepsilon - Bz)\, dx + \frac{1}{2}\int_\Omega Lz\cdot z\, dx\,. \end{aligned} \tag{3.26}$$

Note that the positive definiteness of the matrix L implies that the right-hand side of (3.26) induces in the space $L^2(\Omega; \mathcal{S}^3 \times \mathbb{R}^N)$ a norm equivalent to the standard norm. This norm will be denoted by $\|(\varepsilon, z)\|_\psi$. Moreover, let us denote by $T^{(0)}(x)$ the initial stress obtained as the solution of the problem

$$\operatorname{div}_x T^{(0)}(x) = -b(x, 0)\,, \tag{3.27}$$

$$T^{(0)}(x) = \mathcal{D}(\varepsilon(u^{(0)}(x)) - \varepsilon_p(0))\,, \tag{3.28}$$

$$u^{(0)}(x)_{|\Gamma_1} = \gamma_D(x, 0)\,, \quad T^{(0)}(x)\, n(x)_{|\Gamma_2} = \gamma_N(x, 0)\,. \tag{3.29}$$

Also, we set

$$\Delta(g) = \{h \in L^2(\Omega; \mathbb{R}^N) \,|\, \text{there is } \xi \in L^2(\Omega; \mathbb{R}^N) \text{ such that } \xi(x) \in g(h(x))\,, \text{a.e.}\}\,.$$

Theorem 3.2. *Suppose that the external force b and the boundary data γ_D, γ_N satisfy*

$$b \in H_2^\infty(0, T_e; L^2(\Omega; \mathbb{R}^3))\,, \tag{3.30}$$

$$\begin{aligned} \gamma_D &\in H_3^\infty(0, T_e; H^2_{1/2}(\Gamma_1; \mathbb{R}^3))\,, \\ \gamma_N &\in H_2^\infty(0, T_e; H^2_{-1/2}(\Gamma_2; \mathbb{R}^3))\,, \end{aligned} \tag{3.31}$$

for all $T_e > 0$. Moreover, assume for the initial data $z^{(0)} \in L^2(\Omega; \mathbb{R}^N)$ that

$$B^T T^{(0)} - Lz^{(0)} \in \Delta(g)\,,$$

where the initial stress $T^{(0)}$ is defined as the solution of the system (3.27)–(3.29). Then for all $T_e > 0$ the solution of the approximate problem (3.21)–(3.25) satisfies the following inequality

$$\sup_{t\in(0,T_e)} \|(\varepsilon_t^\lambda(t), z_t^\lambda(t))\|_\psi \le C(T_e)\,,$$

where the constant $C(T_e)$ does not depend on λ.

Proof: Let us fix $T_e > 0$ and for $h > 0$ let us denote by $(\varepsilon_h^\lambda(t), z_h^\lambda(t))$ the shifted functions $(\varepsilon^\lambda(t+h), z^\lambda(t+h))$. Then for the time differences we obtain

$$\begin{aligned}
&\frac{d}{dt}\frac{1}{2}\|(\varepsilon_h^\lambda - \varepsilon^\lambda, z_h^\lambda - z^\lambda)\|_\psi^2 \\
&\quad = \int_\Omega \mathcal{D}(\varepsilon_h^\lambda - \varepsilon^\lambda - Bz_h^\lambda + Bz^\lambda)\cdot(\varepsilon_{h,t}^\lambda - \varepsilon_t^\lambda - Bz_{h,t}^\lambda + Bz_t^\lambda)\,dx \\
&\qquad + \int_\Omega L(z_h^\lambda - z^\lambda)\cdot(z_{h,t}^\lambda - z_t^\lambda)\,dx \\
&\quad = (\text{using the notation } v^\lambda = u_t^\lambda) \\
&\quad = \int_\Omega (v_h^\lambda - v^\lambda)\cdot(b_h - b)\,dx + \int_{\partial\Omega}(v_h^\lambda - v^\lambda)\cdot(T_h^\lambda - T^\lambda)\,n\,d\sigma \\
&\qquad - \int_\Omega \Big\{G_\lambda(B^T T_h^\lambda - Lz_h^\lambda) - G_\lambda(B^T T^\lambda - Lz^\lambda)\Big\} \\
&\qquad\qquad \cdot((B^T T_h^\lambda - Lz_h^\lambda) - (B^T T^\lambda - Lz^\lambda))\,dx \\
&\quad \le (\text{by the monotonicity of the function } G_\lambda) \\
&\quad \le \int_\Omega (v_h^\lambda - v^\lambda)\cdot(b_h - b)\,dx + \int_{\partial\Omega}(v_h^\lambda - v^\lambda)\cdot(T_h^\lambda - T^\lambda)\,n\,d\sigma\,. \qquad (3.32)
\end{aligned}$$

The boundary integral is defined in the following sense

$$\int_{\partial\Omega}(v_h^\lambda - v^\lambda)\cdot(T_h^\lambda - T^\lambda)\,n\,d\sigma = \langle\,(T_h^\lambda - T^\lambda)\,n\,,\,(v_h^\lambda - v^\lambda)\,\rangle_{\partial\Omega}\,, \qquad (3.33)$$

where the brackets $\langle\,\cdot\,,\,\cdot\,\rangle_{\partial\Omega}$ denote the canonical bilinear form between the spaces $H^2_{1/2}(\partial\Omega;\mathbb{R}^3)$ and $H^2_{-1/2}(\partial\Omega;\mathbb{R}^3)$.

Using the boundary conditions we get

$$\begin{aligned}
&\frac{d}{dt}\frac{1}{2}\|(\varepsilon_h^\lambda - \varepsilon^\lambda, z_h^\lambda - z^\lambda)\|_\psi^2 \\
&\quad \le \langle\,(\gamma_{N,h} - \gamma_N)\,,\,(v_h^\lambda - v^\lambda)\,\rangle_{\Gamma_2} \\
&\qquad + \langle\,(T_h^\lambda - T^\lambda)\,n\,,\,(\partial_t\gamma_{D,h} - \partial_t\gamma_D)\,\rangle_{\Gamma_1} \qquad (3.34) \\
&\qquad + \int_\Omega (v_h^\lambda - v^\lambda)\cdot(b_h - b)\,dx\,.
\end{aligned}$$

Next we integrate (3.34) with respect to t, divide by h^2 and shift the difference operators from the velocity and from the stress to the given data (for more details see [8], Theorem 4). Finally we pass to the limit $h \to 0^+$ and arrive at the inequality

$$\begin{aligned}
&\|(\varepsilon_t^\lambda(t), z_t^\lambda(t))\|_\psi^2 \\
&\leq \|(\varepsilon_t^\lambda(0), z_t^\lambda(0))\|_\psi^2 + 2\int_0^t \|\gamma_{N,tt}(\tau)\|_{-1/2,2,\Gamma_2} \|v^\lambda(\tau)\|_{1/2,2,\partial\Omega}\, d\tau \\
&\quad + 2\int_0^t \|T^\lambda(\tau)\, n\|_{-1/2,2,\partial\Omega} \|\gamma_{D,ttt}(\tau)\|_{1/2,2,\Gamma_1}\, d\tau \\
&\quad + 2\int_0^t \|v^\lambda(\tau)\|_{2,\Omega} \|b_{tt}(\tau)\|_{2,\Omega}\, d\tau \\
&\quad + C\Big\{ \sup_{t\in(0,T_e)} \|\gamma_{N,t}(t)\|_{-1/2,2,\Gamma_2} \sup_{t\in(0,T_e)} \|v^\lambda(t)\|_{1/2,2,\partial\Omega} \\
&\qquad + \sup_{t\in(0,T_e)} \|\gamma_{D,tt}(t)\|_{1/2,2,\Gamma_1} \sup_{t\in(0,T_e)} \|T^\lambda(t)\, n\|_{-1/2,2,\partial\Omega} \\
&\qquad + \sup_{t\in(0,T_e)} \|v^\lambda(t)\|_{2,\Omega} \sup_{t\in(0,T_e)} \|b_t(t)\|_{2,\Omega} \Big\}, \qquad (3.35)
\end{aligned}$$

where the constant $C > 0$ is independent of λ. The boundary norms $\|v^\lambda\|_{1/2,2,\partial\Omega}$ and $\|T^\lambda n\|_{-1/2,2,\partial\Omega}$ appearing in the right-hand side of inequality (3.35) can be estimated using the continuity of the trace operator

$$\|v^\lambda\|_{1/2,2,\partial\Omega} \leq C\, \|v^\lambda\|_{1,2,\Omega}, \qquad (3.36)$$

$$\|T^\lambda n\|_{-1/2,2,\partial\Omega} \leq C\, (\|T^\lambda\|_{2,\Omega} + \|\operatorname{div} T^\lambda\|_{2,\Omega}), \qquad (3.37)$$

where the constant $C > 0$ depends on the domain Ω only. Inserting (3.36) and (3.37) into (3.35) yields

$$\begin{aligned}
\|(\varepsilon_t^\lambda(t), z_t^\lambda(t))\|_\psi^2 &\leq \|(\varepsilon_t^\lambda(0), z_t^\lambda(0))\|_\psi^2 + C(T_e, \alpha) \\
&\quad + \alpha \sup_{t\in(0,T_e)} \|v^\lambda(t)\|_{1,2,\Omega}^2 + \alpha \sup_{t\in(0,T_e)} \|T^\lambda(t)\|_{2,\Omega}^2, \qquad (3.38)
\end{aligned}$$

where $\alpha > 0$ is any positive number and the constant $C(T_e, \alpha)$ does not depend on λ. Thus we have only to estimate the L^2-norm of the stress and the H_1-norm of the velocity. The first one can be estimated by the time derivative of the stress

$$\|T^\lambda(t)\|_{2,\Omega}^2 \leq 2\|T^\lambda(0)\|_{2,\Omega}^2 + 2t\int_0^t \|T_t^\lambda(\tau)\|_{2,\Omega}^2\, d\tau, \qquad (3.39)$$

and the second one can be estimated using the ellipticity of the equation of motion

$$\begin{aligned}
\|v^\lambda(t)\|_{1,2,\Omega} &\leq C\, (\|b_t(t)\|_{-1,2,\Omega} + \|\varepsilon_{p,t}^\lambda(t)\|_{2,\Omega} + \|\gamma_{D,t}(t)\|_{1/2,2,\Gamma_1} \\
&\quad + \|\gamma_{N,t}(t)\|_{-1/2,2,\Gamma_2}), \qquad (3.40)
\end{aligned}$$

where the constant $C > 0$ depends on the set Ω only. Inserting (3.39) and (3.40) into (3.38) and choosing α sufficiently small we arrive at the inequality

$$\|(\varepsilon_t^\lambda(t), z_t^\lambda(t))\|_\psi^2 \leq C(T_e)\left(\|(\varepsilon_t^\lambda(0), z_t^\lambda(0))\|_\psi^2 + 1\right). \tag{3.41}$$

Finally, the assumption $B^T T^{(0)} - Lz^{(0)} \in \Delta(g)$ and the properties of the Yosida approximation (see [2] paragraph 3) imply that the sequence $\|(\varepsilon_t^\lambda(0), z_t^\lambda(0))\|_\psi$ is bounded. The proof is complete.

From Theorem 3.2 we conclude that the sequence $\{(\varepsilon^\lambda, z^\lambda)\}_{\lambda>0}$ is bounded in $H_1^\infty(L^2)$, and this together with (3.22) implies that the sequence $\{T^\lambda\}_{\lambda>0}$ is bounded in $H_1^\infty(L^2)$. Hence, we can pass to the weak limit $\lambda \to 0^+$ in the system (3.21)–(3.24) and obtain that the limit functions satisfy the following system of equations

$$\operatorname{div}_x T(x,t) = -b(x,t), \tag{3.42}$$

$$T(x,t) = \mathcal{D}(\varepsilon(x,t) - Bz(x,t)), \tag{3.43}$$

$$\varepsilon(x,t) = \frac{1}{2}(\nabla_x u(x,t) + \nabla_x^T u(x,t)), \tag{3.44}$$

$$z_t(x,t) = \chi(x,t), \tag{3.45}$$

where $\chi(x,t) = w\text{-}\lim_{\lambda\to 0^+} G_\lambda(B^T T^\lambda - Lz^\lambda)$. To end the proof of existence of global solutions to our system (1.9)–(1.12) it remains to show that

$$\chi(x,t) \in g(B^T T(x,t) - Lz(x,t)) \text{ for a.e. } (x,t) \in Z. \tag{3.46}$$

Note that by the L^2-weak-strong closedness of the graph of the maximal monotone operator g the inclusion (3.46) follows immediately from the strong L^2-convergence of the sequence $\{B^T T^\lambda - Lz^\lambda\}_{\lambda>0}$.

Theorem 3.3. *Suppose that the boundary data γ_D, γ_N, the external force b and the initial data z^0 satisfy the assumptions from Theorem 3.2. Then for all $\lambda, \mu > 0$ and for all $T_e > 0$ the inequality*

$$\|(\varepsilon^\lambda - \varepsilon^\mu, z^\lambda - z^\mu)\|_\psi^2 \leq \frac{1}{2}(\lambda + \mu) C^2(T_e) T_e \tag{3.47}$$

holds, where $C(T_e)$ is the constant from Theorem 3.2.

Proof: The proof of this theorem is standard (see for example the proof of Theorem 3.1, p. 54 in [7]). Nevertheless, for completeness we present it here. Proceeding similarly as in the proof of Theorem 3.2, we obtain for the difference of two approximation steps $(\varepsilon^\lambda - \varepsilon^\mu, z^\lambda - z^\mu)$ that

$$\frac{d}{dt}\frac{1}{2}\|(\varepsilon^\lambda - \varepsilon^\mu, z^\lambda - z^\mu)\|_\psi^2 \tag{3.48}$$

$$= -\int_\Omega \left\{G_\lambda(B^T T^\lambda - Lz^\lambda) - G_\mu(B^T T^\mu - Lz^\mu)\right\}$$

$$\cdot ((B^T T^\lambda - Lz^\lambda) - (B^T T^\mu - Lz^\mu))\, dx.$$

Let us denote by $w^i = B^T T^i - Lz^i$ for $i = \lambda, \mu$. Using $G_\lambda = \lambda^{-1}(I - J_\lambda)$ we obtain

$$\begin{aligned}
&-\int_\Omega (w^\lambda - w^\mu)(G_\lambda(w^\lambda) - G_\mu(w^\mu))\,dx \\
&\quad= -\int_\Omega (\lambda G_\lambda(w^\lambda) + J_\lambda(w^\lambda) - \mu G_\mu(w^\mu) \\
&\qquad - J_\mu(w^\mu)) \cdot (G_\lambda(w^\lambda) - G_\mu(w^\mu))\,dx \\
&\quad\leq (\text{by the inclusion } G_i(w^i) \in g(J_i(w^i))) \qquad\qquad (3.49)\\
&\quad\leq -\int_\Omega (\lambda G_\lambda(w^\lambda) - \mu G_\mu(w^\mu)) \cdot (G_\lambda(w^\lambda) - G_\mu(w^\mu))\,dx \\
&\quad\leq \frac{1}{4}(\lambda + \mu)(\|G_\lambda(w^\lambda)\|_{2,\Omega}^2 + \|G_\mu(w^\mu)\|_{2,\Omega}^2)\,.
\end{aligned}$$

We use Theorem 3.2 and equation (3.24) to see that $\|G_i(w^i)\|_{2,\Omega}$ is uniformly bounded for $0 \leq t \leq T_e$. Insertion of (3.49) into (3.48) and integration with respect to t ends the proof.

Proof of Theorem 1.4: From Theorem 3.3 we conclude that the sequences $\{\varepsilon^\lambda\}_{\lambda>0}$ and $\{z^\lambda\}_{\lambda>0}$ are L^2-Cauchy sequences (note that the energy norm $\|\cdot\|_\psi$ is equivalent to the standard L^2 norm). Consequently, the sequence $\{B^T T^\lambda - Lz^\lambda\}_{\lambda>0}$ is a Cauchy sequence in $L^\infty(L^2)$ which implies that the inclusion (3.46) holds. Therefore a solution with the properties stated in Theorem 1.4 exists. Uniqueness follows as in the proof of Theorem 1.3 in Section 2.

References

[1] H.-D. Alber and K. Chełmiński, *Quasistatic problems in viscoplasticity theory II: Models with general hardening*, submitted to J. Nonlinear Sci., 2003.

[2] J.P. Aubin and A. Cellina, *Differential inclusions*, Springer, Berlin Heidelberg New York, 1984.

[3] H.-D. Alber, *Materials with memory – Initial-boundary value problems for constitutive equations with internal variables*, Lecture Notes in Mathematics, vol. 1682, Springer, Berlin Heidelberg New York, 1998.

[4] H.-D. Alber, *Existence of solutions to a class of quasi-static problems in viscoplasticity theory*, Optimal control and partial differential equations (J. Menaldi, E. Rofman, and A. Sulem, eds.), IOS Press, Amsterdam, 2001, pp. 95–104.

[5] H.-D. Alber, *Justification of homogenized models for viscoplastic bodies with microstructure*, Deformation and failure of metallic continua (K. Hutter and H. Baaser, eds.), Springer, Berlin, 2003, to appear.

[6] V. Barbu, *Nonlinear semigroups and differential equations in Banach spaces*, Editura Academiei, Bucharest; Noordhoff, Leyden, 1976.

[7] H. Brézis, *Opérateurs maximaux monotones*, North Holland, Amsterdam, 1973.

[8] K. Chełmiński and P. Gwiazda, *Nonhomogeneous initial-boundary value problems for coercive and self-controlling models of monotone type*, Continuum Mech. Thermodyn. **12** (2000), 217–234.

[9] K. Chełmiński, *Coercive and self-controlling quasistatic models of the gradient type with composite inelastic constitutive equations*, submitted to J. Nonlinear Sci., 2001.

[10] K. Chełmiński, *Coercive limits for a subclass of monotone constitutive equations in the theory of inelastic material behaviour of metals*, Mat. Stos. **40** (1997), 41–81.

[11] K. Chełmiński, *Coercive approximation of viscoplasticity and plasticity*, Asymptotic Analysis **26** (2001), 105–133.

[12] G. Duvaut and J.L. Lions, *Les inéquations en mécanique et en physique*, Dunod, Paris, 1972.

[13] S. Ebenfeld, *Remarks on the quasistatic problem of viscoelasticity: Existence, uniqueness and homogenization*, Continuum Mech. Thermodyn. **14,6** (2002), 511–526.

[14] K. Gröger, *Zur Theorie des quasi-statischen Verhaltens von elastisch-plastischen Körpern*, Z. Angw. Math. Mech. **58** (1978), 81–88.

[15] K. Gröger, *Initial-value problems for elastoplastic and elasto-viscoplastic systems*, Nonlinear Analysis. Function spaces and applications (Leipzig) (S. Fučik and A. Kufner, eds.), Teubner, 1979, pp. 95–127.

[16] B. Halphen and Nguyen Quoc Son, *Sur les matériaux standards généralisés*, J. Méc. **14** (1975), 39–63.

[17] C. Johnson, *Existence theorems for plasticity problems*, J. Math. Pures Appl. **55** (1976), 431–444.

[18] C. Johnson, *On plasticity with hardening*, J. Math. Anal. Appl. **62** (1978), 325–336.

[19] V.A. Kondratiev and O.A. Oleinik, *Boundary-value problems for the systems of elasticity theory in unbounded domains. Korn's inequalities*, Russian Math. Surveys **43** (1988), 65–119.

[20] P. LeTallec, *Numerical analysis of viscoelastic problems*, Masson, Paris; Springer, Berlin, 1990.

[21] J.J. Moreau, *Sur l'évolution d'un système élasto-visco-plastique*, C. R. Acad. Sci. Paris, Sér. A **273** (1971), 118–121.

[22] C. Pares, *Existence, uniqueness and regularity of solution of the equation of a turbulence model for incompressible fluids*, Applicable Analysis **43** (1992), 245–296.

[23] R. Temam, *Problèmes mathématiques en plasticité*, Gauthier–Villars, Paris, 1983.

[24] E. Zeidler, *Nonlinear functional analysis and its applications IV*, Springer, New York, 1997.

Hans-Dieter Alber
Darmstadt University of Technology, Schlossgartenstrasse 7, D-64289 Darmstadt
e-mail: alber@mathematik.tu-darmstadt.de

Krzysztof Chełmiński
University of Konstanz and Cardinal Stefan Wyszyński University, Warsaw
e-mail: chelminski@fmi.uni-konstanz.de

Operator Theory
Advances and Applications, Vol. 147, 131–149

Wiener-Hopf Determinants with Fisher-Hartwig Symbols

Estelle L. Basor and Harold Widom

Dedicated to the memory of Erhard Meister

Abstract. With localization techniques one can obtain general limit theorems for Toeplitz determinants with Fisher-Hartwig singularities from the asymptotics for any symbol with one singularity of general type. There exists a family of these for which the determinants can be evaluated explicitly and their asymptotics determined. But for the Wiener-Hopf analogue, although there are likely analogous localization techniques, there is not a single example known of a symbol with Fisher-Hartwig singularity for which the determinant can be evaluated explicitly. In this paper we determine the asymptotics of Wiener-Hopf determinants for a symbol with one Fisher-Hartwig singularity of general type. We do this by showing that it is asymptotically equal to a Toeplitz determinant with symbol having the corresponding singularity.

1. Introduction

The strong Szegö limit theorem states that if the symbol φ defined on the unit circle has a sufficiently well-behaved logarithm then the determinant of the Toeplitz matrix

$$T_n(\varphi) = (\varphi_{j-k})_{j,k=0,\ldots,n-1}$$

has the asymptotic behavior

$$\det T_n(\varphi) \sim G(\varphi)^n \, E(\varphi) \quad \text{as } n \to \infty, \tag{1.1}$$

where

$$G(\varphi) = e^{(\log \varphi)_0}, \quad E(\varphi) = \exp\left(\sum_{k=1}^{\infty} k \, (\log \varphi)_k \, (\log \varphi)_{-k}\right).$$

Here subscripts denote Fourier coefficients.

Fisher and Hartwig [15] introduced a family of symbols with singularities and conjectured the form of the asymptotics for these. If

$$\varphi_{\alpha,\beta}(e^{i\theta}) = (2 - 2\cos\theta)^{(\alpha+\beta)/2} \, e^{i(\theta-\pi)(\alpha-\beta)/2}, \quad 0 < \theta < 2\pi$$

(this symbol is said to have a pure Fisher-Hartwig singularity), then their symbols had the form

$$\psi(z) = \varphi(z) \prod_{j=1}^{N} \varphi_{\alpha_j,\beta_j}(z/z_j),$$

where φ satisfies the assumption of Szegö's theorem and $z_1, \ldots, z_N$ are distinct points on the unit circle. They conjectured that for some range of the parameters the asymptotics had the form

$$\det\, T_n(\psi) \sim G(\varphi)^n\, n^{\sum \alpha_j\beta_j}\, E(\varphi,\, \alpha_j,\, \beta_j,\, z_j)$$

where $E(\varphi,\, \alpha_j,\, \beta_j,\, z_j)$ is a constant (whose value they did not conjecture). Due to the work of many mathematicians the conjecture has now been proved, and the constant $E(\varphi,\, \alpha_j,\, \beta_j,\, z_j)$ determined, in great generality. The basic condition is that $|\Re\,(\alpha_j \pm \beta_j)| < 1$. In early work the case of several singularities was tackled directly and the proofs were quite involved. But later it was discovered [1, 3, 9] that one could use localization techniques which made it possible to prove general results if one knew the asymptotics for the symbols $\varphi_{\alpha,\beta}$ with pure singularity. Luckily the Toeplitz determinants in these cases could be evaluated explicitly and their asymptotics were then a straightforward matter [8]. (For a detailed history, and a proof of the general result, see [10]. Stronger results for symbols with one singularity are contained in [14], and a status report for a generalized conjecture can be found in [13].)

For the Wiener-Hopf analogue the symbol $\sigma(\xi)$ is defined on the real line and equals 1 at $\pm\infty$. The finite Wiener-Hopf operator $W_R(\sigma)$ acts on $L^2(0, R)$ and is equal to the identity plus the operator with kernel

$$\frac{1}{2\pi} \int_{-\infty}^{\infty} (\sigma(\xi) - 1)\, e^{-i\,(x-y)\,\xi}\, d\xi.$$

Here there is also a "Szegö theorem" for nicely-behaved symbols:

$$\det W_R(\sigma) \sim G(\sigma)^R\, E(\sigma) \quad \text{as } R \to \infty. \tag{1.2}$$

Now

$$G(\sigma) = e^{\tau(0)}, \quad E(\sigma) = \exp\left(\int_0^{\infty} x\, \tau(x)\, \tau(-x)\, dx \right),$$

where τ is the Fourier transform of $\log \sigma$,

$$\tau(x) = \frac{1}{2\pi} \int_{-\infty}^{\infty} \log \sigma(\xi)\, e^{-ix\xi}\, d\xi.$$

A Wiener-Hopf symbol with pure Fisher-Hartwig singularity could be defined by

$$\sigma_{\alpha,\beta}(\xi) = \left(\frac{\xi - 0i}{\xi - i} \right)^{\alpha} \left(\frac{\xi + 0i}{\xi + i} \right)^{\beta}.$$

(We specify the arguments of $\xi \pm 0i$ and $\xi \pm i$ to be zero or close to it when ξ is large and positive.) This has the behavior

$$\sigma_{\alpha,\beta}(\xi) \sim |\xi|^{\alpha+\beta}\, e^{\frac{1}{2} i\pi(\alpha-\beta)\,\mathrm{sgn}\xi} \quad \text{as } \xi \to 0.$$

In [6] there was stated an analogue of the general Fisher-Hartwig conjecture for symbols which have a finite number of singularities of the above type, with explicit values of the constants. These symbols have the form

$$\sigma_0(\xi) \prod_r \sigma_{\alpha_r,\beta_r}(\xi - \xi_r),$$

where σ_0 is a symbol for which (1.2) holds. For an attempted proof, it seems reasonable to try to do what was done in the Toeplitz case, that is, devise the proper localization techniques and then try to evaluate the determinants in the case of the pure singularity. However, while it is likely that quite general localization techniques can be developed (and, in fact, have been in certain cases) there has never been a single example where the finite Wiener-Hopf determinant for a singular symbol could be evaluated explicitly.

In this paper we shall find the Wiener-Hopf asymptotics for the pure symbols $\sigma_{\alpha,\beta}$. A minor complication encountered for these symbols is that the function $\sigma_{\alpha,\beta} - 1$ is not necessarily in L^1 but is in $L^1 + L^2$. The corresponding operators $W_R(\sigma) - I$ will be Hilbert-Schmidt and we shall use the regularized determinants $\det_2 W_R(\sigma)$. (For a Hilbert-Schmidt operator K the regularized determinant is defined by $\det_2(I + K) = \det(I + K)\, e^{-K}$.)

Previous results concerning the regularized determinants go back to [7] where the asymptotics were determined when all $\alpha_j = 0$ or all $\beta_j = 0$. In the case of jump discontinuities, when all $\alpha_j + \beta_j = 0$, general results were also known. These were obtained by (two different kinds of) discretization which led to Toeplitz problems. For these results see [4], [11] and especially [12] where the regularized determinant asymptotics were, in a sense, settled in the jump case. Also, a conjecture for the $\alpha = \beta$ case was given in [17] and proved in some special case.

We determine the asymptotics for the symbols $\sigma_{\alpha,\beta}$ under the condition $|\Re(\alpha \pm \beta)| < 1$. This is done by showing that after normalization it is asymptotically equal to a regularized Toeplitz determinant.

Theorem. *If* $|\Re(\alpha \pm \beta)| < 1$ *then*

$$det_2\, W_R(\sigma_{\alpha,\beta})/G_2(\sigma_{\alpha,\beta})^R \sim det_2\, T_n(\varphi_{\alpha,\beta})/G_2(\varphi_{\alpha,\beta})^n$$

when $R \sim 2n \to \infty$, *where*

$$G_2(\sigma) = \exp\left(\frac{1}{2\pi}\int_{-\infty}^{\infty}(\log\sigma(\zeta) - \sigma(\zeta) + 1)\, d\zeta\right)$$

and

$$G_2(\varphi) = \exp\big((\log\varphi)_0 - \varphi_0 + 1\big).$$

This will be proved by finding exact formulas for the Toeplitz and Wiener-Hopf determinants and regularized determinants and then showing the above quotients are asymptotically equal when $R \sim 2n$. These formulas are expressed in terms of Fredholm determinants of operators acting on $L^2(0,1)$ and are obtained by using an identity of Borodin and Okounkov [5] for Toeplitz determinants with regular symbol and its Wiener-Hopf analogue [2]. What we do is simply stated: in both cases we introduce a parameter to regularize the symbol, apply the identity, and then take the limit.

We remark that in the Toeplitz case, computing a regularized determinant from an ordinary determinant is easy and it causes no problem to go back and forth between the two in computations. The same is true for the finite Wiener-Hopf operators when $\sigma_{\alpha,\beta} - 1$ is in L^1. This holds exactly when $\beta = \alpha$, and the following corollary will follow from the theorem.

Corollary. *If* $|\Re(\alpha)| < 1/2$ *then*

$$\det W_R(\sigma_{\alpha,\alpha}) \sim e^{-R\alpha} \det T_n(\varphi_{\alpha,\alpha})$$

when $R \sim 2n \to \infty$.

2. The case of $\beta = \alpha$

We do this case first since the general case has extra complications, but the main points are the same and will not have to be repeated.

To state the Borodin-Okounkov identity, let φ be a symbol with sufficiently smooth logarithm and let $\varphi(z) = \varphi^-(z)\,\varphi^+(z)$ be its Wiener-Hopf factorization, so that φ^+ extends to a nonzero analytic function inside the unit circle and φ^- outside. Let K_n be the operator on $\ell^2(\{0, 1, \dots\})$ with matrix entries

$$K_n(i,j) = \sum_{k=0}^{\infty} (\varphi^-/\varphi^+)_{n+i+k+1}\, (\varphi^+/\varphi^-)_{-n-j-k-1}.$$

The identity is

$$\det T_n(\varphi) = G(\varphi)^n\, E(\varphi)\, \det(I - K_n), \tag{2.1}$$

where $G(\varphi)$ and $E(\varphi)$ are the constants appearing in (1.1). For the regularized determinant we have

$$\det\nolimits_2 T_n(\varphi) = G_2(\varphi)^n\, E(\varphi)\, \det(I - K_n). \tag{2.2}$$

This follows immediately from the first identity since

$$\det\nolimits_2 A = \det A\, e^{-\mathrm{tr}(A-I)}$$

whenever $A - I$ is trace class. Notice also that we encounter the same expression in computing either

$$\det T_n(\varphi)/G(\varphi)^n$$

or

$$\det\nolimits_2 T_n(\varphi)/G_2(\varphi)^n,$$

so the crucial computation is to find $E(\varphi)\det(I - K_n)$. This will be true when we do the Wiener-Hopf analogue as well.

If we introduce a parameter $r < 1$ the symbol $\varphi_{\alpha,\alpha}(z) = (1-z)^\alpha(1-z^{-1})^\alpha$ becomes $\varphi_r(z) = (1-rz)^\alpha(1-rz^{-1})^\alpha$, a regular symbol for which both (2.1) and (2.2) hold. The limit of $G_2(\varphi_r)$ as $r \to 1$ is easily seen to be $G_2(\varphi_{\alpha,\alpha})$. The constant $E(\varphi_r)$ equals $(1-r^2)^{-\alpha^2}$. The operator K_n is the product of two Hankel operators on $\ell^2(\{0, 1, \dots\})$ which in this case are the same and have i, j entry

$$\frac{1}{2\pi i}\int_{|z|=1}\left(\frac{1-rz}{1-rz^{-1}}\right)^{\alpha} z^{n+i+j}dz = \frac{\sin\pi\alpha}{\pi}\int_0^r\left(\frac{1-rx}{r-x}\right)^{\alpha} x^{n+i+j+\alpha}dx.$$

If the operator is denoted by H then $\det(I-K_n) = \det(I-H^2)$, and so we are interested in $\det(I \pm H)$. We consider at first only $\det(I-H)$, the Fredholm determinant of H.

If $\Re\,\alpha < 1/2$ we can write $H = UV$, where $U : L^2(0,r) \to \ell^2(\{0, 1, \dots\})$ has kernel $U(i,x) = x^i$ and $V : \ell^2(\{0, 1, \dots\}) \to L^2(0,r)$ has kernel

$$V(x,i) = \frac{\sin\pi\alpha}{\pi}\left(\frac{1-rx}{r-x}\right)^{\alpha} x^{n+i+\alpha}.$$

These are both Hilbert-Schmidt (in fact trace class) and $\det(I-UV) = \det(I-VU)$. (See, for example, Chap. 4 of [16].) It follows that H has the same Fredholm determinant as the kernel

$$\frac{\sin\pi\alpha}{\pi}\left(\frac{1-rx}{r-x}\right)^{\alpha} x^{n+\alpha}\frac{1}{1-xy}$$

on $L^2(0,r)$. If we set $r = (1-\varepsilon)/(1+\varepsilon)$ and make the substitutions

$$x \to \frac{1-x}{1+x},\quad y \to \frac{1-y}{1+y},$$

this becomes the kernel

$$\frac{\sin\pi\alpha}{\pi}\left(\frac{x+\varepsilon}{x-\varepsilon}\right)^{\alpha}\left(\frac{1-x}{1+x}\right)^{n+\alpha}\frac{1}{x+y} \tag{2.3}$$

on $L^2(\varepsilon, 1)$. (By this we mean that the kernels represent unitarily equivalent operators and so have the same Fredholm determinant.) We shall determine the limit of its Fredholm determinant as $\varepsilon \to 0$.

Denote by A the operator with the above kernel and by A_0 the operator with kernel

$$A_0(x,y) = \frac{\sin\pi\alpha}{\pi}\frac{1}{x+y}.$$

Let P be multiplication by $\chi_{[\sqrt{\varepsilon},1]}$ and Q multiplication by $\chi_{[\varepsilon,\sqrt{\varepsilon}]}$. Although our operators act on $L^2(\varepsilon,1)$ they can be thought of in the obvious way as acting on $L^2(0,1)$. For example the kernel of A_0 can be replaced by $\chi_{[\varepsilon,1]}(x)\,A_0(x,y)\,\chi_{[\varepsilon,1]}(y)$.

We shall show that $P(A-A_0)$ converges in trace norm as $\varepsilon \to 0$. For this, and later use, we give an estimate for the trace norm of certain kernels.

Lemma 2.1. *The trace norm of a kernel $f(x)g(y)/(x+y)$ on $L^2(J)$, where $J \subset (0,\infty)$, is at most a constant depending on a times the square root of*

$$\int_J |f(x)|^2\frac{dx}{x^{1+a}} \cdot \int_J |g(x)|^2\frac{dx}{x^{1-a}}.$$

Here a belongs to $(-1,1)$ but is otherwise arbitrary.

Proof. If we write the kernel as

$$\int_0^\infty f(x)s^{a/2}e^{-sx}\,e^{-sy}s^{-a/2}g(y)ds$$

we see that it is the (operator) product of two kernels acting between $L^2(J)$ and $L^2(0,\infty)$. The square of the Hilbert-Schmidt norm of the first equals

$$\int_J\int_0^\infty |f(x)|^2 s^a e^{-2sx} ds\, dx,$$

which is a constant depending on a times $\int_J |f(x)|^2 x^{-1-a}dx$. The second is analogous. □

Lemma 2.2. *The operator $P(A-A_0)$ converges in trace norm to the operator on $L^2(0,1)$ with kernel*

$$\frac{\sin\pi\alpha}{\pi}\left[\left(\frac{1-x}{1+x}\right)^{n+\alpha}-1\right]\frac{1}{x+y}.$$

Proof. The kernel of the operator is the sine factor times

$$\chi_{[\sqrt{\varepsilon},1]}(x)\left[\left(\frac{x+\varepsilon}{x-\varepsilon}\right)^{\alpha}\left(\frac{1-x}{1+x}\right)^{n+\alpha}-1\right]\frac{1}{x+y}$$

on $L^2(\varepsilon,1)$. If we replace the first x-factor by 1 then the error is of the form

$$\chi_{[\sqrt{\varepsilon},1]}(x)\,\varepsilon_1(x)\left(\frac{1-x}{1+x}\right)^{n+\alpha}\frac{1}{x+y}$$

where $\varepsilon_1=O(\sqrt{\varepsilon})$. If we apply Lemma 2.1 with small positive a we find that the operator has trace norm $O(\varepsilon^{1/2-\delta})$ for any $\delta>0$. After this replacement we are left with

$$\chi_{[\sqrt{\varepsilon},1]}(x)\left[\left(\frac{1-x}{1+x}\right)^{n+\alpha}-1\right]\frac{1}{x+y},$$

so it suffices to show that

$$\left[\left(\frac{1-x}{1+x}\right)^{n+\alpha}-1\right]\frac{1}{x+y}$$

is trace class on $L^2(0,1)$. This also follows from Lemma 2.1. □

Lemma 2.3. *If α is sufficiently small and $\Re\alpha\le 0$ we have as $\varepsilon\to 0$*

$$\det(I-K_n)\sim\det(I-{A_0}^2)$$
$$\times\det\Big(I-(I-A_0)^{-1}P(A-A_0)\Big)\,\det\Big(I+(I+A_0)^{-1}P(A-A_0)\Big)$$
$$\times\det\Big(I-((I-A_0)^{-1}Q(A-A_0)\Big)\,\det\Big(I+((I+A_0)^{-1}Q(A-A_0)\Big). \quad (2.4)$$

Proof. Recall that $\det(I-K_n)=\det(I-H^2)$ and that $\det(I\pm H)=\det(I\pm A)$. The operator on $L^2(0,1)$ with kernel $A_0(x,y)$ without the sine factor has norm 1. Hence if α is sufficiently small then $I-A_0$ is invertible for all ε and

$$\det(I-A)=\det(I-A_0)\det\Big(I-(I-A_0)^{-1}[P(A-A_0)+Q(A-A_0)]\Big).$$

The operator in the second determinant can be written

$$\Big(I-(I-A_0)^{-1}Q(A-A_0)\Big)\Big(I-(I-A_0)^{-1}P(A-A_0)\Big)$$
$$-(I-A_0)^{-1}Q(A-A_0)(I-A_0)^{-1}P(A-A_0).$$

Now the operators $I-A_0$ are uniformly invertible (their inverses have bounded norms as $\varepsilon\to 0$) if α is small enough. The same is true of the first product above, because of the sine factor in $A-A_0$. It follows that (2.4) will be established if we can show that the last term above is $o_1(1)$, i.e, its trace norm is $o(1)$. (And if the analogous statement holds for $I+A$, which it will.) We know that $P(A-A_0)$ converges in trace norm. It follows from the uniform invertibility of $(I-A_0)^{-1}$ that it converges strongly as $\varepsilon\to 0$ to the corresponding operator on $L^2(0,1)$, where A_0 does not have the $\chi_{[\varepsilon,1]}$ factors. The same is true of $A-A_0$ when $\Re\,\alpha\le 0$ because then multiplication by $((x+\varepsilon)/(x-\varepsilon))^\alpha\,\chi_{(\sqrt{\varepsilon},\,1)}(x)$ converges strongly to I. Hence, since Q converges strongly to 0, the same is true of $Q(A-A_0)(I-A_0)^{-1}$ and this together with the trace norm convergence of $P(A-A_0)$ implies that the last term above is $o_1(1)$. □

We now have the ingredients necessary to derive our formula for

$$\det\nolimits_2 T_n(\varphi_{\alpha,\alpha})/G_2(\varphi_{\alpha,\alpha})^n.$$

Lemma 2.4. *Assume $|\Re\,\alpha|<1/2$. Let A_0 be the operator with kernel*

$$A_0(x,y)=\frac{\sin\pi\alpha}{\pi}\frac{1}{x+y},$$

A_1 the operator with kernel

$$A_1(x,y)=\frac{\sin\pi\alpha}{\pi}\left[\left(\frac{1-x}{1+x}\right)^{n+\alpha}-1\right]\frac{1}{x+y},$$

and A_2 the operator with kernel

$$A_2(x,y)=\frac{\sin\pi\alpha}{\pi}\left[\left(\frac{1+x}{1-x}\right)^{\alpha}-1\right]\frac{1}{x+y}.$$

All act on $L^2(0,1)$. Then

$$\det\nolimits_2 T_n(\varphi_{\alpha,\alpha})/G_2(\varphi_{\alpha,\alpha})^n=4^{-\alpha^2}E(1-\sin\pi\alpha\,\mathrm{sech}\pi\xi)\,E(1+\sin\pi\alpha\,\mathrm{sech}\pi\xi)$$
$$\times\det(I-(I-A_0)^{-1}A_1)\det(I+(I+A_0)^{-1}A_1)$$
$$\times\det(I-(I-A_0)^{-1}A_2)\det(I+(I+A_0)^{-1}A_2),$$

where the E factors are those of (1.2).

Proof. We compute the asymptotics of the right side of (2.4) as $\varepsilon \to 0$ $(r \to 1)$. If we make the substitutions $x \to e^{-x},\ y \to e^{-y}$ then A_0 becomes the operator on $L^2(0, \log \varepsilon^{-1})$ with kernel

$$\frac{\sin \pi\alpha}{2\pi}\ \operatorname{sech}(x-y)/2.$$

This is a finite Wiener-Hopf operator with symbol $\sin \pi\alpha\ \operatorname{sech}\pi\xi$ so we can use (1.2). We have $G(1-\sin^2 \pi\alpha \operatorname{sech}^2 \pi\xi) = e^{-\alpha^2}$. Since $E(\varphi_r) = (1-r^2)^{-\alpha^2}$ and $\varepsilon \sim (1-r)/2$ we see that $E(\varphi_r)\,\varepsilon^{\alpha^2} \to 4^{-\alpha^2}$, which gives

$$\lim_{r\to 1} E(\varphi_r)\ \det\left(I - {A_0}^2\right) = 4^{-\alpha^2} E(1-\sin\pi\alpha\ \operatorname{sech}\pi\xi)\, E(1+\sin\pi\alpha\ \operatorname{sech}\pi\xi).$$

The limit of the factor

$$\det\Big(I - (I-A_0)^{-1} P(A-A_0)\Big)$$

in (2.4) is the determinant of what we get if we replace $P(A-A_0)$ by the operator A_1 on $L^2(0,1)$, by the established trace norm convergence. As for the factor

$$\det\Big(I - (I-A_0)^{-1} Q(A-A_0)\Big),$$

notice that under the variable changes $x \to \varepsilon/x,\ y \to \varepsilon/y$ the operator $Q(A-A_0)$ on $L^2(\varepsilon, 1)$ becomes the one with kernel

$$\chi_{[\sqrt{\varepsilon},1]}(x)\ \left[\left(\frac{1+x}{1-x}\right)^{\alpha}\left(\frac{x-\varepsilon}{x+\varepsilon}\right)^{n+\alpha} - 1\right]\frac{1}{x+y}.$$

And, in analogy with what went before, this converges in trace norm to the kernel $A_2(x,y)$ on $L^2(0,1)$. Under this substitution the kernel of A_0 is unchanged.

All this can be done with the other factors in (2.4). This establishes the lemma under our assumption that α is small enough. But the identity will hold in any connected α-region containing 0 in which both sides are analytic, so where $\varphi_{\alpha,\alpha} \in L^1$ and where $I \pm A_0$ are invertible and A_1 and A_2 trace class. This holds for $|\Re\,\alpha| < 1/2$. □

Remark. The operators $I \pm (I \pm A_0)^{-1} A_1$, those involving n in the statement of the lemma, are uniformly invertible for large n. It suffices to show uniform invertibility for $I\pm$ the operator with kernel the sine factor times

$$\left(\frac{1-x}{1+x}\right)^{n+\alpha}\ \frac{1}{x+y}.$$

If we drop the α we make an error with operator norm $o(1)$, and then it suffices to prove the uniform invertibility of $I\pm$ the sine factor times

$$\left(\frac{1-x}{1+x}\right)^{n/2}\ \frac{1}{x+y}\ \left(\frac{1-y}{1+y}\right)^{n/2}.$$

Since the spectrum of $1/(x+y)$ is $[0,\pi]$ and the factors have absolute value at most 1, the spectrum of the above also lies in $[0,\pi]$. Since it is self-adjoint the uniform invertibility follows whenever $1/\sin\pi\alpha \notin [-1,1]$, and so when $|\Re\,\alpha| < 1/2$.

Now let us go to Wiener-Hopf. To state the analogue of the Borodin-Okounkov identity here, let σ be a symbol with regular logarithm. Write $\sigma = \sigma^+\sigma^-$ where σ^+ extends to be nonzero, bounded and analytic in the upper half-plane and σ^- in the lower and let K_R be the operator on $L^2(0,\infty)$ with kernel

$$K_R(x,y) = \int_0^\infty \left(\frac{\sigma^-}{\sigma^+} - 1\right)_{R+x+z} \left(\frac{\sigma^+}{\sigma^-} - 1\right)_{-R-z-y} dz.$$

(Here for notational convenience the subscripts denote Fourier transform.) Then for a symbol σ with sufficiently regular logarithm and satisfying $\sigma - 1 \in L^1$ we have

$$\det W_R(\sigma) = G(\sigma)^R\, E(\sigma)\, \det(I - K_R). \tag{2.5}$$

The analogous formula for the regularized determinant is the following: for a symbol with sufficiently regular logarithm and satisfying $\sigma - 1 \in L^2$,

$$\det{}_2\, W_R(\sigma) = G_2(\sigma)^R\, E(\sigma)\, \det(I - K_R). \tag{2.6}$$

The formula in the case that $\sigma - 1 \in L^1$ follows from the results in [2]. The derivation of the formula when $\sigma - 1 \in L^2$ is given in the Appendix.

We introduce a small parameter ε and change $\sigma_{\alpha,\alpha}$ to the regular symbol

$$\sigma_\varepsilon(\xi) = \left(\frac{\xi^2+\varepsilon^2}{\xi^2+1}\right)^\alpha.$$

As in the Toeplitz case we can compute the finite regularized Wiener-Hopf determinant of $\sigma_{\alpha,\alpha}$ for fixed R by letting ε tend to zero. To see that this is so observe that since $\sigma_\varepsilon(\xi) - 1$ tends to $\sigma_{\alpha,\alpha} - 1$ in L^1 its Fourier transforms converge in L^2 on any finite interval. The corresponding kernels $k_\varepsilon(x-y)$ and $k(x-y)$ satisfy

$$\int_0^R \int_0^R |k_\varepsilon(x-y) - k(x-y)|^2 dx dy \le R \int_{-\infty}^\infty |k_\varepsilon(x) - k(x)|^2 dx,$$

which tends to zero. This shows that the regularized determinants are limits of ones with smooth symbols. This also holds for the $G_2(\sigma_{\alpha,\alpha})$ term. (This discussion holds also for $\sigma_{\alpha,\beta}$ when $\alpha \neq \beta$; the regularized symbols then converge in the space $L^1 + L^2$ and their Fourier transforms still converge in L^2 on any finite interval. When $\alpha = \beta$ the above remarks hold for the ordinary determinant as well.)

The constant $E(\sigma_\varepsilon)$ which arises in (2.5) equals $((1+\varepsilon)^2/4\varepsilon)^{\alpha^2}$ and so $E(\sigma_\varepsilon)\,\varepsilon^{\alpha^2}$ has the same limit $4^{-\alpha^2}$ as in the Toeplitz case. If our finite Wiener-Hopf operator acts on $L^2(0,R)$ then the operator K_R in (2.5) equals the square of the Hankel operator acting on $L^2(0,\infty)$ with kernel (in variables s and t)

$$\frac{1}{2\pi}\int_{-\infty}^\infty \left[\left(\frac{\xi+\varepsilon i}{\xi-\varepsilon i}\right)^\alpha \left(\frac{\xi-i}{\xi+i}\right)^\alpha - 1\right] e^{i(R+s+t)\xi}\, d\xi$$

$$= \frac{\sin\pi\alpha}{\pi}\int_\varepsilon^1 \left(\frac{x+\varepsilon}{x-\varepsilon}\right)^\alpha \left(\frac{1-x}{1+x}\right)^\alpha e^{-(R+s+t)x}\, dx.$$

Making a switch $UV \to VU$ as before changes this to the kernel

$$\frac{\sin \pi\alpha}{\pi} \left(\frac{x+\varepsilon}{x-\varepsilon}\right)^{\alpha} \left(\frac{1-x}{1+x}\right)^{\alpha} e^{-Rx} \frac{1}{x+y}$$

on $L^2(\varepsilon, 1)$.

This is almost exactly the same as the kernel (2.3). The only difference is that the expression $((1-x)/(1+x))^{n+\alpha}$ there is replaced here by

$$\left(\frac{1-x}{1+x}\right)^{\alpha} e^{-Rx}.$$

The same argument given above for the Toeplitz case gives the following for the Wiener-Hopf case.

Lemma 2.5. *Assume $|\Re\,\alpha| < 1/2$. Let $\tilde{A}_1$ be the operator on $L^2(0,1)$ with kernel*

$$\tilde{A}_1(x,y) = \frac{\sin \pi\alpha}{\pi} \left[\left(\frac{1-x}{1+x}\right)^{\alpha} e^{-Rx} - 1\right] \frac{1}{x+y}.$$

Then with A_0 and A_2 as in Lemma 2.4 we have

$$\begin{aligned} det_2\, W_R(\sigma_{\alpha,\alpha})/G_2(\sigma_{\alpha,\alpha})^R &= 4^{-\alpha^2}\, E(1 - \sin \pi\alpha\ \mathrm{sech}\pi\xi)\, E(1 + \sin \pi\alpha\ \mathrm{sech}\pi\xi) \\ &\quad \times \det(I - (I - A_0)^{-1}\tilde{A}_1) \det(I + (I + A_0)^{-1}\tilde{A}_1) \\ &\quad \times \det(I - (I - A_0)^{-1}A_2) \det(I + (I + A_0)^{-1}A_2). \end{aligned}$$

We can now establish the theorem in the case $\alpha = \beta$. It follows from Lemmas 2.4 and 2.5 that

$$\frac{\det_2 T_n(\varphi_{\alpha,\alpha})/G_2(\varphi_{\alpha,\alpha})^n}{\det_2 W_R(\varphi_{\alpha,\alpha})/G_2(\sigma_{\alpha,\alpha})^n} = \frac{\det(I - (I - A_0)^{-1}A_1) \det(I + (I + A_0)^{-1}A_1)}{\det(I - (I - A_0)^{-1}\tilde{A}_1) \det(I + (I + A_0)^{-1}\tilde{A}_1)}.$$

If we knew that $A_1 - \tilde{A}_1 = o_1(1)$ then this together with the uniform invertibility described in the remark following Lemma 2.4 would show that the determinants involving A_1 and $\tilde{A}_1$ are asymptotically equal. Also, because when n is replaced by anything asymptotic to it the asymptotics of $T_n(\varphi_{\alpha,\alpha})$ are the same, we may replace the condition on R by the stronger one $R = 2n + O(1)$. We suppose first that $R = 2n$ exactly. If we apply Lemma 2.1 with $a = 1/2$, say, we see that we have to show that

$$\int_0^1 \left| \left(\frac{1-x}{1+x}\right)^{n} - e^{-2nx} \right|^2 \frac{dx}{x^{3/2}} = o(1)$$

as $n \to \infty$. The part of the integral where x is bounded away from 0 is clearly $o(1)$. If x is small, say $x < \delta$, we make a variable change $x \to x/2n$ and that part of the integral becomes a constant times

$$n^{1/2} \int_0^{2\delta n} \left| e^{-x+O(x^2/n)} - e^{-x} \right|^2 \frac{dx}{x^{3/2}}.$$

The integral over $x < \sqrt{n}$ is $O(n^{-2})$ and the integral over $x > \sqrt{n}$ is $O(e^{-\sqrt{n}})$ since the $O(x^2/n)$ term in the exponent is $< x/2$ in the range of integration if δ is small enough. Thus $A_1 - \tilde{A}_1 = o_1(1)$ when $R = 2n$. The error incurred when our actual $R = 2n + O(1)$ is replaced by $2n$ easily seen to be $o(1)$. This completes the proof of the theorem in the case $\alpha = \beta$.

Remark. To prove that the Toeplitz and Wiener-Hopf determinants are asymptotically equal it was clearly not necessary to have the individual identities given in Lemmas 2.4 and 2.5, but only the identity for the ratio given above. We derived the individual identities because there was so little extra work involved.

To derive the corollary, we need only observe that

$$\det{}_2\, T_n(\varphi_{\alpha,\alpha}) = \det\, T_n(\varphi_{\alpha,\alpha})\, e^{-\mathrm{tr}\,(T_n(\varphi_{\alpha,\alpha})-I)},$$
$$\det{}_2\, W_R(\sigma_{\alpha,\alpha}) = \det\, W_R(\sigma_{\alpha,\alpha})\, e^{-\mathrm{tr}\,(W_R(\sigma_{\alpha,\alpha})-I)},$$
$$\mathrm{tr}\,(T_n(\varphi_{\alpha,\alpha}) - I) = n\,((\varphi_{\alpha,\alpha})_0 - 1),$$
$$\mathrm{tr}\,(W_R(\sigma_{\alpha,\alpha}) - I) = R\left(\frac{1}{2\pi}\int_{-\infty}^{\infty} (\sigma_{\alpha,\alpha}(\zeta) - 1)\, d\zeta\right),$$
$$G(\varphi_{\alpha,\alpha}) = 1 \quad \text{and} \quad G(\sigma_{\alpha,\alpha}) = e^{-\alpha}.$$

3. General α and β

Introducing a parameter r now leads to the symbol $\varphi_r(z) = (1-rz)^\alpha(1-rz^{-1})^\beta$. We have $E(\varphi_r) = (1-r^2)^{-\alpha\beta}$. Thus we are left with the contribution from $\det\,(I - K_n)$ just as in the previous case. Proceeding as we did before leads to the Fredholm determinant of the product of two different operators on $L^2(\varepsilon, 1)$. One has kernel

$$2^{\beta-\alpha}\varepsilon^{\alpha-\beta}\,\frac{\sin\pi\beta}{\pi}\,\frac{(x+\varepsilon)^\beta}{(x-\varepsilon)^\alpha}\,\frac{(1-x)^{n+\alpha}}{(1+x)^{n+\beta}}\,\frac{1}{x+y}$$

and the kernel of the other is obtained from this by interchanging α and β. For the product the factors $2^{\beta-\alpha}\varepsilon^{\alpha-\beta}$ and $2^{\alpha-\beta}\varepsilon^{\beta-\alpha}$ cancel, and so we can replace them both by 1. The Fredholm determinant of the product equals the Fredholm determinant of the matrix kernel

$$\begin{pmatrix} 0 & \frac{\sin\pi\beta}{\pi}\frac{(x+\varepsilon)^\beta}{(x-\varepsilon)^\alpha}\frac{(1-x)^{n+\alpha}}{(1+x)^{n+\beta}}\frac{1}{x+y} \\ \frac{\sin\pi\alpha}{\pi}\frac{(x+\varepsilon)^\alpha}{(x-\varepsilon)^\beta}\frac{(1-x)^{n+\beta}}{(1+x)^{n+\alpha}}\frac{1}{x+y} & 0 \end{pmatrix}.$$

We proceed as before, with A_0 having matrix kernel

$$\begin{pmatrix} 0 & \frac{\sin\pi\beta}{\pi}\frac{x^{\beta-\alpha}}{x+y} \\ \frac{\sin\pi\alpha}{\pi}\frac{x^{\alpha-\beta}}{x+y} & 0 \end{pmatrix}.$$

We assume at first that α and β are small and purely imaginary. Then all x-factors which arise will be bounded. The analogue of (2.4) is here

$$\det(I-K_n) \sim \det(I-A_0)$$
$$\times \det\Big(I-(I-A_0)^{-1}P(A-A_0)\Big)\,\det\Big(I-(I-A_0)^{-1}Q(A-A_0)\Big) \qquad (3.1)$$

as $r \to 1$ ($\varepsilon \to 0$). To see this, observe that $P(A-A_0)$ is $o_1(1)$ plus the operator with kernel

$$\begin{pmatrix} 0 & \frac{\sin\pi\beta}{\pi}\left[\frac{(1-x)^{n+\alpha}}{(1+x)^{n+\beta}}-1\right]\frac{x^{\beta-\alpha}}{x+y} \\ \frac{\sin\pi\alpha}{\pi}\left[\frac{(1-x)^{n+\beta}}{(1+x)^{n+\alpha}}-1\right]\frac{x^{\alpha-\beta}}{x+y} & 0 \end{pmatrix}.$$

This is true following the same argument as in Lemma 2.2 since the function

$$\frac{(x+\varepsilon)^\alpha}{(x-\varepsilon)^\beta}-x^{\alpha-\beta},$$

and also the one with α and β interchanged, are $O(\sqrt{\varepsilon})$ for $x \in (\sqrt{\varepsilon},1)$. From this it follows that the analogues of Lemmas 2.2 and 2.3 hold and also (3.1).

Now we proceed to find the analogue of Lemma 2.4. The kernel $Q(A-A_0)$ is $o_1(1)$ plus the operator with kernel

$$\chi_{(\varepsilon,\sqrt{\varepsilon})}(x)\begin{pmatrix} 0 & \frac{\sin\pi\beta}{\pi}\left[\frac{(x+\varepsilon)^\beta}{(x-\varepsilon)^\alpha}-x^{\beta-\alpha}\right]\frac{1}{x+y} \\ \frac{\sin\pi\alpha}{\pi}\left[\frac{(x+\varepsilon)^\alpha}{(x-\varepsilon)^\beta}-x^{\alpha-\beta}\right]\frac{1}{x+y} & 0 \end{pmatrix}.$$

If we make the variable changes $x \to \varepsilon/x,\ y \to \varepsilon/y$ this becomes

$$\chi_{(\sqrt{\varepsilon},1)}(x)\begin{pmatrix} 0 & \varepsilon^{\beta-\alpha}\frac{\sin\pi\beta}{\pi}\left[\frac{(1+x)^\beta}{(1-x)^\alpha}-1\right]\frac{x^{\alpha-\beta}}{x+y} \\ \varepsilon^{\alpha-\beta}\frac{\sin\pi\alpha}{\pi}\left[\frac{(1+x^\alpha}{(1-x)^\beta}-1\right]\frac{x^{\beta-\alpha}}{x+y} & 0 \end{pmatrix},$$

and the kernel of A_0 becomes

$$\begin{pmatrix} 0 & \varepsilon^{\beta-\alpha}\frac{\sin\pi\beta}{\pi}\frac{x^{\alpha-\beta}}{x+y} \\ \varepsilon^{\alpha-\beta}\frac{\sin\pi\alpha}{\pi}\frac{x^{\beta-\alpha}}{x+y} & 0 \end{pmatrix}.$$

Since determinants are unchanged if we multiply all kernels on the left by $\begin{pmatrix} I & 0 \\ 0 & \varepsilon^{\beta-\alpha}I \end{pmatrix}$ and on the right by $\begin{pmatrix} I & 0 \\ 0 & \varepsilon^{\alpha-\beta}I \end{pmatrix}$ we can remove these ε factors from both kernels. Therefore for the limit of the second determinant in (3.1) the operators act on $L^2(0,1)$ and we replace the kernel of A_0 by

$$\begin{pmatrix} 0 & \frac{\sin\pi\beta}{\pi}\frac{x^{\alpha-\beta}}{x+y} \\ \frac{\sin\pi\alpha}{\pi}\frac{x^{\beta-\alpha}}{x+y} & 0 \end{pmatrix},$$

and that of $Q(A - A_0)$ by

$$\begin{pmatrix} 0 & \frac{\sin\pi\beta}{\pi}\left[\frac{(1+x)^{\beta}}{(1-x)^{\alpha}} - 1\right]\frac{x^{\alpha-\beta}}{x+y} \\ \frac{\sin\pi\alpha}{\pi}\left[\frac{(1+x^{\alpha}}{(1-x)^{\beta}} - 1\right]\frac{x^{\beta-\alpha}}{x+y} & 0 \end{pmatrix}.$$

Finally, for the limiting operators arising from the $P(A-A_0)$ term we multiply the matrix kernels on the left by $\begin{pmatrix} x^{(\alpha-\beta)/2} & 0 \\ 0 & x^{(\beta-\alpha)/2} \end{pmatrix}$ and on the right by $\begin{pmatrix} y^{(\beta-\alpha)/2} & 0 \\ 0 & y^{(\alpha-\beta)/2} \end{pmatrix}$. Thus the kernel of A_0 here becomes

$$A_0(x,y) = \begin{pmatrix} 0 & \frac{\sin\pi\beta}{\pi}\frac{(x/y)^{(\beta-\alpha)/2}}{x+y} \\ \frac{\sin\pi\alpha}{\pi}\frac{(x/y)^{(\alpha-\beta)/2}}{x+y} & 0 \end{pmatrix} \tag{3.2}$$

and the kernel coming from $P(A - A_0)$ becomes

$$A_1(x,y) = \begin{pmatrix} 0 & \frac{\sin\pi\beta}{\pi}\left[\frac{(1-x)^{n+\alpha}}{(1+x)^{n+\beta}} - 1\right]\frac{(x/y)^{(\beta-\alpha)/2}}{x+y} \\ \frac{\sin\pi\alpha}{\pi}\left[\frac{(1-x)^{n+\beta}}{(1+x)^{n+\alpha}} - 1\right]\frac{(x/y)^{(\alpha-\beta)/2}}{x+y} & 0 \end{pmatrix}. \tag{3.3}$$

Similarly, for the limiting operators arising from the $Q(A - A_0)$ term we multiply the matrix kernels on the left by $\begin{pmatrix} x^{(\beta-\alpha)/2} & 0 \\ 0 & x^{(\alpha-\beta)/2} \end{pmatrix}$ and on the right by $\begin{pmatrix} y^{(\alpha-\beta)/2} & 0 \\ 0 & y^{(\beta-\alpha)/2} \end{pmatrix}$. The kernel of A_0 here becomes

$$A_0(x,y) = \begin{pmatrix} 0 & \frac{\sin\pi\beta}{\pi}\frac{(x/y)^{(\alpha-\beta)/2}}{x+y} \\ \frac{\sin\pi\alpha}{\pi}\frac{(x/y)^{(\beta-\alpha)/2}}{x+y} & 0 \end{pmatrix} \tag{3.4}$$

and the kernel coming from $Q(A - A_0)$ becomes

$$A_2(x,y) = \begin{pmatrix} 0 & \frac{\sin\pi\beta}{\pi}\left[\frac{(1+x)^{\beta}}{(1-x)^{\alpha}} - 1\right]\frac{(x/y)^{(\alpha-\beta)/2}}{x+y} \\ \frac{\sin\pi\alpha}{\pi}\left[\frac{(1+x^{\alpha}}{(1-x)^{\beta}} - 1\right]\frac{(x/y)^{(\beta-\alpha)/2}}{x+y} & 0 \end{pmatrix}. \tag{3.5}$$

None of these operations change the determinants.

For the asymptotics of det $(I - A_0)$ in (3.1) both versions (3.2) and (3.4) give the same determinant. If we make the variable changes $x \to e^{-x}$, $y \to e^{-y}$ then

(3.2) becomes

$$\begin{pmatrix} 0 & \frac{\sin\pi\beta}{2\pi}\,\frac{e^{(\beta-\alpha)(x-y)/2}}{\cosh(x-y)/2} \\ \frac{\sin\pi\alpha}{2\pi}\,\frac{e^{(\alpha-\beta)(x-y)/2}}{\cosh(x-y)/2} & 0 \end{pmatrix} \tag{3.6}$$

on $L^2(0,\infty)$. This is the kernel of a Wiener-Hopf operator with matrix symbol.

The matrix version of (1.2) is well known. The constant $G(\sigma)$ is defined now by using the logarithm of the determinant of the symbol and $E(\sigma)$ is replaced by $\det W(\sigma)W(\sigma^{-1})$. The symbol of I minus the operator with kernel (3.6) is

$$\tau(\xi) = \begin{pmatrix} 1 & -\frac{\sin\pi\beta}{\cosh\pi(\xi+i(\alpha-\beta)/2)} \\ -\frac{\sin\pi\alpha}{\cosh\pi(\xi-i(\alpha-\beta)2)} & 1 \end{pmatrix},$$

whose determinant equals

$$1-\frac{\sin\pi\alpha\,\sin\pi\beta}{\cosh^2\pi\xi-\sin^2\pi(\alpha-\beta)/2}. \tag{3.7}$$

Using the matrix version of (1.2) we find that the G factor times the Toeplitz $E(\varphi_r)$ has limit $4^{-\alpha\beta}$ while the E factor equals $\det W(\tau)W(\tau^{-1})$.

If we set $C(\alpha,\beta) = 4^{-\alpha\beta}\det W(\tau)W(\tau^{-1})$ we see that we have arrived at the following point.

Lemma 3.1. *If α and β are sufficiently small and purely imaginary then*
$$det_2\, T_n(\varphi_{\alpha,\beta})/G_2(\varphi_{\alpha,\beta})^n = C(\alpha,\beta)\,\det(I-(I-A_0)^{-1}A_1)\,\det(I-(I-A_0)^{-1}A_2).$$
In the first determinant on the right the kernels of A_0 and A_1 are given by (3.2) and (3.3) and in the second determinant the kernels of A_0 and A_2 are given by (3.4) and (3.5).

This can be extended to any connected (α,β) region where A_1 and A_2 are trace class and the $I-A_0$ (both versions) are invertible. Requirements just for boundedness of the operators are $\Re\,\alpha,\ \Re\,\beta < 1/2$ and $|\Re\,(\alpha-\beta)| < 1$. An application of Lemma 2.2 shows that these suffice also for the entries of A_1 and A_2 to be trace class. (Under this condition an a can be found in each case such that the integrals that arise are finite.) As for the invertibility of $I-A_0$, a little trigonometry shows that (3.7) is nonzero if $\cos\pi(\alpha+\beta)/2 \notin (-\infty,-1]$, and this holds if $|\Re\,(\alpha+\beta)| < 1$. Since this is a connected set and the index of the determinant is zero for α and β small it must be zero for all these (α,β). Hence under this condition $I-A_0$ is Fredholm of index zero. But we do not know that it is invertible, so here is what we do. We know that $W(\tau)$ is invertible if $W(\tau)\,W(\tau^{-1})$ is. This operator is of the form I plus trace class and its determinant is an analytic function of α and β for $|\Re\,(\alpha\pm\beta)| < 1$. So we assume temporarily that (α,β) is not in the zero set of this determinant. (Notice that the determinant is nonzero for α and β are small enough.) The same applies to the other version of A_0. Recall that we still have the extra requirement that $\Re\,\alpha,\ \Re\,\beta < 1/2$.

Once we have invertibility of both versions of $I - A_0$ we can proceed as before to the Wiener-Hopf analogue. Now the only change is that the expressions involving $1 \pm x$ appearing in (3.3) are replaced by $(1-x)^{\alpha}(1+x)^{-\beta}\, e^{-Rx}$ and $(1-x)^{\beta}(1+x)^{-\alpha}\, e^{-Rx}$ so we must add the assumption that $\Re\,\alpha,\ \Re\,\beta > -1/2$. Other than this everything is as before until when we come to uniform invertibility. After replacing those expressions by e^{-2nx} (the error in doing this being an operator with norm $o(1)$), the problem becomes that of uniform invertibility of I minus

$$\begin{pmatrix} 0 & \frac{\sin \pi\beta}{\pi} e^{-2nx} \frac{(x/y)^{(\beta-\alpha)/2}}{x+y} \\ \frac{\sin \pi\alpha}{\pi} e^{-2nx} \frac{(x/y)^{(\alpha-\beta)/2}}{x+y} & 0 \end{pmatrix}$$

on $L^2(0,1)$. We see upon making the substitutions $x \to x/2n,\ y \to y/2n$ that this is equivalent to the uniform invertibility of I minus

$$\begin{pmatrix} 0 & \frac{\sin \pi\beta}{\pi} e^{-x} \frac{(x/y)^{(\beta-\alpha)/2}}{x+y} \\ \frac{\sin \pi\alpha}{\pi} e^{-x} \frac{(x/y)^{(\alpha-\beta)/2}}{x+y} & 0 \end{pmatrix}$$

on $L^2(0,2n)$. This is completely equivalent to the invertibility of I minus the operator with the same kernel on $L^2(0,\infty)$, and this in turn is equivalent to the invertibility of I minus the operator L with kernel

$$L(x,y) = \begin{pmatrix} 0 & \frac{\sin \pi\beta}{\pi} e^{-x/2} \frac{(x/y)^{(\beta-\alpha)/2}}{x+y} e^{-y/2} \\ \frac{\sin \pi\alpha}{\pi} e^{-x/2} \frac{(x/y)^{(\alpha-\beta)/2}}{x+y} e^{-y/2} & 0 \end{pmatrix}.$$

Lemma 3.2. *The operator $I - L$ is invertible when $|\Re\,(\alpha \pm \beta)| < 1$ and (α,β) does not lie in the zero set of some analytic function which is nonzero for sufficiently small α and β.*

Proof. Let P denote multiplication by $\chi_{[0,1]}$ and think of the kernel A_0 given by (3.2) as acting on $L^2(0,\infty)$. We know that $I - PA_0P$ is invertible except for (α,β) in the zero set of some analytic function which is nonzero for α and β small. Now we can write

$$I - L = I - PA_0P - P(L - A_0)P - PL(I-P) - (I-P)LP$$

and several applications of Lemma 2.2 show that the operators $P(L - A_0)P$, $PL(I-P)$, and $(I-P)LP$ are all trace class. Hence whenever $I - PA_0P$ is invertible the invertibility of $I - L$ is equivalent to the invertibility of

$$I - (I - PA_0P)^{-1}(P(L - A_0)P + PL(I-P) + (I-P)LP),$$

which in turn is equivalent to the nonvanishing of its determinant. Since this is analytic in α and β and nonzero for small α and β the assertion follows. $\square$

Having established the necessary invertibility and uniform invertibility the asymptotics stated in the theorem follow, as in the case $\alpha = \beta$. But recall that we still have two conditions beyond the hypothesis $|\Re(\alpha \pm \beta)| < 1$ of the theorem:

(i) (α, β) does not lie in the zero set of some analytic function $F(\alpha, \beta)$ which is nonzero for sufficiently small α and β.

(ii) $|\Re\,\alpha|$, $|\Re\,\beta| < 1/2$.

To remove requirement (i) we use the analyticity of the regularized determinants and geometric means, the latter also being nonzero. Suppose α and β satisfy (ii). The set S of α for which $F(\alpha, \beta)$ is identically 0 in β is discrete. Assume $\alpha \notin S$. Then $\{\beta : F(\alpha, \beta) = 0\}$ is discrete. Choose any β with $|\Re\,\beta| < 1/2$. There is a little circle Γ with center β such that $F(\alpha, \beta') \neq 0$ for all $\beta' \in \Gamma$. We know that

$$\frac{\det_2 W_R(\sigma_{\alpha,\beta'})/G_2(\sigma_{\alpha,\beta'})^R}{\det_2 T_n(\varphi_{\alpha,\beta'})/G_2(\varphi_{\alpha,\beta'})^n} \to 1$$

for all $\beta' \in \Gamma$, and the denominator is also nonzero for β' inside Γ by the known asymptotics of $\det T_n(\varphi_{\alpha,\beta'})$. The limit holds uniformly for all $\beta' \in \Gamma$. Therefore the limit holds for $\beta' = \beta$ as well. So the statement of the theorem holds for all α satisfying $|\Re\,\alpha| < 1/2$ except for those lying in S. But now we can repeat the previous argument to show that it holds for $\alpha \in S$ as well.

To remove requirement (ii) we do now what we did not do earlier only because it would have made the formulas yet more complicated. We had these requirements because of the factors involving powers of $1 - x$ in (3.5) and the Wiener-Hopf analogue of (3.3). The exponents had to have real part greater than $-1/2$ for the x-factors to belong to L^2. What we could have done is multiply (3.4) and (3.5) on the left by $\begin{pmatrix} (1-x)^{\alpha/2} & 0 \\ 0 & (1-x)^{\beta/2} \end{pmatrix}$ and on the right by $\begin{pmatrix} (1-y)^{-\alpha/2} & 0 \\ 0 & (1-y)^{-\beta/2} \end{pmatrix}$, and multiply the Wiener-Hopf analogues of (3.2) and (3.3) on the left by $\begin{pmatrix} (1-x)^{-\alpha/2} & 0 \\ 0 & (1-x)^{-\beta/2} \end{pmatrix}$ and on the right by $\begin{pmatrix} (1-y)^{\alpha/2} & 0 \\ 0 & (1-y)^{\beta/2} \end{pmatrix}$. These would not have affected the determinants but the x-factors (and now also the y-factors) belong to L^2 under the weaker conditions $|\Re\,\alpha|$, $|\Re\,\beta| < 1$. These hold under the hypothesis of the theorem and so are not extra conditions. The succeeding argument holds with only minor changes with these replacements.

4. Appendix

The continuous analogue of the Borodin-Okounkov identity for generalized determinants is given by the formula

$$\det_2 W_R(\sigma) = G_2(\sigma)^R\, E(\sigma)\, \det(I - K_R) \tag{4.1}$$

where the terms G_2, E and K_R were defined in Section 2. We shall show the identity is valid if $\sigma - 1 \in L^2$, if the Fourier transform k of $\sigma - 1$ is in L^1 and satisfies

$$\int_{-\infty}^{\infty} |x|\,|k(x)|^2\,dx < \infty,$$

and finally if σ is nonzero and has index zero. As we shall see, the last two assumptions imply that $\sigma = \sigma^+\sigma^-$ where σ^+ extends to be nonzero, bounded and analytic in the upper half-plane and σ^- in the lower. All factors tend to one at $\pm\infty$.

That (4.1) is true if $\sigma - 1 \in L^1 \cap L^2$ follows from the corresponding identity for the ordinary determinant proved in [2] and the formula

$$\det_2 A = \det A\; e^{-\mathrm{tr}(A-I)}.$$

To prove (4.1) under the condition $\sigma - 1 \in L^2$ we shall approximate $\sigma - 1$ by a sequence of functions $\sigma_n - 1 \in L^1 \cap L^2$, apply the identity to each σ_n and then take a limit. In order to guarantee convergence of the terms appearing in the identity, we use a Banach algebra approach. We define the algebra $\mathcal{K}$ as the set of all bounded functions ψ whose distributional Fourier transform restricted to $\mathbf{R} - \{0\}$ is equal to a function $\hat{\psi}$ satisfying

$$\|\psi\| := \int_{-\infty}^{\infty} |\hat{\psi}(\zeta)|d\zeta + \left(\int_{-\infty}^{\infty} |x|\,|\hat{\psi}(\zeta)|^2\,d\zeta\right)^{1/2} < \infty.$$

The Fourier transform may have a delta-function summand $c\,\delta$, and the norm on $\mathcal{K}$ is given by $\|\psi\| + |c|$. This is a subalgebra of L^∞ and it is clear that the function σ is contained in the algebra. In [18] it was proved that if $\psi \in \mathcal{K} \cap L^2$ then there is a sequence of functions ψ_n with compact support converging to ψ in both the norm of L^2 and $\mathcal{K}$. This was actually proved for a slightly larger Banach algebra, but the proof is the same here. If we apply this result to the function $\sigma - 1$ we find a sequence of functions $\sigma_n - 1$ for which (4.1) holds.

It remains to show that each of the various terms has the proper limit. Since $\sigma_n - 1$ converges in L^2 to $\sigma - 1$ the terms $\det_2 W_R(\sigma_n)$ and $G_2(\sigma_n)^R$ converge to $\det_2 W_R(\sigma)$ and $G_2(\sigma)^R$, respectively. The factor $E(\sigma_n)$ is equal to $\det W(\sigma_n)W(\sigma_n^{-1}) = \det(I - H(\sigma_n)H(\tilde{\sigma}_n^{-1}))$, where $H(\sigma)$ is the Hankel operator with kernel $\hat{\sigma}(x+y)$ and $H(\tilde{\sigma})$ is the Hankel operator with kernel $\hat{\sigma}(-x-y)$. Now since σ_n converge in $\mathcal{K}$ to the invertible element σ it follows that the analogous statement is true for σ^{-1} and thus the corresponding Hankel operators converge in the Hilbert Schmidt norm. This follows immediately from the definition of the norm on $\mathcal{K}$. Finally, convergence in $\mathcal{K}$ holds for the sequences σ_n^+ and σ_n^- and their quotients, because the well-known projections used in their definitions are continuous in $\mathcal{K}$. This implies the convergence of the term $\det(I - K_R(\sigma_n))$ to the corresponding $\det(I - K_R)$ and completes the proof of the identity.

Acknowledgment

The authors are grateful to Albrecht Böttcher who corrected an error in the corollary of the main theorem in a previous version of this paper and who also pointed out several previous results and conjectures.

The first author was supported by National Science Foundation grant DMS-0200167 and the second author by grant DMS-9732687.

References

[1] E.L. Basor, A localization theorem for Toeplitz determinants, Indiana Univ. Math. J. **28**:6 (1979) 975–983.

[2] E.L. Basor and Y. Chen, *A note on Wiener-Hopf determinants and the Borodin-Okounkov identity*, to appear in Int. Eqns. Operator Th.

[3] E.L. Basor and J.W. Helton, A new proof of the Szegö limit theorem and new results for Toeplitz operators with discontinuous symbol, J. Operator Th. **3** (1980) 23–29.

[4] E.L. Basor and H. Widom, *Toeplitz and Wiener-Hopf determinants with piecewise continuous symbols*, J. Funct. Anal. **50** (1983) 387–413.

[5] A. Borodin and A. Okounkov, *A Fredholm determinant formula for Toeplitz determinants*. Int. Eqns. Operator Th. **37** (2000) 386–396.

[6] A. Böttcher, *Wiener-Hopf determinants with rational symbols*, Math. Nachr **144** (1989) 39–64.

[7] A. Böttcher and B. Silbermann, *Wiener-Hopf determinants with symbols having zeros of analytic type*, Seminar Analysis, 1982/83, Inst. f. Math., Akad. Wiss. DDR, Berlin 1983, 224–243.

[8] A. Böttcher and B. Silbermann, *Toeplitz determinants generated by symbols with one singularity of Fisher-Hartwig type*, Wiss. Z. Tech. Hochsch. Karl-marx-Stadt **26**:2 (1984) 186–188.

[9] A. Böttcher and B. Silbermann, *Toeplitz determinants with Fisher-Hartwig symbols*, J. Funct. Anal. **63** (1985) 178–214.

[10] A. Böttcher and B. Silbermann, Analysis of Toeplitz Operators, Akademie-Verlag Berlin, 1990.

[11] A. Böttcher, B. Silbermann, and H. Widom *A continuous analogue of the Fisher-Hartwig formula for piecewise continuous symbols*, J. Func. Anal. **122** (1994) 222–246.

[12] A. Böttcher, B. Silbermann, and H. Widom, *Determinants of truncated Wiener-Hopf operators with Hilbert-Schmidt kernels and piecewise continuous symbols* Arch. Math. **63** (1994) 60–71.

[13] T. Ehrhardt, *A status report on the asymptotic behavior of Toeplitz determinants with Fisher-Hartwig singularities*, Operator Th: Advances and App. **124** (2001), 217–241.

[14] T. Ehrhardt, B. Silbermann, *Toeplitz determinants with one Fisher-Hartwig singularity*, J. Func. Anal. **148** (1997), 229–256.

[15] M.E. Fisher and R.E. Hartwig, *Toeplitz determinants: some applications, theorems, and conjectures*, Adv. Chem. Phys. **15** (1968) 333–353.

[16] I.C. Gohberg and M.G. Krein, Introduction to the Theory of Linear Nonselfadjoint Operators, Transl. Math. Monographs **18**, Amer. Math. Soc., Providence, 1969.

[17] L.V. Mikaelyan, *Asymptotics of determinants of truncated Wiener-Hopf operators in a singular case* [Russian]. Akad. Nauk Armyan. SSR Dokl. **82** (1986) 151–155.

[18] H. Widom, *On Wiener-Hopf determinants*, Operator Th.: Adv. Appl. **41** (1989) 519–543.

E. L. Basor
Department of Mathematics
California Polytechnic State University
San Luis Obispo, CA 93407
USA
e-mail: `ebasor@calpoly.edu`

H. Widom
Department of Mathematics
University of California
Santa Cruz, CA 95064
USA
e-mail: `widom@math.ucsc.edu`

Operator Theory
Advances and Applications, Vol. 147, 151–174

Convolution Type Operators with Symbols Generated by Slowly Oscillating and Piecewise Continuous Matrix Functions

Maria Amélia Bastos, António Bravo, and Yuri I. Karlovich

To the memory of Erhard Meister

Abstract. The paper deals with a local study of the Banach algebra $\mathcal{A}_{[SO,PC]}$ generated by the convolution type operators $W_{a,b} = a\mathcal{F}^{-1}b\mathcal{F}$ with data $a \in [SO, PC]^{n\times n}$ and $b \in [SO_p, PC_p]^{n\times n}$ which act on the Lebesgue space $L_p^n(\mathbb{R})$ $(1 < p < \infty,\ n \geq 1)$. Here $[SO, PC]^{n\times n}$ means the C^*-algebra generated by slowly oscillating (SO) and piecewise continuous (PC) $n \times n$ matrix functions, and $[SO_p, PC_p]^{n\times n}$ is a Fourier multiplier analogue on $L_p(\mathbb{R})$ of $[SO, PC]^{n\times n}$. The work is based on the study of Fourier multiplier analogue SO_p of SO, on the characterization of the multiplicative linear functionals of slowly oscillating functions, and on the compactness of the commutators $AW_{a,b} - W_{a,b}A$, where $A \in \mathcal{A}_{[SO,PC]}$, $a \in SO$, and $b \in SO_p$. Making use of the Allan–Douglas local principle we construct homomorphisms of $\mathcal{A}_{[SO,PC]}$ onto local Banach algebras and establish a Fredholm criterion for operators $A \in \mathcal{A}_{[SO,PC]}$ in terms of the invertibility of their images in the local algebras.

1. Introduction

Given $1 < p < \infty$ and $n \geq 1$, let $\mathcal{B}(L_p^n(\mathbb{R}))$ denote the Banach algebra of all bounded linear operators on the Lebesgue space $L_p^n(\mathbb{R})$.

Let $[SO, PC]$ be the smallest C^*-subalgebra of the C^*-algebra $L_\infty(\mathbb{R})$ containing all slowly oscillating (SO) and piecewise continuous (PC) functions, and let $[SO_p, PC_p]$ stand for its Fourier multiplier analogue, which is a Banach subalgebra of $\mathcal{M}_p$, the Banach algebra of Fourier multipliers on $L_p(\mathbb{R})$, $1 < p < \infty$.

The paper deals with a local Fredholm study of convolution type operators in the Banach algebra $\mathcal{A}_{[SO,PC]} \subset \mathcal{B}(L_p^n(\mathbb{R}))$. The generators of the algebra $\mathcal{A}_{[SO,PC]}$

2000 *Mathematics Subject Classification.* Primary 47G10; Secondary 47A53.

Key words and phrases. Convolution type operators, slowly oscillating matrix, compactness of commutators, Banach algebra, local principle, Fredholmness.

Partially supported by FCT (Portugal) project POCTI/34222/MAT/2000-FEDER and by PROMEP (México).

have the form

$$W_{a,b} := a\mathcal{F}^{-1}b\mathcal{F} \in \mathcal{B}(L_p^n(\mathbb{R})) \tag{1.1}$$

where $\mathcal{F}$ is the Fourier transform given by

$$(\mathcal{F}\varphi)(x) = \frac{1}{\sqrt{2\pi}} \int_{\mathbb{R}} e^{ixy}\varphi(y)dy, \quad x \in \mathbb{R}, \tag{1.2}$$

a and b are $n \times n$ matrix functions with entries in $[SO, PC]$ and $[SO_p, PC_p]$, respectively.

The Banach algebra $\mathcal{A}_{PC} \subset \mathcal{B}(L_p^n(\mathbb{R}))$ generated by all convolution type operators $W_{a,b}$ with $a \in PC^{n\times n}$ and $b \in PC_p^{n\times n}$ was studied by R.V. Duduchava in [9], [10]. The algebra $\mathcal{A}_{PC}$ in the case of weighted Lebesgue spaces $L_p^n(\mathbb{R}, w)$ with power weights w was investigated by R. Schneider [23] and S. Roch and B. Silbermann [20]. This algebra in the case of weighted Lebesgue spaces $L_p^n(\mathbb{R}, w)$ with general Muckenhoupt weights w was studied by A. Böttcher and I.M. Spitkovsky in [7] (see also [6]). Its shift-invariant analogue for the weighted Lebesgue spaces on the unit circle was treated in [15].

The paper is organized as follows.

In Section 2 we define the spaces that we need later on, introduce and study a Fourier multiplier analogue $SO_p \subset \mathcal{M}_p$ of the C^*-algebra SO of slowly oscillating functions considered by S. C. Power [18], and define the Banach algebras of convolution type operators on the Lebesgue spaces $L_p^n(\mathbb{R})$, objects of the present study. There we also recall the local principle by Allan and Douglas (see, e.g., [5]).

In Section 3 we study the spaces $M(SO)$ and $M(SO_p)$ of multiplicative linear functionals on SO and SO_p, respectively. We prove that the sets $M(SO)$ and $M(SO_p)$ coincide, which implies the inverse closedness of the Banach algebra SO_p in the C^*-algebra SO. We also observe the inverse closedness of the Banach algebra $[SO_p, PC_p]$ in the C^*-algebra $[SO, PC]$ and the coincidence of their maximal ideal spaces as sets. Using the definition of asymptotically independent algebras (see [18]) we describe the fibers $M_\infty([SO, PC]) = M_\infty([SO_p, PC_p])$ at infinity of the corresponding maximal ideal spaces and introduce some useful homomorphisms defined on the algebras $[SO, PC]$ and $[SO_p, PC_p]$.

Sections 4–5 contain the main results of the paper.

In Section 4 we prove the compactness of the commutators $aW_b - W_b aI$ if either $a \in SO$ and $b \in [SO_p, PC_p]$, or $a \in [SO, PC]$ and $b \in SO_p$, or $a \in [SO, C(\overline{\mathbb{R}})]$ and $b \in [SO_p, C_p(\overline{\mathbb{R}})]$. The proof of the compactness of such commutators for $a \in SO$ and $b \in SO_p$ is based on the theory of pseudodifferential operators presented in [8], [19], [26].

Section 5 deals with a local Fredholm study of operators in the Banach algebra $\mathcal{A}_{[SO,PC]}$. First we prove that the Banach algebra

$$\mathcal{A}^{\pi}_{\mathrm{diag}(SO)} := \Big\{ W_{aI_n, bI_n} + \mathcal{K} : \; a \in SO, \; b \in SO_p \Big\},$$

where I_n is the identity $n \times n$ matrix and $\mathcal{K} := \mathcal{K}(L_p^n(\mathbb{R}))$ is the ideal of compact operators in the Banach algebra $\mathcal{B} := \mathcal{B}(L_p^n(\mathbb{R}))$, is a central subalgebra of the quotient algebra $\mathcal{A}^\pi_{[SO,PC]} = \mathcal{A}_{[SO,PC]}/\mathcal{K}$. Further, describing the maximal ideal space $M(\mathcal{A}^\pi_{\mathrm{diag}(SO)})$ of $\mathcal{A}^\pi_{\mathrm{diag}(SO)}$ and making use of the Allan–Douglas local principle we construct homomorphisms of the Banach algebra $\mathcal{A}_{[SO,PC]}$ onto local Banach algebras associated with the points of $M(\mathcal{A}^\pi_{\mathrm{diag}(SO)})$. As a result, we establish a Fredholm criterion for any operator $A \in \mathcal{A}_{[SO,PC]}$ in terms of the invertibility of its images in the local algebras. Finally, we get explicit sufficient conditions for the Fredholmness of the operators

$$A = \sum_{i=1}^{m} W_{a_i,b_i} \quad (a_i \in [SO,PC]^{n\times n},\ b_i \in [SO_p, PC_p]^{n\times n})$$

on the spaces $L_p^n(\mathbb{R})$, $1 < p < \infty$, $n \geq 1$.

Note that the local algebras mentioned above are generated by two idempotents. Therefore, applying the two projections theorem (see, e.g., [11], [13], [2, Section 8.3]) one can construct a symbol calculus for the Banach algebra $\mathcal{A}_{[SO,PC]}$ and establish a Fredholm criterion for any operator $A \in \mathcal{A}_{[SO,PC]}$ in terms of the invertibility of its symbol. The corresponding Fredholm theory will appear in a forthcoming publication.

2. Preliminaries

We begin with defining the spaces that will be considered in this work.

Let $L_p(\mathbb{R})$ $(1 \leq p \leq \infty)$ denote the usual Lebesgue spaces on $\mathbb{R}$, $C(\overline{\mathbb{R}})$ and $C(\dot{\mathbb{R}})$ denote the spaces of continuous functions, respectively, on the two point compactification and one point compactification of the real line. Let $C_0(\mathbb{R})$ stand for the set of all functions $f \in C(\dot{\mathbb{R}})$ with $f(\infty) = 0$. Put $C_b(\mathbb{R}) := C(\mathbb{R}) \cap L_\infty(\mathbb{R})$. Let PC be the set of all functions $f : \dot{\mathbb{R}} \to \mathbb{C}$ which possess finite left-hand and right-hand limits at every point $x \in \dot{\mathbb{R}}$. The set of discontinuities of each function $f \in PC$ is at most countable.

Let $V_1(\mathbb{R})$ be the set of functions $a : \overline{\mathbb{R}} \to \mathbb{C}$ with the finite total variation

$$V_1(a) = \sup\left\{\sum_{i=1}^{n} |a(t_i) - a(t_{i-1})| :\ -\infty \leq t_0 < t_1 < \cdots < t_n \leq +\infty,\ n \in \mathbb{N}\right\}$$

where the supremum is taken over all finite decompositions of the real line $\mathbb{R}$. Clearly, $V_1(\mathbb{R})$ is a Banach algebra with the norm

$$\|a\|_{V_1} = \|a\|_\infty + V_1(a). \tag{2.1}$$

For a continuous function $f : \mathbb{R} \to \mathbb{C}$ and a set $I \subset \mathbb{R}$, let

$$\operatorname{osc}(f, I) = \sup_{t,s\in I} |f(t) - f(s)|.$$

Following [18] we denote by SO the set of *slowly oscillating* functions,

$$SO := \Big\{ f \in C_b(\mathbb{R}) : \lim_{x \to +\infty} \operatorname{osc}(f, [-2x, -x] \cup [x, 2x]) = 0 \Big\}. \tag{2.2}$$

Clearly, SO is a C^*-subalgebra of $L_\infty(\mathbb{R})$ which contains all functions in $C(\dot{\mathbb{R}})$.

Given a continuous function $a : \mathbb{R} \to \mathbb{C}$ and $\varepsilon > 0$, let

$$cm_{x,\varepsilon}(a) := \max_{h \in [-\varepsilon, \varepsilon]} |a(x+h) - a(x)|$$

be a *local oscillation* of a at a point $x \in \mathbb{R}$.

According to [8, p. 122], we introduce the class

$$\mathrm{CM}(\mathbb{R}) := \Big\{ a \in C_b(\mathbb{R}) : \lim_{|x| \to \infty} cm_{x,\varepsilon}(a) = 0 \Big\} \tag{2.3}$$

where ε is any (equivalently, some) small positive constant. Thus, $\mathrm{CM}(\mathbb{R})$ is the class of bounded continuous functions over $\mathbb{R}$ with local oscillation vanishing at infinity. By [8, Chapter 3, Lemma 10.4], $\mathrm{CM}(\mathbb{R})$ is a C^*-algebra being the closure in $L_\infty(\mathbb{R})$ of the algebra

$$\mathrm{CM}^\infty(\mathbb{R}) := \Big\{ a \in C_b^\infty(\mathbb{R}) : \lim_{|x| \to \infty} \Big[\Big(\frac{d}{dx} \Big)^j a \Big](x) = 0, \quad j = 1, 2, \dots \Big\}, \tag{2.4}$$

where $C_b^\infty(\mathbb{R})$ denotes the set of all infinitely differentiable functions $f : \mathbb{R} \to \mathbb{C}$ which are bounded with all their derivatives. For $a \in \mathrm{CM}(\mathbb{R})$, such approximations $a_\varepsilon \in \mathrm{CM}^\infty(\mathbb{R})$ $(\varepsilon \to 0)$ have the form

$$a_\varepsilon = \varphi_\varepsilon * a, \tag{2.5}$$

where $\varphi_\varepsilon(x) = \varepsilon^{-1} \varphi(x/\varepsilon)$ for $x \in \mathbb{R}$, $\varepsilon > 0$; and

$$\varphi \in C^\infty(\mathbb{R}), \quad \operatorname{supp} \varphi \subset [-1, 1], \quad \varphi \geq 0, \quad \int_{\mathbb{R}} \varphi(x) dx = 1. \tag{2.6}$$

Clearly, $\mathrm{CM}^\infty(\mathbb{R}) \cup C(\overline{\mathbb{R}}) \subset \mathrm{CM}(\mathbb{R})$, and functions $a \in \mathrm{CM}(\mathbb{R})$ in general have different slowly oscillating behavior at $+\infty$ and $-\infty$.

We point out that the definitions of slow oscillation for $a \in SO$ and $a \in \mathrm{CM}(\mathbb{R})$ are different (cf. (2.2) and (2.3)). Let us study the relation between these definitions.

By analogy with (2.4) we introduce the non-closed algebra

$$\widetilde{SO}^\infty := \Big\{ f \in C_b^\infty(\mathbb{R}) : \lim_{|x| \to \infty} (D^j f)(x) = 0, \quad j = 1, 2, \dots \Big\} \tag{2.7}$$

where

$$(Df)(x) = x f'(x), \quad x \in \mathbb{R}. \tag{2.8}$$

Clearly, if $f \in \widetilde{SO}^\infty$, then $D^j f \in C_b(\mathbb{R})$ for all $j = 0, 1, 2, \dots$. Moreover, every function $f \in \widetilde{SO}^\infty$ slowly oscillates at $+\infty$ and $-\infty$. Indeed, if, for example,

$0 < x \le y < z \le 2x$, then

$$|f(z) - f(y)| \le \int_y^z |tf'(t)| \, \frac{dt}{t} \le \log 2\Big(\max_{t\in[x,2x]} |tf'(t)| \Big)$$

and hence $\lim_{x\to\infty} \operatorname{osc}(f, [x, 2x]) = 0$ because $\lim_{x\to\infty} xf'(x) = 0$.

In general $\widetilde{SO}^\infty$ is not contained in SO because its elements can have different slowly oscillating behavior at $\pm\infty$. To exclude such a situation, we define an infinite differentiable analogue of SO as $SO^\infty = \widetilde{SO}^\infty \cap SO$. In particular, the set of even functions in $\widetilde{SO}^\infty$ is contained in SO^∞.

Lemma 2.1. *Every even function $a \in SO$ can be uniformly approximated by even functions $a_n \in \widetilde{SO}^\infty$.*

Proof. Let $a \in SO$ be even. The function a can be uniformly approximated by even functions in SO which are constants in a symmetric neighborhood of zero. Without loss of generality we may assume that the initial function $a \in SO$ has that property.

Consider the restriction of a to $\mathbb{R}_+ = (0, +\infty)$ and introduce the function $b(x) = a(e^x)$. It is easily seen that $b \in \mathrm{CM}(\mathbb{R})$. Hence, b is uniformly approximated on $\mathbb{R}$ by functions $b_n \in \mathrm{CM}^\infty(\mathbb{R})$. Then a is uniformly approximated on $\mathbb{R}_+$ by functions $a_n(x) = b_n(\ln x)$.

Since b is a constant at a neighborhood $U_{-\infty}$ of $-\infty$, the functions b_n also are constants in the neighborhoods $U_{-\infty}^{(n)} = \{x - 1/n : x \in U_{-\infty}\}$ of $-\infty$. Indeed, from (2.5)–(2.6) we get

$$b_n(x) = \int_{\mathbb{R}} \varphi_{1/n}(x-y)b(y)dy, \tag{2.9}$$

$$b_n'(x) = \int_{\mathbb{R}} \varphi_{1/n}'(x-y)b(y)dy = \int_{\mathbb{R}} \varphi_{1/n}'(x-y)(b(y) - b(x))dy.$$

The latter is true because $\int_{\mathbb{R}} \varphi_{1/n}'(x-y)dy = 0$, due to the fact that $\varphi_{1/n}$ vanishes for large $|x|$. Taking into account that b is a constant at $U_{-\infty}$ and that $\operatorname{supp} \varphi_{1/n}' \subset [-1/n, 1/n]$, we conclude that for every $n \in \mathbb{N}$,

$$\int_{\mathbb{R}} \varphi_{1/n}'(x-y)(b(y) - b(x))dy = 0$$

for all x in the neighborhoods $U_{-\infty}^{(n)}$ of $-\infty$. Thus $b_n' = 0$ at $U_{-\infty}^{(n)}$ and hence, $b_n = \text{const}$ there.

From the relation $a_n(x) = b_n(\ln x)$ it follows that

$$\Big[\Big(x\frac{d}{dx}\Big)^j a_n\Big](x) = \Big[\Big(\frac{d}{dx}\Big)^j b_n\Big](\ln x), \quad j = 0, 1, 2, \dots, \tag{2.10}$$

and a_n are constants for all sufficiently small $x > 0$. Consequently, $a_n \in C^\infty(\mathbb{R}_+)$, $a_n'(x) = 0$ for all sufficiently small $x > 0$ and, by (2.10) and (2.4),

$$\lim_{x\to\infty} (D^j a_n)(x) = 0, \quad j = 1, 2, \dots .$$

Extending the functions a_n to the whole line $\mathbb{R}$ as even functions and preserving the notation for these extensions, we get even functions $a_n \in \widetilde{SO}^\infty$, where $\widetilde{SO}^\infty$ is given by (2.7). The functions a_n uniformly approximate the function a on $\mathbb{R}$, which completes the proof. □

Let $C_b^1(\mathbb{R})$ consist of all functions $a \in C_b(\mathbb{R})$ with $a' \in C_b(\mathbb{R})$. From Lemma 2.1 we get the following corollary.

Corollary 2.2. *Every even function $a \in SO$ can be uniformly approximated by even functions in the non-closed algebra*

$$\widetilde{SO}^1 := \Big\{ f \in C_b^1(\mathbb{R}) : \lim_{|x|\to\infty} (Df)(x) = 0 \Big\}. \tag{2.11}$$

Lemma 2.3. *Every function $a \in SO$ can be uniformly approximated by functions in the non-closed algebra $SO^\infty = \widetilde{SO}^\infty \cap SO$.*

Proof. Following the proof of Lemma 2.1 we assume that the function $a \in SO$ is constant in a neighborhood of the point $x = 0$. Uniformly approximating the functions $a^+ = a|_{\mathbb{R}_+}$ and $a^- = (a|_{\mathbb{R}_-})(-x)$, respectively, by functions a_n^+ and a_n^- as in Lemma 2.1, we conclude that the functions

$$a_n(x) = \begin{cases} (a_n^+)(x) & \text{if } x > 0, \\ (a_n^-)(-x) & \text{if } x < 0 \end{cases}$$

are bounded and continuous on $\mathbb{R} \setminus \{0\}$ with a possible discontinuity of the first kind at $x = 0$, all their derivatives $(d/dx)^j a_n$ are continuous on $\mathbb{R}$, equal zero at neighborhoods of $x = 0$, and satisfy the condition $\lim\limits_{|x|\to\infty} (D^j a_n)(x) = 0$.

Since $\lim\limits_{n\to\infty} |a_n(+0) - a_n(-0)| = 0$, there are piecewise continuous functions c_n with compact supports and discontinuities only at 0, which have continuous derivatives $(d/dx)^j c_n$ $(j = 1, 2, \dots)$ being equal to zero at neighborhoods of 0 and ∞, and such that $\|c_n\|_{L^\infty(\mathbb{R})} \to 0$ as $n \to \infty$ and the functions $a_n + c_n$ are continuous on $\mathbb{R}$. Obviously, the functions $f_n = a_n + c_n$ belong to $\widetilde{SO}^\infty$ and have the property

$$\lim_{x\to+\infty} \max_{z,y\in[x,2x]} \Big| f_n(z) - f_n(-y) \Big| = 0 \tag{2.12}$$

only simultaneously with a_n. Therefore, without loss of generality we may assume that the initial functions a_n are continuous at the point $x = 0$ and thus they belong to $\widetilde{SO}^\infty$.

Let us show now that the functions a_n satisfy (2.12). To this end we compare the values $b_n^+(z) = a_n^+(e^z)$ and $b_n^-(y) = a_n^-(e^y)$. From (2.9) it follows that

$$b_n^\pm(x) = \int_{\mathbb{R}} \varphi_{1/n}(x-t)\, b^\pm(t)dt = \int_{\mathbb{R}} \varphi_{1/n}(t)\, b^\pm(x-t)dt,$$

where $b^{\pm}(x) = a^{\pm}(e^x)$. Hence,

$$a_n^{\pm}(x) = \int_{\mathbb{R}_+} \varphi_{1/n}(\log t)\, a^{\pm}\Big(\frac{x}{t}\Big)\, \frac{dt}{t}$$

and therefore, for $z, y \in [x, 2x]$, we get

$$\Big|a_n^+(z) - a_n^-(y)\Big| \le \int_{\mathbb{R}_+} \varphi_{1/n}(\log t)\, \Big|a^+\Big(\frac{z}{t}\Big) - a^-\Big(\frac{y}{t}\Big)\Big|\, \frac{dt}{t}$$

$$\le \max_{z,y\in[x,2x]} \max_{t\in[e^{-1/n},e^{1/n}]} \Big|a^+\Big(\frac{z}{t}\Big) - a^-\Big(\frac{y}{t}\Big)\Big|$$

$$\le \operatorname{osc}\Big(a\Big[-2xe^{1/n}, -xe^{-1/n}\Big] \cup \Big[xe^{-1/n}, 2xe^{1/n}\Big]\Big),$$

which implies that $a_n \in SO$. Thus, the functions a_n belong to SO^∞. □

Corollary 2.4. *Every function $a \in SO$ can be uniformly approximated by functions in the non-closed algebra*

$$SO^1 := \widetilde{SO}^1 \cap SO. \tag{2.13}$$

Every function $a \in SO$ can be represented in the form

$$a = a_+ u_+ + a_- u_- + a_0 \tag{2.14}$$

where $a_\pm$ are even functions in SO, $a_0 \in C_0(\mathbb{R})$, and

$$u_\pm(x) = (1 \pm \tanh x)/2, \quad x \in \overline{\mathbb{R}}. \tag{2.15}$$

Obviously, such representation is not unique.

We denote by $[SO, PC]$ the smallest C^*-algebra of $L_\infty(\mathbb{R})$ that contains SO and PC. Clearly, every function $a \in [SO, C(\overline{\mathbb{R}})]$ can be non-uniquely represented in the form (2.14) where $a_\pm \in SO$, $a_0 \in C_0(\mathbb{R})$, and $u_\pm$ are given by (2.15).

Let $1 < p < \infty$. The Cauchy singular integral operator S is defined on the Lebesgue space $L_p(\mathbb{R})$ by

$$(Sf)(t) = \frac{1}{\pi i} \int_{\mathbb{R}} \frac{f(x)}{x - t}\, dx, \quad t \in \mathbb{R}, \tag{2.16}$$

where the integral is understood in the sense of principal value. As is well known (see, e.g., [5], [2]), $S \in \mathcal{B}(L_p(\mathbb{R}))$. Since $S^2 = I$, the operator S generates the two complementary projections

$$P_\pm = (I \pm S)/2 \in \mathcal{B}(L_p(\mathbb{R})).$$

Given $1 < p < \infty$, we define the operator W_a on $L_2(\mathbb{R}) \cap L_p(\mathbb{R})$ by

$$(W_a\varphi)(t) := (\mathcal{F}^{-1} a \mathcal{F}\varphi)(t), \quad \varphi \in L_2(\mathbb{R}) \cap L_p(\mathbb{R}), \tag{2.17}$$

where $\mathcal{F} : L_2(\mathbb{R}) \to L_2(\mathbb{R})$ denotes the Fourier transform given by (1.2) and $\mathcal{F}^{-1}$ is its inverse. A function $a \in L_\infty(\mathbb{R})$ is called a *Fourier multiplier on $L_p(\mathbb{R})$* if the operator W_a given by (2.17) can be extended to a bounded linear operator on $L_p(\mathbb{R})$, and we denote it again by W_a. Since

$$S = \mathcal{F}^{-1}(-\operatorname{sign} x)\mathcal{F} = W_{-\operatorname{sign} x} \tag{2.18}$$

(see, e.g., [10, Section 2] or [4, Example 1.17]), it follows that $W_{\chi_\pm} = P_\mp$ where $\chi_\pm$ are the characteristic functions of $\mathbb{R}_\pm = \{x \in \mathbb{R} : \pm x \geq 0\}$.

As is known [14], the set $\mathcal{M}_p$ of Fourier multipliers on $L_p(\mathbb{R})$ is a Banach algebra with the norm $\|a\|_{\mathcal{M}_p} := \|W_a\|_{\mathcal{B}(L_p(\mathbb{R}))}$. Moreover, if $1 < r < p < 2$ or $2 < p < r < \infty$, then

$$\mathcal{M}_r \subset \mathcal{M}_p \subset \mathcal{M}_2 = L_\infty(\mathbb{R}). \tag{2.19}$$

According to Stechkin's theorem [25], $\mathcal{M}_p$ contains all functions $a : \mathbb{R} \to \mathbb{C}$ of finite total variation and

$$\|a\|_{\mathcal{M}_p} \leq \|S\|_{\mathcal{B}(L_p(\mathbb{R}))}\|a\|_{V_1}$$

where $\|a\|_{V_1}$ is the norm of a in $V_1(\mathbb{R})$ (see (2.1)) and $\|S\|_{\mathcal{B}(L_p(\mathbb{R}))}$ is the norm of the Cauchy singular integral operator S on $L_p(\mathbb{R})$.

Let $C_p(\dot{\mathbb{R}})$ and $C_p(\overline{\mathbb{R}})$ stand for the closures in $\mathcal{M}_p$ of the sets of all functions with finite total variation in $C(\dot{\mathbb{R}})$ and $C(\overline{\mathbb{R}})$, respectively. We denote by PC_p the closure in $\mathcal{M}_p$ of the set of all piecewise constant functions on $\mathbb{R}$ which have at most finite sets of jumps.

By Mikhlin's theorem [17], if

$$f \in C_b(\mathbb{R}), \quad Df \in C(\mathbb{R}), \quad \text{and} \quad \sup_{x\in\mathbb{R}} |xf'(x)| < \infty, \tag{2.20}$$

then $f \in \mathcal{M}_p$ for every $p \in (1, \infty)$.

Since every function $a \in SO^1$ satisfies (2.20) due to (2.13) and (2.11), it follows from the Mikhlin theorem that $a \in \mathcal{M}_p$ for all $p \in (1, \infty)$. Therefore, taking into account Corollary 2.4, by analogy with the definitions of $C_p(\dot{\mathbb{R}})$ and $C_p(\overline{\mathbb{R}})$, for every $p \in (1, \infty)$ we define the set SO_p of slowly oscillating functions in $\mathcal{M}_p$ as the closure in $\mathcal{M}_p$ of the set SO^1. Clearly, SO_p is a Banach subalgebra of $\mathcal{M}_p$. By virtue of Corollary 2.4, the Banach algebra SO_p is a Fourier multiplier analogue of the C^*-algebra SO.

If $1 < r < p < 2$ or $2 < p < r < \infty$, then from (2.19) it follows that

$$SO_r \subset SO_p \subset SO.$$

If f is a continuously differentiable function in $C_b(\mathbb{R})$, $f(x) = f(-x)$ for all $x \in \mathbb{R}$, and

$$\lim_{x\to\infty} xf'(x) = 0, \tag{2.21}$$

then $f \in SO$ in view of (2.2). Thus, if f is an even function in $C_b^1(\mathbb{R})$ such that (2.21) holds, then f satisfies (2.20) and hence $f \in SO_p \subset SO \cap \mathcal{M}_p$ for all $p \in (1, \infty)$. Let $a \in C_b^\infty(\mathbb{R})$ and

$$f(x) = \begin{cases} e^{-1/x}a[\log(\log(x + e))], & x > 0, \\ e^{1/x}a[\log(\log(-x + e))], & x < 0. \end{cases} \tag{2.22}$$

It is easily seen that the even function (2.22) vanishes at $x = 0$ with all its derivatives, belongs to $C_b^\infty(\mathbb{R}) \cap SO$, and satisfies (2.21). Therefore, this function belongs to all the classes SO_p $(1 < p < \infty)$ and oscillates at ∞ if, for example, $a(x) = \sin x$ for $x \in \mathbb{R}$.

Along with $[SO, PC]$, for every $p \in (1, \infty) \setminus \{2\}$ we introduce the Banach subalgebra $[SO_p, PC_p]$ of $\mathcal{M}_p$ generated by SO_p and PC_p. Clearly, $[SO_p, PC_p]$ is a nonclosed subalgebra of $L_\infty(\mathbb{R})$ if $p \neq 2$.

Let X be a subalgebra of $L_\infty(\mathbb{R})$ and let $[X]_{\rm diag}^{n\times n} := \{fI_n : f \in X\}$ be the set of diagonal matrix functions with equal diagonal entries in X.

Given a Banach space X, we denote by $[X]^{n\times m}$ the Banach space of all $n \times m$ matrices $B = (b_{i,j})$ with entries in X and the norm

$$\|B\| = \left(\sum_{i=1}^{n} \sum_{j=1}^{m} \|b_{i,j}\|_X^2 \right)^{1/2}.$$

If $m = 1$, we will write $[X]^n$ instead of $[X]^{n\times 1}$. In particular, $L_p^n(\mathbb{R}) := [L_p(\mathbb{R})]^n$.

Let $1 < p < \infty$. In this paper we will study the Banach subalgebra

$$\mathcal{A}_{[SO,PC]} := \operatorname{alg}\left\{ W_{a,b} : a \in [SO, PC]^{n\times n},\ b \in [SO_p, PC_p]^{n\times n} \right\} \subset \mathcal{B}(L_p^n(\mathbb{R}))$$

which contains all the operators $W_{a,b}$ of the form (1.1) with $a \in [SO, PC]^{n\times n}$ and $b \in [SO_p, PC_p]^{n\times n}$.

Since the Allan-Douglas principle (see, e.g., [5]) plays a fundamental role in this paper we finish this section by recalling its statement. Let $\mathcal{G}A$ stand for the group of invertible elements of a unital Banach algebra A.

Theorem 2.5 (Allan-Douglas principle). *Let A be a Banach algebra with identity I, B a closed subalgebra of the center of A which contains I, $N \subset B$ a maximal ideal of B, J_N the smallest closed two-sided ideal of A containing N, A_N the quotient algebra A/J_N, and $M(B)$ the maximal ideal space of B.*

(i) *$a \in A$ is left (right, two-sided) invertible in A if and only if for every $N \in M(B)$ the coset $a_N = a + J_N$ is left (right, two-sided) invertible in A_N.*
(ii) *The mapping $\mathcal{M}(B) \to \mathbb{R}_+$, $N \mapsto \|a_N\|$ is upper semi-continuous. If $a \in A$ and $a_{N_0} \in \mathcal{G}\, A_{N_0}$, then $a_N \in \mathcal{G}\, A_N$ for all N in some neighborhood of N_0.*
(iii) *If A is a C^*-algebra, then $\|a\| = \max_{N\in M(B)} \|a_N\|$ for every $a \in A$.*

3. The maximal ideal spaces of SO and SO_p

Let us first determine the maximal ideal space $M(SO)$ of SO, that is, the space of all multiplicative linear functionals of SO. To this end we make use of the space $M(C(\dot{\mathbb{R}}))$ of multiplicative linear functionals of $C(\dot{\mathbb{R}})$. Identifying the points $t \in \dot{\mathbb{R}}$ with the evaluation functionals on $\dot{\mathbb{R}}$, $t(f) = f(t)$, where $f(\infty) = \lim_{x\to\infty} f(x)$, we get $M(C(\dot{\mathbb{R}})) = \dot{\mathbb{R}}$. Let

$$M_t(SO) = \left\{\xi \in M(SO) : \xi|_{C(\dot{\mathbb{R}})} = t\right\}$$

be the fiber of $M(SO)$ over $t \in \dot{\mathbb{R}}$. If $t \in \mathbb{R}$, the fiber $M_t(SO)$ consists of the only evaluation functional at t, that is, $M_t(SO) = \{t\}$. Thus

$$M(SO) = \bigcup_{t\in\dot{\mathbb{R}}} M_t(SO) = \mathbb{R} \cup M_\infty(SO). \tag{3.1}$$

The fiber $M_\infty(SO)$ is characterized by the following proposition which can be proved by analogy with [5, Proposition 3.29] (cf. [3, Proposition 4.1]).

Proposition 3.1. *The fiber $M_\infty(SO)$ has the form $M_\infty(SO) = (\operatorname{clos}_{SO^*}\mathbb{R}) \setminus \mathbb{R}$ where $\operatorname{clos}_{SO^*}\mathbb{R}$ is the weak-star closure of $\mathbb{R}$ in SO^*, the dual space of SO.*

Thus any functional $\xi \in M_\infty(SO)$ is the limit of a net $t_\alpha \in \mathbb{R}$ that tends to ∞ and does not converge to functionals $t \in \mathbb{R}$, that is, $\xi(f) = \lim_\alpha f(t_\alpha)$ for every $f \in SO$. Now we are able to relate the maximal ideal space of SO to the partial limits of the function $a \in SO$ at infinity (see [3, Proposition 4.2 and Corollary 4.3]).

Proposition 3.2. *Let $\{a_k\}_{k=1}^\infty$ be a countable subset of SO. If $\xi \in M_\infty(SO)$ then there exists a sequence $g = \{g_n\} \subset \mathbb{R}$ such that $g_n \to \infty$ as $n \to \infty$ and*

$$\xi(a_k) = \lim_{n\to\infty} a_k(g_n), \quad k \in \mathbb{N}. \tag{3.2}$$

Conversely, if $g_n \in \mathbb{R}$, $g_n \to \infty$ as $n \to \infty$, and the limits $\lim_{n\to\infty} a_k(g_n)$ exist for all k, then there is a $\xi \in M_\infty(SO)$ such that (3.2) *holds.*

Corollary 3.3. *If $\{a_k\}_{k=1}^\infty$ is a countable subset of SO and $\xi \in M_\infty(SO)$, then there exists a sequence $\{g_n\} \subset \mathbb{R}$ such that $g_n > 1$, $g_n \to \infty$ as $n \to \infty$, and for every $t \in \mathbb{R} \setminus \{0\}$,*

$$\lim_{n\to\infty} a_k(g_n t) = \xi(a_k), \quad k \in \mathbb{N}. \tag{3.3}$$

Proof. By Proposition 3.2, there is a sequence $\{g_n\} \subset \mathbb{R}$ such that $g_n \to \infty$ as $n \to \infty$ and (3.2) holds. Since in (2.2) we can replace 2 by an arbitrary $t > 1$, that is, equivalently define SO as

$$SO := \left\{f \in C_b(\mathbb{R}) : \lim_{x\to\infty} \operatorname{osc}(f, [-tx, -x] \cup [x, tx]) = 0\right\}, \tag{3.4}$$

we derive from (3.4) and (3.2) that

$$\lim_{n\to\infty} a_k(tg_n) = \lim_{n\to\infty} a_k(g_n) = \xi(a_k). \tag{3.5}$$

Obviously, (3.5) remains valid if $t \in (0,1)$. Hence, (3.5) holds for all $t > 0$. Finally, since due to (2.2),

$$\lim_{n\to\infty} a_k(tg_n) = \lim_{n\to\infty} a_k(-tg_n), \tag{3.6}$$

we get (3.3) for all $t \in \mathbb{R} \setminus \{0\}$. Moreover, in view of (3.6) we may replace g_n by $|g_n|$ in (3.3), that is, (3.3) is fulfilled for a sequence $\{g_n\}$ such that $g_n > 1$ and $g_n \to \infty$ as $n \to \infty$. □

Let us study the maximal ideal space $\mathcal{M}(SO_p)$ of the Banach algebra $SO_p \subset \mathcal{M}_p$. To this end we consider the set SO^1 given by (2.13). One can prove that SO^1 is a commutative Banach algebra with the norm

$$\|f\| = \max\left\{\|f\|_{C_b(\mathbb{R})},\ \|Df\|_{C_b(\mathbb{R})}\right\}$$

where $(Df)(x) = xf'(x)$, $x \in \mathbb{R}$.

Consider the following three unital commutative Banach algebras with the same unit which are homomorphically inclosed one into another:

$$SO^1 \subset SO_p \subset SO.$$

To study the relations between their maximal ideal spaces we use the following result (see [24, Theorem 3.10]).

Theorem 3.4. *Let B_i $(i = 1, 2, 3)$ be commutative Banach algebras with the same unit which are homomorphically inclosed one into another, $B_1 \subset B_2 \subset B_3$. Suppose that B_1 is dense in B_2 and every multiplicative linear functional defined on B_1 extends to a multiplicative linear functional on B_3. Then every multiplicative linear functional on B_2 also extends to a multiplicative linear functional on B_3.*

Lemma 3.5. *Every multiplicative linear functional defined on SO^1 extends to a multiplicative linear functional on SO.*

Proof. As $SO^1 \subset SO$, it follows that $M(SO) \subset M(SO^1)$. Indeed, let $\eta \in M(SO)$ and $a \in SO^1$. Then $a \in SO$ and $\|a\|_\infty \le \|a\|_{SO^1}$. Hence

$$|\eta(a)| \le \|\eta\|_{SO^*}\|a\|_\infty \le \|\eta\|_{SO^*}\|a\|_{SO^1} = \|a\|_{SO^1},$$

and thus η is a (non-zero) multiplicative linear functional on SO^1.

Assume for a moment that $\xi \in M(SO^1)$ is not extendable to a multiplicative linear functional on SO, that is, $\xi \in M(SO^1) \setminus M(SO)$. Clearly, there exists a function $a \in SO^1$ such that $\eta(a) \neq 0$ for all $\eta \in M(SO)$ and $\xi(a) = 0$. Since $a \in SO$ and SO is a C^*-algebra, we infer that the function a is invertible in SO.

But the inverse function $1/a$ automatically belongs to SO^1 because it satisfies (2.21), that is, SO^1 is inverse closed in SO. Then from the Gelfand theory (see, e.g., [21, Chapter 11]) it follows that $\xi(a) \neq 0$, and we arrive at a contradiction, which completes the proof. □

Choosing

$$B_1 = SO^1, \quad B_2 = SO_p, \quad B_3 = SO$$

and taking into that account that SO^1 is dense in SO_p in the norm of $\mathcal{M}_p$, we deduce from Theorem 3.4 and Lemma 3.5 that every multiplicative linear functional on SO_p extends to a multiplicative linear functional on SO, that is, $M(SO_p) \subset M(SO)$. On the other hand, $M(SO) \subset M(SO_p)$ because $SO_p \subset SO$. Thus we have proved the following.

Lemma 3.6. *The maximal ideal spaces of SO_p and SO coincide as sets, that is, $M(SO_p) = M(SO)$.*

Lemma 3.6 and the Gelfand theory immediately give the following assertion.

Corollary 3.7. *The Banach algebra SO_p is inverse closed in the C^*-algebra SO.*

Lemma 3.8. *If $b \in C_p(\dot{\mathbb{R}})$, b vanishes at infinity, and $a \in SO_p$, then the product ba belongs to $C_p(\dot{\mathbb{R}})$ and $(ba)(\infty) = 0$. Hence $[SO_p, C_p(\overline{\mathbb{R}})]$ is a Banach subalgebra of the Banach algebra $\widetilde{SO}_p$, where $\widetilde{SO}_p$ is the closure of $\widetilde{SO}^1$ in $\mathcal{M}_p$.*

Proof. Taking into account that every function $b \in C_p(\dot{\mathbb{R}})$ with $b(\infty) = 0$ can be approximated in $\mathcal{M}_p$ by functions $b_n \in C(\dot{\mathbb{R}}) \cap V_1(\mathbb{R})$ with compact support (see [10, Remark 2.12]) and that every function $a \in SO_p$ can be approximated in $\mathcal{M}_p$ by functions $a_n \in SO^1$, we infer that the product ba of the function $b \in C_p(\dot{\mathbb{R}})$ vanishing at infinity and of the function $a \in SO_p$ belongs to $C_p(\dot{\mathbb{R}})$ because all the functions $b_n a_n$ with compact support belong to $C(\dot{\mathbb{R}}) \cap V_1(\mathbb{R})$.

Since every function $c \in C(\overline{\mathbb{R}}) \cap V_1(\mathbb{R})$ can be approximated in $\mathcal{M}_p$ by continuous piecewise linear functions (see again [10, Remark 2.12]), it is clear that the function c can be also approximated in $\mathcal{M}_p$ by continuously differentiable functions $c_n \in C_b(\overline{\mathbb{R}})$ whose derivative c_n' have compact supports. Hence, c_n belong to $\widetilde{SO}^1$ and thus $c \in \widetilde{SO}_p$, where $\widetilde{SO}_p$ is the closure of $\widetilde{SO}^1$ in $\mathcal{M}_p$ and, consequently, is a Banach subalgebra of $\mathcal{M}_p$. Since $C_p(\overline{\mathbb{R}})$ is the closure of $C(\overline{\mathbb{R}}) \cap V_1(\mathbb{R})$ in $\mathcal{M}_p$, we conclude that $C_p(\overline{\mathbb{R}}) \subset \widetilde{SO}_p$. Obviously, $SO_p \subset \widetilde{SO}_p$. These last two inclusions imply that $[SO_p, C_p(\overline{\mathbb{R}})]$ is a Banach subalgebra of $\widetilde{SO}_p$. □

Taking into account Lemma 3.8 one can easily prove the following analogue of (2.14).

Lemma 3.9. *Let $1 < p < \infty$. Every function a in the Banach algebra $[SO_p, C_p(\overline{\mathbb{R}})]$ is of the form*

$$a = a_+ u_+ + a_- u_- + a_0, \tag{3.7}$$

with $a_\pm \in SO_p$, $u_\pm \in C_p(\overline{\mathbb{R}})$, $a_0 \in C_p(\dot{\mathbb{R}})$ satisfying the conditions

$$u_+ + u_- = 1, \quad u_+(+\infty) = 1, \quad u_-(-\infty) = 1, \quad a_0(\infty) = 0.$$

Consider the homomorphic inclusions of the commutative Banach algebras

$$[SO^1, C(\mathbb{R}) \cap V_1(\mathbb{R})] \subset [SO_p, C_p(\overline{\mathbb{R}})] \subset [SO, C(\overline{\mathbb{R}})],$$

$$[SO^1, PC \cap V_1(\mathbb{R})] \subset [SO_p, PC_p] \subset [SO, PC],$$

where $[SO^1, C(\mathbb{R}) \cap V_1(\mathbb{R})]$ and $[SO^1, PC \cap V_1(\mathbb{R})]$ are Banach algebras with the norm

$$\|f\| = \max\Big\{\|f\|_{L_\infty(\mathbb{R})},\ \|Df\|_{L_\infty(\mathbb{R})},\ V_1(f)\Big\}.$$

Analogously to Lemma 3.6 and Corollary 3.7 one can prove the following result.

Theorem 3.10. *The Banach algebras $[SO_p, C_p(\overline{\mathbb{R}})]$ and $[SO_p, PC_p]$ are inverse closed in the C^*-algebras $[SO, C(\overline{\mathbb{R}})]$ and $[SO, PC]$, respectively, and*

$$M([SO_p, C_p(\overline{\mathbb{R}})]) = M([SO, C(\overline{\mathbb{R}})]), \quad M([SO_p, PC_p]) = M([SO, PC]).$$

Let A and B be C^*-algebras of bounded continuous functions on $\mathbb{R}$ that contain $C(\dot{\mathbb{R}})$, and $[A, B]$ the smallest C^*-algebra that contains A and B. By [18], the algebras A and B are called *asymptotically independent* if the sets $M_\infty([A, B])$ and $M_\infty(A) \times M_\infty(B)$ are naturally homeomorphic.

According to [18] we have the following.

Proposition 3.11. *The fiber $M_\infty([SO, C(\overline{\mathbb{R}})])$ is homeomorphic to*

$$M_\infty(SO) \times \{\pm\infty\}.$$

The maximal ideal spaces $M(PC)$ of PC and $M(PC_p)$ of PC_p can be identified with $\dot{\mathbb{R}} \times \{0, 1\}$ in the following sense: for $a \in PC$ or $a \in PC_p$,

$$a(t, 0) = a(t-0),\ a(t, 1) = a(t+0)\ \ (t \in \mathbb{R}),\ \ a(\infty, 0) = a(+\infty),\ a(\infty, 1) = a(-\infty).$$

Since $M_\infty(C(\overline{\mathbb{R}})) = M_\infty(PC) = \{\pm\infty\}$, from the previous proposition it follows [18] that for every $\zeta \in M_\infty(SO)$ there is a homomorphism

$$\alpha_\zeta : [SO, PC] \to PC|_{M_\infty(PC)}, \quad (\alpha_\zeta \varphi)(\tau) = (\zeta, \tau)\varphi, \quad \tau \in M_\infty(PC). \tag{3.8}$$

Consequently there exists a function $\varphi_\zeta \in PC$ such that

$$\alpha_\zeta \varphi = \varphi_\zeta|_{M_\infty(PC)}. \tag{3.9}$$

As the values $\varphi_\zeta(\pm\infty)$ are defined uniquely, we get an injective homomorphism

$$\beta_\zeta :\ PC|_{M_\infty(PC)} \to PC, \quad \varphi_\zeta|_{M_\infty(PC)} \mapsto \varphi_\zeta(-\infty)\chi_- + \varphi_\zeta(+\infty)\chi_+$$

where $\chi_\pm$ are the characteristic functions of $\mathbb{R}_\pm$. Thus for every $\zeta \in M_\infty(SO)$ we have the homomorphism

$$\gamma_\zeta := \beta_\zeta \circ \alpha_\zeta :\ [SO, PC] \to PC, \quad \gamma_\zeta \varphi = \varphi_\zeta(-\infty)\chi_- + \varphi_\zeta(+\infty)\chi_+, \tag{3.10}$$

where $\varphi_\zeta(\pm\infty)$ are defined by (3.8)–(3.9).

As $M_\infty(PC_p) = M_\infty(PC) = \{\pm\infty\}$ for every $p \in (1, \infty)$, we infer that the restriction of the homomorphism α_ζ to $[SO_p, PC_p]$ sends this set to $PC_p|_{M_\infty(PC_p)} = PC|_{M_\infty(PC)}$ according to (3.8). Therefore for all $\zeta \in M_\infty(SO)$ the homomorphisms $\gamma_\zeta = \beta_\zeta \circ \alpha_\zeta$ given by (3.10) map $[SO_p, PC_p]$ into PC_p.

4. The compactness of commutators

Let $1 < p < \infty$ and let $\mathcal{K} = \mathcal{K}(L_p(\mathbb{R}))$ be the closed two-sided ideal of all compact operators in the Banach algebra $\mathcal{B} = \mathcal{B}(L_p(\mathbb{R}))$.

Lemma 4.1. ([10, Lemmas 7.1–7.4]).

(a) *If $a \in PC$, $b \in PC_p$, and $a(\pm\infty) = b(\pm\infty) = 0$, then aW_b, $W_b aI \in \mathcal{K}$.*

(b) *If $a \in C(\dot{\mathbb{R}})$, $b \in PC_p$ or if $a \in PC$, $b \in C_p(\dot{\mathbb{R}})$, then $aW_b - W_b aI \in \mathcal{K}$.*

(c) *If $a \in C(\overline{\mathbb{R}})$ and $b \in C_p(\overline{\mathbb{R}})$, then $aW_b - W_b aI \in \mathcal{K}$.*

The following is an analogue of Lemma 4.1(b) for slowly oscillating functions.

Theorem 4.2. *If $a \in [SO, PC]$, $b \in SO_p$ or if $a \in SO$, $b \in [SO_p, PC_p]$, then $aW_b - W_b aI \in \mathcal{K}$.*

Proof. If $a \in SO$, then, by [18, Section 4],

$$a \in QC := (H^\infty + C(\dot{\mathbb{R}})) \cap (\overline{H^\infty} + C(\dot{\mathbb{R}}))$$

where QC is the C^*-algebra of quasicontinuous functions (see, e.g., [22]). Hence from the Hartman theorem (see, e.g., [4, Theorem 2.18]) it follows that

$$aS - SaI \in \mathcal{K} \quad \text{for all} \quad a \in SO, \tag{4.1}$$

where S is the Cauchy singular integral operator on $\mathbb{R}$ given by (2.16). Since every function $b \in PC_p$ can be approximated in $\mathcal{M}_p$ by piecewise constant functions, we deduce from (4.1) and (2.18) that

$$aW_b - W_b aI \in \mathcal{K} \quad \text{for all } a \in SO \text{ and } b \in PC_p.$$

Let $a \in PC$ and $b \in SO_p$. By the definition of SO_p, the function $b \in SO_p$ can be approximated in $\mathcal{M}_p$ by functions $b_n \in SO^1$ where the algebra $SO^1 \subset SO \cap \mathcal{M}_p$ is given by (2.13). Hence, by the part already proved, the operators

$$aW_{b_n} - W_{b_n} aI = \mathcal{F}^{-1}\Big(\mathcal{F}a\mathcal{F}^{-1} b_n I - b_n \mathcal{F}a\mathcal{F}^{-1}\Big)\mathcal{F} = \mathcal{F}^{-1}(W_{a(-x)} b_n I - b_n W_{a(-x)})\mathcal{F}$$

are compact on the space $L_2(\mathbb{R})$. Since the operators $aW_{b_n} - W_{b_n} aI$ are bounded on all the spaces $L_p(\mathbb{R})$ $(1 < p < \infty)$, it follows from the Krasnoselskii interpolation theorem (see [16, Theorem 3.10]) that for every $n \in \mathbb{N}$,

$$aW_{b_n} - W_{b_n} aI \in \mathcal{K}(L_p(\mathbb{R})) \quad \text{for all} \quad p \in (1, \infty).$$

Therefore the operator

$$aW_b - W_b aI = \lim_{n \to \infty} (aW_{b_n} - W_{b_n} aI)$$

is also compact on the spaces $L_p(\mathbb{R})$, $1 < p < \infty$. Thus

$$aW_b - W_b aI \in \mathcal{K} \quad \text{for all} \quad a \in PC \quad \text{and} \quad b \in SO_p.$$

It only remains to prove that

$$aW_b - W_b aI \in \mathcal{K} \quad \text{for all} \quad a \in SO \quad \text{and} \quad b \in SO_p.$$

To this end we again approximate $b \in SO_p$ in $\mathcal{M}_p$ by functions $b_n \in SO^1$. Taking into account that the operators $aW_{b_n} - W_{b_n} aI$ are bounded on all the spaces $L_p(\mathbb{R})$, $1 < p < \infty$, we conclude on the basis of the Krasnoselskii interpolation theorem [16] that it is sufficient to prove the compactness of the commutators $aW_{b_n} - W_{b_n} aI$ only on the space $L_2(\mathbb{R})$. Since $a, b_n \in SO$, we have reduced the problem to proving the compactness of the commutator $aW_b - W_b aI$ $(a, b \in SO)$ on the space $L_2(\mathbb{R})$.

By Lemma 2.3, the functions $a, b \in SO$ can be uniformly approximated in $L_\infty(\mathbb{R})$ by functions $a_n, b_n \in \widetilde{SO}^\infty$, where the set $\widetilde{SO}^\infty$ is defined by (2.7). Since $a_n, b_n \in \widetilde{SO}^\infty$, the functions $a_n(x) b_n(\xi)$ belong to the Hörmander class

$S^0_{0,0} = C^\infty_b(\mathbb{R} \times \mathbb{R})$ consisting of all infinitely differentiable functions on $\mathbb{R} \times \mathbb{R}$ which are bounded with all their partial derivatives. By the Calderon–Vaillancourt theorem (see [26, Chapter 13, Theorem 1.3] or [19, Theorem 5.2.1]), if $c \in S^0_{0,0}$, then the pseudodifferential operator $OP(c)$, defined on infinitely differentiable functions $f \in L_2(\mathbb{R})$ with compact support by the iterated integral

$$[OP(c)f](x) = \frac{1}{2\pi} \int_{\mathbb{R}} d\xi \int_{\mathbb{R}} c(x,\xi) e^{i(y-x)\xi} f(y)dy, \quad x \in \mathbb{R},$$

extends to a bounded linear operator on $L_2(\mathbb{R})$.

We write the commutator $a_n W_{b_n} - W_{b_n} a_n I$ as a pseudodifferential operator $OP(c_n) \in \mathcal{B}(L_2(\mathbb{R}))$ where

$$c_n(x,\xi) = a_n(x) b_n(\xi) - \sigma_n(x,\xi), \quad (x,\xi) \in \mathbb{R} \times \mathbb{R}, \tag{4.2}$$

$\sigma_n(x,\xi)$ is the symbol of the pseudodifferential operator

$$OP(\sigma_n) = OP(b_n) OP(a_n),$$

and the pseudodifferential operators $OP(a_n)$ and $OP(b_n)$ act on functions $f \in C^\infty(\mathbb{R})$ with compact support by

$$\begin{aligned} [OP(a_n)f](x) &= \frac{1}{2\pi} \int_{\mathbb{R}} d\xi \int_{\mathbb{R}} a_n(x) e^{i(y-x)\xi} f(y)dy, \quad x \in \mathbb{R}, \\ [OP(b_n)f](x) &= \frac{1}{2\pi} \int_{\mathbb{R}} d\xi \int_{\mathbb{R}} b_n(\xi) e^{i(y-x)\xi} f(y)dy, \quad x \in \mathbb{R}. \end{aligned}$$

From [26, Chapter 2, Theorem 4.4] (see also [19, Theorem 6.2.1]) it follows that

$$\sigma_n(x,\xi) = b_n(\xi) a_n(x) - q_n(x,\xi), \quad (x,\xi) \in \mathbb{R} \times \mathbb{R}, \tag{4.3}$$

where $q_n \in S^0_{0,0}$ and

$$\lim_{|x|+|\xi| \to \infty} q_n(x,\xi) = 0. \tag{4.4}$$

Indeed, in view of [26, Chapter 2, Theorem 4.4],

$$q(x,\xi) \sim \sum_{k=1}^{\infty} \frac{i^{-k}}{k!} (D^k b_n)(\xi)(D^k a_n)(x), \tag{4.5}$$

where the operator D is given by (2.8). Since

$$\lim_{|\xi| \to \infty} (D^k b_n)(\xi) = \lim_{|x| \to \infty} (D^k a_n)(x) = 0 \quad \text{for all} \quad k \geq 1$$

because $a_n, b_n \in \widetilde{SO}^\infty$, from (4.5) we immediately get (4.4). By [19, Theorem 5.8.3], if $q_n \in S^0_{0,0}$ satisfies (4.4), then the pseudodifferential operator $OP(q_n)$ is compact on the space $L_2(\mathbb{R})$. Thus, taking into account that $c_n = q_n$ due to (4.2) and (4.3), we deduce that

$$a_n W_{b_n} - W_{b_n} a_n I = OP(c_n) = OP(q_n) \in \mathcal{K}(L_2(\mathbb{R})). \tag{4.6}$$

Hence, for $a, b \in SO$ we infer from (4.6) that the operator

$$aW_b - W_b aI = \lim_{n\to\infty} (a_n W_{b_n} - W_{b_n} a_n I)$$

also is compact on the space $L_2(\mathbb{R})$, which completes the proof. □

Lemma 4.1(c) and Theorem 4.2 immediately imply the following.

Corollary 4.3. *If $a \in [SO, C(\overline{\mathbb{R}})]$ and $b \in [SO_p, C_p(\overline{\mathbb{R}})]$, then $aW_b - W_b aI \in \mathcal{K}$.*

5. Localization and Fredholmness

Let $C_0^\infty(\mathbb{R})$ stand for the algebra of all infinitely differentiable functions in $C(\mathbb{R})$ with compact support. Obviously, $C_0^\infty(\mathbb{R}) \subset SO^1 \subset SO_p \subset SO$.

By analogy with [12] (see also the proof of [1, Lemma 3]) one can show that

$$\mathcal{K} := \mathcal{K}(L_p^n(\mathbb{R})) \subset \operatorname{alg}\{S, aI : a \in [C(\dot{\mathbb{R}})]^{n\times n}\} \subset \mathcal{A}_{[SO,PC]}.$$

In what follows let A^π denote the coset $A + \mathcal{K}$. Consider the quotient algebra $\mathcal{A}^\pi_{[SO,PC]} := \mathcal{A}_{[SO,PC]}/\mathcal{K}$ and the Banach algebra

$$\mathcal{A}_{\mathrm{diag}(SO)} := \operatorname{alg}\Big\{aW_b : a \in [SO]^{n\times n}_{\mathrm{diag}},\ b \in [SO_p]^{n\times n}_{\mathrm{diag}}\Big\}.$$

By Theorem 4.2, the commutative Banach algebra

$$\mathcal{A}^\pi_{\mathrm{diag}(SO)} := \Big\{aW_b + \mathcal{K} : a \in [SO]^{n\times n}_{\mathrm{diag}},\ b \in [SO_p]^{n\times n}_{\mathrm{diag}}\Big\}$$

is a central subalgebra of the Banach algebra $\mathcal{A}^\pi_{[SO,PC]}$. With every point $(\xi, \eta) \in M(SO)\times M(SO)$ we associate the closed two-sided ideal $I^\pi_{\xi,\eta}$ of $\mathcal{A}^\pi_{\mathrm{diag}(SO)}$ generated by all the cosets

$$\begin{aligned} (a_\xi I)^\pi &= a_\xi I + \mathcal{K} \quad \big(a_\xi \in [SO]^{n\times n}_{\mathrm{diag}},\ a_\xi(\xi) = 0\big), \\ (W_{b_\eta})^\pi &= W_{b_\eta} + \mathcal{K} \quad \big(b_\eta \in [SO_p]^{n\times n}_{\mathrm{diag}},\ b_\eta(\eta) = 0\big). \end{aligned} \tag{5.1}$$

Lemma 5.1. *The algebra $\mathcal{A}^\pi_{\mathrm{diag}(SO)}$ is a closed subalgebra of the center of $\mathcal{A}^\pi_{[SO,PC]}$ and its space $M\big(\mathcal{A}^\pi_{\mathrm{diag}(SO)}\big)$ of maximal ideals is homeomorphic to*

$$\Omega := \big(\mathbb{R} \times M_\infty(SO)\big) \cup \big(M_\infty(SO) \times \mathbb{R}\big) \cup \big(M_\infty(SO) \times M_\infty(SO)\big). \tag{5.2}$$

Proof. First, if $(W_{a,b})^\pi \in \mathcal{A}^\pi_{[SO,PC]}$ and $(W_{a_1,b_1})^\pi \in \mathcal{A}^\pi_{\mathrm{diag}(SO)}$, then from Theorem 4.2 and the equality

$$(aW_b)(a_1 W_{b_1}) - (a_1 W_{b_1})(aW_b) = a(W_b a_1 I - a_1 W_b)W_{b_1} + a_1(aW_{b_1} - W_{b_1} aI)W_b$$

we get

$$(W_{a,b})^\pi (W_{a_1,b_1})^\pi = (W_{a_1,b_1})^\pi (W_{a,b})^\pi.$$

Thus $\mathcal{A}^\pi_{\mathrm{diag}(SO)}$ is a Banach subalgebra of the center of $\mathcal{A}^\pi_{[SO,PC]}$.

Further we partially follow the proof of [20, Proposition 14.1]. Note that if J is a maximal ideal of $\mathcal{A}^{\pi}_{\mathrm{diag}(SO)}$, then

$$J \cap \mathrm{alg}\left\{aI + \mathcal{K} :\ a \in [SO]^{n\times n}_{\mathrm{diag}}\right\} \quad \text{and} \quad J \cap \mathrm{alg}\left\{W_b + \mathcal{K} :\ b \in [SO_p]^{n\times n}_{\mathrm{diag}}\right\}$$

are maximal ideals of the Banach algebras

$$\mathrm{alg}\left\{aI + \mathcal{K} :\ a \in [SO]^{n\times n}_{\mathrm{diag}}\right\} \quad \text{and} \quad \mathrm{alg}\left\{W_b + \mathcal{K} :\ b \in [SO_p]^{n\times n}_{\mathrm{diag}}\right\},$$

respectively (see [5, Lemma 1.33]). Therefore, taking into account that

$$\begin{aligned} M\left(\mathrm{alg}\left\{aI + \mathcal{K} :\ a \in [SO]^{n\times n}_{\mathrm{diag}}\right\}\right) &= M(SO), \\ M\left(\mathrm{alg}\left\{W_b + \mathcal{K} :\ b \in [SO_p]^{n\times n}_{\mathrm{diag}}\right\}\right) &= M(SO_p), \end{aligned}$$

and that $M(SO_p) = M(SO)$ by Lemma 3.6, we conclude that there exists a point $(\xi, \eta) \in M(SO) \times M(SO)$ such that the value of the Gelfand transform of the coset $(aW_b)^{\pi} = aW_b + \mathcal{K}$ at the point J equals $\widetilde{a}(\xi)\widetilde{b}(\eta)$ for each choice of $a \in [SO]^{n\times n}_{\mathrm{diag}}$ and $b \in [SO_p]^{n\times n}_{\mathrm{diag}}$ where $\widetilde{a}$ and $\widetilde{b}$ are the diagonal entries of a and b, respectively. Thus we can identify the ideal J with the point $(\xi, \eta) \in M(SO) \times M(SO)$ and hence, by virtue of (3.1), the maximal ideal space of $\mathcal{A}^{\pi}_{\mathrm{diag}(SO)}$ can be identified with a subset of

$$M(SO) \times M(SO) = \big(\mathbb{R} \cup M_{\infty}(SO)\big) \times \big(\mathbb{R} \cup M_{\infty}(SO)\big).$$

Fix $(t, x) \in \mathbb{R} \times \mathbb{R}$. Given $a \in [SO]^{n\times n}_{\mathrm{diag}}$ and $b \in [SO_p]^{n\times n}_{\mathrm{diag}}$, choose matrix functions $a_1, b_1 \in [C_0^{\infty}(\mathbb{R})]^{n\times n}_{\mathrm{diag}}$ such that $a(t) = a_1(t)$, $b(x) = b_1(x)$. Then

$$aW_b = T_1 + T_2 + T_3 + T_4, \tag{5.3}$$

where

$$T_1 = (a - a_1)W_{(b-b_1)},\ T_2 = (a - a_1)W_{b_1},\ T_3 = a_1 W_{(b-b_1)},\ T_4 = a_1 W_{b_1}.$$

The operator T_4 is compact by Lemma 4.1(a), and the operators T_i ($i = 1, 2, 3$) belong to the ideal $J = (t, x)$. Thus, the smallest closed two-sided ideal of $\mathcal{A}^{\pi}_{\mathrm{diag}(SO)}$ which corresponds to the point $(t, x) \in \mathbb{R} \times \mathbb{R}$ equals the whole algebra. So, the maximal ideals of the algebra $\mathcal{A}^{\pi}_{\mathrm{diag}(SO)}$ can only correspond to points $(t, x) \in \Omega$, where Ω is given by (5.2).

On the other hand, each point of Ω gives a maximal ideal of the commutative Banach algebra $\mathcal{A}^{\pi}_{\mathrm{diag}(SO)}$. Indeed, consider the closed two-sided ideals $I^{\pi}_{\xi,\eta}$ $\big((\xi, \eta) \in \Omega\big)$ of $\mathcal{A}^{\pi}_{\mathrm{diag}(SO)}$ generated by the cosets (5.1). We claim that these ideals are maximal. Since they do not contain the coset $I^{\pi} = I + \mathcal{K}$, these ideals are proper. Suppose, contrary to our claim, that for a point $(\xi, \eta) \in \Omega$ there is a proper closed two-sided ideal $\widetilde{I}^{\pi}_{\xi,\eta}$ of the algebra $\mathcal{A}^{\pi}_{\mathrm{diag}(SO)}$ which properly contains the ideal $I^{\pi}_{\xi,\eta}$. Then there is a coset $A^{\pi} \in \mathcal{A}^{\pi}_{\mathrm{diag}(SO)}$ which belongs to $\widetilde{I}^{\pi}_{\xi,\eta} \setminus I^{\pi}_{\xi,\eta}$. Since, in view of (5.3),

$$(aW_b)^{\pi} = (a(\xi)W_{b(\eta)})^{\pi} + I^{\pi}_{\xi,\eta} = (a(\xi)b(\eta)I)^{\pi} + I^{\pi}_{\xi,\eta}$$

for all $a \in [SO]^{n\times n}_{\rm diag}$ and all $b \in [SO_p]^{n\times n}_{\rm diag}$, there exists a complex number $c \neq 0$ such that $A^\pi = \big((cI_n)I\big)^\pi + \mathcal{K}$. But the coset $\big((cI_n)I\big)^\pi + \mathcal{K}$ is invertible in the algebra $\mathcal{A}^\pi_{{\rm diag}(SO)}$, which implies that the ideal $\widetilde{I}^\pi_{\xi,\eta}$ coincides with the whole algebra $\mathcal{A}^\pi_{{\rm diag}(SO)}$. Thus the ideal $\widetilde{I}^\pi_{\xi,\eta}$ is not proper, a contradiction. Hence all the ideals $I^\pi_{\xi,\eta}$ $\big((\xi,\eta)\in\Omega\big)$ are maximal. □

Let Λ be the set of all operators of $\mathcal{B}(L^n_p(\mathbb{R}))$ that commute modulo compact operators with all operators $A \in \mathcal{A}_{{\rm diag}(SO)}$. Clearly, Λ is a Banach algebra. Due to Theorem 4.2, Λ contains $\mathcal{A}_{[SO,PC]}$ and thus the quotient algebra $\Lambda^\pi := \Lambda + \mathcal{K}$ contains $\mathcal{A}^\pi_{[SO,PC]}$. The algebra Λ^π is just the commutant of $\mathcal{A}^\pi_{{\rm diag}(SO)}$ and hence Λ^π is inverse closed in the Calkin algebra $\mathcal{B}(L^n_p(\mathbb{R}))/\mathcal{K}(L^n_p(\mathbb{R}))$.

Given $(\xi,\eta)\in\Omega$, let $J^\pi_{\xi,\eta}$ be the smallest closed two-sided ideal of Λ^π that contains the maximal ideal $I^\pi_{\xi,\eta}$ and let $\phi_{\xi,\eta} : \Lambda^\pi \to \Lambda^\pi_{\xi,\eta}$ be the canonical homomorphism of Λ^π onto the quotient algebra $\Lambda^\pi_{\xi,\eta} := \Lambda^\pi / J^\pi_{\xi,\eta}$.

The canonical projection $\phi_{\xi,\eta}$ sends $\mathcal{A}^\pi_{[SO,PC]}$ onto

$$\mathcal{A}^\pi_{\xi,\eta} := \big(\mathcal{A}^\pi_{[SO,PC]}\big)_{\xi,\eta} := \phi_{\xi,\eta}\big(\mathcal{A}^\pi_{[SO,PC]}\big) \subset \Lambda^\pi_{\xi,\eta}.$$

By the Allan–Douglas local principle, the coset $A^\pi \in \Lambda^\pi$ is invertible in Λ^π if and only if the cosets $A^\pi_{\xi,\eta} := A^\pi + J^\pi_{\xi,\eta}$ are invertible in $\Lambda^\pi_{\xi,\eta}$ for all $(\xi,\eta)\in\Omega$.

Lemma 5.2. *Let $a \in [SO,PC]^{n\times n}$, $b\in[SO_p,PC_p]^{n\times n}$, $\zeta \in M_\infty(SO)$, and let α_ζ be the homomorphism defined by* (3.8). *If $\xi \in M_\infty(SO)$ and*

$$(\alpha_\xi a)(\tau) = 0 \quad \textit{for all} \quad \tau \in M_\infty(PC), \tag{5.4}$$

then $(aI)^\pi_{\xi,\eta} = 0^\pi_{\xi,\eta} = J^\pi_{\xi,\eta}$ for all $\eta\in\mathbb{R}\cup M_\infty(SO)$. If $\eta\in M_\infty(SO)$ and

$$(\alpha_\eta b)(\tau) = 0 \quad \textit{for all} \quad \tau \in M_\infty(PC), \tag{5.5}$$

then $(W_b)^\pi_{\xi,\eta} = 0^\pi_{\xi,\eta} = J^\pi_{\xi,\eta}$ for all $\xi\in\mathbb{R}\cup M_\infty(SO)$.

Proof. It is sufficient to prove the lemma only in the scalar case $n=1$.

Let $\xi \in M_\infty(SO)$ and let (5.4) hold. Then from Lemma 3.9 and Proposition 3.11 it follows that the restrictions $a_\pm|_{M_\infty(SO)}$ of the functions $a_\pm \in SO$ in the non-unique representation (3.7) are defined uniquely and $a_\pm(\xi) = 0$ due to (5.4). Hence $(a_\pm I)^\pi \in I^\pi_{\xi,\eta}$ for every $\eta\in M(SO)$. Finally, (3.7) implies that $(aI)^\pi \in J^\pi_{\xi,\eta}$ for every $\eta \in M(SO)$, which means that

$$(aI)^\pi_{\xi,\eta} = 0^\pi_{\xi,\eta} = J^\pi_{\xi,\eta} \quad \text{for all} \quad (\xi,\eta)\in M_\infty(SO)\times M(SO)$$

under the condition (5.4).

Analogously, if $\eta\in M_\infty(SO)$ and (5.5) holds, then, by Lemma 3.9 and Proposition 3.11, the functions $b_\pm \in SO_p$ in the representation

$$b = b_+u_+ + b_-u_- + b_0$$

of the form (3.7), with $u_\pm \in C_p(\overline{\mathbb{R}})$ and $b_0 \in C_p(\dot{\mathbb{R}})$ satisfying the conditions

$$u_+ + u_- = 1, \quad u_+(+\infty) = 1, \quad u_-(-\infty) = 1, \quad b_0(\infty) = 0,$$

have the property $b_\pm(\eta)=0$. Hence $(W_{b_\pm})^\pi \in I^\pi_{\xi,\eta}$ for every $\xi \in M(SO)$ and thus, by (3.7), $(W_b)^\pi \in J^\pi_{\xi,\eta}$ for every $\xi \in M(SO)$. This means that

$$(W_b)^\pi_{\xi,\eta} = 0^\pi_{\xi,\eta} = J^\pi_{\xi,\eta} \quad \text{for all} \quad (\xi,\eta) \in M(SO) \times M_\infty(SO)$$

under the condition (5.5). □

Given $y \in \mathbb{R}$, let χ_y^- and χ_y^+ denote the characteristic functions of $(-\infty, y)$ and $(y,+\infty)$, respectively. Obviously, $\chi_0^\pm = \chi_\pm$. For every $p \in (1,\infty)$, every $\zeta \in M_\infty(SO)$, and every $a \in [SO_p, PC_p]^{n\times n}$, we also introduce the matrix function $\widetilde{a}_\zeta \in [C_p(\overline{\mathbb{R}})]^{n\times n}$ by

$$\widetilde{a}_\zeta := a_\zeta(-\infty)u_- + a_\zeta(+\infty)u_+, \tag{5.6}$$

where the matrices $a_\zeta(\pm\infty)$ are uniquely defined by (3.8)–(3.9) with φ replaced by a, and the functions $u_\pm \in C_p(\overline{\mathbb{R}})$ are given by (2.15).

Theorem 5.3. *For every $(\xi,\eta) \in \Omega$, the mapping $\delta_{\xi,\eta} : A \mapsto A^\pi_{\xi,\eta}$ defined on the generators aI ($a \in [SO,PC]^{n\times n}$) and W_b ($b \in [SO_p,PC_p]^{n\times n}$) of the Banach algebra $\mathcal{A}_{[SO,PC]}$ by*

$$\delta_{\xi,\eta}(aI) := \begin{cases} [(a(\xi-0)\chi_\xi^- + a(\xi+0)\chi_\xi^+)I]^\pi_{\xi,\eta}, & (\xi,\eta) \in \mathbb{R} \times M_\infty(SO), \\ [(a_\xi(-\infty)\chi_- + a_\xi(+\infty)\chi_+)I]^\pi_{\xi,\eta}, & (\xi,\eta) \in M_\infty(SO) \times \mathbb{R}, \\ [\widetilde{a}_\xi I]^\pi_{\xi,\eta}, & (\xi,\eta) \in M_\infty(SO) \times M_\infty(SO) \end{cases} \tag{5.7}$$

and

$$\delta_{\xi,\eta}(W_b) := \begin{cases} [W_{b_\eta(-\infty)\chi_- + b_\eta(+\infty)\chi_+}]^\pi_{\xi,\eta}, & (\xi,\eta) \in \mathbb{R} \times M_\infty(SO), \\ [W_{b(\eta-0)\chi_\eta^- + b(\eta+0)\chi_\eta^+}]^\pi_{\xi,\eta}, & (\xi,\eta) \in M_\infty(SO) \times \mathbb{R}, \\ [W_{\widetilde{b}_\eta}]^\pi_{\xi,\eta}, & (\xi,\eta) \in M_\infty(SO) \times M_\infty(SO), \end{cases} \tag{5.8}$$

where the matrix functions $\widetilde{a}_\xi$ and $\widetilde{b}_\eta$ are of the form (5.6), extends to a surjective Banach algebra homomorphism

$$\delta_{\xi,\eta} : \mathcal{A}_{[SO,PC]} \to \mathcal{A}^\pi_{\xi,\eta} = \left(\mathcal{A}^\pi_{[SO,PC]}\right)_{\xi,\eta},$$

and for all $A \in \mathcal{A}_{[SO,PC]}$,

$$\sup_{(\xi,\eta)\in\Omega} \|\delta_{\xi,\eta}(A)\|_{\mathcal{A}^\pi_{\xi,\eta}} \le \|A^\pi\| := \inf_{K\in\mathcal{K}} \|A+K\|.$$

Proof. Let $(\xi,\eta) \in \Omega$. Consider the natural Banach algebra homomorphisms

$$\delta_{\xi,\eta} = \phi_{\xi,\eta} \circ \pi : \mathcal{A}_{[SO,PC]} \to \mathcal{A}^\pi_{[SO,PC]} \to \left(\mathcal{A}^\pi_{[SO,PC]}\right)_{\xi,\eta}, \quad A \mapsto A^\pi \mapsto A^\pi_{\xi,\eta}.$$

Obviously, for every $A \in \mathcal{A}_{[SO,PC]}$,

$$\sup_{(\xi,\eta)\in\Omega} \|\delta_{\xi,\eta}(A)\|_{\mathcal{A}^\pi_{\xi,\eta}} \le \|A^\pi\| \le \|A\|.$$

It only remains to prove the equalities (5.7) and (5.8) for $a \in [SO,PC]^{n\times n}$ and $b \in [SO_p,PC_p]^{n\times n}$.

1) First, let $(\xi,\eta)\in\mathbb{R}\times M_\infty(SO)$. Obviously, the matrix function

$$a-a(\xi-0)\chi_\xi^- - a(\xi+0)\chi_\xi^+$$

belongs to $[SO,PC]^{n\times n}$ and, moreover, it is continuous and vanishing at the point ξ. Hence the coset

$$[(a-a(\xi-0)\chi_\xi^- - a(\xi+0)\chi_\xi^+)I]^\pi$$

belongs to the ideal $J^\pi_{\xi,\eta}$ for each $\eta\in M_\infty(SO)$, and therefore

$$[aI]^\pi_{\xi,\eta}=[(a(\xi-0)\chi_\xi^- + a(\xi+0)\chi_\xi^+)I]^\pi_{\xi,\eta}.$$

On the other hand, from the definitions of the homomorphisms

$$\alpha_\eta:\ [SO,PC]\to PC|_{M_\infty(PC)}\quad\text{and}\quad \gamma_\eta:\ [SO,PC]\to PC$$

given by (3.8) and (3.10), respectively, it follows that

$$\alpha_\eta(b-\gamma_\eta(b))=0,\quad \gamma_\eta(b)=b_\eta(-\infty)\chi_-+b_\eta(+\infty)\chi_+.$$

By Lemma 5.2, the coset $[W_{b-\gamma_\eta(b)}]^\pi$ belongs to the ideal $J^\pi_{\xi,\eta}$ for every $\xi\in\mathbb{R}$. Hence

$$[W_b]^\pi_{\xi,\eta}=[W_{b_\eta(-\infty)\chi_-+b_\eta(+\infty)\chi_+}]^\pi_{\xi,\eta}.$$

2) Secondly, let $(\xi,\eta)\in M_\infty(SO)\times\mathbb{R}$. The matrix function

$$b-b(\eta-0)\chi_\eta^- - b(\eta+0)\chi_\eta^+$$

belongs to $[SO_p,PC_p]^{n\times n}$ and, moreover, it is continuous and vanishing at the point η. Hence the coset

$$[W_{b-b(\eta-0)\chi_\eta^- - b(\eta+0)\chi_\eta^+}]^\pi$$

belongs to the ideal $J^\pi_{\xi,\eta}$ for each $\xi\in M_\infty(SO)$, and therefore

$$[W_b]^\pi_{\xi,\eta}=[W_{b(\eta-0)\chi_\eta^- + b(\eta+0)\chi_\eta^+}]^\pi_{\xi,\eta}.$$

On the other hand, from (3.9) and (3.10) it follows that

$$\alpha_\xi(a-\gamma_\xi(a))=0,\quad \gamma_\xi(a)=a_\xi(-\infty)\chi_-+a_\xi(+\infty)\chi_+.$$

By Lemma 5.2, the coset $[(a-\gamma_\xi(a))I]^\pi$ belongs to the ideal $J^\pi_{\xi,\eta}$ for every $\eta\in\mathbb{R}$. Consequently,

$$[aI]^\pi_{\xi,\eta}=[(a_\xi(-\infty)\chi_-+a_\xi(+\infty)\chi_+)I]^\pi_{\xi,\eta}.$$

3) Finally, let $(\xi,\eta)\in M_\infty(SO)\times M_\infty(SO)$. Since by (3.9) and (5.6),

$$\alpha_\xi(a-\widetilde{a}_\xi)=0,\quad \alpha_\eta(b-\widetilde{b}_\eta)=0,$$

from Lemma 5.2 it follows immediately that

$$[aI]^\pi_{\xi,\eta}=[\widetilde{a}_\xi I]^\pi_{\xi,\eta},\quad [W_b]^\pi_{\xi,\eta}=[W_{\widetilde{b}_\eta}]^\pi_{\xi,\eta}.$$

Thus, the formulas (5.7) and (5.8) are proved. □

Theorem 5.3 and Theorem 2.5 immediately imply the following.

Corollary 5.4. *An operator $A \in \mathcal{A}_{[SO,PC]}$ is Fredholm on the space $L_p^n(\mathbb{R})$ if and only if the cosets $A_{\xi,\eta}^{\pi} = \delta_{\xi,\eta}(A) \in \big(\mathcal{A}_{[SO,PC]}^{\pi}\big)_{\xi,\eta}$ are invertible in the quotient algebras $\Lambda_{\xi,\eta}^{\pi}$ for all $(\xi,\eta) \in \Omega$.*

Let $\nu(\mu)$ be the branch of the function $\sqrt{\mu(1-\mu)}$ in $\mathbb{C} \setminus \big((-\infty,0) \cup (1,+\infty)\big)$ which satisfies the condition $\nu(1/2) = -1/2$ and let

$$\mathcal{L}_p := \Big\{\mu = \big(1 + \coth[\pi(x + i/p)]\big)/2 : \; x \in \overline{\mathbb{R}}\Big\}.$$

Theorem 5.3, Corollary 5.4, and [12] or [24] yield the following explicit result.

Theorem 5.5. *Let $a^{(i)} \in [SO,PC]^{n\times n}$, $b^{(i)} \in [SO_p,PC_p]^{n\times n}$, $i = 1,2,\dots,m$, $1 < p < \infty$, $n \geq 1$. Then the operator*

$$A := \sum_{i=1}^{m} a^{(i)} \mathcal{F}^{-1} b^{(i)} \mathcal{F} \tag{5.9}$$

is Fredholm on the space $L_p^n(\mathbb{R})$ if:

(i) *for $(\xi,\eta,\mu) \in \mathbb{R} \times M_\infty(SO) \times \mathcal{L}_p$,*

$$\det\begin{pmatrix} c_\eta^-(\xi+0)\mu + c_\eta^+(\xi+0)(1-\mu) & \big(c_\eta^-(\xi+0) - c_\eta^+(\xi+0)\big)\nu(\mu) \\ \big(c_\eta^-(\xi-0) - c_\eta^+(\xi-0)\big)\nu(\mu) & c_\eta^-(\xi-0)(1-\mu) + c_\eta^+(\xi-0)\mu \end{pmatrix} \neq 0;$$

(ii) *for $(\xi,\eta,\mu) \in M_\infty(SO) \times \mathbb{R} \times \mathcal{L}_p$,*

$$\det\begin{pmatrix} d_\xi^-(\eta+0)\mu + d_\xi^-(\eta-0)(1-\mu) & \big(d_\xi^-(\eta+0) - d_\xi^-(\eta-0)\big)\nu(\mu) \\ \big(d_\xi^+(\eta+0) - d_\xi^+(\eta-0)\big)\nu(\mu) & d_\xi^+(\eta+0)(1-\mu) + d_\xi^+(\eta-0)\mu \end{pmatrix} \neq 0;$$

(iii) *for $(\xi,\eta) \in M_\infty(SO) \times M_\infty(SO)$,*

$$\det\Big(\sum_{i=1}^{m} a_\xi^{(i)}(\pm\infty)\, b_\eta^{(i)}(+\infty)\Big) \neq 0, \quad \det\Big(\sum_{i=1}^{m} a_\xi^{(i)}(\pm\infty)\, b_\eta^{(i)}(-\infty)\Big) \neq 0;$$

where

$$c_\eta^{\pm}(\xi) = \sum_{i=1}^{m} a^{(i)}(\xi)\, b_\eta^{(i)}(\pm\infty), \quad d_\xi^{\pm}(\eta) = \sum_{i=1}^{m} a_\xi^{(i)}(\pm\infty)\, b^{(i)}(\eta).$$

Proof. Since $W_{\chi_\pm} = P_\mp = (I \mp S)/2$, we get

$$V_\xi^{-1}\chi_\pm V_\xi = \chi_\xi^{\pm}, \quad V_\xi^{-1} P_\mp V_\xi = W_{\chi_\pm}, \quad e_{-\eta} P_\mp e_\eta I = W_{\chi_\eta^{\pm}}, \tag{5.10}$$

where

$$(V_\xi f)(x) = f(x+\xi), \quad e_\eta(x) = e^{i\eta x} \quad (\xi,\eta,x \in \mathbb{R}).$$

Taking into account (5.10), in view of Theorem 5.3 we get

$$A^\pi_{\xi,\eta} = \begin{cases} [V_\xi^{-1} B_{\xi,\eta} V_\xi]^\pi_{\xi,\eta}, & (\xi,\eta) \in \mathbb{R} \times M_\infty(SO), \\ [e_{-\eta} B_{\xi,\eta} e_\eta I]^\pi_{\xi,\eta}, & (\xi,\eta) \in M_\infty(SO) \times \mathbb{R}, \\ [B_{\xi,\eta}]^\pi_{\xi,\eta}, & (\xi,\eta) \in M_\infty(SO) \times M_\infty(SO), \end{cases} \tag{5.11}$$

where

$$B_{\xi,\eta} = \big(c_\eta^-(\xi-0)\chi_- + c_\eta^-(\xi+0)\chi_+\big)P_+ + \big(c_\eta^+(\xi-0)\chi_- + c_\eta^+(\xi+0)\chi_+\big)P_-$$

if $(\xi,\eta) \in \mathbb{R} \times M_\infty(SO)$,

$$B_{\xi,\eta} = \big(d_\xi^-(\eta-0)\chi_- + d_\xi^+(\eta-0)\chi_+\big)P_+ + \big(d_\xi^-(\eta+0)\chi_- + d_\xi^+(\eta+0)\chi_+\big)P_-$$

if $(\xi,\eta) \in M_\infty(SO) \times \mathbb{R}$,

$$B_{\xi,\eta} = c^{++}_{\xi,\eta} u_+ W_{u_+} + c^{-+}_{\xi,\eta} u_- W_{u_+} + c^{+-}_{\xi,\eta} u_+ W_{u_-} + c^{--}_{\xi,\eta} u_- W_{u_-}$$

if $(\xi,\eta) \in M_\infty(SO) \times M_\infty(SO)$, and

$$c^{\pm+}_{\xi,\eta} = \sum_{i=1}^m a_\xi^{(i)}(\pm\infty) b_\eta^{(i)}(+\infty), \quad c^{\pm-}_{\xi,\eta} = \sum_{i=1}^m a_\xi^{(i)}(\pm\infty) b_\eta^{(i)}(-\infty).$$

If $(\xi,\eta) \in M_\infty(SO) \times M_\infty(SO)$, then from (2.15) and Lemma 4.1(c) it follows that the operator $B_{\xi,\eta}$ is Fredholm on $L_p^n(\mathbb{R})$ if and only if

$$\det c^{\pm+}_{\xi,\eta} \neq 0, \quad \det c^{\pm-}_{\xi,\eta} \neq 0.$$

Consequently, (iii) implies the invertibility of the coset $A^\pi_{\xi,\eta}$ in $\Lambda^\pi_{\xi,\eta}$ for every $(\xi,\eta) \in M_\infty(SO) \times M_\infty(SO)$.

On the other hand, if $(\xi,\eta) \in \Omega \setminus \big(M_\infty(SO) \times M_\infty(SO)\big)$, then according to [5], [12], [24, Theorem 1.4] the operator $B_{\xi,\eta}$ is invertible on $L_p^n(\mathbb{R})$ if and only if condition (i) or (ii) is fulfilled. Therefore, conditions (i) and (ii) also imply the invertibility of $A^\pi_{\xi,\eta}$ in $\Lambda^\pi_{\xi,\eta}$ in view of (5.11).

Thus, the cosets $A^\pi_{\xi,\eta}$ are invertible in $\Lambda^\pi_{\xi,\eta}$ for all $(\xi,\eta) \in \Omega$, which implies the Fredholmness of the operator (5.9) on the space $L_p^n(\mathbb{R})$ by Corollary 5.4. □

Note that actually Theorem 5.5 is a Fredholm criterion. Moreover, taking into account that the local algebras are generated by two idempotents and making use of the two projections theorem (see, e.g., [11], [13], [2, Section 8.3]) one can construct a symbol calculus for the Banach algebra $\mathcal{A}_{[SO,PC]}$ and establish a Fredholm criterion for any operator $A \in \mathcal{A}_{[SO,PC]}$. The authors worked out all these extensions already (after submission of this article) and will publish the results in a forthcoming paper.

References

[1] M.A. Bastos, Yu.I. Karlovich, and B. Silbermann, Toeplitz operators with symbols generated by slowly oscillating and semi-almost periodic matrix functions. Preprint DMIST 31/2002, Lisboa 2002, submitted for publication.

[2] A. Böttcher and Yu.I. Karlovich, *Carleson Curves, Muckenhoupt Weights, and Toeplitz Operators.* Progress in Mathematics, vol. 154, Birkhäuser Verlag, Basel, Boston, Berlin, 1997.

[3] A. Böttcher, Yu.I. Karlovich, and V.S. Rabinovich, The method of limit operators for one-dimensional singular integrals with slowly oscillating data. *J. Operator Theory* **43** (2000), 171–198.

[4] A. Böttcher, Yu.I. Karlovich, and I.M. Spitkovsky. *Convolution Operators and Factorization of Almost Periodic Matrix Functions.* OT 131, Birkhäuser Verlag, Basel, Boston, Berlin, 2002.

[5] A. Böttcher and B. Silbermann, *Analysis of Toeplitz Operators.* Akademie-Verlag, Berlin 1989 and Springer-Verlag, Berlin, Heidelberg, New York, 1990.

[6] A. Böttcher and I.M. Spitkovsky, Wiener-Hopf integral operators with *PC* symbols on spaces with Muckenhoupt weight. *Revista Matemática Iberoamericana* **9** (1993), 257–279.

[7] A. Böttcher and I.M. Spitkovsky, Pseudodifferential operators with heavy spectrum. *Integral Equations and Operator Theory* **19** (1994), 251–269.

[8] H.O. Cordes, *Elliptic Pseudo-Differential Operators – An Abstract Theory.* Lect. Notes in Math. **756**, Springer-Verlag, Berlin, Heideberg, New York, 1979.

[9] R.V. Duduchava, Integral equations of convolution type with discontinuous coefficients. *Soobshch. Akad. Nauk Gruz. SSR* **92** (1978), 281–284 [Russian].

[10] R.V. Duduchava, *Integral Equations with Fixed Singularities.* Teubner, Leipzig, 1979.

[11] T. Finck, S. Roch, and B. Silbermann, Two projection theorems and symbol calculus for operators with massive local spectra. *Math. Nachrichten* **162** (1993), 167–185.

[12] I. Gohberg and N. Krupnik, Singular integral operators with piecewise continuous coefficients and their symbols. *Math. USSR - Izv.* **5** (1971), 955–979.

[13] I. Gohberg and N. Krupnik, Extension theorems for invertibility symbols in Banach algebras. *Integral Equations and Operator Theory* **15** (1992), 991–1010.

[14] L. Hörmander, Estimates for translation invariant operators in L^p spaces. *Acta Math.* **104** (1960), 93–140.

[15] Yu.I. Karlovich and E. Ramírez de Arellano, A shift-invariant algebra of singular integral operators with oscillating coefficients. *Integral Equations and Operator Theory* **39** (2001), 441–474.

[16] M.A. Krasnoselskii, P.P. Zabreiko, E.I. Pustylnik, and P.E. Sobolevskii, *Integral Operators in Spaces of Summable Functions.* Nauka, Moscow, 1966 [Russian]. English transl.: I. P. Noordhoff, Leyden, 1976.

[17] S.G. Mikhlin, *Multidimensional Singular Integrals and Integral Equations.* Fizmatgiz, Moscow, 1962 [Russian]. English transl.: Pergamon Press, New York, 1965.

[18] S.C. Power, Fredholm Toeplitz operators and slow oscillation. *Can. J. Math.* **32** (1980), 1058–1071.

[19] V.S. Rabinovich, *An Introductory Course on Pseudodifferential Operators.* Textos de Matemática-1, Centro de Matemática Aplicada, Instituto Superior Técnico, Lisboa, 1998.

[20] S. Roch and B. Silbermann, *Algebras of Convolution Operators and Their Image in the Calkin Algebra.* Report R-Math-05/90, Karl-Weierstrass-Inst. f. Math., Berlin, 1990.

[21] W. Rudin, *Functional Analysis.* Second ed., McGraw-Hill Inc., New York 1991.

[22] D. Sarason, Toeplitz operators with piecewise quasicontinuous symbols. *Indiana Univ. J.* **26** (1977), 817–838.

[23] R. Schneider, Integral equations with piecewise continuous coefficients in L^p-spaces with weight. *J. Integral Equations* **9** (1985), 135–152.

[24] I.B. Simonenko and Chin Ngok Min, *Local Method in the Theory of One-Dimensional Singular Integral Equations with Piecewise Continuous Coefficients. Noetherity.* University Press, Rostov on Don, 1986 [Russian].

[25] S.B. Stechkin, On bilinear forms. *Dokl. Akad. Nauk SSSR* **71** (1950), 237–240 [Russian].

[26] M.E. Taylor: *Pseudodifferential Operators.* Princeton Univ. Press, Princeton, New Jersey, 1981.

M.A. Bastos
Departamento de Matemática
Instituto Superior Técnico
Av. Rovisco Pais
1049 - 001 Lisboa
Portugal
e-mail: `abastos@math.ist.utl.pt`

A. Bravo
Departamento de Matemática
Instituto Superior Técnico
Av. Rovisco Pais
1049 - 001 Lisboa
Portugal
e-mail: `abravo@math.ist.utl.pt`

Yu.I. Karlovich
Facultad de Ciencias
Universidad Autónoma del Estado de Morelos
Av. Universidad 1001, Col. Chamilpa
Cuernavaca, Morelos, C.P. 62210
México
e-mail: `karlovich@buzon.uaem.mx`

Operator Theory
Advances and Applications, Vol. 147, 175–188

Asymptotically Good Pseudomodes for Toeplitz Matrices and Wiener-Hopf Operators

Albrecht Böttcher and Sergei Grudsky

To the memory of Erhard Meister

Abstract. We describe the structure of asymptotically good pseudomodes for Toeplitz matrices and their circulant analogues as well as for Wiener-Hopf integral operators and a continuous analogue of banded circulant matrices. The pseudomodes of circulant matrices and their continuous analogues are extended, while those of Toeplitz matrices or Wiener-Hopf operators are typically strongly localized in the endpoints.

1. Introduction

Let A be a bounded linear operator on a complex Hilbert space $\mathcal{H}$. A point λ in $\mathbf{C}$ is said to be an *ε-pseudoeigenvalue* of A if $\|(A-\lambda I)^{-1}\| \geq 1/\varepsilon$ (with the convention that $\|(A-\lambda I)^{-1}\| := \infty$ in case $A-\lambda I$ is not invertible). If λ is an ε-pseudoeigenvalue, then there exists a nonzero $x \in \mathcal{H}$ such that $\|(A-\lambda I)x\| \leq \varepsilon\|x\|$. Each such x is called an *ε-pseudomode* (or ε-pseudoeigenvector) for A at λ. Papers [10], [12], [13], the web site [5], and the book [4] contain detailed information about these concepts.

Now suppose we are given a sequence $\{A_n\}_{n=1}^{\infty}$ of matrices $A_n \in \mathbf{C}^{n\times n}$. We think of A_n as an operator on $\mathbf{C}^n$ with the ℓ^2 norm. We call a point $\lambda \in \mathbf{C}$ an *asymptotically good pseudoeigenvalue* for $\{A_n\}$ if $\|(A_n-\lambda I)^{-1}\| \to \infty$ as $n \to \infty$. In that case we can find nonzero vectors $x_n \in \mathbf{C}^n$ satisfying

$$\|(A_n-\lambda I)x_n\|/\|x_n\| \to 0 \text{ as } n \to \infty,$$

and each sequence $\{x_n\}$ with this property will be called an *asymptotically good pseudomode* for $\{A_n\}$ at λ. Our terminology is motivated by the papers [10] and [14]: there it is shown that $\|(A_n-\lambda I)^{-1}\|$ increases exponentially for certain classes of matrices, and the corresponding pseudomodes are called *exponentially good*.

S. Grudsky acknowledges financial support by CONACYT grant, Cátedra Patrimonal, Nivel II, No. 010286 (México).

This paper is devoted to the structure of asymptotically good pseudomodes for sequences of Toeplitz matrices. We also embark on Wiener-Hopf integral operators and on the circulant cousins of Toeplitz band matrices (called α-matrices in theoretical chemistry [16]) and their continual analogues.

2. Banded circulant matrices

Given a subset J_n of $\{1, 2, \ldots, n\}$, we denote by P_{J_n} the projection on $\mathbf{C}^n$ defined by

$$(P_{J_n} y)_j = \begin{cases} y_j & \text{for} \quad j \in J_n, \\ 0 & \text{for} \quad j \notin J_n. \end{cases}$$

The number of elements in J_n will be denoted by $|J_n|$. Let $\{y_n\}_{n=1}^{\infty}$ be a sequence of nonzero vectors $y_n \in \mathbf{C}^n$. We say that $\{y_n\}$ is *asymptotically localized* if there exists a sequence $\{J_n\}_{n=1}^{\infty}$ of sets $J_n \subset \{1, \ldots, n\}$ such that

$$\lim_{n\to\infty} \frac{|J_n|}{n} = 0 \quad \text{and} \quad \lim_{n\to\infty} \frac{\|P_{J_n} y_n\|}{\|y_n\|} = 1.$$

We denote by $F_n \in \mathbf{C}^{n\times n}$ the Fourier matrix,

$$F_n = \frac{1}{\sqrt{n}} \left(\omega_n^{jk}\right)_{j,k=0}^{n-1}, \qquad \omega_n := e^{2\pi i/n}.$$

A sequence $\{y_n\}_{n=1}^{\infty}$ of nonzero vectors $y_n \in \mathbf{C}^n$ will be called *asymptotically extended* if $\{F_n y_n\}$ is asymptotically localized. Problems concerning the question whether eigenvectors or pseudoeigenvectors are localized or extended have been extensively studied for many decades, especially for randomly perturbed Toeplitz matrices and their differential operators analogues, and the literature on this topic is vast. A few exemplary works are [1], [7], [8], [9], [11], [14].

Let a be a complex-valued L^∞ function on the complex unit circle $\mathbf{T}$. The $n \times n$ Toeplitz matrix $T_n(a)$ and the infinite Toeplitz matrix $T(a)$ are defined by $T_n(a) = (a_{j-k})_{j,k=1}^{n}$ and $T(a) = (a_{j-k})_{j,k=1}^{\infty}$, where

$$a_\ell = \frac{1}{2\pi} \int_0^{2\pi} a(e^{i\theta}) e^{-i\ell\theta} d\theta, \quad \ell \in \mathbf{Z}.$$

As $a \in L^\infty(\mathbf{T})$, the matrix $T(a)$ generates a bounded linear operator on $\ell^2(\mathbf{N})$. Now suppose that a is actually a trigonometric polynomial,

$$a(t) = \sum_{\ell=-r}^{s} a_\ell t^\ell, \quad t \in \mathbf{T}.$$

Then $T_n(a)$ is a banded matrix. For n large enough, we can add entries in the lower-left and upper-right corners of $T_n(a)$ in order to get a circulant matrix $C_n(a)$. For example, if $a(t) = a_{-1}t^{-1} + a_0 + a_1 t + a_2 t^2$, then

$$C_6(a) = \begin{pmatrix} a_0 & a_{-1} & 0 & 0 & a_2 & a_1 \\ a_1 & a_0 & a_{-1} & 0 & 0 & a_2 \\ a_2 & a_1 & a_0 & a_{-1} & 0 & 0 \\ 0 & a_2 & a_1 & a_0 & a_{-1} & 0 \\ 0 & 0 & a_2 & a_1 & a_0 & a_{-1} \\ a_{-1} & 0 & 0 & a_2 & a_1 & a_0 \end{pmatrix}.$$

Theorem 2.1. *Let a be a trigonometric polynomial. A point $\lambda \in \mathbf{C}$ is an asymptotically good pseudoeigenvalue for $\{C_n(a)\}$ if and only if $\lambda \in a(\mathbf{T})$, in which case every asymptotically good pseudomode for $\{C_n(a)\}$ is asymptotically extended.*

Proof. Clearly, $C_n(a) - \lambda I = C_n(a - \lambda)$. It is well known that

$$C_n(a-\lambda) = F_n^* \operatorname{diag}(a(\omega_n^j) - \lambda)_{j=0}^{n-1} F_n =: F_n^* D_n F_n.$$

Since F_n is unitary, it follows that

$$\|C_n^{-1}(a-\lambda)\| = \frac{1}{\min\limits_{0 \le j \le n-1} |a(\omega_n^j) - \lambda|},$$

which shows that $\|C_n^{-1}(a-\lambda)\| \to \infty$ if and only if $\lambda \in a(\mathbf{T})$.

Now pick $\lambda \in a(\mathbf{T})$ and suppose $\{x_n\}$ is an asymptotically good pseudomode for $\{C_n(a)\}$ at λ. We may without loss of generality assume that $\|x_n\| = 1$. Put $y_n = (y_j^{(n)})_{j=1}^n = F_n x_n$. We have

$$\|C_n(a-\lambda)x_n\| = \|F_n^* D_n F_n x_n\| = \|D_n y_n\|. \tag{2.1}$$

Fix an $\varepsilon > 0$. For $\delta > 0$, we put

$$\begin{aligned} G_n(\delta) &= \{j \in \{1,\dots,n\} : |a(\omega_n^{j-1}) - \lambda| \le \delta\}, \\ E(\delta) &= \{\theta \in [0, 2\pi) : |a(e^{i\theta}) - \lambda| < \delta\}. \end{aligned}$$

Since a is analytic in $\mathbf{C} \setminus \{0\}$, the set $E(\delta)$ is a finite union of intervals. Hence $|G_n(\delta)|/n \to |E(\delta)|/(2\pi)$ as $n \to \infty$, where $|E(\delta)|$ denotes the (length) measure of $E(\delta)$. Because $|E(\delta)| \to 0$ as $\delta \to 0$, there exist $\delta(\varepsilon) > 0$ and $N_1(\varepsilon) \ge 1$ such that $|G_n(\delta(\varepsilon))|/n < \varepsilon$ for all $n \ge N_1(\varepsilon)$. From (2.1) we infer that $\|D_n y_n\|^2 \to 0$ as $n \to \infty$. Consequently, $\|D_n y_n\|^2 < \varepsilon\delta(\varepsilon)^2$ for all $n \ge N_2(\varepsilon)$. Since

$$\|D_n y_n\|^2 = \sum_{j=1}^n |a(\omega_n^{j-1}) - \lambda|^2 |y_j^{(n)}|^2 \ge \delta(\varepsilon)^2 \sum_{j \notin G_n(\delta(\varepsilon))} |y_j^{(n)}|^2,$$

it follows that $\sum_{j \notin G_n(\delta(\varepsilon))} |y_j^{(n)}|^2 < \varepsilon$ for $n \ge N_2(\varepsilon)$. Thus, $\|P_{G_n(\delta(\varepsilon))} y_n\|^2 > 1 - \varepsilon$ for all $n \ge N_2(\varepsilon)$. Put $n(\varepsilon) = \max(N_1(\varepsilon), N_2(\varepsilon))$.

Now let $\varepsilon_k = 1/k$ $(k \ge 2)$. With $\delta_k := \delta(\varepsilon_k)$ and $n_k := n(\varepsilon_k)$ we then have

$$\frac{|G_n(\delta_k)|}{n} < \frac{1}{k} \quad \text{and} \quad \|P_{G_n(\delta_k)} y_n\|^2 > 1 - \frac{1}{k} \quad \text{for } n \ge n_k. \tag{2.2}$$

We may without loss of generality assume that $1 < n_2 < n_3 < \ldots$. For $1 \le n < n_2$, we let J_n denote an arbitrary subset of $\{1, \ldots, n\}$. For $n \ge n_2$, we define the sets $J_n \subset \{1, \ldots, n\}$ by

$$J_{n_2} = G_{n_2}(\delta_2), J_{n_2+1} = G_{n_2+1}(\delta_2), \ldots, J_{n_3-1} = G_{n_3-1}(\delta_2),$$
$$J_{n_3} = G_{n_3}(\delta_3), J_{n_3+1} = G_{n_3+1}(\delta_3), \ldots, J_{n_4-1} = G_{n_4-1}(\delta_3), \quad \ldots.$$

From (2.2) we see that

$$\frac{|J_{n_2}|}{n_2} < \frac{1}{2}, \ldots, \frac{|J_{n_3-1}|}{n_3-1} < \frac{1}{2},$$
$$\frac{|J_{n_3}|}{n_3} < \frac{1}{3}, \ldots, \frac{|J_{n_4-1}|}{n_4-1} < \frac{1}{3}, \quad \ldots,$$

which shows that $|J_n|/n \to 0$ as $n \to \infty$. Also by (2.2),

$$\|P_{J_{n_2}} y_{n_2}\|^2 > 1 - \frac{1}{2}, \ldots, \|P_{J_{n_3-1}} y_{n_3-1}\|^2 > 1 - \frac{1}{2},$$
$$\|P_{J_{n_3}} y_{n_3}\|^2 > 1 - \frac{1}{3}, \ldots, \|P_{J_{n_4-1}} y_{n_4-1}\|^2 > 1 - \frac{1}{3}, \quad \ldots,$$

and hence $\|P_{J_n} y_n\| \to 1$ as $n \to \infty$. Since $\|y_n\| = 1$ for all n, it results that $\{y_n\}$ is asymptotically localized. Consequently, $\{x_n\}$ is asymptotically extended. $\square$

3. Toeplitz matrices

Let $a \in L^\infty(\mathbf{T})$ be a piecewise continuous function, that is, suppose the one-sided limits $a(t-0)$ and $a(t+0)$ exist for each t on the counter-clockwise oriented unit circle. We denote by $a^\#(\mathbf{T})$ the closed and continuous curve that results from the (essential) range of a by filling in the line segments $[a(t-0), a(t+0)]$ at each jump of a. The counter-clockwise orientation of $\mathbf{T}$ induces an orientation of $a^\#(\mathbf{T})$ in the natural manner. For $\lambda \in \mathbf{C} \setminus a^\#(\mathbf{T})$, we let wind (a, λ) denote the winding number of the curve $a^\#(\mathbf{T})$ about λ. It is well known that the spectrum of $T(a)$ on $\ell^2(\mathbf{N})$ is the union of $a^\#(\mathbf{T})$ and all points $\lambda \in \mathbf{C} \setminus a^\#(\mathbf{T})$ with wind $(a, \lambda) \neq 0$. If $\lambda \notin a^\#(\mathbf{T})$ and wind $(a, \lambda) = -m < 0$, then the kernel (= null space) $\operatorname{Ker} T(a-\lambda)$ has the dimension m, while if $\lambda \notin a^\#(\mathbf{T})$ and wind $(a, \lambda) = m > 0$, then the kernel of the adjoint of $T(a-\lambda)$ is m-dimensional. All these facts can be found in [4] or [6], for example.

Suppose that $\lambda \notin a^\#(\mathbf{T})$ and wind $(a, \lambda) = -m < 0$. We then can write $a - \lambda = b\chi_{-m}$, where b is piecewise continuous, $0 \notin b^\#(\mathbf{T})$, wind $(b, 0) = 0$, and χ_k is defined by $\chi_k(t) = t^k$ $(t \in \mathbf{T})$. The operator $T(b)$ is invertible on $\ell^2(\mathbf{N})$ and, moreover, the matrices $T_n(b)$ are invertible for all sufficiently large n,

$$\lim_{n\to\infty} \|T_n^{-1}(b)\| = \|T^{-1}(b)\| \quad \text{and} \quad T_n^{-1}(b) P_n \to T^{-1}(b) \text{ strongly} \tag{3.1}$$

(see, e.g., [2], [4], [6]). Here P_n is the projection on $\ell^2(\mathbf{N})$ given by $(P_n y)_j = y_j$ for $1 \le j \le n$ and $(P_n y)_j = 0$ for $j \ge n+1$. We will frequently identify the image of

P_n with $\mathbf{C}^n$. It is also well known (and easily verified) that the m elements

$$u_j := T^{-1}(b)e_j \quad (j = 1, \ldots, m) \tag{3.2}$$

form a basis in $\operatorname{Ker} T(a - \lambda)$, where $e_j \in \ell^2(\mathbf{N})$ is the sequence whose jth term is 1 and the remaining terms of which are zero. We finally remark that, obviously, $T(a) - \lambda I = T(a - \lambda)$ and $T_n(a) - \lambda I = T_n(a - \lambda)$.

Each point $\lambda \in \mathbf{C} \setminus a^{\#}(\mathbf{T})$ with $\operatorname{wind}(a, \lambda) \neq 0$ is an asymptotically good pseudoeigenvalue for $\{T_n(a)\}$ (see [4], [6], [10]). The following theorem provides us with a complete description of the structure of asymptotically good pseudomodes.

Theorem 3.1. *Suppose* $\lambda \notin a^{\#}(\mathbf{T})$ *and* $\operatorname{wind}(a, \lambda) = -m < 0$. *Let* $x_n \in \mathbf{C}^n$ *be unit vectors. The sequence* $\{x_n\}$ *is an asymptotically good pseudomode for* $\{T_n(a)\}$ *at* λ *if and only if there exist* $c_1^{(n)}, \ldots, c_m^{(n)} \in \mathbf{C}$ *and* $z_n \in \mathbf{C}^n$ *such that*

$$x_n = c_1^{(n)} P_n u_1 + \cdots + c_m^{(n)} P_n u_m + z_n, \tag{3.3}$$

$$\sup_{n \geq 1,\, 1 \leq j \leq m} |c_j^{(n)}| < \infty, \qquad \lim_{n \to \infty} \|z_n\| = 0, \tag{3.4}$$

where $u_1, \ldots, u_m$ *are given by* (3.2).

Proof. Assume that (3.3) and (3.4) hold. Since $\|T_n(a - \lambda)P_n\| \leq \|a - \lambda\|_\infty$, we see that $T_n(a - \lambda)z_n \to 0$. As the numbers $|c_j^{(n)}|$ are bounded by a constant independent of n and as $P_n \to I$ strongly (= pointwise) and $T(a - \lambda)u_j = 0$, we obtain that

$$\lim_{n \to \infty} T_n(a - \lambda)x_n = \sum_{j=1}^{m} \lim_{n \to \infty} c_j^{(n)} T_n(a - \lambda)u_j = 0.$$

Thus, $\{x_n\}$ is an asymptotically good pseudomode.

Conversely, suppose $\|T_n(a - \lambda)x_n\| \to 0$. Put $y_n = T_n(a - \lambda)x_n$. With $Q_n = I - P_n$, we have

$$\begin{aligned} T_n(a - \lambda) &= T_n(\chi_{-m} b) = P_n T(\chi_{-m} b) P_n = P_n T(\chi_{-m}) T(b) P_n \\ &= P_n T(\chi_{-m}) P_n T(b) P_n + P_n T(\chi_{-m}) Q_n T(b) P_n =: A_n + B_n. \end{aligned}$$

Since $T(\chi_{-m})$ is nothing but the shift operator $(\xi_1, \xi_2, \ldots) \mapsto (\xi_{m+1}, \xi_{m+2}, \ldots)$, it follows that

$$\operatorname{Im} A_n \subset \operatorname{Im} P_{n-m}, \quad \operatorname{Im} B_n \subset \operatorname{Im} P_{\{n-m+1, \ldots, n\}}, \tag{3.5}$$

where $\operatorname{Im} C$ refers to the image (= range) of the operator C. This implies that

$$\|y_n\|^2 = \|A_n x_n + B_n x_n\|^2 = \|A_n x_n\|^2 + \|B_n x_n\|^2,$$

and hence $\|A_n x_n\| \to 0$ because $\|y_n\| \to 0$. The equality $T_n(\chi_{-m}) T_n(b) x_n = A_n x_n$ gives

$$T_n(b)x_n = c_1^{(n)} e_1 + \cdots + c_m^{(n)} e_m + T_n(\chi_m) A_n x_n$$

with certain complex numbers $c_1^{(n)}, \dots, c_m^{(n)}$. Since

$$\left(\sum_{j=1}^{m} |c_j^{(n)}|^2\right)^{1/2} \le \|T_n(b)x_n\| + \|T(\chi_m)\| \, \|A_n x_n\| \le \|b\|_\infty + \|A_n x_n\|,$$

we conclude that there is an $M < \infty$ such that $|c_j^{(n)}| \le M$ for all n and j. Finally, from (3.1), (3.2) and the equality

$$x_n = c_1^{(n)} T_n^{-1}(b)e_1 + \dots + c_m^{(n)} T_n^{-1}(b)e_m + T_n^{-1}(b)T_n(\chi_m)A_n x_n$$

we get (3.3) and (3.4) with

$$z_n = T_n^{-1}(b)T_n(\chi_m)A_n x_n + \sum_{j=1}^{m} c_j^{(n)} (T_n^{-1}(b)e_j - P_n T^{-1}(b)e_j).$$

This completes the proof. □

Let a be as in Theorem 3.1, but in addition suppose now that a belongs to the Wiener algebra $W(\mathbf{T})$, i.e., that the Fourier series of a converges absolutely. We write $a - \lambda = b\chi_{-m}$ as above. Clearly, b is also in $W(\mathbf{T})$. Since $0 \notin b(\mathbf{T})$ and $\operatorname{wind}(b, 0) = 0$, the function b has a Wiener-Hopf factorization $b = b_- b_+$. The factors $b_\pm$ can be given by

$$b_-(t) = \exp\left(\sum_{\ell=0}^{\infty} (\log b)_{-\ell} t^{-\ell}\right), \quad b_+(t) = \exp\left(\sum_{\ell=1}^{\infty} (\log b)_\ell t^\ell\right),$$

where $\log b$ is any logarithm of b in $W(\mathbf{T})$. The Wiener-Hopf factorization $b = b_- b_+$ yields the representation $T^{-1}(b) = T(b_+^{-1})T(b_-^{-1})$ (see, e.g., [4] or [6]), or written down in detail, $T^{-1}(b)$ equals

$$\begin{pmatrix} (b_+^{-1})_0 & & & \\ (b_+^{-1})_1 & (b_+^{-1})_0 & & \\ (b_+^{-1})_2 & (b_+^{-1})_1 & (b_+^{-1})_0 & \\ \dots & \dots & \dots & \dots \end{pmatrix} \begin{pmatrix} (b_-^{-1})_0 & (b_-^{-1})_{-1} & (b_-^{-1})_{-2} & \dots \\ & (b_-^{-1})_0 & (b_-^{-1})_{-1} & \dots \\ & & (b_-^{-1})_0 & \dots \\ & & & \dots \end{pmatrix}.$$

If a is even rational, then the sequences $\{(b_+^{-1})_n\}_{n=0}^\infty$ and $\{(b_-^{-1})_{-n}\}_{n=0}^\infty$ decay exponentially, and from (3.2) we deduce that $u_1, \dots, u_m$ are also exponentially decaying. Thus, Theorem 3.1 implies that, up to the $o(1)$ term z_n, *all* asymptotically good pseudomodes are exponentially decaying. We remark that in the case where a is a trigonometric polynomial (which is equivalent to the requirement that $T(a)$ be a banded matrix) the *existence* of exponentially decaying pseudomodes was already proved in [10] and [14].

We now sharpen the definition of an asymptotically localized sequence. We say that a sequence $\{y_n\}$ of vectors $y_n \in \mathbf{C}^n$ is *asymptotically strongly localized in the beginning part* if

$$\lim_{n\to\infty} \frac{\|P_{\{1,\dots,j_n\}} y_n\|}{\|y_n\|} = 1 \tag{3.6}$$

for every sequence $\{j_n\}_{n=1}^{\infty}$ such that $j_n \to \infty$ and $1 \le j_n \le n$. Asymptotic strong localization in the beginning part implies, for example, that (3.6) is true with $j_n = \log\log n$ $(n \ge 3)$.

Theorem 3.2. *Suppose* $\lambda \notin a^{\#}(\mathbf{T})$ *and* $\operatorname{wind}(a, \lambda) = -m < 0$. *Then every asymptotically good pseudomode for* $\{T_n(a)\}$ *at* λ *is asymptotically strongly localized in the beginning part.*

Proof. Let $\{x_n\}$ be an asymptotically good pseudomode for $\{T_n(a)\}$ at λ. We may without loss of generality assume that $\|x_n\| = 1$ for all n. By Theorem 3.1,

$$x_n = x_1^{(n)} P_n u_1 + \cdots + c_m^{(n)} P_n u_m + z_n =: w_n + z_n,$$

where $u_1, \ldots, u_m$ are given by (3.2) and $c_1^{(n)}, \ldots, c_m^{(n)}, z_n$ satisfy (3.4). Choose $M < \infty$ so that $|c_i^{(n)}| \le M$ for all i and n. Let $\{j_n\}$ be any sequence such that $j_n \to \infty$ and $1 \le j_n \le n$. Put $J_n = \{1, \ldots, j_n\}$ and $J_n^c = \{j_n + 1, \ldots, n\}$. From (3.2) we infer that $u_1, \ldots, u_m \in \ell^2(\mathbf{N})$. We have $\|P_{J_n^c} w_n\| \le M \sum_{i=1}^{m} \|P_{J_n^c} u_i\|$. Since $u_i = (u_k^{(i)})_{k=1}^{\infty}$ is in $\ell^2(\mathbf{N})$ and hence

$$\|P_{J_n^c} u_i\|^2 = \sum_{k=j_n+1}^{n} |u_k^{(i)}|^2 \le \sum_{k=j_n+1}^{\infty} |u_k^{(i)}|^2 = o(1) \quad \text{as} \quad j_n \to \infty,$$

it follows that $\|P_{J_n^c} w_n\| \to 0$ as $n \to \infty$. Finally,

$$\begin{aligned} 1 \ge \|P_{J_n} x_n\|^2 &= 1 - \|P_{J_n^c} x_n\|^2 = 1 - \|P_{J_n^c}(w_n + z_n)\|^2 \\ &\ge 1 - \Big(\|P_{J_n^c} w_n\| + \|P_{J_n^c} z_n\|\Big)^2, \end{aligned}$$

and because $\|P_{J_n^c} w_n\| \to 0$ and $\|P_{J_n^c} z_n\| \to 0$ as $n \to \infty$, we arrive at the conclusion that $\|P_{J_n} x_n\| \to 1$. □

To conclude this section, suppose that $\lambda \in \mathbf{C} \setminus a^{\#}(\mathbf{T})$ and that $\operatorname{wind}(a, \lambda) = m > 0$. Following [15], we define $\widetilde{a}$ by $\widetilde{a}(t) := a(1/t)$ $(t \in \mathbf{T})$ and we let W_n be the operator that is P_n followed by reversal of the coordinates. We have $\lambda \notin \widetilde{a}^{\#}(\mathbf{T})$ and $\operatorname{wind}(\widetilde{a}, \lambda) = -m < 0$. Moreover, $W_n T_n(a - \lambda) W_n = T_n(\widetilde{a} - \lambda)$ and hence $\|T_n(a - \lambda)x_n\| = \|T_n(\widetilde{a} - \lambda) W_n x_n\|$. Consequently, by Theorem 3.1, a sequence $\{x_n\}$ of unit vectors is an asymptotically good pseudomode of $\{T_n(a)\}$ at λ if and only if

$$W_n x_n = c_1^{(n)} P_n \widetilde{u}_1 + \cdots + c_m^{(n)} P_n \widetilde{u}_m + z_n, \tag{3.7}$$

where $|c_j^{(n)}| \le M < \infty$ for all j and n, $\|z_n\| \to 0$ as $n \to \infty$, and $\widetilde{u}_1, \ldots, \widetilde{u}_m$ are given by $\widetilde{u}_j = T^{-1}(\widetilde{b}) e_j$. Clearly, (3.7) can be rewritten in the form

$$x_n = c_1^{(n)} W_n \widetilde{u}_1 + \cdots + c_m^{(n)} W_n \widetilde{u}_m + \widetilde{z}_n$$

with $\widetilde{z}_n = W_n z_n$. The analogue of Theorem 3.2 says that every asymptotically good pseudomode $\{x_n\}$ for $\{T_n(a)\}$ at λ is asymptotically strongly localized in the terminating part, that is, the sequence $\{W_n x_n\}$ is asymptotically strongly localized in the beginning part.

4. A continuous analogue of banded circulant matrices

Let k be a function in $L^1(\mathbf{R})$ and suppose $k(x) = 0$ for $|x| \geq r$. For $\tau > 2r$, there is a unique continuation of k to a τ-periodic function k_τ on all of $\mathbf{R}$. A continuous analogue of the operator $C_n(a)$ considered in Section 2 is the operator on $L^2(0,\tau)$ that is defined by

$$(C_\tau(k)f)(x) = \gamma f(x) + \int_0^\tau k_\tau(x-t)f(t)dt, \quad x \in (0,\tau),$$

where γ is a fixed number in $\mathbf{C}$. We put

$$\widehat{k}(\xi) = \gamma + \int_{-r}^{r} k(x)e^{-i\xi x}dx, \quad \xi \in \mathbf{R}.$$

We call a point $\lambda \in \mathbf{C}$ an *asymptotically good pseudoeigenvalue* for $\{C_\tau(k)\}$ if $\|(C_\tau(k) - \lambda I)^{-1}\| \to \infty$ as $\tau \to \infty$, and a family $\{f_\tau\}_{\tau>0}$ of nonzero functions $f_\tau \in L^2(0,\tau)$ is said to be an *asymptotically good pseudomode* for $\{C_\tau(k)\}$ at λ if $\|(C_\tau(k) - \lambda I)f_\tau\|/\|f_\tau\| \to 0$ as $\tau \to \infty$.

Let $\{g_\tau\}_{\tau>0}$ be a family of elements $g_\tau = (g_j^{(\tau)})_{j\in\mathbf{Z}} \in \ell^2(\mathbf{Z})$. We say that $\{g_\tau\}$ is *asymptotically localized* if there exists a family $\{J_\tau\}_{\tau>\tau_0}$ of finite subsets $J_\tau \in \mathbf{Z}$ such that

$$\lim_{\tau\to\infty} \frac{|J_\tau|}{\tau} = 0 \quad \text{and} \quad \lim_{\tau\to\infty} \frac{\|P_{J_\tau} g_\tau\|}{\|g_\tau\|} = 1.$$

Put $\varphi_j(x) = (1/\sqrt{\tau})e^{2\pi i j x/\tau}$. The system $\{\varphi_j\}_{j\in\mathbf{Z}}$ is an orthonormal basis in $L^2(0,\tau)$. Thus, the map

$$\Phi_\tau : L^2(0,\tau) \to \ell^2(\mathbf{Z}), \ f_\tau \mapsto \big((f_\tau, \varphi_j)\big)_{j\in\mathbf{Z}}$$

is a unitary operator. A family $\{f_\tau\}_{\tau>0}$ of functions $f_\tau \in L^2(0,\tau)$ will be called *asymptotically extended* if the family $\{\Phi_\tau f_\tau\}_{\tau>0}$ is asymptotically localized.

Theorem 4.1. *A number $\lambda \in \mathbf{C}$ is an asymptotically good pseudoeigenvalue for $\{C_\tau(k)\}$ if and only if $\lambda \in \gamma + \widehat{k}(\dot{\mathbf{R}})$, where $\dot{\mathbf{R}} := \mathbf{R} \cup \{\infty\}$. If $\lambda + \gamma \in \widehat{k}(\dot{\mathbf{R}})$ and, in addition, $\lambda \neq \gamma$, then every asymptotically good pseudomode for $\{C_\tau(k)\}$ at λ is asymptotically extended.*

Proof. For $f_\tau \in L^2(0,\tau)$ and $x \in (0,\tau)$,

$$\begin{aligned}(C_\tau(k)f_\tau)(x) &= \gamma f(x) + \sum_{j\in\mathbf{Z}} (f_\tau, \varphi_j) \int_0^\tau k_\tau(x-t)\varphi_j(t)dt \\ &= \gamma f(x) + \sum_{j\in\mathbf{Z}} (f_\tau, \varphi_j) \frac{1}{\sqrt{\tau}} e^{2\pi i j x/\tau} \int_{x-\tau}^{x} k_\tau(s)e^{-2\pi i j s/\tau} ds\end{aligned}$$

$$\begin{aligned} &= \gamma f(x) + \sum_{j\in\mathbf{Z}} (f_\tau, \varphi_j) \frac{1}{\sqrt{\tau}}\, e^{2\pi i j x/\tau} \int_{-\tau/2}^{\tau/2} k_\tau(s) e^{-2\pi i j s/\tau} ds \\ &= \gamma f(x) + \sum_{j\in\mathbf{Z}} (f_\tau, \varphi_j) \frac{1}{\sqrt{\tau}}\, e^{2\pi i j x/\tau} \int_{-\tau}^{\tau} k_\tau(s) e^{-2\pi i j s/\tau} ds \\ &= \sum_{j\in\mathbf{Z}} \left(\gamma + \widehat{k}\left(\frac{2\pi j}{\tau}\right)\right) (f_\tau, \varphi_j)\varphi_j(x). \end{aligned}$$

Thus, $C_\tau(k)$ is unitarily equivalent to the diagonal operator

$$\operatorname{diag}\left(\gamma + \widehat{k}\left(\frac{2\pi j}{\tau}\right)\right)_{j\in\mathbf{Z}} : \ell^2(\mathbf{Z}) \to \ell^2(\mathbf{Z}).$$

This shows that λ is an asymptotically good pseudoeigenvalue for $\{C_\tau(k)\}$ if and only if $\lambda \in \gamma + \widehat{k}(\dot{\mathbf{R}})$.

Now suppose $\lambda = \gamma + \widehat{k}(\xi_0)$ with $\xi_0 \in \dot{\mathbf{R}}$ and let $\{f_\tau\}$ be an asymptotically good pseudomode for $\{C_\tau(k)\}$ at λ. Without loss of generality we may assume that $\|f_\tau\| = 1$ for all τ. Thus,

$$\|C_\tau(k-\lambda)f_\tau\|^2 = \sum_{j\in\mathbf{Z}} \left|\widehat{k}\left(\frac{2\pi j}{\tau}\right) - \widehat{k}(\xi_0)\right|^2 |(f_\tau, \varphi_j)|^2 \to 0. \tag{4.1}$$

Fix $\varepsilon > 0$. For $\delta > 0$, consider the sets

$$\begin{aligned} G_\tau(\delta) &= \{j \in \mathbf{Z} : |\widehat{k}(2\pi j/\tau) - \widehat{k}(\xi_0)| < \delta\}, \\ E(\delta) &= \{\xi \in \mathbf{R} : |\widehat{k}(\xi) - \widehat{k}(\xi_0)| < \delta\}. \end{aligned}$$

Since $\widehat{k}(\xi_0) \neq 0$ and since $\widehat{k}$ is an entire function, the set $E(\delta)$ is a finite union of intervals and $|E(\delta)| \to 0$ as $\delta \to 0$. As $|G_\tau(\delta)|/\tau \to |E(\delta)|/(2\pi)$, there are $\delta(\varepsilon) > 0$ and $t_1(\varepsilon) > 0$ such that $|G_\tau(\delta(\varepsilon))|/\tau < \varepsilon$ whenever $\tau \geq t_1(\varepsilon)$. We have

$$\sum_{j\in\mathbf{Z}} \left|\widehat{k}\left(\frac{2\pi j}{\tau}\right) - \widehat{k}(\xi_0)\right|^2 |(f_\tau, \varphi_j)|^2 \geq \delta(\varepsilon)^2 \sum_{j\notin G_\tau(\delta(\varepsilon))} |(f_\tau, \varphi_j)|^2,$$

which in conjunction with (4.1) gives $\|P_{G_\tau(\delta(\varepsilon))}\Phi_\tau f_\tau\|^2 > 1 - \varepsilon$ for all $\tau \geq \tau_2(\varepsilon)$. Let $\tau(\varepsilon) = \max(t_1(\varepsilon), t_2(\varepsilon))$.

Choose $\varepsilon_\ell = 1/\ell$ $(\ell \geq 2)$ and put $\delta_\ell := \delta(\varepsilon_\ell)$ and $\tau_\ell := \tau(\varepsilon_\ell)$. We have proved that

$$\frac{|G_\tau(\delta_\ell)|}{\tau} < \frac{1}{\ell} \quad \text{and} \quad \|P_{G_\tau(\delta_\ell)}\Phi_\tau f_\tau\|^2 > 1 - \frac{1}{\ell} \quad \text{for } \tau \geq \tau_\ell, \tag{4.2}$$

and we may assume that $0 < \tau_2 < \tau_3 < \cdots$. Let J_τ be an arbitrary subset of $\mathbf{Z}$ for $0 < \tau < \tau_2$ and define $J_\tau = G_\tau(\delta_\ell)$ for $\tau_\ell \leq \tau < \tau_{\ell+1}$. Then, by (4.2),

$$\frac{J_\tau}{\tau} < \frac{1}{\ell} \quad \text{and} \quad \|P_{J_\tau}\Phi_\tau f_\tau\|^2 > 1 - \frac{1}{\ell} \quad \text{for } \tau_\ell \leq \tau < \tau_{\ell+1},$$

which shows that $|J_\tau|/\tau \to 0$ and $\|P_{j_\tau}\Phi_\tau f_\tau\| \to 1$ as $\tau \to \infty$. Thus, $\{\Phi_\tau f_\tau\}$ is asymptotically localized. This means that $\{f_\tau\}$ is asymptotically extended. □

Things are different for $\lambda = \gamma$. In the following, $P_{(-\mu\tau,\mu\tau)}$ denotes the canonical projection of $\ell^2(\mathbf{Z})$ onto the subspace of all sequences whose support is contained in the interval $(-\mu\tau, \mu\tau)$.

Theorem 4.2. *Let $\{f_\tau\}_{\tau>0}$ be an asymptotically good pseudomode for $\{C_\tau(k)\}$ at γ and assume without loss of generality that $\|f_\tau\| = 1$ for all τ. If $\widehat{k}(0) \neq \gamma$, then there exists a $\mu > 0$ such that*

$$\|P_{(-\mu\tau,\mu\tau)}\Phi_\tau f_\tau\| \to 0 \quad as \quad \tau \to \infty. \tag{4.3}$$

If $\widehat{k}(0) = \gamma$, then there are μ and ν such that $0 < \nu < \mu$ and

$$\|P_{(-\mu\tau,-\nu\tau)\cup(\nu\tau,\mu\tau)}\Phi_\tau f_\tau\| \to 0 \quad as \quad \tau \to \infty. \tag{4.4}$$

Proof. Suppose first that $\widehat{k}(0) \neq \gamma$. Then $|\widehat{k}(\xi) - \gamma| \geq \delta > 0$ for all $|\xi| \leq 2\pi\mu$ with some $\mu > 0$. Thus, $|\widehat{k}(2\pi j/\tau) - \gamma| \geq \delta > 0$ whenever $|j| < \mu\tau$. Since

$$\begin{aligned} \|C_\tau(k-\gamma)f_\tau\|^2 &= \sum_{j\in\mathbf{Z}} \left|\widehat{k}\left(\frac{2\pi j}{\tau}\right) - \gamma\right|^2 |(f_\tau, \varphi_j)|^2 \\ &\geq \delta^2 \sum_{|j|<\mu\tau} |(f_\tau, \varphi_j)|^2 = \delta^2 \|P_{(-\mu\tau,\mu\tau)}\Phi_\tau f_\tau\|^2, \end{aligned}$$

we obtain (4.3) from (4.1). Now suppose $\widehat{k}(0) = \gamma$. Then the entire $\widehat{k} - \gamma$ has no other zeros than 0 in some open neighborhood of 0, and hence we can find numbers μ and ν such that $0 < \nu < \mu$ and $|\widehat{k}(\xi) - \gamma| \geq \delta > 0$ for $2\pi\nu < |\xi| < 2\pi\mu$. It follows that $|\widehat{k}(2\pi j/\tau) - \gamma| \geq \delta > 0$ for $\nu\tau < |j| < \mu\tau$ and hence

$$\|C_\tau(k-\gamma)f_\tau\|^2 \geq \delta^2 \sum_{\nu\tau<|j|<\mu\tau} |(f_\tau, \varphi_j)|^2 = \delta^2 \|P_{(-\mu\tau,-\nu\tau)\cup(\nu\tau,\mu,\tau)}\Phi_\tau f_\tau\|^2.$$

It is again (4.1) that implies (4.4). □

Remark 4.3. Theorem 4.2 can be improved by taking into account the orders of the zeros of $\widehat{k} - \gamma$ at the origin and at infinity. We will not embark on this question here. We rather wish to point out that Theorem 4.2 is best possible in general.

Let $k(x) = 0$ for $|x| \geq 1$ and $k(x) = 1/2$ for $|x| < 1$. Then $\widehat{k}(\xi) = \dfrac{\sin\xi}{\xi}$ and, for $f_\tau \in L^2(0,\tau)$,

$$\|C_\tau(k)f_\tau\|^2 = \sum_{j\in\mathbf{Z}} \frac{\tau^2}{4\pi^2 j^2} \sin^2\frac{2\pi j}{\tau} |(f_\tau, \varphi_j)|^2.$$

We have $\widehat{k}(0) = 1 \neq 0$. Let $\{f_\tau\}$ by any asymptotically good pseudomode for $\{C_\tau(k)\}$ at 0. Thus, $\|f_\tau\| = 1$ for all $\tau > 0$ and $\|C_\tau(k)f_\tau\| \to 0$ as $\tau \to \infty$. Change f_τ to φ_τ for $\tau \in \mathbf{N}$. Since

$$\|C_\tau(k)\varphi_\tau\|^2 = \frac{\tau^2}{4\pi^2 j^2} \sin^2\frac{2\pi\tau}{\tau} = 0,$$

it follows that the family $\{h_\tau\}_{\tau>0}$ defined by $h_\tau = f_\tau$ for $\tau \notin \mathbf{N}$ and $h_\tau = \varphi_\tau$ for $\tau \in \mathbf{N}$ is also an asymptotically good pseudomode for $\{C_\tau(k)\}$ at the origin. If $\beta : (0,\infty) \to (0,\infty)$ is any superlinear function, that is, any function that increases faster than every linear function, then

$$\|P_{(-\beta(\tau),\beta(\tau))}\Phi_\tau h_\tau\|^2 = \|P_{(-\beta(\tau),\beta(\tau))}e_\tau\|^2 = 1$$

whenever $\tau \in \mathbf{N}$ and $\beta(\tau) > \tau$. This shows that (4.3) is not in general true with $\mu\tau$ replaced by $\beta(\tau)$.

Now let $k(x) = 0$ for $|x| \geq 1$, $k(x) = -i/2$ for $-1 < x < 0$, and $k(x) = i/2$ for $0 \leq x < 1$. Then $\widehat{k}(\xi) = (1-\cos\xi)/\xi$. This time $\widehat{k}(0) = 0$. Since $C_\tau(k)\varphi_\tau = 0$ for $\tau \in \mathbf{N}$, we can argue as above to see that (4.4) is not in general valid with $\mu\tau$ and $\nu\tau$ replaced by a superlinear and sublinear function, respectively. □

5. Wiener-Hopf integral operators

The Wiener-Hopf integral operator $W(a)$ generated by a function $a \in L^\infty(\mathbf{R})$ is the bounded linear operator on $L^2(0,\infty)$ that is defined by $W(a)f = PF^{-1}M(a)Ff$, where F is the Fourier transform,

$$(Ff)(\xi) = \int_{-\infty}^{\infty} f(x)e^{i\xi x}dx, \quad \xi \in \mathbf{R},$$

$M(a)$ stands for the operator of multiplication by a, and P is the orthogonal projection of $L^2(\mathbf{R})$ onto $L^2(0,\infty)$. If a is of the form $a = \gamma + Fk$ with $\gamma \in \mathbf{C}$ and $k \in L^1(\mathbf{R})$, then $W(a)$ can be written as

$$(W(a)f)(x) = \gamma f(x) + \int_0^\infty k(x-t)f(t)dt, \quad x > 0.$$

The Cauchy singular integral operator on the half-line,

$$(Sf)(x) = \frac{1}{\pi i}\,\text{v.p.}\int_0^\infty \frac{f(t)}{t-x}\,dt, \quad x > 0,$$

is $W(\sigma)$ with the piecewise continuous function $\sigma(\xi) := -\operatorname{sign}\xi$. The subject of this section is truncated Wiener-Hopf operators: for $\tau > 0$, the truncated Wiener-Hopf operator $W_\tau(a)$ is the compression of $W(a)$ to $L^2(0,\tau)$.

If $a \in L^\infty(\mathbf{R})$ is piecewise continuous, we define $a^\#(\mathbf{R})$ as the closed, continuous, and naturally oriented curve that results from the (essential) range of a by filling in the line segments $[a(x-0), a(x+0)]$ for the jumps at $x \in \mathbf{R}$ and the line segment $[a(+\infty), a(-\infty)]$ if a has a jump at infinity. We let $\operatorname{wind}(a,\lambda)$ denote the winding number of $a^\#(\mathbf{R})$ about $\lambda \in \mathbf{C} \setminus a^\#(\mathbf{R})$.

Let L_j be the normalized Laguerre polynomial of degree $j-1$,

$$L_j(x) = \frac{e^x}{(j-1)!}\frac{d^{j-1}}{dx^{j-1}}(e^{-x}x^{j-1}), \quad j = 1,2,\ldots.$$

The system $\{e_j\}_{j=1}^{\infty}$ given by $e_j(x) = \sqrt{2}\, L_j(2x)e^{-x}$ is an orthonormal basis in $L^2(0,\infty)$. Notice that

$$(Fe_j)(\xi) = i\sqrt{2}\left(\frac{\xi - i}{\xi + i}\right)^{j-1} \frac{1}{\xi + i}, \quad \xi \in \mathbf{R}.$$

For $k \in \mathbf{Z}$, we put

$$\chi_k(\xi) = \left(\frac{\xi - i}{\xi + i}\right)^{k}, \quad \xi \in \mathbf{R}.$$

Suppose $a \in L^\infty(\mathbf{R})$ is piecewise continuous, $\lambda \notin a^{\#}(\mathbf{R})$, and $\operatorname{wind}(a,\lambda) = -m < 0$. Then $a - \lambda = b\chi_{-m}$, where b is piecewise continuous, $0 \notin b^{\#}(\mathbf{R})$, and $\operatorname{wind}(b,\lambda) = 0$. It follows that $W(b)$ is invertible, that $W_\tau(b)$ is invertible for all sufficiently large τ, that

$$\lim_{\tau\to\infty} \|W_\tau^{-1}(b)\| = \|W^{-1}(b)\| \quad \text{and} \quad W_\tau^{-1}(b)P_\tau \to W^{-1}(b) \text{ strongly},$$

where P_τ is the natural projection of $L^2(0,\infty)$ onto $L^2(0,\tau)$, and that the functions $u_j = W^{-1}(b)e_j$ $(j = 1,\ldots,m)$ form a basis in $\operatorname{Ker} W(a-\lambda)$ (see [2], [3], [6]).

Theorem 5.1. *Let $\lambda \notin a^{\#}(\mathbf{R})$ and* $\operatorname{wind}(a,\lambda) = -m < 0$. *Let further $\{f_\tau\}_{\tau>0}$ be a family of functions $f_\tau \in L^2(0,\tau)$ of norm 1. Then the point λ is an asymptotically good pseudoeigenvalue of $\{W_\tau(a)\}_{\tau>0}$. The family $\{f_\tau\}$ is an asymptotically good pseudomode for $\{W_\tau(a)\}_{\tau>0}$ at λ if and only if there exist complex numbers $c_1^{(\tau)},\ldots,c_m^{(\tau)}$ and functions $z_\tau \in L^2(0,\tau)$ such that*

$$f_\tau = c_1^{(\tau)} P_\tau u_1 + \cdots + c_m^{(\tau)} P_\tau u_m + z_\tau,$$
$$\limsup_{\tau\to\infty} \max_{1\le j\le m} |c_j^{(\tau)}| < \infty, \quad \lim_{\tau\to\infty} \|z_\tau\| = 0,$$

Proof. That λ is an asymptotically good pseudoeigenvalue was established in [6] (also see [2]). The rest of the proof is analogous to the proof of Theorem 3.1, except for a modification of (3.5). Thus, suppose $W_\tau(a-\lambda)f_\tau \to 0$. Let $Q_\tau = I - P_\tau$. Then

$$W_\tau(a-\lambda)f_\tau = P_\tau W(\chi_{-m})P_\tau W(b)P_\tau f_\tau + P_\tau W(\chi_{-m})Q_\tau W(b)P_\tau f_\tau.$$

Put $A_\tau = P_\tau W(\chi_{-m})P_\tau W(b)P_\tau$, $B_\tau = P_\tau W(\chi_{-m})Q_\tau W(b)P_\tau$, $h_\tau = W(b)P_\tau f_\tau$. We denote by $\mathcal{P}_m$ the orthogonal projection of $L^2(0,\infty)$ onto the linear hull of $e_1,\ldots,e_m$ and we set $\mathcal{Q}_m = I - \mathcal{P}_m$. We have $P_\tau W(\chi_{-m})P_\tau = W(\chi_{-m})P_\tau$ and hence, $(\cdot,\cdot)$ denoting the inner product in $L^2(0,\infty)$,

$$\begin{aligned}
(A_\tau f_\tau, B_\tau f_\tau) &= (P_\tau W(\chi_{-m})P_\tau h_\tau, P_\tau W(\chi_{-m})Q_\tau h_\tau)\\
&= (W(\chi_{-m})P_\tau h_\tau, P_\tau W(\chi_{-m})Q_\tau h_\tau)\\
&= (P_\tau W(\chi_{-m})P_\tau h_\tau, W(\chi_{-m})Q_\tau h_\tau)\\
&= (W(\chi_{-m})P_\tau h_\tau, W(\chi_{-m})Q_\tau h_\tau)\\
&= (P_\tau h_\tau, W^*(\chi_{-m})W(\chi_{-m})Q_\tau h_\tau).
\end{aligned}$$

Since $W^*(\chi_{-m}) = W(\chi_m)$ and $W(\chi_m)W(\chi_{-m}) = \mathcal{Q}_m$, it follows that

$$\begin{aligned}(A_\tau f_\tau, B_\tau f_\tau) &= (P_\tau h_\tau, \mathcal{Q}_m Q_\tau h_\tau)\\ &= (P_\tau h_\tau, Q_\tau h_\tau) - (P_\tau h_\tau, \mathcal{P}_m Q_\tau h_\tau) = -(P_\tau h_\tau, \mathcal{P}_m Q_\tau h_\tau).\end{aligned}$$

Thus,

$$|(A_\tau f_\tau, B_\tau f_\tau)| \le \|P_\tau h_\tau\| \, \|\mathcal{P}_m Q_\tau\| \, \|h_\tau\| \le C\|\mathcal{P}_m Q_\tau\|$$

with some constant $C < \infty$ independent of τ. As $\mathcal{P}_m$ is compact and $Q_\tau^* = Q_\tau$ converges strongly to zero, we may conclude that $\|\mathcal{P}_m Q_\tau\| \to 0$. Consequently, $(A_\tau f_\tau, B_\tau f_\tau) \to 0$ as $\tau \to \infty$. Finally, because $\|A_\tau f_\tau + B_\tau f_\tau\| \to 0$ and

$$\|A_\tau f_\tau + B_\tau f_\tau\|^2 = \|A_\tau f_\tau\|^2 + \|B_\tau f_\tau\|^2 + 2\,\mathrm{Re}\,(A_\tau f_\tau, B_\tau f_\tau),$$

we obtain that $\|A_\tau f_\tau\| \to 0$. The rest is as in the proof of Theorem 3.1. □

We call a family $\{f_\tau\}_{\tau>0}$ of nonzero functions $f_\tau \in L^2(0,\tau)$ *asymptotically strongly localized in the beginning part* if

$$\lim_{\tau\to\infty} \frac{\|P_{(0,s_\tau)} f_\tau\|}{\|f_\tau\|} = 1$$

whenever $s_\tau \to \infty$ as $\tau \to \infty$ and $0 < s_\tau < \tau$ for all τ. Again notice that s_τ is allowed to increase as slowly as desired (or required). For example, the choice $s_\tau = \log\log\log\tau$ (τ large enough) is admitted.

Theorem 5.2. *If $\lambda \notin a^\#(\mathbf{R})$ and* wind $(a,\lambda) = -m < 0$, *then every asymptotically good pseudomode for $\{W_\tau(a)\}$ at λ is asymptotically strongly localized in the beginning part.*

This can be proved by the same arguments as in the proof of Theorem 3.2.

References

[1] P. Anderson, Absence of diffusion in certain random lattices. *Phys. Rev.* **109** (1958), 1492–1505.

[2] A. Böttcher, Pseudospectra and singular values of large convolution operators. *J. Integral Equations Appl.* **6** (1994), 267–301.

[3] A. Böttcher and B. Silbermann, *Analysis of Toeplitz Operators.* Springer-Verlag, Berlin 1990.

[4] A. Böttcher and B. Silbermann, *Introduction to Large Truncated Toeplitz Matrices.* Universitext, Springer-Verlag, New York 1999.

[5] M. Embree and L.N. Trefethen, *Pseudospectra Gateway.* Web site: http://www.comlab.ox.ac.uk/pseudospectra.

[6] I. Gohberg and I.A. Feldman, *Convolution Equations and Projection Methods for Their Solution.* Amer. Math. Soc., Providence, RI 1974.

[7] I.Ya. Goldsheid and B.A. Khoruzhenko, Eigenvalue curves of asymmetric tridiagonal random matrices. *Electronic J. Probab.* **5** (2000), paper no. 16, 28 pp.

[8] N. Hatano and D.R. Nelson, Vortex pinning and non-Hermitian quantum mechanics. *Phys. Rev. B* **56** (1997), 8651–8673.

[9] H. Landau, The notion of approximate eigenvalues applied to an integral equation of laser theory. *Quart. Appl. Math.* April 1977, 165–171.

[10] L. Reichel and L.N. Trefethen, Eigenvalues and pseudo-eigenvalues of Toeplitz matrices. *Linear Algebra Appl.* **162** (1992), 153–185.

[11] P. Stollmann, *Caught by Disorder.* Birkhäuser, Boston 2001.

[12] L.N. Trefethen, Pseudospectra of matrices. In: D.F. Griffiths and G.A. Watson, eds., Numerical Analysis 1991 (Dundee, 1991), 234–266. Longman Sci. Tech, Harlow, Essex, UK 1992.

[13] L.N. Trefethen, Pseudospectra of linear operators. *SIAM Review* **39** (1997), 383–406.

[14] L.N. Trefethen and S.J. Chapman, Wave packet pseudomodes of twisted Toeplitz matrices, Oxford Numerical Analysis Group Report 02/22, December 2002.

[15] H. Widom, Asymptotic behavior of block Toeplitz matrices and determinants. II. *Adv. Math.* **21** (1976), 1–29.

[16] P. Zizler, R.A. Zuidwijk, K.F. Taylor, and S. Arimoto, A finer aspect of eigenvalue distribution of selfadjoint band Toeplitz matrices. *SIAM J. Matrix Anal. Appl.* **24** (2002), 59–67.

A. Böttcher
Fakultät für Mathematik
TU Chemnitz
D-09107 Chemnitz
Germany
e-mail: aboettch@mathematik.tu-chemnitz.de

S. Grudsky
Departamento de Matematicas
CINVESTAV del I.P.N.
Apartado Postal 14-740
07000 México, D.F.
México
e-mail: grudsky@math.cinvestav.mx, grudsky@aaanet.ru

Operator Theory
Advances and Applications, Vol. 147, 189–212

Crack-Type Boundary Value Problems of Electro-Elasticity

Tengiz Buchukuri, Otar Chkadua, and Roland Duduchava

Dedicated to the memory of Erhard Meister

Abstract. Dirichlet, Neumann and mixed crack-type boundary value problems of statics are considered in three-dimensional bounded domains filled with a homogeneous anisotropic electro-elastic medium. Applying the method of the potential theory and the theory of pseudodifferential equations, we prove the existence and uniqueness theorems in Besov and Bessel potential spaces, and derive full asymptotic expansion of solutions near the crack edge.

Introduction

The recent years have shown an ever-growing interest in the investigation of models of an anisotropic elastic medium which take into account the influence of various physical fields such as thermal, electric, magnetic etc. A rather strong motivation for such studies is the creation of new artificial materials which possess non-standard properties. Among them are piezoelectric materials that form the core of modern structures and instruments.

Mathematical models of piezoelectric (electro-elastic) bodies and relevant boundary value problems have been studied with sufficient completeness (see, e.g., [BG2, No1, Pa1, To1] and the references therein). Of special interest is the case where the considered body contains cracks or cuts with an edge having a bihedral angle 2π (cuspidal edge) (see [Ch2, CD3]). In that case the presence of an electric field essentially changes the pattern of stress distribution near the cut or crack edge (see [DNS1, Pa1]).

In this work, Dirichlet, Neumann and mixed boundary value problems of statics are considered for a homogeneous anisotropic piezoelectric body with a crack. The existence and uniqueness of solutions of the considered problems are proved in Bessel potential $\mathbb{H}^s_p$ (and Besov $\mathbb{B}^s_{p,q}$) spaces. Complete asymptotic expansion of a solution near the crack edge is obtained. These results are important in the analysis of a stress field in electro-elastic bodies with cracks.

1. Formulation of boundary value problems

Let $\Omega = \Omega_0$ and Ω_1 ($\overline{\Omega}_1 \subset \Omega$) be bounded domains in the three-dimensional Euclidean space $\mathbb{R}^3$ with infinitely smooth boundaries $\partial\Omega = \partial\Omega_0$ and $\partial\Omega_1$ respectively (we use the alternative notation $\Omega_0 = \Omega$ for conciseness of forthcoming formulae). The boundary $\partial\Omega_1$ of the domain Ω_1 (called interface) is divided in two parts: $\partial\Omega_1 = \overline{S}_0 \cup S_1$ with a smooth common boundary $\mathcal{E} := \partial S_0 = \partial S_1$. Let $\Omega_2 = \Omega \backslash \overline{\Omega}_1$.

We assume, that the domain $\Omega \backslash S_1$ is filled with homogeneous anisotropic electro-elastic material, having a crack at S_1.

We use the following notation for function spaces: $\mathbb{W}_p^s(\Omega)$, $\mathbb{W}_p^s(\partial\Omega_i)$ for the Sobolev-Slobodetskij spaces; $\mathbb{H}_p^s(\Omega)$, $\mathbb{H}_p^s(\partial\Omega_i)$ for the Bessel potential spaces; $\mathbb{B}_{p,q}^s(\Omega)$, $\mathbb{B}_{p,q}^s(\partial\Omega_i)$ for the Besov spaces ($i = 1, 2$, $s \in \mathbb{R}$, $1 < p < \infty$, $1 \leq q \leq \infty$), (see [Tr1, Tr2] for the definitions and properties of these spaces). We use the common simplified notation $\mathbb{H}^s(\partial\Omega_i) = \mathbb{H}_2^s(\partial\Omega_i)$. Let further,

$$\mathbb{H}_p^s(S_j) = \{r_{S_j} u : \ u \in \mathbb{H}_p^s(\partial\Omega_1)\},$$
$$\widetilde{\mathbb{H}}_p^s(S_j) = \{u \in \mathbb{H}_p^s(\partial\Omega_1) : \ \operatorname{supp} u \in \overline{S_j}\}, \qquad j = 0, 1,$$

where $r_{S_j}\varphi := \varphi\Big|_{S_j}$ denotes the restriction operator onto the subset S_j. Similarly the Sobolev-Slobodetskii spaces $\mathbb{W}_p^s(S_j)$, $\widetilde{\mathbb{W}}_p^s(S_0)$, $\mathbb{B}_{p,q}^s(S_j)$ and $\widetilde{\mathbb{B}}_{p,q}^s(S_0)$ are defined.

We consider the system of static equations of electro-elasticity for a homogeneous anisotropic medium [No1]

$$\mathbf{A}(D)u(x) + F(x) = 0, \qquad x \in \Omega \backslash S_1, \tag{1.1}$$

where $u = (u_1, u_2, u_3, u_4)$; u_1, u_2, u_3 are displacement vector components, u_4 is an electric potential, F is a mass force. $\mathbf{A}(D)$ is a differential operator of the form

$$\begin{aligned} \mathbf{A}(D) &= \|\mathbf{A}_{jk}(D)\|_{4\times4}, \\ \mathbf{A}_{jk}(D) &= c_{ijlk}\partial_i\partial_l, \quad j,k = 1,2,3, \\ \mathbf{A}_{j4}(D) &= e_{kjl}\partial_k\partial_l, \quad j = 1,2,3, \\ \mathbf{A}_{4k}(D) &= -e_{ikl}\partial_i\partial_l, \quad k = 1,2,3, \\ \mathbf{A}_{44}(D) &= \varepsilon_{il}\partial_i\partial_l, \end{aligned} \tag{1.2}$$

where c_{ijlk}, e_{ikl}, ε_{ik} are the elastic, piezoelectric and dielectric constants, respectively.

Here and in what follows we use the standard convention: the summation is carried out over the repeated indices.

The constants in (1.2) satisfy the symmetry conditions

$$c_{ijlk} = c_{jilk} = c_{lkij}, \quad e_{kjl} = e_{klj}, \quad \varepsilon_{ik} = \varepsilon_{ki}, \quad i,j,k,l = 1,2,3 \tag{1.3}$$

and the condition which provides positiveness of the internal energy:

$$\forall \xi_{ij},\ \eta_i,\ \xi_{ij}=\xi_{ji},\ \exists c_0>0\ \ c_{ijkl}\xi_{ij}\xi_{kl}\geq c_0\xi_{ij}\xi_{ij},\quad \varepsilon_{ij}\eta_i\eta_j\geq c_0\eta_i\eta_i. \tag{1.4}$$

The operator $\mathbf{A}(D)$ has a strongly elliptic symbol:

$$\mathbf{A}(\xi):=\|\mathbf{A}_{jk}(\xi)\|_{4\times 4},$$

$$\begin{aligned}\mathbf{A}_{jk}(\xi)&=-c_{ijkl}\xi_i\xi_l,\quad j,k=1,2,3,\\ \mathbf{A}_{j4}(\xi)&=-e_{kjl}\xi_k\xi_l,\quad j=1,2,3,\\ \mathbf{A}_{4k}(\xi)&=e_{ikl}\xi_i\xi_l,\quad j=1,2,3,\\ \mathbf{A}_{44}(\xi)&=-\varepsilon_{il}\xi_i\xi_l,\end{aligned} \tag{1.5}$$

i.e., it satisfies the inequality

$$-\operatorname{Re}\sum_{i,k=1}^{4}\mathbf{A}_{ik}(\xi)\eta_k\overline{\eta}_i\geq c_0|\xi|^2|\eta|^2 \tag{1.6}$$

for some constant $c_0>0$ and arbitrary $\xi\in\mathbb{R}^3$ and $\eta\in\mathbb{C}^4$. Here the overbar $\overline{\eta}_i$ denotes the complex conjugate to η_i.

Inequality (1.6) follows from (1.3) and (1.4). Note that, in contrast to an analogous operator of classical elasticity [Fi1], the operator $\mathbf{A}(D)$ is neither a formally self-adjoint nor a positive definite operator.

Let Ω be a bounded domain with a piecewise-smooth boundary, $u,v\in C^2(\overline{\Omega})$. Then the following Green formulae are valid:

$$\int_\Omega\big[\overline{v}(x)\mathbf{A}(D)u(x)+E(u,v)\big]dx=\int_{\partial\Omega}\overline{v}(y)T(\partial_y,n(y))u(y)dS, \tag{1.7}$$

$$\int_\Omega\big[\overline{v}(x)\mathbf{A}(D)u(x)-u(x)\mathbf{A}^\top(D)\overline{v}(x)\big]dx$$

$$=\int_{\partial\Omega}\big[\overline{v}(y)T(\partial_y,n(y))u(y)-u(y)\widetilde{T}(\partial_y,n(y))\overline{v}(y)\big]dS, \tag{1.8}$$

where $\mathbf{A}^\top$ is the transposed matrix to $\mathbf{A}$, $n(y)=(n_1(y),n_2(y),n_3(y))$ is the outward unit normal vector to Ω_2 at the point $y\in\partial\Omega\cup\partial\Omega_1$,

$$\begin{aligned}T(\partial_y,n)&=\big\|T_{jk}(\partial_y,n)\big\|_{4\times 4},\\ T_{jk}(\partial_y,n)&=c_{ijlk}n_l\partial_i,\quad j,k=1,2,3,\\ T_{j4}(\partial_y,n)&=e_{kjl}n_l\partial_k,\quad j=1,2,3,\\ T_{4k}(\partial_y,n)&=-e_{ikl}n_i\partial_l,\quad k=1,2,3,\\ T_{44}(\partial_y,n)&=\varepsilon_{ij}n_j\partial_i,\end{aligned}$$

the operator $\widetilde{T}(\partial_y,n(y))$ is obtained from $T(\partial_y,n(y))$ by substituting e_{ijk} with $-e_{ijk}$, and $E(u,v)$ denotes the sesquilinear form

$$E(u,v)\ =c_{ijkl}\partial_i\overline{v}_j\partial_l u_k+e_{kjl}\partial_l\overline{v}_j\partial_k u_4-e_{ikl}\partial_i\overline{v}_4\partial_l u_k\ \ +\varepsilon_{il}\partial_i\overline{v}_4\partial_l u_4. \tag{1.9}$$

Note, that the Green formulae (1.7) and (1.8) are also valid for unbounded domains with a compact boundary provided u and v meet the following constraints at infinity:

$$v_i(y)\partial_k u_j(y) = o(|y|^{-2}) \text{ and } u_i(y)\partial_k v_j(y) = o(|y|^{-2}), \qquad (1.10)$$
$$i,j = 1,2,3,4, \quad k = 1,2,3.$$

Let $r^k u := r_{\Omega_k} u$ denote the restriction operator to the domain Ω_k, $k = 1,2$, and γ^i -the trace operator from the domain Ω_i to the boundary $\partial\Omega_i$, $i = 0,1,2$. We remind that if $u \in \mathbb{W}^1_p(\Omega_i)$ then the trace $\gamma^i u \in \mathbb{W}^{1/p'}_p(\partial\Omega_1) = \mathbb{B}^{1/p'}_{p,p}(\partial\Omega_1)$ ($\gamma^i u \in \mathbb{H}^{1/2}(\partial\Omega_1)$ when $p = 2$), $i = 1,2$, $p' = p/(p-1)$ (see [Tr1]). Moreover, if $\mathbf{A}(D)u \in \mathbb{L}_p(\Omega)$ the trace $\gamma^i T(\partial_y, n)(r^i u)$ exists and is defined through the Green formula (1.7) (see [DNS1, NCS1]). In particular, if $u \in \mathbb{W}^1_p(\Omega)$ is a solution of equation (1.1) with $F \in \mathbb{L}_q(\Omega)$, $q \geq np/(n+p)$, then $\mathbf{A}(D)u \in \mathbb{L}_q(\Omega)$ and the traces

$$\gamma^i u = r_{S_1}\{\gamma^i(r^i u)\},$$
$$\gamma^i T(\partial_y, n)u = r_{S_1}\{\gamma^i T(\partial_y, n)(r^i u)\}, \quad i = 1,2,$$

are defined correctly. The trace $\gamma_{\partial\Omega} T(\partial_y, n)u$ on $\partial\Omega$ is defined similarly.

Based on the above arguments, we will consider the following boundary value problems (BVPs) in the domain $\Omega \backslash S_1$: we look for a function $u \in \mathbb{W}^1(\Omega)$ [1] which solves the following BVPs: :

the Dirichlet BVP:

$$\begin{cases} \mathbf{A}(D)u = 0 & \text{in } \Omega\backslash S_1, \\ \gamma^0 T(\partial_y, n)u = \psi & \text{on } \partial\Omega, \\ \gamma^i u = \varphi_i, \quad i = 1,2, & \text{on } S_1, \end{cases} \qquad (1.11)$$

where $\varphi_i \in \mathbb{H}^{1/2}(S_1)$, $i = 1,2$, $\psi \in \mathbb{H}^{-1/2}(\partial\Omega)$;

the Neumann BVP:

$$\begin{cases} \mathbf{A}(D)u = 0 & \text{in } \Omega\backslash S_1, \\ \gamma^0 T(\partial_y, n)u = \psi & \text{on } \partial\Omega, \\ \gamma^i T(\partial_y, n)u = \psi_i, \quad i = 1,2, & \text{on } S_1, \end{cases} \qquad (1.12)$$

where $\psi_i \in \mathbb{H}^{-1/2}(S_1)$, $i = 1,2$, $\psi \in \mathbb{H}^{-1/2}(\partial\Omega)$.

the mixed BVP:

$$\begin{cases} \mathbf{A}(D)u = 0 & \text{in } \Omega\backslash S_1, \\ \gamma^0 T(\partial_y, n)u = \psi & \text{on } \partial\Omega, \\ \gamma^1 u = \varphi_1, & \text{on } S_1, \\ \gamma^2 T(\partial_y, n)u = \psi_2, & \text{on } S_1, \end{cases} \qquad (1.13)$$

where $\varphi_1 \in \mathbb{H}^{1/2}(S_1)$, $\psi_2 \in \mathbb{H}^{-1/2}(S_1)$, $\psi \in \mathbb{H}^{-1/2}(\partial\Omega)$.

Note, that in the above BVPs (1.11)–(1.13) boundary conditions on the surface S_1 differ, while they are the same on the boundary $\partial\Omega$.

[1] For simplicity we drop the case $p \neq 2$, i.e., when $u \in \mathbb{W}^1_p(\Omega)$); these cases can be considered as in [DNS1, NCS1, Ch2].

Theorem 1.1. *The Dirichlet and the mixed BVPs have a unique solution* $u \in \mathbb{W}^1(\Omega\backslash S_1)$ *, and the solution of the Neumann boundary value problem is defined up to a summand* $u = (u_1, u_2, u_3, u_4)$ *:*

$$u_i = \varepsilon_{ijk} a_j x_k + b_i, \quad i = 1,2,3, \quad v_4 = b_4 . \tag{1.14}$$

Here a_j *,* $j = 1,2,3$ *,* b_i *,* $i = 1,2,3,4$ *, are arbitrary constants,* ε_{ijk} *are the* ε *-tensor components (the Levi-Civita symbol) .*

Proof. The proof is based on the Green formula (1.6) and is standard (see similar proofs in [BG1, DNS1, NCS1, Fi1] etc.). □

2. Properties of potentials

Denote by H a fundamental solution of the operator $\mathbf{A}(D)$ and consider the simple layer potentials

$$V^{(k)}(g)(x) = \int\limits_{\partial\Omega_k} H(x-y)g(y)d_yS, \quad x \in \Omega_1 \cup \Omega_2, \quad k = 0,1 .$$

Theorem 2.1. *Let* $\partial\Omega, \partial\Omega_1$ *be* C^β *-smooth,* $\beta \geq |s| + 1 + 1/p$ *,* $1 < p < \infty$ *,* $1 \leq q < \infty$ *. Then the operators* $V^{(1)}$ *,* $V^{(0)}$ *extend to the continuous operators*

$$V^{(1)} \quad : \quad \mathbb{B}^s_{p,q}(\partial\Omega_1) \to \mathbb{B}^{s+1+1/p}_{p,q}(\Omega_i) \bigcap \mathbb{H}^{s+1+1/p}_p(\Omega_i), \quad i = 1,2,$$

$$V^{(0)} \quad : \quad \mathbb{B}^s_{p,q}(\partial\Omega) \to \mathbb{B}^{s+1+1/p}_{p,q}(\Omega) \bigcap \mathbb{H}^{s+1+1/p}_p(\Omega).$$

We consider the following integral operators on the surfaces

$$V^{(1)}_{-1}(g)(z) = \int\limits_{\partial\Omega_1} H(z-y)g(y)d_yS,$$

$$V^{(1)}_0(g)(z) = \int\limits_{\partial\Omega_1} T(\partial_z, n(z))H(z-y)g(y)d_yS, \quad z \in \partial\Omega_1 = \partial\Omega_2,$$

$$V^{(0)}_{-1}(h)(z) = \int\limits_{\partial\Omega} H(z-y)h(y)d_yS,$$

$$V^{(0)}_0(h)(z) = \int\limits_{\partial\Omega} T(\partial_z, n(z))H(z-y)h(y)d_yS, \quad z \in \partial\Omega,$$

which are the direct values of the corresponding layer potentials. Then for the traces of the layer potentials we have the following Plemelj formulae.

Theorem 2.2. *Let $\partial\Omega, \partial\Omega_1$ be C^β-smooth, $\beta \geq 3$, $g \in C^3(\partial\Omega_1)$, $h \in C^3(\partial\Omega_1)$. Then*

$$\gamma^i V^{(1)}(g)(z) = V_{-1}^{(1)}(g)(z), \quad z \in \partial\Omega_1;$$

$$\gamma^0 V^{(0)}(h)(z) = V_{-1}^{(0)}(h)(z), \quad z \in \partial\Omega.$$

$$\gamma^i\big(T(\partial_z, n(z))V^{(1)}(g)\big)(z) = \frac{(-1)^{i+1}}{2}g(z) + V_0^{(1)}(g)(z), \qquad z \in \partial\Omega_1,$$

$$\gamma^0\big(T(\partial_z, n(z))V^{(0)}(h)\big)(z) = -\frac{1}{2}h(z) + V_0^{(0)}(h)(z), \qquad z \in \partial\Omega,$$

Theorem 2.3. *Let $\partial\Omega, \partial\Omega_1$ be C^β-smooth, $\beta \geq |s| + 3$, $1 < p < \infty$, $1 \leq q \leq \infty$. Let $\mathbb{X}^s = \mathbb{H}_p^s$ or $\mathbb{X}^s = \mathbb{B}_{p,q}^s$. Then the operators $V_{-1}^{(i)}$, $V_0^{(i)}$ extend to the following continuous operators:*

$$\begin{aligned} &V_k^{(1)} : \mathbb{X}^s(\partial\Omega_1) \to \mathbb{X}^{s-k}(\partial\Omega_1), \\ &V_k^{(0)} : \mathbb{X}^s(\partial\Omega) \to \mathbb{X}^{s-k}(\partial\Omega) \qquad \text{for} \quad k = 0, -1. \end{aligned} \tag{2.1}$$

Theorem 2.4. *The operators in (2.1) are strongly elliptic*

$$\begin{aligned} &\mathrm{Re}\langle V_{-1}^{(1)} g, g\rangle \leq 0 \qquad \forall g \in \mathbb{H}_2^{-1/2}(\partial\Omega_1), \\ &\mathrm{Re}\langle V_{-1}^{(0)} h, h\rangle \leq 0 \qquad \forall h \in \mathbb{H}_2^{-1/2}(\partial\Omega) \end{aligned} \tag{2.2}$$

and the equalities in (2.3) are achieved only for $g = 0, h = 0$.

Moreover, the operators $V_{-1}^{(1)}$ and $V_{-1}^{(0)}$ in (2.1) are invertible.

Although the operators $V_{-1}^{(i)}$, $i = 0, 1$ are not self-adjoint as in the theory of elasticity, the proofs of Theorems 2.1–2.4 does not differ from those proved in [DNS1, Sh2] for the classical elasticity case.

The operator $r_{S_0}\mathbf{A} = r_{S_0}\big(V_{-1}^{(1)}\big)^{-1}$ is strongly elliptic and the following is true (cf. [DNS1, CD1]).

Theorem 2.5. *Let $1 < p < \infty$, $1 \leq q \leq \infty$. Then the operator*

$$\begin{aligned} r_{S_0}\mathbf{A} \;&:\; \widetilde{\mathbb{H}}_p^s(S_0) \to \mathbb{H}_p^{s-1}(S_0) \\ &:\; \widetilde{\mathbb{B}}_{p,q}^s(S_0) \to \mathbb{B}_{p,q}^{s-1}(S_0) \end{aligned} \tag{2.3}$$

has the Fredholm property if and only if the conditions

$$\frac{1}{p} - \frac{1}{2} < s < \frac{1}{p} + \frac{1}{2} \tag{2.4}$$

hold. Moreover, if (2.4) holds, the operator (2.3) is invertible.

Theorem 2.6. (see [BG2, NCS1]). *Let* $1 < p < \infty$, $1 \le q \le \infty$, $s \in \mathbb{R}$. *The singular integral operator*

$$\begin{aligned} -\frac{1}{2}I + V_0^{(0)} \quad &: \quad \widetilde{\mathbb{H}}_p^s(\partial\Omega) \to \mathbb{H}_p^s(\partial\Omega) \\ &: \quad \widetilde{\mathbb{B}}_{p,q}^s(\partial\Omega) \to \mathbb{B}_{p,q}^s(\partial\Omega) \end{aligned} \tag{2.5}$$

is Fredholm with index 0.

Theorem 2.7. *Let* $1 < p < \infty$, $1 \le q \le \infty$, $s \in \mathbb{R}$. *Then the pseudodifferential operator*

$$\begin{aligned} \mathbf{B}_N^{(i)} &:= (-V_{-1}^{(1)})^N + \Big(-\frac{1}{2}I + (-1)^i V_0^{(1)}\Big)(V_{-1}^{(1)})^{-1} \\ \mathbf{B}_N^{(i)} \quad &: \quad \widetilde{\mathbb{H}}_p^s(\partial\Omega_1) \to \mathbb{H}_p^{s-1}(\partial\Omega_1) \\ &: \quad \widetilde{\mathbb{B}}_{p,q}^s(\partial\Omega_1) \to \mathbb{B}_{p,q}^{s-1}(\partial\Omega_1) \end{aligned} \tag{2.6}$$

is invertible for $N = 0, 1, \dots$ *and* $i = 1, 2$. *Moreover under the conditions*

$$\frac{1}{p} - \frac{1}{2} < s < \frac{1}{p} + \frac{1}{2} \tag{2.7}$$

the following operator is invertible:

$$\begin{aligned} r_{S_0}\Big[\big(\mathbf{B}_N^{(1)}\big)^{-1} + \big(\mathbf{B}_N^{(2)}\big)^{-1}\Big] \quad &: \quad \widetilde{\mathbb{H}}_p^{s-1}(S_0) \to \mathbb{H}_p^s(S_0) \\ &: \quad \widetilde{\mathbb{B}}_{p,q}^{s-1}(S_0) \to \mathbb{B}_{p,q}^s(S_0) \,. \end{aligned} \tag{2.8}$$

Proof. The proof of the first part is similar to Lemmata 5.2 and 6.2, in [Ch2].

The second part follows because the pseudodifferential operator $r_{S_0}\big(\mathbf{B}_N^{(i)}\big)^{-1}$, $i = 1, 2$, is strongly elliptic. □

Let us consider a $N \times N$ system of pseudodifferential equations on $\mathbb{R}_+^n$

$$\mathbf{R}(x, D)\chi = \Psi \tag{2.9}$$

with a matrix symbol $\sigma_R(x, \xi)$, $x \in \overline{\mathbb{R}}_+^n$, $\xi \in \mathbb{R}^n$ from the Hörmander class $\mathbb{S}_{1,0}^r(\mathbb{R}^n \times \mathbb{R}^n)$. Let $\lambda_j(x')$, $j = 1, \dots, N$ be the eigenvalues of the matrix

$$\big(\sigma_R(x', 0, 0, +1)\big)^{-1}\sigma_R(x', 0, 0, -1), \quad x' \in \mathbb{R}^{n-1} = \partial\mathbb{R}_+^n .$$

Lemma 2.8. *Let the symbol* $\sigma_R(x, \xi)$ *be strongly elliptic and*

$$\delta(x') = \sup_{1 \le j \le N} \frac{1}{2\pi}\big|\arg \lambda_j(x')\big|$$

$$1 < p < \infty, \qquad 1 \le q \le \infty, \qquad \frac{1}{p} - 1 + \frac{r}{2} + \delta(x') < s < \frac{1}{p} + \frac{r}{2} - \delta(x') .$$

Then the operator

$$\begin{aligned} \mathbf{R}(x, D) \quad &: \quad \widetilde{\mathbb{H}}_p^s(\mathbb{R}_+^n) \oplus \widetilde{\mathbb{H}}_p^s(\mathbb{R}_+^n) \to \mathbb{H}_p^{s-r}(\mathbb{R}_+^n) \oplus \mathbb{H}_p^{s-r}(\mathbb{R}_+^n) \\ &: \quad \widetilde{\mathbb{B}}_{p,q}^s(\mathbb{R}_+^n) \oplus \widetilde{\mathbb{B}}_{p,q}^s(\mathbb{R}_+^n) \to \mathbb{B}_{p,q}^{s-r}(\mathbb{R}_+^n) \oplus \mathbb{B}_{p,q}^{s-r}(\mathbb{R}_+^n) \end{aligned}$$

is invertible.

Proof. The result follows from the general theory of pseudodifferential equations on manifolds with boundary in [CD1, Du1, Sh1, Sh2]. □

3. The Dirichlet problem

Let $\Phi_0^{(i)} \in \mathbb{H}^{1/2}(\partial\Omega_1)$ be some extension of the function $\varphi_i \in \mathbb{H}^{1/2}(S_1)$ on $\partial\Omega_1$, $i = 1, 2$. Then any extension $\Phi^{(i)}$ of the function φ_i on $\partial\Omega_1$ has the form

$$\Phi^{(i)} = \Phi_0^{(i)} + \varphi_0^{(i)}, \quad \varphi_0^{(i)} \in \widetilde{\mathbb{H}}^{1/2}(S_0).$$

We will seek a solution of all considered problems in the following form:

$$\begin{array}{lll} r^1 u = V^{(1)} g_1 \ \text{ in } \ \Omega_1, & g_1 \in \mathbb{H}^{1/2}(\partial\Omega_1) & \\ r^2 u = V^{(1)} g_2 + V^{(0)} h \ \text{ in } \ \Omega_2, & g_2 \in \mathbb{H}^{1/2}(\partial\Omega_1), & h \in \mathbb{H}^{1/2}(\partial\Omega). \end{array} \tag{3.1}$$

Since the operator $\mathbf{A}(D)$ is elliptic, any generalized solution of the homogeneous equation (1.1) is an analytic function in the domain $\Omega \backslash S_1$. Then

$$\begin{cases} \gamma^1\{r^1 u\} - \gamma^2\{r^2 u\} = 0 & \text{on } S_0, \\ \gamma^1\{T(r^1 u)\} - \gamma^2\{T(r^2 u)\} = 0 & \text{on } S_0, . \end{cases} \tag{3.2}$$

Due to the boundary conditions of the Dirichlet problem we have

$$\begin{cases} \left(-\frac{1}{2}I + V_0^{(0)}\right)h + \gamma_{\partial\Omega}\left(TV^{(1)}\right)g_2 = \psi & \text{on } \partial\Omega, \\ r_{S_1} V_{-1}^{(1)} g_1 = \varphi_1 & \text{on } S_1, \\ r_{S_1} V_{-1}^{(1)} g_2 + \gamma_{S_1}\left(V^{(0)}\right)h = \varphi_2 & \text{on } S_1, \end{cases}$$

Taking into account (3.2) we get the system of pseudodifferential equations with respect to $(h, g_1, g_2, \varphi_0^{(1)}, \varphi_0^{(2)})$:

$$\begin{cases} \left(-\frac{1}{2}I + V_0^{(0)}\right)h + \gamma_{\partial\Omega}\left(TV^{(1)}\right)g_2 = \psi & \text{on } \partial\Omega, \\ V_{-1}^{(1)} g_1 - \varphi_0^{(1)} = \Phi_0^{(1)} & \text{on } \partial\Omega_1, \\ V_{-1}^{(1)} g_2 - \varphi_0^{(2)} + \gamma_{\partial\Omega_1}\left(V^{(0)}\right)h = \Phi_0^{(2)} & \text{on } \partial\Omega_1, \\ \varphi_0^{(1)} - \varphi_0^{(2)} = -r_{S_0}\Phi_0^{(1)} + r_{S_0}\Phi_0^{(2)} & \text{on } S_0, \\ r_{S_0}\left(\frac{1}{2}I + V_0^{(1)}\right)g_1 - r_{S_0}\left(-\frac{1}{2}I + V_0^{(1)}\right)g_2 - \gamma_{S_0}\left(TV^{(0)}\right)h = 0 & \text{on } S_0, \end{cases} \tag{3.3}$$

where the integral operators $\gamma_{\partial\Omega}\left(TV^{(1)}\right)$, $\gamma_{\partial\Omega_1}\left(V^{(0)}\right)$ and $\gamma_{S_0}\left(TV^{(0)}\right)$ have infinitely smooth kernels (the integration and the outer variable vary on disjoint sets) and are therefore compact. Now we formulate the basic theorems about the existence and uniqueness of a solution to the Dirichlet problem.

Theorem 3.1. *The Dirichlet problem has a unique solution in the class* $\mathbb{W}^1(\Omega\backslash S_1)$, *written as follows*

$$r^1u = V^{(1)}\big((V_{-1}^{(1)})^{-1}(\Phi_0^{(1)} + \varphi_0^{(1)})\big),$$
$$r^2u = V^{(1)}\big((V_{-1}^{(1)})^{-1}(\Phi_0^{(2)} + \varphi_0^{(2)} - r_{\partial\Omega_1}V^{(0)}h\big) + V^{(0)}h\,. \tag{3.4}$$

Here $\Phi_0^{(i)} \in \mathbb{H}^{1/2}(\partial\Omega_1)$ *is a fixed extension of the function* φ_i *on* $\partial\Omega_1$, $h \in \mathbb{H}^{-1/2}(\partial\Omega)$ *and* $\varphi_0^{(1)}, \varphi_0^{(2)} \in \widetilde{\mathbb{H}}^{1/2}(S_0)$ *can be found from the system* (3.3), *which is uniquely solvable.*

Moreover, let $1 < r < \infty$, $1 \le q \le \infty$, $\frac{1}{r} - \frac{1}{2} < s < \frac{1}{r} + \frac{1}{2}$ *and* $u \in \mathbb{W}^1(\Omega\backslash S_1)$ *be the solution of the Dirichlet problem. Then*

$$u \in \mathbb{H}_r^{s+1/r}(\Omega\backslash S_1) \text{ if } \psi \in \mathbb{B}_{r,r}^{s-1}(\partial\Omega),\ \ \varphi_i \in \mathbb{B}_{r,r}^{s}(S_1),\ \ i = 1,2,$$
$$u \in \mathbb{B}_{r,q}^{s+1/r}(\Omega\backslash S_1) \text{ if } \psi \in \mathbb{B}_{r,q}^{s-1}(\partial\Omega),\ \ \varphi_i \in \mathbb{B}_{r,q}^{s}(S_1),\ \ i = 1,2\,. \tag{3.5}$$

Proof. The system (3.3) can be rewritten in the matrix form

$$\mathbf{M}\begin{bmatrix} h \\ g_1 \\ g_2 \\ \varphi_0^{(1)} \\ \varphi_0^{(2)} \end{bmatrix} = \begin{bmatrix} \Psi \\ \Phi_0^{(1)} \\ \Phi_0^{(2)} \\ -r_{S_0}\Phi_0^{(1)} + r_{S_0}\Phi_0^{(2)} \\ 0 \end{bmatrix},$$

where

$$\mathbf{M} = \begin{bmatrix} -\frac{1}{2}I + V_0^{(0)} & 0 & \gamma_{\partial\Omega}\Big(TV^{(1)}\Big) & 0 & 0 \\ 0 & V_{-1}^{(1)} & 0 & -I & 0 \\ \gamma_{\partial\Omega_1}\Big(V^{(0)}\Big) & 0 & V_{-1}^{(1)} & 0 & -I \\ 0 & 0 & 0 & I & -I \\ -\gamma_{S_0}\Big(TV^{(0)}\Big) & r_{S_0}\Big(\frac{1}{2}I + V_0^{(1)}\Big) & -r_{S_0}\Big(-\frac{1}{2}I + V_0^{(1)}\Big) & 0 & 0 \end{bmatrix}.$$

Obviously,

$$\mathbf{M} = \mathbf{Q} + \mathbf{T}_{-\infty}, \qquad \mathbf{Q} = \begin{bmatrix} -\frac{1}{2}I + V_0^{(0)} & 0 \\ 0 & \mathbf{P} \end{bmatrix}, \tag{3.6}$$

where $\mathbf{T}_{-\infty}$, comprised by the operators $\gamma_{\partial\Omega}\Big(TV^{(1)}\Big)$, $\gamma_{\partial\Omega_1}\Big(V^{(0)}\Big)$ and $\gamma_{S_0}\Big(TV^{(0)}\Big)$, is an infinitely smoothing compact operator, while

$$\mathbf{P} = \begin{bmatrix} V_{-1}^{(1)} & 0 & -I & 0 \\ 0 & V_{-1}^{(1)} & 0 & -I \\ 0 & 0 & I & -I \\ r_{S_0}\Big(\frac{1}{2}I + V_0^{(1)}\Big) & -r_{S_0}\Big(-\frac{1}{2}I + V_0^{(1)}\Big) & 0 & 0 \end{bmatrix}.$$

The entry $-\frac{1}{2}I + V_0^{(0)}$ of the operator $\mathbf{Q}$ is a Fredholm singular operator with index 0 (see Theorem 2.6).

Consider the system corresponding to the operator $\mathbf{P}$

$$\begin{cases} V_{-1}^{(1)}\widetilde{g}_i - \widetilde{\varphi}_0^{(i)} = F^{(i)},\ i = 1,2, & \text{on } \partial\Omega_1, \\ \widetilde{\varphi}_0^{(1)} - \widetilde{\varphi}_0^{(2)} = g & \text{on } S_0, \\ r_{S_0}\Big(\frac{1}{2}I + V_0^{(1)}\Big)\widetilde{g}_1 - r_{S_0}\Big(-\frac{1}{2}I + V_0^{(1)}\Big)\widetilde{g}_2 = f & \text{on } S_0 \end{cases}$$

with respect to the unknowns $\widetilde{g}_i$, $\widetilde{\varphi}_0^{(i)}$, $i = 1, 2$. Note that they differ from g_i, $\varphi_0^{(i)}$ by infinitely differentiable functions.

Since the operator $V_{-1}^{(1)}$ is invertible,

$$\widetilde{g}_i = \big(V_{-1}^{(1)}\big)^{-1}\big(F^{(i)} + \widetilde{\varphi}_0^{(i)}\big),\quad i = 1,2.$$

On substituting these equalities in the last two equations in the foregoing system, we obtain a system of pseudodifferential equations on S_0 with respect to $\widetilde{\varphi}_0^{(1)}$ and $\widetilde{\varphi}_0^{(2)}$

$$\begin{cases} \widetilde{\varphi}_0^{(1)} - \widetilde{\varphi}_0^{(2)} = g, \\ r_{S_0}\Big(\frac{1}{2}I + V_0^{(1)}\Big)\big(V_{-1}^{(1)}\big)^{-1}\widetilde{\varphi}_0^{(1)} - r_{S_0}\Big(-\frac{1}{2}I + V_0^{(1)}\Big)\big(V_{-1}^{(1)}\big)^{-1}\widetilde{\varphi}_0^{(2)} = G, \end{cases} \tag{3.7}$$

where

$$G = f + r_{S_0}\Big(-\frac{1}{2}I + V_0^{(1)}\Big)\big(V_{-1}^{(1)}\big)^{-1}F^{(2)} - r_{S_0}\Big(\frac{1}{2}I + V_0^{(1)}\Big)\big(V_{-1}^{(1)}\big)^{-1}F^{(1)}.$$

The system (3.7) is thus reduced to a pseudodifferential equation on the open manifold

$$r_{S_0}\mathbf{A}\widetilde{\varphi}_0^{(1)} = \Psi \quad \text{on } S_0,$$

where $\Psi \in \mathbb{H}_p^{s-1}(S_0)$, $\big(\Psi \in \mathbb{B}_{p,q}^{s-1}(S_0),\big)$ and

$$\mathbf{A} = \Big(\frac{1}{2}I + V_0^{(1)}\Big)\big(V_{-1}^{(1)}\big)^{-1} - \Big(-\frac{1}{2}I + V_0^{(1)}\Big)\big(V_{-1}^{(1)}\big)^{-1} = \big(V_{-1}^{(1)}\big)^{-1}.$$

The operator $r_{S_0}\mathbf{A} = r_{S_0}\big(V_{-1}^{(1)}\big)^{-1}$ is strongly elliptic and, due to Theorem 2.5, the first part of the theorem is proved.

The second part, the solvability properties (3.5), is a consequence of the first part and mapping properties of the potential operators (see [DNS1, NCS1, Ch2] for similar considerations in elasticity). □

Let us look at the asymptotics of the solution to the Dirichlet problem near the boundary $\mathcal{E} = \partial S_1$. We assume that the boundary conditions of the Dirichlet problem are sufficiently smooth, namely: $\varphi_i \in \mathbb{H}_r^{(\infty,s+2N+1),\infty}(S_1)$, $i = 1, 2$ (see [CD1] for the definition and details).

The principal symbol $\sigma_{\mathbf{A}}(x', \xi')$ of the pseudodifferential operator $\mathbf{A}$ in (2.3) is written as (see [CD1])

$$\sigma_{\mathbf{A}}(x', \xi') = \sigma^{-1}_{-V_{-1}^{(1)}}(x', \xi'),\quad x' \in S_0.$$

Moreover, the principal symbol $\sigma_{-V_{-1}^{(1)}}(x', \xi')$ of the operator $-V_{-1}^{(1)}$ is even with respect to the variable ξ' and, therefore, all eigenvalues of the matrix

$$(\sigma_{\mathbf{A}}(x', 0, 0, +1))^{-1}\sigma_{\mathbf{A}}(x', 0, 0, -1) = I$$

are trivial:

$$\lambda_j(x') = 1, \quad j = 1, 2, 3, 4, \quad x' \in \mathcal{E}.$$

In the normal plane $\Pi_{x'}$ to $\mathcal{E}$, containing $x' \in \mathcal{E}$, we consider the polar coordinates (r, θ), where $r \geq 0$ denotes the distance from $x = (x', r, \theta)$ to the boundary $\mathcal{E}$ and $-\pi \leq \theta \leq \pi$ is the angular parameter. Then the points $(x', r, \pm\pi) \in S_1^{\pm}$ belong to the different faces of the surface S_1 and $x' = (x', 0, \theta)$ belongs to the boundary $x' \in \mathcal{E}$ for all $\theta \in [-\pi, \pi]$.

Applying Theorem 2.1 from [CD1] and taking into account the first equality in (3.7), we obtain the asymptotic expansion of the function $\varphi_0^{(i)}$, $i = 1, 2$,

$$\varphi_0^{(i)}(x', r) = c_0(x')r^{\frac{1}{2}} + \sum_{k=1}^{N} c_k(x')r^{\frac{1}{2}+k} + \varphi_{N+1}^{(i)}(x', r), \tag{3.8}$$

where $c_k \in C^\infty(\mathcal{E})$, $k = 0, 1, \ldots, N$, and the remainder term $\varphi_{N+1}^{(i)}$ belongs to the space $\mathbb{H}_r^{(\infty, s+N+1), \infty}(\mathcal{E}_\varepsilon^+) \subset C^{s+N}([0, \varepsilon], C^\infty(\mathcal{E}))$, $\mathcal{E}_\varepsilon^+ = \mathcal{E} \times [0, \varepsilon]$.

As we can see from (3.8) logarithms are absent in the entire asymptotic due to the properties of the symbol $\sigma_{\mathbf{A}}(x', \xi')$ (see [CDD1]).

From Theorem 3.1 it follows that the solution of a Dirichlet problem can be written as a simple-layer potential.

For any $x' \in \mathcal{E}$, let $\tau_1(x'), \ldots, \tau_\ell(x')$ be all different roots of the polynomial equation

$$\det \mathbf{A}((\mathcal{J}_\varkappa^\top(x'))^{-1}(0, 1, \tau)^\top) = 0, \qquad x' \in \mathcal{E}, \quad \operatorname{Im} \tau < 0. \tag{3.9}$$

We recall that $(0, 1, \tau)$ represents the value of the dual variable ξ and that $\mathcal{I}_\varkappa(x')$ is the Jacobian of the local coordinate diffeomorphisms $\varkappa$ (see [CD1]).

We assume that it is possible to enumerate $\tau_1(x'), \ldots, \tau_\ell(x')$ so that the multiplicities $n_1, \ldots, n_\ell$ of $\tau_1(x'), \ldots, \tau_\ell(x')$ are constant on $\mathcal{E}$. Therefore the functions $\tau_1, \ldots, \tau_\ell$ can be chosen smoothly $\tau_m \in C^\infty(\mathcal{E})$.

Since $\mathbf{A}$ is a 4×4 elliptic system of order two, $n_1 + \cdots + n_\ell = 4$ and $\overline{\tau_m(x')}$ are the roots of (3.9) with $\operatorname{Im} \tau > 0$. Let us define the following functions

$$\psi_{m,-1}(x', \theta) := \cos\theta + \tau_m(x')\sin\theta,$$

$$\psi_{m,+1}(x', \theta) := \cos\theta + \overline{\tau}_m(x')\sin\theta, \qquad x' \in \mathcal{E}, \ m = 1, \ldots, \ell. \tag{3.10}$$

Under the assumption $\Phi_0^{(i)} \in \mathbb{H}_r^{(\infty, s+2N+1), \infty}(\partial\Omega_1)$ the solutions of the Dirichlet problem has the following asymptotic form in the vicinity of the crack front $\mathcal{E}$ (see [CD2, Theorems 2.2 and 2.3] and [CDD1, Theorem B.8.1]):

$$\boldsymbol{u} = r^{\frac{1}{2}} \sum_{m=1}^{\ell} \sum_{\omega=\pm 1} \chi(r) \Bigg[\sum_{j=0}^{n_m-1} \sin^j\theta\, \psi_{m,\omega}^{\frac{1}{2}-j}(x',\theta)\, \boldsymbol{d}_{m,\omega}^{j}(x') \tag{3.11}$$
$$+ \sum_{k=1}^{N-1} \sum_{j=0}^{p(m,k)} \sum_{|\alpha|\le N(m,k)} r^k\, \psi_{m,\omega}^{\frac{1}{2}-j+k}(x',\theta) \sin^{\alpha_1}\theta \cos^{\alpha_2}\theta\, \boldsymbol{d}_{m,\omega}^{k,j,\alpha}(x') \Bigg] + \boldsymbol{u}_{\mathrm{rem},N}$$

where $\boldsymbol{u}_{\mathrm{rem},N} \in \mathbb{H}_{\mathrm{loc}}^{\frac{1}{2}+N}(\mathbb{R}^3)$ and the coefficients $\boldsymbol{d}_{m,\omega}^{j}$ and $\boldsymbol{d}_{m,\omega}^{j,k,\alpha}$ are $C^\infty(\mathcal{E})$. $p(m,k)$ and $N(m,k)$ are some positive integers.

Moreover, the explicit formulae expressing the coefficients $\boldsymbol{d}_{m,\omega}^{j}(x')$ by the first coefficients $c_0(x')$, $m = 1,\dots l$, $j = 0,\dots,n_m - 1$, of the surface expansion (3.8) are available as well (see [CD2]).

4. The Neumann problem

Let $\Psi_0^{(i)} \in \mathbb{H}^{-1/2}(\partial\Omega_1)$ be some fixed extension of the function $\psi_i \in \mathbb{H}^{-1/2}(S_1)$ on $\partial\Omega_1 = S_1 \cup \overline{S}_0$. Then any extension $\psi_i \in \mathbb{H}^{-1/2}(S_1)$ of the function ψ_i on $\partial\Omega_1$ has the form

$$\Psi^{(i)} = \Psi_0^{(i)} + \psi_0^{(i)},$$

where $\psi_0^{(i)} \in \widetilde{\mathbb{H}}^{-1/2}(S_0)$, $i = 1,2$. Solutions of the Neumann boundary value problem will be sought in the form (3.1). Due to the boundary conditions (1.12) we have

$$\begin{cases} \Big(-\dfrac{1}{2}I + V_0^{(0)}\Big)h + r_{\partial\Omega}\Big(TV^{(1)}\Big)g_2 = \psi & \text{on } \partial\Omega, \\ r_{S_1}\Big(\frac{1}{2}I + V_0^{(1)}\Big)g_1 = \varphi_1 & \text{on } S_1, \\ r_{S_1}\Big(-\dfrac{1}{2}I + V_0^{(1)}\Big)g_2 + r_{S_1}\Big(TV^{(1)}\Big)h = \varphi_2 & \text{on } S_1, \end{cases} \tag{4.1}$$

Taking into account (3.2) we obtain the following system of equations

$$\begin{cases} \Big(-\dfrac{1}{2}I + V_0^{(0)}\Big)h + r_{\partial\Omega}(TV^{(1)})g_2 = \psi & \text{on } \partial\Omega, \\ \Big(\dfrac{1}{2}I + V_0^{(1)}\Big)g_1 - \psi_0^{(1)} = \Psi_0^{(1)} & \text{on } \partial\Omega_1, \\ \Big(-\dfrac{1}{2}I + V_0^{(1)}\Big)g_2 - \psi_0^{(2)} + r_{\partial\Omega_1}(TV^{(0)})h = \Psi_0^{(2)} & \text{on } \partial\Omega_1, \\ r_{S_0}V_{-1}^{(1)}g_1 - r_{S_0}V_{-1}^{(1)}g_2 - r_{S_0}V^{(0)}h = 0 & \text{on } S_0, \\ \psi_0^{(1)} - \psi_0^{(2)} = -r_{S_0}\Psi_0^{(1)} + r_{S_0}\Psi_0^{(2)} & \text{on } S_0 \end{cases} \tag{4.2}$$

with respect to the known and unknown vector-function

$$(\psi, \Psi_0^{(1)}, \Psi_0^{(2)}) \in \mathbb{H}^{-1/2}(\partial\Omega) \times \mathbb{H}^{-1/2}(\partial\Omega_1) \times \mathbb{H}^{-1/2}(\partial\Omega_1)$$

$$(h, g_1, g_2, \psi_0^{(1)}, \psi_0^{(2)}) \in \mathbb{H}^{1/2}(\partial\Omega) \times \mathbb{H}^{1/2}(\partial\Omega_1) \times \mathbb{H}^{1/2}(\partial\Omega_1) \tag{4.3}$$
$$\times \widetilde{\mathbb{H}}^{-1/2}(S_0) \times \widetilde{\mathbb{H}}^{-1/2}(S_0).$$

Now we can formulate the basic theorem about the existence and uniqueness of a solution of the Neumann problem.

Theorem 4.1. *The Neumann problem has a solution of the class* $\mathbb{W}^1(\Omega\backslash S_1)$ *in the bounded domain* Ω *if and only if the equality*

$$\int_{\partial\Omega} \Big[\varepsilon_{ijk}\psi_i a_j z_k + \sum_{k=1}^{4} b_k\psi_k\Big] d_z S = 0 \tag{4.4}$$

is fulfilled for any constant vectors $a=(a_1,a_2,a_3)$ *and* $b=(b_1,b_2,b_3,b_4)$.

Solutions of the Neumann problem are represented as

$$\begin{cases} r^1 u = V^{(1)}\big(V_{-1}^{(1)}\big)^{-1} g_1 \ \text{ in } \ \Omega_1, \\ r^2 u = V^{(1)}\big(V_{-1}^{(1)}\big)^{-1} g_1 + V^{(0)} h \ \text{ in } \ \Omega_2, \end{cases} \tag{4.5}$$

where h, g_1, g_2 *are found from system* (4.1).

Moreover, let

$$1<r<\infty, \qquad 1\le q\le\infty, \qquad \frac{1}{r}-\frac{1}{2}<s<\frac{1}{r}+\frac{1}{2}. \tag{4.6}$$

If $u\in\mathbb{W}^1(\Omega\backslash S_1)$ *is a solution of the Neumann problem then*

$$\begin{aligned} &u\in\mathbb{H}_r^{s+\frac{1}{r}}(\Omega\backslash S_1) \ \text{ if } \ \psi\in\mathbb{W}_r^{s-1}(\partial\Omega), \ \ \psi_i\in\mathbb{W}_r^{s-1}(S_1), \ \ i=1,2, \\ &u\in\mathbb{B}_{r,q}^{s+\frac{1}{r}}(\Omega\backslash S_1) \ \text{ if } \ \psi\in\mathbb{B}_{r,q}^{s-1}(\partial\Omega), \ \ \psi_i\in\mathbb{B}_{r,q}^{s-1}(S_1), \ \ i=1,2. \end{aligned} \tag{4.7}$$

Proof. The system (4.2) can be rewritten in the matrix form

$$\mathbf{M}\begin{bmatrix} h \\ g_1 \\ g_2 \\ \psi_0^{(1)} \\ \psi_0^{(2)} \end{bmatrix} = \begin{bmatrix} \Psi \\ \Psi_0^{(1)} \\ \Psi_0^{(2)} \\ 0 \\ r_{S_0}\Psi_0^{(2)} - r_{S_0}\Psi_0^{(1)} \end{bmatrix}, \tag{4.8}$$

where

$$\mathbf{M} = \begin{bmatrix} -\frac{1}{2}I + V_0^{(0)} & 0 & r_{\partial\Omega}(TV^{(1)}) & 0 & 0 \\ 0 & \frac{1}{2}I + V_0^{(1)} & 0 & -I & 0 \\ r_{\partial\Omega_1}(TV^{(0)}) & 0 & -\frac{1}{2}I + V_0^{(1)} & 0 & -I \\ -r_{S_0}V^{(0)} & r_{S_0}V_{-1}^{(1)} & -r_{S_0}V_{-1}^{(1)} & 0 & 0 \\ 0 & 0 & 0 & I & -I \end{bmatrix}.$$

The operator $\mathbf{M}$ has the same representation (3.6) with a compact and smoothing $T_{-\infty}$, but yet different $\mathbf{P}$:

$$\mathbf{P} = \begin{bmatrix} \frac{1}{2} I + V_0^{(1)} & 0 & -I & 0 \\ 0 & -\frac{1}{2} I + V_0^{(1)} & 0 & -I \\ r_{S_0} V_{-1}^{(1)} & -r_{S_0} V_{-1}^{(1)} & 0 & 0 \\ 0 & 0 & I & -I \end{bmatrix}.$$

Let us introduce the operators

$$\mathbf{P}_N = \begin{bmatrix} (-V_{-1}^{(1)})^N + \mathcal{M}_+ (V_{-1}^{(1)})^{-1} & 0 & -I & 0 \\ 0 & (-V_{-1}^{(1)})^N + \mathcal{M}_- (V_{-1}^{(1)})^{-1} & 0 & -I \\ -r_{S_0} I & -r_{S_0} I & 0 & 0 \\ 0 & 0 & I & -I \end{bmatrix},$$

$$\mathcal{M}_+ := \Big(\frac{1}{2} I + V_0^{(1)}\Big), \qquad \mathcal{M}_- := \Big(-\frac{1}{2} I + V_0^{(1)}\Big), \qquad N = 1, 2, \dots ,$$

and

$$\mathbf{D} = \begin{bmatrix} -(V_{-1}^{(1)})^{-1} & 0 & 0 & 0 \\ 0 & (V_{-1}^{(1)})^{-1} & 0 & 0 \\ 0 & 0 & I & 0 \\ 0 & 0 & 0 & I \end{bmatrix}.$$

Then $\mathbf{P} \circ D$ differs from $\mathbf{P}_N$ only modulo a compact operator $\mathbf{T}_{-N}$:

$$\mathbf{P} \circ \mathbf{D} - \mathbf{P}_N = \mathbf{T}_{-N}.$$

Consider the following system

$$\begin{cases} \mathbf{B}_N^{(i)} \widetilde{g}_i - \widetilde{\psi}_0^{(i)} = \widetilde{F}^{(i)}, \ i = 1, 2, & \text{on } \partial\Omega_1, \\ -r_{S_0} \widetilde{g}_1 - r_{S_0} \widetilde{g}_2 = \widetilde{G}_1 & \text{on } S_0, \\ \widetilde{\psi}_0^{(1)} - \widetilde{\psi}_0^{(2)} = \widetilde{G}_2 & \text{on } S_0, \end{cases} \tag{4.9}$$

where

$$\mathbf{B}_N^{(i)} = (-V_{-1}^{(1)})^N + \Big(-\frac{1}{2} I + (-1)^i V_0^{(1)}\Big)(V_{-1}^{(1)})^{-1}, \ i = 1, 2.$$

$$\widetilde{g}_i = (\mathbf{B}_N^{(i)})^{-1} \widetilde{\psi}_0^{(i)} + (\mathbf{B}_N^{(i)})^{-1} F^{(i)}, \ i = 1, 2, \qquad N = 1, 2, \dots .$$

The systems (4.1) (i.e., the system (4.2)) and (4.9) are Fredholm-equivalent: are Fredholm or are not Fredholm only simultaneously and have equal indices. Due to Theorem 2.7 the pseudodifferential operator

$$\mathbf{B}_N^{(i)} \ : \ \widetilde{\mathbb{H}}_p^s(S_0) \to \mathbb{H}_p^{s-1}(S_0)$$

$$: \ \widetilde{\mathbb{B}}_{p,q}^s(S_0) \to \mathbb{B}_{p,q}^{s-1}(S_0)$$

is invertible.

Let us define $\widetilde{g}_1$ and $\widetilde{g}_2$ from the first two equations of system (4.9) and insert them into the third equation. We obtain the system of differential equations on the open manifold S_0

$$\begin{cases} r_{S_0}(\mathbf{B}_N^{(1)})^{-1}\widetilde{\psi}_0^{(1)} + r_{S_0}(\mathbf{B}_N^{(2)})^{-1}\widetilde{\psi}_0^{(2)} = F, \\ \widetilde{\psi}_0^{(1)} - \widetilde{\psi}_0^{(2)} = G_2, \end{cases}$$

or

$$\begin{cases} r_{S_0}\mathbf{B}\widetilde{\psi}_0^{(1)} = \widetilde{\Psi}, \\ \widetilde{\psi}_0^{(2)} = \widetilde{\psi}_0^{(1)} - G_2, \end{cases} \tag{4.10}$$

where

$$\mathbf{B} = (\mathbf{B}_N^{(1)})^{-1} + (\mathbf{B}_N^{(2)})^{-1}.$$

The pseudodifferential operator $r_{S_0}\mathbf{B}$ is strongly elliptic and, due to Theorem 2.7 the operator

$$r_{S_0}\mathbf{B} : \widetilde{\mathbb{H}}_p^{s-1}(S_0) \to \mathbb{H}_p^s(S_0) \quad (\widetilde{\mathbb{B}}_{p,q}^{s-1}(S_0) \to \mathbb{B}_{p,q}^s(S_0))$$

is invertible provided the conditions (4.6) hold.

Summarizing we find that if the conditions (4.6) hold, the operator

$$\mathbf{N} : \begin{matrix} \mathbb{H}_p^{s-1}(\partial\Omega) \\ \oplus \\ \mathbb{H}_p^{s-1}(\partial\Omega_1) \\ \oplus \\ \mathbb{H}_p^{s-1}(\partial\Omega_1) \end{matrix} \to \begin{matrix} \mathbb{H}_p^{s-1}(\partial\Omega) \\ \oplus \\ \mathbb{H}_p^{s-1}(S_1) \\ \oplus \\ \mathbb{H}_p^{s-1}(S_1) \end{matrix} \quad \left(\begin{matrix} \mathbb{B}_{p,q}^{s-1}(\partial\Omega) \\ \oplus \\ \mathbb{B}_{p,q}^{s-1}(\partial\Omega_1) \\ \oplus \\ \mathbb{B}_{p,q}^{s-1}(\partial\Omega_1) \end{matrix} \to \begin{matrix} \mathbb{B}_{p,q}^{s-1}(\partial\Omega) \\ \oplus \\ \mathbb{B}_{p,q}^{s-1}(S_1) \\ \oplus \\ \mathbb{B}_{p,q}^{s-1}(S_1) \end{matrix} \right)$$

defined by the left-hand side of (4.1) is Fredholm and Ind $\mathbf{N} = 0$. It is well known that the kernel, the cokernel and, therefore, the index of the operator $\mathbf{N}$ are independent of the parameters of the spaces where it is Fredholm (see [Ag1, DNS1, Ka1]). Therefore it suffices to find Coker $\mathbf{N}$ for $s = 1/2$ and $p = 2$. Thus, we consider the dual (adjoint) operator in the spaces:

$$\mathbf{N}^* : \begin{matrix} \mathbb{H}_2^{1/2}(\partial\Omega) \\ \oplus \\ \widetilde{\mathbb{H}}_2^{1/2}(S_1) \\ \oplus \\ \widetilde{\mathbb{H}}_2^{1/2}(S_1) \end{matrix} \to \begin{matrix} \mathbb{H}_2^{1/2}(\partial\Omega) \\ \oplus \\ \mathbb{H}_2^{1/2}(\partial\Omega_1) \\ \oplus \\ \mathbb{H}_2^{1/2}(\partial\Omega_1) \end{matrix}.$$

If $(\widetilde{h}, \chi_0^{(1)}, \chi_0^{(2)})$ is the solution of homogeneous equation

$$\mathbf{N}^*(\widetilde{h}, \chi_0^{(1)}, \chi_0^{(2)}) = 0, \qquad \operatorname{supp}\chi_0^{(1)} \subset \overline{S}_1, \quad \operatorname{supp}\chi_0^{(2)} \subset \overline{S}_1.$$

Then we have

$$\begin{cases} \left(-\frac{1}{2}I + \widetilde{V}_0^{(0)}\right)\widetilde{h} + r_{\partial\Omega}(\widetilde{U}^{(1)})\chi_0^{(2)} = 0 & \text{on } \partial\Omega, \\ \left(\frac{1}{2}I + \widetilde{V}_0^{(1)}\right)\chi_0^{(1)} = 0 & \text{on } \partial\Omega_1, \\ r_{\partial\Omega_1}(\widetilde{U}^{(0)})\widetilde{h} + \left(-\frac{1}{2}I + \widetilde{V}_0^{(1)}\right)\chi_0^{(2)} = 0 & \text{on } \partial\Omega_1. \end{cases} \tag{4.11}$$

Here $\widetilde{U}^{(i)}$ are the double-layer potentials:

$$\widetilde{U}^{(i)}(g)(x) = \int\limits_{\partial\Omega_i} \left[\widetilde{T}(\partial_y, n(y))H^T(x-y)\right]^T g(y)d_yS\,, \qquad i=0,1\,.$$

Let

$$v^{(2)} = \widetilde{U}^{(0)}\widetilde{h} + \widetilde{U}^{(1)}\chi_0^{(2)},$$
$$v^{(1)} = \widetilde{U}^{(1)}\chi_0^{(1)}.$$

Due to (4.11) $v^{(i)}$, $i=1,2$, satisfy the boundary conditions:

$$\begin{cases} \gamma^- v^{(2)} = 0 \text{ on } \partial\Omega, \\ \gamma^1 v^{(1)} = 0 \text{ on } \partial\Omega_1, \\ \gamma^2 v^{(2)} = 0 \text{ on } \partial\Omega_1\,. \end{cases}$$

Here γ^- denotes the trace on $\partial\Omega$ from the outer domain $\mathbb{R}^3\setminus\overline{\Omega}$. Therefore $v^{(2)}$ is the solution of the following BVP:

$$\begin{cases} \mathbf{A}^T(D)v^{(2)} = 0 \text{ in } \mathbb{R}^3\backslash\overline{\Omega}_2, \\ \gamma^- v^{(2)} = 0 \text{ on } \partial\Omega, \\ \gamma^1 v^{(2)} = 0 \text{ on } \partial\Omega_1, \end{cases}$$

which have only the trivial solution in $\mathbb{R}^3\backslash\overline{\Omega}_2$. Consequently, $v^{(2)} \equiv 0$ in $\mathbb{R}^3\backslash\overline{\Omega_2}$ and from boundary condition

$$\gamma^0\left(\widetilde{T}v^{(2)}\right) - \gamma^-\widetilde{T}v^{(2)} = 0 \text{ on } \partial\Omega;$$
$$\gamma^2\left(\widetilde{T}v^{(2)}\right) - \gamma^1\left(\widetilde{T}v^{(2)}\right) = 0 \text{ on } \partial\Omega_1;$$

it follows that

$$\gamma^0\left(\widetilde{T}v^{(2)}\right) = 0 \text{ on } \partial\Omega, \ \gamma^2\left(\widetilde{T}v^{(2)}\right) = 0 \text{ on } \partial\Omega_1.$$

Thus, $v^{(2)}$ is the solution of the next BVP in Ω_2 :

$$\begin{cases} \mathbf{A}^T(D)v^{(2)} = 0 \text{ in } \Omega_2, \\ \gamma^0\left(\widetilde{T}v^{(2)}\right) = 0 \text{ on } \partial\Omega, \\ \gamma^2\left(\widetilde{T}v^{(2)}\right) = 0 \text{ on } \partial\Omega_1, \end{cases}$$

which have only solution of the form $v^{(2)} = a\cdot x + b$, where a is an antisymmetric matrix and b is an arbitrary constant vector. Due to the boundary conditions:

$$\gamma^0 v^{(2)} - \gamma^- v^{(2)} = \widetilde{h} \text{ on } \partial\Omega,$$
$$\gamma^2 v^{(2)} - \gamma^1 v^{(2)} = \chi_0^{(2)} \text{ on } \partial\Omega_1$$

it follows that

$$\widetilde{h} = a\cdot x + b, \ \ \chi_0^{(2)} = a\cdot x + b,$$

Since $\operatorname{supp}\chi_0^{(2)} \subset \overline{S}_1$, we get $\chi_0^{(2)} = 0$. Analogously we obtain $\chi_0^{(1)} = 0$.

The second part, the solvability properties (4.6), are consequences of the first part and the mapping properties of the potential operators (see [DNS1, NCS1, Ch2] for similar considerations in elasticity). □

Let us investigate the asymptotic behavior of the solution to the Neumann problem in the vicinity of the edge $\mathcal{E}$. Assume that the corresponding boundary data are sufficiently smooth. Namely, $\psi_i \in \mathbb{H}_r^{(\infty,s+2N),\infty}(S_1)$.

We can rewrite the symbol $\sigma_B(x',\xi')$ as

$$\begin{aligned}\sigma_B(x',\xi') &= \Big[\Big(-\frac{1}{2}I+\sigma_{V_0^{(1)}}(x',\xi')\Big)(\sigma_{V_{-1}^{(1)}}(x',\xi'))^{-1}\Big]^{-1}\\ &-\Big[\Big(\frac{1}{2}I+\sigma_{V_0^{(1)}}(x',\xi')\Big)(\sigma_{V_{-1}^{(1)}}(x',\xi'))^{-1}\Big]^{-1}.\end{aligned}$$

The principal symbol $\sigma_{V_0^{(1)}}(x',\xi')$ is odd with respect to the variable ξ', while the symbol $\sigma_{V_{-1}^{(1)}}(x',\xi')$ is even; therefore, $\sigma_B(x',\xi')$ is even with respect to ξ':

$$\sigma_B(x',-\xi')=\sigma_B(x',\xi'),\quad x'\in S_0$$

and all eigenvalues of the matrix $(\sigma_B(x',0,0,+1))^{-1}\sigma_B(x',0,0,-1)=I$ are trivial: $\lambda_j=1$, $j=1,2,3,4$, $x'\in\mathcal{E}$.

Consider the above-described local coordinate system $x=(x',r)\in S_0$. Using the theory of strongly elliptic pseudodifferential equations (see [CD1, Theorem 2.1]) we get the following asymptotic expansion of solutions of equation (4.8):

$$\psi_0^{(i)}(x',r)=c_0(x')r^{-\frac{1}{2}}+\sum_{k=1}^{N}c_k(x')r^{-\frac{1}{2}+k}+\psi_{N+1}^{(i)}(x',r),\qquad i=1,2,\quad (4.12)$$

$$x'\in\mathcal{E},\quad 0<r<\varepsilon,\quad \psi_{N+1}^{(i)}\in\mathbb{H}_r^{(\infty,s+N),\infty}(\mathcal{E}_\varepsilon^+)\subset C^{s+N}([0,\varepsilon],C^\infty(\mathcal{E}),$$

where $c_k\in C^\infty(\mathcal{E})$, $k=0,1,\dots,N$.

And again, logarithms are absent in the entire asymptotic representation (4.12).

Let $(h,g_1,g_2,\psi_0^{(1)},\psi_0^{(2)})$ be a solution of the system (4.2):

$$\begin{aligned}&\mathbf{M}(h,g_1,g_2,\psi_0^{(1)},\psi_0^{(2)})=\Psi,\\ &\Psi=(\psi,\Psi_0^{(1)},\Psi_0^{(2)},0,-(r_{S_0}\Phi_0^{(1)}-r_{S_0}\Phi_0^{(2)})),\end{aligned}$$

then $(g_1,g_2,\psi_0^{(1)},\psi_0^{(2)})$ satisfies the equation

$$\mathbf{P_N}\big(-V_{-1}^{(1)}g_1,V_{-1}^{(1)}g_2,\psi_0^{(1)},\psi_0^{(2)}\big)=\overset{*}{\Psi},\tag{4.13}$$

where

$$\begin{aligned}\overset{*}{\Psi}=\big(\Psi_0^{(1)}+(V_{-1}^{(1)})^{2N}g_1,\Psi_0^{(2)}+(V_{-1}^{(1)})^{2N}g_2-r_{\partial\Omega_1}(TV^{(0)})h,\\ r_{S_0}Vh,-r_{S_0}\Phi_0^{(1)}+r_{S_0}\Phi_0^{(2)}\big).\end{aligned}$$

Equation (4.8) (i.e., equation (4.2)) can be written in the form

$$\begin{cases} \mathbf{B}^{(1)}_{\mathbf{2N}}(-V^{(1)}_{-1})g_1 - \psi^{(1)}_0 = \Psi^{(1)}_0 + (V^{(1)}_{-1})^{2N} g_1, \\ \mathbf{B}^{(1)}_{\mathbf{2N}}(V^{(1)}_{-1})g_2 - \psi^{(2)}_0 = \Psi^{(2)}_0 + (V^{(1)}_{-1})^{2N} g_1 - r_{\partial\Omega_1}(TV^{(0)})h, \\ -r_{S_0} g_1 - r_{S_0} g_2 = r_{S_0} V^{(0)} h, \\ \psi^{(1)}_0 - \psi^{(2)}_0 = -r_{S_0}\Phi^{(1)}_0 + r_{S_0}\Phi^{(2)}_0. \end{cases} \tag{4.14}$$

where

$$\mathbf{B}^{(1)}_{\mathbf{2N}} = \big(V^{(1)}_{-1}\big)^{2N} - \Big(\tfrac{1}{2}I + V^{(1)}_0\Big)\big(V^{(1)}_{-1}\big)^{-1},$$

$$\mathbf{B}^{(2)}_{\mathbf{2N}} = \big(V^{(1)}_{-1}\big)^{2N} + \Big(-\tfrac{1}{2}I + V^{(1)}_0\Big)\big(V^{(1)}_{-1}\big)^{-1}.$$

As we have seen before, equation (4.14) can be reduced to a pseudodifferential equation with a strongly elliptic operator.

Using the first two equations of system (4.14), we find

$$g_1 = \big(-V^{(1)}_{-1}\big)^{-1}\big(\mathbf{B}^{(1)}_{\mathbf{2N}}\big)^{-\mathbf{1}}\psi^{(\mathbf{1})}_{\mathbf{0}} + \mathbf{F_1},$$

$$g_2 = \big(V^{(1)}_{-1}\big)^{-1}\big(\mathbf{B}^{(2)}_{\mathbf{2N}}\big)^{-\mathbf{1}}\psi^{(\mathbf{2})}_{\mathbf{0}} + \mathbf{F_2}.$$

Note that

$$F_i \in \mathbb{H}^{(\infty,s+2N+1),\infty}_r(\partial\Omega_i), \quad i = 1,2.$$

Hence we obtain a representation of the solutions of the Neumann BVP by potential-type functions:

$$r^1 u = V^{(1)}\big(-V^{(1)}_{-1}\big)^{-1}\big(\mathbb{B}^{(1)}_{2N}\big)^{-1}\psi^{(1)}_0 + G_1,$$

$$r^2 u = V^{(1)}\big(V^{(1)}_{-1}\big)^{-1}\big(\mathbb{B}^{(2)}_{2N}\big)^{-1}\psi^{(2)}_0 + G_2,$$

where $G_i \in C^{N+1}(\overline{\Omega}_i)$ $i = 1,2$.

Thus, using the asymptotic expansion (4.12) of the function $\psi^{(i)}_0$, $i = 1,2$ and the asymptotic expansion of functions that can be represented by potential operators (see [CD2, Theorems 2.2 and 2.3]) we obtain the asymptotics of solutions to the Neumann problem in terms of the local coordinates (see § 3) in a neighborhood of the crack boundary:

$$\boldsymbol{u} = \sum_{m=1}^{\ell}\sum_{\omega=\pm1}\chi(r)\Bigg[\sum_{j=0}^{n_m-1} r^{\frac{1}{2}}\sin^j\theta\,\psi^{\frac{1}{2}-j}_{m,\omega}(x',\theta)\,\boldsymbol{d}^j_{m,\omega}(x') \tag{4.15}$$

$$+\sum_{k=1}^{N-1}\sum_{j=0}^{p(m,k)}\sum_{|\alpha|\le N(m,k)} r^{\frac{1}{2}+k}\,\psi^{\frac{1}{2}-j+k}_{m,\omega}(x',\theta)\sin^{\alpha_1}\theta\cos^{\alpha_2}\theta\,\boldsymbol{d}^{k,j,\alpha}_{m,\omega}(x')\Bigg] + \boldsymbol{u}_{\mathrm{rem},N},$$

where $\boldsymbol{u}_{\mathrm{rem},N} \in \mathbb{H}^{\frac{1}{2}+N}_{\mathrm{loc}}(\mathbb{R}^3)$ and the coefficients $\boldsymbol{d}^j_{m,\omega}$ and $\boldsymbol{d}^{j,k,\alpha}_{m,\omega}$ are $C^\infty(\mathcal{E})$ (see (3.11) for a similar notation).

As in § 3, the coefficients $\boldsymbol{d}^j_{m,\omega} \in C^\infty(\mathcal{E})$ of the asymptotic formula (4.15) can be expressed by the first coefficients c_0 of the surface expansion (4.12) (see [CD2]).

5. The mixed problem

Any extension $\Phi^{(1)} \in \mathbb{H}^{1/2}(\partial\Omega_1)$ of the function φ_1 to the whole boundary $\partial\Omega_1 = S_1 \cup \overline{S}_0$ has the form

$$\Phi^{(1)} = \Phi_0^{(1)} + \varphi_0^{(1)},$$

where $\Phi_0^{(1)}$ is a fixed extension of the function φ_1, and $\varphi_0^{(1)} \in \widetilde{\mathbb{H}}^{1/2}(S_0)$.

Any extension $\Psi^{(2)} \in \mathbb{H}^{-1/2}(\partial\Omega_1)$ of the function ψ_2 to the whole boundary $\partial\Omega_1 = S_1 \cup \overline{S}_0$ has the form

$$\Psi^{(2)} = \Psi_0^{(2)} + \psi_0^{(2)},$$

where $\Psi_0^{(2)}$ is a fixed extension of the function ψ_2, and $\psi_0^{(2)} \in \widetilde{\mathbb{H}}^{-1/2}(S_0)$.

Solutions of the mixed boundary value problem will be sought in the form (3.1). Bearing in mind boundary conditions of the mixed problem and equalities (3.2) for the function u, we get a system of equations with respect to h, g_1, g_2, $\varphi_0^{(1)}$, $\varphi_0^{(2)}$:

$$\begin{cases} \Big(-\frac{1}{2}I + V_0^{(0)}\Big)h + r_{\partial\Omega}\Big(TV^{(1)}\Big)g_2 = \psi & \text{on } \partial\Omega, \\ V_{-1}^{(1)}g_1 - \varphi_0^{(1)} = \Phi_0^{(1)} & \text{on } \partial\Omega_1, \\ \Big(-\frac{1}{2}I + V_0^{(1)}\Big)g_2 - \psi_0^{(2)} + r_{\partial\Omega_1}(TV^{(0)})h = \Psi_0^{(2)} & \text{on } \partial\Omega_1, \\ -r_{S_0}V^{(0)}h - r_{S_0}V_{-1}^{(1)}g_2 + \varphi_0^{(1)} = -r_{S_0}\Phi_0^{(1)} & \text{on } S_0, \\ r_{S_0}\Big(\frac{1}{2}I + V_0^{(1)}\Big)g_1 - \psi_0^{(2)} = r_{S_0}\Psi_0^{(2)} & \text{on } S_0, \end{cases} \tag{5.1}$$

Now we can formulate the basic theorem of the existence and uniqueness of a solution of the mixed problem.

Theorem 5.1. *The mixed boundary value problem has a unique solution in the space* $\mathbb{W}^1(\Omega\backslash S_1)$, *which is given by the potential-type functions*

$$u^{(1)} = V^{(1)}(V_{-1}^{(1)})^{-1}\varphi_0^{(1)} + R_1,$$
$$u^{(2)} = V^{(1)}(\mathbf{B}_{2N+1}^{(1)})^{-1}V_{-1}^{(1)}\psi_0^{(2)} + R_2,$$

where

$$R_i \in C^{N+1}(\overline{\Omega}_i), \quad i = 1, 2,$$
$$\mathbf{B}_{2N+1}^{(1)} = -(V_{-1}^{(1)})^{2N+1} + V_{-1}^{(1)}\Big(-\frac{1}{2}I + V_0^{(1)}\Big)$$

and $(-L_+^{-1}\psi_0^{(2)}, \varphi_0^{(1)})$ *is the solution of the strongly elliptic pseudodifferential equation* (5.3).

Moreover, let

$$\delta := \sup_{\substack{1\le j\le 8 \\ x'\in\mathcal{E}}} \frac{1}{2\pi}\left|\arg\lambda_j(x')\right|, \tag{5.2}$$
$$1 < t < \infty, \qquad 1 \le r \le \infty, \qquad \frac{1}{t} - \frac{1}{2} + \delta < s < \frac{1}{t} + \frac{1}{2} - \delta.$$

and $u \in \mathbb{W}^1(\Omega\backslash S_1)$ *be the solution of the mixed BVP. Then:*

If $\varphi_1 \in \mathbb{B}^s_{t,t}(S_1)$, $\psi_2 \in \mathbb{W}^{s-1}_t(S_1)$, $\psi \in \mathbb{W}^{s-1}_t(\partial\Omega)$, *we have* $u \in \mathbb{H}^{s+1/t}_t(\Omega\backslash S_1)$.

If $\varphi_1 \in \mathbb{B}^s_{t,r}(S_1)$, $\psi_2 \in \mathbb{B}^{s-1}_{t,r}(S_1)$, $\psi \in \mathbb{B}^{s-1}_{t,r}(\partial\Omega)$, *we have* $u \in \mathbb{B}^{s+1/t}_{t,r}(\Omega\backslash S_1)$.

Proof. The operator corresponding to the system (5.1) will be denoted by $\mathbf{M}$. It has the form

$$\mathbf{M} = \begin{bmatrix} \left(-\frac{1}{2}I + V_0^{(0)}\right) & 0 & r_{\partial\Omega}(TV^{(1)}) & 0 & 0 \\ 0 & V_{-1}^{(1)} & 0 & -I & 0 \\ r_{\partial\Omega_1}(TV^{(0)}) & 0 & -\frac{1}{2}I + V_0^{(1)} & 0 & -I \\ -r_{S_0}V^{(0)} & 0 & -r_{S_0}V_{-1}^{(1)} & I & 0 \\ 0 & r_{S_0}\left(\frac{1}{2}I + V_0^{(1)}\right) & 0 & 0 & -I \end{bmatrix}.$$

And again, operator $\mathbf{M}$ is decomposed in the form (3.6) with a compact and smoothing $T_{-\infty}$, but yet different P :

$$\mathbf{P} = \begin{bmatrix} V_{-1}^{(1)} & 0 & -I & 0 \\ 0 & -\frac{1}{2}I + V_0^{(1)} & 0 & -I \\ 0 & -r_{S_0}V_{-1}^{(1)} & I & 0 \\ r_{S_0}\left(\frac{1}{2}I + V_0^{(1)}\right) & 0 & 0 & -I \end{bmatrix}.$$

We consider the composition $\mathbf{D}\circ\mathbf{P}$ of $\mathbf{P}$ with the invertible operator

$$\mathbf{D} = \begin{pmatrix} I & 0 & 0 & 0 \\ 0 & V_{-1}^{(1)} & 0 & 0 \\ 0 & 0 & I & 0 \\ 0 & 0 & 0 & I \end{pmatrix}.$$

For the invertibility of the pseudodifferential operator $V_{-1}^{(1)}$ see Theorem 2.4.

Note that, as in previous cases, the difference

$$\mathbf{T}_N = \mathbf{D}\circ\mathbf{P} - \mathbf{P}_N, \quad N = 2, 3, \ldots, ,$$

where

$$\mathbf{P}_N = \begin{pmatrix} V_{-1}^{(1)} & 0 & -I & 0 \\ 0 & (-V_{-1}^{(1)})^N + V_{-1}^{(1)}(-\frac{1}{2}I + V_0^{(1)}) & 0 & -V_{-1}^{(1)} \\ 0 & -r_{S_0}V_{-1}^{(1)} & I & 0 \\ r_{S_0}(-\frac{1}{2}I + V_0^{(1)}) & 0 & 0 & I \end{pmatrix}$$

is a compact operator. So it is sufficient for us to investigate the operator $\mathbf{P}_N$ which acts in the following spaces:

$$\mathbf{P}_N : \begin{array}{c} \mathbb{H}_p^{s-1}(\partial\Omega_1) \\ \oplus \\ \mathbb{H}_p^{s-1}(\partial\Omega_1) \\ \oplus \\ \widetilde{\mathbb{H}}_p^{s}(S_0) \\ \oplus \\ \widetilde{\mathbb{H}}_p^{s-1}(S_0) \end{array} \longrightarrow \begin{array}{c} \mathbb{H}_p^{s}(\partial\Omega_1) \\ \oplus \\ \mathbb{H}_p^{s}(\partial\Omega_1) \\ \oplus \\ \mathbb{H}_p^{s}(S_0) \\ \oplus \\ \mathbb{H}_p^{s-1}(S_0) \end{array} \quad \left(\begin{array}{c} \mathbb{B}_{p,r}^{s-1}(\partial\Omega_1) \\ \oplus \\ \mathbb{B}_{p,r}^{s-1}(\partial\Omega_1) \\ \oplus \\ \widetilde{\mathbb{B}}_{p,r}^{s}(S_0) \\ \oplus \\ \widetilde{\mathbb{B}}_{p,r}^{s-1}(S_0) \end{array} \longrightarrow \begin{array}{c} \mathbb{B}_{p,r}^{s}(\partial\Omega_1) \\ \oplus \\ \mathbb{B}_{p,r}^{s}(\partial\Omega_1) \\ \oplus \\ \mathbb{B}_{p,r}^{s}(S_0) \\ \oplus \\ \mathbb{B}_{p,r}^{s-1}(S_0) \end{array} \right)$$

$$(\beta \geq [|s|] + 3, \quad 1 < p < \infty, \quad 1 \leq t \leq \infty).$$

Now let us consider a system of equations that corresponds to the operator $\mathbf{P}_N$ with respect to h_1, h_2, $\psi_0^{(1)}$, $\psi_0^{(2)}$

$$\left\{ \begin{array}{l} V_{-1}^{(1)}\widetilde{g}_1 - \widetilde{\varphi}_0^{(1)} = \widetilde{\Phi}_0^{(1)}, \\ \left[(-V_{-1}^{(1)})^N + V_{-1}^{(1)}\left(-\frac{1}{2}I + V_0^{(1)}\right)\right]\widetilde{g}_2 - V_{-1}^{(1)}\widetilde{\psi}_0^{(2)} = \widetilde{\Psi}_0^{(2)}, \\ -r_{S_0}V_{-1}^{(2)}\widetilde{g}_2 + \widetilde{\varphi}_0^{(1)} = F_1, \\ r_{S_0}\left(-\frac{1}{2}I + V_0^{(1)}\right)\widetilde{g}_1 + \widetilde{\psi}_0^{(2)} = F_2 . \end{array} \right. \tag{5.3}$$

Similarly as it has been done in [Ch2] system (5.3) is reduced equivalently to a strongly elliptic pseudodifferential equation on S_0

$$\mathbf{R}(x', D')\chi = \Psi , \tag{5.4}$$

where

$$\mathbf{R}(x', D') = \left[\begin{array}{cc} L_- \circ r_{S_0}\mathbf{A}_2(x', D') \circ L_+ & L_- \\ -L_+ & r_{S_0}\mathbf{A}_1(x', D') \end{array} \right],$$

$$\mathbf{A}_1 = \left(-\frac{1}{2}I + V_0^{(1)}\right)(V_{-1}^{(1)})^{-1},$$

$$\mathbf{A}_2 = \left[-(V_{-1}^{(1)})^{2N+1} - \left(\frac{1}{2}I + V_0^{(1)}\right)(V_{-1}^{(1)})^{-1}\right]^{-1},$$

$$L_- = r^+ \mathrm{diag}\Lambda_-\ell, \quad L_+ = \mathrm{diag}\Lambda_+ .$$

Here $\Lambda_\pm$ are the pseudodifferential operators with symbol $\Lambda_\pm(\xi') = \xi_2 \pm i \pm i|\xi_1|$, r^+ denotes the operator of restriction on $\mathbb{R}_+^2$ and ℓ is an extension operator.

Since the operator has the same kernel and cokernel in all spaces where it is Fredholm and thus it has the same index (see [Ag1, Ka1, DNS2]), from Lemma 2.8 and Theorem 1.1 it follows that the operator $\mathbf{N}$ is invertible in the corresponding Besov and Bessel potential spaces (cf. similar proofs in [DNS1, DNS2]).

The second part, the solvability properties, are consequences of the first part and the mapping properties of the potential operators (see [DNS1, NCS1, Ch2] for similar considerations in elasticity). □

We drop the detailed asymptotic expansion of a solution to mixed BVP, because we cannot suggest any simplification in contrast to the Dirichlet and Neumann BVPs considered above. Asymptotic formulae for general PsDOs on a smooth surface with a smooth boundary are exposed in detail in [CD1, CDD1]. Spatial asymptotic representations of the solution to a BVP, based on surface asymptotics of a solution to the corresponding boundary pseudodifferential equation is exposed in [CD2, CDD1].

6. An example: media with cubic symmetry

Consider an electro-elastic medium with cubic symmetry [No1, To1]. In this case the entries of the corresponding operator (1.2) are of the form:

$$\begin{aligned}
\mathbf{A}_{11}(D) &= c_{11}\partial_1^2 + c_{44}\partial_2^2 + c_{44}\partial_3^2,\\
\mathbf{A}_{22}(D) &= c_{44}\partial_1^2 + c_{11}\partial_2^2 + c_{44}\partial_3^2,\\
\mathbf{A}_{33}(D) &= c_{44}\partial_1^2 + c_{44}\partial_2^2 + c_{11}\partial_3^2,\\
\mathbf{A}_{ik}(D) &= (c_{12}+c_{44})\partial_i\partial_k, \quad i,k=1,2,3, \quad i\neq k,\\
\mathbf{A}_{14}(D) &= -\mathbf{A}_{41}(D) = 2e_{14}\partial_2\partial_3,\\
\mathbf{A}_{24}(D) &= -\mathbf{A}_{42}(D) = 2e_{14}\partial_1\partial_3,\\
\mathbf{A}_{34}(D) &= -\mathbf{A}_{43}(D) = 2e_{14}\partial_1\partial_2,\\
\mathbf{A}_{44}(D) &= \varepsilon_{11}\Delta.
\end{aligned}$$

Condition (1.4) of internal energy positiveness imposes the following restrictions on the coefficients c_{ij} :

$$c_{11} > 0, \quad c_{44} > 0, \quad \varepsilon_{11} > 0, \quad -\frac{1}{2} < \frac{c_{12}}{c_{11}} < 1.$$

We can calculate the eigenvalues $\lambda_1, \lambda_2, \dots, \lambda_8$ with the help of the following theorem (see [Ch2, Theorem 6.6]):

Theorem 6.1. *Let* $\lambda_k(x')$, $k = 1,\dots,8,$ *be the eigenvalues of the matrix* $b_{\mathcal{R}}$. *Then*

$$\lambda_k(x') = \begin{cases} i\sqrt{\dfrac{1-2\beta_k(x')}{1+2\beta_k(x')}}, & \text{if } k=1,\dots,4,\\[2mm] -i\sqrt{\dfrac{1-2\beta_{k-4}(x')}{1+2\beta_{k-4}(x')}}, & \text{if } k=5,\dots,8, \ x'\in\mathcal{E}, \end{cases}$$

where $\beta_k \in \left]-\frac{1}{2};\frac{1}{2}\right[$ *are the eigenvalues of the matrix* $\sigma_{\overset{*}{\mathbf{V}}_0}$.

After calculating the matrix $\sigma_{V_0^{(1)}}$ we obtain

$$\sigma_{V_0^{(1)}} = \begin{bmatrix} 0 & 0 & 0 & 0\\ 0 & 0 & i\,\dfrac{(c_{11}-c_{12})}{2c_{11}\sqrt{2(\widetilde{c}+1)}} & 0\\ 0 & -i\,\dfrac{(c_{11}-c_{12})}{2c_{11}\sqrt{2(\widetilde{c}+1)}} & 0 & 0\\ 0 & 0 & 0 & 0 \end{bmatrix},$$

where

$$\widetilde{c} = \frac{c_{11}^2 - c_{12}^2}{2c_{11}c_{44}} - \frac{c_{12}}{c_{44}} > -1.$$

So

$$\beta_1 = \beta_2 = 0, \quad \beta_{3,4} = \pm \frac{c_{11} - c_{12}}{2c_{11}\sqrt{2(\widetilde{c}+1)}}$$

and the eigenvalues λ_j are purely imaginary.

Moreover, in this special case the exponents of the leading term of the asymptotics are $\frac{1}{4}$ and $\frac{1}{4} + i\delta$; exponents of further terms increase by order $\frac{1}{2}$ and not by 1 as it was in (3.8), (3.11), (4.12) and (4.15). A similar asymptotic behavior can be found in mixed problems (see [CD3] and also [Ch2, DN1]). In case of general anisotropy however, these exponents depend on the elastic constants as well as on the geometry of the crack edge.

References

[Agl] M. Agranovich, Elliptic singular integrodifferential operators. *Uspekhi Mat. Nauk* **20** (1965), No. 5, 3–120 (in Russian).

[BG1] T. Buchukuri and T. Gegelia, On the uniqueness of solutions of the basic problems of elasticity for infinite domains. (Russian) *Differentsial'nye Uravneniya* **25** (1989), No. 9, 1556–1565.

[BG2] T. Buchukuri, and T. Gegelia, Some dynamic problems of the theory of electroelasticity. *Mem. Differential Equations Math. Phys.* **10** (1997), 1–53.

[Ch1] O. Chkadua, Singular integro-differential operators on a manifold and the basic boundary contact problems of the theory of elasticity. (Russian) *PhD Dissertation*, A. Razmadze Mathematical Institute, Tbilisi, 1984.

[Ch2] O. Chkadua, The Dirichlet, Neumann and mixed boundary value problems of the theory of elasticity in n-dimensional domains with boundaries containing closed cuspidal edges. *Math. Nachr.* **189** (1998), 61–105.

[CD1] O. Chkadua, and R. Duduchava, Pseudodifferential equations on manifolds with boundary: Fredholm property and asymptotic. *Math. Nachr.* **222** (2001), 79–139.

[CD2] O. Chkadua and R. Duduchava, Asymptotic of functions represented by potentials. *Russian J. Math. Phys.***7** (2000), No. 1, 15–47.

[CD3] O. Chkadua and R. Duduchava, Asymptotic of solutions to some boundary value problems of elasticity for bodies with cuspidal edges. *Mem. Differential Equations Math. Phys.* **15** (1998), 29–58.

[CDD1] M. Costabel, M. Dauge and R. Duduchava, Asymptotics without logarithmic terms for crack problems. *Preprint 01–50, 2001, University of Rennes 1, France.*

[Du1] R. Duduchava, On multidimensional singular integral operators I–II. *Journal of Operator Theory* **11** (1984), 41–76, 199–214.

[DN1] R. Duduchava and D. Natroshvili, Mixed crack type problem in anisotropic elasticity, *Math. Nachr.* **191** (1998), 83–107.

[DNS1] R. Duduchava, D. Natroshvili and E. Shargorodsky, Boundary value problems of the mathematical theory of cracks. (Russian) *Tbiliss. Gos. Univ. Inst. Prikl. Mat. Trudy* **39** (1990), 63–84.

[DNS2] R. Duduchava, D. Natroshvili and E. Shargorodsky, Some basic boundary value problems of thermoelasticity for anisotripic bodies with cuts I, II. *Georgian Math. J.* **2** (1995), No. 2, 123–140; No. 3, 259–276.

[Fi1] G. Fichera, Existence theorems in elasticity. Boundary value problems of elasticity with unilateral constraints. *Handbuch der Physik, Band* 6 a/2. *Springer-Verlag, Berlin*, 1972.

[Ka1] R. Kapanadze, Factorization of matrix–functions and regularization of singular integral operators on manifolds with boundary. *Mem. Differential Equations Math. Phys.* **2** (1994), No. 1, 3–70.

[KP1] R.J. Knops and L.E. Payne, Uniqueness theorems in linear elasticity. *Springer Tracts Nat. Philos.* 19, *Springer-Verlag, Berlin–Heidelberg–New York*, 1971.

[Na1] D. Natroshvili, Investigation of boundary value and initial boundary value problems of the mathematical theory of elasticity and thermoelasticity for homogeneous anisotropic media using the potential methods. (Russian) *Dissertation for Doctor of Sciences Degree, A. Razmadze Mathematical Institute, Tbilisi*, 1984.

[NCS1] D. Natroshvili, O. Chkadua, and E. Shargorodsky, Mixed problems for homogeneous anisotropic elastic media. (Russian) *Tbilisi. Gos. Univ. Inst. Prikl. Mat. Trudy* **39** (1990), 133–181.

[No1] W. Nowacki, Electromagnetic effects in solids. (Russian, translated from Polish) *Mir, Moscow*, 1986; Polish original: Efekty elektromagnetyczne w stalych cialach odksztalcalnych. *Panstwowe Wydawnictwo Naukowe, Warszawa*, 1983.

[Pa1] Y. E. Pak, Linear electro-elastic fracture mechanics of piezoelectric materials. *Internat. J. Fracture* **54** (1992), 79–100.

[Sh1] E. Shargorodsky, Boundary value problems for elliptic pseudodifferential operators on a manifold. *Proc. A. Razmadze Math. Inst.* **104** (1993), 108–132.

[Sh2] E. Shargorodsky, An $\mathbb{L}_p$ -analogue of the Vishik-Eskin theory. *Mem. Diff. Equations Math. Phys.* **2** (1994), 41–146.

[To1] R. A. Toupin, The electric dielectrics. *J. Rat. Mech. Analysis* **5** (1956), 849–915.

[Tr1] H. Triebel, Interpolation Theory. Function spaces. Differential Operators. *North Holland, Amsterdam*, 1978.

[Tr2] H. Triebel, *Theory of Function Spaces*, Birkhäuser-Verlag, Basel 1983.

Tengiz Buchukuri, Otar Chkadua, and Roland Duduchava
A. Razmadze Mathematical Institute
Academy of Sciences of Georgia
M. Alexidze 1
Tbilisi 93
Georgia
e-mail: `buchuk@rmi.acnet.ge`
e-mail: `chkadua@rmi.acnet.ge`
e-mail: `duduch@rmi.acnet.ge`

Operator Theory
Advances and Applications, Vol. 147, 213–240

On a Class of Wedge Diffraction Problems posted by Erhard Meister

Luis P. Castro, Frank-Olme Speck, and Francisco S. Teixeira

Dedicated to the memory of Erhard Meister with gratitude and admiration

Abstract. A class of canonical wedge diffraction problems for Helmholtz equations was formulated by E. Meister in 1986 and subsequently treated by an operator theoretical approach in various publications of his research group including two of the authors. Certain subclasses of those problems, recognized of being unsolved, are subject of the present paper. Some of them are now solved explicitly by refined operator theoretical and analytical methods, others are reduced to systems of equations which contain so-called convolution type operators with symmetry. By a new factorization approach those are proved to be Fredholm in certain (fractional) Sobolev spaces, sometimes with necessary compatibility conditions. Several of the associated operators are therefore explicitly inverted and a number of new problems can be recognized reflecting the challenges of the present state-of-the-art.

1. Introduction

Erhard Meister formulated in his survey papers [10, 11, 12], several classes of boundary value and transmission problems which were of outstanding interest in mathematical physics, particularly for the wave propagation phenomena in acoustics, electrodynamics, elasticity and thermoelasticity theory, so-called *canonical wedge diffraction problems*. The word "canonical" stands (i) for the particular geometrical situations like axi-parallel, rectangular or circular configurations, (ii) for constant coefficients in the corresponding partial differential equations and in the boundary conditions, and (iii) for the importance of the problem to describe basic phenomena in wave propagation. See the conference proceedings [24] of the Workshop on Direct and Inverse Electromagnetic Scattering at Gebze/Istanbul in September 1995, where this question was discussed in a round table, and [13] of the

---

2000 *Mathematics Subject Classification.* Primary 35J25; Secondary 30E25, 45E10, 47A53, 47A68, 47B35.
Key words and phrases. Wedge diffraction problems, Helmholtz equation, boundary pseudodifferential equation, convolution type operator, factorization, Fredholm property, invertibility.

Sommerfeld'96 Workshop in Freudenstadt/Black Forest organized by E. Meister to honor Arnold Sommerfeld for his famous work [25].

In the center of Meister's considerations we meet the following class of problems.

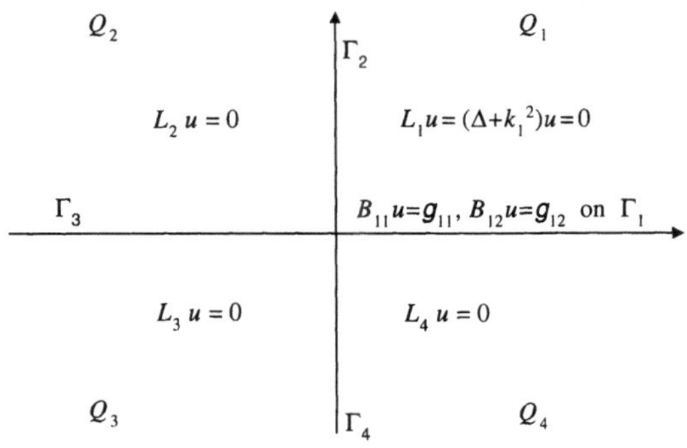

FIGURE 1. Canonical wedge diffraction problem.

Let the real plane $\mathbb{R}^2$ be divided into the four open quadrants Q_j bordered by the (closed) coordinate half-axes Γ_j, $j = 1,2,3,4$, as figured out in Figure 1. Let $k_j \in \mathbb{C}$, $\Im m(k_j) > 0$, $j = 1,2,3,4$, stand for given complex wave numbers due to different media filling the quadrants Q_j. Then we look for a solution of the *Helmholtz equations* (HEs)

$$L_j u = \left(\Delta + k_j^2\right) u = 0 \qquad \text{in} \qquad Q_j \tag{1.1}$$

that satisfies two *boundary/transmission conditions* (BTCs)

$$B_{j1} u = g_{j1}\,, \quad B_{j2} u = g_{j2} \qquad \text{on} \qquad \Gamma_j \tag{1.2}$$

for each $j = 1,2,3,4$ where the boundary operators consist of linear combinations of the Cauchy data and of oblique derivatives as well. They all have the form

$$Bu = \alpha^+ u_0^+ + \beta^+ u_1^+ + \gamma^+ u_\tau^+ + \alpha^- u_0^- + \beta^- u_1^- + \gamma^- u_\tau^- \tag{1.3}$$

with given constant coefficients $\alpha^+, \ldots, \gamma^- \in \mathbb{C}$ which may be different on each Γ_j. Herein $u_0^\pm, u_1^\pm, u_\tau^\pm$ denote the traces of the function u, the normal and tangential first derivatives of u, respectively, on the two banks of Γ_j (due to $x_2 = \pm 0$ on Γ_1 and Γ_3 or $x_1 = \pm 0$ on Γ_2 and Γ_4). The derivatives are always taken in the (positive) direction of the axis x_1 or x_2, respectively.

In the basic setting, we look for the weak solutions u of the HEs (1.1)

$$u_{|Q_j} \in H^1(Q_j)\,, \qquad j = 1,2,3,4. \tag{1.4}$$

Therefore we have to take *boundary data*

$$u_{0\,|\Gamma_1}^+ = T_{0,\Gamma_1} u_{|Q_1} \in H^{1/2}(\Gamma_1) \tag{1.5}$$

etc. due to the trace theorem (we use both notations for convenience) and, via representation formulas [12],

$$\begin{aligned} u^+_{1\,|\Gamma_1} &= T_{1,\Gamma_1} u_{|Q_1} \in H^{-1/2}(\Gamma_1) \\ u^+_{\tau\,|\Gamma_1} &= T_{\tau,\Gamma_1} u_{|Q_1} \in H^{-1/2}(\Gamma_1). \end{aligned} \tag{1.6}$$

Solutions (1.4) of the HEs (1.1) are denoted by $u \in \mathcal{H}^1 = \mathcal{H}^1(\Omega)$ for short, where $\Omega = \cup_{j=1}^4 Q_j = \mathbb{R}^2 \setminus \cup_{j=1}^4 \Gamma_j$ (and analogously for parts of Ω).

Sometimes it is convenient to consider different spaces like $H^{1+\epsilon}(Q_j)$ or $H^{s,p}(Q_j, w)$ with $p \in]1, \infty[$ and a weight function w according to regularity results, edge and radiation conditions, which may be different in each quadrant [10].

It is well known [20] that, if $u \in \mathcal{H}^1$ and $k_1 = k_2$ (e.g.), then the jumps of the Cauchy data across Γ_2 vanish (a.e.):

$$\begin{aligned} u^+_{0\,|\Gamma_2} - u^-_{0\,|\Gamma_2} &= 0 \\ u^+_{1\,|\Gamma_2} - u^-_{1\,|\Gamma_2} &= 0 \end{aligned} \tag{1.7}$$

if and only if the HE holds throughout the upper half-plane

$$\left(\Delta - k_1^2\right) u = 0 \qquad \text{in} \qquad Q_{12} = \text{int clos}\,(Q_1 \cup Q_2) \tag{1.8}$$

(which can be proved with the help of the representation formula for solutions of the HE in the upper half-plane).

Another observation is that the *problem* (1.1)–(1.2) *decomposes* eventually, if some of the BTCs (1.2) are pure boundary conditions. For instance, if

$$\begin{aligned} B_{11} u &= \alpha u_0^+ + \beta u_1^+ + \gamma u_\tau^+ = g_1 \qquad \text{on} \qquad \Gamma_1 \\ B_{21} u &= \alpha' u_0^+ + \beta' u_1^+ + \gamma' u_\tau^+ = g_2 \qquad \text{on} \qquad \Gamma_2 \end{aligned} \tag{1.9}$$

we can obviously "split" a boundary value problem for the first quadrant looking for $u \in \mathcal{H}^1(Q_1)$ that satisfies (1.9). In terms of operator relations, this can be interpreted in the following way: Let L be the operator associated with the full problem (1.1)–(1.2) that maps $u \in \mathcal{H}^1$ into $g \in \times_{j=1}^4 H^{s_{j1}}(\Gamma_j) \times H^{s_{j2}}(\Gamma_j)$ which is the corresponding data space due to (1.2)–(1.6), and let L_1 be the operator associated with the separated problem due to (1.9) that maps $u \in \mathcal{H}^1(Q_1)$ into $g \in H^{s_{11}}(\Gamma_1) \times H^{s_{21}}(\Gamma_2)$, then L can be represented (identifying spaces with their decompositions) as

$$L = \begin{pmatrix} L_1 & * \\ 0 & * \end{pmatrix} \tag{1.10}$$

with obvious consequences for well-posedness, explicit solution, etc. This kind of organization becomes efficient for the extended systems of boundary pseudodifferential equations (BΨDEs) to be treated later.

E. Meister pointed out three subclasses of canonical diffraction problems that are most important from the physical point of view [12, p. 230–232]. These are the following, sketched in Figure 2:

Generalized Sommerfeld half-plane problems: The class of problems with homogeneous transmission conditions of the form (1.7) on Γ_2, Γ_3 and Γ_4 in the case $k_1 = k_2 = k_3 = k_4$ or, if different, with certain weighted transmission conditions for the normal derivatives like

$$\rho_1 u_1^+ - \rho_2 u_1^- = g_{22} \qquad \text{on} \qquad \Gamma_2 \tag{1.11}$$

etc. with $\rho_j > 0$. See also [20].

Generalized (rectangular, exterior) wedge problems: These are characterized by transmission conditions on Γ_3 and Γ_4 and pure boundary conditions on Γ_1 and on Γ_2, i.e., particularly *boundary value problems* (BVPs) for the exterior wedge $Q_{234} = \text{int clos}\,(Q_2 \cup Q_3 \cup Q_4)$.

Mixed two-media half-space BVPs: Here we have transmission conditions on Γ_2 and boundary conditions on Γ_1 and on Γ_3 looking for $u \in \mathcal{H}^1(Q_1 \cup Q_2)$ only.

It turned out [15] that another class is very important and useful for the study of the Meister problems, namely:

Interior (rectangular) wedge problems: They are given by (1.9) looking for $u \in \mathcal{H}^1(Q_1)$ and, in a sense, complementary to the exterior wedge problems [6].

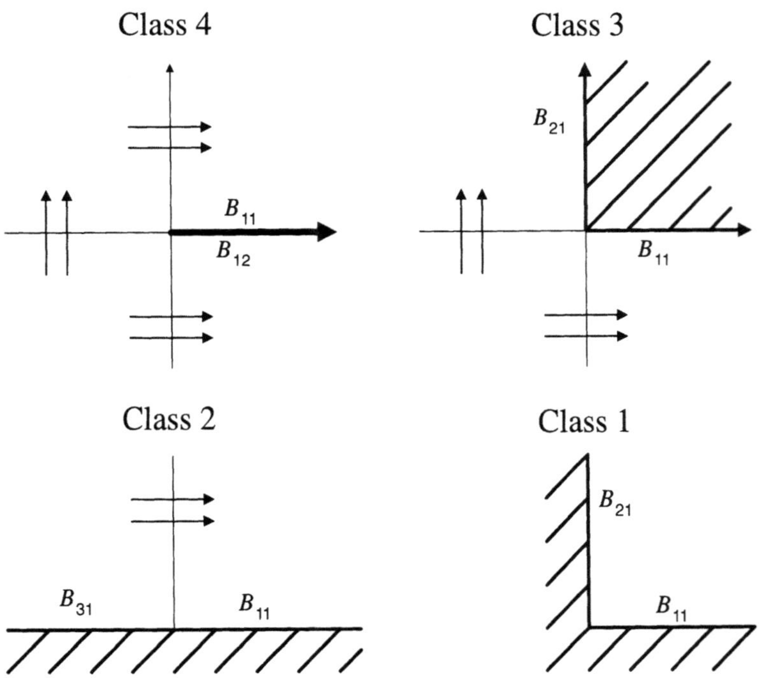

FIGURE 2. Configuration for the three classes of canonical problems posted by E. Meister and for the basic class 1 of interior wedge problems.

We shall number all these classes of problems according to the number of quadrants involved and the type of the two boundary conditions as, for example, $1DD$-the Dirichlet problem for the first quadrant etc., as shown in Table 1. Historically, it might be more reasonable to number them the other way around starting with the Sommerfeld problem as done by E. Meister.

However, the interior wedge problems are of basic interest because they serve to solve the others. Each problem of class 1, which is explicitly solved, gives a new ansatz (or representation formula) to reduce other problems. Thus particular attention is devoted to these problems here and we plan to study further Meister problems that are of particular physical interest in forthcoming publications (it turned out that too many questions have to be answered for class 1 problems).

The following table presents an overview of solved problems in the sense of *explicit analytical solution* (underlined references) and, respectively, partial results like *Fredholm criteria, index formulas, solvability by fixed point arguments* (not underlined items). In general we refer to one-medium problems in weak formulation. Various classical solutions and other previous results are referred to in the cited literature of Table 1. Others do not fall directly into the scheme, e.g., transmission between the first quadrant and its complement [16]. It is evident that studies are much simpler whenever the PDEs hold throughout half-planes (wedges play no role) as possible in class 2, class 4 and partly in class 3.

Class	DD	NN	DN	$\mathcal{II}$	OO	$G1$	HL
1	[15] 1	[15] 1	[15] 0	[23] 1			
2	[14] 1,2	[14] 1,2	[14] 0,3	[3] 6	[3] 6	[3] 6	
3	[15] 1	[15] 1	[15] 0				
4	[26] 1,3	[26] 1,3	[26] 0	[18] 4	[21, 9]	[20, 27, 28] 5	[17]

Legend

- DD *Dirichlet conditions on both parts of the boundary*
- NN *Neumann conditions*
- DN *mixed type (Dirichlet/Neumann)*
- $\mathcal{II}$ *impedance conditions*
- OO *oblique derivatives*
- $G1$ *general third kind (1st order)*
- HL *higher order Leontovich type*

Footnotes

0 *a compatibility condition is not needed*
1 *compatibility conditions are necessary*
2 *also solved explicitly for two media*
3 *partial results for two media*
4 *partial results for two different impedance numbers*
5 *no oblique derivative problems included*
6 *strong solution of so-called piecewise elliptic BVPs for PDEs of higher order*

TABLE 1. Solved and unsolved problems of class 1–4.

This article presents

- an operator theoretical approach which relates the diffraction problems by operator matrix identities to classes of convolution type operators with symmetry (CTOS), replacing the classical ideas to work with potentials and BΨDEs,
- the identification of further problems (mainly of class 1) which are well posed, Fredholm, or normally solvable,
- the image normalization of corresponding operators with non-closed images, discovery and description of the adequate function(al) spaces by compatibility conditions, and
- the explicit analytical solution of various problems by new factorizations methods for CTOS.

2. Basic representation formulas and operator equivalence with BΨDOs

Let us first introduce some notation to formulate the basic problems of class 1 that we like to reduce to systems of BΨDEs by the help of representation formulas derived in [15] where more details are described. We make use of the Fourier transformation in $\mathcal{S}, L^2, \mathcal{S}'$ denoted by

$$\begin{aligned} \mathcal{F}u(\vec{\xi}) &= \widehat{u}(\vec{\xi}) = \textstyle\int_{\mathbb{R}^n} u(\vec{x}) \exp[i\vec{x}\cdot\vec{\xi}]\, dx_1 \cdots dx_n, && \vec{\xi} \in \mathbb{R}^n \\ \mathcal{F}_{x_1 \mapsto \xi} u(x_1, x_2) &= \textstyle\int_{\mathbb{R}} u(x_1, x_2) \exp[i\xi x_1]\, dx_1, && \xi, x_2 \in \mathbb{R} \end{aligned} \tag{2.1}$$

and the usual Sobolev spaces $H^s = H^s(\mathbb{R}^n) \subset \mathcal{S}'(\mathbb{R}^n)$ normed by

$$\|u\|_{H^s}^2 = \int_{\mathbb{R}^n} |\widehat{u}(\vec{\xi})|^2 (1 + |\vec{\xi}|)^2\, d\xi_1 \cdots d\xi_n < \infty. \tag{2.2}$$

For open sets $\Omega \subset \mathbb{R}^n$, let $u \in H^s(\Omega)$ if u is the restriction to Ω of a distribution in $H^s(\mathbb{R}^n)$, and let

$$\|u\|_{H^s(\Omega)} = \inf_{\ell} \|\ell u\|_{H^s(\mathbb{R}^n)} \tag{2.3}$$

taking the infimum with respect to all extensions in H^s; we write $u \in \widetilde{H}^s(\Omega)$ if the extension by zero, $\ell_0 u \in \mathcal{S}'(\mathbb{R}^n)$, belongs to H^s. As pointed out earlier, cf. [15] e.g., for the half-axes $\Omega = \mathbb{R}_\pm = \{x \in \mathbb{R} : \pm x > 0\}$, this space can be identified with the subspace $H^s_{\overline{\Omega}}$ of H^s distributions supported on $\overline{\Omega}$ if and only if $s \geq -1/2$, and it coincides with $H^s(\Omega)$ exactly for $s \in]-1/2, 1/2[$, i.e., in brief

$$\begin{aligned} \widetilde{H}^s(\Omega) &= r_+ H^s_{\overline{\Omega}}, \quad H^s_{\overline{\Omega}} = \ell_0 \widetilde{H}^s(\Omega) && \text{for} \quad s \geq -1/2 \\ \widetilde{H}^s(\Omega) &= H^s(\Omega), \quad H^s_{\overline{\Omega}} = \ell_0 H^s(\Omega) && \text{for} \quad |s| < 1/2 \end{aligned} \tag{2.4}$$

and the same holds for (special) Lipschitz domains $\Omega \subset \mathbb{R}^n$ [29].

The spaces of particular interest in (1.6) admit proper, dense embeddings

$$\widetilde{H}^{1/2}(\mathbb{R}_\pm) \subset H^{1/2}(\mathbb{R}_\pm), \quad \widetilde{H}^{-1/2}(\mathbb{R}_\pm) \subset H^{-1/2}(\mathbb{R}_\pm) \tag{2.5}$$

and the even and odd extension operators from $\mathbb{R}_+$ (e.g.)

$$\begin{aligned} \ell^e &: H^{1/2}(\mathbb{R}_+) \to H^{1/2}(\mathbb{R}) \\ \ell^o &: H^{-1/2}(\mathbb{R}_+) \to H^{-1/2}(\mathbb{R}) \end{aligned} \tag{2.6}$$

are bounded. Conversely, an element $u \in H^{1/2}(\mathbb{R}_+)$ admits an odd extension if and only if $u \in \widetilde{H}^{1/2}(\mathbb{R}_+)$, i.e., it is extendable by zero, and $u \in H^{-1/2}(\mathbb{R}_+)$ has an even extension if and only if $u \in \widetilde{H}^{-1/2}(\mathbb{R}_+)$ [19].

As pointed out in Figure 1, Q_j denote the four open quadrants in $\mathbb{R}^2$ and Γ_j the closed half-axes for $j = 1, 2, 3, 4$. For convenience we write further $\partial Q_j = \Gamma_{jk} (= \Gamma_{kj})$ where $k = j + 1$ for $j = 1, 2, 3$ and $jk = 41$ for $j = 4$; the open half-planes are denoted by $Q_{jk} = \operatorname{int} \operatorname{clos}(Q_j \cup Q_k)$ with analogous numeration, the full axes by Γ_{jk} where $j + k$ is even and *exterior wedges* by $Q_{124} = \operatorname{int}(\mathbb{R}^2 \setminus Q_3)$ etc.

Weak solutions of the HEs (1.2) are briefly denoted by

$$u \in \mathcal{H}^1(\Omega) \tag{2.7}$$

which means that u is a weak solution, i.e., $u \in H^1(\Omega)$, of the Helmholtz equation in Ω provided Ω is a domain (open and connected) such that $\operatorname{int} \operatorname{clos} \Omega = \Omega$ or it is a weak solution in each component of a finite union of such domains.

We shall frequently use the trace operators T_{ι,Γ_j} and $T_{\iota,\Gamma_{jk}}$ introduced in (1.5)–(1.6), convolution operators on the lines Γ_{13} and Γ_{24} (identified with $\mathbb{R}$) given by

$$A_\phi \varphi(x) = \mathcal{F}^{-1}_{\xi \mapsto x} \phi(\xi) \cdot \widehat{\varphi}(\xi), \qquad x \in \mathbb{R} \tag{2.8}$$

in Sobolev spaces for certain $\phi \in L^\infty_{loc}(\mathbb{R})$ and surface (or layer) potential operators which appear in the following representation formulas.

Proposition 2.1. *There is a toplinear isomorphism*

$$\mathcal{K}_{DN,Q_1} : X_1 = H^{1/2}(\Gamma_1) \times H^{-1/2}(\Gamma_2) \to \mathcal{H}^1(Q_1) \tag{2.9}$$

given by the following formulas: If $(f, g) \in X_1$ *then*

$$\begin{aligned} u &= \mathcal{K}_{DN,Q_1}(f,g)^* = \mathcal{K}_{D,Q_{12}} \ell^e f + \mathcal{K}_{N,Q_{14}} \ell^o g \qquad (2.10) \\ &= \mathcal{F}^{-1}_{\xi \mapsto x_1} \left\{ \exp[-t(\xi) x_2] \widehat{\ell^e f}(\xi) \right\} - \mathcal{F}^{-1}_{\xi \mapsto x_2} \left\{ \exp[-t(\xi) x_1] \cdot t^{-1}(\xi) \, \widehat{\ell^o g}(\xi) \right\} \end{aligned}$$

where $t(\xi) = \left(\xi^2 - k_1^2\right)^{1/2}$, $\xi \in \mathbb{R}$ *(with the usual vertical branch cut from* k_1 *to* $-k_1$ *over infinity). Moreover we have*

$$\begin{aligned} (T_{0,\Gamma_1}, T_{1,\Gamma_2})^* \, \mathcal{K}_{DN,Q_1} &= I_{X_1} \\ \mathcal{K}_{DN,Q_1} \, (T_{0,\Gamma_1}, T_{1,\Gamma_2}) &= I_{\mathcal{H}^1(Q_1)} \end{aligned} \tag{2.11}$$

where $*$ *denotes transposition writing vectors in column form.*

Proof. This is another interpretation of the fact that the weak mixed (Dirichlet/Neumann) boundary value problem for the HE in the first quadrant is well posed and explicitly solved by (2.10), see [15, Section 4]. □

For any other half-planes Q_{jk} let $\mathcal{K}_{D,Q_{jk}}$ and $\mathcal{K}_{N,Q_{jk}}$ be defined by analogy (reflection), further $\mathcal{K}_{ND,Q_1}$ by symmetry (exchanging the roles of x_1 and x_2) and $\mathcal{K}_{DN,Q_j}$, $\mathcal{K}_{ND,Q_j}$ as well by analogy.

Thus we obtain a collection of representation formulas for any $u \in \mathcal{H}^1(\Omega)$ which can be enlarged by various other formulas according to explicit solutions mentioned in Table 1. However, taking the *complementary Cauchy data* in (2.10) we arrive at the most important *around the corner operators* (ACOs) in the following:

Proposition 2.2. *The composed operators*

$$\begin{aligned} C_0\ell^e &= T_{0,\Gamma_2}\mathcal{K}_{D,Q_{12}}\ell^e : H^{1/2}(\Gamma_1) \to H^{1/2}(\Gamma_2) \\ C_1\ell^o &= -T_{1,\Gamma_2}\mathcal{K}_{N,Q_{12}}\ell^o : H^{-1/2}(\Gamma_1) \to H^{-1/2}(\Gamma_2) \end{aligned} \tag{2.12}$$

are bounded linear operators where the ACOs are given by

$$\begin{aligned} C_0 f(x) &= (2\pi)^{-1} \textstyle\int_{\mathbb{R}} \exp[-t(\xi)x]\widehat{f}(\xi)\, d\xi, \quad x > 0 \\ C_1 g(x) &= i(2\pi)^{-1} \textstyle\int_{\mathbb{R}} \exp[-t(\xi)x]\frac{\xi}{t(\xi)}\, \widehat{g}(\xi)\, d\xi, \quad x > 0. \end{aligned} \tag{2.13}$$

Moreover, $C_j : H^s \to H^s(\mathbb{R}_+)$ are continuous transformations for any $s \in \mathbb{R}$.

Proof. (2.12) is a consequence of Proposition 2.1 and the symmetry of the HE with respect to the two variables x_1 and x_2, cf. [15, Section 6]. The exponential decrease of the kernels in (2.4) gives us the last conclusion. □

Now we are in the position to derive a system of ΨDEs for the BVP in weak formulation governed by the HE and the general boundary conditions of first kind (1.9). We formulate this equivalence in its strongest form as follows.

Theorem 2.3. *Let $\alpha, \beta, \gamma, \alpha', \beta', \gamma' \in \mathbb{C}$. Then the operator*

$$L_1 : \mathcal{H}^1(Q_1) \to H^{-1/2}(\Gamma_1) \times H^{-1/2}(\Gamma_2) = Y_1 \tag{2.14}$$

that maps the weak solutions of the HE in Q_1 into the data (1.9), is toplinear equivalent *to the operator*

$$\begin{aligned} T &= \begin{pmatrix} r_+(\alpha I - \beta A_t + \gamma A_\vartheta)\,\ell^e & \beta C_1 \ell^o \\ C_0(\alpha' I - \gamma' A_t)\,\ell^e & r_+(-\alpha' A_{t^{-1}} + \beta' I + \gamma' A_\theta)\,\ell^o \end{pmatrix} \\ &: X_1 = H^{1/2}(\Gamma_1) \times H^{-1/2}(\Gamma_2) \to Y_1 \end{aligned} \tag{2.15}$$

where $t(\xi) = (\xi^2 - k^2)^{1/2}$, $\vartheta(\xi) = -i\xi$, $\theta(\xi) = -\vartheta(\xi)/t(\xi) = i\xi(\xi^2 - k^2)^{-1/2}$, $\xi \in \mathbb{R}$. I.e., the two operators coincide up to homeomorphisms being simply related by:

$$\begin{aligned} L_1 &= T\,(T_{0,\Gamma_1}, T_{1,\Gamma_2})^* \\ T &= L_1\,\mathcal{K}_{DN,Q_1}. \end{aligned} \tag{2.16}$$

Proof. Classically speaking: If we put the representation (2.10) of an arbitrary $\mathcal{H}^1(Q_1)$ element in terms of the layer potentials f, g on Γ_1 and Γ_2, respectively, into the BCs (1.9), we obtain directly

$$T(f,g)^* = (g_1, g_2)^* \tag{2.17}$$

because the $D/N/T$ data in terms of (f,g) read on Γ_1:

$$\begin{aligned}
u_0^+(x_1) &= f(x_1) - (2\pi)^{-1}\textstyle\int_{\mathbb{R}} \exp[-i\xi x_2 - t(\xi)x_1]t^{-1}(\xi)\widehat{\ell^o g}(\xi)\,d\xi_{|x_2=0} \\
&= f(x_1) - C_0 A_{t^{-1}}\ell^o g(x_1) = f(x_1) \\
u_1^+(x_1) &= (2\pi)^{-1}\textstyle\int_{\mathbb{R}} \exp[-i\xi x_1 - t(\xi)x_2]\,(-t(\xi))\,\widehat{\ell^e f}(\xi)\,d\xi \\
&\quad +(2\pi)^{-1}\textstyle\int_{\mathbb{R}} i\xi \exp[-i\xi x_2 - t(\xi)x_1]t^{-1}(\xi)\widehat{\ell^o g}(\xi)\,d\xi_{|x_2=0} \\
&= -r_+ A_t \ell^e f(x_1) + C_1 \ell^o g(x_1) \\
u_\tau^+(x_1) &= (2\pi)^{-1}\textstyle\int_{\mathbb{R}} (-i\xi)\exp[-i\xi x_1 - t(\xi)x_2]\widehat{\ell^e f}(\xi)\,d\xi \\
&\quad +(2\pi)^{-1}\textstyle\int_{\mathbb{R}} \exp[-i\xi x_2 - t(\xi)x_1]\widehat{\ell^o g}(\xi)\,d\xi_{|x_2=0} \\
&= f'(x_1) + C_0\ell^o g(x_1) = f'(x_1)
\end{aligned} \tag{2.18}$$

where we used that $C_0\varphi = 0$ for odd functions $\varphi \in H^{\pm 1/2}$ and that $f' = r_+ A_\vartheta \ell^e f \in H^{-1/2}(\mathbb{R}_+)$. Further we have on Γ_2:

$$\begin{aligned}
u_0^-(x_2) &= (2\pi)^{-1}\textstyle\int_{\mathbb{R}} \exp[-i\xi x_1 - t(\xi)x_2]\widehat{\ell^e f}(\xi)\,d\xi \\
&\quad -(2\pi)^{-1}\textstyle\int_{\mathbb{R}} \exp[-i\xi x_2 - t(\xi)x_1]t^{-1}(\xi)\widehat{\ell^0 g}(\xi)\,d\xi_{|x_1=0} \\
&= C_0\ell^e f(x_2) - r_+ A_{t^{-1}}\ell^o g(x_2) \\
u_1^-(x_2) &= (2\pi)^{-1}\textstyle\int_{\mathbb{R}} (-i\xi)\exp[-i\xi x_1 - t(\xi)x_2]\widehat{\ell^e f}(\xi)\,d\xi \\
&\quad +(2\pi)^{-1}\textstyle\int_{\mathbb{R}} \exp[-i\xi x_2 - t(\xi)x_1]\widehat{\ell^o g}(\xi)\,d\xi_{|x_1=0} \\
&= C_0 A_\vartheta \ell^e f(x_2) + g(x_2) = g(x_2) \\
u_\tau^-(x_2) &= (2\pi)^{-1}\textstyle\int_{\mathbb{R}} \exp[-i\xi x_1 - t(\xi)x_2]\,(-t(\xi))\,\widehat{\ell^e f}(\xi)\,d\xi \\
&\quad +(2\pi)^{-1}\textstyle\int_{\mathbb{R}} i\xi \exp[-i\xi x_2 - t(\xi)x_1]t^{-1}(\xi)\widehat{\ell^o g}(\xi)\,d\xi_{|x_1=0} \\
&= -C_0 A_t \ell^e f(x_2) + r_+ A_\theta \ell^o g(x_2).
\end{aligned} \tag{2.19}$$

Note that $A_\vartheta \ell^e f$ is odd and $C_0 : H^{-1/2} \to H^{-1/2}(\mathbb{R}_+)$ is here continuously extended from $H^{1/2}$. Together with Proposition 2.1 we obtain the formulas (2.15)–(2.16). □

Remark 2.4. In other words: L_1 can be factorized into a bounded pseudodifferential operator T and the mixed trace operator $(T_{0,\Gamma_1}, T_{1,\Gamma_2})^*$ which is (boundedly) invertible by $\mathcal{K}_{DN,Q_1}$. The variety of all representations of $H^1(Q_1)$ solutions of the HE in terms of such layer potentials corresponds with such operator factorizations and with the class of all well-posed BVPs for the HE in Q_1 in weak formulation.

Corollary 2.5. *Let $\alpha, \beta, \gamma, \alpha', \beta', \gamma' \in \mathbb{C}$ and $(g_1, g_2) \in H^{-1/2}(\mathbb{R}_+)^2$ be given and consider the BVP 1G1 to look for the solutions $u \in \mathcal{H}^1(Q_1)$ that satisfy* (1.9)*. This BVP is well posed if and only if the operator T given by* (2.15) *is invertible. In this case, the unique solution is given by*

$$u = L_1^{-1}(g_1, g_2)^* = \mathcal{K}_{DN,Q_1} T^{-1}(g_1, g_2)^*. \tag{2.20}$$

Analogously we have results for the Fredholm property, representation of generalized inverses, etc., which are not formulated explicitly here.

3. Interior wedge problems (part 1)

Now we study systematically the BVPs for the first quadrant, particularly the questions: Is the operator T in (2.15) bounded invertible (thus problem $1\mathcal{XY}$ is well posed)? Can we construct an explicit analytical representation of T^{-1} (and consequently of the solution of an interior wedge problem)? In the boundary condition on Γ_1

$$\alpha u_0^+ + \beta u_1^+ + \gamma u_\tau^+ = g_1 \tag{3.1}$$

we distinguish between the following cases:

$$\begin{array}{lllll}
\mathcal{D} & : & \alpha = 1, & \beta = \gamma = 0 & \text{(Dirichlet)} \\
\mathcal{N} & : & \beta = 1, & \alpha = \gamma = 0 & \text{(Neumann)} \\
\mathcal{I} & : & \alpha = ip, & \beta = 1, \gamma = 0 & \text{(impedance)} \\
\mathcal{T} & : & \beta = 0 & & \text{(tangential)} \\
\mathcal{O} & : & \alpha = 0 & & \text{(oblique derivative)} \\
\mathcal{G} & : & \text{———} & & \text{(general)}
\end{array}$$

and so we do in the condition on Γ_2 with coefficients α', β', γ'. Consequently there are $6 \times 6 = 36$ cases to be classified, which reduce to 21 cases by symmetry exchanging Γ_1 and Γ_2. The first question is if a problem reduces directly to a scalar equation (the system of BΨDEs "decomposes"), i.e., if T is triangular or not, respectively. In many cases we obtain a positive answer just looking at the operator matrix (2.15), see the result in Table 2.

$\Gamma_1 \setminus \Gamma_2$	$\mathcal{D}$	$\mathcal{N}$	$\mathcal{I}$	$\mathcal{T}$	$\mathcal{O}$	$\mathcal{G}$
$\mathcal{D}$	▲	▲	▲	▲	▲	▲
$\mathcal{N}$	*	▲	*	*	*	*
$\mathcal{I}$	*	▲	nt	*	nt	nt
$\mathcal{T}$	▲	▲	▲	▲	▲	▲
$\mathcal{O}$	*	▲	nt	*	nt	nt
$\mathcal{G}$	*	▲	nt	*	nt	nt

TABLE 2. Cases of problems $1\mathcal{XY}$ where T is triangular (▲), eventually after exchanging $\mathcal{X}$ and $\mathcal{Y}$ (*) or where T is not triangular (nt), respectively.

Note that only the solution of the problems $1DD$, $1NN$, $1DN$ (see Proposition 2.1) and $1ND$ (by symmetry) are known from [15]. We shall treat various other cases and see that the following properties of T are most important:

- T is triangular (or not, respectively),
- T is *normal type*, i.e., the operators

$$\begin{aligned} A_{\sigma_1} &= \mathcal{F}^{-1}\sigma_1 \cdot \mathcal{F} = \alpha I - \beta A_t + \gamma A_\vartheta &&: H^{1/2} \to H^{s_1} \\ A_{\sigma_2} &= \mathcal{F}^{-1}\sigma_2 \cdot \mathcal{F} = -\alpha' A_{t^{-1}} + \beta' I + \gamma' A_\theta &&: H^{-1/2} \to H^{s_2} \end{aligned} \tag{3.2}$$

  are bijective ($s_j = 1/2$ for the D case, $s_j = -1/2$ else),
- the *pre-symbols* σ_j are even functions or *minus type* i.e., $\sigma_j^{\pm 1}$ have holomorphic extensions from $\mathbb{R}$ into the lower half-plane,
- the images of the main diagonal elements T_1 and T_2 of T are closed,
- $\operatorname{ind} T_j = \dim\ker T_j - \operatorname{codim}\operatorname{im} T_j = 0$.

In the subsequent studies we shall include moreover:

- σ_j admits a certain factorization, see Section 4,
- C_0 and C_1 satisfy certain composition formulas, see [15] and forthcoming research.

Proposition 3.1.

(a) *T is normal type if and only if*

$$\begin{aligned} \sigma_1(\xi) &= \alpha - \beta t(\xi) + \gamma\vartheta(\xi) \neq 0\,, && \xi \in \mathbb{R} \\ \sigma_2(\xi) &= -\alpha' t^{-1}(\xi) + \beta' + \gamma'\theta(\xi) \neq 0\,, && \xi \in \mathbb{R} \\ \sigma_1^{-1}(\xi) &= \mathcal{O}\left(|\xi|^{s_1 - 1/2}\right), && as \quad |\xi| \to \infty \\ \sigma_2^{-1}(\xi) &= \mathcal{O}\left(|\xi|^{s_2 + 1/2}\right), && as \quad |\xi| \to \infty. \end{aligned} \tag{3.3}$$

(b) *σ_j is an even function if and only if the corresponding boundary condition has no tangential derivative term, i.e., is of D, N or $\mathcal{I}$ type.*

(c) *σ_1 is minus type if and only if* (i) *the corresponding boundary condition is tangential ($\beta = 0$) and* (ii) *$\gamma = 0$ or $\Re e(\alpha/\beta) < 0$.*

Proof. It is obvious. □

Theorem 3.2. *Let T be triangular and normal type, let σ_1 be even or minus type and σ_2 be even. Then exactly one of the following statements is fulfilled:*

I. *T is invertible in the given space setting.*

II. *T is injective, $\operatorname{im} T$ is not closed but dense.*

In both cases, the inverse is obtained in closed analytic form by elementary inversion of the main diagonal and usual modification to a triangular operator matrix.

Proof. The various cases are treated (up to analogy) in the following sequence of three reference problems.

Example 3.3. Problem $1D\mathcal{I}$ leads to

$$T = \begin{pmatrix} r_+\ell^e & 0 \\ -C_0 A_t \ell^e & r_+(-ipA_{t^{-1}} + I)\,\ell^o \end{pmatrix} \tag{3.4}$$

where $\sigma_2 = 1 - ipt^{-1}$ is even. Obviously

$$T_2^{-1}T_2 = r_+ A_{\sigma_2^{-1}} \ell^o r_+ A_{\sigma_2} \ell^o = I : H^{-1/2}(\mathbb{R}_+) \to H^{-1/2}(\mathbb{R}_+) \tag{3.5}$$

and $\ell^o r_+ A_{\sigma_2} \ell^o = A_{\sigma_2} \ell^o$ since $A_{\sigma_2} = \mathcal{F}^{-1}\sigma_2 \cdot \mathcal{F}$ transforms odd functionals into odd functionals. The same holds for $T_2 T_2^{-1}$.

Now the inverse of T results from its triangular form:

$$T^{-1} = \begin{pmatrix} I & 0 \\ -C_0 A_t \ell^e & T_2 \end{pmatrix}^{-1} = \begin{pmatrix} I & 0 \\ T_2^{-1} C_0 A_t \ell^e & T_2^{-1} \end{pmatrix} \tag{3.6}$$

as a bounded linear operator in $H^{1/2}(\mathbb{R}_+) \times H^{-1/2}(\mathbb{R}_+)$.

Example 3.4. Problem $1DD$ explains most clearly what happens in case II, based on the analytical solution in [15]. Due to the fact that the trace of a solution $u \in \mathcal{H}^1(Q_1)$ belongs to $H^{1/2}(\Gamma_{12})$, i.e., the two functions $\varphi_1 = T_{0,\Gamma_1}u$, $\varphi_2 = T_{0,\Gamma_2}u$ are extensions of each other within the space $H^{1/2}$, we may think of the data space

$$Y_1 = \left\{ (\varphi_1, \varphi_2) \in H^{1/2}(\mathbb{R}_+)^2 : \varphi_1 - \varphi_2 \in \widetilde{H}^{1/2}(\mathbb{R}_+) \right\} \tag{3.7}$$

identifying Γ_j with $\mathbb{R}_+$.

Formally we have from (2.15)

$$\begin{aligned} T &= \begin{pmatrix} I & 0 \\ C_0 \ell^e & -r_+ A_{t^{-1}} \ell^o \end{pmatrix} \\ T^{-1} &= \begin{pmatrix} I & 0 \\ r_+ A_t \ell^o C_0 \ell^e & -r_+ A_t \ell^o \end{pmatrix}. \end{aligned} \tag{3.8}$$

It is obvious that T is bounded as an operator

$$T : H^{1/2}(\mathbb{R}_+) \times H^{-1/2}(\mathbb{R}_+) \to H^{1/2}(\mathbb{R}_+) \times H^{1/2}(\mathbb{R}_+) \tag{3.9}$$

but the two operators in the last line of (3.8) are not bounded on $H^{1/2}(\mathbb{R}_+)$ if considered separately since ℓ^o is not defined from $H^{1/2}(\mathbb{R}_+)$ into $H^{1/2}(\mathbb{R}_+)$, cf. (2.6). This difficulty disappears by incorporating the *compatibility condition* in the image space of T and considering

$$r_+ A_{t^{-1}} \ell^o : H^{-1/2}(\mathbb{R}_+) \to \widetilde{H}^{1/2}(\mathbb{R}_+) \tag{3.10}$$

since $A_{t^{-1}}$ maps $H^{-1/2,o}$ onto $H^{1/2,o}$ and $r_+ : H^{1/2,o} \to \widetilde{H}^{1/2}(\mathbb{R}_+)$ is bounded invertible by ℓ^o. Thus T in (3.9) is normalized if we replace the last component space by the dense subspace $\widetilde{H}^{1/2}(\mathbb{R}_+)$ and it becomes bijective.

Note that $\widetilde{H}^{1/2}(\mathbb{R}_+)$ is invariant for $C_0\ell^e$ [15] and the operator T is (image) normalized [22] if the adapted data space is re-normed by

$$\|(\varphi_1,\varphi_2)\|_{Y_1} = \|\varphi_1\|_{H^{1/2}(\mathbb{R}_+)} + \|\varphi_1 - \varphi_2\|_{\widetilde{H}^{1/2}(\mathbb{R}_+)}. \tag{3.11}$$

Example 3.5. Problem $1\mathcal{TI}$ leads directly from (2.15) to

$$T = \begin{pmatrix} r_+ (\alpha I + \gamma A_\vartheta)\, \ell^e & 0 \\ \alpha' C_0 \ell^e & r_+ (-\alpha' A_{t^{-1}} + \beta' I)\, \ell^o \end{pmatrix}. \tag{3.12}$$

Now let the pre-symbols

$$\begin{aligned} \sigma_1(\xi) &= \alpha - i\gamma\xi &&= -i\gamma\left(\xi + \frac{\alpha}{\gamma} i\right) \\ \sigma_2(\xi) &= \beta' - \alpha' t^{-1} &&= \beta'\left(1 - ipt^{-1}\right) \end{aligned} \tag{3.13}$$

satisfy (3.3) and σ_1 be minus type, which is the case if $\Re e\,(\alpha/\gamma) < 0$ and, e.g., the physically relevant case $\Im m(p) > 0$, see [18] (for simplicity we may choose $-i\gamma = \beta' = 1$). Then $T_1^{-1} = r_+ A_{\sigma_1^{-1}} \ell$ since

$$T_1^{-1} T_1 = r_+ A_{\sigma_1^{-1}} \ell\, r_+ A_{\sigma_1} \ell^e = I \qquad \text{on} \qquad H^{1/2}(\mathbb{R}_+) \tag{3.14}$$

which does not depend on the extension $\ell\varphi$ of φ in $H^{-1/2}$, see [7, Lemma 4.4], and $T_1 T_1^{-1} = I$ on $H^{-1/2}(\mathbb{R}_+)$ by analogy. Also T_2 is invertible in the given space setting, see Example 3.3, and $\alpha' C_0 \ell^e : H^{1/2}(\mathbb{R}_+) \to H^{-1/2}(\mathbb{R}_+)$ is bounded. Thus T is invertible as in (3.6). No compatibility conditions appear although both boundary operators have order one.

To complete the proof of Theorem 3.2 we observe that all the other cases mentioned in Theorem 3.2 can be treated by analogy to the three given examples as follows:

- $1DD$ – see Example 3.4; belongs to case II.
- $1DN$ – see Theorem 2.3; case I.
- $1D\mathcal{I}$ – see Example 3.3; case I.
- $1D\mathcal{T}$ – σ_2 is not even, nor minus type, see $1\mathcal{T}D$; case I.
- $1ND$ – not triangular, but see $1DN$; case I.
- $1NN$ – analogous to $1DD$, see [15]; case II.
- $1N\mathcal{I}$ – not triangular, but see $1\mathcal{I}N$, like $1NN$; case II.
- $1N\mathcal{T}$ – not triangular, σ_2 not even, but see $1\mathcal{T}N$; case I.
- $1\mathcal{I}D$ – not triangular, but see $1D\mathcal{I}$, case I.
- $1\mathcal{I}N$ – like $1NN$; case II.
- $1\mathcal{I}\mathcal{I}$ – not triangular – still open.
- $1\mathcal{I}\mathcal{T}$ – not triangular, σ_2 not even, but see $\mathcal{T}\mathcal{I}$; case I.
- $1\mathcal{T}D$ – combination of Example 3.5 for T_1 and Example 3.4 for T_2; case II.
- $1\mathcal{T}N$ – like $1\mathcal{T}\mathcal{I}$, Example 3.5; case I.
- $1\mathcal{T}\mathcal{I}$ – see Example 3.5; case I.
- $1\mathcal{T}\mathcal{T}$ – triangular, but σ_2 not even, see Section 5.

This completes the proof of Theorem 3.2 (an overview is given later in Table 3). □

Remark 3.6. 1. Of all the decomposing (triangular) cases of normal type in the first 4×4 block of Table 2 we did not (yet) solve the problems $1\mathcal{T}\mathcal{X}$ when σ_1 is plus type which turns out to be quite subtle (see Section 5).
2. The compatibility condition in $1NN$, $1N\mathcal{I}$, $1\mathcal{I}N$ reads, compare (3.7) and (3.11), as found in [15]:

$$g_1 + g_2 \in \widetilde{H}^{-1/2}(\mathbb{R}_+). \tag{3.15}$$

If we want to invert further operators T_j where the pre-symbol σ_j is not even nor minus type, we need more knowledge about the factorization of σ_j into such functions, provided in the next section.

4. A direct approach to convolution type operators with symmetry

We have seen that various wedge problems can be reduced (in the sense of toplinear equivalence) to the study of *convolution type operators with symmetry*, which have the form

$$T = r_+ A \ell^c : H^r(\mathbb{R}_+) \to H^s(\mathbb{R}_+). \tag{4.1}$$

In the scalar case we had $|r| = |s| = 1/2$, ℓ^c was a special continuous extension operator: even or odd for $r = \pm 1/2$, respectively. $A = A_\phi = \mathcal{F}^{-1}\phi \cdot \mathcal{F} : H^r \to H^s$ is translation invariant (a distributional convolution) and supposed to be bounded invertible (normal type).

Let us outline briefly a most direct constructive factorization method for the (generalized) inversion of (4.1) which will be published in detail elsewhere [5]. Thus we present here only the main features without proofs. An alternative approach via Wiener-Hopf plus Hankel operators can be found in [2].

Lemma 4.1. *Let $H^{r,e}$, $H^{r,o}$ denote the subspaces of H^r elements which are even or odd, respectively. The operator of even or odd extension, respectively,*

$$\ell^c : H^r(\mathbb{R}_+) \to H^r \tag{4.2}$$

is left invertible (by r_+) with image $H^{r,e}$ or $H^{r,o}$ if and only if

$$r \in \begin{cases}]-1/2,\ 3/2[& \text{for} \quad \ell^c = \ell^e \\ ]-3/2,\ 1/2[& \text{for} \quad \ell^c = \ell^o . \end{cases} \tag{4.3}$$

Proposition 4.2. *In case of* (4.3), *the operator* (4.1) *can be* lifted into $L^2(\mathbb{R}_+)$, *i.e., there are toplinear isomorphisms E, F such that*

$$\begin{aligned} T &= E T_0 F \\ T_0 &= r_+ A_0 \ell^c \ : \ L^2(\mathbb{R}_+) \to L^2(\mathbb{R}_+) \\ A_0 &= \mathcal{F}^{-1}\phi_0 \cdot \mathcal{F} \end{aligned} \tag{4.4}$$

where $\phi_0 \in \mathcal{G}L^\infty(\mathbb{R})$ *provided* A *is bijective (normal type). More precisely*

$$\begin{aligned} A_0 &= A_{\lambda_-^s} A A_{\lambda^{-r}} \\ \phi_0 &= \lambda_-^s \phi \lambda^{-r} \\ E &= r_+ A_{\lambda_-^{-s}} \ell, \qquad F = r_+ A_{\lambda^r} \ell^c \end{aligned} \tag{4.5}$$

where $\lambda(\xi) = (\xi^2+1)^{1/2}$, $\lambda_-(\xi) = \xi - i$ *for* $\xi \in \mathbb{R}$.

Remark 4.3. So we reduced, by *toplinear equivalence*, the operators (4.1) to the case $r = s = 0$.

It may be useful to replace $\lambda(\xi)$ by $t(\xi) = (\xi^2 - k^2)^{1/2}$ or $t_{k_j}(\xi) = (\xi^2 - k_j^2)^{1/2}$ and $\lambda_-(\xi)$ by $t_-(\xi) = \xi - k$, in order to obtain simpler pre-symbols ϕ_0 in certain applications provided $\Im m(k) > 0$, $\Im m(k_j) > 0$. We shall see that further symbols are helpful (rational with the same zero/pole characteristics and their square roots) due to certain BTCs.

Definition 4.4. Let $A : H^r \to H^s$ be bounded invertible and one of the cases of (4.3) be satisfied. Then A admits an *asymmetric factorization through an intermediate space* $Z = H^{\tilde{r}}$ (AFIS), if

$$A = A_- C A_e : H^r \to H^{\tilde{r}} \to H^{\tilde{r}} \to H^s \tag{4.6}$$

holds as a composition of bounded invertible linear operators and

$$\begin{aligned} & r_+ A_-^{\pm 1} \ell r_+ = r_+ A_-^{\pm 1} \\ & \ell^c r_+ A_e^{\pm 1} \ell^c = A_e^{\pm 1} \ell^c \\ & C = A_{\zeta^\kappa} = \mathcal{F}^{-1} \zeta^\kappa \cdot \mathcal{F} \\ & \zeta^\kappa(\xi) = (\lambda_-(\xi)/\lambda_+(\xi))^\kappa = \left(\frac{\xi - i}{\xi + i}\right)^\kappa, \quad \xi \in \mathbb{R} \end{aligned} \tag{4.7}$$

where ℓ denotes any extension in the corresponding space to the full line and ℓ^c stands for the bounded extension operator ℓ^e or ℓ^o, respectively, i.e., $\tilde{r}$ satisfies the same condition as r does, see (4.3), and $\kappa \in \mathbb{Z}$.

As before, a useful modification is to choose, instead of ζ^κ, any other rational function with the same number of zeros and poles in the upper/lower complex half-plane and no zeros and poles on $\dot{\mathbb{R}}$ (cf. Proposition 5.3).

Theorem 4.5. *Let* $A = A_- C A_e$ *be an AFIS. Then the operator* T *in* (4.1) *is Fredholm and one-sided invertible. A left or right inverse is given by*

$$T^- = r_+ A_e^{-1} \ell^c r_+ C^{-1} \ell^c r_+ A_-^{-1} \ell \tag{4.8}$$

and

$$\begin{aligned} \alpha(T) &= \dim\ker T &&= \max\{0, -\kappa\} \\ \beta(T) &= \operatorname{codim} \operatorname{im} T &&= \max\{0, \kappa\} \end{aligned} \tag{4.9}$$

i.e., $\operatorname{ind} T = -\kappa$.

Theorem 4.6. *Let $A = \mathcal{F}^{-1}\phi \cdot \mathcal{F} : L^2 \to L^2$ where $\phi \in \mathcal{G}C^\nu(\ddot{\mathbb{R}})$ and $\nu \in]0,1[$, further let*

$$\begin{aligned} \omega &= (2\pi i)^{-1} \int_{\mathbb{R}} d \log \phi \\ \Re e(\omega) &\pm 1/4 \notin \mathbb{Z} \end{aligned} \tag{4.10}$$

where the sign depends on the case $\ell^c = \ell^e$ or ℓ^o, respectively. Then A admits an AFIS given by the following formulas:

$$\begin{aligned} \kappa &= \mathcal{I}nt\,(\Re e(\omega) \pm 1/4) = \max\{z \in \mathbb{Z} : z \leq \Re e(\omega) \pm 1/4\} \\ \psi &= \zeta^{-\omega}\,\phi^{-1}(+\infty)\,\phi \\ \phi_- &= \lambda_-^{2(\omega-\kappa)} \exp\left\{P_- \log\left(\psi\widetilde{\psi}^{-1}\right)\right\} \\ \phi_e &= \zeta^{-\kappa}\phi_-^{-1}\phi \\ A_- &= \mathcal{F}\phi_- \cdot \mathcal{F} \\ A_e &= \mathcal{F}^{-1}\phi_e \cdot \mathcal{F} \end{aligned} \tag{4.11}$$

where $P_- = (I - S_{\mathbb{R}})/2$ (with $S_{\mathbb{R}}$ denoting the Hilbert transformation) and $\widetilde{\psi} = J\psi$ for the reflection operator J.

Corollary 4.7. *Let T be given by (4.1), r satisfying (4.3), $s \in \mathbb{R}$, $A = \mathcal{F}^{-1}\phi \cdot \mathcal{F}$, $\phi_0 = \lambda_-^s \phi \lambda^{-r} \in \mathcal{G}C^\nu(\ddot{\mathbb{R}})$ and ω be given by (4.10). Then the following conditions are equivalent:*

(j) *$\Re e(\omega) \pm 1/4 \notin \mathbb{Z}$ in case of ℓ^e or ℓ^o, respectively,*

(jj) *A admits an AFIS,*

(jjj) *T is Fredholm.*

In this case $Z = H^{2\eta}$ where $\eta = \Re e(\omega) - \kappa \in]-1/4,\ 3/4[$ for $\ell^c = \ell^e$ or $\eta \in]-3/4,\ 1/4[$ for $\ell^c = \ell^o$, respectively. The kernels of T and T^- can be computed easily, see [5].

Corollary 4.8. *If A and A_0 are related as in Proposition 4.2 and if ϕ_0 is factored as in Theorem 4.6 (there denoted by ϕ), then*

$$A = A_{\lambda_-^{-s}} A_0 A_{\lambda^r} = (A_{\lambda_-^{-s}} A_-)\, C (A_e A_{\lambda^r}) \tag{4.12}$$

represents an AFIS of A according to (4.5) and (4.11).

5. Interior wedge problems solved by factorization

Now we are able to complete the discussion of class 1 problems with triangular T in (2.15). Let us first study the scalar operators that appear in this context (lifted main diagonal elements of T). There are two types.

Proposition 5.1. *Let*

$$S_\mu = r_+ A_{\phi_\mu} \ell^o : L^2(\mathbb{R}_+) \to L^2(\mathbb{R}_+) \tag{5.1}$$

where

$$\begin{aligned}\phi_\mu &= \left(\alpha t^{-1} - \beta - \gamma\theta\right)\zeta^\mu \\ \phi_0(\xi) &= \frac{\alpha}{\left(\xi^2 - k^2\right)^{1/2}} - \beta - \gamma\frac{i\xi}{\left(\xi^2 - k^2\right)^{1/2}} \neq 0, \quad \textit{for} \quad \xi \in \ddot{\mathbb{R}}\end{aligned} \tag{5.2}$$

and $\alpha, \beta, \gamma \in \mathbb{C}$. *Then* S_μ *is Fredholm if and only if*

$$\exp\left[2\pi i\left(\mu - 1/4\right)\right]\frac{\beta + i\gamma}{\beta - i\gamma} \notin \mathbb{R}_+. \tag{5.3}$$

In this case, an AFIS of A_{ϕ_μ} *and a one-sided inverse* S_μ^- *of* S_μ *are given by* (4.8)–(4.11) *(substitute* $A = A_{\phi_\mu}$ *and* $T = S_\mu$*). Furthermore*

$$\begin{aligned}|\operatorname{ind} S_0| &= |\mathcal{I}nt\left(\Re e(\omega_0) + 3/4\right)| \leq 3 \\ \operatorname{ind} S_\mu &= -\mathcal{I}nt\left(\Re e(\omega_\mu) + 3/4\right) = -\mathcal{I}nt\left(\Re e(\omega_0) + \mu + 3/4\right)\end{aligned} \tag{5.4}$$

for any choice of admissible parameters. In particular S_μ *is invertible if and only if*

$$\Re e(\omega_\mu) = \Re e\left(\left(2\pi i\right)^{-1}\int_{\mathbb{R}} d\log\phi_\mu\right) \in \left]-3/4,\ 1/4\right[. \tag{5.5}$$

Proof. Firstly we consider the case $\mu = 0$. Obviously $\phi_0 \in \mathcal{G}C^\nu(\ddot{\mathbb{R}})$ because of (5.2), so that Theorem 4.5, Theorem 4.6 and Corollary 4.7 apply. Thus the Fredholm property of S_0 is equivalent to each of the following conditions:

$$\begin{aligned}&\Re e(\omega_0) - 1/4 \notin \mathbb{Z} \\ &\left(2\pi\right)^{-1}\arg\phi_0(+\infty) - \left(2\pi\right)^{-1}\arg\phi_0(-\infty) \neq 1/4 \qquad \operatorname{mod}\ \mathbb{Z} \\ &\phi_0(+\infty) \neq iy\,\phi_0(-\infty) \quad \text{for} \quad y \in \mathbb{R}_+ \\ &\frac{-\beta - i\gamma}{-\beta + i\gamma} \neq i\gamma \quad \text{for} \quad y \in \mathbb{R}_+\end{aligned} \tag{5.6}$$

that is (5.3) where $\phi_0(\pm\infty) = -\beta \mp \gamma i \neq 0$ follows from (5.2).

Therefore all statements result from Section 4 up to the index estimate, $|\operatorname{ind} S_0| \leq 3$, which is a consequence of the discussion of the graphs of t^{-1} and $\sigma = \beta + \xi t^{-1}$ (putting $i\gamma = 1$ for simplicity), combining considerations from [18] and [21].

Now consider the case $\mu \in \mathbb{R}$. (5.2) remains unchanged since $\zeta^\mu \in \mathcal{G}C^\nu(\ddot{\mathbb{R}})$. The Fredholm property is equivalent to $\Re e(\omega_\mu) - 1/4 \notin \mathbb{Z}$ which results in (5.3) due to the modification of (5.6). In this case, we have an AFIS, the invertibility criterion (5.5), and an index formula (5.4) by the fact that $\omega_\mu = \omega_0 + \mu$. □

Remark 5.2. Condition (5.2) can be checked easily by computing the roots of $\phi_0(\xi) = 0$, which is a quadratic equation in $\mathbb{C}$. We obtain immediately the essential spectrum of S_μ from (5.3) and the spectrum from (5.5), since $\kappa = 0$ corresponds with the invertibility of S_μ.

Proposition 5.3. *Let*

$$S = r_+ A_\phi \ell^e : H^{1/2}(\mathbb{R}_+) \to H^{-1/2}(\mathbb{R}_+) \tag{5.7}$$

where

$$\phi(\xi) = \alpha + \gamma\vartheta(\xi) = \alpha - i\gamma\xi \tag{5.8}$$

and $\alpha, \gamma \in \mathbb{C} \setminus \{0\}$, $\Re e(\alpha/\gamma) > 0$ *(i.e.,* ϕ *is plus type). Then* S *is Fredholm with* $\alpha(S) = 1$, $\beta(S) = 0$ *and right invertible by*

$$S^- = r_+ C^{-1} \ell^e r_+ A_-^{-1} \ell \tag{5.9}$$

where $S = A_- C$ *stands for a (modified) AFIS with Fourier symbols*

$$\phi = (\alpha - \gamma\vartheta) \frac{\alpha + \gamma\vartheta}{\alpha - \gamma\vartheta} \tag{5.10}$$

Proof. Obviously, $A_- = \mathcal{F}^{-1}(\alpha - \gamma\vartheta) \cdot \mathcal{F}$ and $C = \mathcal{F}^{-1} \frac{\alpha + \gamma\vartheta}{\alpha - \gamma\vartheta} \cdot \mathcal{F}$ are minus type and C^{-1} is plus type. Therefore

$$\begin{aligned} SS^- &= r_+ A_- C \ell^e r_+ C^{-1} \ell^e r_+ A_-^{-1} \ell \\ &= r_+ A_- \ell r_+ C \ell^e r_+ C^{-1} \ell^e r_+ A_-^{-1} \ell \\ &= r_+ A_- \ell r_+ A_-^{-1} \ell = I \end{aligned} \tag{5.11}$$

because

$$\begin{aligned} r_+ A_- &= r_+ A_- \ell r_+ \ : \ H^{1/2} \to H^{-1/2}(\mathbb{R}_+) \\ r_+ C &= r_+ C \ell r_+ \ : \ H^{1/2} \to H^{1/2}(\mathbb{R}_+). \end{aligned} \tag{5.12}$$

For obtaining $\alpha(S) = 1$, we lift the operators by applying Proposition 4.2 and take profit of the properties of the present AFIS in order to work with the kernel of S^-. Note that the even extension operator ℓ^e acts here always into $H^{1/2}$. □

Theorem 5.4. *Let* T *be the operator* (2.15) *due to Problem* 1*DG and let* T *be of normal type. Then (dropping the dashes of the coefficients)* T *has the form*

$$T = \begin{pmatrix} I & 0 \\ K_0 & T_2 \end{pmatrix} = \begin{pmatrix} I & 0 \\ K_0 & r_+ A_{\lambda_-^{1/2}} \ell S_\mu r_+ A_{\lambda^{-1/2}} \ell^o \end{pmatrix} \tag{5.13}$$

where S_μ *satisfies (5.1)–(5.2),* $\mu = -1/4$, *and* $K_0 = C_0(\alpha I - \gamma A_t)\ell^e : H^{1/2}(\mathbb{R}_+) \to H^{-1/2}(\mathbb{R}_+)$ *is bounded. The pre-symbol* ϕ_μ *of* S_μ *can be factored as*

$$\begin{aligned} &\phi_\mu = \phi_- \zeta^{\omega_\mu} \phi_e \\ &\omega_\mu = \kappa + \eta + i\tau \\ &\kappa \in \mathbb{Z}, \ \ \eta \in [-3/4, \ 1/4[\, , \ \ \tau \in \mathbb{R} \end{aligned} \tag{5.14}$$

and we have the following alternative: Either

IIIa. *Case $\eta \neq -3/4$, i.e., $\Re e(\omega_\mu - 1/4) \notin \mathbb{Z}$: T is Fredholm with the same characteristics as S_μ,* $\operatorname{ind} T = -\kappa$, *and one-sided invertible by*

$$T^- = \begin{pmatrix} I & 0 \\ T_2^- K_0 & T_2^- \end{pmatrix} \tag{5.15}$$

$$T_2^- = r_+ A_{\lambda^{1/2}} \ell^o S_\mu^- r_+ A_{\lambda_-^{-1/2}} \ell$$

where S_μ^- is given by Proposition 5.1 *and* (4.8)–(4.11) – *or*

IIIb. *Case $\eta = -3/4$: The images of T and S_μ are not closed but*

$$\kappa = \begin{cases} -\dim\ker S_\mu & \text{if } \kappa \leq 0 \\ \dim L^2(\mathbb{R}_+)/\overline{\operatorname{im} S_\mu} & \text{if } \kappa \geq 0. \end{cases} \tag{5.16}$$

In this case T_2 (and thus T) is image normalized *if we replace the original image space $Y_2 = H^{-1/2}(\mathbb{R}_+)$ of T_2 by a dense subspace, namely by*

$$\overset{\prec}{Y}_2 = \overset{\prec}{H}^{-1/2-i\tau}(\mathbb{R}_+) = r_+ A_{\lambda_-^{i\tau}} H_+^{-1/2} \tag{5.17}$$

with the induced norm of $H_+^{-1/2}$. I.e., the operators with smaller image spaces written as

$$\begin{aligned} \overset{\prec}{T}_2 &= \operatorname{Rst} T_2 : H^{-1/2}(\mathbb{R}_+) \to \overset{\prec}{Y}_2 \\ \overset{\prec}{T} &= \operatorname{Rst} T : H^{1/2}(\mathbb{R}_+) \times H^{-1/2}(\mathbb{R}_+) \to H^{1/2}(\mathbb{R}_+) \times \overset{\prec}{Y}_2 \end{aligned} \tag{5.18}$$

are Fredholm with the characteristics of (5.16). *They have one-sided (bounded) inverses of the same form as before in the case* IIIa *in the sense of continuous extension of $T_2^-{}_{|H^{-1/2+\epsilon}(\mathbb{R}_+)}$ to $\overset{\prec}{Y}_2$, where $\epsilon \in]0,1[$.*

Proof. Formula (5.13) is evident as well as the boundedness of S_μ in $L^2(\mathbb{R}_+)$ (by lifting) and of K_0 in the trace spaces. Further (5.13) implies that T and S are (toplinear) equivalent after extension and thus have isomorphic kernels and cokernels [1, 4], which yields that the Fredholm characteristics coincide.

In the first case IIIa, the existence and representation of a one-sided inverse S_μ^- follows from Proposition 5.1 and formula (5.15) is easily verified. In the second case, when S is not Fredholm, the conclusions for image normalization are completely analogous to the corresponding results for Wiener-Hopf operators with $\mathcal{G}C^\nu(\ddot{\mathbb{R}})$ symbols and known from [22]. □

Theorem 5.5. *Let T be the operator* (2.15) *due to Problem* $1GN$ *and let T be of normal type. Then*

$$T = \begin{pmatrix} T_1 & K_1 \\ 0 & I \end{pmatrix} = \begin{pmatrix} r_+ A_{t_-^{1/2}} \ell S_\mu^e r_+ A_{t^{1/2}} \ell^e & K_1 \\ 0 & I \end{pmatrix} \tag{5.19}$$

$$: H^{1/2}(\mathbb{R}_+) \times H^{-1/2}(\mathbb{R}_+) \to H^{-1/2}(\mathbb{R}_+) \times H^{-1/2}(\mathbb{R}_+)$$

where $K_1 = \beta C_1 \ell^o$ *is bounded in* $H^{-1/2}(\mathbb{R}_+)$,

$$S^e_\mu = r_+ A_{\phi_\mu} \ell^e : L^2(\mathbb{R}_+) \to L^2(\mathbb{R}_+) \tag{5.20}$$

and ϕ_μ *as given in* (5.2), $\mu = -1/4$. *Writing now*

$$\begin{aligned} &\omega_\mu = (2\pi i)^{-1} \textstyle\int_{\mathbb{R}} d \log \phi_\mu = \kappa + \eta + i\tau \\ &\kappa \in \mathbb{Z}, \quad \eta \in [-1/4, 3/4[, \quad \tau \in \mathbb{R} \end{aligned} \tag{5.21}$$

we have the following alternative: Either

IIIa′. *Case* $\eta \neq -1/4$, *i.e.* $\Re e(\omega + 1/4) \notin \mathbb{Z}$: T *is Fredholm and one-sided invertible by*

$$\begin{aligned} T^- &= \begin{pmatrix} T_1^- & -T_1^- K_1 \\ 0 & I \end{pmatrix} \\ T_1^- &= r_+ A_{t^{-1/2}} \ell^e S^{e-}_\mu r_+ A_{t_-^{-1/2}} \ell \end{aligned} \tag{5.22}$$

where S^{e-}_μ *is constructed as in Theorem* 4.5 *and Theorem* 4.6 – *or:*

IIIb′. *Case* $\eta = -1/4$. *The images of* T, T_1 *and* S^e_μ *are not closed, image normalization is possible by putting*

$$\begin{aligned} \overset{\prec}{Y}_1 &= \overset{\prec}{H}{}^{1/2-i\tau}(\mathbb{R}_+) = r_+ A_{\lambda_-^{i\tau}} H_+^{1/2} \\ \overset{\prec}{T}_1 &= \operatorname{Rst} T_1 : H^{1/2}(\mathbb{R}_+) \to \overset{\prec}{Y}_1. \end{aligned} \tag{5.23}$$

A one-sided inverse of $\overset{\prec}{T}$ *is obtained as before in* IIIb.

In all cases, $-\kappa$ *represents the index of* T *or* $\overset{\prec}{T}$, *respectively, and* $|\kappa| \leq 4$.

Proof. By analogy to the previous proof, noting that the pre-symbol ϕ_1 of T_1 is lifted by

$$\begin{aligned} \phi_0 &= t_-^{-1/2} \phi_1 t^{-1/2} \\ &= t_-^{-3/4} t_+^{-1/4} (\alpha - \beta t + \gamma \vartheta) \\ &= \left(\frac{t_-}{t_+}\right)^{-1/4} \left(\frac{\alpha}{t} - \beta - \gamma\theta\right) \end{aligned} \tag{5.24}$$

and the first factor can be changed into $\zeta^{-1/4}$ (see Proposition 4.2). Modifying Proposition 5.1 for S^e_μ with even extension in (5.1), we need to replace the representation (5.14) for ω_μ by (5.21) because of (4.10). □

Corollary 5.6. *Consider the operator* T *of* (2.15) *due to Problem* $1\mathcal{TG}$

$$T = \begin{pmatrix} T_1 & 0 \\ K & T_2 \end{pmatrix} = \begin{pmatrix} r_+(\alpha I + \gamma A_\vartheta)\ell^e & 0 \\ C_0(\alpha' I - \gamma' A_t)\ell^e & r_+(-\alpha' A_{t^{-1}} + \beta' I + \gamma' A_\theta)\ell^o \end{pmatrix} \tag{5.25}$$

and assume that T *is of normal type, i.e.,*

$$\begin{aligned} \phi_1(\xi) &= \alpha + \gamma\vartheta(\xi) \neq 0, \quad \xi \in \mathbb{R} \\ \phi_2(\xi) &= -\alpha' t^{-1}(\xi) + \beta' + \gamma'\theta(\xi) \neq 0, \quad \xi \in \ddot{\mathbb{R}} \end{aligned} \tag{5.26}$$

as well as $\gamma \neq 0$. *Then we have the following different cases of solvability:*

IVa,r. T_2 *is Fredholm and right invertible (see Proposition* 5.1 *and Theorem* 5.4*). Then* T *is also Fredholm and right invertible by*

$$T^- = \begin{pmatrix} T_1^- & 0 \\ -T_2^- K T_1^- & T_2^- \end{pmatrix} \tag{5.27}$$

where T_1^- *is given by Example* 3.5 *or Proposition* 5.3 *and* T_2^- *by Theorem* 5.4*,*

IVa,l. T_1 *is right invertible, not invertible (i.e.,* ϕ_1 *is plus type) and* T_2 *is Fredholm and not right invertible (i.e., only left invertible, not invertible,* $\kappa > 0$*), then* T *is Fredholm (in general not one-sided invertible) and* (5.27) *is a two-sided regularizer up to operators of finite characteristics.*

IVb,r *and* **IVb,l.** T_2 *is not Fredholm. Then we have the two cases as before replacing* T_2 *by* $\overset{\prec}{T}_2$ *(see Theorem* 5.4*).*

Theorem 5.7. *The following classes of interior wedge problems can be explicitly solved by generalized inversion of* T *provided* T *is of normal type:*

$\Gamma_1 \setminus \Gamma_2$	D	N	$\mathcal{I}$	$\mathcal{T}$	O	G
D	**II**	**O**	**I**	$\text{II}^*_\pm$	III	**III**
N	O*	II	II*	I*	III*	III*
$\mathcal{I}$	I*	II		$\text{I}^*_\pm$		
$\mathcal{T}$	$\mathbf{II_\pm}$	I	$\mathbf{I_\pm}$	IV	IV	**IV**
O	III*	III		IV*		
G	III*	**III**		IV*		

Legend:

bold – *reference problems discussed in detail.*

* – *belongs to the corresponding class where the two variables are exchanged, see* (2.10).

O – *invertible by representation formulas* (2.10).

I – *direct inversion by Theorem* 3.2 *(even pre-symbol), Example* 3.3.

I_- – *dito (minus type), Example* 3.5.

I_+ – *right inversion via AFIS, see Proposition* 5.3*, cf. Corollary* 5.6.

II, $\text{II}_\pm$ – *like* I, $\text{I}_\pm$ *after image normalization, Example 3.4.*

III – *Fredholm and one-sided invertible via AFIS, eventually after normalization, see Theorem* 5.4*, Theorem* 5.5*, only one scalar factorization needed.*

IV – *similar with two scalar operators, see Corollary* 5.6

empty spaces – *correspond with open problems, not decomposing (triangular) in the preceding sense.*

TABLE 3. Solvability classes of normal type operators T in (2.15) due to interior wedge problems $1\mathcal{X}\mathcal{Y}$.

Remark 5.8. In order to treat the remainder open problems (with non-triangular T), it seems to be necessary to know more about matrix factorization [8] and about composition formulas for the involved ACOs, extending the ideas of [15].

6. A particular example from class 2: The two-media impedance problem

Coming back to the classes of Meister diffraction problems we like to give an example of how the present approach can be employed to solve more complicated boundary/transmission problems. As pointed out by E. Meister [12, 18], the case $2\mathcal{I}\mathcal{I}$ is of particular physical interest (cf. Figure 2):

$$
\begin{aligned}
&u \in \mathcal{H}^1(Q_{12}) = \{u \in L^2(Q_1 \cup Q_2) : \\
&\qquad\qquad u_{|Q_j} \in H^1(Q_j),\ (\Delta + k_j^2)u = 0 \text{ in } Q_j, j = 1,2\} \\
&B_1 u = \alpha u_0^+ + \beta u_1^+ = g_1 \quad \text{on} \quad \Gamma_1 \\
&\left.\begin{aligned} u_0^+ - u_0^- &= f_2 \\ \rho_1 u_1^+ - \rho_2 u_1^- &= g_2 \end{aligned}\right\} \quad \text{on} \quad \Gamma_2 \\
&B_3 u = \alpha' u_0^+ + \beta' u_1^+ = g_3 \quad \text{on} \quad \Gamma_3.
\end{aligned}
\tag{6.1}
$$

Herein the coefficients $\alpha, \beta, \alpha', \beta', \rho_1, \rho_2$ are given and physically reasonable, e.g., $\beta = \beta' = 1$, $\alpha = ip_1$, $\alpha' = ip_3$, where the impedance numbers are such that $\Re e(p_j) > 0$, $\Im m(p_j) > 0$, the densities $\rho_j > 0$ [12], $\rho_1 \neq \rho_2$ and $\Im m(k_j) > 0$, $k_1 \neq k_2$, i.e., the two quadrants are filled with different lossy media. As before we assume the data to be given as

$$
\begin{aligned}
&f_2 \in H^{1/2}(\mathbb{R}_+) \\
&g_1, g_2, Jg_3 \in H^{-1/2}(\mathbb{R}_+)
\end{aligned}
\tag{6.2}
$$

with the common identification of Γ_j with $\mathbb{R}_\pm$, and we denote the associated operator by $L : u \mapsto (g_1, f_2, g_2, g_3)$.

The spontaneous idea to represent u in Q_1 and Q_2 by DN- or ND-data as done in Proposition 2.1 yields rather complicated systems of BΨDEs. Thus we modify the classical trick from potential theory to present u by a combined single/double layer potential ansatz admitting also tangential data. The operator calculus for class 1 problems leads to a rapid optimal choice of representation formulas (see Table 3) in brief, by the idea to solve $1\mathcal{I}\mathcal{I}_*$ where the "partner boundary operator" $\mathcal{I}_*$ yields a simple structure of the corresponding BΨDE.

Proposition 6.1. *Let* $\alpha, \beta \in \mathbb{C} \setminus \{0\}$, $\Re e(\alpha/\beta) < 0$,

$$
\left.\begin{aligned}
\lambda_1(\xi) &= \alpha - i\beta\xi \\
\sigma_1(\xi) &= \alpha - \beta t_1(\xi) \neq 0
\end{aligned}\right\} \quad \xi \in \mathbb{R}
\tag{6.3}
$$

where $t_1(\xi) = (\xi^2 - k_1^2)^{1/2}$, *with* $\Im m(k_1) > 0$, *is chosen as before. Then we have a toplinear isomorphism*

$$\begin{aligned} &\mathcal{K}_{\mathcal{II}_*,Q_1} : H^{-1/2}(\mathbb{R}_+)^2 \to \mathcal{H}^1(Q_1) \\ &u = \mathcal{K}_{\mathcal{II}_*,Q_1}(\varphi_1,\varphi_2)^* = \mathcal{K}_1\varphi_1 + \mathcal{K}_2\varphi_2 \\ &u(x) = \mathcal{F}^{-1}_{\xi\mapsto x_1} \exp[-x_2 t(\xi)]\sigma_1^{-1}(\xi)\cdot \widehat{\ell^o\varphi_1}(\xi) \\ &\qquad + \mathcal{F}^{-1}_{\xi\mapsto x_2} \exp[-x_1 t(\xi)]\lambda_1^{-1}(\xi)\cdot \widehat{\ell^o\varphi_2}(\xi) \end{aligned} \tag{6.4}$$

such that

$$\begin{aligned} B_1 u &= (\alpha T_{0,\Gamma_1} + \beta T_{1,\Gamma_1})\, u = \varphi_1 \\ B_2 u &= (\alpha T_{0,\Gamma_2} + \beta T_{\tau,\Gamma_2})\, u = \varphi_2 \end{aligned} \tag{6.5}$$

i.e.,

$$\begin{aligned} (B_1,B_2)^*\,(\mathcal{K}_1,\mathcal{K}_2) &= I_{H^{-1/2}(\mathbb{R}_+)^2} \\ (\mathcal{K}_1,\mathcal{K}_2)\,(B_1,B_2)^* &= I_{\mathcal{H}^1(Q_1)} \end{aligned} \tag{6.6}$$

Proof. The formulas can be either verified directly as in the proof of Theorem 2.3 or considered as a consequence of the solution of Problem $1\mathcal{IT}$ (see $1\mathcal{TI}$ in Example 3.5 and exchange the variables) together with the representation formulas of Proposition 2.1. For the bijectivity of the operators in (6.6) we need that λ_1^{-1} is minus type. Otherwise, if λ_1^{-1} is plus type, the operator $(B_1,B_2)^*$ has a kernel of dimension 1, see Proposition 5.3 and Corollary 5.6. Note that no compatibility condition is needed. □

Theorem 6.2. *Under the same assumptions* (6.3) *and analogous assumptions for the* Q_2 *parameters (with index* 2*), the operator* L *associated to Problem* $2\mathcal{II}$ *given by* (6.1)–(6.2) *can be factored as follows*

$$Lu = \mathbf{T}\mathbf{K}u = \left(\begin{array}{c|cc|c} I & 0 & 0 & 0 \\ \hline 0 & \multicolumn{2}{c|}{\multirow{2}{*}{T}} & 0 \\ T_{31} & & & T_{34} \\ \hline 0 & 0 & 0 & I \end{array}\right) \begin{pmatrix} B_1 & 0 \\ B_2 & 0 \\ 0 & B_2' \\ 0 & B_3 \end{pmatrix} \begin{pmatrix} u_{|Q_1} \\ u_{|Q_2} \end{pmatrix} = \begin{pmatrix} g_1 \\ f_2 \\ g_2 \\ g_3 \end{pmatrix} \tag{6.7}$$

where

$$\begin{aligned} T &= \begin{pmatrix} r_+ A_{\lambda_1^{-1}}\ell^o & -r_+ A_{\lambda_2^{-1}}\ell^o \\ -r_+\rho_1 A_{t_1\lambda_1^{-1}}\ell^o & -r_+\rho_2 A_{t_2\lambda_2^{-1}}\ell^o \end{pmatrix} \\ &\quad : H^{-1/2}(\mathbb{R}_+)^2 \to H^{1/2}(\mathbb{R}_+)\times H^{-1/2}(\mathbb{R}_+) \\ T_{31} &= 2\rho_1 C_0 A_{\vartheta\sigma_1^{-1}}\ell^o \\ T_{34} &= -2\rho_2 C_0 A_{\vartheta\sigma_2^{-1}}\ell^o \\ B_2' &= \alpha' T_{0,\Gamma_2} + \beta' T_{\tau,\Gamma_2} : \mathcal{H}^1(Q_2) \to H^{-1/2}(\Gamma_2). \end{aligned} \tag{6.8}$$

Therefore, L *and* T *are toplinear equivalent after extension.*

Proof. Let $u \in \mathcal{H}^1(Q_{12})$ and let $u_{|Q_1}$ be represented by formula (6.4). Take the corresponding representation of u in the second quadrant

$$\begin{aligned} u_{|Q_2} &= \mathcal{K}_3\varphi_3 + \mathcal{K}_2'\varphi_2' \\ &= \mathcal{F}^{-1}_{\xi\mapsto x_1} \exp[-x_2 t_2(\xi)]\sigma_2^{-1}(\xi)\widehat{\ell^o\varphi_3}(\xi) + \mathcal{F}^{-1}_{\xi\mapsto x_2} \exp[x_1 t_2(\xi)]\lambda_2^{-1}(\xi)\widehat{\ell^o\varphi_2'}(\xi) \end{aligned} \tag{6.9}$$

with $\sigma_2(\xi) = \alpha' - \beta' t_2(\xi)$, $\lambda_2(\xi) = \alpha' - i\beta'\xi$ and

$$\begin{aligned} B_3 u_{|\Gamma_3} &= (\alpha' T_{0,\Gamma_3} + \beta' T_{1,\Gamma_3})\, u = \varphi_3 \\ B_2' u_{|\Gamma_2} &= (\alpha' T_{0,\Gamma_2} + \beta' T_{\tau,\Gamma_2})\, u = \varphi_2'. \end{aligned} \tag{6.10}$$

Substituted into the transmission conditions on Γ_2 (see (6.1)), the following jump relations hold there:

$$\begin{aligned} u_0^+ - u_0^- &= r_+ A_{\lambda_1^{-1}} \ell^o \varphi_2 - r_+ A_{\lambda_2^{-1}} \ell^o \varphi_2' = f_2 \\ \rho_1 u_1^+ - \rho_2 u_1^- &= \rho_1 \left(C_0 A_{\vartheta\sigma_1^{-1}} \ell^o \varphi_1 - r_+ A_{t_1\lambda_1^{-1}} \ell^o \varphi_2 \right) \\ &\quad - \rho_2 \left(C_0 A_{\vartheta\sigma_2^{-1}} \ell^o \varphi_3 + r_+ A_{t_2\lambda_2^{-1}} \ell^o \varphi_2' \right) = g_2 \end{aligned} \tag{6.11}$$

where the C_0 terms in the first formula vanished (C_0 acts on odd functions, see (2.13)). The C_0 terms in the last formula represent known data ($\varphi_1 = g_1$, $\varphi_3 = g_3$). So Problem $2\mathcal{II}$ is (by substitution) "reduced to"

$$T \begin{pmatrix} \varphi_2 \\ \varphi_2' \end{pmatrix} = \begin{pmatrix} f_2 \\ g_2 \end{pmatrix} + \begin{pmatrix} 0 \\ -\rho_1 C_0 A_{\vartheta\sigma_1^{-1}} \ell^o g_1 + \rho_2 C_0 A_{\vartheta\sigma_2^{-1}} \ell^o g_3 \end{pmatrix} \tag{6.12}$$

where T is given by (6.8), (φ_2, φ_2') stands for the unknown "$\mathcal{I}_*$ potentials" on the two banks of Γ_2, and the rest is given. Formula (6.7) summarizes the preceding formulas and the mapping properties are obvious from the orders of the operators and from the invertibility of the Poisson operators by trace operators, see (6.4)–(6.5). □

Corollary 6.3. *If T is invertible by T^{-1}, the boundary/transmission problem $2\mathcal{II}$ is well posed and explicitly solved by*

$$u = \left(\begin{array}{cc|cc} \mathcal{K}_1 & \mathcal{K}_2 & 0 & 0 \\ \hline 0 & 0 & J\mathcal{K}_2 & -\mathcal{K}_1 J \end{array} \right) \left(\begin{array}{c|cc|c} I & 0 & 0 & 0 \\ \hline S_{11} & \multicolumn{2}{c|}{\multirow{2}{*}{T^{-1}}} & S_{21} \\ S_{12} & & & S_{22} \\ \hline 0 & 0 & 0 & I \end{array} \right) \begin{pmatrix} g_1 \\ f_2 \\ g_2 \\ g_3 \end{pmatrix} \tag{6.13}$$

where

$$S = (S_{ij}) = -T^{-1} \begin{pmatrix} 0 & 0 \\ T_{31} & T_{34} \end{pmatrix}. \tag{6.14}$$

Proof. This is a straightforward computation based on the formulas (6.7)–(6.10). □

Proposition 6.4. *The operator T of (6.8) can be factored into bounded operators by*

$$T = \begin{pmatrix} r_+ A_{\lambda_1^{-1}} \ell^o & 0 \\ -r_+ \rho_1 A_{t_1 \lambda_1^{-1}} \ell^o & U \end{pmatrix} \begin{pmatrix} I & -r_+ A_{\lambda_1 \lambda_2^{-1}} \ell^o \\ 0 & I \end{pmatrix}$$
$$U = -r_+ \rho_1 A_{t_1 \lambda_1^{-1}} \ell^o r_+ A_{\lambda_1 \lambda_2^{-1}} \ell^o - r_+ \rho_2 A_{t_2 \lambda_2^{-1}} \ell^o$$
$$: H^{-1/2}(\mathbb{R}_+) \to H^{-1/2}(\mathbb{R}_+) \tag{6.15}$$

Proof. The result is evident, but note that λ_1^{-1} is minus type due to $\Re e(\alpha/\beta) < 0$ as assumed in Theorem 6.2. □

Theorem 6.5. *Let again* (6.3) *be satisfied and the corresponding conditions for $\alpha', \beta', \lambda_2, \sigma_2, t_2, k_2$ hold as well. Let $\lambda_2 = \lambda_1$, i.e., we consider $2\mathcal{I}\mathcal{I}$ with the same impedance condition on Γ_1 and Γ_3, but (in general) different media in Q_1 and Q_2.*

Then the corresponding operators $L, \mathbf{T}, T, U$ are not normally solvable. Image normalization of U by

$$\overset{\prec}{U} = \mathrm{Rst}\, U : H^{-1/2}(\mathbb{R}_+) \to \widetilde{H}^{-1/2}(\mathbb{R}_+) \tag{6.16}$$

and resulting normalizations of the other operators yield bounded invertibility of all of them and the fact that Problem $2\mathcal{I}\mathcal{I}$ is well posed. The resulting compatibility condition for the given data, which is necessary and sufficient for solution in $\mathcal{H}(Q_{12})$, reads

$$g_2 + \rho_2 C_0 A_{\vartheta\sigma_2^{-1}} \ell^o g_3 - \rho_1 C_0 A_{\vartheta\sigma_1^{-1}} \ell^o g_1 + r_+ \rho_1 A_{t_1 \lambda_1^{-1}} \ell^o r_+ A_{\lambda_1} \ell f_2 \in \widetilde{H}^{-1/2}(\mathbb{R}_+). \tag{6.17}$$

Proof. Consider the operator U of (6.15) for $\lambda_2 = \lambda_1$:

$$U = -r_+ A_\sigma \ell^o : H^{-1/2}(\mathbb{R}_+) \to H^{-1/2}(\mathbb{R}_+)$$
$$\sigma = \lambda_1^{-1} (\rho_1 t_1 + \rho_2 t_2) = \sigma_- \sigma_e \tag{6.18}$$

which gives us a factorization of $A_\sigma = A_- A_e$ through the space $Z = H^{-3/2}$ that is forbidden in Definition 4.4. Corollary 4.7 yields that U is not Fredholm. As in the proof of Theorem 5.4 we obtain the image normalization (6.16) and corresponding results for $L, \mathbf{T}, T$ which are equivalent after extension to U, thus all have isomorphic kernels and co-kernels. Formula (6.16) is then a direct consequence of (6.12) and (6.15) resolving

$$\begin{pmatrix} r_+ A_{\lambda_1^{-1}} \ell^o & 0 \\ -r_+ \rho_1 A_{t_1 \lambda_1^{-1}} \ell^o & U \end{pmatrix} \begin{pmatrix} \psi_1 \\ \psi_2 \end{pmatrix} = T \begin{pmatrix} \varphi_2 \\ \varphi_2' \end{pmatrix} \tag{6.19}$$
$$= \begin{pmatrix} f_2 \\ g_2 - \rho_1 C_0 A_{\vartheta\sigma_1^{-1}} \ell^o g_1 + \rho_2 C_0 A_{\vartheta\sigma_2^{-1}} \ell^o g_3 \end{pmatrix}$$

with respect to $U\psi_2 \in \widetilde{H}^{-1/2}(\mathbb{R}_+)$. □

Remark 6.6. It is possible to regard Problem $2\mathcal{II}$ with different face impedances as a perturbation of the previous case provided $\alpha'/\beta' \approx \alpha/\beta$.

Acknowledgment

The authors would like to acknowledge the financial support of *Centro de Matemática e Aplicações* of Instituto Superior Técnico and *Unidade de Investigação Matemática e Aplicações* of Universidade de Aveiro, through *Programa Operacional "Ciência, Tecnologia, Inovação"* (POCTI) of the *Fundação para a Ciência e a Tecnologia* (FCT), cofinanced by the European Community fund FEDER.

References

[1] H. Bart, V.E. Tsekanovskii, *Matricial coupling and equivalence after extension.* T. Ando et al. (eds.), Operator Theory and Complex Analysis. Birkhäuser, Basel. Oper. Theory, Adv. Appl. **59** (1992), 143–160.

[2] E.L. Basor, T. Ehrhardt, *On a class of Toeplitz + Hankel operators.* New York J. Math. **5** (1999), 1–16.

[3] L.P. Castro, R. Duduchava, F.-O. Speck, *Localization and minimal normalization of some basic mixed boundary value problems.* S. Samko et al. (eds.), Factorization, Singular Operators and Related Problems, Kluwer Academic Publishers, Dordrecht, 2003, 73–100.

[4] L.P. Castro, F.-O. Speck, *Regularity properties and generalized inverses of delta-related operators.* Z. Anal. Anwend. **17** (1998), 577–598.

[5] L.P. Castro, F.-O. Speck, F.S. Teixeira, *A direct approach to convolution type operators with symmetry.* To appear.

[6] M. Costabel, E. Stephan, *A direct boundary integral equation method for transmission problems.* J. Math. Anal. Appl. **106** (1985), 367–413.

[7] G.I. Èskin, *Boundary value problems for elliptic pseudodifferential equations.* Translations of Mathematical Monographs 52. AMS, Providence, R.I., 1981.

[8] I. Gohberg, M.A. Kaashoek, I.M. Spitkovsky, *An overview of matrix factorization theory and operator applications.* I. Gohberg et al. (eds.), Factorization and Integrable Systems, Proceedings of the Summer School, Faro, Portugal, 2000. Operator Theory: Advances and Applications, **141**. Birkhäuser, Basel, 2003, 1–102.

[9] A.B. Lebre, A. Moura Santos, F.-O. Speck, *Factorization of a class of matrices generated by Sommerfeld diffraction problems with oblique derivatives.* Math. Meth. Appl. Sci. **20** (1997), 1185–1198.

[10] E. Meister, *Some solved and unsolved canonical problems of diffraction theory.* Differential Equations and their Applications, Equadiff 6, Lect. Notes Math. **1192** (1986), 393–398.

[11] E. Meister, *Some solved and unsolved canonical problems of diffraction theory.* Differential Equations and Mathematical Physics, Lect. Notes Math. **1285** (1987), 320–336.

[12] E. Meister, *Einige gelöste und ungelöste kanonische Probleme der mathematischen Beugungstheorie.* Expo. Math. **5** (1987), 193–237.

[13] E. Meister (ed.), *Modern mathematical methods in diffraction theory and its applications in engineering.* Proceedings of the Sommerfeld'96 workshop. Freudenstadt, Germany. Methoden und Verfahren der Mathematischen Physik. 42. Peter Lang, Frankfurt, Europ. Verlag der Wissenschaften, 1997.

[14] E. Meister, F. Penzel, F.-O. Speck, F.S. Teixeira, *Two-media scattering problems in a half-space.* H. Begehr et al. (eds.), Partial Differential Equations with Real Analysis. Longman Scientific & Technical, Harlow. Pitman Res. Notes Math. Ser. **263** (1992), 122–146.

[15] E. Meister, F. Penzel, F.-O. Speck, F.S. Teixeira, *Some interior and exterior boundary-value problems for the Helmholtz equation in a quadrant.* Proc. R. Soc. Edinb., Sect. A **123** (1993), 275–294.

[16] E. Meister, F. Penzel, F.-O. Speck, F.S. Teixeira, *Two canonical wedge problems for the Helmholtz equation.* Math. Methods Appl. Sci. **17** (1994), 877–899.

[17] E. Meister, P.A. Santos, F.S. Teixeira, *A Sommerfeld-type diffraction problem with second-order boundary conditions.* Z. Angew. Math. Mech. **72** (1992), 621–630.

[18] E. Meister, F.-O. Speck, *Diffraction problems with impedance conditions.* Appl. Anal. **22** (1986), 193–211.

[19] E. Meister, F.-O. Speck, *A contribution to the quarter-plane problem in diffraction theory.* J. Math. Anal. Appl. **130** (1988), 223–236.

[20] E. Meister, F.-O. Speck, *Modern Wiener-Hopf methods in diffraction theory.* Ordinary and Partial Differential Equations, Vol. II, Proc. 10th Dundee Conf., Pitman Res. Notes Math. Ser. **216** (1989), 130–171.

[21] A. Moura Santos, F.-O. Speck, *Sommerfeld diffraction problems with oblique derivatives.* Math. Methods Appl. Sci. **20** (1997), 635–652.

[22] A. Moura Santos, F.-O. Speck, F.S. Teixeira, *Minimal normalization of Wiener-Hopf operators in spaces of Bessel potentials.* J. Math. Anal. Appl. **225** (1998), 501–531.

[23] F. Penzel, F.S. Teixeira, *The Helmholtz equation in a quadrant with Robin's conditions.* Math. Methods Appl. Sci. **22** (1999), 201–216.

[24] A.H. Serbest (ed.), S.R. Cloude (ed.), *Direct and inverse electromagnetic scattering.* Proceedings of the Workshop in Gebze/Istanbul, Turkey. Pitman Research Notes in Mathematics Series **361**. Longman, Harlow, 1996.

[25] A. Sommerfeld, *Mathematische Theorie der Diffraction*, Math. Ann. **47** (1896), 317–374.

[26] F.-O. Speck, *Mixed boundary value problems of the type of Sommerfeld's half-plane problem.* Proc. R. Soc. Edinb., Sect. A **104** (1986), 261–277.

[27] F.-O. Speck, *Sommerfeld diffraction problems with first and second kind boundary conditions.* SIAM J. Math. Anal. **20** (1989), 396–407.

[28] F.-O. Speck, R.A. Hurd, E. Meister, Sommerfeld diffraction problems with third kind boundary conditions. *SIAM J. Math. Anal.* **20** (1989), 589–607.

[29] H. Triebel, *Theory of function spaces II.* Monographs in Mathematics 84. Birkhäuser Verlag, Basel, 1992.

Luis P. Castro
Department of Mathematics
and R&D Unit *Mathematics and Applications*
University of Aveiro
3810–193 Aveiro
Portugal
e-mail: `lcastro@mat.ua.pt`

Frank-Olme Speck
Department of Mathematics
Instituto Superior Técnico
Technical University of Lisbon
Av. Rovisco Pais
1049-001 Lisboa
Portugal
e-mail: `fspeck@math.ist.utl.pt`

Francisco S. Teixeira
Department of Mathematics
Instituto Superior Técnico
Technical University of Lisbon
Av. Rovisco Pais
1049-001 Lisboa
Portugal
e-mail: `fteix@math.ist.utl.pt`

Operator Theory
Advances and Applications, Vol. 147, 241–256

Corner Singularities of Maxwell Interface and Eddy Current Problems

Martin Costabel, Monique Dauge, and Serge Nicaise

Dedicated to the memory of Erhard Meister

Abstract. The aim of this work is to provide a description of the corner asymptotics for the solutions of Maxwell equations in and outside a conductor body and to investigate the limit as the ratio permittivity/conductivity tends to zero (the eddy current limit). Corner singularities of the Maxwell transmission problem and also of the eddy current model have been described elsewhere [6, 7]. Here we concentrate on the uniform behavior with respect to the small parameter describing the eddy current limit – analyticity of the singular functions and stability of the decomposition of the fields into regular and singular parts.

Introduction

We consider the time-harmonic Maxwell equations in a medium with a high conductivity in one part and isolating in the other part. We are not interested in scattering aspects here, but will study the local regularity of the fields, in particular near corners of the conductor. Since the questions are local, we can assume from the outset that the domain is bounded. Let therefore $\Omega \subset \mathbb{R}^3$ be a bounded domain decomposed into the two subdomains Ω_C, the conductor, and Ω_E, which corresponds to the exterior isolator. For the sake of simplicity, we assume that $B := \partial\Omega_C$ is connected, and that $\partial\Omega_E = \partial\Omega \cup B$. The conductivity σ is a constant $\sigma_C > 0$ in Ω_C and vanishes in Ω_E. The electric permittivity ε and the magnetic permeability μ are supposed to be positive constants on Ω_C and on Ω_E. The frequency ω is a fixed positive constant. The eddy current model describes the situation where the quotient $\omega\varepsilon/\sigma_C$ is very small.

We consider the case where B has a conical singularity. As is well known, in the neighborhood of this corner point the electrical field will, in general, be unbounded. All possible singularities for the solutions of the time-harmonic Maxwell equations near conical corner points have been described in [5] for the case of perfect conductor boundary conditions, in [6] for interface conditions, and in [7] for

2000 *Mathematics Subject Classification.* Primary 35J; Secondary 00A00.
Key words and phrases. Maxwell, Eddy current, corners, singularities.

the eddy current model. In all these cases, it has been shown that the singular functions can be obtained from corresponding scalar problems for the Laplace equation. Therefore the analytical tools for singularity analysis, mainly Kondrat'ev's Mellin transform based technique, are applied to the well-studied boundary value and interface problems for the Laplacian.

If, however, we want to describe the behavior of the singularities as the full Maxwell problem tends towards the eddy current problem, we need to use tools that have been developed for the situation of singularity problems depending on a parameter ([4, 14, 16]). Since the coefficients of the operator are complex, we cannot expect any simplification, but have to take into account all the possible complications that may appear in such situations, such as "crossing" and "branching" of the exponents. In order to get a stable description of the singular behavior, we no longer can reduce everything to the Laplace operator. Instead we have to use Mellin transformation directly for the Maxwell system. The corresponding constructions of spaces and operators do not seem to exist in the literature.

In the eddy current limit, the PDE problem itself changes its type from a pure transmission problem for a strongly elliptic second order system to a mixture of one-sided boundary conditions and transmission conditions for a system of Maxwell type in the conductor and of Laplace type with a divergence-free constraint in the isolator. On the other hand, from the point of view of the description of singularities, the eddy current limit is a regular perturbation problem.

Thus it is not hard to show that in this limit, not only the solution of the Maxwell problem converges in the energy norm to the solution of the eddy current problem, but also the singularity exponents converge at the same time. This has been proved in [7].

The continuity of the solution and of the singularity exponents does not imply, however, that in any decomposition of the solution into regular and singular parts, all the terms – regular part, singular functions, and coefficients of the singular functions – will also depend continuously on the small parameter δ that characterizes the eddy current limit. This is the problem we are studying here.

In general, when the singularity exponent in the eddy current limit is of multiplicity higher than one, a "naïve" decomposition into a regular part and individual singular functions will produce coefficients tending to infinity as $\delta \to 0$. We show how to choose bases of singular functions that lead to coefficients continuous as $\delta \to 0$, and we prove stability for the decomposition into regular and singular parts in this case.

For the Mellin analysis, the "Mellin symbol", i.e., the angular part of the Maxwell transmission problem, has to be considered in a strong form, that is, on a level of regularity above the level of the energy norm related to the variational formulation. This means in particular that the natural transmission conditions now appear explicitly in the formulation of the Mellin symbol. We indicate in Section 5 how they are obtained. We also show how the classification of the singularities into types 1, 2, and 3 is obtained from the Mellin analysis.

1. Maxwell and eddy current problems

Let $\omega > 0$ be a fixed frequency. The time harmonic Maxwell equations are

$$\begin{cases} \mathbf{curl}\,\mathbf{E} = -i\omega\mu\mathbf{H} & \text{in } \Omega, \\ \mathbf{curl}\,\mathbf{H} = (i\omega\varepsilon + \sigma)\mathbf{E} + \mathbf{j}_0 & \text{in } \Omega, \\ \mathbf{E}\times\mathbf{n} = 0 \quad \text{and} \quad \mathbf{H}\cdot\mathbf{n} = 0 & \text{on } \partial\Omega, \end{cases} \tag{1.1}$$

$\mathbf{E}$ (resp. $\mathbf{H}$) is the electric (resp. magnetic) field and $\mathbf{j}_0$ is the source current density which is supposed to be a $L^2(\Omega)$ field with support in Ω_C and to be divergence free, i.e., $\operatorname{div}\mathbf{j}_0 = 0$ in Ω. Note that the assumption on $\operatorname{div}\mathbf{j}_0$ is equivalent to

$$\operatorname{div}\mathbf{j}_0 = 0 \text{ in } \Omega_C \quad \text{and} \quad \mathbf{j}_0\cdot\mathbf{n} = 0 \text{ on } B.$$

Thus, taking the divergence of the second equation of (1.1), we obtain the following equation on the divergence of $\mathbf{E}$:

$$\operatorname{div}(i\omega\varepsilon + \sigma)\mathbf{E} = 0 \quad \text{in} \quad \Omega. \tag{1.2}$$

The time-harmonic eddy current problem [2, 3, 1, 9] reads

$$\begin{cases} \mathbf{curl}\,\mathbf{E} = -i\omega\mu\mathbf{H} & \text{in } \Omega, \\ \mathbf{curl}\,\mathbf{H} = \sigma\mathbf{E} + \mathbf{j}_0 & \text{in } \Omega. \\ \mathbf{E}\times\mathbf{n} = 0 \quad \text{and} \quad \mathbf{H}\cdot\mathbf{n} = 0 & \text{on } \partial\Omega, \end{cases} \tag{1.3}$$

Let us write $\mathbf{E}_C = \mathbf{E}|_{\Omega_C}$ and $\mathbf{E}_E = \mathbf{E}|_{\Omega_E}$.

Taking the divergence of the second equation of (1.3), instead of condition (1.2) we only obtain $\operatorname{div}\mathbf{E}_C = 0$ in Ω_C and $\mathbf{E}_C\cdot\mathbf{n} = 0$ on B. These conditions *have to be completed by the gauge conditions*:

$$\operatorname{div}\mathbf{E}_E = 0 \ \text{in} \ \Omega_E \quad \text{and} \quad \int_B \mathbf{E}_E\cdot\mathbf{n}\,\mathrm{d}S = 0. \tag{1.4}$$

1.1. Eddy current limit

Following [3, Ch. 4], we consider the eddy current limit as the limiting situation when the quantities $\omega\varepsilon_C/\sigma_C$ and $\omega\varepsilon_E/\sigma_C$ are small. For a conducting material, the permittivity ε_C is of the same order of magnitude as ε_E (also denoted ε_0), but ε_C/σ_C is very small. For moderate frequencies ω the quantities $\omega\varepsilon_C/\sigma_C$ and $\omega\varepsilon_E/\sigma_C$ are still small. Let us fix two numbers $\hat\varepsilon_C$ and $\hat\varepsilon_E$ which are of the same order as σ_C and such that there exists $\delta > 0$ (thus δ is small)

$$\varepsilon_C = \delta\hat\varepsilon_C \quad \text{and} \quad \varepsilon_E = \delta\hat\varepsilon_E. \tag{1.5}$$

Thus we can write (defining by the same token the complex electric transmission coefficient α)

$$i\omega\alpha := i\omega\varepsilon + \sigma = \begin{cases} i\omega\delta\hat\varepsilon_C + \sigma_C & \text{in } \Omega_C \\ i\omega\delta\hat\varepsilon_E & \text{in } \Omega_E. \end{cases} \tag{1.6}$$

We fix σ_C, ω, $\hat\varepsilon_C$ and $\hat\varepsilon_E$. The eddy current limit is the limit $\delta \to 0$.

1.2. Strong electric formulations

We use the electric approach consisting in eliminating $\mathbf{H}$ from equations (1.1)–(1.2) and (1.3)–(1.4). We denote by $\mathbf{E}^\delta$ for $\delta > 0$ according to (1.6), the solution of equations (1.1)–(1.2) and by $\mathbf{E}^0$ the solution of (1.3)-(1.4). We note that we can write the equations satisfied by $\mathbf{E}^\delta$ in a unified way for $\delta > 0$ and $\delta = 0$:

$$\left\{\begin{array}{lrcll}
\text{(i)} & \mathbf{curl}\,\mu_C^{-1}\,\mathbf{curl}\,\mathbf{E}_C^\delta + i\omega\sigma_C\mathbf{E}_C^\delta - \delta\omega^2\hat{\varepsilon}_C\mathbf{E}_C^\delta & = & -i\omega\mathbf{j}_0 & \text{in } \Omega_C,\\
\text{(ii)} & \operatorname{div}\mathbf{E}_C^\delta & = & 0 & \text{in } \Omega_C,\\
\text{(iii)} & \mathbf{curl}\,\mu_E^{-1}\,\mathbf{curl}\,\mathbf{E}_E^\delta - \delta\omega^2\hat{\varepsilon}_E\mathbf{E}_E^\delta & = & 0 & \text{in } \Omega_E,\\
\text{(iv)} & \operatorname{div}\mathbf{E}_E^\delta & = & 0 & \text{in } \Omega_E,\\
\text{(v)} & \int_B \mathbf{E}_E^\delta\cdot\mathbf{n}\,\mathrm{d}S & = & 0 & \\
\text{(vi)} & [\mathbf{E}^\delta\times\mathbf{n}] & = & 0 & \text{on } B,\\
\text{(vii)} & i\delta\omega[\hat{\varepsilon}\mathbf{E}^\delta\cdot\mathbf{n}] + \sigma_C\mathbf{E}_C^\delta\cdot\mathbf{n} & = & 0 & \text{on } B,\\
\text{(viii)} & \mathbf{E}^\delta\times\mathbf{n} & = & 0 & \text{on } \partial\Omega.
\end{array}\right. \tag{1.7}$$

1.3. Variational formulations

The variational space which we will use is *independent of* δ, i.e., suitable for both the Maxwell and the eddy current problem. Let $H_0(\mathbf{curl}\,,\Omega)$ be the standard space

$$H_0(\mathbf{curl}\,,\Omega) = \{\mathbf{u}\in L^2(\Omega)^3 : \mathbf{curl}\ \mathbf{u}\in L^2(\Omega)^3, \quad \mathbf{u}\times\mathbf{n} = 0 \text{ on } \partial\Omega\}.$$

Our variational space $\mathbf{Y}(\Omega)$ is defined as

$$\mathbf{Y}(\Omega) = \Big\{\mathbf{u}\in H_0(\mathbf{curl}\,,\Omega) : \operatorname{div}\mathbf{u}_C\in L^2(\Omega_C),\ \operatorname{div}\mathbf{u}_E\in L^2(\Omega_E),\ \int_B \mathbf{E}\cdot\mathbf{n} = 0\Big\}$$

equipped with the norm

$$\|\mathbf{u}\|^2_{\mathbf{Y}(\Omega)} = \|\mathbf{u}\|^2_{0,\Omega} + \|\,\mathbf{curl}\ \mathbf{u}\|^2_{0,\Omega} + \|\operatorname{div}\mathbf{u}_C\|^2_{0,\Omega_C} + \|\operatorname{div}\mathbf{u}_E\|^2_{0,\Omega_E}.$$

There is a full family of sesquilinear forms a^δ on $\mathbf{Y}(\Omega)$ adapted to a *regularized variational formulation* of the problem (1.7) for $\delta \geq 0$: We arbitrarily fix some positive parameter γ (possibly different in Ω_C and Ω_E) and we define a^δ as follows: For $\mathbf{u}$, $\mathbf{v}\in\mathbf{Y}(\Omega)$:

$$\begin{aligned}
a^\delta(\mathbf{u},\mathbf{v}) \;=\; & \int_\Omega \big(\mu^{-1}\,\mathbf{curl}\ \mathbf{u}\cdot\mathbf{curl}\ \overline{\mathbf{v}} - \delta\omega^2\hat{\varepsilon}\mathbf{u}\cdot\overline{\mathbf{v}}\big)\,\mathrm{d}x + i\omega\int_{\Omega_C}\sigma_C\mathbf{u}\cdot\overline{\mathbf{v}}\,\mathrm{d}x\\
& + \int_{\Omega_C}\gamma_C\operatorname{div}\mathbf{u}_C\operatorname{div}\overline{\mathbf{v}}_C\,\mathrm{d}x + \int_{\Omega_E}\gamma_E\operatorname{div}\mathbf{u}_E\operatorname{div}\overline{\mathbf{v}}_E\,\mathrm{d}x.
\end{aligned}$$

Lemma 1.1. *Let the positive constants μ_C, μ_E, $\hat{\varepsilon}_C$, $\hat{\varepsilon}_E$, σ_C and ω be fixed. Then there exists $\delta_0 > 0$ such that for all $\delta\in[0,\delta_0]$, a^δ is strongly coercive on $\mathbf{Y}(\Omega)$: $\exists c_0 > 0$, $\forall\delta\in[0,\delta_0]$, $\forall\mathbf{u}\in\mathbf{Y}(\Omega)$*

$$\operatorname{Re}\big((1-i)a^\delta(\mathbf{u},\mathbf{u})\big) \geq c_0\|\mathbf{u}\|^2_{\mathbf{Y}(\Omega)}.$$

For all $\delta \in [0, \delta_0]$, we consider the variational problem:

$$\textit{Find } \mathbf{E} \in \mathbf{Y}(\Omega) \ \ s.\ t.\ \ a^\delta(\mathbf{E}, \mathbf{v}) = -i\omega(\mathbf{j}_0, \mathbf{v})_{\Omega_C}, \ \ \forall \mathbf{v} \in \mathbf{Y}(\Omega), \tag{1.8}$$

where $(\cdot,\cdot)_D$ is the $L^2(D)^3$ hermitian inner product.

The following result is proved in [7]:

Theorem 1.2. *Under the assumptions of Lemma* 1.1, *there holds for all* $\delta \in [0, \delta_0]$*:*

(i) *There exists a unique solution* $\mathbf{E}^\delta$ *to problem* (1.8).

(ii) *The solution* $\mathbf{E}^\delta$ *satisfies all equations in* (1.7).

(iii) *The norms of the* $\mathbf{E}^\delta$ *in* $\mathbf{Y}(\Omega)$ *are uniformly bounded:*

$$\exists C > 0, \quad \forall \delta \in [0, \delta_0], \quad \|\mathbf{E}^\delta\|_{\mathbf{Y}(\Omega)} \le C.$$

(iv) *As* $\delta \to 0$, $\mathbf{E}^\delta \to \mathbf{E}^0$ *and we have the convergence estimate*

$$\exists C > 0, \quad \forall \delta \in [0, \delta_0], \quad \|\mathbf{E}^\delta - \mathbf{E}^0\|_{\mathbf{Y}(\Omega)} \le C\delta.$$

2. Localization at a conical point

Let us assume that Ω_C has a smooth boundary except at one point, say O, where it coincides with the tip of a cone Γ_C. The solid angle $G_C := \Gamma_C \cap \mathbb{S}^2$ is a smooth domain in $\mathbb{S}^2$. In a neighborhood of O, the exterior domain Ω_E coincides with the open cone Γ_E such that $\overline{\Gamma}_C \cup \overline{\Gamma}_E = \mathbb{R}^3$ and $\Gamma_C \cap \Gamma_E = \emptyset$. Let $G_E := \Gamma_E \cap \mathbb{S}^2$. Let (ρ, ϑ) be spherical coordinates centered in O.

As far as elliptic boundary value or transmission problems on domains with conical points are concerned, the standard tool for the investigation of the structure of their solution is the Mellin transform defined for all $\mathbf{u} \in \mathcal{C}_0^\infty(\mathbb{R}^3 \setminus \{0\})$ and all $\lambda \in \mathbb{C}$ by

$$\mathbf{u} = \mathbf{u}(\rho, \vartheta) \quad \longmapsto \quad \mathfrak{M}\mathbf{u}(\lambda) = \mathbf{U}(\lambda, \vartheta) := \int_0^\infty \rho^{-\lambda}\mathbf{u}(r, \vartheta)\, \frac{\mathrm{d}\rho}{\rho}.$$

The argument in [5, 6, 7] is that the equivalent regularized variational formulation (1.8) provides an equivalent elliptic transmission problem where the standard Kontrat'ev approach [11] applies. This is the reason why in these works only the structure of singularities is investigated.

If we want to investigate the possibility of a *stable asymptotics* with respect to the parameter δ, we have to revisit the Mellin approach, in order to look for a Mellin symbol with meromorphic inverse which should depend analytically on δ (and acting between a couple of spaces independent from δ).

2.1. Ellipticity

We integrate by parts in the variational problem (1.8) and find (with $\mathbf{u} = \mathbf{E}^\delta$)

$$\begin{cases} \text{(i) } \mathbf{curl}\, \mu_C^{-1}\, \mathbf{curl}\, \mathbf{u}_C - \nabla\gamma_C \operatorname{div} \mathbf{u}_C - \omega^2\alpha_C \mathbf{u}_C = -i\omega\mathbf{j}_0 & \text{in } \Omega_C, \\ \text{(ii) } \mathbf{curl}\, \mu_E^{-1}\, \mathbf{curl}\, \mathbf{u}_E - \nabla\gamma_E \operatorname{div} \mathbf{u}_E - \omega^2\alpha_E \mathbf{u}_E = 0 & \text{in } \Omega_E, \end{cases} \tag{2.1}$$

with the essential boundary conditions, – we recall that α is defined in (1.6):

$$\begin{cases} \text{(i)}\ [\mathbf{u}\times\mathbf{n}] = 0 & \text{on } B, \\ \text{(ii)}\ [\alpha\mathbf{u}\cdot\mathbf{n}] = 0 & \text{on } B, \end{cases} \tag{2.2}$$

which we complement by the Neumann type transmission conditions

$$\begin{cases} \text{(i)}\ [\mu^{-1}\,\mathbf{curl}\,\mathbf{u}\times\mathbf{n}] = 0 & \text{on } B, \\ \text{(ii)}\ [\operatorname{div}\alpha\mathbf{u}] = 0 & \text{on } B. \end{cases} \tag{2.3}$$

Proposition 2.1. *Let L_C and L_E be the principal parts of the operators in* (2.1). *There exists $\delta_0 > 0$ such that for all $\delta \in [0, \delta_0]$ the set of transmission conditions* (2.2) *and* (2.3) *covers*[1] *the couple of operators (L_C, L_E) at any smooth point of the interface B.*

Proof. Let us take $\delta = 0$. We check that the 6×6 determinant obtained after partial Fourier transform at any point of the interface and reduction to the interface by the interior equations $L_C\mathbf{u} = 0$, $L_E\mathbf{u} = 0$ (symbol at the interface – cf. standard covering boundary conditions) is non-zero for any real $\xi = (\xi_1, \xi_2) \neq 0$.

The corresponding determinant is therefore non-zero for δ small enough. □

2.2. Local regularity

We deduce that our solution $\mathbf{u}$ corresponding to a charge density $\mathbf{j}_0$ in $L^2(\Omega_C)$ has an optimal local regularity up to the interface outside the corner O: for any neighborhood $\mathcal{V}$ such that $O \notin \overline{\mathcal{V}}$:

$$\mathbf{u}_C \in H^2(\Omega_C \cap \mathcal{V}) \quad \text{and} \quad \mathbf{u}_E \in H^2(\Omega_E \cap \mathcal{V}).$$

Let χ be a smooth cut-off function which is $\equiv 1$ in a neighborhood of O and has its support in the region where Ω_C and Ω_E coincide with the cones Γ_C and Γ_E respectively. Still denoting $\chi\mathbf{u}$ by $\mathbf{u}$, we are left with the following problem, instead of (2.1)–(2.3):

$$\begin{cases} \text{(i)}\ \mathbf{curl}\,\mu_C^{-1}\,\mathbf{curl}\,\mathbf{u}_C - \nabla\gamma_C\operatorname{div}\mathbf{u}_C\ (= L_C\mathbf{u}_C) = \mathbf{j}_C & \text{in } \Gamma_C, \\ \text{(ii)}\ \mathbf{curl}\,\mu_E^{-1}\,\mathbf{curl}\,\mathbf{u}_E - \nabla\gamma_E\operatorname{div}\mathbf{u}_E\ (= L_E\mathbf{u}_E) = \mathbf{j}_E & \text{in } \Gamma_E, \end{cases} \tag{2.4}$$

where $\mathbf{j}_C$ and $\mathbf{j}_E$ belong to $L^2(\Gamma_C)^3$ and $L^2(\Gamma_E)^3$ respectively (we have used the fact that $\mathbf{u}$ belongs to $L^2(\mathbb{R}^3)^3$ to put the term of order zero into the RHS),

$$\begin{cases} \text{(i)}\ [\mathbf{u}\times\mathbf{n}] = 0 & \text{on } I, \\ \text{(ii)}\ [\alpha\mathbf{u}\cdot\mathbf{n}] = 0 & \text{on } I, \end{cases} \tag{2.5}$$

where I denotes the interface $\partial\Gamma_C = \partial\Gamma_E$, and

$$\begin{cases} \text{(i)}\ [\mu^{-1}\,\mathbf{curl}\,\mathbf{u}\times\mathbf{n}] = \mathbf{h}_\top & \text{on } I, \\ \text{(ii)}\ [\operatorname{div}\alpha\mathbf{u}] = h_n & \text{on } I, \end{cases} \tag{2.6}$$

[1] In a sense extended from the classical Shapiro-Lopatinskii condition, [12, 15].

where $\mathbf{h}_\top \in H^{1/2}(I)^2$ and $h_n \in H^{1/2}(I)$. Since the support of $\mathbf{h}_\top$ and h_n is away from zero and infinity there also holds:

$$\rho^{-1/2}\mathbf{h}_\top \in L^2(I)^2 \quad \text{and} \quad \rho^{-1/2}h_n \in L^2(I). \tag{2.7}$$

A standard homogeneity argument based on the a priori estimate between the two nested annuli $\mathcal{A}_j = \{x : 2^{-j} < \rho < 2^j\}$, $j = 1, 2$:

$$\begin{aligned}\|\mathbf{u}_C\|_{H^2(\Gamma_C\cap\mathcal{A}_1)} + \|\mathbf{u}_E\|_{H^2(\Gamma_E\cap\mathcal{A}_1)} \leq C\big(&\|\mathbf{j}_C\|_{L^2(\Gamma_C\cap\mathcal{A}_2)} + \|\mathbf{j}_E\|_{L^2(\Gamma_E\cap\mathcal{A}_2)}\\ &+ \|\mathbf{h}_\top\|_{H^{1/2}(I\cap\mathcal{A}_2)} + \|h_n\|_{H^{1/2}(I\cap\mathcal{A}_2)}\big)\end{aligned}$$

yields the weighted regularity for $\mathbf{u}_C$ and $\mathbf{u}_E$:

$$\rho^{|\alpha|}\partial^\alpha \mathbf{u}_C \in L^2(\Gamma_C) \quad \text{and} \quad \rho^{|\alpha|}\partial^\alpha \mathbf{u}_E \in L^2(\Gamma_E), \quad \forall \alpha, |\alpha| \leq 2. \tag{2.8}$$

3. Mellin transform

For each fixed $\delta \geq 0$, we can apply to problem (2.4)–(2.6) the standard tools of the Mellin transform and residue formula. Let us recall that the Mellin symbol of an operator A homogeneous of degree m with constant coefficients is $\mathbb{C} \ni \lambda \mapsto \mathfrak{A}(\lambda)$ where

$$A(\partial_x) = \rho^{-m}\mathfrak{A}(\vartheta; \rho\partial_\rho, \partial_\vartheta) \quad \text{and} \quad \mathfrak{A}(\lambda) := \mathfrak{A}(\vartheta; \lambda, \partial_\vartheta).$$

If $\mathbf{U}(\lambda)$ is the Mellin transform of $\mathbf{u}$, then the Mellin transform $\mathbf{F}(\lambda)$ of $\rho^m A\mathbf{u}$ is $\mathfrak{A}(\lambda)\mathbf{U}(\lambda)$. Practically, we have the relation

$$A(\partial_x)\big(\rho^\lambda \mathbf{U}(\vartheta)\big) = \rho^{\lambda-m}\mathfrak{A}(\lambda)\mathbf{U}(\lambda).$$

3.1. Mellin symbol

In the case of our problem, we define a Mellin symbol acting between spaces which do not depend on λ of course, but also not on δ. We take as source space

$$\mathbf{Y}^+(G_C, G_E) = \big\{\mathbf{U} \in L^2(\mathbb{S}^2) : \mathbf{U}_C \in H^2(G_C),\ \mathbf{U}_E \in H^2(G_E),\ [\mathbf{U}\times\mathbf{n}] = 0\big\}$$

and as target space

$$\mathbf{Z}(G_C, G_E) = L^2(G_C)^3 \times L^2(G_E)^3 \times H^{3/2}(J) \times H^{1/2}(J)^2 \times H^{1/2}(J),$$

where we recall that $G_C = \Gamma_C \cap \mathbb{S}^2$ and $G_E = \Gamma_C \cap \mathbb{S}^2$, and J denotes $\partial G_C \cap \partial G_E$. Let us fix $\lambda \in \mathbb{C}$. Our Mellin symbol $\mathfrak{L}(\lambda)$ is defined as follows

$$\begin{aligned}\mathfrak{L}(\lambda): \ \mathbf{Y}^+(G_C, G_E) &\longrightarrow \mathbf{Z}(G_C, G_E)\\ (\mathbf{U}_C, \mathbf{U}_E) &\longmapsto \mathbf{F} = (\mathbf{J}_C, \mathbf{J}_E, G, \mathbf{H}_\top, H_n),\end{aligned} \tag{3.1}$$

with

$$L_C(\partial_x)\big(\rho^\lambda \mathbf{U}_C(\vartheta)\big) = \rho^{\lambda-2}\mathbf{J}_C(\vartheta), \quad L_E(\partial_x)\big(\rho^\lambda \mathbf{U}_E(\vartheta)\big) = \rho^{\lambda-2}\mathbf{J}_E(\vartheta)$$

$$\big[\alpha\rho^\lambda \mathbf{U}(\vartheta)\cdot\mathbf{n}\big] = \rho^\lambda G(\vartheta)$$

$$\big[\mu^{-1}\,\mathbf{curl}\,\rho^\lambda\mathbf{U}(\vartheta)\times\mathbf{n}\big] = \rho^{\lambda-1}\mathbf{H}_\top(\vartheta), \quad \big[\operatorname{div}\alpha\rho^\lambda\mathbf{U}(\vartheta)\big] = \rho^{\lambda-1}H_n(\vartheta).$$

As a standard consequence of the ellipticity (Proposition 2.1), we obtain that

- $\mathfrak{L}(\lambda)$ is invertible except for λ in a discrete set $\mathfrak{S} \subset \mathbb{C}$,
- $\lambda \mapsto \mathfrak{L}(\lambda)^{-1}$ is meromorphic,
- at each pole $\lambda \in \mathfrak{S}$, the range of the polar part of $\mathfrak{L}^{-1}$ is finite-dimensional,
- in each strip of the form $\xi_1 \leq \operatorname{Re}\lambda \leq \xi_2$ there are at most a finite number of elements of $\mathfrak{S}$.

The elements λ of $\mathfrak{S}$ are the poles of $\mathfrak{L}(\lambda)^{-1}$ and they are also called *singular exponents* because they are the possible degrees of homogeneity of the singular parts in the conical asymptotics of solutions.

3.2. Splitting in regular and singular parts

As a consequence of the regularity and support properties of the data, cf. (2.7), the Mellin transform $\lambda \mapsto \mathbf{F}(\lambda)$ of

$$(\rho^2 \mathbf{f}_C, \rho^2 \mathbf{f}_E, 0, \rho \mathbf{h}_\top, \rho h_n)$$

is holomorphic for λ in the half-plane $\operatorname{Re}\lambda < \frac{1}{2}$ with values in the space $\mathbf{Z}(G_C, G_E)$, whereas, thanks to (2.8), the Mellin transform $\lambda \mapsto \mathbf{U}(\lambda)$ of $\mathbf{u}$ is holomorphic for λ in the half-plane $\operatorname{Re}\lambda < -\frac{3}{2}$ with values in the space $\mathbf{Y}^+(G_C, G_E)$. Moreover, there holds

$$\mathfrak{L}(\lambda)\mathbf{U}(\lambda) = \mathbf{F}(\lambda), \quad \forall \lambda, \operatorname{Re}\lambda \leq -\frac{3}{2}. \tag{3.2}$$

The function $\lambda \mapsto \mathfrak{L}(\lambda)^{-1}\mathbf{F}(\lambda)$ is a meromorphic extension of $\mathbf{U}(\lambda)$ to the strip $-\frac{3}{2} \leq \operatorname{Re}\lambda < \frac{1}{2}$. As standard in the Kondrat'ev Cauchy residue analysis we obtain

$$\mathbf{u} = \mathbf{u}_{\mathsf{reg}} - \frac{1}{2i\pi} \int_\gamma \rho^\lambda \mathfrak{L}(\lambda)^{-1} \mathbf{F}(\lambda)\, \mathrm{d}\lambda \tag{3.3}$$

where γ is a simple curve surrounding all the poles of $\mathfrak{L}(\lambda)^{-1}$ in the strip $-\frac{3}{2} \leq \operatorname{Re}\lambda < \frac{1}{2}$, and $\mathbf{u}_{\mathsf{reg}}$ is the regular part satisfying, if $\mathfrak{S} \cap \{\lambda : \operatorname{Re}\lambda = \frac{1}{2}\}$ is empty:

$$\rho^{|\alpha|-2} \partial^\alpha \mathbf{u}_{\mathsf{reg},C} \in L^2(\Gamma_C) \quad \text{and} \quad \rho^{|\alpha|-2} \partial^\alpha \mathbf{u}_{\mathsf{reg},E} \in L^2(\Gamma_E), \quad \forall \alpha, |\alpha| \leq 2. \tag{3.4}$$

Note that, in particular, the regular part $\mathbf{u}_{\mathsf{reg}}$ is H^2 in any neighborhood of zero inside Γ_C and Γ_E.

The properties of the polar part of $\mathfrak{L}^{-1}$ inherited from the ellipticity imply that the residue in (3.3) (the singular part) spans a finite-dimensional space.

4. Stable asymptotics with respect to δ

We now trace the dependency with respect to δ as $\delta \to 0$ in the above decomposition (3.3). We fix the right-hand side $\mathbf{j}_0$ and consider $\mathbf{u}^\delta := \chi \mathbf{E}^\delta$ the localized solution of problem (1.8). Since the form a^δ depends analytically on δ, the solution $\mathbf{E}^\delta$ also depends analytically on δ in $\mathbf{Y}(\Omega)$ for δ in a neighborhood of $\delta = 0$. Therefore, the localized solutions $\mathbf{u}^\delta$ and the associated right-hand sides $(\mathbf{f}_C^\delta, \mathbf{f}_E^\delta, \mathbf{h}_\top^\delta, h_n^\delta)$ depend analytically on δ.

Thus, with $\lambda \mapsto \mathbf{F}^\delta(\lambda)$ the Mellin transform of

$$(\rho^2 \mathbf{f}_C^\delta, \rho^2 \mathbf{f}_E^\delta, 0, \rho \mathbf{h}_\mathrm{T}^\delta, \rho h_n^\delta)$$

we finally have the splitting of $\mathbf{u}^\delta$ for each δ according to (3.3):

$$\mathbf{u}^\delta = \mathbf{u}^\delta_{\mathrm{reg}} - \frac{1}{2i\pi}\int_\gamma \rho^\lambda \mathfrak{L}^\delta(\lambda)^{-1}\mathbf{F}^\delta(\lambda)\,\mathrm{d}\lambda, \tag{4.1}$$

where $\mathfrak{L}^\delta$ is the Mellin symbol of problem (2.4)–(2.6) with α depending on δ via formula (1.6).

4.1. Spaces of singularities and characteristic polynomials

Definition 4.1. For a $\delta \geq 0$ and a contour γ which does not meet any pole of $\mathfrak{L}^\delta(\lambda)^{-1}$, we denote by $\mathbf{S}^\delta_\gamma$ the finite-dimensional space spanned by the residues

$$\frac{1}{2i\pi}\int_\gamma \rho^\lambda \mathfrak{L}^\delta(\lambda)^{-1}\mathbf{F}(\lambda)\,\mathrm{d}\lambda$$

for all holomorphic $\lambda \mapsto \mathbf{F}(\lambda)$ with values in $\mathbf{Z}(G_C, G_E)$. We denote by m^δ_γ its dimension.

We have the classical result, see [13] for example:

Lemma 4.2. *For δ and γ as in the definition above, let D be the interior region such that $\partial D = \gamma$. Then*

$$m^\delta_\gamma = \sum_{\nu \in D \cap \mathfrak{S}^\delta} m^\delta(\nu)$$

where $\mathfrak{S}^\delta$ is the set of poles of $\mathfrak{L}^\delta(\lambda)^{-1}$ and $m^\delta(\nu)$ is the sum of the lengths of all Keldysh chains associated with the pole ν.

Definition 4.3. We call characteristic polynomial of $\mathfrak{L}^\delta$ inside γ:

$$P^\delta_\gamma(\lambda) = \prod_{\nu \in D \cap \mathfrak{S}^\delta} (\lambda - \nu)^{m^\delta(\nu)}.$$

The operator-valued function $\lambda \mapsto P^\delta_\gamma(\lambda)\,\mathfrak{L}^\delta(\lambda)^{-1}$ is holomorphic in $\overline{D}$.

From the "operator valued Rouché formula" of Gohberg-Sigal [8], we find:

Lemma 4.4. *The following representation holds for $P^\delta_\gamma(\mu)$ when $\mu \notin \overline{D}$:*

$$P^\delta_\gamma(\mu) = \exp\Big\{\mathrm{tr}\,\frac{1}{2i\pi}\int_\gamma \mathfrak{L}^\delta(\lambda)'\,\mathfrak{L}^\delta(\lambda)^{-1}\log(\mu - \lambda)\,d\lambda\Big\}.$$

Here we fix the contour γ so that $\mathfrak{L}^0(\lambda)^{-1}$ has no pole on γ. Therefore, for $0 \leq \delta \leq \delta_0$ with $\delta_0 > 0$ small enough, $\mathfrak{L}^\delta(\lambda)^{-1}$ has still no poles on γ. The analytic dependency of $\mathfrak{L}^\delta(\lambda)^{-1}$ on δ when $\lambda \in \gamma$ implies that there holds:

Corollary 4.5. *The coefficients of the characteristic polynomial P^δ_γ depend analytically on δ for $0 \leq \delta \leq \delta_0$. In particular its degree does not depend on δ for $0 \leq \delta \leq \delta_0$: we denote it by m_γ.*

4.2. Stable asymptotics: general result

Let us recall the notion of "distance" between two subspaces E and F of the same Hilbert space H, cf. [10]:

$$\operatorname{dist}(E,F) = \max_{\mathbf{u}\in E}\min_{\mathbf{v}\in F} \frac{\|\mathbf{u}-\mathbf{v}\|_H}{\|\mathbf{u}\|_H}\,. \tag{4.2}$$

In general this distance is not symmetric. However, if E and F are finite-dimensional subspaces satisfying $\operatorname{dist}(E,F) < 1$ and $\operatorname{dist}(F,E) < 1$, then $\dim E = \dim F$ and $\operatorname{dist}(E,F) = \operatorname{dist}(F,E)$.

Again as a consequence of the analytic dependency of $\mathfrak{L}^\delta(\lambda)^{-1}$ on δ we find that

$$\operatorname{dist}(\mathbf{S}^0_\gamma\,,\,\mathbf{S}^\delta_\gamma) = \operatorname{dist}(\mathbf{S}^\delta_\gamma\,,\,\mathbf{S}^0_\gamma) = \mathcal{O}(\delta)$$

and that there holds

Theorem 4.6. *We can choose a basis $\mathbf{s}^\delta_m$, $m = 1,\dots,m_\gamma$, of $\mathbf{S}^\delta_\gamma$ depending analytically on δ. The decomposition* (4.1) *of the localized solution $\mathbf{u}^\delta = \chi\mathbf{E}^\delta$ of problem* (1.8) *can be written*

$$\mathbf{u}^\delta = \mathbf{u}^\delta_{\mathsf{reg}} + \sum_{m=1}^{m_\gamma} c^\delta_m \mathbf{s}^\delta_m, \quad \textit{with} \quad [0,\delta_0] \ni \delta \mapsto c^\delta_m \in \mathbb{C} \textit{ analytic}. \tag{4.3}$$

with $\mathbf{u}^\delta_{\mathsf{reg},C}$ in $H^2(\Omega_C)$ and $\mathbf{u}^\delta_{\mathsf{reg},E}$ in $H^2(\Omega_E)$.

Let $\nu^0_1,\dots,\nu^0_d$ be the distinct elements of $\mathfrak{S}^0 \cap D$. Let $\gamma_1,\dots,\gamma_d$ be contours around each of them and such that their interior regions $D_1,\dots,D_d$ are pairwise disjoint. For $0 \le \delta \le \delta_0$ with $\delta_0 > 0$ small enough, $\gamma_1,\dots,\gamma_d$ do not intersect $\mathfrak{S}^\delta$. We clearly have

$$\mathbf{S}^\delta_\gamma = \bigoplus_{j=1}^d \mathbf{S}^\delta_{\gamma_j}. \tag{4.4}$$

According to Corollary 4.5, each characteristic polynomial $P^\delta_{\gamma_j}$ has coefficients depending analytically on δ. The roots of $P^\delta_{\gamma_j}$ form the *cluster* of singular exponents ν^δ_i which tend to ν^0_j as $\delta \to 0$.

4.3. Stable asymptotics: cases of degrees 1 and 2

In particular, if the degree of $P^\delta_{\gamma_j}$ is one, the situation is "simple" as described in the following proposition.

Proposition 4.7. *Let $\nu^0 = \nu^0_j \in \mathfrak{S}^0 \cap D$. If its multiplicity $m^0(\nu^0)$ is 1, then there exists a unique element ν^δ of $\mathfrak{S}^\delta \cap D$ which tends to ν^0 as $\delta \to 0$. Moreover, $\delta \mapsto \nu^\delta$ is analytic and a basis $\mathbf{s}^\delta$ of $\mathbf{S}^\delta_{\gamma_j}$ can be chosen of the form*

$$\mathbf{s}^\delta(\rho,\vartheta) = \rho^{\nu^\delta}\mathbf{U}^\delta(\vartheta)$$

where $\delta \mapsto \mathbf{U}^\delta$ is analytic in δ with values in $\mathbf{Y}^+(G_C,G_E)$.

Although unpredictable in general, the simplicity of limiting exponents of singularity is generic. Nevertheless, the general question of *stable* asymptotics with respect to δ keeps its interest.

If the characteristic polynomial $P^{\delta}_{\gamma_j}$ is of degree 2, the situation is more complicated, but still possible to describe:

Proposition 4.8. *Let $\nu^0 = \nu^0_j \in \mathfrak{S}^0 \cap D$. If its multiplicity $m^0(\nu^0)$ is 2, then there exists at least one and at most two elements ν^δ_i, $i = 1, 2$, of $\mathfrak{S}^\delta \cap D$ which tend to ν^0 as $\delta \to 0$.*

Let us assume that the pole of $\mathfrak{L}(\lambda)^{-1}$ in ν^0 is of order 2 and that $\nu^\delta_1 \neq \nu^\delta_2$ for all $\delta \neq 0$ small enough. Then we can choose a basis of $\mathbf{S}^0_{\gamma_j}$ of the form

$$\mathbf{s}^0(\rho,\vartheta) = \rho^{\nu^0}\mathbf{U}^0(\vartheta) \quad \textit{and} \quad \mathbf{t}^0(\rho,\vartheta) = \mathbf{s}^0(\rho,\vartheta)\log\rho + \rho^{\nu^0}\mathbf{V}^0(\vartheta).$$

Moreover, for all $\delta \neq 0$ small enough there exist two singular functions $\mathbf{s}^\delta_i$, $i = 1, 2$ of the form

$$\mathbf{s}^\delta_1(\rho,\vartheta) = \rho^{\nu^\delta_1}\mathbf{U}^\delta_1(\vartheta) \quad \textit{and} \quad \mathbf{s}^\delta_2(\rho,\vartheta) = \rho^{\nu^\delta_2}\mathbf{U}^\delta_2(\vartheta)$$

such that

(i) *$\mathbf{s}^\delta_i$ tends to $\mathbf{s}^0$ as $\delta \to 0$, $i = 1, 2$,*

(ii) *the two functions*

$$\frac{\mathbf{s}^\delta_1 - \mathbf{s}^\delta_2}{\nu^\delta_1 - \nu^\delta_2} \quad \textit{and} \quad \mathbf{s}^\delta_1 + \mathbf{s}^\delta_2$$

are a basis of $\mathbf{S}^\delta_{\gamma_j}$ depending analytically on δ. The first function tends to $\mathbf{t}^0$ and the second one to $2\mathbf{s}^0$ as $\delta \to 0$.

Proof. The first assertions are classical.

Let us prove the existence of $\mathbf{s}^\delta_i$ satisfying the next assertions of the proposition. Let us denote for short the "small" contour surrounding ν^0 by γ instead of γ_j. By construction $P^0_\gamma(\lambda)\mathfrak{L}^0(\lambda)^{-1}$ is holomorphic in the region $\overline{D}$ surrounded by γ. As the pole ν^0 is double, there exists $\mathbf{F}(\lambda)$ holomorphic in $\overline{D}$ such that $(\lambda - \nu^0)\mathfrak{L}^0(\lambda)^{-1}\mathbf{F}(\lambda)$ has a pole of order 1 in ν^0. Let us define

$$\mathbf{U}^\delta(\lambda) := P^\delta_\gamma(\lambda)\mathfrak{L}^\delta(\lambda)^{-1}\mathbf{F}(\lambda).$$

Thus $\mathbf{U}^\delta(\nu^0)$ is not zero and the residue of $\rho^\lambda(\lambda - \nu^0)P^\delta_\gamma(\lambda)^{-1}\mathbf{U}^\delta(\lambda)$ is an element of $\mathbf{S}^0_\gamma$, which we denote by $\mathbf{s}^0$:

$$\mathbf{s}^0(\rho,\vartheta) = \rho^{\nu^\delta_0}\mathbf{U}^\delta(\nu^\delta_0). \tag{4.5}$$

The function $\lambda \mapsto \mathbf{U}^\delta(\lambda)$ is holomorphic in $\overline{D}$ and analytic in δ. The residues of

$$\lambda \longmapsto \rho^\lambda \frac{\mathbf{U}^\delta(\lambda)}{P^\delta_\gamma(\lambda)} \quad \text{in} \quad \nu^\delta_1 \text{ and } \nu^\delta_2$$

are

$$(\nu^\delta_1 - \nu^\delta_2)^{-1}\rho^{\nu^\delta_1}\mathbf{U}^\delta(\nu^\delta_1) \quad \text{and} \quad (\nu^\delta_2 - \nu^\delta_1)^{-1}\rho^{\nu^\delta_2}\mathbf{U}^\delta(\nu^\delta_2)$$

respectively, and define elements of $\mathbf{S}^\delta_\gamma$. Let us set

$$\mathbf{s}_i^\delta := \rho^{\nu_i^\delta}\,\mathbf{U}^\delta(\nu_i^\delta), \quad i = 1, 2. \tag{4.6}$$

From (4.5), (4.6), the continuity of $\delta \mapsto \nu_i^\delta$ as $\delta \to 0$ and the regularity of $(\delta, \lambda) \mapsto \mathbf{U}^\delta(\lambda)$, we deduce that $\mathbf{s}_i^\delta$ tend to $\mathbf{s}^0$ as $\delta \to 0$. We check that

$$\frac{1}{2i\pi}\int_\gamma \rho^\lambda P^\delta_\gamma(\lambda)^{-1}\mathbf{U}^\delta(\lambda)\,\mathrm{d}\lambda = \frac{\mathbf{s}_1^\delta - \mathbf{s}_2^\delta}{\nu_1^\delta - \nu_2^\delta}$$

and

$$\frac{1}{2i\pi}\int_\gamma \rho^\lambda P^\delta_\gamma(\lambda)^{-1}(2\lambda - \nu_1^\delta - \nu_2^\delta)\mathbf{U}^\delta(\lambda)\,\mathrm{d}\lambda = \mathbf{s}_1^\delta + \mathbf{s}_2^\delta.$$

They both belong to $\mathbf{S}^\delta_\gamma$ and, since $\nu_1^\delta + \nu_2^\delta$ depends analytically on δ, they both depend analytically on δ too. □

General stable behaviors with respect to ρ for more general characteristic polynomials are investigated in [4] and [16].

5. Maxwell singularities

In [5] it is proved that the singularities of the electric fields at corners and edges of a perfectly conducting polyhedral body all derive from scalar potentials, namely via three different types:

Type 1 are gradients of singularities of the Dirichlet Laplacian.

Type 2 are such that their curl is equal to the gradient of singularities of the Neumann Laplacian.

Type 3 are such that their divergence is a singularity of the Dirichlet Laplacian.

In [6, 7] we have simply applied the same procedure as in [5] to describe the three different types in the case of transmission and eddy current problems. In this section, we are going to develop the arguments leading to such a description in these cases.

We investigate the structure of the elements of $\mathbf{S}^\delta_\gamma$. Let $\mathbf{s}^\delta$ belong to $\mathbf{S}^\delta_\gamma$: There exists $\mathbf{F}$ holomorphic such that

$$\mathbf{s}^\delta = \int_\gamma \rho^\lambda \mathfrak{L}^\delta(\lambda)^{-1}\mathbf{F}(\lambda)\,\mathrm{d}\lambda.$$

Let $\mathfrak{L}^\delta = \mathfrak{L}^\delta(\rho\partial_\rho)$ be the collection of interior and interface operators

$$\mathbf{u} \mapsto \mathfrak{L}^\delta\mathbf{u} = (\rho^2 L_C\mathbf{u}_C,\ \rho^2 L_E\mathbf{u}_E,\ [\alpha\mathbf{u}\cdot\mathbf{n}],\ \rho[\mu^{-1}\,\mathbf{curl}\,\mathbf{u}\times\mathbf{n}],\ \rho[\operatorname{div}\alpha\mathbf{u}]),$$

with α defined in (1.6), compare with (2.4)–(2.6). We have

$$\mathfrak{L}^\delta\mathbf{s}^\delta = \int_\gamma \rho^\lambda \mathfrak{L}^\delta(\lambda)\mathfrak{L}^\delta(\lambda)^{-1}\mathbf{F}(\lambda)\,\mathrm{d}\lambda = \int_\gamma \rho^\lambda\mathbf{F}(\lambda)\,\mathrm{d}\lambda = 0.$$

Let us fix $\delta \in [0, \delta_0]$, and thus α which is given by (1.6). From now on, we drop the exponent δ.

In other words, $\mathbf{s}$ is a solution of the elliptic interface problem (2.4)–(2.6) with zero right-hand side:

$$\begin{cases} \text{(i) } \mathbf{curl}\,\mu_C^{-1}\,\mathbf{curl}\,\mathbf{s}_C - \nabla\gamma_C \operatorname{div}\mathbf{s}_C = 0 & \text{in } \Gamma_C, \\ \text{(ii) } \mathbf{curl}\,\mu_E^{-1}\,\mathbf{curl}\,\mathbf{s}_E - \nabla\gamma_E \operatorname{div}\mathbf{s}_E = 0 & \text{in } \Gamma_E, \end{cases} \tag{5.1}$$

$$\begin{cases} \text{(i) } [\mathbf{s}\times\mathbf{n}] = 0 & \text{on } I, \\ \text{(ii) } [\alpha\mathbf{s}\cdot\mathbf{n}] = 0 & \text{on } I, \end{cases} \tag{5.2}$$

$$\begin{cases} \text{(i) } [\mu^{-1}\,\mathbf{curl}\,\mathbf{s}\times\mathbf{n}] = 0 & \text{on } I, \\ \text{(ii) } [\operatorname{div}\alpha\mathbf{s}] = 0 & \text{on } I. \end{cases} \tag{5.3}$$

5.1. Uncoupled systems

The key of the investigation of solutions $\mathbf{s}$ of (5.1)–(5.3) is the introduction of the auxiliary unknowns

$$q \text{ s.t. } \begin{cases} q_C = \operatorname{div}\mathbf{s}_C & \text{in } \Gamma_C, \\ q_E = \operatorname{div}\mathbf{s}_E & \text{in } \Gamma_E, \end{cases} \quad \text{and} \quad \boldsymbol{\psi} \text{ s.t. } \begin{cases} \boldsymbol{\psi}_C = \mu_C^{-1}\,\mathbf{curl}\,\mathbf{s}_C & \text{in } \Gamma_C, \\ \boldsymbol{\psi}_E = \mu_E^{-1}\,\mathbf{curl}\,\mathbf{s}_E & \text{in } \Gamma_E. \end{cases}$$

Taking the divergence of equations (5.1), we find that q_C and q_E are harmonic. Condition (5.3)(ii) gives that $[\alpha q] = 0$. Taking the normal component (with respect to the interface I) of equations (5.1), we find that

$$\gamma_C\partial_n \operatorname{div}\mathbf{s}_C = \mathbf{curl}_{\mathsf{T}}(\mu_C^{-1}\,\mathbf{curl}\,\mathbf{s}_C)_{\mathsf{T}}$$

$$\gamma_E\partial_n \operatorname{div}\mathbf{s}_E = \mathbf{curl}_{\mathsf{T}}(\mu_E^{-1}\,\mathbf{curl}\,\mathbf{s}_E)_{\mathsf{T}}$$

Thanks to (5.3)(i), we obtain that the jump of $\mathbf{curl}_{\mathsf{T}}(\mu^{-1}\,\mathbf{curl}\,\mathbf{s})_{\mathsf{T}}$ is zero, whence $[\gamma\partial_n \operatorname{div}\mathbf{s}] = 0$, i.e., $[\gamma\partial_n q] = 0$. When $\delta \neq 0$, we can write equivalently that $[\alpha^{-1}\gamma\partial_n(\alpha q)] = 0$. Summing up, we have obtained if $\delta \neq 0$

$$\begin{cases} \Delta q_C = 0 & \text{in } \Gamma_C, \\ \Delta q_E = 0 & \text{in } \Gamma_E, \\ [\alpha q] = 0, \ [\alpha^{-1}\gamma\partial_n \alpha q] = 0 & \text{on } I, \end{cases} \tag{5.4}$$

and if $\delta = 0$:

$$\begin{cases} \Delta q_C = 0 & \text{in } \Gamma_C, \\ \Delta q_E = 0 & \text{in } \Gamma_E, \\ q_C = 0, \ [\gamma\partial_n q] = 0 & \text{on } I. \end{cases} \tag{5.5}$$

Both problems (5.4) and (5.5) are scalar elliptic transmission problems. Besides the trivial zero solution, their solutions are the singularities of these problems.

For $\boldsymbol{\psi}$, we obtain from (5.1), (5.2)(i) and (5.3)(i):

$$\begin{cases} \mathbf{curl}\,\boldsymbol{\psi}_C = \nabla q_C, \ \operatorname{div}(\mu_C\boldsymbol{\psi}_C) = 0 & \text{in } \Gamma_C, \\ \mathbf{curl}\,\boldsymbol{\psi}_E = \nabla q_E, \ \operatorname{div}(\mu_E\boldsymbol{\psi}_E) = 0 & \text{in } \Gamma_E, \\ [\boldsymbol{\psi}\times\mathbf{n}] = 0, \ [\mu\boldsymbol{\psi}\cdot\mathbf{n}] = 0 & \text{on } I. \end{cases} \tag{5.6}$$

Finally for $\mathbf{s}$ we have:

$$\begin{cases} \mathbf{curl}\,\mathbf{s}_C = \mu_C \boldsymbol{\psi}_C, \ \operatorname{div} \mathbf{s}_C = q_C & \text{in } \Gamma_C, \\ \mathbf{curl}\,\mathbf{s}_E = \mu_E \boldsymbol{\psi}_E, \ \operatorname{div} \mathbf{s}_E = q_E & \text{in } \Gamma_E, \\ [\mathbf{s} \times \mathbf{n}] = 0, \ [\alpha \mathbf{s} \cdot \mathbf{n}] = 0 & \text{on } I, \end{cases} \tag{5.7}$$

the last transmission condition becoming $\mathbf{s}_C \cdot \mathbf{n} = 0$ on I if $\delta = 0$.

5.2. The three types of Maxwell singularities

The Maxwell or eddy current singularities of Type 1 are the solutions of (5.7) with $\boldsymbol{\psi} = 0$ and $q = 0$. We can see that they are the gradient of scalar singularities: $\mathbf{s}_C = \nabla \Phi_C$ and $\mathbf{s}_E = \nabla \Phi_E$ with

$$\begin{cases} \Delta \Phi_C = 0 & \text{in } \Gamma_C, \\ \Delta \Phi_E = 0 & \text{in } \Gamma_E, \\ [\Phi] = 0, \ [\alpha \partial_n \Phi] = 0 & \text{on } I, \end{cases} \tag{5.8}$$

the last transmission condition becoming $\partial_n \Phi_C = 0$ on I if $\delta = 0$.

The Maxwell or eddy current singularities of Type 2 are particular solutions of (5.7) with $\boldsymbol{\psi}$ solution of (5.6) and $q = 0$. We find that $\boldsymbol{\psi}$ is a gradient: $\boldsymbol{\psi}_C = \nabla \Psi_C$ and $\boldsymbol{\psi}_E = \nabla \Psi_E$ with

$$\begin{cases} \Delta \Psi_C = 0 & \text{in } \Gamma_C, \\ \Delta \Psi_E = 0 & \text{in } \Gamma_E, \\ [\Psi] = 0, \ [\mu \partial_n \Psi] = 0 & \text{on } I. \end{cases}$$

The solutions of the above transmission problem with *real coefficients* are linear combinations of terms of the form $\Psi = \rho^\lambda \psi(\vartheta)$. Then a particular solution $\mathbf{s}$ associated with such a Ψ is given by (for $\lambda \neq -1$)

$$\mathbf{s} = (\lambda + 1)^{-1} (\mu \nabla \Psi \times \mathbf{x} - \nabla r)$$

with r solution of

$$\begin{cases} \Delta r_C = 0 & \text{in } \Gamma_C, \\ \Delta r_E = 0 & \text{in } \Gamma_E, \\ [r] = 0, \ [\alpha \partial_n r] = [\alpha \mu] (\nabla \Psi \times \mathbf{x}) \cdot \mathbf{n} & \text{on } I, \end{cases}$$

the last condition being replaced with $\partial_n r_C = \mu_C (\nabla \Psi_C \times \mathbf{x}) \cdot \mathbf{n}$ on I when $\delta = 0$.

The singularities of Type 3 are particular solutions of (5.7), with $\boldsymbol{\psi}$ particular solution of (5.6) and q general solution of (5.5).

In our case, the singularities of Type 3 can be discarded because their divergences form the asymptotic part of the divergence of the solution. Since our solutions are divergence-free, these singularities do not appear in the asymptotics (4.1).

For any singular exponent $\nu^0 \in \mathfrak{S}^0$ of the eddy current problem which is not at the same time an exponent of type 1 and an exponent of type 2, we have as $\delta \to 0$ a stable cluster of singular exponents ν_i^δ of the same type tending to ν^0.

Let us develop that point for singularities of type 1 (which are the most important ones, since they are the only possible non-H^1 singularities). Let $\nu^0 \in \mathfrak{S}^0$ be of type 1, and let γ be a simple contour isolating ν^0 from the other elements of $\mathfrak{S}^0$. There exists $\delta_0 > 0$ such that for all $\delta \in [0, \delta_0]$, $\gamma \cap \mathfrak{S}^\delta = \emptyset$.

Let $\lambda \mapsto \mathfrak{M}^\delta(\lambda)$ be the Mellin symbol of problem (5.8) and let $\mathbf{T}^\delta_{\gamma+1}$ be the singularity space for potentials Φ: $\mathbf{T}^\delta_{\gamma+1}$ is spanned by the residues

$$\frac{1}{2i\pi} \int_{\gamma+1} \rho^\lambda \mathfrak{M}^\delta(\lambda)^{-1} \mathbf{G}(\lambda) \, \mathrm{d}\lambda$$

for all holomorphic $\lambda \mapsto \mathbf{G}(\lambda)$. Then $\mathbf{S}^\delta_\gamma = \nabla \mathbf{T}^\delta_{\gamma+1}$.

References

[1] H. Ammari, A. Buffa and J.-C. Nédélec, *A justification of eddy currents model for the Maxwell equations*, SIAM J. Appl. Math. **60** (2000), 1805–1823.

[2] A. Bossavit, *Two dual formulations of the 3D eddy-current problem*, COMPEL **4** (1985), 103–116.

[3] A. Bossavit, *Electromagnétisme en vue de la modélisation*, Springer Verlag, 1993.

[4] M. Costabel and M. Dauge, *Singularités d'arêtes pour les problèmes aux limites elliptiques*. In *Journées "Équations aux Dérivées Partielles" (Saint-Jean-de-Monts, 1992)*. École Polytech., Palaiseau, 1992, Exp. No. IV, 12p.

[5] M. Costabel and M. Dauge, *Singularities of Maxwell's equations on polyhedral domains*, Arch. Rational Mech. Anal. **151** (2000), 221–276.

[6] M. Costabel, M. Dauge and S. Nicaise, *Singularities of Maxwell interface problems*, RAIRO Modél. Math. Anal. Numér. **33** (1999), 627–649.

[7] M. Costabel, M. Dauge and S. Nicaise, *Singularities of eddy current problems*, Preprint of the Newton Institute **03019**, Cambridge (2003).

[8] I. Gohberg and E. Sigal, *An operator generalization of the logarithmic residue theorem and the theorem of Rouché*, Math. USSR Sbornik **13 (4)** (1971), 603–625.

[9] R. Hiptmair, *Symmetric coupling for eddy currents problems*, SIAM J. Numer. Anal. **40** (2002), 41–65.

[10] T. Kato, *Perturbation theory for nullity, deficiency and other quantities of linear operators*, J. Analyse Math. **6** (1958), 261–322.

[11] V.A. Kondrat'ev, *Boundary-value problems for elliptic equations in domains with conical or angular points*, Trans. Moscow Math. Soc. **16** (1967), 227–313.

[12] J.-L. Lions and E. Magenes, *Problèmes aux limites non homogènes et applications*, Vol. 1. Dunod, Paris, 1968.

[13] V.G. Maz'ya, B.A. Plamenevskii, *On the coefficients in the asymptotic of solutions of the elliptic boundary problem in domains with conical points,* Amer. Math. Soc. Trans. (2) **123** (1984), 57–88.

[14] V.G. Maz'ya and J. Rossmann, *On a problem of Babuška (Stable asymptotics of the solution to the Dirichlet problem for elliptic equations of second order in domains with angular points)*, Math. Nachr. **155** (1992), 199–220.

[15] S. Nicaise, *Polygonal interface problems*, Peter Lang, Berlin, 1993.

[16] B. Schmutzler, *Branching asymptotics for elliptic boundary value problems in a wedge*, in "Boundary value problems and integral equations in nonsmooth domains (Luminy, 1993)", M. Costabel, M. Dauge, and S. Nicaise, eds., Dekker, New York, (1995), 255–267.

Martin Costabel
IRMAR
Université de Rennes 1
Campus de Beaulieu
F-35042 Rennes
France
e-mail: costabel@univ-rennes1.fr

Monique Dauge
IRMAR
Université de Rennes 1
Campus de Beaulieu
F-35042 Rennes
France
e-mail: dauge@univ-rennes1.fr

Serge Nicaise
Université de Valenciennes et du Hainaut Cambrésis
MACS, Le Mont Houy
F-59313 Valenciennes Cedex 9
France
e-mail: snicaise@univ-valenciennes.fr

Operator Theory
Advances and Applications, Vol. 147, 257–276

On Some Aspects of Analysis of Incompressible Flow through Cascades of Profiles

Miloslav Feistauer and Tomáš Neustupa

Dedicated to the memory of Professor Dr. Erhard Meister.

Abstract. The paper is concerned with the analysis of the model of incompressible, viscous, stationary flow through a plane cascade of profiles. We give a classical as well as weak formulation of the problem and discuss the existence of a weak solution.

1. Introduction

One of the most important mathematical models used in the investigation of aerodynamical properties of blade rows of turbines, compressors and pumps is the flow through cascades of profiles. A series of works are concerned with various types of cascade flows. For example, the potential theory and the complex function theory were extensively used ([17], [2]). The well-known Martensen's method applies the integral equation method for the solution of inviscid, irrotational, incompressible flow through a plane cascade of profiles. See, for example, [16], [18].

Modern computers allow the numerical simulation of the cascade flow in general situations (e.g., viscous compressible flow, see [3]), but a mathematical theory of such flows is still missing even under simplified assumptions.

In this paper we shall be concerned with the theoretical analysis of incompressible, viscous, stationary flow through a plane cascade of profiles. Our main goal is to prove the existence of a weak solution of this problem. As we shall see, it will be necessary to overcome some obstacles. As a result we shall get the solvability of the problem in the case of small data.

The theory presented here will be a starting point for the finite element analysis of the cascade flow problem, which is of a great importance for engineers and fluid dynamists. This is the reason that we try to make the presented analysis sufficiently accessible and, therefore, the treatment is more detailed than would be necessary for specialists in nonlinear partial differential equations.

2000 *Mathematics Subject Classification.* 35Q30, 76D05.
Key words and phrases. Incompressible Navier-Stokes equations, cascade flow, periodic boundary conditions, classical and weak solutions, Galerkin method.

2. Classical formulation

Let C_1 be a simple, closed, piecewise C^2 curve in $\mathbb{R}^2$, whose interior as well as exterior are domains with a Lipschitz-continuous boundary. Let $\mathbb{N}$ denote the set of positive integers. For $k \in \mathbb{N}$ we define

$$\begin{aligned} C_k &= \{(x_1, x_2 + (k-1)\tau);\ (x_1, x_2) \in C_1\}, \\ C_{-k} &= \{(x_1, x_2 - k\tau);\ (x_1, x_2) \in C_1\}, \end{aligned}$$

where $\tau > 0$ is a constant called the period of the cascade of profiles. We assume that τ is so large that the sets C_k are mutually disjoint. By $\operatorname{Int} C_k$ we denote the interior of the curve C_k. The set

$$M = \bigcup_{k=1}^{+\infty} \left(\overline{\operatorname{Int} C_k} \cup \overline{\operatorname{Int} C_{-k}} \right)$$

is called a *cascade of profiles* and each of its component $\overline{\operatorname{Int} C_k}$ is called a *profile.*

Let $d_i < d_o$ be such real numbers that all profiles lie inside the strip

$$\{(x_1, x_2) \in \mathbb{R}^2;\ d_i < x_1 < d_o\}.$$

Then the domain

$$O = \{(x_1, x_2) \in \mathbb{R}^2;\ d_i < x_1 < d_o\} \setminus M$$

represents the region occupied by the fluid. The line G_i given by $x_1 = d_i$ represents the inlet through which the fluid enters the domain O and the line G_o with the equation $x_1 = d_o$ is the outlet through which the fluid leaves O. Further, we set $G_w = \cup_{k=1}^{+\infty}(C_{-k} \cup C_k)$ (surfaces of all profiles).

We assume the existence of a function $\gamma \in C^1([d_i, d_o])$ such that its graph Γ_- on the interval (d_i, d_o) is a part of O and cuts O into two parts, O^+ and O^-, containing the profiles $C_k, k = 1, 2, \ldots$, and $C_k, k = -1, -2, \ldots$, respectively. We set $A_0 = (d_i, \gamma(d_i))$ and $B_0 = (d_o, \gamma(d_o))$ (end-points of Γ_-), $\Gamma_w = C_1$, $\Gamma_+ = \{(x_1, x_2 + \tau); (x_1, x_2) \in \Gamma_-\}$, $A_1 = (d_i, \gamma((d_i) + \tau)$, $B_1 = (d_o, \gamma(d_o) + \tau)$ (end-points of Γ_+), and denote by Γ_i and Γ_o the straight segments with end-points A_0, A_1 and B_0, B_1, respectively.

Our goal is the solution of viscous, incompressible, stationary flow in the domain O. It is described by the continuity equation, the Navier-Stokes equations and suitable boundary conditions representing the inlet and outlet flow and the flow past the cascade of fixed impermeable profiles. The domain O is unbounded in the direction of the axis x_2, but periodic in this direction with period τ. Therefore, we shall assume that the functions describing the flow are also periodic in the direction x_2 with period τ. It is clear that the velocity is zero on the profiles C_k. On the inlet we prescribe the distribution of the velocity. However, from the physical point of view, it is not reasonable to prescribe the velocity on the outlet. Here we use a "soft" natural boundary condition.

Due to the periodicity, we shall be able to reformulate the problem in a bounded domain Ω of the form of one period only. The boundary of this domain has the form

$$\partial\Omega = \Gamma_i \cup \Gamma_o \cup \Gamma_- \cup \Gamma_+ \cup \Gamma_w \,. \tag{2.1}$$

2.1. Governing equations

The flow in consideration is described by the *Navier-Stokes equations*

$$\sum_{j=1}^{2} u_j \frac{\partial u_i}{\partial x_j} = f_i - \frac{\partial p}{\partial x_i} + \nu\,\Delta u_i \quad i = 1, 2, \tag{2.2}$$

and the *continuity equation*

$$\operatorname{div} \boldsymbol{u} = 0. \tag{2.3}$$

We denote by $\boldsymbol{u} = (u_1, u_2)$ the velocity vector, $\boldsymbol{f} = (f_1, f_2)$ the vector of volume force, p the pressure and $\nabla = \big(\partial/\partial x_1,\, \partial/\partial x_2\big)$. System (2.2) can be written in the form

$$(\boldsymbol{u} \cdot \nabla)\boldsymbol{u} = \boldsymbol{f} - \nabla p + \nu\,\Delta \boldsymbol{u}. \tag{2.4}$$

2.2. Boundary conditions

We consider the following boundary conditions on the boundary of the domain O:

$$\boldsymbol{u}\,|_{G_i} = \boldsymbol{g}, \tag{2.5}$$

$$\boldsymbol{u}\,|_{G_w} = 0. \tag{2.6}$$

On the outlet the following natural boundary condition is prescribed:

$$-\nu \frac{\partial \boldsymbol{u}}{\partial \boldsymbol{n}} + p\boldsymbol{n} - \frac{1}{2}(\boldsymbol{u} \cdot \boldsymbol{n})^- \boldsymbol{u} = \boldsymbol{h} \qquad \text{for } (x_1, x_2) \in G_o. \tag{2.7}$$

Here $\boldsymbol{g}$ and $\boldsymbol{h}$ are given functions, periodic with period τ. Moreover, we also assume that the volume force $\boldsymbol{f}$ is periodic in the direction x_2 with period τ. For $a \in \mathbb{R}$ we set $a^+ = (|a| + a)/2$ and $a^- = (|a| - a)/2$.

2.3. Classical formulation of the problem in the domain O

Let us prescribe continuous functions $\boldsymbol{f} = (f_1, f_2)$ (in O), $\boldsymbol{g} = (g_1, g_2)$ (on G_i) and $\boldsymbol{h} = (h_1, h_2)$ (on G_o), periodic with respect to x_2 with period τ. Find $\boldsymbol{u} = (u_1, u_2) \in C^2(\overline{O})^2$ and $p \in C^1(\overline{O})$, periodic with respect to x_2 with period τ, such that they satisfy equations (2.4), (2.3) in the domain O and the boundary conditions (2.5) (on G_i), (2.6) (on G_w) and (2.7) (on G_o).

2.4. Classical formulation of the problem in the domain Ω

The domain O is unbounded, which is not convenient for the numerical solution. However, due to the periodicity, it is possible to reformulate the problem in an equivalent way in the bounded domain Ω.

Let continuous functions $\boldsymbol{f} = (f_1, f_2)$ (in $\overline{\Omega}$), $\boldsymbol{g} = (g_1, g_2)$ (on Γ_i) and $\boldsymbol{h} = (h_1, h_2)$ (on Γ_o) satisfy the conditions

$$\boldsymbol{f}(x_1, x_2) = \boldsymbol{f}(x_1, x_2 + \tau) \qquad \text{for } (x_1, x_2) \in \Gamma_-, \tag{2.8}$$

$$\boldsymbol{g}(A_0) = \boldsymbol{g}(A_1), \tag{2.9}$$

$$\boldsymbol{h}(B_0) = \boldsymbol{h}(B_1). \tag{2.10}$$

Find $\boldsymbol{u} = (u_1, u_2)$ and p such that

a) $\boldsymbol{u} \in C^2(\overline{\Omega})^2$, b) $p \in C^1(\overline{\Omega})$,

c) $\boldsymbol{u}$ and p satisfy equations (2.4), (2.3) in Ω and boundary conditions (2.5), (2.6) and (2.7) restricted to the boundary of the domain Ω:

$$\boldsymbol{u}\,|_{\Gamma_i} = \boldsymbol{g}, \tag{2.11}$$

$$\boldsymbol{u}\,|_{\Gamma_w} = 0, \tag{2.12}$$

$$-\nu \frac{\partial \boldsymbol{u}}{\partial \boldsymbol{n}} + p\boldsymbol{n} - \frac{1}{2}(\boldsymbol{u}\cdot\boldsymbol{n})^- \boldsymbol{u} = \boldsymbol{h} \quad \text{on } \Gamma_o. \tag{2.13}$$

Moreover, we consider the periodicity conditions

$$\boldsymbol{u}(x_1, x_2 + \tau) = \boldsymbol{u}(x_1, x_2), \qquad (x_1, x_2) \in \Gamma_-, \tag{2.14}$$

$$\frac{\partial \boldsymbol{u}}{\partial \boldsymbol{n}}(x_1, x_2 + \tau) = -\frac{\partial \boldsymbol{u}}{\partial \boldsymbol{n}}(x_1, x_2), \qquad (x_1, x_2) \in \Gamma_-, \tag{2.15}$$

$$p(x_1, x_2 + \tau) = p(x_1, x_2), \qquad (x_1, x_2) \in \Gamma_-. \tag{2.16}$$

Remark. In various models we meet the problem how to choose suitable outlet conditions. The Dirichlet condition imposed on the velocity is usually rather restrictive and does not lead to physical results. Therefore, the goal is to prescribe some "soft" natural outlet boundary conditions. One possibility often used is the so-called "do-nothing" condition

$$-\nu \frac{\partial \boldsymbol{u}}{\partial \boldsymbol{n}} + p\boldsymbol{n} = \boldsymbol{h}, \tag{2.17}$$

used, for example, in [7] or [8]. However, this condition does not allow to control the flux of the kinetic energy through the outlet and to prove the existence of a solution. In [9] and [10], there is an attempt to avoid this drawback by the formulation of the viscous channel flow, equipped by the "do-nothing" outlet condition, with the aid of variational inequalities. This problem was also investigated in [11] and [12]. Moreover, sometimes condition (2.17) yields nonphysical solutions. Therefore, we apply condition (2.7) proposed in [1].

3. Weak formulation

We shall work with the following function spaces. By $H^1(\Omega)$ we denote the well-known Sobolev space. It is a Hilbert space with the scalar product

$$(u,v)_{H^1(\Omega)} = \int_\Omega (uv + \nabla u\cdot\nabla v)\,dx.$$

We also consider the space $H^1(\Omega)^2 = H^1(\Omega)\times H^1(\Omega)$ of vector functions, equipped with the scalar product

$$(\boldsymbol{u},\boldsymbol{v})_{H^1(\Omega)^2} = \sum_{i=1}^{2}(u_i,v_i)_{H^1(\Omega)},$$

where $\boldsymbol{u} = (u_1,u_2), \boldsymbol{v} = (v_1,u_2) \in H^1(\Omega)^2$. Further, we define the following spaces:

$$\begin{aligned}
\mathcal{X} &= \Big\{\boldsymbol{v}\in C^\infty(\overline{\Omega})^2;\ \boldsymbol{v}=0 \text{ on } \Gamma_i\cup\Gamma_w,\\
&\qquad \boldsymbol{v}(x_1,x_2+\tau) = \boldsymbol{v}(x_1,x_2)\ \forall\,(x_1,x_2)\in\Gamma_-\Big\},\\
\mathcal{V} &= \Big\{\boldsymbol{v}\in\mathcal{X};\ \operatorname{div}\boldsymbol{v} = 0 \text{ in } \Omega\Big\},\\
&X \ \text{ is the closure of } \mathcal{X} \text{ in } H^1(\Omega)^2,\\
&V \ \text{ is the closure of } \mathcal{V} \text{ in } H^1(\Omega)^2.
\end{aligned}$$

Using the same process as in Theorem 6.6.4 from [13], we can show that

$$\begin{aligned}
X \;=\; &\Big\{\boldsymbol{v}\in H^1(\Omega)^2;\ \boldsymbol{v}=0 \text{ in } \Gamma_i\cup\Gamma_w,\ \boldsymbol{v}(x_1,x_2+\tau)=\boldsymbol{v}(x_1,x_2)\\
&\text{for } (x_1,x_2)\in\Gamma_-\Big\}.
\end{aligned}$$

The identities on Γ_i, Γ_w and Γ_- are interpreted in the sense of traces. Similarly as in [4], pages 142–143, we can prove that

$$V = \Big\{\boldsymbol{v}\in X;\ \operatorname{div}\boldsymbol{v}=0 \text{ in } \Omega\Big\}.$$

In the space V we can use the norm defined by

$$|||\boldsymbol{v}||| = \left(\int_\Omega \sum_{i,j=1}^{2}\frac{\partial v_i}{\partial x_j}\frac{\partial v_i}{\partial x_j}\,dx\right)^{1/2} \tag{3.1}$$

and induced by the scalar product

$$(\boldsymbol{u},\boldsymbol{v})_V = \int_\Omega \sum_{i,j=1}^{2}\frac{\partial u_i}{\partial x_j}\frac{\partial v_i}{\partial x_j}\,dx. \tag{3.2}$$

It is possible to prove that $|||\,.\,|||$ is a norm on V equivalent with the norm $\|\,.\,\|_{H^1(\Omega)^2}$: There exists a constant $c_1 > 0$ such that

$$\|\boldsymbol{v}\|_{H^1(\Omega)^2} \le c_1\,|||\boldsymbol{v}||| \le c_1\,\|\boldsymbol{v}\|_{H^1(\Omega)^2}. \tag{3.3}$$

3.1. The derivation of a weak formulation

Let us multiply equation (2.4) by an arbitrary test function $\boldsymbol{v} = (v_1, v_2) \in V$, integrate over Ω and use Green's theorem. We get

$$\begin{aligned}\int_\Omega \boldsymbol{f}\cdot\boldsymbol{v}\,dx &= \int_\Omega\Big(-\nu\,\Delta\boldsymbol{u} + \sum_{j=1}^2 u_j\,\frac{\partial\boldsymbol{u}}{\partial x_j} + \nabla p\Big)\cdot\boldsymbol{v}\,dx\\ &= \nu\int_\Omega\sum_{i,j=1}^2 \frac{\partial u_i}{\partial x_j}\,\frac{\partial v_i}{\partial x_j}\,dx - \nu\int_{\partial\Omega}\frac{\partial\boldsymbol{u}}{\partial\boldsymbol{n}}\cdot\boldsymbol{v}\,dS + \int_\Omega\sum_{i,j=1}^2 u_j\,\frac{\partial u_i}{\partial x_j}\,v_i\,dx\\ &\quad - \int_\Omega p\,\operatorname{div}\boldsymbol{v}\,dx + \int_{\partial\Omega} p\,\boldsymbol{v}\cdot\boldsymbol{n}\,dS. \end{aligned} \tag{3.4}$$

For integrals along $\partial\Omega$ we have

$$\int_{\partial\Omega} = \int_{\Gamma_i} + \int_{\Gamma_o} + \int_{\Gamma_-} + \int_{\Gamma_+} + \int_{\Gamma_w}.$$

Since the function $\boldsymbol{v}\in V$ has a zero trace on $\Gamma_i\cup\Gamma_w$, the integrals along Γ_i and Γ_w vanish. From the relations

$$\boldsymbol{n}(x_1,x_2) = -\boldsymbol{n}(x_1,x_2+\tau) \qquad \text{for } (x_1,x_2)\in\Gamma_-, \tag{3.5}$$

and (2.15) and properties of the test function $\boldsymbol{v}$ we get

$$\int_{\Gamma_-\cup\Gamma_+}\Big(-\nu\,\frac{\partial\boldsymbol{u}}{\partial\boldsymbol{n}}\cdot\boldsymbol{v} + p\,\boldsymbol{v}\cdot\boldsymbol{n}\Big)\,dS = 0. \tag{3.6}$$

The integral along Γ_o can be transformed with the aid of (2.13):

$$\begin{aligned}&\int_{\Gamma_o}\Big(-\nu\,\frac{\partial\boldsymbol{u}}{\partial\boldsymbol{n}}\cdot\boldsymbol{v} + p\,\boldsymbol{v}\cdot\boldsymbol{n}\Big)\,dS\\ &\qquad = \int_{\Gamma_o}\frac{1}{2}\,(\boldsymbol{u}\cdot\boldsymbol{n})^-\,\boldsymbol{u}\cdot\boldsymbol{v}\,dS + \int_{\Gamma_o}\boldsymbol{h}\cdot\boldsymbol{v}\,dS.\end{aligned} \tag{3.7}$$

Further, taking into account that $\operatorname{div}\boldsymbol{v} = 0$ in Ω and substituting (3.5)–(3.7) into (3.4), we obtain

$$\begin{aligned}\int_\Omega \boldsymbol{f}\cdot\boldsymbol{v}\,dx &= \nu\int_\Omega\sum_{i,j=1}^2 \frac{\partial u_i}{\partial x_j}\,\frac{\partial v_i}{\partial x_j}\,dx + \int_\Omega\sum_{i,j=1}^2 u_j\,\frac{\partial u_i}{\partial x_j}\,v_i\,dx\\ &\quad + \int_{\Gamma_o}\frac{1}{2}\,(\boldsymbol{u}\cdot\boldsymbol{n})^-\,\boldsymbol{u}\cdot\boldsymbol{v}\,dS + \int_{\Gamma_o}\boldsymbol{h}\cdot\boldsymbol{v}\,dS.\end{aligned} \tag{3.8}$$

For $\boldsymbol{u} = (u_1, u_2)$, $\boldsymbol{v} = (v_1, v_2)$, $\boldsymbol{w} = (w_1, w_2) \in V$ we introduce the following notation:

$$
\begin{aligned}
a_1(\boldsymbol{u}, \boldsymbol{v}) &= \nu \int_\Omega \sum_{i,j=1}^{2} \frac{\partial u_i}{\partial x_j} \frac{\partial v_i}{\partial x_j}\, dx, \\
a_2(\boldsymbol{u}, \boldsymbol{v}, \boldsymbol{w}) &= \int_\Omega \sum_{i,j=1}^{2} u_j \frac{\partial v_i}{\partial x_j} w_i\, dx, \\
a_3(\boldsymbol{u}, \boldsymbol{v}, \boldsymbol{w}) &= \int_{\Gamma_o} \frac{1}{2} (\boldsymbol{u} \cdot \boldsymbol{n})^- \boldsymbol{v} \cdot \boldsymbol{w}\, dS, \\
a(\boldsymbol{u}, \boldsymbol{v}) &= a_1(\boldsymbol{u}, \boldsymbol{v}) + a_2(\boldsymbol{u}, \boldsymbol{u}, \boldsymbol{v}) + a_3(\boldsymbol{u}, \boldsymbol{u}, \boldsymbol{v}), \\
(\boldsymbol{f}, \boldsymbol{v}) &= \int_\Omega \boldsymbol{f} \cdot \boldsymbol{v}\, dx, \\
b(\boldsymbol{h}, \boldsymbol{v}) &= -\int_{\Gamma_o} \boldsymbol{h} \cdot \boldsymbol{v}\, dS.
\end{aligned}
$$

It is possible to show that all these forms make sense for $\boldsymbol{u}, \boldsymbol{v}, \boldsymbol{w} \in V$, $\boldsymbol{f} \in L^2(\Omega)$ and $\boldsymbol{h} \in L^2(\Gamma_o)$.

Now the integral identity (3.8) can be written in the form

$$a(\boldsymbol{u}, \boldsymbol{v}) = (\boldsymbol{f}, \boldsymbol{v}) + b(\boldsymbol{h}, \boldsymbol{v}). \tag{3.9}$$

3.1.1. Definition of a weak solution Let $\boldsymbol{g} \in C(\Gamma_i)^2$ satisfy the condition $\boldsymbol{g}(A_1) = \boldsymbol{g}(A_0)$. (Let us recall that A_0 and A_1 are the end-point of Γ_i.) Let $\boldsymbol{f} \in L^2(\Omega)^2$ and $\boldsymbol{h} \in L^2(\Gamma_o)^2$. We define a *weak solution* of the problem as a vector function $\boldsymbol{u} \in H^1(\Omega)^2$ satisfying identity (3.9) for all test functions $\boldsymbol{v} \in V$, the continuity equation (2.3) in Ω and the boundary conditions

$$
\begin{aligned}
\boldsymbol{u}|_{\Gamma_i} &= \boldsymbol{g}, && (3.10)\\
\boldsymbol{u}|_{\Gamma_w} &= 0, && (3.11)\\
\boldsymbol{u}(x_1, x_2 + \tau) &= \boldsymbol{u}(x_1, x_2) \quad \text{for } (x_1, x_2) \in \Gamma_- && (3.12)
\end{aligned}
$$

in the sense of traces.

In what follows we shall need to specify the properties of the function $\boldsymbol{g}$ in a more precise way. To this end we shall assume that $\boldsymbol{g} \in H^s(\Gamma_i)^2$, where $s \in (1/2, 1]$ and $H^s(\Gamma_i)$ is the Sobolev-Slobodetskii space with the norm

$$\|\boldsymbol{g}\|_{H^s(\Gamma_i)^2} = \|\boldsymbol{g}\|_{L^2(\Gamma_i)^2} + \left(\int_{\Gamma_i} \int_{\Gamma_i} \frac{|\boldsymbol{g}(d_i, y_2) - \boldsymbol{g}(d_i, x_2)|^2}{|y_2 - x_2|^{1+2s}}\, dy_2\, dx_2 \right)^{1/2}. \tag{3.13}$$

See, for example, [15], Section 9.1.

In view of Theorem 9.8 from [15], the function $\boldsymbol{g} \in H^s(\Gamma_i)^2$ is continuous and, thus, the condition $\boldsymbol{g}(A_0) = \boldsymbol{g}(A_1)$ makes sense. Moreover, the embedding of

$H^s(\Gamma_i)^2$ into $C(\Gamma_i)^2$ is continuous. Thus, there exists a constant $c_2 > 0$ independent of $\boldsymbol{g}$ such that

$$\|\boldsymbol{g}\|_{C(\Gamma_i)^2} \leq c_2 \|\boldsymbol{g}\|_{H^s(\Gamma_i)^2} . \tag{3.14}$$

For our further considerations we shall put $\gamma_k = \gamma(d_i) + k\tau$. Hence, we have $A_0 = (d_i, \gamma_0)$ and $A_1 = (d_i, \gamma_1)$.

It follows from Theorem 11.3 in [15] that there exists a constant $c_3 > 0$ such that for each function $\boldsymbol{g} \in H^s(\Gamma_i)^2$ satisfying the condition $\boldsymbol{g}(A_0) = \boldsymbol{g}(A_1)$ (i.e., $\boldsymbol{g}(d_i, \gamma_0) = \boldsymbol{g}(d_i, \gamma_1)$) it holds

$$\int_{\gamma_0}^{\gamma_1} \frac{|\boldsymbol{g}(d_i, x_2) - \boldsymbol{g}(d_i, \gamma_0)|^2}{(x_2 - \gamma_0)^{2s}} \, dx_2 \;\leq\; c_3 \|\boldsymbol{g}\|^2_{H^s(\Gamma_i)^2} , \tag{3.15}$$

$$\int_{\gamma_0}^{\gamma_1} \frac{|\boldsymbol{g}(d_i, \gamma_1) - \boldsymbol{g}(d_i, x_2)|^2}{(\gamma_1 - x_2)^{2s}} \, dx_2 \;\leq\; c_3 \|\boldsymbol{g}\|^2_{H^s(\Gamma_i)^2} . \tag{3.16}$$

3.2. Extension of function $\boldsymbol{g}$ on the whole boundary $\partial\Omega$

In the proof of the existence of a weak solution we shall need a suitable prolongation of function $\boldsymbol{g}$ from Γ_i on the domain Ω. First we prolong $\boldsymbol{g}$ on the whole boundary $\partial\Omega$ in the following way:

$$\boldsymbol{g}(x_1, x_2) = \begin{cases} \boldsymbol{g}(A_1) & \text{for } (x_1, x_2) \in \Gamma_+ , \\ \boldsymbol{g}(A_0) & \text{for } (x_1, x_2) \in \Gamma_- , \\ \boldsymbol{g}(d_i, \gamma(d_i) + x_2 - \gamma(d_o)) & \text{for } (x_1, x_2) = (d_o, x_2) \in \Gamma_o , \\ 0 & \text{for } (x_1, x_2) \in \Gamma_w . \end{cases} \tag{3.17}$$

Hence, the prolonged function $\boldsymbol{g}$ is equal to the constant $\boldsymbol{g}(A_0) = \boldsymbol{g}(A_1)$ on $\Gamma_+ \cup \Gamma_-$, is equal to zero on Γ_w and has a similar distribution on Γ_o as on Γ_i.

Lemma 3.1. *The extension* (3.17) *of function $\boldsymbol{g}$ from Γ_i onto $\partial\Omega$ is an element of the space $H^{1/2}(\partial\Omega)^2$. Moreover, there exists a constant $c_4 > 0$ independent of $\boldsymbol{g}$ such that*

$$\|\boldsymbol{g}\|_{H^{1/2}(\partial\Omega)^2} \leq c_4 \|\boldsymbol{g}\|_{H^s(\Gamma_i)^2} . \tag{3.18}$$

Proof. The norm of function $\boldsymbol{g}$ in the space $H^{1/2}(\partial\Omega)^2$ is defined by

$$\|\boldsymbol{g}\|_{H^{1/2}(\partial\Omega)^2} = \|\boldsymbol{g}\|_{L^2(\partial\Omega)^2} + \left(\int_{\partial\Omega} \int_{\partial\Omega} \frac{|\boldsymbol{g}(y) - \boldsymbol{g}(x)|^2}{|x - y|^2} \, dS_y \, dS_x \right)^{1/2} . \tag{3.19}$$

We want to show that the expression on the right-hand side of (3.19) is less than or equal to the right-hand side of (3.13). From the definition of $\boldsymbol{g}$ on $\partial\Omega$ and inequality (3.14) we deduce that there exists a constant $c_5 > 0$ independent of $\boldsymbol{g}$ such that

$$\|\boldsymbol{g}\|_{L^2(\partial\Omega)^2} \leq c_5 \|\boldsymbol{g}\|_{H^s(\Gamma_i)^2} . \tag{3.20}$$

The integral over $\partial\Omega \times \partial\Omega$ on the right-hand side of (3.19) can be written as the sum of integrals over $\Gamma_i \times \Gamma_i$, $\Gamma_i \times \Gamma_+$, $\Gamma_i \times \Gamma_o$, $\Gamma_i \times \Gamma_-$, $\Gamma_i \times \Gamma_w$, $\Gamma_+ \times \Gamma_+$, $\Gamma_+ \times \Gamma_o$, $\Gamma_+ \times \Gamma_-$, $\Gamma_+ \times \Gamma_w$, $\Gamma_o \times \Gamma_o$, $\Gamma_o \times \Gamma_-$, $\Gamma_o \times \Gamma_w$, $\Gamma_- \times \Gamma_-$, $\Gamma_- \times \Gamma_w$ and $\Gamma_w \times \Gamma_w$, and the integrals over the Cartesian products of nonequal sets has to be considered twice. Integrals over $\Gamma_i \times \Gamma_i$ and $\Gamma_o \times \Gamma_o$ can be estimated by $\|\boldsymbol{g}\|^2_{H^s(\Gamma_i)^2}$. Integrals over $\Gamma_+ \times \Gamma_+$, $\Gamma_+ \times \Gamma_-$, $\Gamma_- \times \Gamma_-$ and $\Gamma_w \times \Gamma_w$ obviously vanish. Integrals over $\Gamma_i \times \Gamma_w$, $\Gamma_+ \times \Gamma_w$, $\Gamma_o \times \Gamma_w$ and $\Gamma_- \times \Gamma_w$ can be estimated by the expression $c_6 \, \|\boldsymbol{g}\|_{L^2(\Gamma_i)^2}$ (where c_6 is a suitable constant), since Γ_w has a positive distance from Γ_i, Γ_+, Γ_o and Γ_-. It remains to estimate the integrals over $\Gamma_i \times \Gamma_+$, $\Gamma_i \times \Gamma_-$, $\Gamma_+ \times \Gamma_o$ and $\Gamma_o \times \Gamma_-$. Let us prove the estimate of the integral over $\Gamma_i \times \Gamma_+$. (Other integrals can be bounded similarly.)

We can write

$$\int_{\Gamma_i}\int_{\Gamma_+} \frac{|\boldsymbol{g}(y) - \boldsymbol{g}(x)|^2}{|y-x|^2}\, dS_y\, dS_x = \int_{\Gamma_i}\int_{\Gamma_+} \frac{|\boldsymbol{g}(d_i,\gamma_1) - \boldsymbol{g}(d_i,x_2)|^2}{|y-x|^2}\, dS_y\, dS_x.$$

Using the relations

$$x = (d_i, x_2),\ dS_x = dx_2,\ y = (y_1, \gamma(y_1)),\ dS_y = \sqrt{1+\gamma'(y_1)^2}\, dy_1,$$

we find that

$$\begin{aligned}&\int_{\Gamma_i}\int_{\Gamma_+} \frac{|\boldsymbol{g}(y) - \boldsymbol{g}(x)|^2}{|y-x|^2}\, dS_y dS_x \\ &= \int_{\gamma_0}^{\gamma_1}\int_{d_i}^{d_o} \frac{|\boldsymbol{g}(d_i,\gamma_1) - \boldsymbol{g}(d_i,x_2)|^2}{(y_1-d_i)^2 + (\gamma(y_1)-x_2)^2}\sqrt{1+\gamma'(y_1)^2}\, dy_1 dx_2.\end{aligned}$$

Since function γ is smooth, there exists a constant $c_7 > 0$ such that

$$\sqrt{1+\gamma'(y_1)^2} \le c_7 \quad \forall\, y_1 \in (d_i, d_o).$$

Further, there exists a constant $c_8 > 0$ such that

$$|y-x| \ge c_8 \left[\mathrm{dist}(y; A_1) + \mathrm{dist}(A_1; x)\right] \ge c_8 \left[(y_1 - d_i) + (\gamma_1 - x_2)\right]$$

for all $x = (d_i, x_2) \in \Gamma_i$ and $y = (y_1, \gamma(y_1)) \in \Gamma_+$. This and inequality (3.16) imply that the last integral can be estimated from above by

$$\begin{aligned}&\frac{c_7}{c_8}\int_{\gamma_0}^{\gamma_1}\int_{d_i}^{d_o} \frac{|\boldsymbol{g}(d_i,\gamma_1) - \boldsymbol{g}(d_i,x_2)|^2}{[(y_1-d_i) + (\gamma_1 - x_2)]^2}\, dy_1\, dx_2 \\ &= \frac{c_7}{c_8}\int_{\gamma_0}^{\gamma_1} |\boldsymbol{g}(d_i,\gamma_1) - \boldsymbol{g}(d_i,x_2)|^2 \left(\frac{1}{\gamma_1 - x_2} - \frac{1}{(d_o - d_i) + (\gamma_1 - x_2)}\right) dx_2 \\ &\le \frac{c_7}{c_8}\int_{\gamma_0}^{\gamma_1} \frac{|\boldsymbol{g}(d_i,\gamma_1) - \boldsymbol{g}(d_i,x_2)|^2}{\gamma_1 - x_2}\, dx_2 \le c_9 \int_{\gamma_0}^{\gamma_1} \frac{|\boldsymbol{g}(d_i,\gamma_1) - \boldsymbol{g}(d_i,x_2)|^2}{(\gamma_1 - x_2)^{2s}}\, dx_2 \\ &\le c_9\, c_3\, \|\boldsymbol{g}\|^2_{H^s(\Gamma_1)^2},\end{aligned}$$

which we wanted to prove. □

Lemma 3.2. *The prolongation* (3.17) *of function* $\boldsymbol{g}$ *satisfies the condition*

$$\int_{\partial\Omega} \boldsymbol{g} \cdot \boldsymbol{n} \, dS = 0. \tag{3.21}$$

Proof. The proof follows immediately from the definition of the prolongation of function $\boldsymbol{g}$. □

3.3. Extension of function $\boldsymbol{g}$ into the domain Ω

Lemma 3.3. *Let* $\boldsymbol{g} \in H^{1/2}(\partial\Omega)^2$, *satisfy* (3.21). *Then there exists its extension* $\boldsymbol{g}^* \in H^1(\Omega)^2$ *such that*

$$\boldsymbol{g}^* \Big|_{\partial\Omega} = \boldsymbol{g} \quad \textit{(in the sense of traces)}, \tag{3.22}$$

$$\operatorname{div} \boldsymbol{g}^* = 0 \quad \mathrm{v}\ \Omega, \tag{3.23}$$

$$\|\boldsymbol{g}^*\|_{H^1(\Omega)^2} \leq c_{10} \, \|\boldsymbol{g}\|_{H^{1/2}(\partial\Omega)^2}, \tag{3.24}$$

where the constant $c_{10} > 0$ *is independent of* $\boldsymbol{g}$.

Proof. The proof follows from Lemma 4.1 in [5]. □

Now we shall seek the weak solution $\boldsymbol{u}$ in the form $\boldsymbol{u} = \boldsymbol{g}^* + \boldsymbol{z}$, where $\boldsymbol{z} \in V$. Then $\boldsymbol{u}$ is an element of the space $H^1(\Omega)^2$ and satisfies the conditions

$$\boldsymbol{u}|_{\Gamma_i} = \boldsymbol{g}, \tag{3.25}$$

$$\boldsymbol{u}|_{\Gamma_w} = 0, \tag{3.26}$$

$$\boldsymbol{u}(x_1, x_2 + \tau) = \boldsymbol{u}(x_1, x_2) \quad \text{for } (x_1, x_2) \in \Gamma_-, \tag{3.27}$$

$$\operatorname{div} \boldsymbol{u} = 0 \quad \text{in } \Omega. \tag{3.28}$$

Substituting $\boldsymbol{u} = \boldsymbol{g}^* + \boldsymbol{z}$ into equation (3.9), we get a problem for a new unknown function $\boldsymbol{z} \in V$:

$$a(\boldsymbol{g}^* + \boldsymbol{z}, \boldsymbol{v}) = (\boldsymbol{f}, \boldsymbol{v}) + b(\boldsymbol{h}, \boldsymbol{v}), \quad \forall\, \boldsymbol{v} \in V. \tag{3.29}$$

The following result will play an important role in the proof of the existence of a weak solution.

Theorem 3.1. *There exist positive constants* c_{11} *and* c_{12} *such that*

$$a(\boldsymbol{g}^* + \boldsymbol{z}, \boldsymbol{z}) \geq |||\boldsymbol{z}||| \Big(\nu \, |||\boldsymbol{z}||| - \nu \, c_{10} \, c_4 \, \|\boldsymbol{g}\|_{H^s(\Gamma_i)^2} - c_{11} \, \|\boldsymbol{g}\|^2_{H^s(\Gamma_i)^2} - c_{12} \, \|\boldsymbol{g}\|_{H^s(\Gamma_i)^2} \, |||\boldsymbol{z}||| \Big) \tag{3.30}$$

for all $\boldsymbol{z} \in V$.

Proof. Using the definitions of the forms a, a_2 and a_3, we find that

$$\begin{aligned} a(\boldsymbol{g}^* + \boldsymbol{z}, \boldsymbol{z}) = {} & a_1(\boldsymbol{z}, \boldsymbol{z}) + a_1(\boldsymbol{g}^*, \boldsymbol{z}) \\ & + a_2(\boldsymbol{g}^*, \boldsymbol{g}^*, \boldsymbol{z}) + a_2(\boldsymbol{g}^*, \boldsymbol{z}, \boldsymbol{z}) + a_2(\boldsymbol{z}, \boldsymbol{g}^*, \boldsymbol{z}) + a_2(\boldsymbol{z}, \boldsymbol{z}, \boldsymbol{z}) \\ & + a_3(\boldsymbol{g}^*, \boldsymbol{g}^*, \boldsymbol{z}) + a_3(\boldsymbol{g}^*, \boldsymbol{z}, \boldsymbol{z}) + a_3(\boldsymbol{z}, \boldsymbol{g}^*, \boldsymbol{z}) + a_3(\boldsymbol{z}, \boldsymbol{z}, \boldsymbol{z}). \end{aligned}$$

This implies that

$$\begin{aligned} a(\boldsymbol{g}^* + \boldsymbol{z}, \boldsymbol{z}) \;\geq\; & a_1(\boldsymbol{z}, \boldsymbol{z}) - |a_1(\boldsymbol{g}^*, \boldsymbol{z})| \\ & - |a_2(\boldsymbol{g}^*, \boldsymbol{g}^*, \boldsymbol{z})| - |a_2(\boldsymbol{g}^*, \boldsymbol{z}, \boldsymbol{z})| - |a_2(\boldsymbol{z}, \boldsymbol{g}^*, \boldsymbol{z})| + a_2(\boldsymbol{z}, \boldsymbol{z}, \boldsymbol{z}) \\ & - |a_3(\boldsymbol{g}^*, \boldsymbol{g}^*, \boldsymbol{z})| - |a_3(\boldsymbol{g}^*, \boldsymbol{z}, \boldsymbol{z})| - |a_3(\boldsymbol{z}, \boldsymbol{g}^*, \boldsymbol{z})| + a_3(\boldsymbol{z}, \boldsymbol{z}, \boldsymbol{z}). \end{aligned} \tag{3.31}$$

Let us estimate the terms on the right-hand side of (3.31). If we use the Cauchy inequality, the continuous embedding of $H^1(\Omega)$ into $L^4(\Omega)$, inequalities (3.3), Green's theorem and theorem on traces, we successively obtain

$$a_1(\boldsymbol{z}, \boldsymbol{z}) \;=\; \nu \int_\Omega \sum_{i,j=1}^{2} \frac{\partial z_i}{\partial x_j} \frac{\partial z_i}{\partial x_j}\, dx \;\geq\; \nu\, |||\boldsymbol{z}|||^2, \tag{3.32}$$

$$\begin{aligned} |a_1(\boldsymbol{g}^*, \boldsymbol{z})| \;=\; & \left| \nu \int_\Omega \sum_{i,j=1}^{2} \frac{\partial g_i^*}{\partial x_j} \frac{\partial z_i}{\partial x_j}\, dx \right| \\ \leq\; & \nu\, \|\boldsymbol{g}^*\|_{H^1(\Omega)^2}\; |||\boldsymbol{z}|||, \end{aligned} \tag{3.33}$$

$$\begin{aligned} |a_2(\boldsymbol{g}^*, \boldsymbol{g}^*, \boldsymbol{z})| \;=\; & \left| \int_\Omega \sum_{i,j=1}^{2} g_j^* \frac{\partial g_i^*}{\partial x_j} z_i\, dx \right| \\ \leq\; & c_{13}\, \|\boldsymbol{g}^*\|_{H^1(\Omega)^2} \sum_{i,j=1}^{2} \left(\int_\Omega g_j^{*4}\, dx \right)^{1/4} \left(\int_\Omega z_i^4\, dx \right)^{1/4} \\ \leq\; & c_{14}\, \|\boldsymbol{g}^*\|^2_{H^1(\Omega)^2}\; |||\boldsymbol{z}|||, \end{aligned} \tag{3.34}$$

$$\begin{aligned} |a_2(\boldsymbol{g}^*, \boldsymbol{z}, \boldsymbol{z})| \;=\; & \left| \int_\Omega \sum_{i,j=1}^{2} g_j^* \frac{\partial z_i}{\partial x_j} z_i\, dx \right| \\ \leq\; & \sum_{i,j=1}^{2} \left(\int_\Omega \left|\frac{\partial z_i}{\partial x_j}\right|^2 dx \right)^{1/2} \left(\int_\Omega (g_j^*)^4\, dx \right)^{1/4} \left(\int_\Omega z_i^4\, dx \right)^{1/4} \\ \leq\; & c_{15} |||\boldsymbol{z}||| \|\boldsymbol{g}^*\|_{L^4(\Omega)} \|\boldsymbol{z}\|_{L^4(\Omega)} \;\leq\; c_{16} |||\boldsymbol{z}|||^2 \|\boldsymbol{g}^*\|_{H^1(\Omega)}. \end{aligned} \tag{3.35}$$

Moreover,

$$\begin{aligned} |a_2(\boldsymbol{z}, \boldsymbol{g}^*, \boldsymbol{z})| \;=\; & \left| \int_\Omega \sum_{i,j=1}^{2} z_j \frac{\partial g_i^*}{\partial x_j} z_i\, dx \right| \;\leq\; \left(\int_\Omega |\boldsymbol{z}|^4\, dx \right)^{1/2} \left(\int_\Omega |\nabla \boldsymbol{g}^*|^2\, dx \right)^{1/2} \\ & \leq \|\boldsymbol{z}\|^2_{L^4(\Omega)^2}\; \|\boldsymbol{g}^*\|_{H^1(\Omega)^2} \;\leq\; c_{17}\, |||\boldsymbol{z}|||^2\; \|\boldsymbol{g}^*\|_{H^1(\Omega)^2}, \end{aligned} \tag{3.36}$$

$$a_2(\boldsymbol{z}, \boldsymbol{z}, \boldsymbol{z}) + a_3(\boldsymbol{z}, \boldsymbol{z}, \boldsymbol{z}) \;=\; \int_\Omega \sum_{i,j=1}^{2} z_j \frac{\partial z_i}{\partial x_j} z_i\, dx \;+\; \frac{1}{2} \int_{\Gamma_o} (\boldsymbol{z} \cdot \boldsymbol{n})^- \, |\boldsymbol{z}|^2\, dS$$

$$
\begin{aligned}
&= \frac{1}{2}\int_\Omega \sum_{i,j=1}^{2} z_j \frac{\partial z_i^2}{\partial x_j}\, dx + \frac{1}{2}\int_{\Gamma_o} (\boldsymbol{z}\cdot\boldsymbol{n})^- \, |\boldsymbol{z}|^2 \, dS \\
&= \frac{1}{2}\int_{\partial\Omega} (\boldsymbol{z}\cdot\boldsymbol{n})\, |\boldsymbol{z}|^2 \, dS - \frac{1}{2}\int_\Omega \operatorname{div}\boldsymbol{z}\, |\boldsymbol{z}|^2 \, dx + \frac{1}{2}\int_{\Gamma_o} (\boldsymbol{z}\cdot\boldsymbol{n})^- \, |\boldsymbol{z}|^2 \, dS \\
&= \frac{1}{2}\int_{\Gamma_o} \left[(\boldsymbol{z}\cdot\boldsymbol{n}) + (\boldsymbol{z}\cdot\boldsymbol{n})^-\right] |\boldsymbol{z}|^2 \, dS = \frac{1}{2}\int_{\Gamma_o} (\boldsymbol{z}\cdot\boldsymbol{n})^+ \, |\boldsymbol{z}|^2 \, dS \geq 0, \qquad (3.37)
\end{aligned}
$$

$$
\begin{aligned}
|a_3(\boldsymbol{g}^*, \boldsymbol{g}^*, \boldsymbol{z})| &= \left| \frac{1}{2}\int_{\Gamma_o} (\boldsymbol{g}^*\cdot\boldsymbol{n})^- \, (\boldsymbol{g}^*\cdot\boldsymbol{z}) \, dS \right| \leq \frac{1}{2}\int_{\Gamma_o} |\boldsymbol{g}^*|^2 \, |\boldsymbol{z}| \, dS \\
&\leq \frac{1}{2}\left(\int_{\Gamma_o} |\boldsymbol{g}^*|^4 \, dS\right)^{1/2} \left(\int_{\Gamma_o} |\boldsymbol{z}|^2 \, dS\right)^{1/2} \leq c_{18}\, \|\boldsymbol{g}^*\|^2_{H^1(\Omega)^2} \, \|\boldsymbol{z}\|_{H^1(\Omega)^2} \\
&\leq c_{19}\, \|\boldsymbol{g}^*\|^2_{H^1(\Omega)^2} \, |||\boldsymbol{z}|||, \qquad (3.38)
\end{aligned}
$$

$$
\begin{aligned}
|a_3(\boldsymbol{g}^*, \boldsymbol{z}, \boldsymbol{z})| &= \left| \frac{1}{2}\int_{\Gamma_o} (\boldsymbol{g}^*\cdot\boldsymbol{n})^- \, |\boldsymbol{z}|^2 \, dS \right| \leq \frac{1}{2}\int_{\Gamma_o} |\boldsymbol{g}^*| \, |\boldsymbol{z}|^2 \, dS \\
&\leq \frac{1}{2}\left(\int_{\Gamma_o} |\boldsymbol{g}^*|^2 \, dS\right)^{1/2} \left(\int_{\Gamma_o} |\boldsymbol{z}|^4 \, dS\right)^{1/2} \leq c_{20}\, \|\boldsymbol{g}^*\|_{H^1(\Omega)^2} \, \|\boldsymbol{z}\|^2_{H^1(\Omega)^2} \\
&\leq c_{21}\, \|\boldsymbol{g}^*\|_{H^1(\Omega)^2} \, |||\boldsymbol{z}|||^2, \qquad (3.39)
\end{aligned}
$$

$$
\begin{aligned}
|a_3(\boldsymbol{z}, \boldsymbol{g}^*, \boldsymbol{z})| &= \left| \frac{1}{2}\int_{\Gamma_o} (\boldsymbol{z}\cdot\boldsymbol{n})^- \, \boldsymbol{g}^* \cdot \boldsymbol{z} \, dS \right| \leq \frac{1}{2}\int_{\Gamma_o} |\boldsymbol{g}^*| \, |\boldsymbol{z}|^2 \, dS \\
&\leq \frac{1}{2}\left(\int_{\Gamma_o} |\boldsymbol{g}^*|^2 \, dS\right)^{1/2} \left(\int_{\Gamma_o} |\boldsymbol{z}|^4 \, dS\right)^{1/2} \leq c_{22}\, \|\boldsymbol{g}^*\|_{H^1(\Omega)^2} \, \|\boldsymbol{z}\|^2_{H^1(\Omega)^2} \\
&\leq c_{23}\, \|\boldsymbol{g}^*\|_{H^1(\Omega)^2} \, |||\boldsymbol{z}|||^2. \qquad (3.40)
\end{aligned}
$$

Substituting (3.32)–(3.40) into (3.31), using (3.18) and (3.24), we get

$$
\begin{aligned}
a(\boldsymbol{g}^* + \boldsymbol{z}, \boldsymbol{z}) &\geq \nu\, |||\boldsymbol{z}|||^2 - \nu\, \|\boldsymbol{g}^*\|_{H^1(\Omega)^2} \, |||\boldsymbol{z}||| - (c_{14} + c_{19})\, \|\boldsymbol{g}^*\|^2_{H^1(\Omega)^2} \, |||\boldsymbol{z}||| \\
&\quad - (c_{16} + c_{17} + c_{21} + c_{23})\, \|\boldsymbol{g}^*\|_{H^1(\Omega)^2} \, |||\boldsymbol{z}|||^2 \\
&\geq |||\boldsymbol{z}||| \Big(\nu\, |||\boldsymbol{z}||| - \nu\, \|\boldsymbol{g}^*\|_{H^1(\Omega)^2} - (c_{14} + c_{19})\, \|\boldsymbol{g}^*\|^2_{H^1(\Omega)^2} \\
&\quad - (c_{16} + c_{17} + c_{21} + c_{23})\, \|\boldsymbol{g}^*\|_{H^1(\Omega)^2} \, |||\boldsymbol{z}||| \Big).
\end{aligned}
$$

$$
\begin{aligned}
&\geq |||\boldsymbol{z}||| \Big(\nu\, |||\boldsymbol{z}||| - \nu\, c_{10}\, c_4\, \|\boldsymbol{g}\|_{H^s(\Gamma_i)^2} - (c_{14} + c_{19})\, c_{10}^2\, c_4^2\, \|\boldsymbol{g}\|^2_{H^s(\Gamma_i)^2} \\
&\quad - (c_{16} + c_{17} + c_{21} + c_{23})\, c_{10}\, c_4\, \|\boldsymbol{g}\|_{H^s(\Gamma_i)^2} \, |||\boldsymbol{z}||| \Big). \qquad (3.41)
\end{aligned}
$$

Finally putting $c_{11} = (c_{14} + c_{19})\, c_{10}^2\, c_4^2$ and $c_{12} = (c_{16} + c_{17} + c_{21} + c_{23})\, c_{10}\, c_4$, we arrive at the inequality (3.30). □

Theorem 3.2. *There exists $\epsilon > 0$ such that if*

$$\|\boldsymbol{g}\|_{H^s(\Gamma_i)^2} < \epsilon, \tag{3.42}$$

then the form $a(\boldsymbol{g}^* + \boldsymbol{z}, \boldsymbol{z})$ is coercive on the space V:

$$\lim_{|||\boldsymbol{z}||| \to +\infty} a(\boldsymbol{g}^* + \boldsymbol{z}, \boldsymbol{z}) = +\infty. \tag{3.43}$$

Proof. Inequality (3.30) implies that it is sufficient to choose $\epsilon = \frac{\nu}{c_{12}}$. □

4. Existence of a weak solution

On the basis of the above results we can prove the existence of a weak solution to problem (3.29). We shall use the Galerkin method which will be applied in several steps.

We start with the citation of a well-known lemma.

Lemma 4.1. *Let $A : \mathbb{R}^n \to \mathbb{R}^n$ be a continuous mapping. If there exists $R > 0$ such that*

$$A(\vartheta) \cdot \vartheta \geq 0 \tag{4.1}$$

for all $\vartheta \in \mathbb{R}^n$ satisfying the condition $|\vartheta| = R$, then the equation

$$A(\vartheta) = 0 \tag{4.2}$$

has at least one solution $\vartheta \in \mathbb{R}^n$ such that $|\vartheta| \leq R$.

For the proof, see Lemma 4.1.53 from [2], or Lemma II.1.4 from [19].

4.1. Approximate solutions

The space V is a separable Hilbert space. Similarly as in [2], Chapter 8 or [19] it is possible to construct a basis in V formed by elements from $\mathcal{V}$. With the aid of the Schmidt orthogonalization process we obtain a basis $\{\boldsymbol{e}_i\}_{i=1}^{\infty}$ with $\boldsymbol{e}_i \in \mathcal{V}, i = 1, 2, \ldots,$ orthonormal with respect to the scalar product $(.,.)_V$. For any $n \in \mathbb{N}$ let us set

$$V_n = \mathcal{L}\{\boldsymbol{e}_1, \boldsymbol{e}_2, \ldots, \boldsymbol{e}_n\},$$

which is the linear space spanned on the functions $\boldsymbol{e}_1, \boldsymbol{e}_2, \ldots, \boldsymbol{e}_n$. In this space we shall define an approximate solution of problem (3.29):

$$\boldsymbol{z}_n = \sum_{k=1}^{n} \vartheta_k\, \boldsymbol{e}_k\,. \tag{4.3}$$

If we set $\vartheta = (\vartheta_1, \ldots, \vartheta_n)$ and

$$|\vartheta| = \left(\sum_{k=1}^{n} \vartheta_k^2 \right)^{1/2},$$

then

$$|||z_n||| = \sqrt{(z_n, z_n)_V} = \left(\sum_{k,l=1}^{n} \vartheta_k \vartheta_l (e_k, e_l)_V \right)^{1/2} = |\vartheta|. \tag{4.4}$$

The approximate solution $z_n \in V_n$ is defined by

$$a(g^* + z_n, v) = (f, v) + b(h, v) \quad \forall\, v \in V_n , \tag{4.5}$$

which is equivalent to the system

$$a(g^* + z_n, e_k) = (f, e_k) + b(h, e_k) \quad \text{for } k = 1, 2, \ldots, n. \tag{4.6}$$

Expressing the form a with the aid of a_1, a_2 and a_3, we can rewrite (4.6) in the form

$$\begin{aligned} a_1(g^* + z_n, e_k) + a_2(g^* + z_n, g^* + z_n, e_k) + a_3(g^* + z_n, g^* + z_n, e_k) \\ = (f, e_k) + b(h, e_k) \quad \text{for } k = 1, 2, \ldots, n. \end{aligned} \tag{4.7}$$

Substituting (4.3) into (4.7), we get

$$\begin{aligned} & a_1(g^*, e_k) + \sum_{l=1}^{n} \vartheta_l\, a_1(e_l, e_k) + a_2(g^*, g^*, e_k) \\ & + \sum_{l=1}^{n} \vartheta_l \left[a_2(g^*, e_l, e_k) + a_2(e_l, g^*, e_k) \right] \\ & + \sum_{l,m=1}^{n} \vartheta_l \vartheta_m\, a_2(e_l, e_m, e_k) + a_3(g^*, g^*, e_k) \\ & + \sum_{l=1}^{n} \vartheta_l \left[a_3(g^*, e_l, e_k) + a_3(e_l, g^*, e_k) \right] \\ & + \sum_{l,m=1}^{n} \vartheta_l \vartheta_m\, a_3(e_l, e_m, e_k) - (f, e_k) - b(h, e_k) = 0, \quad k = 1, 2, \ldots, n. \end{aligned} \tag{4.8}$$

This is a system of n quadratic equations for unknown coefficients $\vartheta_1, \vartheta_2, \ldots, \vartheta_n$. If we denote by $A_k(\vartheta)$ the left-hand side of the kth equation and set

$$A(\vartheta) = (A_1(\vartheta), \ldots, A_n(\vartheta)),$$

system (4.8) can be written as an equation

$$A(\vartheta) = 0 \tag{4.9}$$

in $\mathbb{R}^n$. Obviously, $A : \mathbb{R}^n \to \mathbb{R}^n$ is a continuous mapping. Using inequalities (3.30) and (3.3), representation (4.3) and theorem on traces, we successively deduce that

$$\begin{aligned}
A(\vartheta)\cdot\vartheta &= \sum_{k=1}^{n} A_k(\vartheta)\,\vartheta_k = \sum_{k=1}^{n} \vartheta_k\, a_1(\boldsymbol{g}^*, \boldsymbol{e}_k) + \sum_{k,l=1}^{n} \vartheta_k\,\vartheta_l\, a_1(\boldsymbol{e}_l, \boldsymbol{e}_k) \\
&\quad + \sum_{k=1}^{n} \vartheta_k\, a_2(\boldsymbol{g}^*, \boldsymbol{g}^*, \boldsymbol{e}_k) + \sum_{k,l=1}^{n} \vartheta_k\,\vartheta_l\,[a_2(\boldsymbol{g}^*, \boldsymbol{e}_l, \boldsymbol{e}_k) + a_2(\boldsymbol{e}_l, \boldsymbol{g}^*, \boldsymbol{e}_k)] \\
&\quad + \sum_{k,l,m=1}^{n} \vartheta_k\,\vartheta_l\,\vartheta_m\, a_2(\boldsymbol{e}_l, \boldsymbol{e}_m, \boldsymbol{e}_k) + \sum_{k=1}^{n} \vartheta_k\, a_3(\boldsymbol{g}^*, \boldsymbol{g}^*, \boldsymbol{e}_k) \\
&\quad + \sum_{k,l=1}^{n} \vartheta_k\,\vartheta_l\,[a_3(\boldsymbol{g}^*, \boldsymbol{e}_l, \boldsymbol{e}_k) + a_3(\boldsymbol{e}_l, \boldsymbol{g}^*, \boldsymbol{e}_k)] + \sum_{k,l,m=1}^{n} \vartheta_k\,\vartheta_l\,\vartheta_m\, a_3(\boldsymbol{e}_l, \boldsymbol{e}_m, \boldsymbol{e}_k) \\
&\quad - \sum_{k=1}^{n} \vartheta_k\,(\boldsymbol{f}, \boldsymbol{e}_k) - \sum_{k=1}^{n} \vartheta_k\, b(\boldsymbol{h}, \boldsymbol{e}_k) \\
&= a_1(\boldsymbol{g}^*, \boldsymbol{z}_n) + a_1(\boldsymbol{z}_n, \boldsymbol{z}_n) + a_2(\boldsymbol{g}^*, \boldsymbol{g}^*, \boldsymbol{z}_n) + a_2(\boldsymbol{g}^*, \boldsymbol{z}_n, \boldsymbol{z}_n) + a_2(\boldsymbol{z}_n, \boldsymbol{g}^*, \boldsymbol{z}_n) \\
&\quad + a_3(\boldsymbol{g}^*, \boldsymbol{g}^*, \boldsymbol{z}_n) + a_3(\boldsymbol{g}^*, \boldsymbol{z}_n, \boldsymbol{z}_n) + a_3(\boldsymbol{z}_n, \boldsymbol{g}^*, \boldsymbol{z}_n) - (\boldsymbol{f}, \boldsymbol{z}_n) - b(\boldsymbol{h}, \boldsymbol{z}_n) \\
&= a(\boldsymbol{g}^* + \boldsymbol{z}_n, \boldsymbol{z}_n) - (\boldsymbol{f}, \boldsymbol{z}_n) - b(\boldsymbol{h}, \boldsymbol{z}_n) \\
&\geq |||\boldsymbol{z}_n||| \Big(\nu\, |||\boldsymbol{z}_n||| - \nu\, c_{10}\, c_4\, \|\boldsymbol{g}\|_{H^s(\Gamma_i)^2} - c_{11}\, \|\boldsymbol{g}\|^2_{H^s(\Gamma_i)^2} \\
&\quad - c_{12}\, \|\boldsymbol{g}\|_{H^s(\Gamma_i)^2}\, |||\boldsymbol{z}_n|||\Big) - c_1\, \|\boldsymbol{f}\|_{L^2(\Omega)^2}\, |||\boldsymbol{z}_n||| - c_{24}\, \|\boldsymbol{h}\|_{L^2(\Gamma_o)^2}\, |||\boldsymbol{z}_n|||. \qquad (4.10)
\end{aligned}$$

Let us assume that the norm $\|\boldsymbol{g}\|_{H^s(\Gamma_i)^2}$ is so small that (3.42) holds with $\varepsilon = \nu/c_{12}$. Hence,

$$\|\boldsymbol{g}\|_{H^s(\Gamma_i)^2} < \nu/c_{12}. \qquad (4.11)$$

Now we see that $A(\vartheta)\cdot\vartheta \geq 0$, if ϑ satisfies the condition $|\vartheta| = R_0$, with

$$R_0 = \frac{\nu\, c_{10}\, c_4\, \|\boldsymbol{g}\|_{H^s(\Gamma_i)^2} + c_{11}\, \|\boldsymbol{g}\|^2_{H^s(\Gamma_i)^2} + c_1\, \|\boldsymbol{f}\|_{L^2(\Omega)^2} + c_{24}\, \|\boldsymbol{h}\|_{L^2(\Gamma_o)^2}}{\nu - c_{12}\, \|\boldsymbol{g}\|_{H^s(\Gamma_i)^2}}. \qquad (4.12)$$

It follows from Lemma 4.1 that equation (4.9) has at least one solution ϑ with $|\vartheta| \leq R_0$. The function $\boldsymbol{z}_n$ given by (4.3), is a sought approximate solution. In view of (4.4),

$$|||\boldsymbol{z}_n||| \leq R_0. \qquad (4.13)$$

Thus, we have proved the following theorem.

Theorem 4.1. *Let us assume that condition* (4.11) *is satisfied. Let* $n \in \mathbb{N}$. *Then problem* (4.5) *has a solution* $\boldsymbol{z}_n \in V_n$ *satisfying estimate* (4.13).

4.2. Convergence of approximate solutions

Let us assume that the norm $\|\boldsymbol{g}\|_{H^s(\Gamma_i)^2}$ satisfies condition (4.11). Since the space V is reflexive, the boundedness of the sequence $\{\boldsymbol{z}_n\}$ implies that there exist its subsequence, for the sake of simplicity denoted again by $\{\boldsymbol{z}_n\}$, and an element $\boldsymbol{z} \in V$ such that

$$\boldsymbol{z}_n \longrightarrow \boldsymbol{z} \quad \text{for } n \to +\infty \text{ weakly in } V. \tag{4.14}$$

Hence,

$$\boldsymbol{z}_n \longrightarrow \boldsymbol{z} \quad \text{for } n \to +\infty \text{ weakly in } H^1(\Omega)^2. \tag{4.15}$$

The space $H^1(\Omega)^2$ is compactly embedded into $L^q(\Omega)^2$ for each $q \geq 1$. (See, for example, [6], Theorem 1.3.) This implies that

$$\boldsymbol{z}_n \longrightarrow \boldsymbol{z} \quad \text{for } n \to +\infty \text{ strongly in } L^q(\Omega)^2 \tag{4.16}$$

for each $q \geq 1$. The trace operator from $H^1(\Omega)^2$ into $L^q(\partial\Omega)^2$ is compact for each $q \geq 1$. See, e.g., [13], Theorem 6.4.2. Hence, it follows from (4.14) that

$$\boldsymbol{z}_n \longrightarrow \boldsymbol{z} \quad \text{for } n \to +\infty \text{ strongly in } L^q(\partial\Omega)^2$$

and, in particular,

$$\boldsymbol{z}_n \longrightarrow \boldsymbol{z} \quad \text{for } n \to +\infty \text{ strongly in } L^q(\Gamma_o)^2 \tag{4.17}$$

for each $q \geq 1$. This implies that

$$\begin{aligned} \boldsymbol{z}_n \cdot \boldsymbol{n} \quad &\longrightarrow \quad \boldsymbol{z} \cdot \boldsymbol{n} \quad \text{for } n \to +\infty \text{ strongly in } L^q(\Gamma_o), \\ (\boldsymbol{z}_n \cdot \boldsymbol{n})^- \quad &\longrightarrow \quad (\boldsymbol{z} \cdot \boldsymbol{n})^- \quad \text{for } n \to +\infty \text{ strongly in } L^q(\Gamma_o)^2. \end{aligned} \tag{4.18}$$

In virtue of (4.17) and (4.18), the sequences $\{\boldsymbol{z}_n\}$ a $\{(\boldsymbol{z}_n \cdot \boldsymbol{n})^-\}$ are bounded in $L^q(\Gamma_o)^2$ and $L^q(\Gamma_o))$, respectively for each $q \geq 1$.

4.3. Limit process in (4.5)

By the definition of the form a, equation (4.5) can be written in the form

$$\begin{aligned} a_1(\boldsymbol{g}^*, \boldsymbol{v}) + a_1(\boldsymbol{z}_n, \boldsymbol{v}) + a_2(\boldsymbol{g}^*, \boldsymbol{g}^*, \boldsymbol{v}) + a_2(\boldsymbol{z}_n, \boldsymbol{g}^*, \boldsymbol{v}) + a_2(\boldsymbol{g}^*, \boldsymbol{z}_n, \boldsymbol{v}) \\ +a_2(\boldsymbol{z}_n, \boldsymbol{z}_n, \boldsymbol{v}) + a_3(\boldsymbol{g}^*, \boldsymbol{g}^*, \boldsymbol{v}) + a_3(\boldsymbol{z}_n, \boldsymbol{g}^*, \boldsymbol{v}) + a_3(\boldsymbol{g}^*, \boldsymbol{z}_n, \boldsymbol{v}) \\ +a_3(\boldsymbol{z}_n, \boldsymbol{z}_n, \boldsymbol{v}) \;=\; (\boldsymbol{f}, \boldsymbol{v}) + b(\boldsymbol{h}, \boldsymbol{v}). \end{aligned} \tag{4.19}$$

Let $\boldsymbol{v} \in V_m$ for a fixed $m \in \mathbb{N}$. (Then $\boldsymbol{v}$ is infinitely differentiable in $\overline{\Omega}$.) Further, let us assume that $n \in \mathbb{N}$, $n \geq m$, and write $\boldsymbol{z}_n = (z_{n1}, z_{n2})$, $\boldsymbol{z} = (z_1, z_2)$, $\boldsymbol{g}^* = (g_1^*, g_2^*)$ a $\boldsymbol{v} = (v_1, v_2)$. By (4.15),

$$a_1(\boldsymbol{z}_n, \boldsymbol{v}) \longrightarrow a_1(\boldsymbol{z}, \boldsymbol{v}) \quad \text{for } n \to +\infty. \tag{4.20}$$

Further,

$$a_2(\boldsymbol{z}_n, \boldsymbol{g}^*, \boldsymbol{v}) = \int_\Omega \sum_{i,j=1}^2 z_{nj} \frac{\partial g_i^*}{\partial x_j} v_i \, dx.$$

The expressions $\partial g_i^* / \partial x_j \, v_i$ $(i,j = 1,2)$ are elements of $L^2(\Omega)$. Thus, in view of (4.16),

$$|a_2(\boldsymbol{z}_n, \boldsymbol{g}^*, \boldsymbol{v}) - a_2(\boldsymbol{z}, \boldsymbol{g}^*, \boldsymbol{v})| \\ \leq c_{25} \left(\int_\Omega \sum_{i,j=1}^{2} \left| \frac{\partial g_i^*}{\partial x_j} \right|^2 |v_i|^2 \, dx \right)^{1/2} \left(\int_\Omega |\boldsymbol{z}_n - \boldsymbol{z}|^2 \, dx \right)^{1/2},$$

where the constant c_{25} is independent of n. This implies that

$$a_2(\boldsymbol{z}_n, \boldsymbol{g}^*, \boldsymbol{v}) \longrightarrow a_2(\boldsymbol{z}, \boldsymbol{g}^*, \boldsymbol{v}) \quad \text{for } n \to +\infty. \tag{4.21}$$

Further,

$$a_2(\boldsymbol{g}^*, \boldsymbol{z}_n, \boldsymbol{v}) = \int_\Omega \sum_{i,j=1}^{2} g_j^* \frac{\partial z_{ni}}{\partial x_j} v_i \, dx.$$

Obviously, $g_j^* \, v_i \in H^1(\Omega)^2$ $(i,j = 1,2)$ and, hence, it follows from (4.15) that

$$a_2(\boldsymbol{g}^*, \boldsymbol{z}_n, \boldsymbol{v}) \longrightarrow a_2(\boldsymbol{g}^*, \boldsymbol{z}, \boldsymbol{v}) \quad \text{for } n \to +\infty. \tag{4.22}$$

We have

$$a_2(\boldsymbol{z}_n, \boldsymbol{z}_n, \boldsymbol{v}) = \int_\Omega \sum_{i,j=1}^{2} z_{nj} \frac{\partial z_{ni}}{\partial x_j} v_i \, dx,$$

$$\begin{aligned}
&|a_2(\boldsymbol{z}_n, \boldsymbol{z}_n, \boldsymbol{v}) - a_2(\boldsymbol{z}, \boldsymbol{z}, \boldsymbol{v})| \\
&\quad \leq |a_2(\boldsymbol{z}_n, \boldsymbol{z}_n, \boldsymbol{v}) - a_2(\boldsymbol{z}, \boldsymbol{z}_n, \boldsymbol{v})| + |a_2(\boldsymbol{z}, \boldsymbol{z}_n, \boldsymbol{v}) - a_2(\boldsymbol{z}, \boldsymbol{z}, \boldsymbol{v})| \\
&\quad = \left| \int_\Omega \sum_{i,j=1}^{2} (z_{nj} - z_j) \frac{\partial z_{ni}}{\partial x_j} v_i \, dx \right| + \left| \int_\Omega \sum_{i,j=1}^{2} z_j \left(\frac{\partial z_{ni}}{\partial x_j} - \frac{\partial z_i}{\partial x_j} \right) v_i \, dx \right| \\
&\quad \leq c_{26} \, \|\boldsymbol{z}_n - \boldsymbol{z}\|_{L^2(\Omega)^2} \, |||\boldsymbol{z}_n||| + \left| \int_\Omega \sum_{i,j=1}^{2} z_j \left(\frac{\partial z_{ni}}{\partial x_j} - \frac{\partial z_i}{\partial x_j} \right) v_i \, dx \right|
\end{aligned}$$

where c_{26} is a constant independent of n. The first term on the right-hand side converges to zero as $n \to +\infty$ due to (4.16) and the boundedness of the sequence $\{\boldsymbol{z}_n\}$ in V. The second term tends to zero for $n \to +\infty$ due to the weak convergence (4.15). Therefore,

$$a_2(\boldsymbol{z}_n, \boldsymbol{z}_n, \boldsymbol{v}) \longrightarrow a_2(\boldsymbol{z}, \boldsymbol{z}, \boldsymbol{v}) \quad \text{for } n \to +\infty. \tag{4.23}$$

Taking into account that

$$a_3(\boldsymbol{g}^*, \boldsymbol{z}_n, \boldsymbol{v}) = \frac{1}{2} \int_{\Gamma_o} (\boldsymbol{g}^* \cdot \boldsymbol{n})^- \, \boldsymbol{z}_n \cdot \boldsymbol{v} \, dS,$$

$(\boldsymbol{g}^* \cdot \boldsymbol{n})^- \, \boldsymbol{v} \in L^q(\Gamma_o)^2$ for each $q \geq 1$ and using (4.17), we find that

$$a_3(\boldsymbol{g}^*, \boldsymbol{z}_n, \boldsymbol{v}) \longrightarrow a_3(\boldsymbol{g}^*, \boldsymbol{z}, \boldsymbol{v}) \quad \text{for } n \to +\infty. \tag{4.24}$$

Similarly we get

$$a_3(\boldsymbol{z}_n, \boldsymbol{g}^*, \boldsymbol{v}) = \frac{1}{2}\int_{\Gamma_o} (\boldsymbol{z}_n \cdot \boldsymbol{n})^- \, \boldsymbol{g}^* \cdot \boldsymbol{v} \, dS,$$

$\boldsymbol{g}^* \cdot \boldsymbol{v} \in L^q(\Gamma_o)$ for each $q \geq 1$. This and (4.18) imply that

$$a_3(\boldsymbol{z}_n, \boldsymbol{g}^*, \boldsymbol{v}) \longrightarrow a_3(\boldsymbol{z}, \boldsymbol{g}^*, \boldsymbol{v}) \quad \text{for } n \to +\infty. \tag{4.25}$$

Finally, we have

$$a_3(\boldsymbol{z}_n, \boldsymbol{z}_n, \boldsymbol{v}) = \frac{1}{2}\int_{\Gamma_o} (\boldsymbol{z}_n \cdot \boldsymbol{n})^- \, \boldsymbol{z}_n \cdot \boldsymbol{v} \, dS,$$

$$\begin{aligned}
&|a_3(\boldsymbol{z}_n, \boldsymbol{z}_n, \boldsymbol{v}) - a_3(\boldsymbol{z}, \boldsymbol{z}, \boldsymbol{v})| \\
&\quad \leq |a_3(\boldsymbol{z}_n, \boldsymbol{z}_n, \boldsymbol{v}) - a_3(\boldsymbol{z}_n, \boldsymbol{z}, \boldsymbol{v})| + |a_3(\boldsymbol{z}_n, \boldsymbol{z}, \boldsymbol{v}) - a_3(\boldsymbol{z}, \boldsymbol{z}, \boldsymbol{v})| \\
&\quad = \left|\frac{1}{2}\int_{\Gamma_o} (\boldsymbol{z}_n \cdot \boldsymbol{n})^- \, (\boldsymbol{z}_n - \boldsymbol{z}) \cdot \boldsymbol{v} \, dS\right| + \left|\frac{1}{2}\int_{\Gamma_o} \Big((\boldsymbol{z}_n - \boldsymbol{z}) \cdot \boldsymbol{n}\Big)^- \, \boldsymbol{z} \cdot \boldsymbol{v} \, dS\right| \\
&\quad \leq c_{27} \left(\int_{\Gamma_o} ((\boldsymbol{z}_n \cdot \boldsymbol{n})^-)^2 \, dS\right)^{1/2} \left(\int_{\Gamma_o} |\boldsymbol{z}_n - \boldsymbol{z}|^2 \, dS\right)^{1/2} \\
&\qquad + c_{28} \left(\int_{\Gamma_o} \Big(\Big((\boldsymbol{z}_n - \boldsymbol{z}) \cdot \boldsymbol{n}\Big)^-\Big)^2 \, dS\right)^{1/2} \left(\int_{\Gamma_o} |\boldsymbol{z}|^2 \, dS\right)^{1/2}
\end{aligned}$$

where c_{27} and c_{28} are constants depending on $\boldsymbol{v} \in V_m$ and, thus, on m, but independent of n. Using the boundedness of the sequence $\{(\boldsymbol{z}_n \cdot \boldsymbol{n})^-\}$ in $L^2(\Gamma_o)$, (4.17) and (4.18), we get

$$a_3(\boldsymbol{z}_n, \boldsymbol{z}_n, \boldsymbol{v}) \longrightarrow a_3(\boldsymbol{z}, \boldsymbol{z}, \boldsymbol{v}) \quad \text{for } n \to +\infty. \tag{4.26}$$

Hence, by the limit process in (4.19) we obtain the identity

$$a(\boldsymbol{g}^* + \boldsymbol{z}, \boldsymbol{v}) = (\boldsymbol{f}, \boldsymbol{v}) + b(\boldsymbol{h}, \boldsymbol{v}), \quad \forall\, \boldsymbol{v} \in V_m \tag{4.27}$$

and for each $m \in \mathbb{N}$. Therefore, (4.27) is satisfied for all test functions $\boldsymbol{v} \in \bigcup_{m=1}^{+\infty} V_m$. This union is dense in V and the individual terms in (4.27) considered as functions of $\boldsymbol{v} \in V$ are continuous linear functionals. From this we conclude that (4.27) holds for all $\boldsymbol{v} \in V$. As we see, function $\boldsymbol{z}$ is a solution of problem (3.29) and, thus, $\boldsymbol{u} = \boldsymbol{g}^* + \boldsymbol{z}$ is a weak solution of the cascade flow problem, defined in 3.1.1.

The above results can formulated as the following theorem.

Theorem 4.2. (On the existence of a weak solution.) *Under the assumption that the norm* $\|\boldsymbol{g}\|_{H^s(\Gamma_i)^2}$ *satisfied the bound* (4.11), *there exists a weak solution of the cascade flow problem defined in* 3.1.1.

Remark. In the monographs [6], [14], [19] and [2], the existence of a weak solution of the boundary value problem for the Navier-Stokes equations in a bounded domain Ω with the Dirichlet boundary condition

$$\boldsymbol{u}\Big|_{\partial\Omega} = \boldsymbol{g}, \tag{4.28}$$

prescribed on the whole boundary $\partial\Omega$ is established under the assumption that $\boldsymbol{g} \in H^{1/2}(\partial\Omega)^k$ (where $k = 2$ or $k = 3$) and assumptions ensuring the condition

$$\int_\Gamma \boldsymbol{g} \cdot \boldsymbol{n}\, dS = 0 \tag{4.29}$$

for each component Γ of the boundary $\partial\Omega$. In this case, it is not necessary to assume the smallness of function $\boldsymbol{g}$ in a suitable norm.

However, in applications we can meet the necessity to solve viscous flow in a domain Ω whose boundary has several components, the Dirichlet boundary condition (4.28) is prescribed on $\partial\Omega$ and function $\boldsymbol{g}$ satisfies the condition

$$\int_{\partial\Omega} \boldsymbol{g} \cdot \boldsymbol{n}\, dS = 0, \tag{4.30}$$

but the integrals of $\boldsymbol{g} \cdot \boldsymbol{n}$ over individual components of the boundary are nonzero. This general problem is studied in [5], but it is necessary to assume that function $\boldsymbol{g}$ is sufficiently small. The existence of a solution in the case, when this assumption is not satisfied, remains open. Our problem has a similar character and, therefore, we are able to prove the existence of a weak solution under the restrictive assumption (4.11). There is a hope that this assumption will not be necessary for the existence of a weak solution of the nonstationary problem, which will be the subject of a further analysis. Another direction of a future research will be the finite element approximation of the problem.

Acknowledgement This research was supported under the Grant No. 201/01/1200 of the Grant Agency of the Czech Republic and the Grant No. MSM 113200007 of the Ministry of Education of the Czech Republic. The authors acknowledge the support of these institutions.

References

[1] Bruneau C.H., Fabrie P.: New efficient boundary conditions for incompressible Navier–Stokes equations: A well-posedness result. *Mathematical Modelling and Numerical Analysis* **30**, No. 7, 1996, 815–840.

[2] Feistauer M.: *Mathematical Methods in Fluid Dynamics.* Pitman Monographs and Surveys in Pure and Applied Mathematics 67, Longman Scientific & Technical, Harlow, 1993.

[3] Dolejší V., Feistauer M., Felcman J., Kliková A.: Error estimates for barycentric finite volumes combined with nonconforming finite elements applied to nonlinear convection-diffusion problems. *Appl. Math.* **47**, No. 4, 2002, 301–340.

[4] Galdi G.P.: *An Introduction to the Mathematical Theory of the Navier–Stokes Equations, Vol. I: Linearized Steady Problems.* Springer Tracts in Natural Philosophy, Vol. 38, Springer-Verlag, New York-Berlin-Heidelberg, 1994.

[5] Galdi G.P.: *An Introduction to the Mathematical Theory of the Navier–Stokes Equations, Vol. II: Nonlinear Steady Problems.* Springer Tracts in Natural Philosohy, Vol. 39, Springer-Verlag, New York-Berlin-Heidelberg, 1994.

[6] Girault V., Raviart P.-A.: *Finite Elements Methods for Navier–Stokes Equations.* Springer-Verlag, Berlin-Heidelberg-New York-Tokyo, 1986.

[7] Glowinski R.: *Numerical Methods for Nonlinear Variational Problems.* Springer-Verlag, New York-Berlin-Heidelberg-Tokyo, 1984.

[8] Heywood J.G., Rannacher R., Turek S.: Artificial boundaries and flux and pressure conditions for the incompressible Navier-Stokes equations. *Int. J. for Numerical Methods in Fluids* **22**, 1996, 325–352.

[9] Kračmar S., Neustupa J.: Global existence of weak solutions of a nonsteady variational inequality of the Navier–Stokes type with mixed boundary conditions. *Proc. of the "International Symposium on Numerical Analysis"*, Charles University Prague, 1993, 156–177.

[10] Kračmar S., Neustupa J.: A weak solvability of steady variational inequality of the Navier-Stokes type with mixed boundary conditions. *Nonlinear Analysis* **47**, 2001, 4169–4180.

[11] Kučera P., Skalák Z.: Local solutions to the Navier-Stokes equations with mixed boundary conditions. *Acta Applicandae Mathematicae* **54**, No. 3, 1998, 275–288.

[12] Kučera P.: Solutions to the stationary Navier-Stokes equations with mixed boundary conditions in a bounded domain. *Proc. of the conference "Analysius, Numerics and Applications of Differential and Integral Equations"*, editors M. Bach, C. Constanda, G.C. Hsiao, A.M. Sändig & P. Werner, Pitman Research Notes in Mathematical Series 379, 1998, 127–132.

[13] Kufner A., Fučík S., John O.: *Function Spaces.* Academia, Praha, 1977.

[14] Ladyzhenskaya O.A.: The Mathematical Theory of Viscous Incompressible Flow. Gordon and Breach, New York 1969.

[15] Lions J.L.: *Non-Homogeneous Boundary Value Problems and Applications I.* Springer-Verlag, Berlin-Heidelberg-New York, 1972.

[16] Martensen E.: Berechnung der Druckverteilung an Gitterprofilen in ebener Potentialströmung mit einer Fredholmschen Integralgleichung. *Arch. Rat. Mech. Anal.* **3**, 1959, 253–270.

[17] Meister E.: *Randwertaufgaben der Funktionentheorie.* Teubner, Stuttgart, 1983.

[18] Polášek J., Vlášek Z.: Berechnung der ebenen Potentialströmung von rotierenden radialen Profilgittern. *Appl. Math.* **17**, 1972, 295–308.

[19] Temam R.: *Navier-Stokes Equations.* North-Holland, Amsterdam-New York-Oxford, 1977.

Miloslav Feistauer and Tomáš Neustupa
Faculty of Mathematics and Physics
Charles University, Sokolovská 83
186 75 Praha 8, Czech Republic
e-mail: feist@karlin.mff.cuni.cz
e-mail: tneu@centrum.cz

Operator Theory
Advances and Applications, Vol. 147, 277–285

On Inversion of Convolution Integral Operators on a Finite Interval

Israel Gohberg, Marinus A. Kaashoek, and Freek van Schagen

Dedicated to the memory of Erhard Meister.
We remember him with respect and admiration.

Abstract. The Gohberg-Heinig inversion formula for convolution operators on a finite interval is computed explicitly for the case when the kernel function is given by $k(t) = Ce^{tA}B$, where A, B, and C are matrices of appropriate sizes.

1. Introduction

Let $k \in L^1_{m\times m}[-\tau, \tau]$, where τ is a positive real and m is a positive integer, and let T be the convolution operator on $L^2_m[0, \tau]$ given by

$$(Tf)(t) = f(t) - \int_0^\tau k(t-s)f(s)\mathrm{d}s, \quad 0 \le t \le \tau. \tag{1.1}$$

This paper concerns the Gohberg-Heinig inversion theorem [2] for T. To state this result we need the following integral equations in $m \times m$ matrix functions associated with T:

$$x(t) - \int_0^\tau k(t-s)x(s)\mathrm{d}s = k(t), \quad 0 \le t \le \tau, \tag{1.2}$$

$$z(t) - \int_{-\tau}^0 k(t-s)z(s)\mathrm{d}s = k(t), \quad -\tau \le t \le 0, \tag{1.3}$$

$$w(t) - \int_0^\tau w(s)k(t-s)\mathrm{d}s = k(t), \quad 0 \le t \le \tau, \tag{1.4}$$

$$y(t) - \int_{-\tau}^0 y(s)k(t-s)\mathrm{d}s = k(t), \quad -\tau \le t \le 0. \tag{1.5}$$

In the sequel we follow the convention that solutions of the equations (1.2)–(1.5) will be considered on the full interval $[-\tau, \tau]$ by defining them to be zero on the appropriate complementary intervals.

2000 *Mathematics Subject Classification.* Primary 45E10; Secondary 45B05.
Key words and phrases. Convolution integral operators on a finite interval, inversion formula.

Theorem 1. ([2]) *The following statements are equivalent:*

(i) *the operator* T *is invertible on* $L^2_m[0,\tau]$;

(ii) *equations* (1.2) *and* (1.3) *have solutions in* $L^1_{m\times m}[0,\tau]$ *and* $L^1_{m\times m}[-\tau,0]$, *respectively;*

(iii) *equations* (1.4) *and* (1.5) *have solutions in* $L^1_{m\times m}[0,\tau]$ *and* $L^1_{m\times m}[-\tau,0]$, *respectively.*

Moreover in this case T^{-1} *is given by*

$$(T^{-1}f)(t) = f(t) + \int_0^\tau \gamma(t,s)f(s)\mathrm{d}s, \quad 0 \le t \le \tau, \tag{1.6}$$

with the kernel function γ *being given by*

$$\gamma(t,s) = x(t-s)+y(t-s)+\int_0^{\min\{t,s\}} [x(t-r)y(r-s)-z(t-r-\tau)w(r-s+\tau)]\mathrm{d}r, \tag{1.7}$$

where x, w, z *and* y *are the (unique) solutions of* (1.2)–(1.5).

In [2] (see also Chapter 7 in [1]) the inversion formula (1.7) is proved in two steps. The first step establishes this formula for kernel functions of the form

$$k(t) = \sum_{j=-n}^{n} \mathrm{e}^{i\pi jt/\tau} A_j, \tag{1.8}$$

where A_j, $-n \le j \le n$, is an $m \times m$ complex matrix. The second step uses the result of the first step and an approximation argument based on the fact that functions of the form (1.8) are dense in $L^1_{m\times m}[-\tau,\tau]$.

In this paper we concentrate on the first of these two steps. Instead of functions of the form (1.8) we work with kernel functions k of the form

$$k(t) = C\mathrm{e}^{tA}B, \quad -\tau \le t \le \tau, \tag{1.9}$$

where the matrices A, B, and C are of size $n\times n$, $n\times m$, and $m\times n$, respectively. The class of functions defined by (1.9) contains the functions of the form (1.8), and hence this second class is also dense in $L^1_{m\times m}[-\tau,\tau]$. We refer to a function of the form (1.9) as a *kernel of exponential type*. Notice that these kernels are of finite rank.

For kernels of exponential type we give a direct and simple proof of Theorem 1. Our proof uses elements of system theory and has a linear algebraic flavor. It yields explicit formulas for all functions involved. The following theorem is our main result.

Theorem 2. *Let* T *be the integral operator* (1.1) *on* $L^2_m[0,\tau]$ *with kernel* k *given by* (1.9). *Then* T *is invertible if and only if the* $n\times n$ *matrix*

$$M = I_n - \int_0^\tau \mathrm{e}^{-sA}BC\mathrm{e}^{sA}\mathrm{d}s \tag{1.10}$$

is invertible. In that case the resolvent kernel γ *and the solutions* x, w, z *and* y *of the equations* (1.2–1.5) *are given by*

$$\gamma(t,s) = Ce^{tA}M^{-1}e^{-sA}B, \quad 0 \le s \le \tau, 0 \le t \le \tau, \tag{1.11}$$

$$x(t) = Ce^{tA}M^{-1}B = \gamma(t,0), \quad 0 \le t \le \tau, \tag{1.12}$$

$$z(t) = Ce^{(\tau+t)A}M^{-1}e^{-\tau A}B = \gamma(\tau+t,\tau), \quad -\tau \le t \le 0, \tag{1.13}$$

$$w(t) = Ce^{\tau A}M^{-1}e^{-(\tau-t)A}B = \gamma(\tau,\tau-t), \quad 0 \le t \le \tau, \tag{1.14}$$

$$y(t) = CM^{-1}e^{tA}B = \gamma(0,-t), \quad -\tau \le t \le 0. \tag{1.15}$$

Furthermore, the functions γ *and* x, y, z *and* w *satisfy the relation* (1.7).

In what follows we refer to the matrix M given by (1.10) as the *indicator* of T corresponding to the representation (1.9).

In the next section we prove Theorem 2. The final section presents a direct proof of the equivalence of statements (i)–(iii) in Theorem 1 for the case when k is given by (1.9).

2. Kernels of exponential type

Throughout this section k is given (1.9), and K is the integral operator on $L^2_m[0,\tau]$ given by

$$(Kf)(t) = \int_0^\tau k(t-s)f(s)\mathrm{d}s, \quad 0 \le t \le \tau. \tag{2.1}$$

Thus $T = I - K$. We first prove Theorem 2.

Proof of Theorem 2. We divide the proof into three parts. In the first part we prove that T is invertible if and only if M is invertible, and derive formula (1.11). In the two other parts we assume that T (or equivalently the indicator M) is invertible.

Part 1. Notice that $K = FG$, where the operators $F : \mathbb{C}^n \to L^2_m[0,\tau]$ and $G : L^2_m[0,\tau] \to \mathbb{C}^n$ are given by

$$(Fx)(t) = Ce^{tA}x \quad (x \in \mathbb{C}^n), \quad Gf = \int_0^\tau e^{-sA}Bf(s)\mathrm{d}s.$$

From the equality

$$\begin{pmatrix} I & -F \\ 0 & I_n \end{pmatrix}\begin{pmatrix} I & 0 \\ G & I_n - GF \end{pmatrix}\begin{pmatrix} I & F \\ 0 & I_n \end{pmatrix} = \begin{pmatrix} I - FG & 0 \\ G & I_n \end{pmatrix}, \tag{2.2}$$

it follows that $T = I - K$ is invertible if and only if $M = I - GF$ is invertible. Moreover, in that case $T^{-1} = I + FM^{-1}G$. Hence

$$(T^{-1}f)(t) = f(t) + \int_0^\tau Ce^{tA}M^{-1}e^{-sA}Bf(s)\mathrm{d}s,$$

which shows that the resolvent kernel γ is indeed given by (1.11).

Part 2. We prove the formulas (1.12)–(1.15). First (1.12). Since T is invertible, we have $x = T^{-1}k$. But we know that $k = FB$. Using $T^{-1} = I + FM^{-1}G$ we obtain the equalities

$$x = T^{-1}FB = (I + FM^{-1}G)FB = FM^{-1}(M + GF)B = FM^{-1}B$$

Thus $x = FM^{-1}B$, and hence $x(t) = Ce^{tA}M^{-1}B = \gamma(t,0)$.

Next, observe that equation (1.3) can be transformed into an equation of the form (1.2). Indeed, put $\tilde{z}(t) = z(-t)$ and $\tilde{k}(t) = k(-t)$. Then equation (1.3) can be rewritten in the following equivalent form:

$$\tilde{z}(t) - \int_0^\tau \tilde{k}(t-s)\tilde{z}(s)ds = \tilde{k}(t), \quad 0 \le t \le \tau. \tag{2.3}$$

Notice that $\tilde{k}(t) = Ce^{-tA}B$, and hence we can apply the results of the previous paragraphs with A being replaced by $-A$. For the corresponding indicator $\tilde{M}$ we have

$$\tilde{M} := I_n - \int_0^\tau e^{sA}BCe^{-sA}ds = I_n - e^{\tau A}\int_0^\tau e^{-sA}BCe^{-sA}dse^{-\tau A} = e^{\tau A}Me^{-\tau A}.$$

Hence $\tilde{M}$ is invertible, and we conclude that (2.3) is uniquely solvable, and by the result of the previous paragraph $\tilde{z}(t) = Ce^{-tA}\tilde{M}^{-1}B$. Therefore

$$z(t) = Ce^{(\tau+t)A}M^{-1}e^{-\tau A}B = \gamma(\tau+t,\tau),$$

and (1.13) is proved.

To prove (1.14), we consider the transposed of (1.4) which is an equation of the same type as (1.2). Indeed, put $\hat{w}(t) = w(t)^T$ and $\hat{k}(t) = k(t)^T$, and consider

$$\hat{w}(t) - \int_0^\tau \hat{k}(t-s)\hat{w}(s)ds = \hat{k}(t), \quad 0 \le t \le \tau. \tag{2.4}$$

Notice that $\hat{k}(t) = B^Te^{-tA^T}C^T$, and hence $\hat{k}$ is again a linear combination of exponentials. For the corresponding indicator $\hat{M}$ we have

$$\hat{M} = I_n - \int_0^\tau e^{-sA^T}C^TB^Te^{sA^T}ds = (I_n - \int_0^\tau e^{sA}BCe^{-sA}ds)^T = (e^{\tau A}Me^{-\tau A})^T.$$

Thus $\hat{M}$ is invertible, and $\hat{w}(t) = B^Te^{tA^T}\hat{M}^{-1}C^T$ is the unique solution of (2.4). This shows that $w(t) = Ce^{\tau A}M^{-1}e^{-(\tau-t)A}B = \gamma(\tau,\tau-t)$ is the solution of (1.4), and (1.14) is proved.

Finally, in order to solve equation (1.5) put $\tilde{y}(t) = y(-t)$. Then the equation reduces to the equation (1.4) with $-A$ replacing A. Therefore the solution is

$$\tilde{y}(t) = Ce^{-\tau A}\hat{M}^{-1}e^{(\tau-t)A}B = CM^{-1}e^{-tA}B = \gamma(0,t).$$

Since $y(t) = \tilde{y}(-t)$, the preceding calculation yields (1.15).

Part 3. In this part we verify that the functions γ and x, y, z and w satisfy the relation (1.7). For this purpose we first note that

$$MA - AM = \int_0^\tau (Ae^{-sA}BCe^{sA} - e^{-sA}BCe^{sA}A)\,ds = BC - e^{-\tau A}BCe^{\tau A}. \tag{2.5}$$

Hence,

$$M^{-1}(BC - e^{-\tau A}BCe^{\tau A})M^{-1} = AM^{-1} - M^{-1}A.$$

By integration we obtain

$$\int_0^s e^{-rA}M^{-1}(BC - e^{-\tau A}BCe^{\tau A})M^{-1}e^{rA}dr = M^{-1} - e^{-sA}M^{-1}e^{sA}. \tag{2.6}$$

Multiplying (2.6) to the left by Ce^{tA} and to the right by $e^{-sA}B$ we obtain

$$\begin{aligned} &Ce^{tA}M^{-1}e^{-sA}B - Ce^{(t-s)A}M^{-1}B \\ &\quad = \int_0^s Ce^{(t-r)A}M^{-1}BCM^{-1}e^{(r-s)A}B \\ &\qquad -Ce^{(t-r)A}M^{-1}e^{-\tau A}BCe^{\tau A}M^{-1}e^{(r-s)A}Bdr. \end{aligned}$$

For $s < t$ this is equivalent to

$$\gamma(t,s) = x(t-s) + \int_0^{\min\{t,s\}} [x(t-r)y(r-s) - z(t-r-\tau)w(r-s+\tau)]dr,$$

which proves (1.7) for $s < t$.

Next we replace s by t in formula (2.6) and multiply to the left by Ce^{tA} and to the right by $e^{-sA}B$ to obtain

$$\begin{aligned} &Ce^{tA}M^{-1}e^{-sA}B - CM^{-1}e^{(t-s)A}B \\ &\quad = \int_0^t Ce^{(t-r)A}M^{-1}BCM^{-1}e^{(r-s)A}B \\ &\qquad -Ce^{(t-r)A}M^{-1}e^{-\tau A}BCe^{\tau A}M^{-1}e^{(r-s)A}Bdr. \end{aligned}$$

which is for the case when $s > t$ the equality

$$\gamma(t,s) = y(t-s) + \int_0^{\min\{t,s\}} [x(t-r)y(r-s) - z(t-r-\tau)w(r-s+\tau)]dr.$$

We finished the proof of (1.7). □

Notice that the matrix M depends on the choice of A, B and C, but that γ, x, w, z and y do not. In particular Theorem 2 implies that the matrix M is invertible for all choices of the representation of k if it is invertible for one.

3. The equivalence of (i)–(iii) in Theorem 1

In this section we present a direct proof of the equivalence of statements (i)–(iii) in Theorem 1 for the case when k is a kernel of exponential type. In this case these equivalences may be derived as a corollary from the following proposition.

Proposition 3. *Let A, B, and C be matrices of sizes $n \times n$, $n \times m$, and $m \times n$, respectively. Consider the matrix*

$$M = I_n - \int_0^\tau e^{-sA} B C e^{sA} ds. \tag{3.1}$$

The following statements are equivalent:

(i) *the matrix M is invertible;*

(ii) *the matrix equations $MX = B$ and $MZ = e^{-\tau A}B$ have solutions;*

(iii) *the matrix equations $YM = C$ and $WM = Ce^{\tau A}$ have solutions.*

Furthermore, in this case

$$M^{-1} = I_n + \int_0^\tau e^{-tA} B C e^{tA}\, dt + \int_0^\tau \int_0^\tau e^{-tA} B \gamma(t,s) C e^{sA}\, ds\, dt. \tag{3.2}$$

Here

$$\gamma(t,s) = \begin{cases} Ce^{(t-s)A}X + Ce^{tA}R(s)e^{-sA}B, & s < t, \\ Ye^{(t-s)A}B + Ce^{tA}R(t)e^{-sA}B, & s > t, \end{cases} \tag{3.3}$$

with the function $R(\cdot)$ being given by

$$R(t) = \int_0^t e^{-rA}(XY - ZW)e^{rA}\, dr. \tag{3.4}$$

Proof. Notice that the equivalence of (i) and (iii) follows from the equivalence of (i) and (ii) by transposing. Obviously, (i) implies (ii). So we have to prove that (ii) implies (i). In order to do this we identify A, B, and C with the linear operators of which the action relative to the standard bases in $\mathbb{C}^n$ and $\mathbb{C}^m$ are given by the respective matrices. First we show that without loss of generality we may assume that

$$\operatorname{Im} \begin{bmatrix} B & AB & \cdots & A^{n-1}B \end{bmatrix} = \mathbb{C}^n. \tag{3.5}$$

Let $\mathcal{H}_0$ be the subspace of $\mathbb{C}^n$ defined by the left-hand side of (3.5), and let $\mathcal{H}_1$ be the orthogonal complement of $\mathcal{H}_0$ in $\mathbb{C}^n$. Notice that $\mathcal{H}_0$ is invariant under A, and the range of B is contained in $\mathcal{H}_0$. Thus relative to the decomposition $\mathbb{C}^n = \mathcal{H}_0 \oplus \mathcal{H}_1$, the operators A, B, and C admit the following partitionings:

$$A = \begin{bmatrix} A_0 & * \\ 0 & * \end{bmatrix}, \quad B = \begin{bmatrix} B_0 \\ 0 \end{bmatrix}, \quad C = \begin{bmatrix} C_0 & * \end{bmatrix}. \tag{3.6}$$

Here the stars denote operators which we will not specify further. A straightforward computation using the partitionings in (3.6) shows that relative to $\mathbb{C}^n = \mathcal{H}_0 \oplus \mathcal{H}_1$ the operator M in (3.1) can be partitioned as:

$$M = \begin{bmatrix} M_0 & * \\ 0 & I_{\mathcal{H}_1} \end{bmatrix}, \tag{3.7}$$

where

$$M_0 = I_{\mathcal{H}_0} - \int_0^\tau e^{-sA_0} B_0 C_0 e^{sA_0} ds. \tag{3.8}$$

Here $I_{\mathcal{H}_0}$ and $I_{\mathcal{H}_1}$ are the identity operators on $\mathcal{H}_0$ and $\mathcal{H}_1$, respectively. From (3.7) we see that M is invertible if and only if M_0 is invertible.

Next, notice that the equations $MX = B$ and $MZ = \mathrm{e}^{-\tau A}B$ can be written as

$$\begin{bmatrix} M_0 & * \\ 0 & I_{\mathcal{H}_1} \end{bmatrix}\begin{bmatrix} X_0 \\ X_1 \end{bmatrix} = \begin{bmatrix} B_0 \\ 0 \end{bmatrix}, \quad \begin{bmatrix} M_0 & * \\ 0 & I_{\mathcal{H}_1} \end{bmatrix}\begin{bmatrix} Z_0 \\ Z_1 \end{bmatrix} = \begin{bmatrix} \mathrm{e}^{-\tau A_0}B_0 \\ 0 \end{bmatrix}.$$

This implies that $MX = B$ and $MZ = \mathrm{e}^{-\tau A}B$ are solvable if and only if $M_0X_0 = B_0$ and $M_0Z_0 = \mathrm{e}^{-\tau A_0}B_0$ are solvable. Thus in order to prove (ii)⇒(i) we may without loss of generality assume that (3.5) holds.

Assume (ii) holds, and let (3.5) be satisfied. Let x be a vector in $\mathbb{C}^n$ such that $x^TM = 0$, where x^T is the transpose of x. Since the equations $MX = B$ and $MZ = \mathrm{e}^{-\tau A}B$ are solvable, $x^TM = 0$ implies that $x^TB = 0$ and $x^T\mathrm{e}^{-\tau A}B = 0$. As we have seen in the previous section, M satisfies the following intertwining relation:

$$MA - AM = BC - \mathrm{e}^{-\tau A}BC\mathrm{e}^{\tau A}. \tag{3.9}$$

We conclude that $x^TAM = 0$. By repeating the argument with x^TA in place of x^T we see that $x^TA^2M = 0$. By continuing in this way we show that $x^TA^kM = 0$ for $k = 0, 1, \ldots, n-1$. Recall that $MX = B$ for some X. Hence $x^TA^kB = 0$ for $k = 0, 1, \ldots, n-1$. But then we see from (3.5) that $x^T = 0$. Thus $x^T = 0$ implies $x = 0$, and hence M is invertible. This completes the proof of (ii)⇒(i).

It remains to prove the inversion formula (3.2). Since M is invertible, we see from the intertwining relation (3.9) that

$$AM^{-1} - M^{-1}A = XY - ZW.$$

Thus R in (3.4) is also given by

$$\begin{aligned} R(t) &= \int_0^t \mathrm{e}^{-rA}(AM^{-1} - M^{-1}A)\mathrm{e}^{rA}\,\mathrm{d}r \\ &= -\int_0^t \frac{\mathrm{d}}{\mathrm{d}r}(\mathrm{e}^{-rA}M^{-1}\mathrm{e}^{rA})\,\mathrm{d}r = M^{-1} - \mathrm{e}^{-tA}M^{-1}\mathrm{e}^{tA}. \end{aligned}$$

Inserting this expression of R into (3.3) and using that $X = M^{-1}B$ and $Y = CM^{-1}$ yields

$$\gamma(t,s) = C\mathrm{e}^{tA}M^{-1}\mathrm{e}^{-sA}B,$$

and hence

$$\begin{aligned} &\int_0^\tau\int_0^\tau \mathrm{e}^{-tA}B\gamma(t,s)C\mathrm{e}^{sA}\,\mathrm{d}t\,\mathrm{d}s \\ &\qquad = \Big(\int_0^\tau \mathrm{e}^{-tA}BC\mathrm{e}^{tA}\,\mathrm{d}t\Big)M^{-1}\Big(\int_0^\tau \mathrm{e}^{-sA}BC\mathrm{e}^{sA}\,\mathrm{d}s\Big) \\ &\qquad = (I_n - M)M^{-1}(I_n - M) = M^{-1} - 2I_n + M. \end{aligned}$$

This yields (3.2) and completes the proof. □

Lemma 4. *Let k be given by (1.9), and let M be the corresponding indicator. Then the following statements are true.*

(i) *equation (1.2) has a solution in $L^1_{m\times m}[0,\tau]$ if and only if $MX = B$ has a solution.*

(ii) *equation (1.3) has a solution in $L^1_{m\times m}[-\tau,0]$ if and only if $MZ = \mathrm{e}^{-\tau A}B$ has a solution.*

(iii) *equation (1.4) has a solution in $L^1_{m\times m}[0,\tau]$ if and only if $WM = C\mathrm{e}^{\tau A}$ has a solution.*

(iv) *equation (1.5) has a solution in $L^1_{m\times m}[-\tau,0]$ if and only if $YM = C$ has a solution.*

Proof. By transposing we see that (i) follows from (iii), and (ii) from (iv). Furthermore, (iv) follows from (iii) by passing from t to $-t$. Thus it suffices to prove (iii).

Assume the matrix equation $WM = C\mathrm{e}^{\tau A}$ has a solution. Consider the function $w(t) = W\mathrm{e}^{(t-\tau)A}B$ on $0 \le t \le \tau$. We claim that w is a solution of (1.4). Indeed

$$\begin{aligned} w(t) - \int_0^\tau w(s)k(t-s)\mathrm{d}s &= W\mathrm{e}^{(t-\tau)A}B - \int_0^\tau W\mathrm{e}^{(s-\tau)A}BC\mathrm{e}^{(t-s)A}B\,\mathrm{d}s \\ &= \big[W\mathrm{e}^{-\tau A} - W\mathrm{e}^{-\tau A}\big(\int_0^\tau \mathrm{e}^{sA}BC\mathrm{e}^{-sA}\,\mathrm{d}s\big)\big]\mathrm{e}^{tA}B \\ &= \big[W\mathrm{e}^{-\tau A} - W\big(\int_0^\tau \mathrm{e}^{-sA}BC\mathrm{e}^{sA}\,\mathrm{d}s\big)\mathrm{e}^{-\tau A}\big]\mathrm{e}^{tA}B \\ &= \big[WM\mathrm{e}^{-\tau A}\big]\mathrm{e}^{tA}B = C\mathrm{e}^{tA}B = k(t), \quad 0 \le t \le \tau. \end{aligned}$$

Thus w is a solution of (1.4).

Conversely, assume (1.4) has a solution w.Then

$$\begin{aligned} C\mathrm{e}^{tA}B &= w(t) - \int_0^\tau w(s)C\mathrm{e}^{(t-s)A}B\,\mathrm{d}s \\ &= w(t) - \big(\int_0^\tau w(s)C\mathrm{e}^{-sA}\,\mathrm{d}s\big)\mathrm{e}^{tA}B, \end{aligned}$$

and hence w is of the form $w(t) = \tilde{W}\mathrm{e}^{tA}B$. Put $W = \tilde{W}\mathrm{e}^{\tau A}$. Then $(WM - C\mathrm{e}^{\tau A})\mathrm{e}^{tA}B = 0$ for each $0 \le t \le \tau$. We claim that this implies that $WM = C\mathrm{e}^{\tau A}$. When (3.5) is satisfied this is trivial, and if not one uses the same arguments as in the second paragraph of the proof of Proposition 3. □

From Proposition 3 and the above lemma the equivalence of (i), (ii), and (iii) in Theorem 1 is clear for the case when k is of the form (1.9).

In conclusion, let us remark that in system theory formula (3.5) is known as a condition of controllability (see, e.g., [3], Section 6.2). The transformation of the pair A, B into the controllable pair A_0, B_0 described in the second paragraph of the proof of Proposition 3 is also standard in system theory.

References

[1] R.L. Ellis, I. Gohberg, *Orthogonal Polynomials and Convolution Equations*, Operator Theory: Advances and Applications, Vol. **140**, Birkhäuser Verlag, Basel, 2003.

[2] I. Gohberg, G. Heinig, On matrix valued integral operators on a finite interval with matrix kernels that depend on the difference of arguments, *Rev. Roumaine Math. Pures Appl.*, **20** (1975), 1. 55–73. (in Russian)

[3] T. Kailath, *Linear systems*, Prentice Hall, Englewood Cliffs, N.J., 1980.

I. Gohberg
School of Mathematical Sciences
Ryamond and Beverly Sackler Faculty of Exact Sciences
Tel-Aviv University
69978 Tel Aviv
Israel
e-mail: gohberg@post.tau.ac.il

M.A. Kaashoek
Afdeling Wiskunde
Faculteit der Exacte Wetenschappen
Vrije Universiteit
De Boelelaan 1081a
1081 HV Amsterdam
The Netherlands
e-mail: kaash@few.vu.nl

F. van Schagen
Afdeling Wiskunde
Faculteit der Exacte Wetenschappen
Vrije Universiteit
De Boelelaan 1081a
1081 HV Amsterdam
The Netherlands
e-mail: freek@cs.vu.nl

Operator Theory
Advances and Applications, Vol. 147, 287–296

Some Algebras of Functions with Fourier Coefficients in Weighted Orlicz Sequence Spaces

Alexei Yu. Karlovich

To the memory of Erhard Meister

Abstract. In this paper, the author proves that the set of all integrable functions whose sequences of negative (resp. nonnegative) Fourier coefficients belong to $\ell^1 \cap \ell^\Phi_{\varphi,w}$ (resp. to $\ell^1 \cap \ell^\Psi_{\psi,\varrho}$), where $\ell^\Phi_{\varphi,w}$ and $\ell^\Psi_{\psi,\varrho}$ are two-weighted Orlicz sequence spaces, forms an algebra under pointwise multiplication whenever the weight sequences

$$\varphi = \{\varphi_n\}, \quad \psi = \{\psi_n\}, \quad w = \{w_n\}, \quad \varrho = \{\varrho_n\}$$

increase and satisfy the Δ_2-condition.

1. Introduction

Let $\mathbb{T}$ be the unit circle. For a complex-valued function $f \in L^1(\mathbb{T})$, let $\{f_n\}_{n\in\mathbb{Z}}$ be the sequence of the Fourier coefficients of f,

$$f_n := \frac{1}{2\pi} \int_0^{2\pi} f(e^{i\theta}) e^{-in\theta}\, d\theta.$$

Let W be the Wiener algebra of all functions f on $\mathbb{T}$ for which

$$\|f\|_W := \sum_{k=-\infty}^{\infty} |f_k| < \infty.$$

Let $F\ell^{p,r}_{\alpha,\beta}$, where $1 \le p, r < \infty$ and $0 \le \alpha, \beta < \infty$, denote the set of all functions f on $\mathbb{T}$ for which

$$\|f\|_{F\ell^{p,r}_{\alpha,\beta}} := \left(\sum_{k=1}^{\infty} |f_{-k}|^p (k+1)^{\alpha p} \right)^{1/p} + \left(\sum_{k=0}^{\infty} |f_k|^r (k+1)^{\beta r} \right)^{1/r} < \infty.$$

2000 *Mathematics Subject Classification.* Primary 46J10; Secondary 46B45.
Key words and phrases. Wiener algebra, weighted Orlicz sequence space.
The author is supported by F.C.T. (Portugal) grant SFRH/BPD/11619/2002.

The following result was conjectured by A. Böttcher and B. Silbermann and proved by Detlef Horbach. The proof is in [1, Section 6.54].

Theorem 1.1. *If* $1 \le p, r < \infty$ *and* $0 \le \alpha, \beta < \infty$, *then* $W \cap F\ell^{p,r}_{\alpha,\beta}$ *is an algebra under pointwise multiplication.*

Note that a stronger fact is actually proved in [1, Section 6.54]: the inequality $\|f^2\|_{F\ell^{p,r}_{\alpha,\beta}} \le C_1 \|f\|_W \|f\|_{F\ell^{p,r}_{\alpha,\beta}}$ is fulfilled for some $C_1 > 0$ and any $f \in W \cap F\ell^{p,r}_{\alpha,\beta}$. From the latter inequality and $\|fg\|_W \le \|f\|_W \|g\|_W$ one can easily get

$$\|fg\|_W + \|fg\|_{F\ell^{p,r}_{\alpha,\beta}} \le C\Big(\|f\|_W + \|f\|_{F\ell^{p,r}_{\alpha,\beta}}\Big)\Big(\|g\|_W + \|g\|_{F\ell^{p,r}_{\alpha,\beta}}\Big)$$

for some $C > 0$ and all $f, g \in W \cap F\ell^{p,r}_{\alpha,\beta}$, that is, $W \cap F\ell^{p,r}_{\alpha,\beta}$ is a Banach algebra. This result has important applications in the theory of Toeplitz determinants and the theory of Toeplitz operators on Lebesgue spaces $l^p_\varrho (1 < p < \infty)$ with Khvedelidze weights ϱ (see [1, Ch. 7 and Ch. 10]). The aim of the present note is to generalize this result by replacing in the definition of $F\ell^{p,r}_{\alpha,\beta}$:

1. norms in weighted Lebesgue sequence spaces by norms in two-weighted Orlicz sequence spaces;
2. canonical weight sequences $\{(n+1)^\lambda\}_{n=0}^\infty$ with $\lambda \in \{\alpha, \beta\}$ by general increasing weight sequences satisfying the Δ_2-condition.

The author hopes that such an extension will be useful because the scale of Orlicz spaces is much wider and sensitive than the scale of Lebesgue spaces.

Note also that similar questions were considered by P.L. Ul'yanov [5] for the set E of functions $f \in C(\mathbb{T})$ with

$$\|f\|_E := \sum_{k \in \mathbb{Z}} \omega(|f_k|)\tau_n < \infty,$$

where ω is a function with certain properties (a so-called modulus of continuity) and $\{\tau_n\}_{n\in\mathbb{Z}}$ is a weight sequence. Under some natural assumptions he proved that E is an algebra under pointwise multiplication.

This note is organized as follows. In Section 2 we remind the notion of two-weighted Orlicz sequence spaces and define a related class $F\ell^{\Phi,\Psi}_{\varphi,w;\psi,\varrho}$ that generalizes the class $F\ell^{p,r}_{\alpha,\beta}$. Further we formulate the main result: the set $W \cap F\ell^{\Phi,\Psi}_{\varphi,w;\psi,\varrho}$ is a Banach algebra under pointwise multiplication whenever the weight sequences $\varphi, w, \psi, \varrho$ increase and satisfy the Δ_2-condition. Finally we state a corollary about factorization of nondegenerate functions in this algebra. Section 3 contains the proofs.

2. Preliminaries and the main result

2.1. Weighted Orlicz sequence spaces

An *Orlicz function* Φ is a continuous non-decreasing and convex function defined for $t \ge 0$ such that $\Phi(0) = 0$ and $\lim_{t\to\infty} \Phi(t) = \infty$ (see [3, Definition 4.a.1]). Let $\mathbb{I}$ be either $\mathbb{N} := \{1, 2, \dots\}$ or $\mathbb{Z}_+ := \mathbb{N} \cup \{0\}$ and let $\{\Phi_n\}_{n\in\mathbb{I}}$ be a sequence of Orlicz functions. The set $\ell^{\{\Phi_n\}}(\mathbb{I})$ of all sequences of complex numbers $c = \{c_n\}_{n\in\mathbb{I}}$ with

$$\sum_{n\in\mathbb{I}} \Phi_n\left(\frac{|c_n|}{\lambda}\right) < \infty$$

for some $\lambda = \lambda(c) > 0$ is a Banach space when equipped with the norm

$$\|c\|_{\ell^{\{\Phi_n\}}(\mathbb{I})} = \inf\left\{\lambda > 0 : \quad \sum_{n\in\mathbb{I}} \Phi_n\left(\frac{|c_n|}{\lambda}\right) \le 1\right\}.$$

The space $\ell^{\{\Phi_n\}}(\mathbb{I})$ is called a *modular sequence space* (see [3, Definition 4.d.1]) or a *Musielak-Orlicz sequence space* (see, e.g., [4]).

Any sequence $\{\nu_n\}_{n\in\mathbb{I}}$ of positive numbers is called a *weight sequence.* Let Φ be an Orlicz function and $\varphi = \{\varphi_n\}_{n\in\mathbb{I}}$ and $w = \{w_n\}_{n\in\mathbb{I}}$ be weight sequences. Clearly

$$\Phi_n(x) := \Phi(x\varphi_n)w_n \quad (x \ge 0) \tag{2.1}$$

is an Orlicz function for every $n \in \mathbb{I}$. The special case of a Musielak-Orlicz sequence space generated by the sequence of Orlicz functions (2.1) is denoted by $\ell^{\Phi}_{\varphi,w}(\mathbb{I})$ and is called the *two-weighted Orlicz space* generated by the Orlicz function Φ and the weight sequences $\varphi = \{\varphi_n\}_{n\in\mathbb{I}}$ and $w = \{w_n\}_{n\in\mathbb{I}}$.

In particular, if $\Phi(x) = x^p, 1 \le p < \infty$, then $\ell^{\Phi}_{\varphi,w}(\mathbb{I})$ is the weighted Lebesgue sequence space $\ell^p_\rho(\mathbb{I})$ and

$$\|c\|_{\ell^{\Phi}_{\varphi,w}(\mathbb{I})} = \left(\sum_{n\in\mathbb{I}} |c_n|^p \rho_n\right)^{1/p} \qquad \Big(\rho_n = \varphi_n^p w_n, \quad n \in \mathbb{I}\Big).$$

So, in this case, $\ell^{\Phi}_{\varphi,1}(\mathbb{I}) = \ell^{\Phi}_{1,\varphi^p}(\mathbb{I})$, where $\varphi^p := \{\varphi_n^p\}_{n\in\mathbb{I}}$. But in the general case an Orlicz function Φ is not homogeneous. Hence the weighted Orlicz sequence spaces $\ell^{\Phi}_{\varphi,1}(\mathbb{I})$ and $\ell^{\Phi}_{1,w}(\mathbb{I})$ are essentially different.

Applying criteria of coincidence of Musielak-Orlicz sequence spaces and Musielak-Orlicz sequence classes (see [4, Theorem 8.13(b)]) to the case of the sequence of Orlicz functions (2.1), one can get the following.

Proposition 2.1. *Suppose Φ is an Orlicz function, $\Phi(x) > 0$ whenever $x > 0$, and $\varphi = \{\varphi_n\}_{n\in\mathbb{I}}, w = \{w_n\}_{n\in\mathbb{I}}$ are weight sequences. Then*

$$\ell^{\Phi}_{\varphi,w}(\mathbb{I}) = \left\{c = \{c_n\}_{n\in\mathbb{I}} : \quad \sum_{n\in\mathbb{I}} \Phi(|c_n|\varphi_n)w_n < \infty\right\}$$

if and only if there exist positive numbers δ, K and a sequence of nonnegative numbers $\{d_n\}_{n\in\mathbb{I}}$ such that for all $x \ge 0$ and $n \in \mathbb{I}$,

$$\Phi(x\varphi_n)w_n < \delta \quad \Longrightarrow \quad \Phi(2x\varphi_n)w_n \le K\Phi(x\varphi_n)w_n + d_n$$

and $\sum_{n\in\mathbb{I}} d_n < \infty$.

2.2. The main result

Let Φ and Ψ be Orlicz functions and let

$$\varphi = \{\varphi_n\}_{n=1}^{\infty}, \quad w = \{w_n\}_{n=1}^{\infty}, \quad \psi = \{\psi_n\}_{n=0}^{\infty}, \quad \varrho = \{\varrho_n\}_{n=0}^{\infty}$$

be weight sequences. We denote by $F\ell^{\Phi,\Psi}_{\varphi,w;\psi,\varrho}$ the set of all functions $f \in L^1(\mathbb{T})$ such that the sequence $\{f_{-n}\}_{n\in\mathbb{N}}$ of all negative Fourier coefficients of f belongs to the two-weighted Orlicz space $\ell^{\Phi}_{\varphi,w}(\mathbb{N})$ and the sequence $\{f_n\}_{n\in\mathbb{Z}_+}$ of all nonnegative Fourier coefficients of f belongs to the two-weighted Orlicz space $\ell^{\Psi}_{\psi,\varrho}(\mathbb{Z}_+)$. The set $W \cap F\ell^{\Phi,\Psi}_{\varphi,w;\psi,\varrho}$ is a Banach space with respect to the norm

$$\|f\|_{W\cap F} := \|f\|_W + \|f\|_- + \|f\|_+,$$

where

$$\|f\|_- := \left\|\{f_{-k}\}_{k\in\mathbb{N}}\right\|_{\ell^{\Phi}_{\varphi,w}(\mathbb{N})} = \inf\left\{\lambda > 0 : \sum_{k=1}^{\infty} \Phi\left(\frac{|f_{-k}|\varphi_k}{\lambda}\right) w_k \le 1\right\},$$

$$\|f\|_+ := \left\|\{f_k\}_{k\in\mathbb{Z}_+}\right\|_{\ell^{\Psi}_{\psi,\varrho}(\mathbb{Z}_+)} = \inf\left\{\mu > 0 : \sum_{k=0}^{\infty} \Psi\left(\frac{|f_k|\psi_k}{\mu}\right) \varrho_k \le 1\right\}.$$

We denote by $\mathcal{W}^+$ (resp. by $\mathcal{W}^-$) the collection of all weight sequences $\{\nu_n\}_{n=0}^{\infty}$ (resp. $\{\nu_n\}_{n=1}^{\infty}$) such that

(i) $\nu_0 > 0$ (resp. $\nu_1 > 0$);
(ii) $\nu_n \le \nu_{n+1}$ for $n \in \mathbb{Z}_+$ (resp. for $n \in \mathbb{N}$);
(iii) $\{\nu_n\}_{n=0}^{\infty}$ (resp. $\{\nu_n\}_{n=1}^{\infty}$) satisfies the Δ_2-condition, that is, there exists a constant $C_\nu \in (0,\infty)$ such that $\nu_{2n} \le C_\nu \nu_n$ for $n \in \mathbb{N}$.

From (ii) and (iii) it follows that $C_\nu \ge 1$.

Theorem 2.2. *If Φ, Ψ are arbitrary Orlicz functions, $\varphi = \{\varphi_n\}_{n=1}^{\infty}, w = \{w_n\}_{n=1}^{\infty}$ are weight sequences in $\mathcal{W}^-$, and $\psi = \{\psi_n\}_{n=0}^{\infty}, \varrho = \{\varrho_n\}_{n=0}^{\infty}$ are weight sequences in $\mathcal{W}^+$, then for every $f, g \in W \cap F\ell^{\Phi,\Psi}_{\varphi,w;\psi,\varrho}$,*

$$\|fg\|_{W\cap F} \le C\|f\|_{W\cap F}\|g\|_{W\cap F} \tag{2.2}$$

where $C := 1 + (1 + C_w)C_\varphi + (1 + C_\varrho)C_\psi$.

This theorem will be proved in Sections 3.1–3.2.

Clearly, the weight sequences $\widetilde{\varphi} = \{(n+1)^\alpha\}_{n=1}^{\infty}$ and $\widetilde{\psi} = \{(n+1)^\beta\}_{n=0}^{\infty}$ belong to $\mathcal{W}^-$ and $\mathcal{W}^+$, respectively, whenever $0 \le \alpha, \beta < \infty$. If

$$\widetilde{\Phi}(x) := x^p, \quad \widetilde{\Psi}(x) := x^r, \quad x \ge 0, \quad 1 \le p, r < \infty,$$

then $F\ell^{\widetilde{\Phi},\widetilde{\Psi}}_{\widetilde{\varphi},1;\widetilde{\psi},1}$ with the norm $\|f\|_F := \|f\|_- + \|f\|_+$ is isometrically isomorphic to $F\ell^{p,r}_{\alpha,\beta}$. Hence Theorem 1.1 follows from Theorem 2.2.

Corollary 2.3. *Under the assumptions of Theorem 2.2 we have the following.*

(a) *$W \cap F\ell^{\Phi,\Psi}_{\varphi,w;\psi,\varrho}$ is a commutative Banach algebra under pointwise multiplication. The maximal ideal space of this algebra coincides with $\mathbb{T}$.*

(b) *If $b \in W \cap F\ell^{\Phi,\Psi}_{\varphi,w;\psi,\varrho}$ does not vanish on $\mathbb{T}$ and the Cauchy index of b vanishes, then b has a logarithm in $W \cap F\ell^{\Phi,\Psi}_{\varphi,w;\psi,\varrho}$.*

(c) *If we let*

$$G(b) := \exp\Big((\log b)_0\Big), \qquad b_\pm(t) := \exp\left(\sum_{n=1}^{\infty}(\log b)_{\pm n} t^{\pm n}\right) \qquad (t \in \mathbb{T}),$$

then $b = G(b)b_- b_+$ and $b_\pm^{\pm 1} \in W \cap F\ell^{\Phi,\Psi}_{\varphi,w;\psi,\varrho}$.

This corollary will be proved in Section 3.3.

3. Proofs

3.1. Auxiliary results

Proposition 3.1. *If $\{\nu_n\}_{n=0}^\infty \in \mathcal{W}^+$ (resp. $\{\nu_n\}_{n=1}^\infty \in \mathcal{W}^-$) and $k \in \mathbb{Z}_+$ (resp. $k \in \mathbb{N}$), then*

$$\nu_k \le C_\nu \nu_j \quad \text{for} \quad j \ge k - [k/2]. \tag{3.1}$$

Proof. If $k < 2$, then $k - [k/2] = k$. Hence from (ii) and $C_\nu \ge 1$ we get (3.1).

If $k \ge 2$, then $2[k/2] \le k$. Therefore $k \le 2(k - [k/2])$. In that case from (ii) and (iii) it follows that

$$\nu_k \le \nu_{2(k-[k/2])} \le C_\nu \nu_{k-[k/2]} \le C_\nu \nu_j$$

for $j \ge k - [k/2]$, i.e., we obtain (3.1). □

Let $\{f_n\}_{n=-\infty}^\infty$ and $\{g_n\}_{n=-\infty}^\infty$ be the Fourier coefficient sequences of functions $f \in L^1(\mathbb{T})$ and $g \in L^1(\mathbb{T})$, respectively. Put $a_n := |f_n|$ and $b_n := |g_n|$.

Proposition 3.2. (a) *If $k \in \mathbb{N}$, then*

$$|(fg)_{-k}| \le \sum_{j=0}^{\infty} a_j b_{-k-j} + \sum_{j=1}^{[k/2]} a_{-j} b_{-k+j} + \sum_{j=0}^{\infty} b_j a_{-k-j} + \sum_{j=1}^{[k/2]} b_{-j} a_{-k+j}. \tag{3.2}$$

(b) *If $k \in \mathbb{Z}_+$, then*

$$|(fg)_k| \le \sum_{j=1}^{\infty} a_{-j} b_{k+j} + \sum_{j=0}^{[k/2]} a_j b_{k-j} + \sum_{j=1}^{\infty} b_{-j} a_{k+j} + \sum_{j=0}^{[k/2]} b_j a_{k-j}.$$

Proof. (a) For $k \in \mathbb{N}$,

$$\begin{aligned} |(fg)_{-k}| &= \left|\sum_{j=-\infty}^{\infty} f_j g_{-k-j}\right| \le \sum_{j=-\infty}^{\infty} a_j b_{-k-j} \qquad (3.3)\\ &= \sum_{j=-\infty}^{-k} a_j b_{-k-j} + \sum_{j=-k+1}^{-1} a_j b_{-k-j} + \sum_{j=0}^{\infty} a_j b_{-k-j} \\ &=: \sigma_1 + \sigma_2 + \sigma_3. \end{aligned}$$

Changing variables in σ_1 ($r = -k - j$), we obtain

$$\sigma_1 := \sum_{j=-\infty}^{-k} a_j b_{-k-j} = \sum_{r=0}^{\infty} a_{-k-r} b_r. \tag{3.4}$$

Obviously, $\sigma_2 = 0$ if $k = 1$. Hence (3.2) follows from (3.3) and (3.4) for $k = 1$.

If $k > 1$ then changing variables in σ_2 ($r = -j$) we get

$$\sigma_2 := \sum_{j=-k+1}^{-1} a_j b_{-k-j} = \sum_{r=1}^{k-1} a_{-r} b_{-k+r}.$$

If $k = 2m$ and $m \in \mathbb{N}$, then

$$\sigma_2 = \sum_{j=1}^{2m-1} a_{-j} b_{-2m+j} = \sum_{j=1}^{m} a_{-j} b_{-2m+j} + \sum_{j=m+1}^{2m-1} a_{-j} b_{-2m+j}. \tag{3.5}$$

Changing variables in the second sum ($r = 2m - j$), we obtain

$$\sum_{j=m+1}^{2m-1} a_{-j} b_{-2m+j} = \sum_{r=1}^{m-1} a_{-2m+r} b_{-r} \le \sum_{j=1}^{m} b_{-j} a_{-2m+j}. \tag{3.6}$$

Since $[k/2] = [2m/2] = m$, from (3.3)–(3.6) we deduce that (3.2) holds for $k = 2m$ and $m \in \mathbb{N}$.

If $k = 2m + 1$ and $m \in \mathbb{N}$, then

$$\sigma_2 = \sum_{j=1}^{2m} a_{-j} b_{-(2m+1)+j} = \sum_{j=1}^{m} a_{-j} b_{-(2m+1)+j} + \sum_{j=m+1}^{2m} a_{-j} b_{-(2m+1)+j}. \tag{3.7}$$

Changing variables in the second sum ($r = 2m + 1 - j$), we infer that

$$\sum_{j=m+1}^{2m} a_{-j} b_{-(2m+1)+j} = \sum_{r=1}^{m} a_{-(2m+1)+r} b_{-r}. \tag{3.8}$$

Since $[k/2] = [(2m+1)/2] = m$, from (3.3)–(3.4) and (3.7)–(3.8) we conclude that (3.2) is satisfied for $k = 2m + 1$ and $m \in \mathbb{N}$. Part (a) is proved. Part (b) is proved analogously to Part (a). □

3.2. Proof of Theorem 2.2

Proof. The idea of this proof is borrowed from [1, Theorem 6.54]. Let us show that

$$\|fg\|_- \le C_-(\|f\|_W \|g\|_- + \|g\|_W \|f\|_-), \tag{3.9}$$

where $C_- := (1 + C_w)C_\varphi$, C_φ and C_w are the constants in the Δ_2-condition for the sequences $\{\varphi_n\}_{n=1}^{\infty}$ and $\{w_n\}_{n=1}^{\infty}$, respectively. If $\|f\|_W \|g\|_- + \|g\|_W \|f\|_- = 0$, then (3.9) is obvious.

Assume that $\|f\|_W\|g\|_- + \|g\|_W\|f\|_- > 0$. Since Φ is increasing, from Proposition 3.2(a) it follows that for $k \in \mathbb{N}$,

$$\Phi\left(\frac{|(fg)_{-k}|\varphi_k}{C_-(\|f\|_W\|g\|_- + \|g\|_W\|f\|_-)}\right) \le \Phi\left(\sum_{i=1}^{6}\Sigma_i \Big/ \sum_{i=1}^{6}\sigma_i\right), \tag{3.10}$$

where

$$\Sigma_1 := \sum_{j=0}^{\infty}(a_j\|g\|_-)\frac{b_{-k-j}\varphi_k}{C_-\|g\|_-}, \qquad \sigma_1 := \sum_{j=0}^{\infty} a_j\|g\|_-,$$

$$\Sigma_2 := \sum_{j=1}^{[k/2]}(a_{-j}\|g\|_-)\frac{b_{-k+j}\varphi_k}{C_-\|g\|_-}, \qquad \sigma_2 := \sum_{j=1}^{[k/2]} a_{-j}\|g\|_-,$$

$$\Sigma_3 := \sum_{j<-[k/2]}(a_j\|g\|_-)\cdot 0 = 0, \qquad \sigma_3 := \sum_{j<-[k/2]} a_j\|g\|_-,$$

$$\Sigma_4 := \sum_{j=0}^{\infty}(b_j\|f\|_-)\frac{a_{-k-j}\varphi_k}{C_-\|f\|_-}, \qquad \sigma_4 := \sum_{j=0}^{\infty} b_j\|f\|_-,$$

$$\Sigma_5 := \sum_{j=1}^{[k/2]}(b_{-j}\|f\|_-)\frac{a_{-k+j}\varphi_k}{C_-\|f\|_-}, \qquad \sigma_5 := \sum_{j=1}^{[k/2]} b_{-j}\|f\|_-,$$

$$\Sigma_6 := \sum_{j<-[k/2]}(b_j\|a\|_-)\cdot 0 = 0, \qquad \sigma_6 := \sum_{j<-[k/2]} b_j\|f\|_-.$$

Since Φ is convex and $\Phi(0) = 0$, from (3.10) and Jensen's inequality (see, e.g., [2, Theorem 90]) it follows that

$$\sum_{k=1}^{N}\Phi\left(\frac{|(fg)_{-k}|\varphi_k}{C_-(\|f\|_W\|g\|_- + \|g\|_W\|f\|_-)}\right)w_k \tag{3.11}$$

$$\le \frac{\|g\|_-\Big(\sigma_1(N) + \sigma_2(N)\Big) + \|f\|_-\Big(\sigma_3(N) + \sigma_4(N)\Big)}{\|f\|_W\|g\|_- + \|g\|_W\|f\|_-},$$

where

$$\sigma_1(N) := \sum_{k=1}^{N}\sum_{j=0}^{\infty} a_j\Phi\left(\frac{b_{-k-j}\varphi_k}{C_-\|g\|_-}\right)w_k, \quad \sigma_2(N) := \sum_{k=1}^{N}\sum_{j=1}^{[k/2]} a_{-j}\Phi\left(\frac{b_{-k+j}\varphi_k}{C_-\|g\|_-}\right)w_k,$$

$$\sigma_3(N) := \sum_{k=1}^{N}\sum_{j=0}^{\infty} b_j\Phi\left(\frac{a_{-k-j}\varphi_k}{C_-\|f\|_-}\right)w_k, \quad \sigma_4(N) := \sum_{k=1}^{N}\sum_{j=1}^{[k/2]} b_{-j}\Phi\left(\frac{a_{-k+j}\varphi_k}{C_-\|f\|_-}\right)w_k.$$

Taking into account (i), (ii), and $C_\varphi \geq 1$, we have

$$\sigma_1(N) = \sum_{j=0}^{\infty} a_j \sum_{k=1}^{N} \Phi\left(\frac{b_{-k-j}\varphi_k}{C_-\|g\|_-}\right) w_k \leq \sum_{j=0}^{\infty} a_j \sum_{k=1}^{N} \Phi\left(\frac{b_{-k-j}\varphi_{k+j}}{C_-\|g\|_-}\right) w_{k+j}$$

$$\leq \sum_{j=0}^{\infty} a_j \sum_{k=1}^{\infty} \Phi\left(\frac{b_{-k}\varphi_k}{C_-\|g\|_-}\right) w_k \leq \|f\|_W \sum_{k=1}^{\infty} \Phi\left(\frac{C_\varphi b_{-k}\varphi_k}{C_-\|g\|_-}\right) w_k. \tag{3.12}$$

Changing variables in the second sum in $\sigma_2(N)$ $(r = k - j)$, we obtain

$$\sigma_2(N) := \sum_{k=1}^{N} \sum_{j=1}^{[k/2]} a_{-j} \Phi\left(\frac{b_{-k+j}\varphi_k}{C_-\|g\|_-}\right) w_k = \sum_{k=1}^{N} \sum_{r=k-[k/2]}^{k-1} a_{r-k} \Phi\left(\frac{b_{-r}\varphi_k}{C_-\|g\|_-}\right) w_k.$$

Since Φ is increasing, it follows from the latter equality and Proposition 3.1 that

$$\sigma_2(N) \leq C_w \sum_{k=1}^{N} \sum_{r=k-[k/2]}^{k-1} a_{-k+r} \Phi\left(\frac{C_\varphi b_{-r}\varphi_r}{C_-\|g\|_-}\right) w_r \tag{3.13}$$

$$\leq C_w \sum_{k=1}^{N} \sum_{r=1}^{\infty} a_{-k+r} \Phi\left(\frac{C_\varphi b_{-r}\varphi_r}{C_-\|g\|_-}\right) w_r = C_w \sum_{r=1}^{\infty} \Phi\left(\frac{C_\varphi b_{-r}\varphi_r}{C_-\|g\|_-}\right) w_r \sum_{k=1}^{N} a_{-k+r}$$

$$\leq C_w \sum_{r=1}^{\infty} \Phi\left(\frac{C_\varphi b_{-r}\varphi_r}{C_-\|g\|_-}\right) w_r \sum_{k=-\infty}^{\infty} a_{-k} = C_w \|f\|_W \sum_{k=1}^{\infty} \Phi\left(\frac{C_\varphi b_{-k}\varphi_k}{C_-\|g\|_-}\right) w_k.$$

Combining (3.12)–(3.13), we arrive at

$$\sigma_1(N) + \sigma_2(N) \leq \|f\|_W (1 + C_w) \sum_{k=1}^{\infty} \Phi\left(\frac{C_\varphi b_{-k}\varphi_k}{C_-\|g\|_-}\right) w_k. \tag{3.14}$$

Since Φ is an Orlicz function, $\Phi(x)/x$ is a non-decreasing function (see, e.g., [3, p. 139]). Thus,

$$(1 + C_w)\Phi(x) \leq \Phi((1 + C_w)x), \quad x \geq 0.$$

Applying this inequality to (3.14), we obtain for $N \geq 1$,

$$\sigma_1(N) + \sigma_2(N) \leq \|f\|_W \sum_{k=1}^{\infty} \Phi\left(\frac{b_{-k}\varphi_k}{\|g\|_-}\right) w_k \leq \|f\|_W \tag{3.15}$$

Analogously one can show that for $N \geq 1$,

$$\sigma_3(N) + \sigma_4(N) \leq \|g\|_W. \tag{3.16}$$

Taking into account that N is arbitrary, from (3.11), (3.15), and (3.16) we get

$$\sum_{k=1}^{\infty} \Phi\left(\frac{|(fg)_{-k}|\varphi_k}{C_-(\|f\|_W\|g\|_- + \|g\|_W\|f\|_-)}\right) w_k \leq 1.$$

Therefore,

$$\|fg\|_- \le C_-(\|f\|_W\|g\|_- + \|g\|_W\|f\|_-) \le C_-\|f\|_{W\cap F}\|g\|_{W\cap F}. \tag{3.17}$$

By using of Proposition 3.2(b), one can similarly prove that

$$\|fg\|_+ \le C_+(\|f\|_W\|g\|_+ + \|g\|_W\|f\|_+) \le C_+\|f\|_{W\cap F}\|g\|_{W\cap F}, \tag{3.18}$$

where $C_+ := (1+C_\varrho)C_\psi$, C_ψ and C_ϱ are the constants in the Δ_2-condition for the sequences $\psi = \{\psi_n\}_{n=0}^\infty$ and $\varrho = \{\varrho_n\}_{n=0}^\infty$, respectively. Combining (3.17), (3.18), and

$$\|fg\|_W \le \|f\|_W\|g\|_W \le \|f\|_{W\cap F}\|g\|_{W\cap F},$$

we arrive at (2.2). □

3.3. Proof of Corollary 2.3

Proof. (a) By Theorem 2.2, $W\cap F\ell^{\Phi,\Psi}_{\varphi,w;\psi,\varrho}$ is a commutative Banach algebra under pointwise multiplication. Clearly, $W\cap F\ell^{\Phi,\Psi}_{\varphi,w;\psi,\varrho} \subset W \subset C(\mathbb{T})$, where $C(\mathbb{T})$ is the Banach algebra of all complex-valued continuous functions on $\mathbb{T}$ with the maximum norm. The description of the maximal ideal space of $W\cap F\ell^{\Phi,\Psi}_{\varphi,w;\psi,\varrho}$ is standard. Part (a) is proved.

Part (b) follows from Part (a) and [1, Section 2.41(e)].

(c) By Part (b), $\log b \in W\cap F\ell^{\Phi,\Psi}_{\varphi,w;\psi,\varrho}$, that is,

$$\log b(t) = \sum_{n=-\infty}^{-1}(\log b)_n t^n + (\log b)_0 + \sum_{n=1}^{\infty}(\log b)_n t^n =: a_-(t) + (\log b)_0 + a_+(t),$$

whenever $t\in\mathbb{T}$, and

$$\{(\log b)_n\}_{n\in\mathbb{Z}_+} \in \ell^1(\mathbb{Z}_+)\cap\ell^{\Psi}_{\psi,\varrho}(\mathbb{Z}_+), \quad \{(\log b)_{-n}\}_{n\in\mathbb{N}} \in \ell^1(\mathbb{N})\cap\ell^{\Phi}_{\varphi,w}(\mathbb{N}).$$

Obviously, $\pm a_-$ and $\pm a_+$ belong to $W\cap F\ell^{\Phi,\Psi}_{\varphi,w;\psi,\varrho}$. Therefore, $b_-^{\pm1} = \exp(\pm a_-)$ and $b_+^{\pm1} = \exp(\pm a_+)$ belong to $W\cap F\ell^{\Phi,\Psi}_{\varphi,w;\psi,\varrho}$ and

$$b = \exp(a_- + (\log b)_0 + a_+) = b_-G(b)b_+.$$

Part (c) and the corollary are proved. □

Acknowledgment

I would like to thank Albrecht Böttcher (Chemnitz Technical University, Germany) for useful remarks on an earlier version of this paper.

References

[1] A. Böttcher, B. Silbermann, *Analysis of Toeplitz Operators*, Springer-Verlag, Berlin, 1990.

[2] G.H. Hardy, J.E. Littlewood, G. Pólya, *Inequalities*, 2d ed. Cambridge University Press, Cambridge, 1952.

[3] J. Lindenstrauss, L. Tzafriri, *Classical Banach Spaces. I. Sequence Spaces*, Ergebnisse der Mathematik und ihrer Grenzgebiete, **92**. Springer-Verlag, Berlin, New York, 1977.

[4] J. Musielak, *Orlicz Spaces and Modular Spaces*, Lecture Notes in Mathematics, **1034**. Springer-Verlag, Berlin, 1983.

[5] P.L. Ul'yanov, Absolute convergence of trigonometric Fourier series, *Dokl. Akad. Nauk SSSR*, **322** (1992), no. 2, 253–258 (in Russian). English translation: *Soviet Math. Dokl.*, **45** (1992), no. 1, 83–88.

Alexei Yu. Karlovich
Departamento de Mathemática
Instituto Superior Técnico
Av. Rovisco Pais
1049-001 Lisboa
Portugal
e-mail: akarlov@math.ist.utl.pt

Operator Theory
Advances and Applications, Vol. 147, 297–325

Invertibility at Infinity of Band-Dominated Operators on the Space of Essentially Bounded Functions

Marko Lindner and Bernd Silbermann

To the memory of Professor Erhard Meister.

Abstract. The topic of this paper is band operators and the norm limits of such – so-called band-dominated operators, both classes acting on $L^\infty(\mathbb{R}^n)$. Invertibility at infinity is closely related to Fredholmness. In fact, in the discrete case $\ell^p(\mathbb{Z}^n)$, $1 \le p \le \infty$, both properties coincide. For many applications, e.g., the question of applicability of certain approximation methods, in the situation at hand, $L^p(\mathbb{R}^n)$, $1 \le p \le \infty$, it has however proved to be useful to study invertibility at infinity rather than Fredholmness.

We will present a criterion for a band-dominated operator's invertibility at infinity in terms of the invertibility of its limit operators. It is the same criterion that was found for $\ell^p(\mathbb{Z}^n)$, $1 < p < \infty$ in [21], and for the C^*-algebra $L^2(\mathbb{R}^n)$ in [22]. Our investigations concentrate on one of the most unloved cases, being $L^\infty(\mathbb{R}^n)$. With the techniques presented here it is clear now how the remaining cases ℓ^1, ℓ^∞ and L^p, $(p \ne 2)$ have to be treated.

1. Introduction and main results

An operator on $\ell^\infty(\mathbb{Z})$ is regarded as a band operator if its matrix representation with respect to the canonical basis is a band matrix – a matrix whose entries a_{ij} vanish if $|i-j|$ exceeds some number which is then called band-width.

By an analogous construction, band operators are defined on $\ell^\infty(\mathbb{Z}^n)$. Every function $f \in L^\infty(\mathbb{R}^n)$ can be identified with a $L^\infty([0,1)^n)$-valued $\ell^\infty(\mathbb{Z}^n)$ sequence, and by this identification an operator on $L^\infty(\mathbb{R}^n)$ can be identified with an operator on $\ell^\infty(\mathbb{Z}^n, L^\infty([0,1)^n))$, and thus we can carry over the notion of band operators to operators on $L^\infty(\mathbb{R}^n)$. A norm-limit of a sequence of band operators is then called *band-dominated operator.*

Band and band-dominated operators can be found in many fields of mathematics and physics. Prominent examples are acoustic scattering problems and discretizations of partial differential equations.

2000 *Mathematics Subject Classification.* Primary 47B38; Secondary 47L80.

The study of concrete classes of band and band-dominated operators (such as convolution, Wiener-Hopf, and Toeplitz operators) goes back to the 1930's starting with [28] by WIENER and HOPF, continued in the 1950's for example in [9] and [5] by GOHBERG and KREIN, and was culminating in the 1970/80's with the huge monographs [4] by GOHBERG/FELDMAN and [1] by BÖTTCHER and the second author. A general theory for this class of operators was initiated by SIMONENKO [25], [26] whose constructions (after some appropriate modifications) can be found in the proof of some essential ingredient to the proof of Theorem 1.1 of this paper.

The property we are interested in is called *invertibility at infinity*. For an operator A of a special quality (see Proposition 2.22 for the exact formulation), this means that there are two operators B_1, B_2 (of the same quality) and a bounded and measurable subset U of $\mathbb{R}^n$ such that

$$Q_U A B_1 = Q_U = B_2 A Q_U$$

holds, where Q_U is the operator of multiplication by the function that is 0 on and 1 outside U. This property is closely related with Fredholmness. Indeed, in the discrete cases ℓ^p, $1 < p < \infty$, both properties even coincide. In the cases L^p, $1 < p < \infty$, invertibility at infinity is still necessary for Fredholmness (where in many cases the missing facts for sufficiency are readily checked). In Section 4 we will demonstrate why it is however convenient to follow the notion of invertibility at infinity rather than Fredholmness, without losing contact to the latter property.

Every convolution operator with an absolutely summable kernel function is band-dominated. Moreover, every operator in the algebra generated by such convolution operators and bounded multiplication operators is band-dominated. So we will be able to give a criterion on invertibility at infinity for such operators, provided the multiplication operators involved do not behave "too strange" at infinity. We will specify clearly what kind of behavior we do expect there, but it is important to note that we do not need convergence at infinity, as many of our fore-runners did. The reason for that is the special tool we are using: *limit operators*

With every direction of $\mathbb{R}^n$ in which infinity can be approached, one or more limit operators of a given operator A can be associated. We collect them in the so-called operator spectrum which is denoted by $\sigma^{\mathsf{op}}(A)$. If a band-dominated operator A has sufficiently many limit operators – we write $A \in \mathcal{B}_\$$ and regard A as a *rich* operator in this case – we can prove the following theorem which is the main outcome of this paper.

Theorem 1.1. *An operator $A \in \mathcal{B}_\$$ is invertible at infinity if and only if its set of limit operators $\sigma^{\mathsf{op}}(A)$ is uniformly invertible.*

... where "uniformly invertible" means that all elements are invertible, and their inverses are collectively bounded. $\mathcal{B}_\$$ turns out to be a Banach subalgebra of $\mathcal{L}(L^\infty)$, and we know some of its ingredients, including convolution operators and operators of multiplication by: uniformly continuous functions, slowly oscillating functions, periodic functions (with rational period), step functions (with rational step length) and of course, convergent functions.

Limit operators are supposed to be introduced in the 1920's by FAVARD [3] for studying ordinary differential equations with almost-periodic coefficients. Since that time limit operators have been used in context with partial differential and pseudo-differential operators and in many other fields of numerical analysis (see [7]). The Fredholmness of pseudo-differential operators in spaces of functions on $\mathbb{R}^n$ has been studied by MUHAMADIEV in [16], [17], and by LANGE and RABINOVICH in [13] and [19]. The first time limit operator techniques were applied to the general case of band-dominated operators for the study of their Fredholm property was in 1985 by LANGE and RABINOVICH in [11] and [12], continued in 1992 by RABINOVICH in [20].

During the last years RABINOVICH, ROCH and the second author proved Theorem 1.1 in the discrete case ℓ^p with $1 < p < \infty$ [21] and in the case L^2 [22]. Our paper is heavily inspired by these two papers, and we demonstrate in the example of the most unloved case, L^∞, how the remaining cases ℓ^1, ℓ^∞ and L^p, $p \neq 2$ have to be treated. In either case, Theorem 1.1 proves to be valid.

Section 2 is written to introduce tools and language, and to prove some basic properties. In Section 3 we will motivate and prove Theorem 1.1, and in Section 4 we will give some outlook to an application of Theorem 1.1 in the theory of approximation methods, and we will show how invertibility at infinity can be used to derive a sufficient Fredholm criterion in a subalgebra of $\mathcal{L}(L^\infty)$ containing sufficiently many interesting operators.

2. Some basic theory

2.1. Basic agreements

Throughout the following, n is some integer used as dimension in $\mathbb{Z}^n$ and $\mathbb{R}^n$. For $x=(x_1,\dots,x_n)\in\mathbb{R}^n$, we put $|x|:=\max(|x_1|,\dots,|x_n|)$, and for two sets $U,V\subset\mathbb{R}^n$, their distance $\operatorname{dist}(U,V)$ shall be the infimum of $|u-v|$, taken over all $u\in U, v\in V$.

Let $\mathbf{X}$ be a Banach space, T be some measurable subset of $\mathbb{R}^n$, and let $p \in [1,\infty)$. Then we refer to the Banach space of all (equivalence classes of) functions $f : T \to \mathbb{C}$ for which the pth power of f is Lebesgue integrable by $L^p(T)$. By $L^\infty(T)$ we denote the Banach space of all (equivalence classes of) essentially bounded functions $f : T \to \mathbb{C}$. $L^p(\mathbb{R}^n)$ and $L^\infty(\mathbb{R}^n)$ will be abbreviated by L^p and L^∞, respectively. Analogously, the Banach spaces $\ell^p(S,\mathbf{X})$ and $\ell^\infty(S,\mathbf{X})$ of $\mathbf{X}$-valued sequences on $S \subseteq \mathbb{Z}^n$ are defined, where we abbreviate $\ell^p(\mathbb{Z}^n,\mathbb{C})$ and $\ell^\infty(\mathbb{Z}^n,\mathbb{C})$ by ℓ^p and ℓ^∞, respectively.

By $\mathcal{L}(\mathbf{X})$ we denote the unital Banach algebra of linear and bounded operators $A : \mathbf{X} \to \mathbf{X}$, equipped with the usual operations and norm. For operators $A, A_1, A_2, \dots \in \mathcal{L}(\mathbf{X})$, we talk about *strong convergence*, and write $A_m \to A$, if $\|A_m x - Ax\|_{\mathbf{X}} \to 0$ as $m \to \infty$ for every $x \in \mathbf{X}$ and about *norm-convergence*, and write $A_m \rightrightarrows A$, if $\|A_m - A\|_{\mathcal{L}(\mathbf{X})} \to 0$ as $m \to \infty$.

2.2. Operators of shift and multiplication

Definition 2.1. *For every function $b \in L^\infty$, let $M_b \in \mathcal{L}(L^\infty)$ denote the operator of multiplication by b, acting on every $u \in L^\infty$ by $(M_b u)(x) = b(x)u(x)\ \forall x \in \mathbb{R}^n$. Frequently, we will call b the symbol of M_b.*

Fix some Banach space $\mathbf{X}$. If $b = (B_\alpha)_{\alpha\in\mathbb{Z}^n}$ is a bounded sequence of operators $B_\alpha \in \mathcal{L}(\mathbf{X})$, i.e., $b \in \ell^\infty(\mathbb{Z}^n, \mathcal{L}(\mathbf{X}))$, then by $\hat{M}_b$ we will denote the generalized multiplication operator $(\hat{M}_b u)_\alpha = B_\alpha u_\alpha\ \forall \alpha \in \mathbb{Z}^n$ on the space $\ell^\infty(\mathbb{Z}^n, \mathbf{X})$.

Definition 2.2. *For $\alpha \in \mathbb{Z}^n$, let V_α denote the so-called shift operator on $\ell^\infty(\mathbb{Z}^n, \mathbf{X})$, acting by the rule $(V_\alpha u)_\beta = u_{\beta-\alpha}$, i.e., shifting the whole sequence $u \in \ell^\infty(\mathbb{Z}^n, \mathbf{X})$ forward by α components. Without introducing a new symbol, we will say that V_τ is the shift operator on L^∞, shifting by $\tau \in \mathbb{R}^n$, i.e., $(V_\tau u)(x) = u(x-\tau)$, $u \in L^\infty$.*

Definition 2.3. *For every measurable set $U \subset \mathbb{R}^n$, put $P_U := M_{\chi_U}$, where χ_U is the characteristic function of U. Clearly, P_U is a projector. We will refer to its complementary projector $I - P_U$ by Q_U.*

2.3. Band- and band-dominated operators

For a quite natural introduction to these classes of operators on $\ell^\infty(\mathbb{Z}^n, \mathbf{X})$ with some $n \in \mathbb{N}$ and some Banach space $\mathbf{X}$, we will have a short recourse to the case $\ell^\infty(\mathbb{Z})$ ($n = 1$ and $\mathbf{X} = \mathbb{C}$).

One usually regards a linear bounded operator A on $\ell^\infty(\mathbb{Z})$ as *band operator* if its matrix representation $(a_{ij})_{i,j\in\mathbb{Z}}$ with respect to the canonical basis in $\ell^\infty(\mathbb{Z})$ is a band matrix, which means the entries a_{ij} vanish if $|i-j|$ is larger than some number B which is then called *band-width* of A. A little thought shows that this are exactly those operators A which can be written as a finite sum $\sum_{|i|\le B} \hat{M}_{b^{(i)}} V_i$ with $b^{(i)} \in \ell^\infty(\mathbb{Z})$ and V_i being the operator of shifting forward by i components.

With $\alpha \in \mathbb{Z}^n$ and $b^{(\alpha)} = (B_\beta^{(\alpha)})_{\beta\in\mathbb{Z}^n} \in \ell^\infty(\mathbb{Z}^n, \mathcal{L}(\mathbf{X}))$ for every α, we can carry over this notation to $\ell^\infty(\mathbb{Z}^n, \mathbf{X})$ by saying that every finite sum of the form

$$\sum_{|\alpha|\le B} \hat{M}_{b^{(\alpha)}} V_\alpha \tag{2.1}$$

be a *band operator* of *band-width* B on $\ell^\infty(\mathbb{Z}^n, \mathbf{X})$. Moreover, every norm limit of a sequence of band operators is referred to as a *band-dominated operator* on $\ell^\infty(\mathbb{Z}^n, \mathbf{X})$.

This was the discrete case ℓ^∞. Since we will finally deal with the function case, we will export the notion of band- and band-dominated operators to L^∞. This can be done by a rather easy and natural construction:

Let therefore H denote the (hyper-)cube $[0,1)^n$, and put $H_\alpha := \alpha + H$ for every $\alpha \in \mathbb{Z}^n$. Clearly, $\{H_\alpha\}_{\alpha\in\mathbb{Z}_n}$ is a disjoint decomposition of $\mathbb{R}^n$. Let G denote the operator that assigns to every function $f \in L^\infty$ the sequence of its restrictions to the cubes H_α, $G : f \mapsto (f|_{H_\alpha})_{\alpha\in\mathbb{Z}^n}$. By identifying $L^\infty(H_\alpha)$ with $L^\infty(H)$, we get that G is an isometrical isomorphism from L^∞ to $\ell^\infty(\mathbb{Z}^n, L^\infty(H))$. Consequently, every operator A on L^∞ can be identified with the operator $A^G := GAG^{-1}$ on $\ell^\infty(\mathbb{Z}^n, L^\infty(H))$.

If $M_f \in \mathcal{L}(L^\infty)$ is the operator of multiplication by $f \in L^\infty$, then its discretization $(M_f)^G$ is a generalized multiplication operator $\hat{M}_b$, where the components of the sequence b are the (usual) multiplication operators $B_\alpha = M_{f|_{H_\alpha}}$ on $L^\infty(H)$. In this case we will write $\hat{f}$ for the sequence b, i.e., $\hat{M}_{\hat{f}} := \hat{M}_b = (M_f)^G$.

Sometimes we will also pass from the discrete case ℓ^∞ to L^∞, and we therefore introduce the abbreviation B^{-G} for the operator $G^{-1}BG \in \mathcal{L}(L^\infty)$ with $(B^{-G})^G = B \in \mathcal{L}(\ell^\infty(\mathbb{Z}^n, L^\infty(H)))$. And conversely, we have $(A^G)^{-G} = A$.

Definition 2.4. *We say that an operator $A \in \mathcal{L}(L^\infty)$ is a band operator with band-width B if A^G is band operator with band-width B on $\ell^\infty(\mathbb{Z}^n, L^\infty(H))$, and we call A band-dominated if A^G is band-dominated. The classes of band- and band-dominated operators on L^∞ will be denoted by* BO *and* BDO*, respectively.*

We will now prove two lemmas which will lead us to a nice and very useful characterization of both sets, BO and BDO. The first lemma merely says that A is a band operator with band-width B if and only if all entries $A^G_{\gamma,\delta}$ of the matrix representation of A^G vanish for $|\gamma - \delta| > B$.

Lemma 2.5.

a) *If $A \in$ BO with band-width B, i.e., A^G is of the form (2.1), and $\gamma, \delta \in \mathbb{Z}^n$ are arbitrary, then*

$$P_{H_\gamma} A P_{H_\delta} = \begin{cases} V_\gamma B_\gamma^{(\gamma-\delta)} V_{-\delta} & \text{if } |\gamma-\delta| \le B, \\ 0 & \text{if } |\gamma-\delta| > B. \end{cases} \tag{2.2}$$

b) *Conversely, if A is an operator with $P_{H_\gamma} A P_{H_\delta} = 0$ for all $\gamma, \delta \in \mathbb{Z}^n$ with $|\gamma - \delta|$ being larger than some B, then A^G is of the form (2.1) with*

$$B_\beta^{(\alpha)} = V_{-\beta} P_{H_\beta} A P_{H_{\beta-\alpha}} V_{\beta-\alpha}, \tag{2.3}$$

i.e., $A \in$ BO with band-width B.

Proof. **Part a)** We start with an easy computation:

$$\begin{aligned} (P_{H_\gamma} A P_{H_\delta})^G &= P^G_{H_\gamma} A^G P^G_{H_\delta} = P^G_{H_\gamma} \Big(\sum_{|\alpha|\le B} \hat{M}_{b^{(\alpha)}} V_\alpha \Big) P^G_{H_\delta} \\ &= \sum_{|\alpha|\le B} P^G_{H_\gamma} \hat{M}_{b^{(\alpha)}} V_\alpha P^G_{H_\delta} = \sum_{|\alpha|\le B} \hat{M}_{b^{(\alpha)}} P^G_{H_\gamma} P^G_{H_{\delta+\alpha}} V_\alpha \end{aligned}$$

Clearly, $P^G_{H_\gamma \cap H_{\delta+\alpha}} = P^G_{H_\gamma}$ if $\alpha = \gamma - \delta$, and $P^G_{H_\gamma \cap H_{\delta+\alpha}} = 0$ otherwise. Moreover, $\hat{M}_{b^{(\alpha)}} P^G_{H_\gamma} V_\alpha = V_\gamma (B_\gamma^{(\alpha)})^G V_{-\gamma} V_\alpha$, which is equal to $V_\gamma (B_\gamma^{(\gamma-\delta)})^G V_{-\delta}$ if $\alpha = \gamma - \delta$. Putting this together, proves our claim.

Part b) Clearly, equation (2.3) arises from (2.2) with $\alpha := \gamma - \delta$ and $\beta := \gamma$. So the proof of **b)** merely is verifying the proof of **a)** in reverse order (note that $B_\beta^{(\alpha)}$ are allowed to be arbitrary operators on $L^\infty(H)$; cf. Definition 2.1). □

The following lemma will be needed again at some later point in Section 3. For our purposes in the current section, we remark that the characteristic function χ_H of H is subject to the conditions on φ in Lemma 2.6 with $N_\varphi = 1$ and $M_{\varphi_\alpha} = P_{H_\alpha}$ for all $\alpha \in \mathbb{Z}^n$.

Lemma 2.6. *Let $\{A_\alpha\}_{\alpha\in\mathbb{Z}^n}$ be some family of operators on L^∞ with $\|A_\alpha\| \le M < \infty$ $\forall\alpha \in \mathbb{Z}^n$, and let $\varphi \in L^\infty$ be some function with $\sum_{\alpha\in\mathbb{Z}^n} |\varphi_\alpha(x)| \le N_\varphi < \infty$ $\forall x \in \mathbb{R}^n$, where $\varphi_\alpha(x) = \varphi(x-\alpha)$. Then by $\sum_{\alpha\in\mathbb{Z}^n} M_{\varphi_\alpha} A_\alpha$ a unique operator A on L^∞ is given whose norm does not exceed MN_φ.*

Proof. Take some arbitrary $u \in L^\infty$, and put $u_\alpha := A_\alpha u$ $\forall\alpha \in \mathbb{Z}^n$. Then

$$|u_\alpha(x)| \le \|u_\alpha\|_\infty \le \|A_\alpha\|\|u\|_\infty \le M\|u\|_\infty \qquad \forall\alpha \in \mathbb{Z}^n, \text{ almost all } x \in \mathbb{R}^n.$$

Evaluating the action of our operator A on u at (almost) every $x \in \mathbb{R}^n$ gives

$$\left|\Big(\sum M_{\varphi_\alpha} A_\alpha u\Big)(x)\right| \le \sum |\varphi_\alpha(x)|\,|u_\alpha(x)| \le M\|u\|_\infty \sum |\varphi_\alpha(x)| \le MN_\varphi\|u\|_\infty,$$

and consequently, $\|Au\|_\infty = \sup_{x\in\mathbb{R}^n} |(Au)(x)| \le MN_\varphi\|u\|_\infty$. □

Besides the (hyper-)cube $H = [0,1)^n$, we will introduce the cube $C := [-1,1]^n$ which will be used to abbreviate many notations in what follows. Now we are ready to prove the mentioned characterizations of BO and BDO.

Proposition 2.7.

a) $A \in$ BO *with band-width B if and only if*

$$P_U A P_V = 0 \qquad \forall U, V \subset \mathbb{R}^n : \ \operatorname{dist}(U,V) \ge B. \tag{2.4}$$

b) $A \in$ BDO *if and only if*

$$P_U A P_V \rightrightarrows 0 \qquad \text{as} \quad \operatorname{dist}(U,V) \to \infty, \tag{2.5}$$

precisely, $\forall\varepsilon > 0$ there is a $B > 0$ such that $\|P_U A P_V\| < \varepsilon$ for **all** *measurable subsets $U, V \subset \mathbb{R}^n$ with* $\operatorname{dist}(U,V) \ge B$.

Proof. **Part a)** ($\Longrightarrow$) Suppose A^G is of the form (2.1), and take $U, V \subset \mathbb{R}^n$ arbitrarily with $\operatorname{dist}(U,V) \ge B$. Doing the same computations as in the proof of part a) of Lemma 2.5, we get

$$(P_U A P_V)^G = \sum_{|\alpha|\le B} \hat{M}_{b(\alpha)} P^G_{U\cap(\alpha+V)} V_\alpha = 0$$

since $U \cap (\alpha + V) = \emptyset$ for all α with $|\alpha| \le B$.

($\Longleftarrow$) Suppose $P_U A P_V = 0$ for all $U, V \subset \mathbb{R}^n$ with distance $\ge B$. Then this is true for all $U = H_\gamma$ and $V = H_\delta$ with $|\gamma - \delta| > B$, and from Lemma 2.5 b) we get that $A \in$ BO with band-width B.

Part b) ($\Longrightarrow$) Take a sequence $(A_m) \subset$ BO with $A_m \rightrightarrows A \in$ BDO as $m \to \infty$. Given an arbitrary $\varepsilon > 0$, there is some m such that $\|A - A_m\| < \varepsilon$. From part a) we know that there is a B such that A_m is subject to (2.4). Then for all sets U, V with distance $\geq B$ one has

$$\|P_U A P_V\| \leq \|P_U A_m P_V\| + \|P_U (A - A_m) P_V\| < 0 + \varepsilon.$$

($\Longleftarrow$) For every $m \in \mathbb{N}$, put $A_m := \sum_{\alpha \in \mathbb{Z}^n} P_{H_\alpha} A P_{\alpha + mC}$ which is well defined by Lemma 2.6. Due to Lemma 2.5 b), it is an easy observation that $A_m \in \mathrm{BO}$ with band-width $B = m$. Moreover, by Lemma 2.6,

$$\|A - A_m\| = \|\sum_{\alpha \in \mathbb{Z}^n} P_{H_\alpha} A Q_{\alpha + mC}\| \leq \sup_{\alpha \in \mathbb{Z}^n} \|P_{H_\alpha} A Q_{\alpha + mC}\| \to 0$$

as $m = \mathrm{dist}\left(H_\alpha \,,\, \mathbb{R}^n \setminus (\alpha + mC)\right) + 1 \to \infty$ by property (2.5). So $A \in \mathrm{BDO}$. □

Remark 2.8 It is important to note that the norm convergence in (2.5) has to be understood as uniformly with respect to the distance B, exactly as it is stated in "precisely,..."! For instance, the operator of contraction, $(Au)(x) = u(2x)$, also has the property that $\|P_U A P_V\|$ tends to zero as U and V drift apart, but no B is large enough that for **all** measurable sets U, V with distance $\geq B$ this norm is as small as desired.

Operators $A \in \mathcal{L}(L^\infty)$ that are subject to (2.5) are referred to as *operators of local type* in [18], and the set of all these operators is denoted by OLT. It turns out (see Theorems 4.37 and 4.39 in [18]) that OLT is an inverse closed Banach subalgebra of $\mathcal{L}(L^\infty)$, and by Proposition 2.7 b) we have BDO = OLT. □

2.4. The algebra $\mathcal{S}$

Of course, L^1 and L^∞ are two closely related extreme cases of L^p, each at an other end of the scale, and both resist to most of the usual L^p arguments and techniques. We will mainly deal with L^∞, but will frequently pass to L^1 by several duality arguments. Here some caution is required:

Given a bounded linear operator A on L^∞, its adjoint operator A^* clearly acts on the dual space $(L^\infty)'$ of L^∞ which **contains** L^1 as a proper subspace. But on the other hand, the dual of L^1 **equals** L^∞ (irreflexivity of L^∞). So one cannot pass from L^∞ to L^1 and back by repeatedly taking adjoints, as one usually does when $p \in (1, \infty)$. We will therefore restrict ourselves to operators $A \in \mathcal{L}(L^\infty)$ whose adjoint operator A^* maps L^1-functions to L^1-functions, i.e., $A^*(L^1) = \mathrm{im}\,(A^*|_{L^1}) \subseteq L^1$.

$$\mathcal{S} := \left\{ \; A \in \mathcal{L}(L^\infty) \;:\; A^*(L^1) \subseteq L^1 \; \right\}.$$

Now, if $A \in \mathcal{S}$, the restriction $B := A^*|_{L^1}$, seen as operator in $\mathcal{L}(L^1)$, has the property $B^* = A$, whence we regard B as the *pre-adjoint* operator of A – the operator whose adjoint equals A –, and we will frequently denote it by $^*\!A$. So for $A \in \mathcal{S}$, we can pass from $\mathcal{L}(L^\infty)$ to $\mathcal{L}(L^1)$ by $A \mapsto {}^*\!A$, and back to $\mathcal{L}(L^\infty)$ by $B \mapsto B^*$. It is important to note that $^*\!A$ is invertible in $\mathcal{L}(L^1)$ if and only if $(^*\!A)^* = A$ is invertible in $\mathcal{L}(L^\infty)$. Moreover, $\|^*\!A\|_{\mathcal{L}(L^1)} = \|(^*\!A)^*\|_{\mathcal{L}((L^1)')} = \|A\|_{\mathcal{L}(L^\infty)}$.

Proposition 2.9. *$\mathcal{S}$ is an inverse closed Banach subalgebra of $\mathcal{L}(L^\infty)$.*

Proof. If $A, B \in \mathcal{S}$, then also $AB \in \mathcal{S}$. Indeed, $(AB)^* = B^*A^*$ maps L^1 to L^1 since A^* and B^* do so. For the proof of closedness, take some sequence (A_m) in $\mathcal{S}$ with $A_m \rightrightarrows A$ as $m \to \infty$. From $\|{}^*\!A_m\| = \|A_m\|$ we conclude that $({}^*\!A_m)$ is a Cauchy sequence since (A_m) is one. Denote the uniform limit of ${}^*\!A_m$ by B. This and $\operatorname{im} {}^*\!A_m \subseteq L^1\ \forall m$ show that $\operatorname{im} B \subseteq L^1$. But from $A_m = ({}^*\!A_m)^* \rightrightarrows B^*$ as $m \to \infty$ and $A_m \rightrightarrows A$ we conclude $B^* = A$, i.e., B is the pre-adjoint of A. So $A \in \mathcal{S}$ which shows that $\mathcal{S}$ is closed.

To see that $\mathcal{S}$ is inverse closed, take some arbitrary $A \in \mathcal{S}$ being invertible as an operator on L^∞. We have to show that $(A^{-1})^*$, restricted to L^1, only maps to L^1-functions. Since A is invertible, A^* is invertible as well, and $(A^*)^{-1} = (A^{-1})^*$. Also ${}^*\!A$ is invertible, and its inverse $({}^*\!A)^{-1} = (A^*|_{L^1})^{-1}$ (the inverse of the restriction of A^*) clearly coincides with the restriction of the inverse of A^* which is $(A^*)^{-1}|_{L^1} = (A^{-1})^*|_{L^1}$, and hence $A^{-1} \in \mathcal{S}$. □

2.5. The algebras $\mathcal{B}$ and $\mathcal{B}_\mathcal{S}$

Definition 2.10. *By* BC *we denote the set of bounded and continuous complex-valued functions on* $\mathbb{R}^n$, *and let* BUC *refer to those* BC *functions that are even uniformly continuous on* $\mathbb{R}^n$.

Given a function $\varphi \in L^\infty$ and a vector $t = (t_1, \ldots, t_n) \in \mathbb{R}^n$, we define the function φ_t at the point $x = (x_1, \ldots, x_n) \in \mathbb{R}^n$ as $\varphi_t(x) := \varphi(t_1x_1, \ldots, t_nx_n)$. Further let $[A\,,\,B]$ refer to the *commutator* $AB - BA$ of two operators $A, B \in \mathcal{L}(\mathbf{X})$.

Definition 2.11. *By* $\mathcal{B}$ *we denote the set of all operators* $A \in \mathcal{L}(L^\infty)$ *for which*

$$[M_{\varphi_t}\,,\,A] \rightrightarrows 0 \quad \text{as } t \to 0 \tag{2.6}$$

holds for all functions $\varphi \in$ BUC. *By* $\mathcal{B}_\mathcal{S}$ *we will denote the intersection* $\mathcal{B} \cap \mathcal{S}$.

Proposition 2.12. $\mathcal{B}$ *and* $\mathcal{B}_\mathcal{S}$ *are inverse closed Banach algebras in* $\mathcal{L}(L^\infty)$.

Proof. From $[M\,,\,A+B] = [M\,,\,A] + [M\,,\,B]$ and $[M\,,\,AB] = [M\,,\,A]B + A[M\,,\,B]$ we get that $\mathcal{B}$ is an algebra. Its norm-closedness follows from

$$\|\,[M_{\varphi_t}\,,\,A]\,\| \le \|\,[M_{\varphi_t}\,,\,A_m]\,\| + 2\|\varphi\|_\infty\,\|A - A_m\|,$$

and if $A \in \mathcal{B}$ is invertible, then by $[M\,,\,A^{-1}] = A^{-1}[A\,,\,M]A^{-1}$, we get that also $A^{-1} \in \mathcal{B}$. From Proposition 2.9 it follows that the same is true for $\mathcal{B}_\mathcal{S}$. □

Here is the important connection between the two algebras $\mathcal{B}_\mathcal{S}$ and BDO:

Proposition 2.13. *Every* $A \in \mathcal{B}_\mathcal{S}$ *is band-dominated, i.e.,* $\mathcal{B}_\mathcal{S} \subset$ BDO.

Proof. Pick some arbitrary $A \in \mathcal{B}_\mathcal{S}$, and put $B := A^G \in \mathcal{L}(\ell^\infty(\mathbb{Z}^n, L^\infty(H)))$. We have to show that B can be approximated (in the operator norm) by operators of the form (2.1).

For $x = (x_1, \ldots, x_n) \in \mathbb{R}^n$, define

$$\varepsilon(x) := \exp(i(x_1 + \cdots + x_n)). \tag{2.7}$$

Clearly, $\varepsilon \in \mathrm{BUC}$, and consequently, $[M_{\varepsilon_t}, A] \rightrightarrows 0$ as $t \to 0$. Now we will pass to the discretization:

$$[\hat{M}_{\hat{\varepsilon}_t}, B] \rightrightarrows 0 \quad \text{as } t \to 0 \tag{2.8}$$

Since $\varepsilon_t(x) = \exp i(t,x)$ with $(.,.)$ referring to the usual inner product of $\mathbb{R}^n$, one has $\varepsilon_s\varepsilon_t = \varepsilon_{s+t}$ for all $s,t \in \mathbb{R}^n$. Consequently,

$$\|\hat{M}_{\hat{\varepsilon}_t} B \hat{M}_{\hat{\varepsilon}_{-t}} - \hat{M}_{\hat{\varepsilon}_s} B \hat{M}_{\hat{\varepsilon}_{-s}}\| \le \|\hat{M}_{\hat{\varepsilon}_{t-s}} B - B\hat{M}_{\hat{\varepsilon}_{t-s}}\|,$$

which, together with (2.8), implies the continuity of the function

$$\varphi : \mathbb{R}^n \to \mathcal{L}(\ell^\infty(\mathbb{Z}^n, L^\infty(H))), \qquad \varphi : t \mapsto \hat{M}_{\hat{\varepsilon}_t} B \hat{M}_{\hat{\varepsilon}_{-t}}. \tag{2.9}$$

Moreover, the functions $t \mapsto \hat{M}_{\hat{\varepsilon}_t}$ and $t \mapsto \hat{M}_{\hat{\varepsilon}_{-t}}$ are 2π-periodic in each variable t_j, and so is the function φ from (2.9). So we can approximate φ by its Riesz means $\varphi^{(k)}$ with sufficiently large exponent (see [27, VII.2.14f]), i.e.,

$$\sup_{t\in[0,2\pi]^n} \|\varphi(t) - \varphi^{(k)}(t)\| \to 0 \quad \text{as } k \to \infty, \tag{2.10}$$

where $\varphi^{(k)}$ are polynomials of the form

$$\varphi^{(k)}(t) = \sum_{|\alpha|\le k} B^{(\alpha)} e^{i(t,\alpha)}, \quad \alpha \in \mathbb{Z}^n \tag{2.11}$$

$$\text{with} \quad B^{(\alpha)} = c^{(\alpha)} \int_{[0,2\pi]^n} \varphi(\tau) e^{-i(\tau,\alpha)} d\tau \quad \in \mathcal{L}(\ell^\infty(\mathbb{Z}^n, L^\infty(H))) \tag{2.12}$$

with certain constants $c^{(\alpha)}$. Note that these approximation techniques work for functions φ with values in a Banach space as well as for complex-valued functions (see [8]).

The operators $B^{(\alpha)} V_{-\alpha}$ commute with each of the (generalized) multiplication operators $\hat{M}_{\hat{\varepsilon}_s}$ where $s \in \mathbb{R}^n$. Indeed,

$$\hat{M}_{\hat{\varepsilon}_s} B^{(\alpha)} V_{-\alpha} \hat{M}_{\hat{\varepsilon}_{-s}} = c^{(\alpha)} \int_{[0,2\pi]^n} (\hat{M}_{\hat{\varepsilon}_{s+t}} B \hat{M}_{\hat{\varepsilon}_{-s-t}}) e^{-i(s+t,\alpha)} dt\, V_{-\alpha} = B^{(\alpha)} V_{-\alpha}.$$

Since $A \in \mathcal{S}$, B also has the $\mathcal{S}$-property as well as $B^{(\alpha)} V_{-\alpha}$. Theorem 2.1 in [24] shows that such operators are continuous in the (L^∞, L^1) topology. Moreover, Proposition 5.1.11 in [10] says that an (L^∞, L^1)-continuous operator M which commutes with every $\hat{M}_{\hat{\varepsilon}_s}$ even commutes with every scalar multiplicator, that is, $\hat{M}_b$ with $b = (B_\alpha)$, where all components B_α are multiples of the identity on $L^\infty(H)$. Consequently, M also commutes with $\hat{M}_{b^{(\beta)}} = (P_{H_\beta})^G$ for all $\beta \in \mathbb{Z}^n$. The equation $M\hat{M}_{b^{(\beta)}} = \hat{M}_{b^{(\beta)}} M$ shows that M maps sequences $(c_\alpha)_{\alpha\in\mathbb{Z}^n}$, that are only supported in component number β, to exactly such sequences again, i.e., M acts on every component c_α independently from all other components. By Definition 2.1, M is a generalized multiplication operator.

From this discussion it follows that every operator $B^{(\alpha)} V_{-\alpha}$ is of the form $\hat{M}_{a^{(\alpha)}}$ with some $a^{(\alpha)} \in \ell^\infty(\mathbb{Z}^n, L^\infty(H))$, and hence, $B^{(\alpha)} = \hat{M}_{a^{(\alpha)}} V_\alpha$. This shows that the value $\varphi^{(k)}(0) = \sum_{|\alpha|\le k} B^{(\alpha)}$ of the kth Riesz mean polynomial (2.11) is a band operator! From (2.10) we infer that

$$\varphi^{(k)}(0) \rightrightarrows \varphi(0) = B \quad \text{as } k \to \infty,$$

and hence, $B(= A^G)$ is band dominated. And so is A. □

Remark 2.14 The reverse inclusion is not true. To see this take some generalized multiplication operator $\hat{M}_b$ where the sequence $b = (B_\alpha)_{\alpha\in\mathbb{Z}^n}$ contains operators $B_\alpha \in \mathcal{L}(L^\infty(H))$ whose adjoint B_α^* does not map $L^1(H)$ to $L^1(H)$. □

The proof of Proposition 2.13 gives us some nice idea: Since we used property (2.6) only for the function $\varphi = \varepsilon$ from (2.7), we will introduce the algebra $\mathcal{B}'$ consisting of all $A \in \mathcal{L}(L^\infty)$ for which (2.6) holds for $\varphi(x) = \varepsilon(x) = \exp(i(x_1 + \cdots + x_n))$.

Theorem 2.15. *Put* $\mathcal{B}'_\mathcal{S} := \mathcal{B}' \cap \mathcal{S}$ *and* $\mathrm{BDO}_\mathcal{S} := \mathrm{BDO} \cap \mathcal{S}$. *Then*

$$\mathrm{BDO}_\mathcal{S} = \mathcal{B}_\mathcal{S} = \mathcal{B}'_\mathcal{S} \subset \mathrm{BDO} \subset \mathcal{B} \subset \mathcal{B}'.$$

Proof. $\mathcal{B} \subset \mathcal{B}'$ is obvious. The proof of Proposition 2.13 shows that $\mathcal{B}'_\mathcal{S} \subset \mathrm{BDO}$. To see that $\mathrm{BDO} \subset \mathcal{B}$, check the generators of the algebra BDO: It is readily seen that shift operators are in $\mathcal{B}$. Generalized multiplication operators $\hat{M}_b$ commute with operators of multiplication by functions which are constant on every H_α, for instance, with $\dot{\varphi}_t(x_1,\ldots,x_n) := \varphi_t([x_1],\ldots,[x_n])$, where $[x_i]$ is the integer part of x_i. But since for every function $\varphi \in \mathrm{BUC}$ $\|\varphi_t - \dot{\varphi}_t\|_\infty \to 0$ as $t \to 0$, we also have $\hat{M}_b \in \mathcal{B}$. Summarizing this, we get $\mathcal{B}'_\mathcal{S} \subset \mathrm{BDO} \subset \mathcal{B} \subset \mathcal{B}'$, and taking intersections with $\mathcal{S}$ proves the rest of our claim. □

2.6. Invertibility at infinity

Usually in the theory of approximation methods on L^p, the ideal $\mathcal{K}$ of compact operators plays an important role. Invertibility modulo compact perturbations is then called Fredholmness. We will show quite shortly why $\mathcal{K}$ is **not** the appropriate tool in our case at hand, $p = \infty$.

For the (hyper-)cube $C = [-1,1]^n$, the sequence of (outside) cut-off projections $(Q_{rC})_{r=1}^\infty$ strongly converges to 0 on every space L^p, $p < \infty$ (where functions need to vanish more or less slowly towards infinity). But not so in L^∞.

One crucial property of operators $T \in \mathcal{K}$ is that they turn strong convergence into norm convergence if they are applied to the convergent sequence from the right (and, under additional assumptions on the adjoint sequence, also from the left). Consequently, if $1 < p < \infty$, then $TQ_{rC} \rightrightarrows 0$ and $Q_{rC}T \rightrightarrows 0$ as $r \to \infty$ for every compact operator T. But for $p = \infty$ (and the same would be necessary for $p = 1$), we will introduce a new ideal $\mathcal{J}$ which will serve as a substitute of the ideal $\mathcal{K}$ of compact operators.

Definition 2.16. *By* $\mathcal{J}$ *we denote the set of all operators* $T \in \mathcal{L}(L^\infty)$ *with*

$$TQ_{rC} \rightrightarrows 0 \quad \text{and} \quad Q_{rC}T \rightrightarrows 0 \quad \text{as} \quad r \to \infty,$$

and by $\mathcal{J}_\mathcal{S}$ *we will denote the intersection* $\mathcal{J} \cap \mathcal{S}$.

Proposition 2.17. *$\mathcal{J}$ is a closed ideal in $\mathcal{B}$, and $\mathcal{J}_S$ is a closed ideal in $\mathcal{B}_S$.*

Proof. Take arbitrary $T \in \mathcal{J}$ and $\varphi \in \mathrm{BUC}$, and put $\psi := \varphi - \varphi(0) \in \mathrm{BUC}$. Then

$$[M_{\varphi_t}, T] = [M_{\psi_t}, T]. \tag{2.13}$$

For every $r > 0$, the inequality $\|M_{\psi_t}T\| \le \|(M_{\psi_t}P_{rC})T\| + \|M_{\psi_t}(Q_{rC}T)\|$ holds, where the second term on the right can be made as small as desired by choosing r large enough, and the first term evidently tends to 0 as $t \to 0$. The same argumentation for $\|TM_{\psi_t}\|$ shows that

$$M_{\psi_t}T \rightrightarrows 0 \quad \text{and} \quad TM_{\psi_t} \rightrightarrows 0 \quad \text{as} \quad t \to 0. \tag{2.14}$$

Now (2.13) shows that $T \in \mathcal{B}$, and hence $\mathcal{J} \subset \mathcal{B}$. Clearly, $\mathcal{J}$ is closed, and we are left with the proof of the ideal property.

Let therefore $A \in \mathcal{B}$, and choose $\varphi \in \mathrm{BUC}$ such that φ vanishes on C and is equal to 1 outside $2C$. Let $\Phi_r(x) := \varphi_{(1/r,\dots,1/r)}(x) = \varphi(x/r)\ \forall x \in \mathbb{R}^n$. Then

$$\begin{aligned} Q_{2rC}(AT) = Q_{2rC}M_{\Phi_r}AT &= Q_{2rC}([M_{\Phi_r}, A]T + AM_{\Phi_r}T) \\ &= Q_{2rC}([M_{\Phi_r}, A]T + AM_{\Phi_r}Q_{rC}T) \rightrightarrows 0 \end{aligned}$$

as $r \to \infty$. The convergence $(AT)Q_{rC} \rightrightarrows 0$ as $r \to \infty$ is obvious, and hence $AT \in \mathcal{J}$. Analogously one shows that also $TA \in \mathcal{J}$.

Our second claim is then trivial by the definition of $\mathcal{B}_S$ and $\mathcal{J}_S$. □

In analogy to Fredholmness, we will introduce the notion of *invertibility at infinity*, where $\mathcal{K}$ is substituted by $\mathcal{J}$.

Definition 2.18. *An operator $A \in \mathcal{L}(L^\infty)$ is said to be invertible at infinity if there exists an operator $B \in \mathcal{L}(L^\infty)$ such that*

$$AB = I + T_1 \quad \text{and} \quad BA = I + T_2$$

hold with operators $T_1, T_2 \in \mathcal{J}$.

Remark 2.19 Definitions 2.16 and 2.18 can be given in L^p, $1 < p < \infty$ as well, and from what we wrote earlier it is clear that in this setting every compact operator is automatically in $\mathcal{J}$. (The reverse inclusion is not true since, for example, P_C is in $\mathcal{J} \setminus \mathcal{K}$.) Consequently, in L^p every Fredholm operator is invertible at infinity.

In L^∞ neither inclusion holds between $\mathcal{J}$ and $\mathcal{K}$. □

Proposition 2.20. *If $A \in \mathcal{B}$ is invertible at infinity, then the operator B from Definition 2.18 is in $\mathcal{B}$ as well.*

Proof. Take some arbitrary $\varphi \in \mathrm{BUC}$, and define $\psi := \varphi - \varphi(0) \in \mathrm{BUC}$. Then

$$[M_{\varphi_t}, B] = [M_{\psi_t}, B] = B[A, M_{\psi_t}]B + C_t$$

holds with $C_t = T_2M_{\psi_t}B - BM_{\psi_t}T_1 \rightrightarrows 0$ by (2.14). □

Corollary 2.21. *An operator $A \in \mathcal{B}$ is invertible at infinity if and only if the coset $A + \mathcal{J}$ is invertible in the factor algebra $\mathcal{B}/\mathcal{J}$.*

Proposition 2.22. *$A \in \mathcal{B}$ is invertible at infinity if and only if there exist operators $B_1, B_2 \in \mathcal{B}$ such that*

$$Q_{rC}AB_1 \;=\; Q_{rC} \;=\; B_2AQ_{rC}$$

holds for some $r > 0$.

Proof. Take $B \in \mathcal{B}$ such that $AB = I + T_1$ and $BA = I + T_2$ with some $T_1, T_2 \in \mathcal{J}$, and take r large enough that $\|Q_{rC}T_1\| < 1$ and $\|T_2Q_{rC}\| < 1$. Then

$$Q_{rC}AB \;=\; Q_{rC} + Q_{rC}T_1 \;=\; Q_{rC}(I + Q_{rC}T_1),$$

where $D := I + Q_{rC}T_1$ is invertible, and so we have $Q_{rC}AB_1 = Q_{rC}$ with $B_1 := BD^{-1} \in \mathcal{B}$ (note that $D \in \mathcal{B}$, and $\mathcal{B}$ is inverse closed). By the same computations, we see that also the second equality holds with $B_2 := (I + T_2Q_{rC})^{-1}B \in \mathcal{B}$.

Conversely, if $Q_{rC}AB_1 = Q_{rC} = B_2AQ_{rC}$ with $B_1, B_2 \in \mathcal{B}$, then $AB_1 = I - P_{rC} + P_{rC}AB_1 =: I - T_1$. Analogously, $B_2A = I - T_2$ with $T_1, T_2 \in \mathcal{J}$. □

2.7. $\mathcal{P}$-convergence

Having replaced $\mathcal{K}$ by $\mathcal{J}$, we will now replace strong convergence by another type of convergence which is compatible to $\mathcal{J}$ as strong convergence is to $\mathcal{K}$.

2.7.1. Definition Let $\mathcal{P}$ denote the collection of all P_U with U running through all bounded and measurable (frequently abbreviate this by b&m) subsets of $\mathbb{R}^n$.

Definition 2.23. *We say that a sequence $(A_\tau) \subset \mathcal{L}(L^\infty)$ $\mathcal{P}$-converges to $A \in \mathcal{L}(L^\infty)$ and that $(u_\tau) \subset L^\infty$ $\mathcal{P}$-converges (or strictly converges) to $u \in L^\infty$ if for all $P_U \in \mathcal{P}$*

$$\|P_U(A_\tau - A)\| \to 0, \qquad \|(A_\tau - A)P_U\| \to 0,$$
$$(u_\tau) \text{ is bounded} \quad \text{and} \quad \|P_U(u_\tau - u)\| \to 0 \quad \text{as } \tau \to \infty, \text{ respectively.}$$

It is readily seen that a $\mathcal{P}$-limit is unique if it exists. Indeed, suppose for every b&m set $U \subset \mathbb{R}^n$, we have $P_U(A_\tau - A) \rightrightarrows 0$ and $P_U(A_\tau - B) \rightrightarrows 0$; and consequently, $P_U(A - B) \rightrightarrows 0$ as $\tau \to \infty$, i.e., $P_U(A - B) = 0$. Then clearly, $\operatorname{im}(A - B) \subset \cap_U \ker P_U = \{0\}$, i.e., $A - B = 0$. By the same argumentation we see that also $\mathcal{P}\text{-}\lim u_\tau$ is unique.

Remark 2.24 Although this is trivial, it is worth to be mentioned that $\mathcal{P} \subsetneq \mathcal{J}$. □

2.7.2. Characterization and Properties By Definition 2.23, operators $P_U \in \mathcal{P}$ produce a norm-convergent sequence if they are applied to a $\mathcal{P}$-convergent sequence of functions or operators. Not only operators from $\mathcal{P}$ but all operators from $\mathcal{J}$ have this effect on (bounded!) $\mathcal{P}$-convergent sequences:

Proposition 2.25. *For bounded sequences (A_τ) in $\mathcal{L}(L^\infty)$ and (u_τ) in L^∞, we have $A = \mathcal{P}\text{-}\lim A_\tau$ and $u = \mathcal{P}\text{-}\lim u_\tau$ if and only if for all $T \in \mathcal{J}$ and $\tau \to \infty$*

$$\|T(A_\tau - A)\| \to 0 \quad \text{and} \quad \|(A_\tau - A)T\| \to 0 \tag{2.15}$$
$$\text{and} \quad \|T(u_\tau - u)\| \to 0, \text{ respectively.} \tag{2.16}$$

Proof. Suppose $A = \mathcal{P}\text{-}\lim A_\tau$ and $u = \mathcal{P}\text{-}\lim u_\tau$. Then for every $r > 0$, inequality $\|T(A_\tau - A)\| \leq \|T(P_{rC}(A_\tau - A))\| + \|(TQ_{rC})(A_\tau - A)\|$ holds, where the first term tends to zero as $\tau \to \infty$, and the second is as small as desired if r is large enough. The remaining inequalities of (2.15) and (2.16) are shown absolutely analogous. Conversely, if (2.15), (2.16) hold for all $T \in \mathcal{J}$, then also for $T = P_U \in \mathcal{P} \subset \mathcal{J}$. □

At the beginning of Section 2.6 we mentioned that compact operators turn strong convergence into norm-convergence, and we mused that $\mathcal{K}$ is not the appropriate tool in our case, introducing $\mathcal{J}$ instead. Consequently, we started studying invertibility modulo $\mathcal{J}$, called "invertibility at infinity" rather than invertibility modulo $\mathcal{K}$ alias Fredholmness. Now we moreover substituted strong convergence by $\mathcal{P}$-convergence and have seen that operators in $\mathcal{J}$ turn $\mathcal{P}$-convergence into norm-convergence if they are applied from the left or right to the convergent sequence – provided it is bounded.

Summarizing this journey, the pair $(\mathcal{J}, \mathcal{P}\text{-}\lim)$ fits together in the L^∞ setting as the pair $(\mathcal{K}, s - \lim)$ did in earlier cases like $\ell^p(\mathbb{Z}^n)$ with $1 < p < \infty$. It turns out that these pairs actually coincide in the latter case!

Remark 2.26 Note that if one defines $\mathcal{P}$-limits by (2.15), then every $\mathcal{P}$-convergent sequence is automatically bounded, starting from some $\tau_* > 0$ (for a proof use Proposition 2.31 and the Banach-Steinhaus Theorem). We will nevertheless use Definition 2.23 which also allows unbounded sequences to $\mathcal{P}$-converge, for example, $(A_\tau) = (\tau Q_{\tau C})$ is unbounded – but $\mathcal{P}$-convergent to the zero operator. □

Before continuing with $\mathcal{P}$-limits and their properties, we will introduce the appropriate set of operators to work with.

Definition 2.27. *Let $\mathfrak{A}$ refer to the set of all operators $A \in \mathcal{L}(L^\infty)$ for which both sequences*

$$(AQ_{rC}) \qquad \text{and} \qquad (Q_{rC}A) \tag{2.17}$$

$\mathcal{P}$-converge to 0 as $r \to \infty$.

Proposition 2.28.

a) *$\mathfrak{A}$ is a Banach subalgebra of $\mathcal{L}(L^\infty)$,*

b) OLT $\subset \mathfrak{A}$,

c) *$\mathcal{J}$ is a closed two-sided ideal in $\mathfrak{A}$.*

Proof. The only thing that needs a proof in **a)** is that $AB \in \mathfrak{A}$ if $A, B \in \mathfrak{A}$. To see this, write $\|P_U(AB)Q_{rC}\| \leq \|P_UA(P_{sC}BQ_{rC})\| + \|(P_UAQ_{sC})BQ_{rC}\|$ with some $s > 0$, and observe that the first term tends to zero as $r \to \infty$ and the second one can be made as small as desired by choosing s large enough.

Concerning **b)**, we remark that clearly, for every b&m set U, the distance of U and $\mathbb{R}^n \setminus rC$ goes to infinity as $r \to \infty$, and hence, OLT $\subset \mathfrak{A}$ holds.

For **c)** we remark that every norm-convergent sequence is $\mathcal{P}$-convergent, and hence $\mathcal{J} \subset \mathfrak{A}$. To show the ideal property, take arbitrary $A \in \mathfrak{A}$ and $T \in \mathcal{J}$. Then, for every $r > 0$, $\|Q_{rC}(AT)\| \leq \|(Q_{rC}AP_{sC})T\| + \|Q_{rC}A(Q_{sC}T)\|$, where again the first term tends to zero as $r \to \infty$, and the second one becomes small if s

is sufficiently large. Clearly, $(AT)Q_{rC} \rightrightarrows 0$ for $r \to \infty$ as well and the proof of $TA \in \mathcal{J}$ is analogous. □

Remark 2.29 Note that $\mathfrak{A}$ is strictly larger than OLT alias BDO. It also contains operators for which the convergence in (2.5) is not uniformly with respect to the distance of U and V, for instance, the contraction operator $(Au)(x) = u(2x)$ (cf. Remark 2.8). Moreover, $\mathfrak{A}$ even contains operators which do not enjoy property (2.5) in either sense, like the flip operator $(Ju)(x) = u(-x)$.

It turns out that $\mathfrak{A}$ is the set of all operators A for which TA and AT are in $\mathcal{J}$ whenever $T \in \mathcal{J}$, i.e., $\mathfrak{A}$ is the largest algebra where $\mathcal{J}$ is an ideal of. □

So by Propositions 2.13, 2.7 b) and 2.28 b), we have $\mathcal{B}_S \subsetneqq \text{BDO} = \text{OLT} \subsetneqq \mathfrak{A}$.

Remark 2.30 So all operators $A \in \mathcal{B}_S$ have the property that the sequences (2.17) $\mathcal{P}$-converge to zero, and those $A \in \mathcal{B}_S$ for which this $\mathcal{P}$-convergence is even norm-convergence form the ideal $\mathcal{J}_S$. □

As a consequence of Proposition 2.28 c), with every operator $A \in \mathfrak{A}$ one can associate two operators **on** $\mathcal{J}$ by

$$A^l : T \mapsto AT \qquad \text{and} \qquad A^r : T \mapsto TA$$

for every $T \in \mathcal{J}$. This observation as well as the following proposition and its corollary are due to ROCH, RABINOVICH and the second author. For completeness we will give all proofs here.

Proposition 2.31.

a) $\|A^l\|_{\mathcal{L}(\mathcal{J})} = \|A\| = \|A^r\|_{\mathcal{L}(\mathcal{J})}$ *for every* $A \in \mathfrak{A}$.

b) *A bounded sequence* $(A_\tau) \subset \mathfrak{A}$ $\mathcal{P}$*-converges to some* $A \in \mathfrak{A}$ *if and only if* $A^l_\tau \to A^l$ *and* $A^r_\tau \to A^r$ *strongly in* $\mathcal{L}(\mathcal{J})$.

Proof. **Part a)** Take an arbitrary $A \in \mathfrak{A}$. Clearly, $\|A^l\|, \|A^r\| \le \|A\|$. For the reverse inequalities take an arbitrary $\varepsilon > 0$, and choose $x_0 \in L^\infty$ with $\|x_0\| = 1$ such that $\|Ax_0\| \ge \|A\| - \varepsilon$. Moreover, choose a b&m set $U \subset \mathbb{R}^n$ with $\|Ax_0\| - \|P_U Ax_0\| \le \varepsilon$. Then $\|P_U Ax_0\| \ge \|A\| - 2\varepsilon$. Since $P_U \in \mathcal{J}$, we have

$$\|A^r\| = \|A^r\|\|P_U\| \ge \|P_U A\| \ge \|P_U Ax_0\| \ge \|A\| - 2\varepsilon. \tag{2.18}$$

Choosing $r > 0$ large enough that $\|P_U AQ_{rC}\| \le \varepsilon$ and taking the inequality $\|P_U A\| \ge \|A\| - 2\varepsilon$ from (2.18), we conclude from $P_{rC} \in \mathcal{J}$ to

$$\begin{aligned} \|A^l\| &= \|A^l\|\|P_{rC}\| \ge \|AP_{rC}\| \ge \|P_U AP_{rC}\| \\ &\ge \|P_U A\| - \|P_U AQ_{rC}\| \ge \|A\| - 3\varepsilon. \end{aligned} \tag{2.19}$$

Since (2.18) and (2.19) hold for every $\varepsilon > 0$, we are done with **a)**.

Part b) By **a)**, $\mathfrak{A}$ is isometrically embedded into $\mathcal{L}(\mathcal{J})$ by each one of the mappings $A \mapsto A^l$ and $A \mapsto A^r$. Consequently, $\|T(A_\tau - A)\| \to 0$ for all $T \in \mathcal{J}$ is equivalent to the strong convergence of A^r_τ to A^r on $\mathcal{J}$, and the second property in (2.15) is the same for A^l_τ and A^l. So it remains to apply Proposition 2.25 to the bounded sequence (A_τ). □

Corollary 2.32. *Let (A_τ), $(B_\tau) \subset \mathfrak{A}$ be bounded sequences having $\mathcal{P}$-limits $A, B \in \mathfrak{A}$, respectively. Then*

$$\begin{array}{lrcl} \textbf{a)} & \|A\| \leq \liminf \|A_\tau\| & \leq & \sup \|A_\tau\| < \infty, \\ \textbf{b)} & \mathcal{P}\text{-}\lim(A_\tau + B_\tau) & = & A + B, \\ \textbf{c)} & \mathcal{P}\text{-}\lim(A_\tau B_\tau) & = & AB. \end{array}$$

Proof. Using Proposition 2.31, we only have to recall the Banach-Steinhaus Theorem as well as the compatibility of strong limits with addition and composition.

Alternatively, we present elementary proofs for **b)** and **c)** using Definition 2.23. Equality **b)** is trivial, and **c)** can be seen as follows: For every b&m set $U \subset \mathbb{R}^n$ and every $r > 0$,

$$\begin{aligned} \|P_U(A_\tau B_\tau - AB)\| \quad \leq \quad & \|(P_U(A_\tau - A))B_\tau\| + \|P_U A(P_{rC}(B_\tau - B))\| \\ & + \|(P_U A Q_{rC})(B_\tau - B)\| \end{aligned}$$

holds, where the first two terms tend to zero as $\tau \to \infty$, and the third one becomes as small as desired if we choose r large enough (note that (B_τ) is bounded). $(A_\tau B_\tau - AB)P_U \rightrightarrows 0$ is shown analogously, using the boundedness of (A_τ). □

2.8. Limit operators

In Section 2.6 we introduced a way to evaluate an operator's behavior "at infinity". Here we will present a completely different look at this point which will lead us to some additional insight.

Definition 2.33. *Let $A \in \mathcal{L}(L^\infty)$ and $h = (h_m) \subset \mathbb{Z}^n$ tend to infinity. If*

$$V_{-h_m} A V_{h_m}, \qquad m = 1, 2, \ldots$$

$\mathcal{P}$-converges to some operator $B \in \mathcal{L}(L^\infty)$ as $m \to \infty$, then we call B the limit operator of A with respect to the sequence h and denote it by A_h. In this case we will also say that the sequence h leads to a limit operator of A.

If $h \subset \mathbb{Z}^n$ is a sequence tending to infinity such that A_h exists, and if g is an infinite subsequence of h, then also g leads to a limit operator of A, and $A_g = A_h$.

Limit operators are compatible with addition, composition, passing to the norm limit and to the (pre-)adjoint operator.[1] Here we summarize these basic properties.

Proposition 2.34. *Let A, B and $A^{(m)}$ $(m = 1, 2, \ldots)$ be arbitrary operators in $\mathfrak{A}$, and let $h \subset \mathbb{Z}^n$ be some sequence tending to infinity.*

- **a)** *If A_h exists, then $\|A_h\| \leq \|A\|$.*
- **b)** *If A_h and B_h exist, then $(A + B)_h$ exists and is equal to $A_h + B_h$.*
- **c)** *If A_h and B_h exist, then $(AB)_h$ exists and is equal to $A_h B_h$.*
- **d)** *If $A = \lim_{m\to\infty} A^{(m)}$ and the limit operators $(A^{(m)})_h$ exist for all sufficiently large m, then A_h exists and equals $\lim_{m\to\infty}(A^{(m)})_h$.*

This proposition is from [21]. It is a simple consequence of Corollary 2.32.

---

[1] Note that in $\mathcal{L}(L^1)$ limit operators are defined exactly as in Definition 2.33.

Proposition 2.35. *Take an arbitrary $A \in \mathfrak{A}$, and suppose the sequence $h \subset \mathbb{Z}^n$ leads to the limit operator A_h of A.*

a) *If $A \in \mathcal{S}$, then also $A_h \in \mathcal{S}$, and $^*(A_h) = (^*A)_h =: {}^*A_h$.*
b) *If $A \in \mathcal{B}$, then also $A_h \in \mathcal{B}$.*
c) *If $A \in \mathcal{B}_\mathcal{S}$, then also $A_h \in \mathcal{B}_\mathcal{S}$.*
d) *If $A \in$ BO or BDO, then also $A_h \in$ BO or BDO, respectively.*

Proof. Concerning **a)**, note that $^*P_U = P_U$ and $^*V_\alpha = V_{-\alpha}$, and consequently, for every b&m set $U \subset \mathbb{R}^n$ and $m \to \infty$, we have

$$\|P_U(V_{-h_m}{}^*AV_{h_m} - {}^*A)\| = \|^*(\,(V_{-h_m}AV_{h_m} - A)P_U\,)\| \to 0.$$

b) is from [22]: Let $A \in \mathcal{B}$, and take arbitrary $\varphi \in$ BUC and $t \in \mathbb{R}^n$. Then $[M_{\varphi_t}\,,\,V_{-h_m}AV_{h_m}]$ $\mathcal{P}$-converges to $[M_{\varphi_t}\,,\,A_h]$ as $m \to \infty$. By Corollary 2.32 a),

$$\|[M_{\varphi_t}\,,\,A_h]\| \le \liminf \|[M_{\varphi_t}\,,\,V_{-h_m}AV_{h_m}]\| \le 2\|[M_{\varphi_t}\,,\,V_{-h_{m_0}}AV_{h_{m_0}}]\|$$

for some m_0. But from $A, V_{\pm h_{m_0}} \in \mathcal{B}$ and Proposition 2.12 we get that this tends to zero as $t \to 0$. Consequently, $A_h \in \mathcal{B}$. Clearly, **c)** follows from **a)** and **b)**.

For **d)** first note that limit operators of operators A, for which A^G is a generalized multiplicator, are again operators with this property. The claim for BO then follows from $(V_\alpha)_h = V_\alpha \,\forall h$ and Proposition 2.34 b), c). Finally, a simple approximation argument involving Proposition 2.34 d) shows the claim for BDO. □

Definition 2.36. *Let $A \in \mathcal{L}(L^\infty)$. The set of all limit operators of A is denoted by $\sigma^{\mathrm{op}}(A)$, and we refer to it as the operator spectrum of A.*

We will now come to some class of operators which will be of great importance to us in what follows. It is the set of operators which we will refer to as *rich* operators (in terms of: having limit operators everywhere).

Definition 2.37. *By $\mathcal{B}_\$$ we denote the set of all operators $A \in \mathcal{B}_\mathcal{S}$ with the following property: Every sequence $h = (h_m) \subset \mathbb{Z}^n$ with $h_m \to \infty$ possesses a subsequence g leading to a limit operator of A. We will regard operators in $\mathcal{B}_\$$ as rich operators.*

Proposition 2.38. *$\mathcal{B}_\$$ is a Banach subalgebra of $\mathcal{L}(L^\infty)$.*

Proof. Let $A, B \in \mathcal{B}_\$$, and take an arbitrary sequence $h = (h_m) \subset \mathbb{Z}^n$ tending to infinity. Since $A \in \mathcal{B}_\$$, there is a subsequence g of h which leads to a limit operator of A. But since $B \in \mathcal{B}_\$$, we can once more choose a subsequence f of g for which B_f exists, and clearly $A_f = A_g$, and by Proposition 2.34 b) and c), f leads to a limit operator of $A+B$ and of AB. The closedness of $\mathcal{B}_\$$ is verified by an analogous construction of a subsequence of h and finally applying Proposition 2.34 d). □

3. Limit operators vs. invertibility at infinity

Invertibility at infinity is a property that only depends on an operator's behavior in a neighborhood of infinity. For rich operators, all necessary information about this behavior is accurately stored in the operator spectrum $\sigma^{\mathrm{op}}(A)$, and it is the topic of this section to show the interplay between these two concepts.

3.1. The role of the ideal $\mathcal{J}$

The ideal $\mathcal{J}$ is an appropriate item to describe this relationship: On the one hand, invertibility at infinity is nothing but invertibility modulo $\mathcal{J}$. On the other hand, the ideal $\mathcal{J}$ is exactly what becomes "factored out" by considering limit operators:

Proposition 3.1. *Let, for a moment, $\mathcal{L}_\$$ denote the set of all operators $A \in \mathcal{L}(L^\infty)$ for which every sequence $h \subset \mathbb{Z}^n$ tending to infinity has a subsequence that leads to a limit operator of A (so $\mathcal{B}_\$ = \mathcal{L}_\$ \cap \mathcal{B}_S$). Then*

$$\mathcal{J} = \left\{T \in \mathcal{L}_\$: \sigma^{\mathrm{op}}(T) = \{0\}\right\} \quad \textit{and} \quad \mathcal{J}_S = \left\{T \in \mathcal{B}_\$: \sigma^{\mathrm{op}}(T) = \{0\}\right\}.$$

Proof. From $V_\alpha P_U = P_{\alpha+U} V_\alpha$ and V_α being an isometry, we conclude that for a sequence $h = (h_m) \to \infty$ and an operator T, the limit operator T_h equals 0 if and only if for all b&m subsets $U \subset \mathbb{R}^n$ both,

$$\begin{aligned} \|P_U(V_{-h_m} T V_{h_m} - 0)\| &= \|V_{-h_m} P_{h_m+U} T V_{h_m}\| = \|P_{h_m+U} T\| \quad \text{and} \\ \|(V_{-h_m} T V_{h_m} - 0) P_U\| &= \|V_{-h_m} T P_{h_m+U} V_{h_m}\| = \|T P_{h_m+U}\| \end{aligned}$$

tend to zero as $m \to \infty$.

If $T \in \mathcal{J}$, then clearly, $\|P_{h_m+U}T\|$ and $\|TP_{h_m+U}\|$ tend to zero for all sequences $h = (h_m) \to \infty$, so $T \in \mathcal{L}_\$$, and $\sigma^{\mathrm{op}}(T) = \{0\}$.

Conversely, if some $T \in \mathcal{L}_\$$ is not in $\mathcal{J}$, then, by Definition 2.16, $\|Q_{rC}T\|$ or $\|TQ_{rC}\|$ does not tend to zero as $r \to \infty$. A little thought shows that then there is a sequence $h = (h_m)$ tending to infinity and a b&m set U such that $\|P_{h_m+U}T\|$ or $\|TP_{h_m+U}\|$ remains bounded away from zero as m goes to ∞. Since $T \in \mathcal{L}_\$$, we can choose a subsequence g of h which leads to a limit operator of T. But if T_g were 0, then $\|P_{g_m+U}T\|$ and $\|TP_{g_m+U}\|$ would both tend to zero which they do not. So $\sigma^{\mathrm{op}}(T) \neq \{0\}$, and we have shown the first equality. On both sides taking the intersection with $\mathcal{B}_S$, gives the second claim since $\mathcal{J} \subset \mathcal{B}$ and $\mathcal{B}_\$ = \mathcal{L}_\$ \cap \mathcal{B}_S$. □

This observation is the key motivation for – roughly speaking – identifying the coset $A + \mathcal{J}$ with $\sigma^{\mathrm{op}}(A)$ if A is some rich operator. The essence of this identification is Theorem 1.1 which is proven on the following pages.

3.2. Invertibility at infinity in terms of limit operators

In this section we will prove Theorem 1.1. We still need some propositions and the following basic lemma from the theory of Banach algebras.

Lemma 3.2. *Consider some Banach algebra* $\mathbf{B}$ *with unit e and a convergent sequence $(x_n)_{n=1}^\infty$ of invertible elements in* $\mathbf{B}$. *If $\|x_n^{-1}\| < M < \infty$, $\forall n$ for some M, then also the limit $x := \lim\limits_{n\to\infty} x_n$ is invertible, and $\|x^{-1}\| \le M$.*

Proof. From

$$\|x_m^{-1} - x_n^{-1}\| = \|x_m^{-1}(x_n - x_m)x_n^{-1}\| < M^2 \|x_n - x_m\|$$

we conclude that also $(x_n^{-1})_{n=1}^\infty$ is a Cauchy sequence in $\mathbf{B}$. Denote its limit by y. Then we have $x_n \to x$ and $x_n^{-1} \to y$ as $n \to \infty$, and consequently, $e = x_n x_n^{-1} \to xy$ and $e = x_n^{-1} x_n \to yx$ which shows that $y = x^{-1}$. Clearly, $\|y\| \le M$. □

Proposition 3.3. *Let $A \in \mathcal{L}(L^\infty)$, and suppose a limit operator A_h of A is invertible. Then for every function $\xi \in L^\infty$ with bounded support there exists an m_0 such that for every $m > m_0$ there are operators B_m and C_m with $\|B_m\|, \|C_m\| < 2\|A_h^{-1}\|$ and*

$$B_m A M_{V_{h_m}\xi} = M_{V_{h_m}\xi} = M_{V_{h_m}\xi} A C_m.$$

Proof. Take such a function ξ, and put $U := \operatorname{supp}\xi$. Then

$$V_{-h_m} A V_{h_m} P_U = A_h P_U + T_m$$

with some sequence $T_m \rightrightarrows 0$ as $m \to \infty$. Multiplying the latter from the right by $M_\xi V_{-h_m}$ and from the left by A_h^{-1}, gives

$$A_h^{-1} V_{-h_m} A V_{h_m} M_\xi V_{-h_m} = (I + A_h^{-1} T_m) M_\xi V_{-h_m}. \tag{3.1}$$

Now take m_0 large enough that $\|A_h^{-1} T_m\| < 1/2 \ \forall m > m_0$. Then $I + A_h^{-1} T_m$ is invertible with $\|(I + A_h^{-1} T_m)^{-1}\| < 2$. Finally, multiply (3.1) from the left by $V_{h_m}(I + A_h^{-1} T_m)^{-1}$, and consider that $V_{h_m} M_\xi V_{-h_m} = M_{V_{h_m}\xi}$ to see that

$$\underbrace{V_{h_m}(I + A_h^{-1} T_m)^{-1} A_h^{-1} V_{-h_m}}_{=:\, B_m} A M_{V_{h_m}\xi} = M_{V_{h_m}\xi}$$

which is our first claim. The second one is shown completely symmetric. □

We will now define a set of continuous functions, each of them being supported in a small cube only, with the property that their sum is 1 everywhere in $\mathbb{R}^n$. Such a set will be regarded as a partition of unity.

By $\theta : \mathbb{R} \to [0, 1]$ we denote the continuous function

$$\theta(x) := \begin{cases} 3x + 2 & \text{for } x \in [-2/3, -1/3) \\ 1 & \text{for } x \in [-1/3, 1/3] \\ -3x + 2 & \text{for } x \in (1/3, 2/3) \\ 0 & \text{otherwise} \end{cases} \tag{3.2}$$

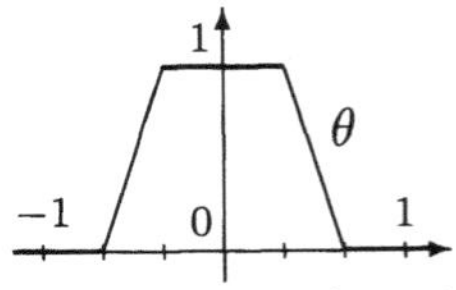

with support in $[-2/3, 2/3]$. Let $\omega : \mathbb{R}^n \to [0, 1]$ be the continuous function with support in $[-2/3, 2/3]^n$ which is given by

$$\omega(x_1, \ldots, x_n) := \sqrt{\theta(|(x_1, \ldots, x_n)|)}.$$

For $\alpha \in \mathbb{Z}^n$, we put $\omega_\alpha(x) := \omega(x - \alpha) \ \forall x \in \mathbb{R}^n$. Note that ω is chosen such that

$$\sum_{\alpha \in \mathbb{Z}^n} \omega_\alpha^2(x) = 1 \qquad \forall x \in \mathbb{R}^n$$

holds. In this sense we will say that $\{\omega_\alpha^2\}_\alpha$ is a partition of unity. Note that for every x, the latter sum is actually a finite sum.

For every $R > 0$, we define the function $\omega_{\alpha,R}(x) := \omega_\alpha(x/R) = \omega(x/R - \alpha)$ $\forall x \in \mathbb{R}^n$. Obviously $\{\omega_{\alpha,R}^2\}_\alpha$ is a partition of unity as well for every $R > 0$.

Finally, let $\psi : \mathbb{R}^n \to [0, 1]$ be some continuous function with bounded support which is equal to 1 on $[-3/4, 3/4]^n$ (and hence on the whole support of ω). In analogy to ω_α and $\omega_{\alpha,R}$ we define ψ_α and $\psi_{\alpha,R}$, respectively.

Note that the functions ω and ψ are subject to the conditions on φ in Lemma 2.6 with $N_\omega, N_\psi < 2^n$ (the number of adjacent cubes in $\mathbb{R}^n$). We will make use of this in the proof of the following proposition.

Proposition 3.4. *Let A be some operator in $\mathcal{B}_{\mathcal{S}}$. If there is some $M > 0$ such that for every $R > 0$ there exists a $\rho(R)$ such that for every $\alpha \in \mathbb{Z}^n$ with $|\alpha| > \rho(R)$ there are two operators $B_{\alpha,R}$ and $C_{\alpha,R}$ with $\|B_{\alpha,R}\|, \|C_{\alpha,R}\| < M$ and*

$$B_{\alpha,R}AM_{\omega_{\alpha,R}} \;=\; M_{\omega_{\alpha,R}} \;=\; M_{\omega_{\alpha,R}}AC_{\alpha,R}, \tag{3.3}$$

then A is invertible at infinity.

Proof. Our proof requires some elementary geometry: We will interpret the $\mathbb{R}^n$ as a puzzle of infinitely many (hyper-)cubes whose size increases with their distance from the origin. We will choose this decomposition such that, by our premise, on every one of these cubes, a left and a right inverse to A exist, and by puzzling these operators together (see Lemma 2.6) we will get some left and right regularizer for A at infinity.

The width of our cubes will run through the set of positive odd numbers $U := \{1, 3, 5, \dots\}$. To provide us with a frame for this "cubic puzzle", we will in a first step think of an infinite sequence of increasing concentric (centered in 0) cubes $C_{-1} \subset C_1 \subset C_3 \subset C_5 \subset \cdots$, reaching every point of $\mathbb{R}^n$.

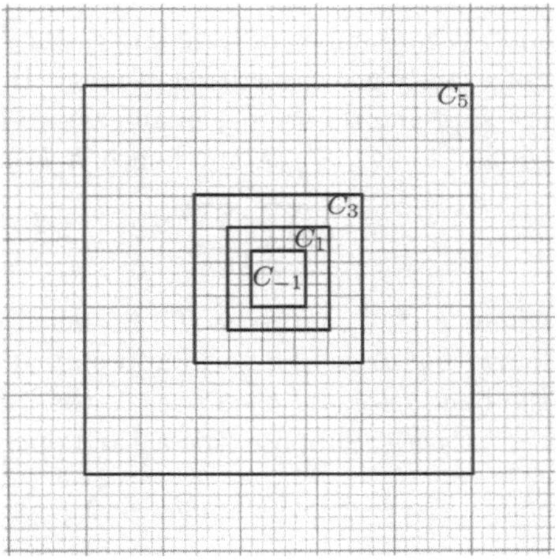

The choice of this sequence shall be such that for every $k \in U$, firstly, the width w_k of cube C_k is an odd multiple of $k(k+2)$, and secondly, the inequality

$$w_{k-2}/2 \geq k\rho(k)$$

holds.

Our first constraint guarantees that for every $k \in U$, both, w_k and w_{k-2} are divisible by k, and hence, both, C_k and C_{k-2} can be puzzled by cubes of width k. But what we actually do then is puzzling **their difference** $C_k \setminus C_{k-2}$ with cubes of width k. Doing this for all $k \in U$, gives us the proposed (disjoint up to sets of measure zero) decomposition of $\mathbb{R}^n \setminus C_{-1}$ into cubes of increasing width.

Our second constraint yields that from $k\alpha \in C_k \setminus C_{k-2}$ for $\alpha \in \mathbb{Z}^n$, we can conclude $|\alpha| > \rho(k)$, and hence, the existence of operators $B_{\alpha,k}$, $C_{\alpha,k}$ with (3.3).

For every $k \in U$, we put $D_k := \{\alpha \in \mathbb{Z}^n : k\alpha \in C_k \setminus C_{k-2}\}$, whence $\{\omega^2_{\alpha,k}\}_{\alpha \in D_k}$ is a partition of unity on $C_k \setminus C_{k-2}$ with exception of some points near the boundary of $C_k \setminus C_{k-2}$. Consequently, the set $\bigcup_{k\in U}\{\omega^2_{\alpha,k}\}_{\alpha\in D_k}$ is a partition of unity on $\mathbb{R}^n \setminus C_{-1}$ with exception of the boundary of every cube C_k and of some points in the near, i.e., where cubes of different width meet one another.

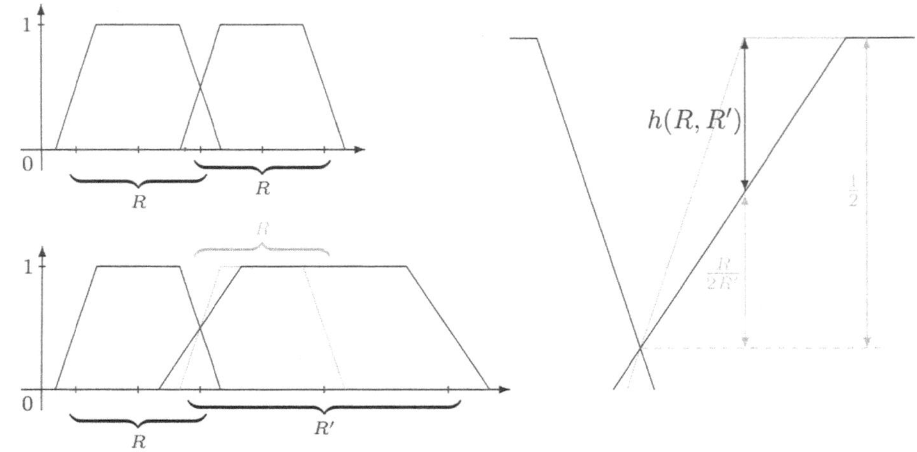

Without loss of generality, we study the situation for the case $n = 1$. Two copies of the function ω^2, stretched and shifted onto adjacent intervals ($\mathbb{R}^1$-cubes) of the same and of different width. A zoom into the second picture shows how much this situation differs from a partition of unity.

Where cubes of different width $R = k$ and $R' = k + 2$ are adjacent, we do not exactly get a partition of unity. But we are only $h(k, k + 2)$ away from it – in the supremum norm. It is an easy computation to see that the distance $h(R, R')$ equals $(1 - R/R')/2$, and hence, $h(k, k+2)$ tends to zero as $k \to \infty$. Consequently,

$$\sum_{k\in U} \sum_{\alpha\in D_k} M_{\omega^2_{\alpha,k}} = I - \sum_{\alpha\in C_{-1}} M_{\omega^2_\alpha} + T = I + T', \tag{3.4}$$

where $T \in \mathcal{J}$, and so is T'. So much about geometry. From $A \in \mathcal{B}_\mathcal{S} \subset$ BDO we know that A can be norm approximated by band operators. Let us first deal with the band case and then pass to the limit.

Case 1. Suppose A is a band operator. We present left and right regularizers:

$$B := \sum_{k\in U} \sum_{\alpha\in D_k} M_{\omega_{\alpha,k}} B_{\alpha,k} M_{\omega_{\alpha,k}}$$

is well defined by Lemma 2.6 with $\|B\| \le N_\omega M$. Then

$$\begin{aligned} BA &= \sum_{k\in U} \sum_{\alpha\in D_k} M_{\omega_{\alpha,k}} B_{\alpha,k} M_{\omega_{\alpha,k}} A \\ &= \sum_{k\in U} \sum_{\alpha\in D_k} M_{\omega_{\alpha,k}} \underbrace{B_{\alpha,k} A M_{\omega_{\alpha,k}}}_{M_{\omega_{\alpha,k}}} + \sum_{k\in U} \underbrace{\sum_{\alpha\in D_k} M_{\omega_{\alpha,R}} B_{\alpha,R} [M_{\omega_{\alpha,R}}, A]}_{=:T_k}. \end{aligned}$$

So $BA = I + T' + \sum_{k\in U} T_k$, where $T' \in \mathcal{J}$ is the operator from equation (3.4), and by Lemma 2.6, we have $\|T_k\| \le N_\omega \sup_{\alpha\in D_k} M\|[M_{\omega_{\alpha,k}}, A]\|$. From $A \in \mathcal{B}$ we conclude that $T_k \rightrightarrows 0$ as $k \to \infty$, and, again thinking of Lemma 2.6, $\sum_{k\in U} T_k \in \mathcal{J}$ which shows that B is indeed a left regularizer for A in terms of invertibility at infinity.

Recall that A is a band operator. By definition of ψ and ω, the distance of $\operatorname{supp}(1-\psi_{\alpha,k})$ and $\operatorname{supp}\omega_{\alpha,k}$ tends to infinity as $k \to \infty$. Proposition 2.7 a) tells us that $M_{1-\psi_{\alpha,k}}AM_{\omega_{\alpha,k}} = 0$ if k is large enough, say $k > k_0$, and hence, $AM_{\omega_{\alpha,k}} = M_{\psi_{\alpha,k}}AM_{\omega_{\alpha,k}}$ if $k > k_0$. Put $U_0 := \{k \in U : k > k_0\}$. Then

$$C := \sum_{k\in U_0} \sum_{\alpha\in D_k} M_{\omega_{\alpha,k}} C_{\alpha,k} M_{\omega_{\alpha,k}}$$

is a right regularizer for A with $\|C\| \leq N_\omega M$. Indeed,

$$\begin{aligned} AC &= \sum_{k\in U_0} \sum_{\alpha\in D_k} M_{\psi_{\alpha,k}} AM_{\omega_{\alpha,k}} C_{\alpha,k} M_{\omega_{\alpha,k}} \\ &= \sum_{k\in U_0} \sum_{\alpha\in D_k} M_{\psi_{\alpha,k}} \underbrace{M_{\omega_{\alpha,k}} AC_{\alpha,k}}_{M_{\omega_{\alpha,k}}} M_{\omega_{\alpha,k}} \\ &\quad + \sum_{k\in U_0} \underbrace{\sum_{\alpha\in D_k} M_{\psi_{\alpha,k}} [A, M_{\omega_{\alpha,k}}] C_{\alpha,k} M_{\omega_{\alpha,k}}}_{=:T_k} . \end{aligned}$$

So $AC = I + T'' + \sum_{k\in U_0} T_k$ with $T'' \in \mathcal{J}$ and $T_k \rightrightarrows 0$. And as in the left-sided part, we see that this right-hand side is of the form $I + T$ with some $T \in \mathcal{J}$.

Case 2. Now suppose A is the norm limit of a sequence of band operators $(A_m)_{m=1}^\infty$ with $\|A - A_m\| < 1/m$. Moreover, suppose A is subject to the condition culminating in equation (3.3). We will show that then every operator A_m with $m > M$ is subject to this condition as well, but with M replaced by $M/(1-M/m)$. Indeed, if $B_{\alpha,R}$ is such that $\|B_{\alpha,R}\| \leq M$ and $B_{\alpha,R}AM_{\omega_{\alpha,R}} = M_{\omega_{\alpha,R}}$, then

$$\begin{aligned} B_{\alpha,R}A_m M_{\omega_{\alpha,R}} &= B_{\alpha,R}AM_{\omega_{\alpha,R}} - B_{\alpha,R}(A - A_m)M_{\omega_{\alpha,R}} \\ &= M_{\omega_{\alpha,R}} - B_{\alpha,R}(A - A_m)M_{\omega_{\alpha,R}} \\ &= (I - B_{\alpha,R}(A - A_m))M_{\omega_{\alpha,R}}, \end{aligned}$$

where $D := I - B_{\alpha,R}(A - A_m)$ is invertible since $\|B_{\alpha,R}(A-A_m)\| \leq M \cdot 1/m < 1$, and $\|D^{-1}\| \leq 1/(1-M/m)$. Now it is easy to see that A_m is indeed subject to (3.3) with $B_{\alpha,R}^{(m)} := D^{-1}B_{\alpha,R}$ and $M/(1 - M/m)$ in place of $B_{\alpha,R}$ and M, respectively. The construction of local right inverses is analogous.

By case 1, we know that then all (band-) operators A_m are invertible at infinity with the norm of their regularizers being not larger than $N_\omega M/(1-M/m)$. If $m > 2M$, all these norms are collectively bounded by $N_\omega M/(1 - M/(2M)) = 2N_\omega M$. By Corollary 2.21, all cosets $A_m + \mathcal{J}$ with $m > 2M$ are invertible in $\mathcal{B}/\mathcal{J}$ with $\|(A_m + \mathcal{J})^{-1}\|_{\mathcal{B}/\mathcal{J}} \leq 2N_\omega M$. Lemma 3.2 now guarantees that the norm limit $A + \mathcal{J}$ is invertible in $\mathcal{B}/\mathcal{J}$, i.e., A is invertible at infinity. □

Now we are ready for the

Proof of Theorem 1.1

($\Longrightarrow$) Suppose A is invertible at infinity. Then there is some $B \in \mathcal{L}(L^\infty)$ such that $I = BA + T_1$ and $I = AB + T_2$ with $T_1, T_2 \in \mathcal{J}$. For arbitrary elements $u \in L^\infty$ and $v \in (L^\infty)'$ (passing to adjoints in the second equation), this yields to

$$\begin{aligned} \|u\|_\infty &\le \|B\|\|Au\|_\infty + \|T_1 u\|_\infty \\ \|v\|_{\infty'} &\le \|B^*\|\|A^*v\|_{\infty'} + \|T_2^* v\|_{\infty'}. \end{aligned}$$

Restricting v to $L^1 \subset (L^\infty)'$ and putting $C := \max(\|B\|, 1)$, we get that the norm inequalities

$$\|u\|_\infty \le C(\|Au\|_\infty + \|T_1 u\|_\infty) \tag{3.5}$$

$$\|v\|_1 \le C(\|{}^*Av\|_1 + \|T_2^* v\|_{\infty'}) \tag{3.6}$$

hold for all $u \in L^\infty$ and $v \in L^1$.

Let $h = (h_m) \subset \mathbb{Z}^n$ be an arbitrary sequence tending to infinity and leading to a limit operator A_h of A. We will show that A_h is invertible.

Since V_α is an isometry, for every $\alpha \in \mathbb{Z}^n$, we get from (3.5) that

$$\begin{aligned} \|u\|_\infty = \|V_{h_m} u\|_\infty &\le C(\|AV_{h_m} u\|_\infty + \|T_1 V_{h_m} u\|_\infty) \\ &= C(\|V_{-h_m} A V_{h_m} u\|_\infty + \|V_{-h_m} T_1 V_{h_m} u\|_\infty) \quad \forall u \in L^\infty. \end{aligned}$$

For the function ψ from above and every $R > 0$, we define the function ψ_R by $\psi_R(x) := \psi(x/R)\ \forall x \in \mathbb{R}^n$.

Replacing u in the last inequality by the function $M_{\psi_R} u$, and doing all the same steps with inequality (3.6), we get that

$$\begin{aligned} \|M_{\psi_R} u\|_\infty &\le C(\|V_{-h_m} A V_{h_m} M_{\psi_R} u\|_\infty + \|V_{-h_m} T_1 V_{h_m} M_{\psi_R} u\|_\infty) \\ \|M_{\psi_R} v\|_1 &\le C(\|V_{-h_m} {}^*A V_{h_m} M_{\psi_R} v\|_1 + \|V_{-h_m} T_2^* V_{h_m} M_{\psi_R} v\|_{\infty'}) \end{aligned}$$

are true for all $R > 0$, all $u \in L^\infty$ and all $v \in L^1$.

From $(V_{-h_m} A V_{h_m} - A_h) M_{\psi_R} \rightrightarrows 0$, Proposition 2.35 b), $V_{-h_m} T_1 V_{h_m} M_{\psi_R} \rightrightarrows 0$ and $\|V_{-h_m} T_2^* V_{h_m} M_{\psi_R}\|_{\infty'} = \|(M_{\psi_R} V_{-h_m} T_2 V_{h_m})^*\|_{\infty'} = \|M_{\psi_R} V_{-h_m} T_2 V_{h_m}\|_\infty \to 0$ as $m \to \infty$ we conclude

$$\begin{aligned} \|M_{\psi_R} u\|_\infty &\le C\|A_h M_{\psi_R} u\|_\infty \le C(\|M_{\psi_R} A_h u\|_\infty + \|[A_h\,,\, M_{\psi_R}]u\|_\infty) \\ \|M_{\psi_R} v\|_1 &\le C\|{}^*A_h M_{\psi_R} v\|_1 \le C(\|M_{\psi_R} {}^*A_h v\|_1 + \|[{}^*A_h\,,\, M_{\psi_R}]v\|_1). \end{aligned}$$

Since $A_h \in \mathcal{B}$ and $[{}^*A_h\,,\, M_{\psi_R}] = {}^*[M_{\psi_R}\,,\, A_h]$ holds, letting $R \to \infty$ leads to

$$\begin{aligned} \|u\|_\infty &\le C\|A_h u\|_\infty \qquad \forall u \in L^\infty \\ \|v\|_1 &\le C\|{}^*A_h v\|_1 \qquad \forall v \in L^1. \end{aligned}$$

Taking this together, we see that ${}^*A_h$ and $A_h = ({}^*A_h)^*$ are injective, whence ${}^*A_h$ is invertible. Consequently, A_h is invertible and $\|A_h^{-1}\| \le C$, where C does not depend on the sequence h.

($\Longleftarrow$) Let $\sigma^{\mathrm{op}}(A)$, the set of limit operators of A, be uniformly invertible, and suppose A were not invertible at infinity.

Then by Proposition 3.4, for every $M > 0$, and hence, for

$$M := 2 \sup_{A_h \in \sigma^{\mathrm{op}}(A)} \|A_h^{-1}\|,$$

there is an $R > 0$ such that for all ρ, there exists an $\alpha \in \mathbb{Z}^n$ with $|\alpha| > \rho$ such that

$$BAM_{\omega_{\alpha,R}} \neq M_{\omega_{\alpha,R}}$$

for all operators B with $\|B\| < M$. This means there is a sequence $(\alpha_k) \subseteq \mathbb{Z}^n$ tending to infinity such that for **no** $k \in \mathbb{N}$ there is an operator B_k with $\|B_k\| < M$ for which

$$B_k AM_{\omega_{\alpha_k,R}} = M_{\omega_{\alpha_k,R}} \tag{3.7}$$

holds.

Since $A \in \mathcal{B}_{\$}$, there is a subsequence g of (α_k) such that A_g exists. But $A_g \in \sigma^{\mathrm{op}}(A)$ is invertible by our premise. And so Proposition 3.3 (putting $\xi := \omega_{0,R}$) says that for infinitely many k, there is an operator B_k with $\|B_k\| < M$ that is subject to equation (3.7). This is clearly a contradiction. □

Remark 3.5 Note that for the proof of the ($\Longrightarrow$) part, we only needed that $A \in \mathcal{B}_S$, while for the reverse direction the richness of A is involved. □

It is sort of "the big question in limit operator business" whether in Theorem 1.1 **uniform** invertibility of $\sigma^{\mathrm{op}}(A)$ can be replaced by **elementwise** invertibility! The answer has proven to be "Yes" for some classes of operators including operators in the Wiener algebra, for example. For new achievements on this question see [15]!

4. An outlook to applications

4.1. A criterion on Fredholmness

As we already mentioned, Fredholmness and invertibility at infinity are two closely related properties. We will prove a Fredholm criterion in a suitable subalgebra of $\mathcal{L}(L^\infty)$, showing that invertibility at infinity plus some other property (which is readily verified) yields Fredholmness.

4.1.1. Localization Let $\dot{\mathbb{R}}^n$ refer to the one-point compactification $\mathbb{R}^n \cup \{\infty\}$ of $\mathbb{R}^n$. By Λ we will denote the set of all operators $A \in \mathcal{L}(L^\infty)$ for which all commutators

$$[A, M_f], \qquad f \in C(\dot{\mathbb{R}}^n)$$

are compact. Clearly, Λ is an inverse closed Banach subalgebra of $\mathcal{L}(L^\infty)$, and the compact operators $\mathcal{K}$ form a closed two-sided ideal in Λ. For simplicity, we abbreviate the factor algebra $\Lambda/\mathcal{K}$ by $\Lambda^{\mathcal{K}}$ and the coset $A + \mathcal{K}$ by $A^{\mathcal{K}}$ for every $A \in \Lambda$.

If $A \in \Lambda$ is Fredholm, then there are $B \in \mathcal{L}(L^\infty)$ and $T_1, T_2 \in \mathcal{K}$ such that

$$AB = I + T_1 \qquad \text{and} \qquad BA = I + T_2$$

hold.

In this case, B is in Λ as well. Indeed, for every $f \in C(\dot{\mathbb{R}}^n)$, we have

$$\begin{aligned}[B\,,\,M_f] &= BM_f(I+T_1) - (I+T_2)M_fB - BM_fT_1 + T_2M_fB \\ &= BM_fAB - BAM_fB - BM_fT_1 + T_2M_fB \\ &= -B[A\,,\,M_f]B - BM_fT_1 + T_2M_fB \in \mathcal{K}.\end{aligned}$$

Consequently, $A \in \Lambda$ is Fredholm if and only if $A^{\mathcal{K}}$ is invertible in $\Lambda^{\mathcal{K}}$. We will study invertibility in $\Lambda^{\mathcal{K}}$ by means of ALLANs local principle.

Therefore, note that $\{M_f^{\mathcal{K}} \,:\, f \in C(\dot{\mathbb{R}}^n)\}$ is a central subalgebra of $\Lambda^{\mathcal{K}}$. Its maximal ideal space can be identified with $\dot{\mathbb{R}}^n$, where the maximal ideal M_x corresponding with $x \in \dot{\mathbb{R}}^n$ consists of all $M_f^{\mathcal{K}}$ for which $f(x) = 0$. Moreover, put $J_x := \text{closid}_{\Lambda^{\mathcal{K}}} M_x$ for every $x \in \dot{\mathbb{R}}^n$. We abbreviate the so-called *local algebras* $\Lambda^{\mathcal{K}}/J_x$ by $\Lambda_x^{\mathcal{K}}$ and the so-called *local representatives* $A^{\mathcal{K}} + J_x$ by $A_x^{\mathcal{K}}$ for every $A \in \Lambda$ and $x \in \dot{\mathbb{R}}^n$. ALLANs local principle [1, Theorem 1.34] now says that

Proposition 4.1. *$A \in \Lambda$ is Fredholm if and only if all local representatives $A_x^{\mathcal{K}}$ are invertible in their local algebra $\Lambda_x^{\mathcal{K}}$ for all $x \in \dot{\mathbb{R}}^n$.*

4.1.2. AN ALGEBRA OF CONVOLUTION AND MULTIPLICATION OPERATORS We will evaluate this Fredholm criterion in a concrete subalgebra of Λ. For a function $k \in L^1$, let Fk refer to its Fourier transform

$$(Fk)(x) = \int_{\mathbb{R}^n} k(t)e^{i(t,x)}\,dt,$$

and denote the set of functions $\{Fk : k \in L^1\}$ by FL^1.

With every function $a = Fk$ we associate a *convolution operator* $\mathring{W}_a \in \mathcal{L}(L^\infty)$

$$(\mathring{W}_a\varphi)(t) := \int_{\mathbb{R}^n} k(t-s)\varphi(s)\,ds,$$

and say that the function a is the *symbol* of the operator $\mathring{W}_a$.

Besides convolution operators we are also interested in operators of multiplication. So let $\mathcal{A}$ denote the smallest closed subalgebra of $\mathcal{L}(L^\infty)$ containing all $\mathring{W}_a$ with $a \in FL^1$ and all M_b with $b \in L^\infty$:

$$\mathcal{A} := \text{alg}_{\mathcal{L}(L^\infty)}\Big\{\mathring{W}_a,\ M_b \;:\; a \in FL^1,\ b \in L^\infty\Big\}.$$

Lemma 4.2. *If $a = Fk$ with some $k \in L^1$ and $U \subset \mathbb{R}^n$ is bounded and measurable, then $P_U\mathring{W}_a$ and $\mathring{W}_aP_U$ are compact operators.*

Proof. Since k can be approximated in L^1 as good as desired by some compactly supported continuous function l, it suffices to prove our claim with l in place of k.

Let L denote some compact which contains U as well as the support of l. Then $P_L\mathring{W}_{Fl}$ is an integral operator with a continuous kernel function supported in $(2L) \times L$. But such operators are compact, and hence,

$$P_U\mathring{W}_{Fl} = P_U(P_L\mathring{W}_{Fl})$$

is compact as well. The proof of the second claim is completely symmetric. □

Corollary 4.3. *If $a \in FL^1$ and $b \in C(\dot{\mathbb{R}}^n)$, then $M_b\mathring{W}_a$ and $\mathring{W}_a M_b$ are compact.*

Proof. Take $U := mC$, and write $M_b\mathring{W}_a = M_b(P_U\mathring{W}_a) + (M_bQ_U)\mathring{W}_a$, where the first term is compact and the second one tends to zero as $m \to \infty$. □

This fact has remarkable consequences as we will see. Every element of $\mathcal{A}$ is of the form $A = \sum_{i=1}^{k} A_i$ or a norm-limit of such, where A_i are finite products of convolution and multiplication operators. Some of these A_i might be pure multiplication operators, say $A_j, \dots, A_k$. Their sum is again a multiplication operator whose symbol shall be denoted by b^A. So

$$A = M_{b^A} + \sum_{i=1}^{j-1} A_i. \tag{4.1}$$

Clearly, for finite sum-products A, this decomposition is unique. But we will show that this is also true for arbitrary $A \in \mathcal{A}$.

Proposition 4.4. $\mathcal{A} = \{M_b : b \in L^\infty\} \dotplus \operatorname{closid}_{\mathcal{A}}\{\mathring{W}_a : a \in FL^1\}$.

Proof. Clearly, we here have two closed subalgebras of $\mathcal{A}$. Suppose A is contained in both. Then, for every b&m set $U \subset \mathbb{R}^n$, the operator AP_U is compact by Lemma 4.2. This implies $b = 0$, and hence, $A = 0$.
Let us now show that the sum is all of $\mathcal{A}$. Every operator in the (dense) subset consisting of all finite sum-products can be written as such a sum by (4.1). It remains to show that the (linear) mapping $A \mapsto b^A$ is continuous.

To see this, take some $\varepsilon > 0$ and a b&m set $U \subset \mathbb{R}^n$ with positive measure, where $|b^A(u)| > \|b^A\|_\infty - \varepsilon$ for all $u \in U$. By (4.1) and Lemma 4.2, we have $AP_U = M_{b^A}P_U + K$ with $K \in \mathcal{K}$. Now

$$\|b^A\|_\infty - \varepsilon \;\le\; \|M_{b^A}P_U\| \;=\; \|(AP_U)^{\mathcal{K}}\|_{\Lambda^{\mathcal{K}}} \;\le\; \|A\|$$

for every $\varepsilon > 0$, and hence, $\|b^A\| \le \|A\|$. □

The continuous extension of the mapping $A \mapsto M_{b^A}$ will be denoted by $S \in \mathcal{L}(\mathcal{A})$. S is clearly a projector from $\mathcal{A}$ onto $\{M_b : b \in L^\infty\}$.

4.1.3. Fredholmness in the algebra $\mathcal{A}$ By Corollary 4.3, we have $\mathcal{A} \subset \Lambda$ since the generators of $\mathcal{A}$ are in Λ. Consequently, Proposition 4.1 holds for all $A \in \mathcal{A}$, leading us to the study of local representatives of the generators of $\mathcal{A}$. So suppose $a \in FL^1$ and $b \in L^\infty$. We start at the finite points x: Take some $x \in \mathbb{R}^n$, and let φ denote a continuous function with compact support and $\varphi(x) = 1$.

By Corollary 4.3, $M_\varphi\mathring{W}_a$ is compact. Consequently,

$$\left(\mathring{W}_a^{\mathcal{K}}\right)_x = \left((M_\varphi\mathring{W}_a)^{\mathcal{K}}\right)_x = \left(0^{\mathcal{K}}\right)_x = 0.$$

Analogously, we see that $(M_b^{\mathcal{K}})_x$ is invertible in $\Lambda_x^{\mathcal{K}}$ if and only if 0 is not contained in the local essential range of b in x. So for finite points x, we can conclude from $A = S(A) + W$ with $W \in \operatorname{closid}_{\mathcal{A}}\{\mathring{W}_a : a \in FL^1\}$ to

$$A_x^{\mathcal{K}} = S(A)_x^{\mathcal{K}} + W_x^{\mathcal{K}} = S(A)_x^{\mathcal{K}} + 0_x^{\mathcal{K}} = S(A)_x^{\mathcal{K}}.$$

Summarizing, we can say that $A_x^{\mathcal{K}}$ is invertible for all $x \in \mathbb{R}^n$ if and only if $S(A) = M_{b^A}$ is invertible, i.e., b^A has no zeros. So this was the simple part of our Fredholm criterion.

But what about the invertibility of $A_\infty^{\mathcal{K}}$?
We will show that this is guaranteed if A is invertible at infinity! Indeed, suppose there is some $B \in \mathcal{L}(L^\infty)$ and a b&m set $U \subset \mathbb{R}^n$ such that $Q_U AB = Q_U$. Write $A = S(A) + W$. Then

$$(Q_U A Q_U)(Q_U B Q_U) = Q_U A B Q_U - Q_U A P_U B Q_U = Q_U - Q_U K Q_U, \qquad (4.2)$$

where $Q_U S(A) P_U B Q_U = 0$, and $K := W P_U B$ is compact by Lemma 4.2. But (4.2) and its symmetric counterpart (derived from $B' A Q_U = Q_U$) say that the compression of A to $\operatorname{im} Q_U$ is Fredholm on $\operatorname{im} Q_U$. Studying this Fredholmness as in Section 4.1.1, we get the invertibility of $A_x^{\mathcal{K}}$ for points $x \in \dot{\mathbb{R}}^n \setminus U$, including that of $A_\infty^{\mathcal{K}}$. Involving Proposition 4.1, we get

Proposition 4.5. *If $A \in \mathcal{A}$ is invertible at infinity and the multiplication operator $S(A)$ is invertible, then A is Fredholm.*

4.2. Approximation methods and applicability

Another interesting application of the presented results is the answer to the question if some approximation method is applicable to an operator $A \in \mathcal{B}_\$$ or not. We restrict ourselves to presenting the idea. For a detailed description see [14].

Suppose the sequence $(A_\tau)_{\tau>0} \subset \mathrm{BDO}$ is $\mathcal{P}$-convergent to $A \in \mathrm{BDO}$. Then Roch and the second author proved in [23] that the approximation method (A_τ) is *applicable* to A, i.e., the sequence of solutions x_τ of

$$A_\tau x_\tau = b$$

$\mathcal{P}$-converges to the solution x of

$$Ax = b$$

as $\tau \to \infty$ for every right-hand side $b \in L^\infty$, if and only if A is invertible and (A_τ) is *stable*, the latter means that $\{A_\tau\}_{\tau>\tau_0}$ is uniformly invertible for some $\tau_0 > 0$.

So the important question is that about the stability of a given sequence of operators A_τ, each of them acting on $L^\infty(\mathbb{R}^n)$. We find the answer to that question by (temporarily) increasing the dimension of the problem, which goes like that:

Clearly, $\mathbb{R}^{n+1}$ is what results from stacking a continuum of copies of the $\mathbb{R}^n$ into the $(n+1)$th dimension. By the same idea, stacking the operators A_τ (put $A_\tau := I$ for $\tau \le 0$) into the $(n+1)$th dimension results in an operator on $L^\infty(\mathbb{R}^{n+1})$ which we will denote by $\oplus A_\tau$.

It can be shown that under fairly general assumptions on the smoothness of the mapping $\tau \mapsto A_\tau$ the sequence (A_τ) is stable if and only if $\oplus A_\tau$ is invertible at infinity! Now Theorem 1.1 links this to the uniform invertibility of $\sigma^{\mathrm{op}}(\oplus A_\tau)$.

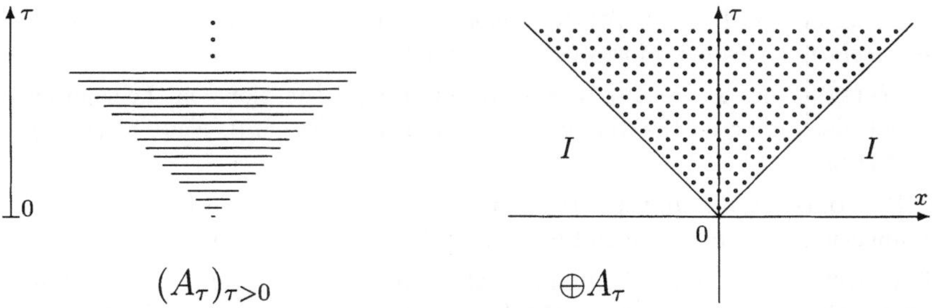

The finite section sequence $(A_\tau)_{\tau>0}$ and its stacked operator $\oplus A_\tau$ for $n = 1$ and $\Omega = [-1, 1]$.

In the discrete case, the observation that the applicability of some approximation method is equivalent to the Fredholm property of an associated operator on a space of functions on $\mathbb{R}^{n+1}$, was first made for the finite section method of Toeplitz operators with homogenous symbol. For $n = 1$ this goes back to DOUGLAS and HOWE [2], and for arbitrary n it was proven by GORODETSKI [6].

Things get much more sophisticated in the function case, when stacking a continuum of operators. The method presented in [22] is connected with some rather drastic restrictions to the operator sequence. We will demonstrate a much more natural idea in [14] by studying what we will call the essential invertibility of sufficiently smooth sequences.

If (A_τ) is the finite section method for A, then we have a very clear description of all limit operators of $\oplus A_\tau$ which also allows us to pass back to the n-dimensional context and to operators that are related to A itself and to limit operators of A.

4.3. The system case

For simplicity, we restricted ourselves to operators on the space $L^\infty := L^\infty(\mathbb{R}^n, \mathbb{C})$ of all essentially bounded complex **scalar**-valued functions on $\mathbb{R}^n$. It is worth to be mentioned that all our investigations and results remain valid if we pass to **vector**-valued functions $L^\infty(\mathbb{R}^n, \mathbb{C}^N)$ with some integer N.

But as every function $f \in L^\infty(\mathbb{R}^n, \mathbb{C}^N)$ can be thought of as an N-vector $(f_1, \ldots, f_N)$ of functions $f_i \in L^\infty$, we can think of every operator A on $L^\infty(\mathbb{R}^n, \mathbb{C}^N)$ as an $N \times N$-matrix whose entries A_{ij} are operators on L^∞.

This is what we usually call "the *system case*", and we would like to emphasize that consequently, also this system case is covered by our results.

References

[1] A. BÖTTCHER and B. SILBERMANN: *Analysis of Toeplitz Operators*, Akademie-Verlag, Berlin, 1989 and Springer Verlag, Berlin, Heidelberg, New York 1990.

[2] R.G. DOUGLAS and R. HOWE: On the C^*-algebra of Toeplitz operators on the quarter plane. *Trans. Amer. Math. Soc.* **158** (1971), 203–217.

[3] J.A. FAVARD: Sur les équations differentiales a coefficients presque périodiques, *Acta Math.* **51** (1927), 31–81.

[4] I. GOHBERG and I.A. FELDMAN: *Convolution equations and projection methods for their solutions*, Nauka, Moskva 1971 (Russian).

[5] I. GOHBERG and M.G. KREIN: Systems of integral equations on the semi-axis with kernels depending on the difference of arguments, *Usp. Mat. Nauk* **13** (1958), no. 5, 3–72 (Russian).

[6] M.B. GORODETSKI: On the Fredholm theory and the finite section method for multidimensional discrete convolutions, *Sov. Math.* **25** (1981), no. 4, 9–12.

[7] R. HAGEN, S. ROCH and B. SILBERMANN: *Spectral Theory of Approximation Methods for Convolution Equations*, Birkhäuser Verlag, Basel, Boston, Berlin 1995.

[8] Y. KATZNELSON: *An Introduction to Harmonic Analysis*, New York - London - Sydney - Toronto, 1968.

[9] M.G. KREIN: Integral equations on the semi-axis with kernels depending on the difference of arguments, *Usp. Mat. Nauk* **13** (1958), no. 2, 3–120 (Russian).

[10] V.G. KURBATOV: *Functional Differential Operators and Equations*, Kluwer Academic Publishers, Dordrecht, Boston, London 1999.

[11] B.V. LANGE and V.S. RABINOVICH: On the Noether property of multidimensional discrete convolutions, *Mat. Zametki* **37** (1985), no. 3, 407–421 (Russian).

[12] B.V. LANGE and V.S. RABINOVICH: On Noetherian multidimensional convolution operators with bounded measurable coefficients, *Izv. VUZ Matemat.* **29** (1985), no. 6, 22–30 (Russian).

[13] B.V. LANGE and V.S. RABINOVICH: Pseudo-differential operators on $\mathbb{R}^n$ and limit operators, *Mat. Sbornik* **129** (1986), no. 2, 175–185 (Russian).

[14] M. LINDNER: The Finite Section Method in the space of essentially bounded functions. An approach using Limit Operators., *(submitted to) Functional Analysis and Optimization* (2003).

[15] M. LINDNER: Contours at the horizon: Limit operators, *(submitted to) Zeitschrift für Analysis und ihre Anwendungen* (2003).

[16] E.M. MUHAMADIEV: On normal solvability and Noether property of elliptic operators in spacesof functions on $\mathbb{R}^n$, Part I: *Zapiski nauchnih sem. LOMI* **110** (1981), 120–140 (Russian).

[17] E.M. MUHAMADIEV: On normal solvability and Noether property of elliptic operators in spacesof functions on $\mathbb{R}^n$, Part II: *Zapiski nauchnih sem. LOMI* **138** (1985), 108–126 (Russian).

[18] S. PRÖSSDORF and B. SILBERMANN: *Numerical Analysis for Integral and Related Operator Equations*, Akademie-Verlag, Berlin, 1991 and Birkhäuser Verlag, Basel, Boston, Berlin 1991.

[19] V.S. RABINOVICH: Fredholmness of pseudo-differential operators on $\mathbb{R}^n$ in the scale of $L_{p,q}$-spaces. *Siberian Math. J.* **29** (1988), no. 4, 635–646 (Russian).

[20] V.S. RABINOVICH: Operator-valued discrete convolutions and some of its applications, *Mat. Zametki* **51** (1992), no. 5, 90–101 (Russian).

[21] V.S. RABINOVICH, S. ROCH and B. SILBERMANN: Fredholm Theory and Finite Section Method for Band-dominated Operators, *Integral Equations Operator Theory* **30** (1998), no. 4, 452–495.

[22] V.S. Rabinovich, S. Roch and B. Silbermann: Band-dominated operators with operator-valued coefficients, their Fredholm properties and finite sections, *Integral Equations Operator Theory* **40** (2001), no. 3, 342–381.

[23] S. Roch and B. Silbermann: Non-strongly converging approximation methods, *Demonstratio Math.* **22** (1989), no. 3, 651–676.

[24] H.H. Schaefer: *Topological vector spaces*, MacMillan Company New York, Collier-MacMillan Ltd., London 1966.

[25] I.B. Simonenko: Operators of convolution type in cones, *Mat. Sb.* **74** (116),(1967), 298–313 (Russian).

[26] I.B. Simonenko: On multidimensional discrete convolutions, *Mat. Issled.* **3** (1968), no. 1, 108–127 (Russian).

[27] E.M. Stein and G. Weiss: *Introduction to Fourier Analysis on Euclidean Spaces*, Princeton University Press, 1971.

[28] N. Wiener and E. Hopf: Über eine Klasse singulärer Integralgleichungen, *Sitzungsberichte Akad. Wiss. Berlin* (1931), 696–706.

Marko Lindner and Bernd Silbermann
Fakultät für Mathematik
TU Chemnitz
D-09107 Chemnitz
Germany
e-mail: Marko.Lindner@Mathematik.TU-Chemnitz.de
e-mail: Bernd.Silbermann@Mathematik.TU-Chemnitz.de

Operator Theory
Advances and Applications, Vol. 147, 327–352

Half-Plane Diffraction Problems Using Alternative Integral Representations

Ernst Lüneburg

Dedicated to the memory of Erhard Meister

Abstract. This tutorial review presents the formulation of some diffraction problems involving half-plane structures as Riemann-Hilbert problems. The formulations of the problems of diffraction by an infinite set of Rawlins's type half-planes, a bifurcated and a trifurcated waveguide and by a special infinite strip grating are considered in detail.

1. Introduction

The scattering of electromagnetic waves by edges and corners of obstacles whose characteristic dimensions are large compared to the wavelength of the illuminating wave can be considered as a local phenomenon that involves a region of several wavelength in the vicinity of the geometrical discontinuities. This implies that the actual shape of the scatterer can be substituted by simplified structures such as half-planes, wedges or cones that locally agree with the discontinuities of complicated object under study but which are easier to solve as a regular boundary value problem of mathematical physics. Material properties of the scatterer are taken into account by properly defining and adapting higher order impedance boundary conditions, see Senior and Volakis [59]. Historical aspects of the theory of diffraction by half-planes are considered in a contribution of the author to the Sommerfeld Memorial Conference [30].

The first systematic method of solution of the so-called Sommerfeld half-plane problem [61] was the application of the Wiener-Hopf method by Schwinger and by Copson in the fortieth. In the following decades this method has been applied to a large variety of physical and engineering problems involving half-plane structures. These developments are summarized in various textbooks as for instance by Noble [49], Weinstein [65], Mittra and Lee [48], Meister [41], Jones [23] and Estrada and Kanwal [13]. Especially Noble's monograph that appeared as early as 1958 became the standard reading text for any newcomer to this field at

2000 *Mathematics Subject Classification.* Primary 78; Secondary 47.
Key words and phrases. Diffraction, Half planes, Wiener-Hopf and Riemann-Hilbert problems.

least from the engineering community. All these textbooks deal with scalar Wiener-Hopf problems or with finite-dimensional vector Wiener-Hopf problems that can be reduced to scalar ones by elementary decoupling, cf. a series of papers by Meister [39, 40] and Igarashi [21] dealing with half-plane structures or for instance a truncated parallel plate waveguide, Weinstein [64]. An immediate extension of the scalar Wiener-Hopf factorization method to the vector case was suggested in 1950 by Heins [14], taking the logarithm of the Fourier symbol matrix function before applying the additive Cauchy integral decomposition. This approach, however, fails if the factor matrices do not commute (Baker-Campbell-Hausdorff formula). This restriction hampered further applications of the vector Wiener-Hopf technique for several decades.

It should be noted that there exist other methods that can be used to solve problems reducing to scalar Wiener-Hopf equations. These are among others the residue calculus technique, see for instance Whitehead [63], Mittra and Lee [48] and Lüneburg [28]), the direct inversion technique (see Mittra [47]) and the Riemann method, see Shestopalov [60]. We also mention a function-theoretic method due to Heins and MacCamy [15] that is applicable to some half-plane structures. This method has not been pursued further; its application to the strip (or slit) diffraction problem by Schmeltzer and Levin [58] is considered to be incorrect.

All these methods are more or less equivalent to the scalar Wiener-Hopf technique in the sense that if a problem could be solved by one method it could be solved also by the other methods (this statement may be valid only for problems in electromagnetic and acoustic scattering and has not been demonstrated in all cases). For an account of the vector Wiener-Hopf technique in electromagnetics reference is made to Meister and Speck [43], Büyükaksoy and Serbest [4] and to the introduction of a paper by Lüneburg and Serbest [34].

Fourier symbol matrices of problems in diffraction theory often involve a term of the form of the square root $\gamma(\alpha) = \sqrt{\alpha^2 - k^2}$ where $k = 2\pi/\lambda$ is the wave number and α is the complex Fourier transform variable. In very many cases it is possible to formulate the diffraction problem as a true vector Wiener-Hopf equation as well as a scalar Riemann-Hilbert problem on the two-sheeted Riemann surface of $\gamma(\alpha)$. This was first demonstrated by Lüneburg and Westpfahl [35] in 1971 for a special strip grating problem, see also Lüneburg [25, 26]. This approach will be explained in some detail in the present contribution. In 1976 Hurd [17, 20] invented the so-called Wiener-Hopf-Hilbert (WHH) method and was able to solve the problem of diffraction by a half-plane with different impedance boundary conditions on both sides. This includes the case of the so-called Rawlins' half-plane with a Dirichlet condition on one side and a Neumann condition on the other that had been solved previously by Rawlins [51] by a heuristic method. For a general formulation of the Wiener-Hopf-Hilbert technique we refer to Spitkovsky and Tashbaev [62].

These methods, however, were not pursued intensively. It become possible to factorize directly an entire class of Fourier symbol matrices that relate to important problems in electromagnetic, acoustic and elastodynamic scattering theory. This class is nowadays called the Daniele-Khrapkov class and goes back to Khrapkov [24]

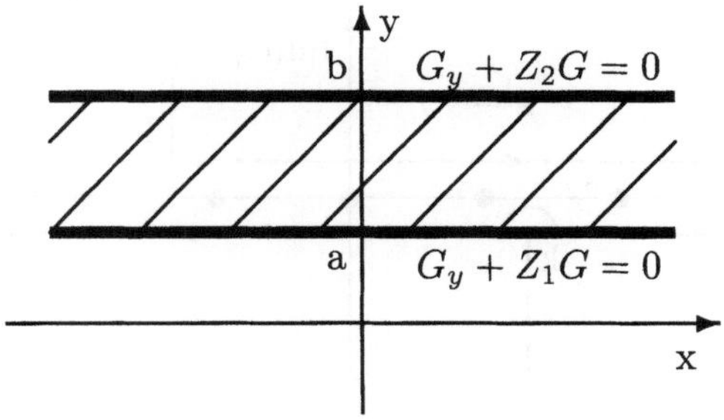

FIGURE 1. The slab geometry of the Green function

using previous results by Chebotarev [5] (a report of the Kazan State University) and to Daniele [7, 8, 9], see also Rawlins [52, 53, 54]. The importance of the work of Khrapkov and Chebotarev became known in the western engineering community only after the publication of Lukyanov's papers [36, 37] in 1980/81, see for instance Lüneburg and Hurd [31, 33] and Lüneburg and Serbest [34]. The Daniele-Khrapkov method and the Riemann-Hilbert approach appear to be equivalent for problems in engineering sciences involving the square root $\gamma(\alpha)$ or a small modification of it. This seems to be true also for the modified residue calculus method as presented by Weinstein and Volman [66].

In the following we consider some examples in diffraction theory that can be solved by the Riemann-Hilbert method. In any case we limit ourselves to indicate the reduction of the problem to the generic matrix form but do not carry out the rather tedious explicit solution for all problems considered. We refer to Lüneburg and Westpfahl [35] and Lüneburg [25, 26, 27] for details of the solution procedure in some special cases.

2. The Helmholtz integral and alternative Green functions representations

Let $G(\mathbf{r}|\mathbf{r}')$ be a fundamental solution (Green's function) of the two-dimensional inhomogeneous reduced wave equation (Helmholtz equation)

$$(\nabla + k^2)G(\mathbf{r}|\mathbf{r}') = -\delta(\mathbf{r} - \mathbf{r}'), \tag{2.1}$$

that satisfies the radiation condition for $r = |\mathbf{r}| \to \infty$. If $\phi(\mathbf{r})$ denotes a solution of the homogeneous Helmholtz equation in the exterior simply connected region R with piecewise smooth boundary ∂R that satisfies the Sommerfeld radiation condition at infinity, then the Helmholtz integral representation reads (cf. Meister [41], Jones [23])

$$\left.\begin{matrix}\phi(\mathbf{r})\\ 0\end{matrix}\right\} = \int_{\partial R}\left\{G(\mathbf{r},\mathbf{r}')\frac{\partial\phi}{\partial n'}(\mathbf{r}') - \frac{\partial G}{\partial n'}(\mathbf{r},\mathbf{r}')\phi(\mathbf{r}')\right\}ds' \text{ for } \left\{\begin{matrix}\mathbf{r}\in R\\ \mathbf{r}\notin R\end{matrix}\right., \tag{2.2}$$

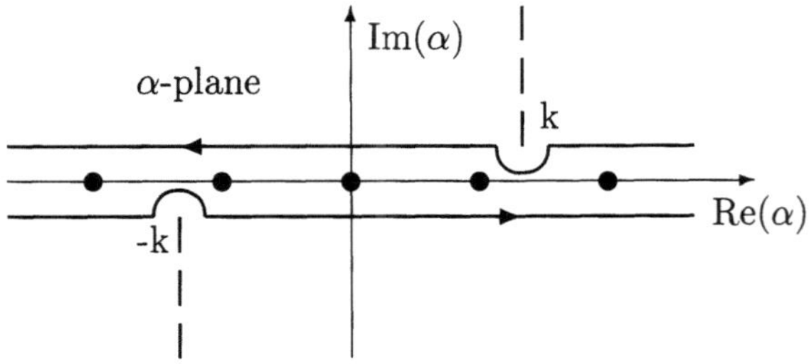

FIGURE 2. Integration contour in the complex α-plane
(-- branch cuts of $\gamma(\alpha)$; • poles)

where $\partial/\partial n'$ denotes differentiation along the outward normal at the point $\mathbf{r}'$ of the boundary ∂R. For $\exp(-i\omega t)$-time dependence, the free space Green's function is given by

$$G(\mathbf{r}|\mathbf{r}') = \frac{i}{4} H_0^{(1)}(k|\mathbf{r}-\mathbf{r}'), \tag{2.3}$$

where $H_0^{(1)}$ is the Hankel function of the first kind of order zero with the Fourier integral representation ($\mathbf{r} = (x, y)$)

$$H_0^{(1)}(k\sqrt{(x-x')^2+(y-y')^2}) = \frac{1}{\pi}\int_{-\infty}^{\infty} e^{i[\alpha(x-x')+\sqrt{k^2-\alpha^2}|y-y'|]}\frac{d\alpha}{\sqrt{k^2-\alpha^2}} \tag{2.4}$$

with $\kappa(\alpha) := \sqrt{k^2-\alpha^2} = i\sqrt{\alpha^2-k^2} =: i\gamma(\alpha)$. A detailed discussion of the branch cuts and branch points of this two-valued analytic function is given by Noble [49].

Referring to Cohen [6] alternative representations of the Green's function of the Helmholtz equation for the slab geometry of Fig. 1 with the boundary conditions

$$\frac{\partial}{\partial y}G(x,y|x',y') + Z_1 G(x,y|x',y') = 0 \quad \text{for } y = a \tag{2.5}$$

$$\frac{\partial}{\partial y}G(x,y|x',y') + Z_2 G(x,y|x',y') = 0 \quad \text{for } y = b \tag{2.6}$$

can be expressed in the form of a loop contour integral

$$G(x,y|x',y') = \frac{1}{4\pi i}\oint \frac{N}{D}\,\frac{e^{i\sqrt{\alpha^2-k^2}|x-x'|}}{\sqrt{\alpha^2-k^2}}d\alpha \tag{2.7}$$

with the nominator

$$N = \{\alpha\cos[\alpha(y'-a)] + Z_1\sin[\alpha(y'-a)]\}\,\{\alpha\cos[\alpha(b-y)] + Z_2\sin[\alpha(b-y)]\} \tag{2.8}$$

and the denominator

$$D = (\alpha^2 + Z_1 Z_2)\sin[(b-a)\alpha] - (Z_2 - Z_1)\cos[(b-a)]. \tag{2.9}$$

The loop integral encircles the poles of the integrand, i.e., the zeros of the denominator D, in the positive sense, passing below/above the branch points at $\pm k$

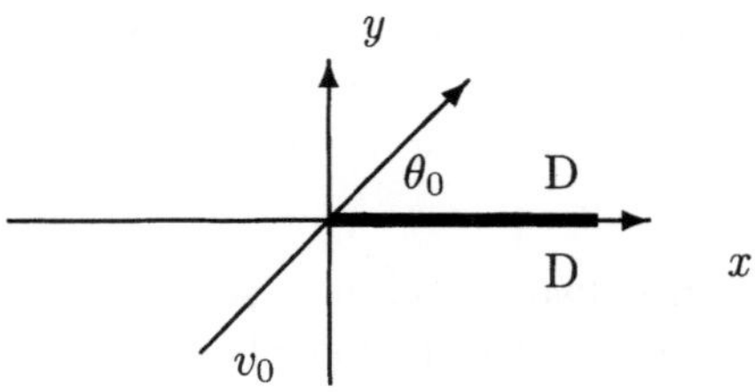

FIGURE 3. Sommerfeld half-plane with Dirichlet boundary conditions (D) and incident plane wave v_0

assuming that k has a small positive imaginary part. The position of the branch cuts and the loop integral are sketched in Fig. 2. The standard representations of the Green's functions assumes the form of a Fourier integral with a difference kernel $\exp[(x-x')\alpha]$. They are used extensively in the Wiener-Hopf formulation of such problems.

Neumann boundary conditions are obtained by setting $Z_1 = 0$ and/or $Z_2 = 0$ and Dirichlet boundary conditions by $Z_1 \to \infty$ and/or $Z_2 \to \infty$.

3. Boundary value problems for the upper half-plane

A detailed account of Sommerfeld's half-plane problem [61] has been given by Nobel [49] and Lüneburg [30]. Noble in his famous textbook considers three boundary value problems connected with the steady-state wave equation (we assume a time factor $\exp(-i\omega t)$)

$$\frac{\partial^2\phi(x,y)}{\partial x^2} + \frac{\partial^2\phi(x,y)}{\partial y^2} + k^2\phi(x,y) = 0, \qquad -\infty < x < \infty,\ y \geq 0 \tag{3.1}$$

where ϕ represents an outgoing wave at infinity and k is the wave number such that the following boundary conditions are satisfied in each of the three separated cases

a) $$\phi(x,0) = f(x) \quad \text{on} \quad y = 0, \quad -\infty < x < \infty;$$

b) $$\frac{\partial\phi(x,0)}{\partial y} = g(x) \quad \text{on} \quad y = 0, \quad -\infty < x < \infty;$$

c) $$\phi(x,0) = f(x) \quad \text{on} \quad y = 0, \quad 0 < x < \infty,$$

$$\frac{\partial\phi(x,0)}{\partial y} = g(x) \quad \text{on} \quad y = 0, \quad -\infty < x < 0.$$

Introducing the Fourier transform pair along the boundary x-axis

$$\Phi(\alpha, y) = \frac{1}{\sqrt{2\pi}} \int_{-\infty}^{\infty} \phi(x, y) e^{i\alpha x} d\alpha \quad \Longleftrightarrow \quad \phi(\alpha, y) = \frac{1}{\sqrt{2\pi}} \int_{-\infty}^{\infty} \Phi(x, y) e^{-i\alpha x} d\alpha \tag{3.2}$$

Noble points out that the first two cases can be solved directly by using the inverse real Fourier transform whereas the third case, however, cannot be solved elementarily and leads to a scalar Wiener-Hopf equation using Fourier transformations in the complex plane. As an example for the third case the classical Sommerfeld half-plane problem with Dirichlet boundary conditions, see Fig. 3, is treated in some detail where $f(x) = 0$ and $g(x)$ is a known function depending on the form of the incident wave.

Taking into account Meixner's edge condition [45, 46], see also Dos Santos and Teixeira [11], the scalar Wiener-Hopf technique leads to the solution

$$v(x, y) = v_0(x, y) - \frac{\sqrt{k + \alpha_0}}{2\pi i} \int_{-\infty}^{\infty} e^{i[\alpha x + \sqrt{k^2 - \alpha^2}|y|]} \frac{1}{\sqrt{k^2 - \alpha^2}} \frac{d\alpha}{\alpha - \alpha_0|_\vee} \tag{3.3}$$

where θ_0 is the angle of incidence of the plane wave $v_0 = \exp[ik(\cos\theta_0 x + \sin\theta_0 y)]$ and $\alpha_0 = k\cos\theta_0$. Here $\alpha - \alpha_0|_\vee$ indicates that the path of integration $-\infty < \alpha < \infty$ passes below the pole α_0. The essential feature of the Wiener-Hopf solution of this problem is the multiplicative factorization

$$\sqrt{k^2 - \alpha^2} = \sqrt{k - \alpha}\,\sqrt{k + \alpha} \tag{3.4}$$

where the factor $\sqrt{k - \alpha}$ is holomorphic in the lower α-half-plane whereas the factor $\sqrt{k + \alpha}$ is holomorphic in the upper α-half-plane due to the appropriate judicious choice of branch cuts, see Noble [49]. Noble refers to three different methods all depending on the Wiener-Hopf technique: (i) dual integral equation method, (ii) Green's function formulation, (iii) Jones' approach [22]. The unknown function is either the jump of the derivatives of the field components on the screen $\{y = \pm 0,\ x > 0\}$ or the field on the supplementary screen $\{y = 0,\ x < 0\}$. It should be pointed out that the unknown function can also be the field or the normal derivative of the field on the half-line $x = 0$, $y > 0$. The corresponding Wiener-Hopf equation is of the same type.

4. Rawlins-type half-plane problems

Rawlins was the first to consider a modification of the Sommerfeld half-plane problem consisting of a half-plane with Dirichlet boundary condition on one side and Neumann boundary condition on the other side, see Fig. 4. The formulation of that problem leads to a vector Wiener-Hopf equation with a 2×2 Fourier symbol matrix that cannot be decoupled elementarily. Rawlins was able to solve this problem by a heuristic method [51]. Only one year later in 1976 Hurd invented the Wiener-Hopf-Hilbert method [17, 20] which he used to solve the more general

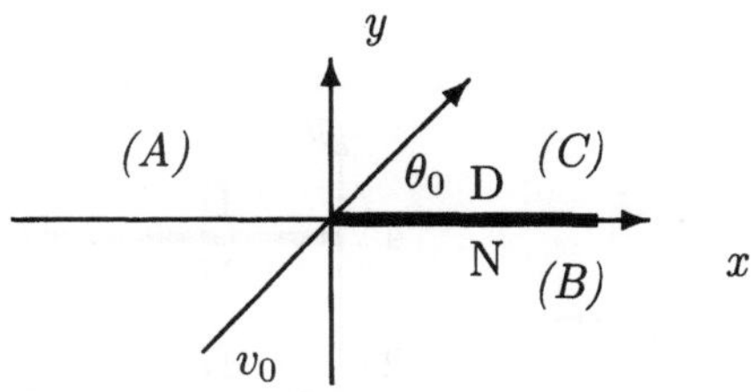

FIGURE 4. Rawlins half-plane with Dirichlet (D) and Neumann (N) boundary conditions and incident plane wave v_0

2-face impedance half-plane problem. Hurd also showed that his solution coincides with Maliuzhinets' results [38]. Maliuzhinets' solution was well known but according to Hurd it was not possible to derive any valuable clues from that solution for the factorization of the corresponding Fourier symbol matrix (private communication). The Fourier symbol matrix function of Rawlins' problem reads

$$M(\alpha) = \begin{bmatrix} 1 & -i\kappa(\alpha) \\ -i/\kappa(\alpha) & 1 \end{bmatrix} = M^+(\alpha)M^-(\alpha) \tag{4.1}$$

with $\kappa(\alpha) = \sqrt{k^2 - \alpha^2}$. This matrix is of the Daniele-Khrapkov type

$$M(\alpha) = a(\alpha)I + b(\alpha)R(\alpha) \tag{4.2}$$

where $a(\alpha)$ and $b(\alpha)$ are scalar functions, I is the identity matrix and $R(\alpha)$ is a polynomial 2×2 matrix function with trace $R(\alpha) = 0$, see Ehrhardt and Speck [12] for a recent review of 2×2 Daniele-Khrapkov type matrix factorizations.

For the canonical factorization of the matrix function (4.1) which is a much more difficult problem than the factorization of the scalar function (3.4) we refer for instance to Rawlins [52, 53, 54], Rawlins and Williams [57], Heins [16], Williams [67], Büyükaksoy and Serbest [4] and Meister [42].

The Wiener-Hopf factorization of (4.1) reads

$$M^+(\alpha) = \sqrt{2} \begin{bmatrix} \sqrt{\dfrac{\sqrt{2k} + \sqrt{k+\alpha}}{2\sqrt{2k}}} & -i\kappa(\alpha)\sqrt{\dfrac{\sqrt{2k} - \sqrt{k+\alpha}}{2\sqrt{2k}}} \\ -\dfrac{i}{\kappa(\alpha)}\sqrt{\dfrac{\sqrt{2k} - \sqrt{k+\alpha}}{2\sqrt{2k}}} & \sqrt{\dfrac{\sqrt{2k} + \sqrt{k+\alpha}}{2\sqrt{2k}}} \end{bmatrix} \tag{4.3}$$

and $M^-(\alpha) = M^+(-\alpha)$ where $M^+(\alpha)$ and $M^-(\alpha)$ commute. The factorization $M(\alpha) = M^+(\alpha)M^-(\alpha)$ is not unique since $M(\alpha) = \bar{M}^+(\alpha)\bar{M}^-(\alpha)$ with $\bar{M}^+(\alpha) = M^+(\alpha)C$ and $\bar{M}^-(\alpha) = C^{-1}M^-(\alpha)$ where C is a nonsingular constant matrix is also an acceptable form where, however, $\bar{M}^+(\alpha)$ and $\bar{M}^-(\alpha)$ may no longer

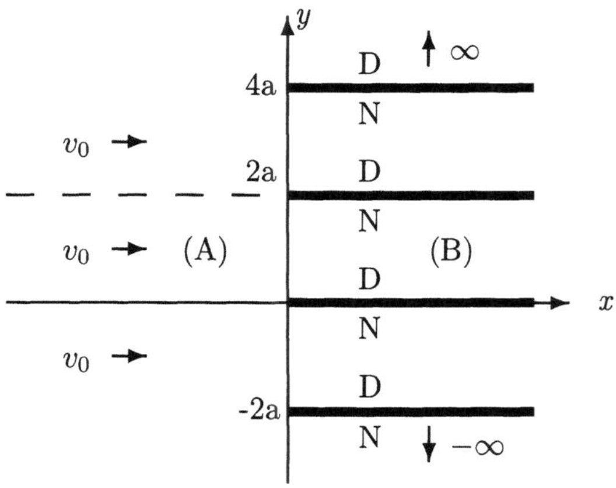

FIGURE 5. An infinite set of Rawlins' half-planes with Dirichlet (D) and Neumann (N) boundary conditions

commute. This is the reason why different factorization expressions occur in the literature.

The Riemann-Hilbert method which in this case is equivalent to the Wiener-Hopf-Hilbert technique leads to the factorization of the matrix $M(\alpha)$, cf. (4.1),

$$H(\alpha) = \begin{bmatrix} 0 & 1/\kappa(\alpha) \\ -\kappa(\alpha) & 0 \end{bmatrix} = H^I(\alpha)H^{II}(\alpha), \quad \alpha \in C \cup \bar{C} \tag{4.4}$$

where $C \cup \bar{C}$ is a contour along the branch cuts of $\kappa(\alpha) = \sqrt{k^2 - \alpha^2}$. The factor matrix $H^I(\alpha)$ is holomorphic in the α-plane cut along $C \cup \bar{C}$, whereas $H^{II}(\alpha)$ is holomorphic in the analytic continuation of $H^I(\alpha)$ through the cut $C \cup \bar{C}$ of the two-sheeted Riemann surface of $\kappa(\alpha)$. Choosing other functions as unknowns different Riemann-Hilbert problems can be set up. For instance the fields or its normal derivatives can be matched at the interface between region (A) and regions (B) and (C) of Fig. 4, see the treatment of the following problem.

We now consider an infinite set of parallel Rawlins' half-planes, see Fig. 5. Rather than using a standard Wiener-Hopf matrix formulation, see Lüneburg and Hurd [27], we use an alternative integral representation. From the Helmholtz integral representation (2.2) and the alternative loop integral representation (2.7) we obtain for region $(A) : 0 < y < 2a,\ x < 0$ of Fig. 5 the expression

$$v^A(x,y) = v_0(x,y) + v_0(-x,y) + \frac{1}{4\pi i}\oint \bar{\psi}(\alpha)\cot(a\alpha)e^{i(y-a)\alpha}e^{\gamma(\alpha)x}\frac{d\alpha}{\gamma(\alpha)} \tag{4.5}$$

with

$$\bar{\psi}(\alpha) = \int_0^{2a} \frac{\partial v}{\partial x}(0,y)e^{-i(y-a)\alpha}dy. \tag{4.6}$$

For region $(B) : 0 < y < 2a,\ x > 0$ we obtain correspondingly

$$v^B(x,y) = -\frac{1}{4\pi}\oint \frac{\bar{\psi}(-\alpha) - e^{-2ia\alpha}\bar{\psi}(\alpha)}{\cos(2a\alpha)} e^{i(y-a)\alpha}\frac{d\alpha}{\gamma(\alpha)}. \tag{4.7}$$

Rather than matching directly the fields v^A and v^B at $x = 0,\ 0 < y < 2a$ we match the field combinations

$$v(0,y) \pm iv(0, 2a - y). \tag{4.8}$$

Field matching at the interface $x = 0$, multiplication by $\exp(-i\alpha y)$ and integration over the unit cell extension $0 \le y \le 2a$ leads to two decoupled singular integral equations

$$\begin{aligned}&\tfrac{1}{2\pi i}\oint \bar{\psi}_\pm(\alpha')\left[\cot(a\alpha') \pm \tan(\tfrac{\pi}{4} \mp a\alpha')\right]\sin[a(\alpha'-\alpha)]\tfrac{1}{\gamma(\alpha')}\tfrac{d\alpha'}{\alpha'-\alpha}\\ &= -4(1\pm i)\tfrac{\sin(a\alpha)}{\alpha}\end{aligned} \tag{4.9}$$

with

$$\bar{\psi}_+(\alpha) = \bar{\psi}(\alpha) + i\bar{\psi}(-\alpha), \quad \bar{\psi}_-(\alpha) = \bar{\psi}(\alpha) - i\bar{\psi}(-\alpha). \tag{4.10}$$

Trigonometric simplifications of these expressions (taking into account the property that $\bar{\psi}_\pm(\alpha)$ are entire functions), shift of the path of integration from the loop integral encircling the simple poles on the real α-axis to the path $C \cup \bar{C}$ along the branch cuts of $\gamma(\alpha)$ and taking principal values yields

$$\frac{1}{2\pi i}PV\int_{C\cup\bar{C}} \bar{\psi}_\pm(\alpha')\left\{\frac{\sin(a\alpha)}{\sin(a\alpha')} + \frac{\cos(\pi/4 \mp a\alpha)}{\cos(\pi/4 \mp a\alpha')}\right\}\frac{1}{\gamma(\alpha')}\frac{d\alpha'}{\alpha'-\alpha} \tag{4.11}$$

$$= 2(1\pm i)\frac{\sin(a\alpha)}{\alpha}, \quad \alpha \in C \cup \bar{C}. \tag{4.12}$$

The solution of these Riemann-Hilbert equations read

$$\bar{\psi}_\pm(\alpha) = 2i(1\pm i)M^I(0)\cos(\pi/4 \mp a\alpha)\left[M^I(\pm\alpha) + M^{II}(\pm\alpha)\right] \tag{4.13}$$

with the factorization

$$M(\alpha) = \frac{\sin(a\alpha)/\alpha}{\cos(\pi/4 - a\alpha)} = M^I(\alpha)M^{II}(\alpha) \tag{4.14}$$

on the two-sheeted Riemann surface of $\gamma(\alpha)$. The factor function $M^I(\alpha)$ is holomorphic and without zeros on the cut sheet I of the Riemann surface of $\gamma(\alpha)$ whereas the other factor $M^{II}(\alpha)$ has corresponding properties on sheet II. Explicit expressions of $M^{I(II)}(\alpha)$ are given in Lüneburg [25]. In matrix notation this procedure can be written as the factorization of two 2×2 Riemann-Hilbert matrix functions

$$M(\alpha) = \begin{bmatrix} 0 & F(\alpha) \\ 1/F(\alpha) & 0 \end{bmatrix} \quad \text{with} \quad F(\alpha) = \frac{\sin(a\alpha)/\alpha}{\cos(\pi/4 \mp a\alpha)}. \tag{4.15}$$

A Wiener-Hopf solution of this problem has been given by Lüneburg and Hurd [27]. The method can easily be extended to Bragg angle incidence where $\{\theta_{0m} : -\pi \le \theta_{0m} \le \pi;\ \sin\theta_{0m} = \pi m/2ka,\ m \text{ integer}\}$. The isolated Rawlins' half-plane solution can be found from the solution of the grating problem letting a tend to infinity yielding the familiar factorization of Rawlins' problem, see Lüneburg [25]. For arbitrary angles of incidence no solution has been found yet; the corresponding Wiener-Hopf matrix function is no longer of the Daniele-Khrapkov type, cf. Lüneburg [28].

5. The bifurcated waveguide

One of the simplest diffraction configurations that has been extensively treated by different modifications of the classical Wiener-Hopf technique is the bifurcated parallel waveguide problem with Dirichlet boundary conditions on the outer walls and on both faces of the inserted half-plane, see Fig. 6. Here the fundamental waveguide mode is assumed to be incident from the wider waveguide on the half-plane but incidence out of one of the smaller waveguide regions or several modes can be treated along the same lines. Also the Dirichlet conditions may be replaced everywhere by Neumann boundary conditions. Scattering of the incident mode at the edge of the inserted half-plane leads to the excitation of propagating and evanescent waveguide modes in all the three regions (A), (B) and (C). It should be noted that the half-plane is positioned asymmetrically inside the parallel-plate duct. A detailed analysis is presented in the Mittra and Lee's textbook [48]. The general case of N equally spaced inserted half-planes has been treated by Igarashi in a series of papers [21]. The case with outer BC's both being of the Dirichlet or the Neumann type but with Dirichlet and Neumann conditions on the inserted half-plane has been solved by Lüneburg and Hurd [33] for the special case that $a = b$.

Replacing the free-space Green's function for the Wiener-Hopf solution of Sommerfeld's classical half-plane problem by the Green's function of a parallel waveguide this problem can be reduced to the solution of a scalar Wiener-Hopf equation for the semi-infinite Fourier transform of the jump of the normal derivative of the field on the inserted half-plane. The problem can also be formulated as a doubly infinite set of linear algebraic equations for the infinite set of modal coefficients inside the three waveguide regions. Imposing continuity of the fields and the normal derivatives at the adjacent regions the latter systems can be solved either by the residue calculus technique or by using a direct matrix inversion technique, cf. Mittra [47]. All these methods are explained in detail for this canonical example in the monograph by Mittra and Lee referred to above, see also the survey article by Lüneburg [28]. This simple example has been chosen primarily to illustrate the applicability of the Riemann-Hilbert method to plane waveguide problems. For region (A) we choose a Green's function that vanishes on the outer walls $y = a$ and $y = -b$ and whose normal derivative (z-derivative) vanishes at the interface $x = 0$ of the waveguide regions (A) with (B) and (C). From the general equation of the previous section follows

$$G^A(x,y|x',y') = -\frac{1}{4\pi i}\oint \frac{\sin[(a-y)\alpha]\sin[(y'+b)\alpha]}{\sin[(a+b)\alpha]}\left\{e^{i\gamma|x-x'|}+e^{i\gamma|x+x'|}\right\}\frac{d\alpha}{\gamma} \tag{5.1}$$

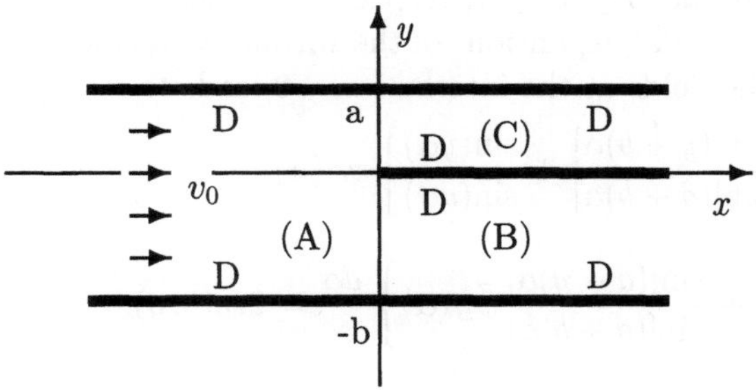

FIGURE 6. The bifurcated waveguide problem with Dirichlet boundary conditions

with $\gamma \equiv \gamma(\alpha) = \sqrt{\alpha^2 - k^2}$ or

$$G^A(x, y|x', y') = -\frac{1}{4\pi i} \oint \frac{\sin[(x+b)\alpha]\sin[(a-y')\alpha]}{\sin[(a+b)\alpha]} \left\{ e^{i\gamma|x-x'|} + e^{i\gamma|x+x'|} \right\} \frac{d\alpha}{\gamma} \tag{5.2}$$

due to the symmetry of the Green's function with respect to the dashed and undashed arguments. The first form will be used for $-b \leq y \leq 0$ and the second form for $0 \leq y \leq a$. This is done in order to introduce the same (unknown) finite Fourier transforms in regions (A) and (B) and (A) and (C). Inside the waveguide region (A) the field can be expressed in term of the Green's function G^A as

$$v_A(x, y) = v_0(x, y) + v_0(-x, y) + \int_{-b}^{a} \frac{\partial}{\partial x'} v(0, y') G^A(x, y|0, y') dy' \tag{5.3}$$

valid for $-b \leq y \leq a, x \leq 0$. v_0 is the incident modal field. For x = 0 it follows in particular

$$v_A(0, y) = 2v_0(0, y) - \frac{1}{2\pi i} \oint \frac{\sin[(a-y)\alpha]\bar{\psi}_B(\alpha) + \sin[(y+b)\alpha]\bar{\psi}_C(\alpha)}{\sin[(a+b)\alpha]} \frac{d\alpha}{\gamma} \tag{5.4}$$

with the finite Fourier transforms

$$\bar{\psi}_B(\alpha) = \int_{-b}^{0} \frac{\partial v(0, y)}{\partial x} \sin[(y+b)\alpha] dy \tag{5.5}$$

$$\bar{\psi}_C(\alpha) = \int_{0}^{a} \frac{\partial v(0, y)}{\partial z} \sin[(a-y)\alpha] dy. \tag{5.6}$$

Introducing the appropriate Green's function for the waveguide regions (B) and (C) yields

$$v_B(0, y) = -\frac{1}{2\pi i} \oint \frac{\sin(y\alpha)}{\sin(b\alpha)} \bar{\psi}_B(\alpha) \frac{d\alpha}{\gamma} \tag{5.7}$$

$$v_C(0, y) = \frac{1}{2\pi i} \oint \frac{\sin(y\alpha)}{\sin(a\alpha)} \bar{\psi}_C(\alpha) \frac{d\alpha}{\gamma} \tag{5.8}$$

valid for $-b \leq y \leq 0$ and $0 \leq y \leq a$, respectively. Closing the contour in these integrals provides the modal expansions of the various waveguides at the interface.

Continuity of the fields at the interface $x = 0$ leads to

$$\frac{1}{2\pi i} \oint \left\{ \left[\frac{\sin[(y+b)\alpha]}{\sin[(a+b)\alpha]} + \frac{\sin(y\alpha)}{\sin(a\alpha)} \right] \bar{\psi}_C(\alpha) + \frac{\sin[(a-y)\alpha]}{\sin[(a+b)\alpha]} \bar{\psi}_B(\alpha) \right\} \frac{d\alpha}{\gamma} = 2v_0(0,y), \quad 0 \leq y \leq a \tag{5.9}$$

and

$$\frac{1}{2\pi i} \oint \left\{ \left[\frac{\sin[(a-y)\alpha]}{\sin[(a+b)\alpha]} - \frac{\sin(y\alpha)}{\sin(b\alpha)} \right] \bar{\psi}_B(\alpha) + \frac{\sin[(y+b)\alpha]}{\sin[(a+b)\alpha]} \bar{\psi}_C(\alpha) \right\} \frac{d\alpha}{\gamma} = 2v_0(0,y), \quad -b \leq y \leq 0. \tag{5.10}$$

Replacing $\alpha \to \alpha'$ in equation (5.9), multiplying the resulting equation by $\sin[(a-y)\alpha]$ and integrating over the domain $0 \leq y \leq a$ yields

$$\frac{1}{2\pi i} \oint \left\{ \left[\frac{\sin(b\alpha' + a\alpha)}{\sin[(a+b)\alpha']} + \frac{\sin(a\alpha)}{\sin(a\alpha')} \right] \bar{\psi}_C(\alpha') + \frac{\sin[a(\alpha'-\alpha)]}{\sin[(a+b)\alpha']} \bar{\psi}_B(\alpha') \right\} \frac{1}{\gamma'} \frac{d\alpha'}{\alpha' - \alpha} = 2\bar{v}_{0C}(\alpha) \tag{5.11}$$

where

$$\bar{v}_{0C}(\alpha) := \int_0^a v_0(0,y) \sin[(a-y)\alpha] dy. \tag{5.12}$$

Similarly, replacing $\alpha \to \alpha'$ in equation (5.10), multiplying the resulting equation by $\sin[(y+b)\alpha]$ and integrating over the interval $-b \leq y \leq 0$ yields

$$\frac{1}{2\pi i} \oint \left\{ \left[\frac{\sin(a\alpha' + b\alpha)}{\sin[(a+b)\alpha']} + \frac{\sin(b\alpha)}{\sin(b\alpha')} \right] \bar{\psi}_B(\alpha') + \frac{\sin[b(\alpha'-\alpha)]}{\sin[(a+b)\alpha']} \bar{\psi}_C(\alpha') \right\} \frac{1}{\gamma'} \frac{d\alpha'}{\alpha' - \alpha} = 2\bar{v}_{0B}(\alpha) \tag{5.13}$$

where

$$\bar{v}_{0B}(\alpha) := \int_{-b}^0 v_0(0,y) \sin[(y+b)\alpha] dy. \tag{5.14}$$

With equations (5.11) and (5.13) we form the expression

$$\sin(b\alpha) \times \text{equation}\,(5.11) + \sin(a\alpha) \times \text{equation}\,(5.13). \tag{5.15}$$

The partial contributions of one and the other equation resembles the procedure used by Mittra in his direct-inversion method [47]. The final standard form reads

$$\frac{1}{2\pi i} \oint \bar{\Psi}(\alpha') \{F(\alpha) + F(\alpha')\} \frac{1}{\gamma'} \frac{d\alpha'}{\alpha' - \alpha} = \bar{e}(\alpha) \tag{5.16}$$

with the unknown function

$$\bar{\Psi}(\alpha) = \frac{\bar{\psi}_{0B}(\alpha)}{\sin(b\alpha)} + \frac{\bar{\psi}_{0C}(\alpha)}{\sin(a\alpha)}, \tag{5.17}$$

the excitation term

$$\bar{e}(\alpha) = 2\frac{\sin(b\alpha)\bar{v}_{0B}(\alpha) + \sin(a\alpha)\bar{v}_{0C}(\alpha)}{\sin[(a+b)\alpha]} \tag{5.18}$$

and the kernel function

$$F(\alpha) = \frac{\sin[(a+b)\alpha]}{\sin(a\alpha)\sin(b\alpha)}. \tag{5.19}$$

Here α lies outside the contour integral $\oint$. $F(\alpha)$ is a meromorphic function with poles on the real axis only, $\Psi(\alpha)$ is the unknown and $\bar{e}(\alpha)$ is the given excitation function. Comparing equations (5.4), (5.7) and (5.8) it should be noted that $F(\alpha)$ is the ratio of the denominator of waveguide region (A) and the product of the denominators of waveguide regions (B) and (C).

The reduction of the integral equation equation (5.16) to a Riemann-Hilbert problem along the banks of the branch cuts of $\gamma(\alpha)$ and its solution is well known and has been carried out for a number of special cases. The shift of the path of integration from the contour $\oint$ of Fig. 2 to the contour Γ along the branch cuts of γ relies on the edge condition. The resulting 2-vector Riemann-Hilbert problem has a matrix kernel M of the special form

$$M(\tau) = \begin{pmatrix} 0 & F^{-1}(\tau) \\ F(\tau) & 0 \end{pmatrix} \quad \text{for } \tau \in \Gamma. \tag{5.20}$$

The meromorphic matrix M has vanishing diagonal elements. This is a direct consequence of the particular standard form of equation (5.16). Details, however, are outside the scope of the present contribution and we refer to Lüneburg and Westpfahl [35] and Lüneburg [25, 26, 27, 28].

6. Rawlins' trifurcated waveguide

Recently Rawlins discovered that the solution of two special trifurcated waveguide problems can be reduced to the solution of a 2×2 Wiener-Hopf equation with a matrix kernel function of the Daniele-Khrapkov form [55, 56]. One of these configurations which are used by Rawlins for modelling exhaust systems is sketched in Fig. 7 for case 1. In the second configuration (case 2) the boundary conditions on the two inner half-planes are interchanged. It should be pointed out that the general trifurcated waveguide problem with arbitrary spacings between the inserted half-planes and the walls and between themselves has not found a closed form solution yet even for Dirichlet boundary conditions everywhere, cf. [50]. Introducing an appropriate Green's function the fields in the different regions can be expressed

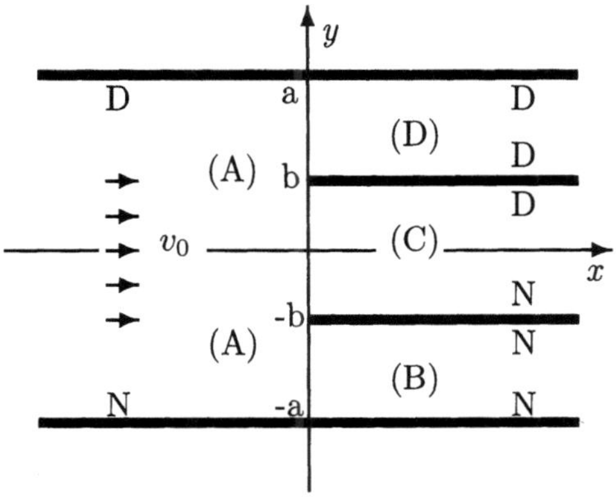

FIGURE 7. Rawlins' trifurcated waveguide with Dirichlet (D) and Neumann (N) boundary conditions (Case 1)

in the form:

$$v_A(0,y) = 2v_0(0,y) - \frac{1}{2\pi i}\oint\left\{\frac{\sin[(a-y)\alpha]}{\cos(2a\alpha)}\bar{\psi}_B(\alpha) + \tan(\frac{\pi}{4}+a\alpha)\cos(\frac{\pi}{4}+a\alpha)\bar{\psi}_C(\alpha) + \frac{\cos[(y+a)\alpha]}{\cos(2a\alpha)}\bar{\psi}_D(\alpha)\right\}\frac{d\alpha}{\gamma} \tag{6.1}$$

with the finite Fourier transform

$$\bar{\psi}_B(\alpha) = \int_{-a}^{-b}\frac{\partial}{\partial x}v(0,y)\cos[(y+a)\alpha]dy \tag{6.2}$$

$$\bar{\psi}_C(\alpha) = \int_{-b}^{b}\frac{\partial}{\partial z}v(0,y)\cos(\frac{\pi}{4}+y\alpha)dy \tag{6.3}$$

$$\bar{\psi}_D(\alpha) = \int_{b}^{a}\frac{\partial}{\partial z}v(0,y)\sin[(a-y)\alpha]dy. \tag{6.4}$$

Furthermore

$$v_B(0,y) = -\frac{1}{2\pi i}\oint\frac{\cos[(b+y)]}{\sin[(a-b)\alpha]}\bar{\psi}_B(\alpha)\frac{d\alpha}{\gamma} \quad (-a \le y \le -b), \tag{6.5}$$

$$v_C(0,y) = \frac{1}{2\pi i}\oint\tan(\frac{\pi}{4}+b\alpha)\cos(\frac{\pi}{4}+y\alpha)\bar{\psi}_C(\alpha)\frac{d\alpha}{\gamma} \quad (-b \le y \le b), \tag{6.6}$$

$$v_D(0,y) = \frac{1}{2\pi i}\oint\frac{\sin[(y-b)]}{\sin[(a-b)\alpha]}\bar{\psi}_D(\alpha)\frac{d\alpha}{\gamma} \quad (b \le y \le a). \tag{6.7}$$

Continuity of the field representations at the interfaces leads to the following y-triplet of equations

$$\begin{aligned}&\frac{1}{2\pi i}\oint\left[\frac{\sin[(a-y)\alpha]}{\cos(2a\alpha)}-\frac{\cos[(b+y)\alpha]}{\sin[(a-b)\alpha]}\right]\bar{\psi}_B(\alpha)\frac{d\alpha}{\gamma}\\&+\frac{1}{2\pi i}\oint\tan(\frac{\pi}{4}+a\alpha)\cos(\frac{\pi}{4}+y\alpha)\bar{\psi}_C(\alpha)\frac{d\alpha}{\gamma}\qquad(-a\le y\le -b)\\&+\frac{1}{2\pi i}\oint\frac{\cos[(y+a)]}{\cos(2a\alpha)}\bar{\psi}_D(\alpha)\frac{d\alpha}{\gamma}=2v_0(0,y),\end{aligned}\tag{6.8}$$

$$\begin{aligned}&\frac{1}{2\pi i}\oint\frac{\sin[(a-y)\alpha]}{\cos(2a\alpha)}\bar{\psi}_B(\alpha)\frac{d\alpha}{\gamma}\\&+\frac{1}{2\pi i}\oint\left[\tan(\frac{\pi}{4}+a\alpha)+\tan(\frac{\pi}{4}+b\alpha)\right]\cos(\frac{\pi}{4}+y\alpha)\bar{\psi}_C(\alpha)\frac{d\alpha}{\gamma}\\&+\frac{1}{2\pi i}\oint\frac{\cos[(y+a)]}{\cos(2a\alpha)}\bar{\psi}_D(\alpha)\frac{d\alpha}{\gamma}=2v_0(0,y)\quad(-b\le y\le b),\end{aligned}\tag{6.9}$$

$$\begin{aligned}&\frac{1}{2\pi i}\oint\{\frac{\sin[(a-y)]}{\cos(2a\alpha)}\bar{\psi}_B(\alpha)\frac{d\alpha}{\gamma}\\&+\frac{1}{2\pi i}\oint\tan(\frac{\pi}{4}+a\alpha)\cos(\frac{\pi}{4}+y\alpha)\bar{\psi}_C(\alpha)\frac{d\alpha}{\gamma},\quad(b\le y\le a),\\&+\frac{1}{2\pi i}\oint\left[\frac{\cos[(y+a)\alpha]}{\cos(2a\alpha)}+\frac{\sin[(y-b)\alpha]}{\sin[(a-b)\alpha]}\right]\bar{\psi}_D(\alpha)\frac{d\alpha}{\gamma}=2v_0(0,y).\end{aligned}\tag{6.10}$$

+Replacing $\alpha\to\alpha'$, multiplication of equation (6.8) by $\cos[(y+a)\alpha]$, of equation (6.9) by $\cos(\frac{\pi}{4}+y\alpha)$, and of equation (6.10) by $\sin[(a-y)\alpha]$ and integrating over the respective y-intervals leads to the following α-triplet of equations

$$\begin{aligned}&\frac{1}{2\pi i}\oint\left[\frac{\cos[(a+b)\alpha'+(a-b)\alpha]}{\cos(2a\alpha')}+\frac{\sin[(a-b)\alpha]}{\sin[(a-b)\alpha']}\right]\bar{\psi}_B(\alpha')\frac{1}{\gamma'}\frac{d\alpha'}{\alpha'-\alpha}\\&+\frac{1}{2\pi i}\oint\tan(\frac{\pi}{4}+a\alpha')\frac{1}{2}\left[\frac{\cos[\frac{\pi}{4}+b\alpha'+(a-b)\alpha]}{\alpha'-\alpha}+\frac{\cos[\frac{\pi}{4}+b\alpha'-(a-b)\alpha]}{\alpha'+\alpha}\right]\\&\bar{\psi}_C(\alpha')\frac{d\alpha'}{\gamma'}+\frac{1}{2\pi i}\oint\frac{\sin[(a-b)(\alpha'-\alpha)]}{\cos(2a\alpha')}\bar{\psi}_D(\alpha')\frac{1}{\gamma'}\frac{d\alpha'}{\alpha'-\alpha}=2\bar{v}_{0B}(\alpha)\end{aligned}\tag{6.11}$$

$$\begin{aligned}&\frac{1}{2\pi i}\oint\{\frac{\sin[b(\alpha'-\alpha)]}{\cos(\frac{\pi}{4}+a\alpha')}\left[\bar{\psi}_B(\alpha')+\bar{\psi}_C(\alpha')\right]\frac{1}{\gamma'}\frac{d\alpha'}{\alpha'-\alpha}\\&+\frac{1}{2\pi i}\oint\left[\frac{\cos(\frac{\pi}{4}+b\alpha)}{\cos(\frac{\pi}{4}+b\alpha')}+\tan(\frac{\pi}{4}+a\alpha')\sin[b(\alpha'-\alpha)]\right]\\&\times\bar{\psi}_C(\alpha')\frac{1}{\gamma'}\frac{d\alpha'}{\alpha'-\alpha}=2\bar{\psi}_{0C}(\alpha),\end{aligned}\tag{6.12}$$

$$\frac{1}{2\pi i}\oint \frac{\sin[(a-b)(\alpha'-\alpha)]}{\cos(2a\alpha')}\bar{\psi}_B(\alpha')\frac{1}{\gamma'}\frac{d\alpha'}{\alpha'-\alpha} + \frac{1}{2\pi i}\oint \tan(\frac{\pi}{4}+a\alpha')$$
$$\times \frac{1}{2}\left[\frac{\cos[\frac{\pi}{4}+b\alpha'+(a-b)\alpha]}{\alpha'-\alpha} - \frac{\cos[\frac{\pi}{4}+b\alpha'-(a-b)\alpha]}{\alpha'+\alpha}\right]\bar{\psi}_C(\alpha')\frac{d\alpha'}{\gamma'} \tag{6.13}$$
$$+\frac{1}{2\pi i}\oint\left[\frac{\cos[(a+b)\alpha'+(a-b)\alpha]}{\cos(2a\alpha')} + \frac{\sin[(a-b)\alpha]}{\sin[(a-b)\alpha']}\right]\frac{\bar{\psi}_D(\alpha')}{\gamma'}\frac{d\alpha'}{\alpha'-\alpha} = 2\bar{v}_{0D}(\alpha)$$

with

$$\bar{v}_{0B}(\alpha) = \int_{-a}^{-b} v_0(0,y)\cos[(y+a)\alpha]dy \tag{6.14}$$

$$\bar{v}_{0C}(\alpha) = \int_{-b}^{b} v_0(0,y)\cos(\frac{\pi}{4}+y\alpha)dy \tag{6.15}$$

$$\bar{v}_{0D}(\alpha) = \int_{b}^{a} v_0(0,y)\sin[(a-y)\alpha]dy. \tag{6.16}$$

Note the occurrence of a term proportional to $1/(\alpha'+\alpha)$ in equations (6.11) and (6.13) with opposite signs. Adding the equations (6.11) and (6.13) these terms cancel and we obtain

$$\frac{1}{2\pi i}\oint\left\{\frac{\cos[\frac{\pi}{4}+a\alpha+b(\alpha'-\alpha)]}{\cos(\frac{\pi}{4}+a\alpha')} + \frac{\sin[(a-b)\alpha]}{\sin[(a-b)\alpha']}\right\}\frac{\bar{\psi}_B(\alpha')+\bar{\psi}_D(\alpha')}{\gamma'}\frac{d\alpha'}{\alpha'-\alpha}$$
$$+\frac{1}{2\pi i}\oint \tan(\frac{\pi}{4}+a\alpha')\cos[\frac{\pi}{4}+b\alpha'+(a-b)\alpha]\bar{\psi}_C(\alpha')\frac{1}{\gamma'}\frac{d\alpha'}{\alpha'-\alpha}$$
$$= 2\{\bar{v}_{0B}(\alpha)+\bar{v}_{0D}(\alpha)\}. \tag{6.17}$$

Similarly to what has been done in the case of the bifurcated half-plane, see equation (5.15), we form the sum

$$\cos(\frac{\pi}{4}+b\alpha)\times \text{equation (6.17)} + \sin[(a-b)\alpha]\times \text{equation (6.12)} \tag{6.18}$$

and find the standard general form of equation (5.16) with the unknown function

$$\bar{\Psi}(\alpha) = \frac{\bar{\psi}_B(\alpha)+\bar{\psi}_D(\alpha)}{\sin[(a-b)\alpha]} + \frac{\bar{\psi}_C(\alpha)}{\cos(\frac{\pi}{4}+b\alpha)}, \tag{6.19}$$

the excitation term

$$2\bar{e}(\alpha) = \frac{\cos(\frac{\pi}{4}+b\alpha)\,[\bar{v}_{0B}(\alpha)+\bar{v}_{0D}(\alpha)] + \sin[(a-b)\alpha]\bar{v}_{0C}(\alpha)}{\cos(\frac{\pi}{4}+a\alpha)} \tag{6.20}$$

and the kernel function

$$F(\alpha) = \frac{\sin[(a-b)\alpha]\cos(\frac{\pi}{4}+b\alpha)}{\cos(\frac{\pi}{4}+a\alpha)}. \tag{6.21}$$

For $b=a$ we have the case of undisturbed propagation with $F(\alpha)=0$ whereas for $b=0$ we obtain the special case of a bifurcated waveguide with a symmetrically inserted Rawlins half-plane

$$F(\alpha) = \frac{1}{\sqrt{2}}\frac{\sin(a\alpha)}{\cos(\frac{\pi}{4}+a\alpha)}. \tag{6.22}$$

The second Rawlins' trifurcated waveguide problem has the inner inserted half-planes interchanged and the pertinent kernel function of this problem assumes the form

$$F(\alpha) = \frac{\cos[(a-b)\alpha]\sin(\frac{\pi}{4}+b\alpha)}{\cos(\frac{\pi}{4}+a\alpha)}. \tag{6.23}$$

For $b = a$ it follows in this case

$$F(\alpha) = \tan(\frac{\pi}{4}+a\alpha). \tag{6.24}$$

This describes the scattering characteristics of a waveguide with changing boundary conditions: $D \to N$ on the upper wall and $N \to D$ on the lower wall. For $b = 0$ we obtain

$$F(\alpha) = \frac{1}{\sqrt{2}}\frac{\sin(a\alpha)}{\cos(\frac{\pi}{4}+a\alpha)} \tag{6.25}$$

which again describes a symmetrically inserted Rawlins half-plane but with reversed boundary conditions with respect to the outer walls. The cases corresponding to equations (6.22) and (6.25) have been solved by the Wiener-Hopf technique by Lüneburg and Hurd [33].

The explicit solution of both trifurcated waveguide problems by the Riemann-Hilbert technique will not be presented here. Here we only wanted to demonstrate that the Riemann-Hilbert method is capable to handle these problems which have been solved by Rawlins using the Wiener-Hopf technique [55, 56]. The Wiener-Hopf technique requires the factorization of the 2×2 Fourier symbol matrices of the Daniele-Khrapkov form ($\kappa(\alpha) = \sqrt{k^2-\alpha^2}$, $0 \le \arg\kappa(\alpha) \le \pi$)

$$G_1(\alpha) = \begin{bmatrix} 1 & -\dfrac{\kappa(\alpha)\sin[\kappa(\alpha)(b+a)]}{\cos[\kappa(\alpha)(b-a)]} \\ -\dfrac{\sin[\kappa(\alpha)(b+a)]}{\kappa(\alpha)\cos[\kappa(\alpha)(b-a)]} & 1 \end{bmatrix} \tag{6.26}$$

for case 1. For case 2 the function sin[...] and cos[...] in equation (6.26) are interchanged.

The foregoing analysis can be extended to include discontinuous boundary conditions on the outer walls of the waveguide, i.e., for $x > 0$ and $x < 0$, respectively, in the present notation. There are essentially three different configurations for combinations of Dirichlet (D) and Neumann (N) boundary conditions. These are listed together with the pertinent kernel functions

$$\begin{aligned} \frac{D-N}{D-N}: &\qquad F(\alpha) = \tan(a\alpha) \\ \frac{D-D}{N-D}: &\qquad F(\alpha) = \tan(2a\alpha) \\ \frac{D-N}{N-D}: &\qquad F(\alpha) = \tan(\frac{\pi}{4}+a\alpha). \end{aligned} \tag{6.27}$$

Here, for instance, $\frac{D-N}{N-D}$ denotes a waveguide with Dirichlet conditions for $x < 0$, $y = a$ and $x > 0$, $y = -a$ and Neumann conditions for $x > 0$, $y = a$ and $x < 0$, $y = -a$.

For one inserted half-plane with either Dirichlet or Neumann conditions on the upper or lower side there are altogether 6 different half-planes involved yielding in principle $2^6 = 64$ different combinations. Furthermore, a distinction has to be made whether the inserted half-plane is placed symmetrically or asymmetrically inside the duct. A number of cases has been investigated so far but the answer is still incomplete. What can be said, however, is the following: whenever such a waveguide problem leads to a scalar or to a 2-by-2 matrix Wiener-Hopf system of the Daniele-Khrapkov form it appears that it can be formulated also as a standard Riemann-Hilbert problem and vice versa. Hurd and Meister [19] investigated the general bifurcated waveguide problem including waveguide discontinuities with the boundary conditions $\partial v/\partial x = Z_i v$ $(i = 1, \ldots, 6)$ on the six sections $x > o$ and $x < 0$ and pointed out those problems that can be solved by the Wiener-Hopf method.

In this context the bifurcated half-plane sketched previously leads to the Wiener-Hopf kernel function

$$F(\alpha) = \frac{\sin[(a+b)\alpha]}{\sin(a\alpha)\sin(b\alpha)}. \tag{6.28}$$

If the Dirichlet boundary condition (D) on the lower wall is replaced by the Neumann boundary condition (N) the kernel changes to

$$F(\alpha) = \frac{\cos[(a+b)\alpha]}{\sin(a\alpha)\cos(b\alpha)}. \tag{6.29}$$

In some cases the analysis can be extended to a finite number of equidistant inserted half-planes, cf. [21] and [33]. The latter papers make use of the fact that the corresponding infinite grating problems can be solved for arbitrary angles of incidence where it should be reminded that for infinite grating problems the angle of incidence appears in the kernel matrix to be factorized. Letting a tend to infinity, $a \to \infty$, or a tend to b, $a \to b$, various other diffraction problems can be derived using the preceding solutions. A survey of waveguide problems with different impedance boundary conditions by the Wiener-Hopf technique is given by Hurd and Meister [19]. Finally it should be pointed out that the corresponding circular waveguide problems can be treated similarly, cf. Weinstein [65] for some of these problems.

7. Diffraction by an infinite strip grating

Diffraction of an incident plane electromagnetic wave $\phi_0(x, z|\theta_0)$ and arbitrary angle of incidence θ_0 by the special infinite grating with strips and gaps of equal width and Dirichlet (or Neumann) boundary conditions on both sides is a classical problem in the mathematical theory of diffraction that has been a test problem for

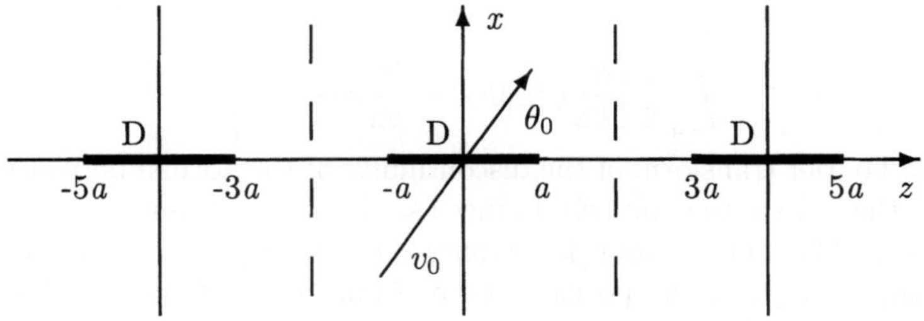

FIGURE 8. Infinite strip grating with Dirichlet boundary conditions

a variety of different solution techniques. The more general problem with strips and gaps of unequal widths has not found a closed-form solution yet. For approximate solutions obtained by truncating an infinite set of algebraic equations reference will be made to Agranovich et al. [1] and Shestopalov [60]. For arbitrary angle of incidence the Wiener-Hopf approach leads to the factorization of two decoupled 2×2 symbol matrices $K_i(\alpha)$ $(i = 1, 2)$ of the form

$$K_1(\alpha) = G_1(\alpha) + G_2(\alpha), \quad K_2(\alpha) = G_1(\alpha) - G_2(\alpha) \tag{7.1}$$

with

$$G_1(\alpha) = \frac{1}{\Delta(\alpha)} \left\{ \sinh(d) I + \frac{\sinh[4a\gamma(\alpha)]}{\gamma(\alpha)} C(\alpha) \right\} \tag{7.2}$$

$$G_2(\alpha) = \frac{2}{\Delta(\alpha)} \left\{ \sinh(d/2)\cosh[2a\gamma(\alpha)] I + \frac{\cosh(d/2)\sinh[2a\gamma(\alpha)]}{\gamma(\alpha)} C(\alpha) \right\} \tag{7.3}$$

and where $d = 4ia\alpha_0$, $\alpha_0 = k\cos\theta_0$, I is the 2×2 unit matrix and

$$\Delta(\alpha) = \cosh[4a\gamma(\alpha)] - \cosh(d), \quad C(\alpha) = \begin{bmatrix} 0 & 1 \\ \alpha^2 - k^2 & 0 \end{bmatrix}. \tag{7.4}$$

The matrices $G_i(\alpha)$ $(i = 1, 2)$ are of the Daniele-Khrapkov form and can be factorized explicitly. For details see Lukyanov [36, 37], a recent reevaluation by Daniele et al. [10] and a survey article by Lüneburg [28].

For normal incidence $\theta_0 = 0$ this problem has been solved by Baldwin and Heins [2] as early as 1954 and independently by Weinstein [64] by scalar Wiener-Hopf technique using a dual integral equation formulation. This analysis has been extended by Weinstein [65] to Bragg angle incidence.

Restricting all considerations to the unit cell $\{-2a \le x \le 2a, -\infty < z < \infty\}$ we obtain by means of alternative integral representations of appropriate Green's function for the total field

$$v(x, z|\phi_i) = v_i(x, z|\phi_i) + \frac{1}{4\pi} \oint \hat{\psi}(\alpha) \cot[2a(\alpha - \alpha_0)] e^{i\alpha x + i\kappa(\alpha)|z|} \frac{d\alpha}{\kappa(\alpha)} \tag{7.5}$$

where

$$\hat{\psi}(\alpha) = \int_{-a}^{a} \frac{1}{2}\left[\frac{\partial v}{\partial z}\psi(x|0-) - \frac{\partial v}{\partial z}\psi(x|0+)\right] e^{-i\alpha x} dx \tag{7.6}$$

is the finite Fourier transform of the discontinuity of the normal derivatives of the field across the strip in the unit cell (induced surface current) and hence is an entire function of α. The loop contour $\oint$ encircles the simple poles of $1/\cos[a(\alpha - \alpha_0)]$ and $1/\sin[a(\alpha - \alpha_0)]$ in the positive sense. Application of the Dirichlet boundary condition $v(x|0\pm) = 0$ for $-a < x < a$, multiplication by $\exp\{-i\alpha x\}$ and integration over x from $-a$ to $+a$ leads to the following functional equation

$$\begin{aligned} \frac{1}{8\pi}\oint \hat{\psi}(\alpha')\left\{\frac{\cos[a(\alpha-\alpha_0)]}{\cos[a(\alpha'-\alpha_0)]} + \frac{\sin[a(\alpha-\alpha_0)]}{\sin[a(\alpha'-\alpha_0)]}\right\}\frac{1}{\kappa(\alpha')}\frac{d\alpha'}{(\alpha'-\alpha)} \\ = \frac{\sin[a(\alpha-\alpha_0)]}{\alpha-\alpha_0} \end{aligned} \tag{7.7}$$

where the points α lie outside the loop contour of integration. Due to the edge condition the functions $\hat{\psi}(\alpha)\exp\{\pm ia\alpha\}$ are holomorphic in the upper/lower α-half-plane and are of the order $O(\alpha^{-\frac{1}{2}})$ as $\alpha \to \infty$.

A shift of the path of integration to a loop C around the branch cuts of $\kappa(\alpha)$, combining the contributions on both sides (opposite signs but opposite directions of integration) along the contour Γ that goes from the branch point k to infinity and then to the other branch point $-k$. Taking into account the residue contribution at $\alpha' = \alpha$ yields

$$\begin{aligned} \frac{\hat{\psi}(\alpha)}{\kappa} + \frac{1}{2\pi i}\int_{\Gamma} \hat{\psi}(\alpha')\left\{\frac{\cos[a(\alpha-\alpha_0)]}{\cos[a(\alpha'-\alpha_0)]} + \frac{\sin[a(\alpha-\alpha_0)]}{\sin[a(\alpha'-\alpha_0)]}\right\}\frac{1}{\kappa(\alpha')}\frac{d\alpha'}{(\alpha'-\alpha)} \\ = 2i\frac{\sin[a(\alpha-\alpha_0)]}{\alpha-\alpha_0}. \end{aligned} \tag{7.8}$$

Letting $\alpha \to \Gamma$ from one side or the other leads to the singular integral equation

$$\begin{aligned} PV\frac{1}{\pi i}\int_{\Gamma} \hat{\psi}(\alpha')\left\{\frac{\cos[a(\alpha-\alpha_0)]}{\cos[a(\alpha'-\alpha_0)]} + \frac{\sin[a(\alpha-\alpha_0)]}{\sin[a(\alpha'-\alpha_0)]}\right\}\frac{1}{\kappa(\alpha')}\frac{d\alpha'}{(\alpha'-\alpha)} \\ = 4i\frac{\sin[a(\alpha-\alpha_0)]}{\alpha-\alpha_0}, \qquad \alpha \in \Gamma \end{aligned} \tag{7.9}$$

on the open arc Γ. This equation can also be considered as a boundary value problem on the two-sheeted Riemann surface of $\kappa(\alpha)$, see Zverovich [68] and Meister and Penzel [44]. Introducing the Cauchy type integrals

$$H_1(\alpha) = \frac{1}{2\pi i}\int_{\Gamma} \hat{\psi}(\alpha')\frac{1}{\cos[a(\alpha'-\alpha)]}\frac{1}{\kappa(\alpha')}\frac{d\alpha'}{\alpha-\alpha'}, \tag{7.10}$$

$$H_2(\alpha) = \frac{1}{2\pi i}\int_{\Gamma} \hat{\psi}(\alpha')\frac{1}{\sin[a(\alpha'-\alpha)]}\frac{1}{\kappa(\alpha')}\frac{d\alpha'}{\alpha-\alpha'} \tag{7.11}$$

the following Riemann-Hilbert problem will be obtained from standard procedures

$$M(\alpha)H^{+}(\alpha)+H^{-}(\alpha)=R(\alpha), \quad \alpha\in\Gamma \tag{7.12}$$

with the 2-vectors

$$H(\alpha)=\begin{bmatrix} H_1(\alpha)\\ H_2(\alpha)\end{bmatrix}, \quad R(\alpha)=\frac{2i}{\alpha-\alpha_0}\begin{bmatrix}\tan[a(\alpha-\alpha_0)]\\ 0\end{bmatrix} \tag{7.13}$$

and the coupling matrix

$$M(\alpha)=\begin{bmatrix} 0 & \tan[a(\alpha-\alpha_0)]\\ \cot[a(\alpha-\alpha_0)] & 0\end{bmatrix}. \tag{7.14}$$

The matrix $M(\alpha)$ has zero elements in the main diagonal and meromorphic functions with product one in the secondary diagonal. The final solution reads

$$\hat{\psi}(\alpha)=2i\frac{\kappa(\alpha_0)}{L(\alpha_0)}\left\{\frac{\sin[a(\alpha-\alpha_0)]}{\alpha-\alpha_0}L(\alpha)+\cos[a(\alpha-a\alpha_0)]\frac{1}{L(\alpha)}\right\} \tag{7.15}$$

where

$$L(\alpha)=\exp\left\{-\frac{\kappa(\alpha)}{2\pi i}\int_\Gamma \ln\left[(\alpha-\alpha_0)\cot[a(\alpha'-\alpha_0)]\right]\frac{1}{\kappa(\alpha')}\frac{d\alpha'}{\alpha'-\alpha}\right\}. \tag{7.16}$$

Identifying $L(\alpha)$ with $L^{I}(\alpha)$, a function that is holomorphic on the upper (physical) sheet of the two-sheeted Riemann surface of $\kappa(\alpha)$ along the branch cut Γ, and denoting by $L^{II}(\alpha)$ its analytical continuation into the second sheet of $\kappa(\alpha)$ we note the relation

$$L^{I}(\alpha)L^{II}(\alpha)=(\alpha-\alpha_0)\cot[a(\alpha-\alpha_0)] \tag{7.17}$$

as another example of the factorization of a meromorphic function on the two-sheeted Riemann surface of $\kappa(\alpha)$. Infinite product expansions and high-frequency asymptotic approximations are provided by Lüneburg and Westpfahl [35].

It is quite interesting that the problem by this special strip grating for arbitrary angles of incidence ϕ_i does not present any additional complications when compared with the case of normal incidence or more generally Bragg angle incidence. This is in remarkable contrast to the Wiener-Hopf method where the generalization from Bragg angle incidence to arbitrary incidence signifies a transition from a scalar Wiener-Hopf equation to a vector Wiener-Hopf equation with a Fourier symbol matrix function of the Daniele-Khrapkov form. The canonical factorization of such matrices became effectively possible on the early 80's after the publication of Daniele's work [7, 8, 9] following basic ideas introduced by Heins [14] in 1950 and after the recognition in the western world of the pioneering work of Chebotarev [5] in 1956 and Khrapkov [24] in 1971.

This special strip grating problem has been reconsidered as a matrix Wiener-Hopf equation by Daniele et al [10] and in terms of convolution operators on a finite interval by Bastos and dos Santos [3].

8. Conclusions

It has been shown that the Riemann-Hilbert method offers an alternative to the conventional Wiener-Hopf technique for solving parallel-plate waveguide problems. Apparently every problem that can be formulated as a 2×2 matrix Wiener-Hopf system of the Daniele-Khrapkov form can also be formulated as a standard Riemann-Hilbert problem. This is demonstrated for two new trifurcated waveguide problems recently solved by Rawlins.

It is interesting that the standard Riemann-Hilbert form comprises scalar as well as 2×2 matrix Wiener-Hopf equations of the Daniele-Khrapkov form. Historically the special strip grating problem (strips and gaps of equal width) with Dirichlet conditions was solved first by the Riemann-Hilbert approach in 1971 [35]. Khrapkov's results [24] first became known in the west at the beginning of the eighties, complementing and extending Daniele's findings [7, 8, 9] and [10].

The formulation of the boundary value problem for parallel-plate waveguides as a matrix Wiener-Hopf equation is usually straightforward and it is readily seen if the resulting kernel matrix is of the solvable Daniele-Khrapkov form or not. The formulation as a Riemann-Hilbert problem is much more tedious. It involves the proper definition of unknowns, the appropriate step from the coupled integral equations in coordinate space to the complex transformation space and the weighted addition of the individual equations. At present these steps are guided by familiarity and experience. Unfortunately this lack of simple comprehensive guiding lines seems to restrict the dissemination of this approach.

References

[1] Agranovich, Z.S., V.A. Marchenko and V.P. Shestopalov, *The diffraction of electromagnetic waves from plane metallic lattices*, Sov. Phys. Tech. Phys., Engl. transl. 7(4) (1962) 277–286.

[2] Baldwin, G.L. and A.E. Heins, *The reflection of electromagnetic waves by an infinite plane grating*, Math. Scand. **2** (1954) 103–118.

[3] Bastos, M.A. and A.F. Dos Santos, *Convolution operators on a finite interval with periodic kernel-Fredholm property and invertibility,* Intgr. Equat. Oper. Theory **16** (1993), 186–223.

[4] Büyükaksoy, A. and A.H. Serbest, *Matrix Wiener-Hopf factorization methods and applications to some diffraction problem*, Chapter 6 in: Analytical and Numerical Methods in Electromagnetic Wave Theory, edited by M. Hashimoto, M. İdemen, and O.A. Tretyakov, Science House Co, Tokyo, 1993, 257–315.

[5] Chebotarev, G.N., *On the closed form solution of the Riemann boundary value problem for a system of n pairs of functions*, Kazan. Gos. Univ. Uchen. Zap. **116**(4) (1956) 31–58 (in Russian).

[6] Cohen, D.S., *Separation of variables and alternative representations for non self-adjoint boundary value problems,* Comm. Pure and Appl. Math. **17** (1964), 1–22.

[7] Daniele, V., *On the factorization of Wiener-Hopf matrices in problems solvable with Hurd's method,* IEEE Trans. Antennas Propagat. **AP-26** (1978), 614–616.

[8] Daniele, V., *On the solution of two coupled Wiener-Hopf equations,* SIAM J. Appl. Math. **44** (1984), 667–680.

[9] Daniele, V., *On the solution of vector Wiener-Hopf equations occurring in scattering problems,* Radio Science **19** (1984), 1173–1178.

[10] Daniele, V., M. Gilli and E. Viterbo, *Diffraction of a plane wave by a strip grating,* Electromagnetics **10** (1990), 245–269.

[11] Dos Santos, A.F. and F.S. Teixeira, *The Sommerfeld problem revisited: Solution spaces and the edge condition,* J. Math. Analysis and Applications, **143**(2) (1989) 341–357.

[12] Ehrhardt, T. and F.-O. Speck, *Transformation techniques towards the factorization of non-rational* 2×2 *matrix functions*, Linear Algebra and its Applications, **353**(1–3) (2002) 53–90.

[13] Estrada, R. and R.P. Kanwal, *Singular Integral Equations*, Birkhäuser, Boston, 2000.

[14] Heins, A.E., *Systems of Wiener-Hopf integral equations and their applications to some boundary value problems in electromagnetic theory*, in Proc. Symp. Appl. Math. II, McGraw-Hill Book Company, New York, 1950, 76–81.

[15] Heins, A.E. and R.C. MacCamy, *A function-theoretic solution of certain integral equations (I), (II),* Quart. J. Math. Oxford (2), **9** (1958) 132–143; dto **10** (1959) 280–293.

[16] Heins, A.E., *The Sommerfeld half-plane problem revisited, II. The factoring of a matrix of analytic functions*, Math. Meth. Appl. Sci. **5** (1983) 14–21.

[17] Hurd, R.A., *The Wiener-Hopf-Hilbert method for diffraction problems*, Can. J. Phys **54** (1976) 775–780

[18] Hurd, R.A. and E. Lüneburg, *Scattering by hard and soft parallel half-planes*, Can. J. Phys. **59** (1981) 1879–1885.

[19] Hurd, R.A. and E. Meister, *Generalized waveguide bifurcation problems*, Quart. J. Mech. Appl. Math. **41** (1981) 127–139.

[20] Hurd, R.A. and S. Przeździecki, *Diffraction by a half-plane with different face impedances – a re-examination,* Can. J. Phys., **59** (1981) 1337–1347.

[21] Igarashi, A., *Simultaneous Wiener-Hopf equations and their application to diffraction problems in electromagnetic theory I, II, III, IV,* J. Phys. Soc. Japan **19** (1964), 1213–1221; ibid. **25** (1968), 260–271; ibid. **25** (1968), 607–615; ibid. **26** (1969), 549–560.

[22] Jones, D.S., *A simplifying technique in the solution of a class of diffraction problems,* Quart. J. Math. Oxford (2), **3** (1952) 189–196.

[23] Jones, D.S., *Acoustic and Electromagnetic Waves*, Clarendon Press, Oxford, 1986.

[24] Khrapkov, A.A., *Certain cases of the elastic equilibrium of an infinite wedge with a non-symmetric notch at the vortex subjected to concentrated forces,* Prikl. Mat. Mekh. **35** (1971), 677–689; Engl. trans. J. Appl. Math. Mech. **35**, 625–637.

[25] Lüneburg, E., *Diffraction by an infinite set of parallel half-planes,* National URSI Conference, Kleinheubacher Berichte **24** (1981), 17–35.

[26] Lüneburg, E., *Field matching and singular integral equations,* Radio Science **16** (1981), 979–981.

[27] Lüneburg, E., *Diffraction by an infinite set of soft/hard parallel half-planes: The Riemann approach,* Can. J. Phys. **60** (1982), 1125–1138.

[28] Lüneburg, E., *Diffraction by an infinite set of parallel half-planes and by an infinite strip grating: Comparison of different methods,* Chapter 7 in: Analytical and Numerical Methods in Electromagnetic Wave Theory, edited by M. Hashimoto, M. İdemen, and O.A. Tretyakov, Science House Co, Tokyo, 1993, 317–372.

[29] Lüneburg, E., *Plane waveguides and the Riemann-Hilbert method: An alternative to the Wiener-Hopf technique*, in: Direct and Inverse Electromagnetic Scattering, H. Serbest and S. Cloude, eds., Pitman Research Notes in Mathematics Series 361, Addison Wesley Longman, Harlow, U.K., 1996, 56–68.

[30] Lüneburg, E., *The Sommerfeld Problem: Methods, Generalizations and Frustrations*, in: Modern Mathematical Methods in Diffraction Theory and ist Applications in Engineering, Methoden und Verfahren der mathematischen Physik, Bd. 42, Peter Lang, Frankfurt, 1997, 145–162.

[31] Lüneburg, E. and R.A. Hurd, *Diffraction by an infinite set of hard and soft parallel half-planes,* Can. J. Phys. **60** (1982), 1–9.

[32] Lüneburg, E. and R.A. Hurd, *Diffraction by an infinite set of soft/hard parallel half-planes,* Radio Science **17** (1982), 453–462.

[33] Lüneburg, E. and R.A. Hurd, *Two waveguide bifurcation problems,* Zeit. angew. Math. Mech. (ZAMM) **65** (1985), 551–559.

[34] Lüneburg, E. and H. Serbest, *Skew incidence half-plane diffraction problems: Particular aspects of the factorization problem*, in Singular Integral Operators, Factorization and Applications, edited by A. Böttcher et al., Vol. 142 of Operator Theory: Advances and Applications, Birkhäuser, Basel, 2003, 223–251.

[35] Lüneburg, E. and K. Westpfahl, *Diffraction of plane waves by an infinite strip grating,* Ann. der Physik **27** (1971), 257–288.

[36] Lukyanov, V.D., *Exact solution of the problem of diffraction of an obliquely incident wave at a grating*, Dokl. Akad. Nauk SSSR **255** (1980), 78–80; Engl. transl. Soviet Phys. Dokl. **25**(11) (1980), 905–906.

[37] Lukyanov, V.D., *Exact solution of the problem of diffraction of plane waves obliquely incident on a grating*, Zh. Tekh. Fiz. **51** (1981), 2001–2006; Engl. transl. Sov. Phys. Tech. Phys. **26**(10) (1981), 1167–1169.

[38] Maliuzhinets, G.D., *Excitation, reflection and emission of surface waves from a wedge with given face impedances*, Sov. Phys. Dokl., Engl. transl.**3**(1958) 752–755.

[39] Meister, E., *Zum Dirichlet-Problem der Helmholtzschen Schwingungsgleichung für ein gestaffeltes Streckengitter*, Arch. Rat. Mech. Anal. **10** (1962), 67–100.

[40] Meister, E., *Zum Neumann-Problem der Helmholtzschen Schwingungsgleichung für ein gestaffeltes Streckengitter*, Arch. Rat. Mech. Anal. **10** (1962), 127–148.

[41] Meister, E., *Randwertaufgaben der Funktionentheorie*, B.G. Teubner, Stuttgart, 1983.

[42] Meister, E., *Some multiple-part Wiener-Hopf problems in mathematical physics*, St. Banach Center, Publ. **15** (1985) 395–407.

[43] Meister, E. and F.-O. Speck, *Modern Wiener-Hopf methods in diffraction theory*, in: Ordinary and Partial Differential Equations, Vol. II, Proc. 10th Dundee Conf.,

Dundee/UK 1988 (Eds. B.D. Slater, R.J. Jarvis), Pitman Res. Notes Math. Ser. **216** (1989) 130–171.

[44] Meister, E. and F. Penzel, *On the reduction of the factorization of matrix functions of Daniele-Khrapkov type to a scalar boundary value problem on a Riemann surface,* Complex Variables, Theory Appl. **18**(1–2) (1992) 63–71.

[45] Meixner, J., *Die Kantenbedingung in der Theorie der Beugung elektromagnetischer Wellen an vollkommen leitenden ebenen Schirmen,* Ann. der Physik **6** (1949), 1–9.

[46] Meixner, J., *The behavior of electromagnetic fields at edges,* IEEE Trans. Antennas Propagat. **AP-20** (1972), 442–446.

[47] Mittra, R., *Relative convergence of the solution of a doubly infinite set of equations,* J. Res. Natl. Bur. Std. **67D** (1963), 245–254.

[48] Mittra, R. and S.W. Lee, *Analytical Techniques in the Theory of Guided waves,* The MacMillan Company, New York, 1971.

[49] Noble, B., *Methods Based on the Wiener-Hopf Technique for the Solution of Partial Differential Equations,* Pergamon Press, London, 1958; second edition, Chelsea Publishing, New York, 1988.

[50] Pace, J.R. and R. Mittra, *The trifurcated waveguide,* Radio Science **1** (1966), 117–121.

[51] Rawlins, A.D., *The solution of a mixed boundary value problem in the theory of diffraction by a semi-infinite plane*, Proc. Roy. Soc. London, Ser. A **346** (1975) 469–484.

[52] Rawlins, A.D., *A note on the factorization of matrices occurring in Wiener-Hopf problems*, IEEE Trans. Ant. Propag. **AP-28** (1980) 933–934.

[53] Rawlins, A.D., *Simultaneous Wiener-Hopf equations*, Can. J. Phys., **58** (1980) 420–428.

[54] Rawlins, A.D., *The explicit Wiener-Hopf factorization of a special matrix*, Z. Angew. Math. Mech. (ZAMM) **61** (1981) 527–528.

[55] Rawlins, A.D., *Two trifurcated waveguide problems solved exactly by matrix Wiener-Hopf factorization*, in: Direct and Inverse Electromagnetic Scattering, edited by H. Serbest and S. Cloude, Pitman Research Notes in Mathematics Series 361, Addison Wesley Longman, Harlow, U.K., 1996, 82–88.

[56] Rawlins, A.D., *Two Waveguide Trifurcation Problems*, Math. Proc. Cambridge Phil. Soc. **121**(3) (1998), 555–573.

[57] Rawlins, A.D. and W.E. Williams, *Matrix Wiener-Hopf factorization*, Quart. J. Mech. Appl. Math., **34**(1) (1981) 1–8.

[58] Schmeltzer, R.A. and M. Levin, *Function-theoretic solution to a class of dual integral equations and an application to diffraction theory,* Quart. Appl. Math., **21** (1963) 269–283, 1963; cf. MacCamy in MR 27, Nr. 5102, 1964.

[59] Senior, T.B.A. and J.L. Volakis, *Approximate Boundary Conditions in Electromagnetics,* Inst. of Electr. Eng., London, 1995.

[60] Shestopalov, V.P., *The Riemann Method in the Theory of Diffraction and Propagation of Electromagnetic Waves*, Isdatelstvo Kharkov Univ., Kharkov, 1971 (in Russian).

[61] Sommerfeld, A., *Zur Theorie der Diffraction,* Math. Ann. **47** (1896) 317–374.

[62] Spitkovsky, I.M. and A.M. Tashbaev, *On the efficient factorization of matrix functions*, Izv. Vyssh. Uchebn. Zaved. Mat., **33**(4) (1989) 69–76 (in Russian); Engl. transl. in: Soviet Math. (Iz VUZ) **33**(4) (1989) 85–93.

[63] Whitehead, E.A.N., *The theory of parallel plate media fore microwave lenses*, Proc. IEE (London) **98**, pt. III (1951), 133–140.

[64] Weinstein, L.A., *Diffraction of electromagnetic waves at gratings consisting of parallel conducting strips*, Zh. Tekh. Fiz. **25**(2) (1955) 847–853.

[65] Weinstein, L.A., *The Theory of Diffraction and the Factorization Method*, The Golem Press, Boulder, 1969.

[66] Weinstein, L.A. and W.I. Volman, *On the rigorous solution of a diffraction problem,* Dokl. Acad. Nauk SSSR **286**(6) (1985), 1360–1364; IX All-Union Sympos. on Diffraction and Propagation of Waves, Tbilisi, 1985 (in Russian).

[67] Williams, W.E., *Recognition of same readily "Wiener-Hopf factorizable" matrices*, IMA J. Appl. Math., **32** (1984) 367–378.

[68] Zverovich, E.I., *Boundary value problems in the theory of analytic functions in Hölder classes on Riemann surfaces*, Russ. Surveys (Uspekhi Mat. Nauk), **26** (1971) 117–192.

Ernst Lüneburg
EML Consultants
Georg Schmid Weg 4
D-82234 Wessling
Germany
e-mail: `e.lueneburg@t-online.de`

Operator Theory
Advances and Applications, Vol. 147, 353–363

The Integral Equation Method for Dipole Equilibrium

Erich Martensen and Stefan Ritter

Dedicated to the memory of Erhard Meister

Abstract. The equations of dipole distributions being in equilibrium in a compact three-dimensional domain are treated by the integral equation method. It is shown that electric or magnetic dipole equilibria are governed by the eigenfunctions of the electrostatic integral operator or its adjoint operator, respectively.

1. Dipole potentials and fields

Let $D \subset \mathbb{R}^3$ denote a compact domain with closed exterior $\hat{D} = \overline{\mathbb{R}^3 \setminus D}$. We assume the boundary S of D consisting of at most a finite number of disjoint continuously curved closed surfaces (the components of S) with unit normal n directed outward of D. Then we will be concerned with dipole distributions $p \in [C^\alpha(D)]^3$ spread over D where the Hölder exponent $\alpha \in (0,1)$ is considered as a priori fixed. For such distribution p the scalar and vector potentials are given by

$$\Phi(x) = \frac{1}{4\pi} \int_D p(y) \cdot \operatorname{grad}_y \frac{1}{|y-x|} dV(y), \quad x \in \mathbb{R}^3, \tag{1.1}$$

$$A(x) = \frac{1}{4\pi} \int_D p(y) \times \operatorname{grad}_y \frac{1}{|y-x|} dV(y), \quad x \in \mathbb{R}^3, \tag{1.2}$$

and equipped with regularities [1,3]

$$\Phi \in C^\alpha(\mathbb{R}^3) \cap C^{1+\alpha}(D) \cap C^{1+\alpha}(\hat{D}),$$

$$A \in [C^\alpha(\mathbb{R}^3)]^3 \cap [C^{1+\alpha}(D)]^3 \cap [C^{1+\alpha}(\hat{D})]^3,$$

respectively.

We first consider $p \in [C^\alpha(D)]^3$ to be *complex*. Since the potentials (1.1) and (1.2) are continuous on S, their tangential derivatives are also continuous. Hence

$$n \times \operatorname{grad} \Phi|_+ = n \times \operatorname{grad} \Phi|_- \quad \text{on } S, \tag{1.3}$$

$$n \cdot \operatorname{curl} A|_+ = n \cdot \operatorname{curl} A|_- \quad \text{on } S, \tag{1.4}$$

with $\cdot|_+$ and $\cdot|_-$ denoting the limiting values on the exterior or interior side of S, respectively. For (1.1) and (1.2) we have

$$\operatorname{grad}\Phi(x) + \operatorname{curl} A(x) = (-\operatorname{grad}\operatorname{div} + \operatorname{curl}\operatorname{curl})\left(\frac{1}{4\pi}\int_D \frac{p(y)}{|y-x|}dV(y)\right),$$

and by using the formula $-\operatorname{grad}\operatorname{div} + \operatorname{curl}\operatorname{curl} = -\Delta$ we get

$$\operatorname{grad}\Phi + \operatorname{curl} A = \begin{cases} p & \text{in } D, \\ 0 & \text{in } \hat{D}. \end{cases} \tag{1.5}$$

Let $p, \overline{p} \in [C^\alpha(D)]^3$ be a complex dipole distribution and its conjugate. Then, by linearity, the corresponding scalar and vector potentials are $\Phi, \overline{\Phi}$ and $A, \overline{A}$, respectively. By integration of

$$\operatorname{grad}\Phi \cdot \operatorname{curl}\overline{A} = \operatorname{div}(\Phi \operatorname{curl}\overline{A}) \quad \text{in } D, \hat{D},$$

where the divergence is to be understood in the weak sense, and application of the (weak) Gauss theorem together with the conjugate of (1.4), it follows that

$$\int_{\mathbb{R}^3} \operatorname{grad}\Phi \cdot \operatorname{curl}\overline{A}\, dV = 0. \tag{1.6}$$

If $|\operatorname{grad}\Phi + \operatorname{curl} A|^2$ is integrated over $\mathbb{R}^3$, then (1.5) and (1.6) lead to

$$\int_{\mathbb{R}^3} |\operatorname{grad}\Phi|^2 dV + \int_{\mathbb{R}^3} |\operatorname{curl} A|^2 dV = \int_D |p|^2 dV. \tag{1.7}$$

For a non-trivial complex dipole distribution $p \in [C^\alpha(D)]^3$ the *relative potential energy* [2] is given by the Rayleigh quotient

$$\Lambda = \frac{\int_{\mathbb{R}^3} |\operatorname{grad}\Phi|^2\, dV}{\int_D |p|^2\, dV} = \frac{\int_D \operatorname{grad}\Phi \cdot \overline{p}\, dV}{\int_D p \cdot \overline{p}\, dV} \tag{1.8}$$

where the transformation of the numerator is based on the conjugate of (1.5) together with (1.6). From (1.7) and (1.8) it is seen that $\Lambda \in [0, 1]$. The dipole distribution p is *in equilibrium (in D)* if (1.8) becomes stationary. From this an isoperimetric variational problem arises. It is solved by all complex vector functions $p \in [C^\alpha(D)]^3$ satisfying for some complex parameter λ one of the two equivalent equations [6, 8, 9]

$$\operatorname{grad}\Phi = \frac{1+\lambda}{2} p, \quad \operatorname{curl} A = \frac{1-\lambda}{2} p \quad \text{in } D; \tag{1.9}$$

note that the equivalence of both equations is immediately clear from (1.5).

Now let $p \in [C^\alpha(D)]^3$ be a non-trivial complex *equilibrium distribution* satisfying (1.9) to complex λ. Then the Rayleigh quotient (1.8) becomes

$$\Lambda = \frac{1+\lambda}{2} \in [0, 1],$$

i.e., λ is *real lying in the interval* $[-1, 1]$. Hence the equilibrium is stable if $\lambda = -1$ and unstable if $\lambda \in (-1, 1]$, the "degree" of instability increasing with λ. Therefore beside the case $\lambda = -1$, equilibria with λ close to -1 are of particular interest.

From the evaluation of (1.6) by means of (1.5) and (1.9) we find that

$$\int_{\hat{D}} |\operatorname{grad}\Phi|^2 dV = \int_{\hat{D}} |\operatorname{curl} A|^2 dV = \frac{1-\lambda^2}{4} \int_D |p|^2 dV. \tag{1.10}$$

Finally observing from (1.9) that the real and imaginary parts of a complex solution are forming real solutions to the same (real) parameter λ, there is no restriction for the further proceeding if we confine ourselves to *real* $p \in [C^\alpha(D)]^3$ solving (1.9) to $\lambda \in [-1, 1]$.

Physically, a dipole distribution p in an appropriate material D may be realized by an *(electric) polarization* (SI unit: As/m^2) or a *magnetization* (SI unit: A/m). In case of equilibrium, p is called a *permanent polarization* or a *permanent magnetization,* respectively. The corresponding electric or magnetic fields are given by

$$E = -\varepsilon_0^{-1} \operatorname{grad}\Phi \quad \text{in } D, \hat{D}, \tag{1.11}$$

$$B = \mu_0 \operatorname{curl} A \quad \text{in } D, \hat{D}, \tag{1.12}$$

respectively, where ε_0 is the permittivity and μ_0 the permeability of vacuum. With regard to (1.3) and (1.4), the tangential components of E and the normal components of B are continuous when passing through S.

Electric or magnetic dipole distributions in equilibrium p in D are governed by the same equations (1.9). The corresponding electric or magnetic fields (1.11) or (1.12), however, differ essentially. Particularly, as seen from (1.9), the electric field (1.11) has the opposite direction whereas the magnetic field (1.12) has the same direction as p in D. With regard to this physical background, the equilibrium equations (1.9) will be treated by different integral equation methods. It results that the electric or magnetic dipole equilibria are represented by *electric single layers* or *magnetic double layers* on the boundary, respectively. An integral equation approach for magnetic dipole equilibrium based on *surface currents* has been given in [6, 9].

Placing behind the limiting cases $\lambda = 1, -1$ (these cases are characterized from (1.10) by vanishing exterior fields), we shall draw our main attention to the case $\lambda \in (-1, 1)$. For such λ, according to (1.10), the exterior fields only vanish for the trivial solution p and, consequently, every possible exterior equilibrium field is obtained from exactly one solution p of (1.9).

For the following it is useful to introduce the linear space $P^\alpha(D)$ of all *flux-free and circulation-free*[1] dipole distributions $p \in [C^\alpha(D)]^3$. Due to the theorem of Koebe-Bôcher [4], such vector functions attain the additional properties that $p \in [C^1(\underline{D})]^3$ with $\operatorname{div} p = 0$, $\operatorname{curl} p = 0$ in $\underline{D}$ valid in the strong sense. Let further $P^\alpha_\lambda(D)$ denote the linear space of all $p \in [C^\alpha(D)]^3$ solving (1.9) to the parameter λ. Then there is obviously the inclusion

$$P^\alpha_\lambda(D) \subseteq P^\alpha(D), \quad \lambda \in (-1, 1). \tag{1.13}$$

[1] Flux-freeness or circulation-freeness of a vector function in $\mathbb{R}^3$ is to be understood with respect to all *closed surfaces* or all *closed curves*, respectively, placed in the domain under consideration.

2. Electric dipole equilibrium

We consider an electric single layer $q \in C^\alpha(S)$ (SI unit: As/m^2) placed on the boundary S. The corresponding scalar potential

$$\Phi(x) = \frac{1}{4\pi} \int_S \frac{q(y)}{|y-x|} dS(y), \quad x \in \mathbb{R}^3, \tag{2.1}$$

has the regularity [1, 3]

$$\Phi \in C^\alpha(\mathbb{R}^3) \cap C^{1+\alpha}(D) \cap C^{1+\alpha}(\hat{D}).$$

Lemma 1. *Let $Q^\alpha(S)$ denote the linear space of all single layers $q \in C^\alpha(S)$ vanishing in the mean over every boundary component of S. Then the linear spaces $P^\alpha(D)$ and $Q^\alpha(S)$ are isomorphic where the mapping is characterized by equal scalar potentials* (1.1) *for $p \in P^\alpha(D)$ and* (2.1) *for $q \in Q^\alpha(S)$. Furthermore, corresponding p and q are due to*

$$n \cdot p = q \quad \text{on } S. \tag{2.2}$$

Proof. The mapping. For $p \in P^\alpha(D)$ integration by parts together with $\operatorname{div} p = 0$ in $\underline{D}$ yields

$$\int_D p(y) \cdot \operatorname{grad}_y \frac{1}{|y-x|} dV(y) = \int_S \frac{(n \cdot p)(y)}{|y-x|} dS(y), \quad x \in \mathbb{R}^3. \tag{2.3}$$

From this the scalar potentials (1.1) and (2.1) coincide if the single layer $q \in C^\alpha(S)$ is defined by (2.2). From (2.2) it further follows that q vanishes in the mean over every component of S since p is flux-free. Hence $q \in Q^\alpha(S)$. – If $q \in Q^\alpha(S)$ is the *difference* of two single layers which have the same scalar potentials as $p \in P^\alpha(D)$, then (2.1) vanishes, and from the jump relation for (14) it results that $q = 0$ on S. Thus for every $p \in P^\alpha(D)$ there is exactly one $q \in Q^\alpha(S)$ with equal scalar potential.

Injectivity. If $p \in P^\alpha(D)$ and $q = 0$ have equal scalar potentials (1.1) and (2.1), so (1.1) must vanish and (2.3) holds with left-hand side 0. Then the jump relation leads to $n \cdot p = 0$ on S. As $\operatorname{div} p = 0$, $\operatorname{curl} p = 0$ in $\underline{D}$, and p in D is circulation-free, it follows that $p = 0$ in D.

Surjectivity. For $q \in Q^\alpha(S)$, by solving a vector Neumann problem for D, there is exactly one circulation-free vector function $p \in [C^\alpha(D)]^3$ satisfying

$$\operatorname{div} p = 0, \quad \operatorname{curl} p = 0 \quad \text{in } \underline{D}, \quad n \cdot p = q \quad \text{on } S; \tag{2.4}$$

note that the necessary and sufficient integrability condition is fulfilled since q vanishes in the mean over S. As q also vanishes in the mean over every component of S, we conclude from (2.4) and the Gauss theorem that p in D is flux-free. Hence $p \in P^\alpha(D)$, and from (2.3) and (2.4) it follows that p and q have equal scalar potentials. So p is the original image of q.

Next we take into consideration the eigenvalue equation for the *electrostatic integral operator* $T : C^\alpha(S) \to C^\alpha(S)$ [1], i.e., the integral equation

$$Tq(x) := \frac{1}{2\pi} \int_S q(y) \frac{\partial}{\partial n_x} \frac{1}{|y-x|} dS(y) = \lambda q(x), \quad x \in S. \tag{2.5}$$

Due to Plemelj [7], all eigenvalues of T are real lying in the interval $[-1, 1]$.

Lemma 2. *Let $Q^\alpha_\lambda(S)$ denote the linear space of all solutions $q \in C^\alpha(S)$ of* (2.5). *Then there is the inclusion*

$$Q^\alpha_\lambda(S) \subseteq Q^\alpha(S), \quad \lambda \in (-1, 1).$$

Proof. For $q \in Q^\alpha_\lambda(S)$, $\lambda \in (-1,1)$, the scalar potential (2.1) is harmonic in $\mathbb{R}^3 \setminus S$. Then from the Gauss theorem, the limiting relations for (2.1), and (2.5) we find that

$$\int_S \frac{\partial \Phi}{\partial n}\Big|_- dS = \frac{1}{2} \int_S (Tq + q) dS = \frac{\lambda + 1}{2} \int_S q\, dS = 0.$$

Analogously, if there is any interior boundary component S_i of S, we get

$$\int_{S_i} \frac{\partial \Phi}{\partial n}\Big|_+ dS = \frac{1}{2} \int_{S_i} (Tq - q) dS = \frac{\lambda - 1}{2} \int_{S_i} q\, dS = 0,$$

i.e., q vanishes in the mean over every component of S and hence $q \in Q^\alpha(S)$.

Theorem 1. *For $\lambda \in (-1,1)$ the linear spaces $P^\alpha_\lambda(D)$ and $Q^\alpha_\lambda(S)$ are isomorphic where the mapping is characterized by equal scalar potentials* (1.1) *and* (2.1).

Proof. With regard to (1.13) and Lemmata 1 and 2, it is sufficient to show that under the mapping mentioned $P^\alpha_\lambda(D)$ has the image $Q^\alpha_\lambda(S)$, $\lambda \in (-1,1)$.

Let first $p \in P^\alpha_\lambda(D)$ have the image $q \in Q^\alpha(S)$. From the coincidence of the scalar potentials (1.1) and (2.1), the limiting relations for (2.1), and (1.9) it follows that

$$\frac{\partial \Phi}{\partial n}\Big|_- = \frac{1}{2}(Tq + q) = \frac{1+\lambda}{2}(n \cdot p) \quad \text{on } S.$$

Hence together with (2.2) the integral equation (2.5) is obtained, i.e., $q \in Q^\alpha_\lambda(S)$.

Let, vice versa, $q \in Q^\alpha_\lambda(S)$ have the original image $p \in P^\alpha(D) \subseteq [C^\alpha(D)]^3 \cap [C^1(\underline{D})]^3$. As p in D is circulation-free, there is the representation

$$p = \operatorname{grad} \varphi \quad \text{in } D \tag{2.6}$$

with $\varphi \in C^{1+\alpha}(D) \cap C^2(\underline{D})$; further from $\operatorname{div} p = 0$ in $\underline{D}$ it follows that φ in $\underline{D}$ is harmonic. If Φ is the common scalar potential of p and q, then with regard to (2.1), (2.2), (2.5), and (2.6),

$$\psi := \Phi - \frac{1+\lambda}{2} \varphi \quad \text{in } D \tag{2.7}$$

belongs to $C^{1+\alpha}(D)$, is harmonic in $\underline{D}$, and due to

$$\frac{\partial \psi}{\partial n} = \frac{1}{2}(Tq+q) - \frac{1+\lambda}{2}(n \cdot p) = 0 \quad \text{on } S.$$

Hence $\psi = \text{const}$ in D, and elimination of φ from (2.6) and (2.7) leads to (1.9). So we have $p \in P^{\alpha}_{\lambda}(D)$.

Remark. For $q \in Q^{\alpha}_{\lambda}(S)$ with corresponding $p \in P^{\alpha}_{\lambda}(D)$ Theorem 1 allows computation of the scalar potential (1.1) – and just so the electric field (1.11) –, *without knowledge of p* by computing the scalar potential (2.1) for q. To compute p from q one can either solve the vector Neumann problem occurring in the proof of Lemma 1 or refer to the scalar potential (2.1) and the first equation (1.9).

Example 1. For a *ball* D: $|x| \leq R$, $R > 0$, the solutions of the integral equation (18) to $\lambda \in (-1,1)$ are well known as

$$\lambda_\nu = -\frac{1}{2\nu+1}, \quad q_\nu(x) = \frac{\nu}{R} K_\nu(x), \quad x \in S, \quad \nu = 1, 2, \ldots,$$

K_ν being a spherical harmonic of order ν. The scalar potentials (2.1) of the electric single layers q_ν on S are computed as

$$\Phi_\nu(x) = \left\{ \begin{array}{ll} \dfrac{\nu}{2\nu+1} K_\nu(x) & , \quad x \in D, \\[2ex] \dfrac{\nu}{2\nu+1} \left(\dfrac{R}{|x|}\right)^{2\nu+1} K_\nu(x), & x \in \hat{D}, \end{array} \right\}, \; \nu = 1, 2, \ldots;$$

in the exterior this corresponds to the potential of a 2^ν-multipole at the origin. The first equation (1.9) shows the underlying permanent polarizations

$$p_\nu(x) = \frac{2}{1+\lambda_\nu} \operatorname{grad} \Phi_\nu(x) = \operatorname{grad} K_\nu(x), \quad x \in D, \quad \nu = 1, 2, \ldots.$$

For the lowest eigenvalue $\lambda_1 = -\frac{1}{3}$ there is a constant permanent polarization p_1 in D. From (1.11) the corresponding electric field is $E = -\varepsilon_0^{-1} \operatorname{grad} \Phi_1 = -\frac{1}{3}\varepsilon_0^{-1} p_1$ in D, completed by a dipole field in $\hat{D}$ with continuous tangential components on S. Hereby a *ball electret* is modelled.

3. Magnetic dipole equilibrium

In this chapter we start from a magnetic double layer $\mu \in C^{1+\alpha}(S)$ (SI unit: A) distributed on the boundary S. The corresponding scalar and vector potentials[2]

$$\Psi(x) = \frac{1}{4\pi} \int_S \mu(y) \frac{\partial}{\partial n_y} \frac{1}{|y-x|} dS(y), \quad x \in D, \hat{D}, \tag{3.1}$$

$$A(x) = \frac{1}{4\pi} \int_S \mu(y) \left(n(y) \times \operatorname{Grad}_y \frac{1}{|y-x|} \right) dS(y), \quad x \in D, \hat{D}, \tag{3.2}$$

[2] We denote by Grad the *surface gradient.*

own the regularities [1, 3]

$$\Psi \in C^{1+\alpha}(D) \cap C^{1+\alpha}(\hat{D}),$$

$$A \in [C^{\alpha}(\mathbb{R}^3)]^3 \cap [C^{1+\alpha}(D)]^3 \cap [C^{1+\alpha}(\hat{D})]^3,$$

respectively[3]. For (3.1) and (3.2), similarly to (1.5), it is obtained that

$$\operatorname{grad} \Psi + \operatorname{curl} A = 0 \quad \text{in } D, \hat{D}. \tag{3.3}$$

Lemma 3. *Let $M^{\alpha}(S)$ denote the linear space of all double layers $\mu \in C^{1+\alpha}(S)$ with properties*

(i) *$\int_S q_0 \mu \, dS = 0$, $q_0 \in C^{\alpha}(S)$ being a non-trivial solution[4] of $Tq_0 = -q_0$;*

(ii) *$\operatorname{grad} \mu$ in D is flux-free, $\mu \in C^{1+\alpha}(D) \cap C^2(\underline{D})$ being the harmonic extension[5] of $\mu \in C^{1+\alpha}(S)$.*

Then the linear spaces $P^{\alpha}(D)$ and $M^{\alpha}(S)$ are isomorphic where the mapping is characterized by equal vector potentials (1.2) for $p \in P^{\alpha}(D)$ and (3.2) for $\mu \in M^{\alpha}(S)$. Furthermore, corresponding p and μ are due to

$$p = \operatorname{grad} \mu \quad \text{in } D. \tag{3.4}$$

Proof. The mapping. Let $p \in P^{\alpha}(D) \subseteq [C^{\alpha}(D)]^3 \cap [C^1(\underline{D})]^3$. As p in D is circulation-free, there exists exactly one $\mu \in C^{1+\alpha}(D) \cap C^2(\underline{D})$ satisfying (3.4) and normed by $\int_S q_0 \mu \, dS = 0$. Since p in D is flux-free, so $\operatorname{grad} \mu$ in D is also, and μ in $\underline{D}$ turns out to be harmonic. Thus the restriction $\mu \in C^{1+\alpha}(S)$ forms a double layer which belongs to $M^{\alpha}(S)$. Carrying out an integration by parts, we have

$$\int_D \operatorname{grad} \mu(y) \times \operatorname{grad}_y \frac{1}{|y-x|} dV(y) = \int_S \mu(y) \left(n(y) \times \operatorname{Grad}_y \frac{1}{|y-x|} \right) dS(y), \quad x \in D, \hat{D}; \tag{3.5}$$

thus with regard to (3.4), the coincidence of the vector potentials (1.2) and (3.2) becomes evident. – If $\mu \in M^{\alpha}(S)$ is the *difference* of two double layers which have the same vector potentials as $p \in P^{\alpha}(D)$, then (3.2) vanishes, and another integration by parts, using that S consists of *closed* surfaces, gives

$$-\int_S \frac{(n \times \operatorname{Grad} \mu)(y)}{|y-x|} dS(y) = 0, \quad x \in \mathbb{R}^3 \setminus S.$$

---

[3] On S the values of Ψ, A are to be understood as one side limits. For A, especially, both limits coincide.

[4] Such solution q_0 on S is uniquely determined up to a non-vanishing constant factor. It has no zeros on the (only) exterior boundary component of S and vanishes on all interior boundary components of S (if there are any) so that, in particular, $\int_S q_0 \, dS \neq 0$ [5]. Physically q_0 describes an electric charge distribution (single layer) on S with constant scalar potential in D.

[5] Such extension, harmonic in $\underline{D}$, is uniquely determined by a Dirichlet problem for D. For the regularity of the extension we refer to [1].

Next the jump relation yields $n \times \operatorname{Grad} \mu = 0$ on S, so $\operatorname{Grad} \mu = 0$ on S, and μ turns out with constant values on every component of S. Because of (ii) the normal derivatives $\frac{\partial \mu}{\partial n}$ on S vanish in the mean over every component of S. Together this leads to $\mu = \text{const}$ in D, and from (i) this constant becomes normed to 0. Thus for every $p \in P^\alpha(D)$ there is exactly one $\mu \in M^\alpha(S)$ with equal vector potential.

Injectivity. If $p \in P^\alpha(D)$ and $\mu = 0$ have equal vector potentials (1.2) and (3.2), then (1.2) must vanish, and partial integration together with $\operatorname{curl} p = 0$ in $\underline{D}$ yields

$$-\int_S \frac{(n \times p)(y)}{|y - x|} dS(y) = 0 \quad \text{in } \mathbb{R}^3.$$

Again by the jump relation we conclude that $n \times p = 0$ on S, and observing that $\operatorname{div} p = 0$, $\operatorname{curl} p = 0$ in $\underline{D}$, and p in D is flux-free we get $p = 0$ in D.

Surjectivity. Let $\mu \in M^\alpha(S)$ have the harmonic extension $\mu \in C^{1+\alpha}(D)$. Then $p \in [C^\alpha(D)]^3$, if defined by (3.4), is flux-free according to (ii) and immediately circulation-free; it thus belongs to $P^\alpha(D)$. From (3.4) and (3.5) it follows that p and μ have equal vector potentials, so p turns out as original image of μ.

Now we consider the eigenvalue equation for the *adjoint electrostatic integral operator* $T' : C^{1+\alpha}(S) \to C^{1+\alpha}(S)$ [1], i.e., the integral equation

$$T'\mu(x) := \frac{1}{2\pi} \int_S \mu(y) \frac{\partial}{\partial n_y} \frac{1}{|y - x|} dS(y) = \lambda \mu(x), \quad x \in S; \tag{3.6}$$

adjointness of the operators T, T' means that

$$\int_S (Tq)\mu \, dS = \int_S q(T'\mu) dS \tag{3.7}$$

for $q \in C^\alpha(S)$, $\mu \in C^{1+\alpha}(S)$. Again according to Plemelj [7], all eigenvalues of T' are real lying in the interval $[-1, 1]$.

Lemma 4. *Let $M_\lambda^\alpha(S)$ denote the linear space of all solutions $\mu \in C^{1+\alpha}(S)$ of* (3.6). *Then there is the inclusion*

$$M_\lambda^\alpha(S) \subseteq M^\alpha(S), \quad \lambda \in (-1, 1).$$

Proof. Let $\mu \in M_\lambda^\alpha(S)$, $\lambda \in (-1, 1)$. Then (3.6), (3.7), and $Tq_0 = -q_0$ yield

$$\int_S q_0(T'\mu) dS - \int_S (Tq_0)\mu \, dS = (\lambda + 1) \int_S q_0 \mu \, dS = 0$$

and thus (i). Let μ have the scalar potential (3.1) which is harmonic in $\underline{D}$. Then the limiting relations for (21) and (3.6) show that

$$\Psi|_- = \frac{1}{2}(T'\mu - \mu) = \frac{\lambda - 1}{2} \mu \quad \text{on } S,$$

and the harmonic extension of μ to D is performed by

$$\Psi = \frac{\lambda - 1}{2} \mu \quad \text{in } D. \tag{3.8}$$

From (3.3) we see that grad Ψ in D is flux-free and from (3.8) we see that grad μ in D is also. Thus (ii) and herewith $\mu \in M^\alpha(S)$ is shown.

Theorem 2. *For $\lambda \in (-1,1)$ the linear spaces $P_\lambda^\alpha(D)$ and $M_\lambda^\alpha(S)$ are isomorphic where the mapping is characterized by equal vector potentials* (1.2) *and* (3.2).

Proof. With regard to (1.13) and Lemmata 3 and 4, we only need to show that under the mapping mentioned $P_\lambda^\alpha(D)$ has the image $M_\lambda^\alpha(S)$, $\lambda \in (-1,1)$.

Let first $p \in P_\lambda^\alpha(D)$ have the image $\mu \in M^\alpha(S)$. From the coincidence of the vector potentials (1.2) and (3.2) together with (1.9), (3.3), and (3.4) the scalar potential (3.1) satisfies

$$\operatorname{grad}\Psi = -\operatorname{curl} A = \frac{\lambda-1}{2}p = \frac{\lambda-1}{2}\operatorname{grad}\mu \quad \text{in } D$$

which is integrated by

$$\Psi = \frac{\lambda-1}{2}\mu + \text{const} \quad \text{in } D.$$

From the limiting relations for (3.1), the adjointness (3.7), and the condition (i) in Lemma 3 it follows that

$$\int_S q_0\left(\Psi|_- - \frac{\lambda-1}{2}\mu\right) dS = \frac{1}{2}\int_S q_0(T'\mu - \lambda\mu)dS = -\frac{1+\lambda}{2}\int_S q_0\mu\, dS = 0$$

by which the constant obtained before is determined to 0. Hence (3.8) arises where the limit to the interior side of S leads to the integral equation (3.6). Thus $\mu \in M_\lambda^\alpha(S)$.

Let, vice versa, $\mu \in M_\lambda^\alpha(S)$ have the original image $p \in P^\alpha(D)$. From the proof of Lemma 4 we see that the harmonic extension of μ is due to (3.8). Now the negative gradient of (3.8) together with (3.3) and (3.4) yields

$$\operatorname{curl} A = \frac{1-\lambda}{2}p \quad \text{in } D$$

where A has the meaning of the vector potential (3.2). But since A coincides with the vector potential (1.2), so p satisfies (1.9), i.e., $p \in P_\lambda^\alpha(D)$.

Remark. For $\mu \in M_\lambda^\alpha(S)$ with corresponding $p \in P_\lambda^\alpha(D)$ Theorem 2 enables computation of the vector potential (1.2) – and just so the magnetic field (1.12) – *without knowledge of* p by computing the vector potential (3.2) for μ. In view of (3.3), the magnetic field (1.12) can be expressed in a more convenient manner by the scalar potential (3.1) for μ like

$$B = -\mu_0 \operatorname{grad}\Psi \quad \text{in } D, \hat{D}. \tag{3.9}$$

Computation of p from μ is done by (3.4) with μ harmonically extended to D. This extension may be obtained from the scalar potential (3.1) together with (3.8).

Example 2. For the *ball D* as before, the solutions of the integral equation (3.6) to $\lambda \in (-1, 1)$ are given by

$$\lambda_\nu = -\frac{1}{2\nu+1}, \quad \mu_\nu(x) = K_\nu(x), \quad x \in S, \quad \nu = 1, 2, \ldots.$$

The scalar potentials (3.1) of the magnetic double layers μ_ν on S are computed as

$$\Psi_\nu(x) = \left\{ \begin{array}{ll} -\dfrac{\nu+1}{2\nu+1} K_\nu(x) & , \quad x \in D, \\ \dfrac{\nu}{2\nu+1} \left(\dfrac{R}{|x|}\right)^{2\nu+1} K_\nu(x), & x \in \hat{D}, \end{array} \right\}, \ \nu = 1, 2, \ldots.$$

Then from (3.4) and (3.8) we find the underlying permanent magnetizations

$$p_\nu(x) = -\frac{2}{1-\lambda_\nu} \operatorname{grad} \Psi_\nu(x) = \operatorname{grad} K_\nu(x), \quad x \in D, \quad \nu = 1, 2, \ldots.$$

For the lowest eigenvalue $\lambda_1 = -\frac{1}{3}$ there is a constant permanent magnetization p_1 in D. From (3.9) the corresponding magnetic field is $B = -\mu_0 \operatorname{grad} \Psi_1 = \frac{2}{3}\mu_0 p_1$ in D, completed by a dipole field in $\hat{D}$ with continuous normal components on S. By this a *ball magnet* is modeled which, for instance, is realized by the *Earth*.

4. The limiting cases $\lambda = 1, -1$

In the limiting cases $\lambda = 1, -1$ the integral equation method is not available for solving the equilibrium equations (1.9). The solutions of (1.9), however, may be easily characterized as follows.

Theorem 3. *Let* $\lambda = 1$. *Then the following two conditions for* $p \in [C^\alpha(D)]^3$ *are equivalent with* (1.9):

(i) *The vector potential* (1.2) *vanishes in* $\mathbb{R}^3$;

(ii) p *in* D *is circulation-free,* $n \times p = 0$ *on* S.

Proof. The second equation (1.9) states that $\operatorname{curl} A = 0$ in D, and from (1.10) we have $\operatorname{curl} A = 0$ in $\hat{D}$. Together with (1.3) and (1.5) we obtain (ii). – From (ii) an integration by parts in the weak sense for (1.2) leads to (i). – The step from (i) to the second equation (1.9) is trivial.

Example 3. Let D be a *solid spherical shell.* Then a radial electric dipole distribution p in D with $|p|$ decreasing with the square of the reciprocal distance to the centre satisfies the condition (ii) in Theorem 3 and is therefore in equilibrium to $\lambda = 1$. Such dipole distribution p in D, circulation-free and divergence-free but not flux-free, models an (extremely unstable) *shell electret.* From (1.9), (1.10), and (1.11) we get the electric field $E = -\varepsilon_0^{-1} p$ in D and $E = 0$ in $\hat{D}$.

Theorem 4. *Let* $\lambda = -1$. *Then the following two conditions for* $p \in [C^\alpha(D)]^3$ *are equivalent with* (1.9):

(i) *The scalar potential* (1.1) *vanishes in* $\mathbb{R}^3$;

(ii) p *in* D *is flux-free,* $n \cdot p = 0$ *on* S.

Proof. The first equation (1.9) states that $\operatorname{grad} \Phi = 0$ in D, and from (1.10) we get $\operatorname{grad} \Phi = 0$ in $\hat{D}$. Together with (1.4) and (1.5) we are led to (ii). – From (ii) an integration by parts in the weak sense for (1.1) yields (i). – The step from (i) to the first equation (1.9) is trivial.

Example 4. Let D be a *solid circular torus.* Then an azimuthal magnetic dipole distribution p in D with $|p|$ decreasing with the reciprocal distance to the axis satisfies the condition (ii) in Theorem 4 and is thus in equilibrium to $\lambda = -1$. Such dipole distribution p in D, flux-free and curl-free but not circulation-free, models a (stable) *ring magnet.* From (1.9), (1.10), and (1.12) the magnetic field is $B = \mu_0 p$ in D and $B = 0$ in $\hat{D}$.

Remark. With the help of condition (ii) in Theorem 4 it is not difficult to construct a non-trivial equilibrium solution to $\lambda = -1$. Hence any dipole equilibrium to $\lambda \in (-1, 1]$ is *confirmed to be unstable.*

References

[1] D. Colton, R. Kress, *Integral Equation Methods in Scattering Theory.* Wiley, New York, 1983.

[2] G. Falk, W. Ruppel, *Energie und Entropie.* Springer, Berlin/Heidelberg/New York, 1976.

[3] D. Gilbarg, N.S. Trudinger, *Elliptic Partial Differential Equations of Second Order.* Springer, Berlin/Heidelberg/New York, 1977.

[4] O.D. Kellogg, *Foundations of Potential Theory.* Dover, New York, 1953.

[5] E. Martensen, *Potentialtheorie.* Teubner, Stuttgart, 1968.

[6] E. Martensen, *Zwei Abhandlungen zu den mathematischen Grundlagen der Geodäsie.* Verlag der Bayerischen Akademie der Wissenschaften, München, 1998.

[7] J. Plemelj, *Potentialtheoretische Untersuchungen.* Preisschriften der Fürstlich Jablonowskischen Gesellschaft zu Leipzig. Teubner, Leipzig, 1911.

[8] S. Ritter, *Dipolverteilungen im Gleichgewicht für das dreiachsige Ellipsoid.* Doctorate thesis. Karlsruhe, 1994.

[9] S. Ritter, *On the Magnetostatic Integral Operator for Ellipsoids.* J. Math. Anal. Appl. **207** (1997), 12–28.

Erich Martensen, Mathematisches Institut II, Universität Karlsruhe
D-76128 Karlsruhe, Germany
e-mail: `mrt@mi2b-serv.mathematik.uni-karlsruhe.de`

Stefan Ritter, DaimlerChrysler AG, Mercedes-Benz-Str. 1
D-76725 Germersheim, Germany
e-mail: `stefan.ritter@daimlerchrysler.com`

Operator Theory
Advances and Applications, Vol. 147, 365–386

Characterization of Multipliers in Pairs of Besov Spaces

Vladimir Maz'ya and Tatyana Shaposhnikova

Dedicated to the memory of Erhard Meister

Abstract. We give necessary and sufficient conditions for a function to be a multiplier from one Besov space $B_p^m(\mathbf{R}^n)$ into another $B_p^l(\mathbf{R}^n)$ where $0 < l \leq m$ and $p \in (1, \infty)$. We also show that the space of multipliers acting from the Sobolev space $W_p^m(\mathbf{R}^n)$ into a distribution Sobolev space $W_p^{-k}(\mathbf{R}^n)$ is isomorphic to $W_{p,\mathrm{unif}}^{-k}(\mathbf{R}^n) \cap W_{p',\mathrm{unif}}^{-m}(\mathbf{R}^n)$ for either $k \geq m > 0$, $k > n/p'$, or $m \geq k > 0$, $m > n/p$, where $p \in (1, \infty)$ and $p + p' = pp'$.

1. Introduction

By a multiplier acting from one Banach function space S_1 into another S_2 we call a function γ such that $\gamma u \in S_2$ for any $u \in S_1$. By $M(S_1 \to S_2)$ we denote the space of multipliers $\gamma : S_1 \to S_2$ with the norm

$$\|\gamma\|_{M(S_1 \to S_2)} = \sup\{\|\gamma u\|_{S_2} : \quad \|u\|_{S_1} \leq 1\}.$$

We write MS instead of $M(S \to S)$.

A theory of pointwise multipliers was developed in our book [MS], where a complete bibliography and description of related results obtained before 1985 can be found. In particular, [MS] contains characterization of the spaces $M(H_p^m(\mathbf{R}^n) \to H_p^l(\mathbf{R}^n))$ with $1 < p < \infty$, where $H_p^k(\mathbf{R}^n)$ is the Bessel potential space. We also described multipliers $M(W_p^m(\mathbf{R}^n) \to W_p^l(\mathbf{R}^n))$ in Sobolev (k integer)-Slobodetskii (k noninteger) spaces with $1 \leq p < \infty$ and both m and l being either integer or noninteger.

We mention known results on multipliers preserving a certain Besov space. Necessary and sufficient conditions for a function to belong to $MB_p^l(\mathbf{R}^n)$, $1 < p < \infty$, $0 < l < \infty$, are given in [MS]. Recently a characterization of $MB_{p,q}^s(\mathbf{R}^n)$ for $1 \leq p \leq q \leq \infty$, $s > n/p$, was obtained by Sickel and Smirnov [SS]. The spaces $MB_{\infty,1}^0(\mathbf{R}^n)$ and $MB_{\infty,\infty}^0(\mathbf{R}^n)$ were described by Koch and Sickel [KS].

2000 *Mathematics Subject Classification.* Primary 26D10, 46E35, 46E25, 42B25.
Key words and phrases. Pointwise multipliers, Besov spaces, distribution Sobolev spaces.
The authors were supported by the Swedish Natural Science Research Council.

The main goal of the present paper is to characterize the space $M(B_p^m(\mathbf{R}^n) \to B_p^l(\mathbf{R}^n))$ for $m \geq l > 0$, $p \in (1,\infty)$.

A sufficient condition for inclusion into the space $M(W_p^m(\mathbf{R}^n) \to W_p^{-k}(\mathbf{R}^n))$ of Sobolev multipliers can be found in Sect. 1.5 [MS]. Maz'ya and Verbitsky [MV2], [MV3] described the spaces $M(W_2^1(\mathbf{R}^n) \to W_2^{-1}(\mathbf{R}^n))$ and $M(W_2^{1/2}(\mathbf{R}^n) \to W_2^{-1/2}(\mathbf{R}^n))$ recently, solving the problem of the form boundedness of the Schrödinger and the relativistic Schrödinger operators (see [MV2] and [MV4] for further results in the same vein). We conclude the present paper by showing that the space $M(W_p^m(\mathbf{R}^n) \to W_p^{-k}(\mathbf{R}^n))$ is isomorphic to $W_{p,\mathrm{unif}}^{-k}(\mathbf{R}^n) \cap W_{p',\mathrm{unif}}^{-m}(\mathbf{R}^n)$ provided $k \geq m > 0$, $k > n/p'$ or $m \geq k > 0$, $m > n/p$, where $p \in (1,\infty)$ and $p + p' = pp'$. This is a straightforward corollary of the above-mentioned sufficient condition from Sect. 1.5 [MS]. However, the result seems to be new even for $n = 1$, except for the case $k = m = 1$ treated in [MV4].

Let $s = k + \alpha$, where $\alpha \in (0,1]$ and k is a nonnegative integer. Further, let

$$\Delta_h^{(2)} u(x) = u(x+2h) - 2u(x+h) + u(x)$$

and

$$(C_{p,s}u)(x) = \Big(\int_{\mathbf{R}^n} |\Delta_h^{(2)} \nabla_k u(x)|^p |h|^{-n-p\alpha} dh\Big)^{1/p},$$

where ∇_k stands for the gradient of order k, i.e., $\nabla_k u = \{\partial_{x_1}^{\alpha_1} \dots \partial_{x_n}^{\alpha_n}\}$, $\alpha_1 + \dots + \alpha_n = k$. The Besov space $B_p^s(\mathbf{R}^n)$ is introduced as the completion of $C_0^\infty(\mathbf{R}^n)$ in the norm

$$\|C_{p,s}u; \mathbf{R}^n\|_{L_p} + \|u; \mathbf{R}^n\|_{L_p}.$$

Let $\{s\}$ and $[s]$ denote the fractional and integer parts of a positive number s and let

$$(D_{p,s}u)(x) = \Big(\int |\Delta_h \nabla_{[s]} u(x)|^p |h|^{-n-p\{s\}} dh\Big)^{1/p},$$

where $\Delta_h v(x) = v(x+h) - v(x)$. The fractional Sobolev space W_p^s is defined as the closure of C_0^∞ in the norm

$$\|D_{p,s}u\|_{L_p} + \|u\|_{L_p}.$$

(Here and in the sequel, we omit $\mathbf{R}^n$ in the notation of norms, spaces, and in the range of integration.) For $\{s\} > 0$ the spaces B_p^s and W_p^s have the same elements and their norms are equivalent since

$$(2 - 2^{\{s\}}) D_{p,s}u \leq C_{p,s}u \leq (2 + 2^{\{s\}}) D_{p,s}u \tag{1.1}$$

which follows directly from the identity

$$2[u(x+h) - u(x)] = -[u(x+2h) - 2u(x+h) + u(x)] + [u(x+2h) - u(x)].$$

In what follows the equivalence $a \sim b$ means that there exist positive constants c_1, c_2 such that $c_1 b \leq a \leq c_2 b$.

With any Banach space S of functions on $\mathbf{R}^n$ one can associate the spaces

$$S_{\mathrm{loc}} = \{u : \eta u \in S \quad \text{for all } \eta \in C_0^\infty\}$$

and

$$S_{\text{unif}} = \{u : \sup_{z\in\mathbf{R}^n} \|\eta_z u\|_S < \infty\},$$

where $\eta_z(x) = \eta(x-z)$, $\eta \in C_0^\infty$, $\eta = 1$ on $\mathcal{B}_1$. Here and in what follows $\mathcal{B}_r(x)$ is the ball $\{y \in \mathbf{R}^n : |y-x| < r\}$ and $\mathcal{B}_r = \mathcal{B}_r(0)$. The space S_{unif} is endowed with the norm

$$\|u\|_{S_{\text{unif}}} = \sup_{z\in\mathbf{R}^n} \|\eta_z u\|_S.$$

The obvious consequence of the definition of the multiplier space $M(S_1 \to S_2)$ is the embedding

$$M(S_1 \to S_2) \subset S_{2,\text{unif}}.$$

Let Λ^μ be the operator defined for any $\mu \in \mathbf{R}$ by

$$\Lambda^\mu = (-\Delta + 1)^{\mu/2} = F^{-1}\big(1 + |\xi|^2\big)^{\mu/2} F,$$

where F is the Fourier transform in $\mathbf{R}^n$ and F^{-1} is the inverse of F. By J_l we denote the Bessel potential of order l, that is the operator Λ^{-l}. Throughout the paper we assume that $m > 0$ and use the notion of the (p,m)-capacity $\text{cap}_{p,m}(e)$ of a compact set $e \subset \mathbf{R}^n$ which is defined by

$$\text{cap}_{p,m}(e) = \inf\{\|f\|_{L_p}^p : \ f \in L_p,\ f \geq 0 \text{ and } J_m f(x) \geq 1 \text{ for all } x \in e\}.$$

For properties of this capacity see [M], Ch. 7 and [AH], Ch. 2 and Sect. 4.4. In particular, it is well known that if $0 < r \leq 1$, then

$$\text{cap}_{p,m}(\mathcal{B}_r) \sim \begin{cases} r^{n-mp}, & \text{for } mp < n, \\ (\log \frac{2}{r})^{1-p}, & \text{for } mp = n, \\ 1, & \text{for } mp > n. \end{cases} \tag{1.2}$$

and if e is a compact set in $\mathbf{R}^n$ with $\text{diam}(e) \leq 1$, then

$$\text{cap}_{p,m}(e) \geq \begin{cases} c(\text{mes}_n\, e)^{(n-mp)/n}, & \text{for } mp < n, \\ \left(\log \dfrac{2^n}{\text{mes}_n\, e}\right)^{1-p}, & \text{for } mp = n. \end{cases} \tag{1.3}$$

The following assertion is the main result of this article.

Theorem 1. *Let $0 < l \leq m$, $p \in (1,\infty)$, and let $\gamma \in B^l_{p,\text{loc}}$. There holds the equivalence relation*

$$\|\gamma\|_{M(B_p^m \to B_p^l)} \sim \sup_e \frac{\|C_{p,l}\gamma; e\|_{L_p}}{[\text{cap}_{p,m}(e)]^{1/p}} + \begin{cases} \|\gamma\|_{L_{1,\text{unif}}}, & m > l, \\ \|\gamma\|_{L_\infty}, & m = l, \end{cases} \tag{1.4}$$

where e is an arbitrary compact set in $\mathbf{R}^n$. The finiteness of the right-hand side in (1.4) *is necessary and sufficient for $\gamma \in M(B_p^m \to B_p^l)$.*

The relation (1.4) *remains valid if one adds the condition* $\text{diam}(e) \leq 1$.

For $mp > n$ the statement of the above theorem simplifies. Namely, the relation (1.4) is equivalent to

$$\|\gamma\|_{M(B_p^m \to B_p^l)} \sim \|\gamma\|_{B^l_{p,\text{unif}}} \quad \text{for } m \geq l, \tag{1.5}$$

and for $lp > n$

$$\|\gamma\|_{MB^l_p} \sim \|C_{p,l}\gamma\|_{L_{p,\mathrm{unif}}} + \|\gamma\|_{L_\infty}. \tag{1.6}$$

From results of Kerman and Saywer [KeS] and Maz'ya and Verbitsky [MV1] it follows that the supremum in the right-hand side of (1.4) is equivalent to each of the suprema

$$\sup_{\{Q\}} \frac{\|J_m \chi_Q (C_{p,l}\gamma)^p; Q\|_{L_{p/(p-1)}}}{\|C_{p,l}\gamma; Q\|^{p-1}_{L_p}}, \tag{1.7}$$

where $\{Q\}$ is the collection of all cubes, χ_Q is the characteristic function of Q, and

$$\sup_{x\in\mathbf{R}^n} \frac{J_m(J_m(C_{p,l}\gamma)^p)^{p/(p-1)}(x)}{J_m(C_{p,l}\gamma)^p(x)}. \tag{1.8}$$

From (1.4), (1.7), and (1.8) one can deduce various precise upper and lower estimates for the norm in $M(B^m_p \to B^l_p)$ formulated in more conventional terms (compare with [MS], Ch. 3).

2. Preliminaries

In this section, we collect some auxiliary assertions used in the sequel.

Lemma 1. (see [St], Sect. 5.1) *There holds the equivalence relation*

$$\|u\|_{B^k_p} \sim \|\Lambda^\alpha u\|_{B^{k-\alpha}_p}, \tag{2.1}$$

where $p \in (1,\infty)$ *and* $\alpha \in (0,k)$.

By H^k_p, $k \geq 0$, $p \in (1,\infty)$, we denote the space of Bessel potentials defined as the completion of C^∞_0 in the norm

$$\|u\|_{H^k_p} = \|\Lambda^k u\|_{L_p}. \tag{2.2}$$

The following relations are well known

$$\|\gamma\|_{M(B^k_p\to L_p)} \sim \|\gamma\|_{M(H^k_p\to L_p)} \sim \sup_e \frac{\|\gamma; e\|_{L_p}}{[\mathrm{cap}_{p,k}(e)]^{1/p}} \sim \sup_{e,\mathrm{diam}(e)\leq 1} \frac{\|\gamma; e\|_{L_p}}{[\mathrm{cap}_{p,k}(e)]^{1/p}} \tag{2.3}$$

(see [MS], Lemma 2.2.2/1, Corollary 3.2.1/1, Remark 3.2.1/1 and [AH], Sect. 4.4).

Using estimates (1.2) for the capacity of a ball, one obtains the following relations from (2.3)

$$\|\gamma\|_{M(B^k_p\to L_p)} \sim \|\gamma\|_{L_{p,\mathrm{unif}}} \quad \text{for } pk > n, \tag{2.4}$$

$$\|\gamma\|_{M(B^k_p\to L_p)} \geq c \sup_{x\in\mathbf{R}^n, r\in(0,1)} r^{k-n/p}\|\gamma; \mathcal{B}_r(x)\|_{L_p} \quad \text{for } pk < n, \tag{2.5}$$

$$\|\gamma\|_{M(B^k_p\to L_p)} \geq c \sup_{x\in\mathbf{R}^n, r\in(0,1)} \Big(\log\frac{2}{r}\Big)^{(p-1)/p}\|\gamma; \mathcal{B}_r(x)\|_{L_p} \quad \text{for } pk = n. \tag{2.6}$$

Lemma 2. *Let γ_ρ denote a mollifier of a function γ which is defined as*

$$\gamma_\rho(x) = \rho^{-n} \int K\big(\rho^{-1}(x-\xi)\big)\gamma(\xi)d\xi,$$

where $K \in C_0^\infty(\mathcal{B}_1)$, $K \geq 0$, and $\|K\|_{L_1} = 1$. The inequalities

$$\|\gamma_\rho\|_{M(B_p^m \to B_p^l)} \leq \|\gamma\|_{M(B_p^m \to B_p^l)} \leq \liminf_{\rho \to 0} \|\gamma_\rho\|_{M(B_p^m \to B_p^l)},$$

$$\|\gamma_\rho\|_{M(B_p^m \to L_p)} \leq \|\gamma\|_{M(B_p^m \to L_p)} \leq \liminf_{\rho \to 0} \|\gamma_\rho\|_{M(B_p^m \to L_p)},$$

and

$$\sup_e \frac{\|C_{p,l}\gamma_\rho; e\|_{L_p}}{[\mathrm{cap}_{p,m}(e)]^{1/p}} \leq \sup_e \frac{\|C_{p,l}\gamma; e\|_{L_p}}{[\mathrm{cap}_{p,m}(e)]^{1/p}}$$

are valid.

Proof. The proof of two-sided estimates is the same as in Lemma 3.2.1/1 [MS]. By Minkowski's inequality

$$\frac{\|C_{p,l}\gamma_\rho; e\|_{L_p}}{[\mathrm{cap}_{p,m}(e)]^{1/p}} \leq \frac{\displaystyle\int K(z)\Big(\int_e \big(C_{p,l}\gamma(x-\rho z)\big)^p dx\Big)^{1/p} dz}{[\mathrm{cap}_{p,m}(e)]^{1/p}}$$

$$\leq \frac{\displaystyle\int_{\mathcal{B}_1} K(z)\Big(\int_E \big(C_{p,l}\gamma(\xi)\big)^p d\xi\Big)^{1/p} dz}{[\mathrm{cap}_{p,m}(E)]^{1/p}} \leq \|K\|_{L_1} \sup_e \frac{\|C_{p,l}\gamma; e\|_{L_p}}{[\mathrm{cap}_{p,m}(e)]^{1/p}}$$

where $E = \{x - \rho z : x \in e, z \in \mathcal{B}_1\}$.

Below we use the interpolation properties

$$B_p^{m-k} = \Big(B_p^m,\, H_p^{m-l}\Big)_{k/l,p} \tag{2.7}$$

and

$$B_p^{m-k} = \Big(B_p^m,\, B_p^{m-l}\Big)_{k/l,p}, \tag{2.8}$$

where $l < k < m$ (see, [Tr], Th. 2.4.2). In particular, (2.8) implies

$$\|\gamma\|_{MB_p^r} \leq c\, \|\gamma\|_{MB_p^\sigma}^{\theta}\, \|\gamma\|_{MB_p^\rho}^{1-\theta}, \tag{2.9}$$

where $p \in (1,\infty)$, $\sigma > \rho > 0$, $0 < \theta < 1$, and $r = \theta\sigma + (1-\theta)\rho$. It follows from (2.3) and (2.8) that $\gamma \in M(B_p^m \to B_p^l) \cap M(B_p^{m-l} \to L_p)$ implies $\gamma \in M(B_p^{m-k} \to B_p^{l-k})$ for $0 < k < l$. Moreover,

$$\|\gamma\|_{M(B_p^{m-k} \to B_p^{l-k})} \leq c\|\gamma\|_{M(B_p^m \to B_p^l)}^{1-k/l} \|\gamma\|_{M(B_p^{m-l} \to L_p)}^{k/l} \tag{2.10}$$

for $0 < k < l < m$ and

$$\|\gamma\|_{MB_p^{l-k}} \leq c\|\gamma\|_{MB_p^l}^{1-k/l} \|\gamma\|_{L_\infty}^{k/l} \tag{2.11}$$

for $0 < k < l$.

In what follows we shall use five following assertions proved in the book [MS].

Lemma 3. (see [MS], Lemma 3.1.2/1) *Let $\mathcal{M}$ be the Hardy-Littlewood maximal operator defined by*

$$\mathcal{M}v(x) = \sup_{t>0} \frac{1}{\mathrm{mes}_n \mathcal{B}_t} \int_{\mathcal{B}_t(x)} |v(y)|dy.$$

Also let $J_r^{(n+s)}$ denote the Bessel potential in $\mathbf{R}^{n+s}$, $s \geq 1$. Then, for any nonnegative function $f \in L_p(\mathbf{R}^{n+s})$

$$(J^{(n+s)}_{r\theta+s/p}f)(x,0) \leq c \left((J^{(n+s)}_{r+s/p}f)(x,0)\right)^{\theta} (\mathcal{M}F(x))^{1-\theta},$$

where $F(x) = \|f(x,\cdot);\mathbf{R}^s\|_{L_p}$ and $0 < \theta < 1$.

Lemma 4. (see [MS], Lemma 3.2.1/3) *For any nonnegative function $\varphi \in L_{p\mu,loc}$, $p \in (1,\infty)$, and $0 < \lambda \leq \mu$, there holds*

$$\sup_e \left(\frac{\int_e \varphi^{\lambda p}(x)dx}{\mathrm{cap}_{p,\lambda}(e)} \right)^{1/\lambda} \leq c \sup_e \left(\frac{\int_e \varphi^{\mu p}(x)dx}{\mathrm{cap}_{p,\mu}(e)} \right)^{1/\mu}. \tag{2.12}$$

Lemma 5. (see [MS], Lemma 3.1.1./1) *For any positive $\alpha > 0$ and $\beta > 0$ there holds inequalities*

$$(C_{p,\alpha}u)(x) \leq \big(J_\beta C_{p,\alpha}\Lambda^\beta u\big)(x), \tag{2.13}$$

$$(D_{p,\alpha}u)(x) \leq \big(J_\beta D_{p,\alpha}\Lambda^\beta u\big)(x). \tag{2.14}$$

Lemma 6. (see Lemma 3.9.1 [MS]). *For $\delta \in (0,1)$ and any $k \geq 1$ there holds*

$$\left(\int\int |\Delta_h\gamma(x)\Delta_h u(x)|^p |h|^{-n-p} dh dx\right)^{1/p} \leq c \sup_e \frac{\|C_{p,\delta}\gamma; e\|_{L_p}}{[\mathrm{cap}_{p,k-1+\delta}(e)]^{1/p}} \|u\|_{B_p^k}. \tag{2.15}$$

Lemma 7. (see Lemma 3.1.1/2 [MS]) *For any α, $\beta > 0$ with $\alpha + \beta < 1$ there holds*

$$\|D_{p,\alpha}D_{p,\beta}u\|_{L_p} \leq c\, \|D_{p,\alpha+\beta}u\|_{L_p}.$$

3. Lower estimates of the norm in $M(B_p^m \to B_p^l)$

The following is the main result of this section.

Lemma 8. *Let $0 < l \leq m$ and $p \in (1,\infty)$. Then*

$$\|\gamma\|_{L_\infty} \leq \|\gamma\|_{MB_p^l} \qquad \text{for } m = l \tag{3.1}$$

and

$$\|\gamma\|_{M(B_p^{m-l}\to L_p)} \leq c\,\|\gamma\|_{M(B_p^m \to B_p^l)} \qquad \text{for } m > l. \tag{3.2}$$

Proof. Let $u \in B_p^l$ and let N be a positive integer. Clearly,

$$\|\gamma^N u\|_{L_p}^{1/N} \leq \|\gamma^N u\|_{B_p^l}^{1/N} \leq \|\gamma\|_{MB_p^l}\, \|u\|_{B_p^l}^{1/N}.$$

Passing to the limit as $N \to \infty$ we arrive at (3.1).

Now suppose $0 < l < m$. Let γ_ρ be the mollification of $\gamma \in M(B_p^m \to B_p^l)$. By Lemma 2, it suffices to prove (3.2) for γ_ρ. To simplify the notation we write γ in place of γ_ρ.

We consider two cases: $m \geq 2l$ and $2l > m > l$. Assume first that $m \geq 2l$. Let $U \in H_p^{m-l+1/p}(\mathbf{R}^{n+1})$ denote an extension of the function $u \in B_p^{m-l}(\mathbf{R}^n)$ to $\mathbf{R}^{n+1}$ such that

$$\|U; \mathbf{R}^{n+1}\|_{H_p^{m-l+1/p}} \leq c\|u; \mathbf{R}^n\|_{B_p^{m-l}}. \tag{3.3}$$

It is standard that the converse estimate

$$\|u; \mathbf{R}^n\|_{B_p^{m-l}} \leq c\|U; \mathbf{R}^{n+1}\|_{H_p^{m-l-1/p}} \tag{3.4}$$

holds for all extensions U. Let us represent U as the Bessel potential $J_{m-l+1/p}^{(n+1)} f$ with density $f \in L_p(\mathbf{R}^{n+1})$. By Lemma 3,

$$|u(x)| \leq c\left((J_{m+1/p}^{(n+1)}|f|)(x,0)\right)^{(m-l)/l} \left(\mathcal{M}F(x)\right)^{l/m},$$

where $F(x) = \|f(x,\cdot); \mathbf{R}\|_{L_p}$. Therefore,

$$\|\gamma u\|_{L_p} \leq c\,\|f; \mathbf{R}^{n+1}\|_{L_p}^{l/m}\, \|\, |\gamma|^{m/(m-l)} \left(J_{m+1/p}^{(n+1)}|f|\right)(\cdot,0)\|_{L_p}^{(m-l)/m}.$$

The right-hand side does not exceed

$$c\,\|f; \mathbf{R}^{n+1}\|_{L_p}^{l/m}\, \|\gamma\left(J_{m+1/p}^{(n+1)}|f|\right)(\cdot,0)\|_{B_p^l}^{(m-l)/m} \sup_e \left(\frac{\displaystyle\int_e |\gamma|^{\frac{pl}{m-l}}\,dx}{\mathrm{cap}_{p,l}(e)}\right)^{(m-l)/mp}. \tag{3.5}$$

Setting $\varphi = |\gamma|^{\frac{1}{m-l}}$, $\lambda = l$, $\mu = m-l$ in Lemma 4, we find that in the case $m \geq 2l$ the supremum in (3.5) is dominated by

$$c\left(\sup_e \frac{\displaystyle\int_e |\gamma|^p dx}{\mathrm{cap}_{p,m-l}(e)}\right)^{l/mp} \leq c\,\|\gamma\|_{M(B_p^{m-l}\to L_p)}^{l/m}.$$

Hence and by (3.5)

$$\|\gamma u\|_{L_p} \leq c\|f; \mathbf{R}^{n+1}\|_{L_p}^{l/m} \|\gamma\|_{M(B_p^m \to B_p^l)}^{(m-l)/m} \|J_{m+1/p}^{(n+1)}|f|(\cdot,0)\|_{B_p^m}^{(m-l)/m} \|\gamma\|_{M(B_p^{m-l}\to L_p)}^{l/m}.$$

Using first (3.4) and then (2.2) and (3.3), we obtain

$$\|J_{m+1/p}^{(n+1)}|f|(\cdot,0)\|_{B_p^m} \leq c\,\|J_{m+1/p}^{(n+1)}|f|; \mathbf{R}^{n+1}\|_{H_p^{m+1/p}} = c\|f; \mathbf{R}^{n+1}\|_{L_p}$$

$$= c\|U; \mathbf{R}^{n+1}\|_{H_p^{m-l+1/p}} \leq c\,\|u; \mathbf{R}^n\|_{B_p^{m-l}}.$$

Thus,

$$\|\gamma u\|_{L_p} \leq c\,\|\gamma\|_{M(B_p^{m-l}\to L_p)}^{l/m} \|\gamma\|_{M(B_p^m\to B_p^l)}^{(m-l)/m} \|u\|_{B_p^{m-l}},$$

which implies (3.2) for $m \geq 2l$.

Suppose $2l > m > l$. Let μ be an arbitrary positive number less than $m-l$. By (2.10) with $k = l - \mu$,

$$\|\gamma\|_{M(B_p^{m-l+\mu}\to B_p^{\mu})} \le c\,\|\gamma\|^{(l-\mu)/l}_{M(B_p^{m-l}\to L_p)}\|\gamma\|^{\mu/l}_{M(B_p^{m}\to B_p^{l})}.$$

Since $m - l + \mu > 2\mu$, it follows from the first part of the proof that there holds inequality (3.2) with m and l replaced by $m - l + \mu$ and μ, respectively, i.e.,

$$\|\gamma\|_{M(B_p^{m-l}\to L_p)} \le c\,\|\gamma\|_{M(B_p^{m-l+\mu}\to B_p^{\mu})}.$$

Consequently,

$$\|\gamma\|_{M(B_p^{m-l}\to L_p)} \le c\,\|\gamma\|^{(l-\mu)/l}_{M(B_p^{m-l}\to L_p)}\|\gamma\|^{\mu/l}_{M(B_p^{m}\to B_p^{l})}$$

and (3.2) is proved for $2l > m > l$ as well.

By Lemma 8 and (2.3), the following assertion holds.

Corollary 1. *Let $\gamma \in M(B_p^m \to B_p^l)$, $0 < l < m$. Then*

$$\sup_e \frac{\|\gamma; e\|_{L_p}}{[\mathrm{cap}_{p,m-l}(e)]^{1/p}} \le c\,\|\gamma\|_{M(B_p^m\to B_p^l)}.$$

Lemma 8 in combination with (2.10) and (2.11) implies

Corollary 2. *Let $\gamma \in M(B_p^m \to B_p^l)$, $0 < l \le m$. Then $\gamma \in M(B_p^{m-k} \to B_p^{l-k})$, $0 < k < l$, and*

$$\|\gamma\|_{M(B_p^{m-k}\to B_p^{l-k})} \le c\|\gamma\|_{M(B_p^m\to B_p^l)}.$$

The following assertion contains an estimate for derivatives of a multiplier.

Lemma 9. *Let $\gamma \in M(B_p^m \to B_p^l)$, $0 < l \le m$. Then $D^\alpha\gamma \in M(B_p^m \to B_p^{l-|\alpha|})$ for any multi-index α of order $|\alpha| \le l$. The inequality holds*

$$\|D^\alpha\gamma\|_{M(B_p^m\to B_p^{l-|\alpha|})} \le c\,\|\gamma\|_{M(B_p^m\to B_p^l)}.$$

Proof. It suffices to consider the case $|\alpha| = 1$, $l \ge 1$. Clearly,

$$\|u\nabla\gamma\|_{B_p^{l-1}} \le \|u\gamma\|_{B_p^l} + \|\gamma\nabla u\|_{B_p^{l-1}}$$

$$\le \left(\|\gamma\|_{M(B_p^m\to B_p^l)} + \|\gamma\|_{M(B_p^{m-1}\to B_p^{l-1})}\right)\|u\|_{B_p^m}.$$

This and Corollary 2 imply

$$\|u\nabla\gamma\|_{B_p^{l-1}} \le c\,\|\gamma\|_{M(B_p^m\to B_p^l)}\|u\|_{B_p^m}$$

which completes the proof.

Lemmas 8 and 9 imply the following

Corollary 3. *Let $\gamma \in M(B_p^m \to B_p^l)$, $0 < l \le m$. Then, for any multi-index α of order $|\alpha| \le l$, $D^\alpha\gamma \in M(B_p^{m-|\alpha|} \to L_p)$. The inequality holds*

$$\|D^\alpha\gamma\|_{M(B_p^{m-l+|\alpha|}\to L_p)} \le c\,\|\gamma\|_{M(B_p^m\to B_p^l)}.$$

4. Proof of necessity in Theorem 1

In this section we derive the inequalities

$$\sup_e \frac{\|C_{p,l}\gamma; e\|_{L_p}}{[\mathrm{cap}_{p,m}(e)]^{1/p}} + \sup_{x\in\mathbf{R}^n} \|\gamma; \mathcal{B}_1(x)\|_{L_p} \le c\|\gamma\|_{M(B_p^m\to B_p^l)}, \quad m>l \tag{4.1}$$

and

$$\sup_e \frac{\|C_{p,l}\gamma; e\|_{L_p}}{[\mathrm{cap}_{p,l}(e)]^{1/p}} + \|\gamma\|_{L_\infty} \le c\|\gamma\|_{MB_p^l}. \tag{4.2}$$

The core of the proof is the following assertion.

Lemma 10. *Let $\gamma \in M(B_p^m \to B_p^l)$, where $0 < l \le m$ and $p \in (1,\infty)$. Then*

$$\sup_e \frac{\|C_{p,l}\gamma; e\|_{L_p}}{[\mathrm{cap}_{p,m}(e)]^{1/p}} \le c\,\|\gamma\|_{M(B_p^m\to B_p^l)}. \tag{4.3}$$

Proof. We use induction in l and start by showing that (4.3) is valid for $l \in (0,1]$.

(i) Let $l \in (0,1)$. We have

$$\|uC_{p,l}\gamma\|_{L_p} \le c\big(\|\gamma u\|_{B_p^l} + \|\gamma C_{p,l}u\|_{L_p}\big)$$
$$\le c\big(\|\gamma\|_{M(B_p^m\to B_p^l)}\|u\|_{B_p^l} + \|\gamma C_{p,l}u\|_{L_p}\big). \tag{4.4}$$

Consider first the case $m = l$. Clearly, $\|\gamma C_{p,l}u\|_{L_p} \le \|\gamma\|_{L_\infty}\|u\|_{B_p^l}$ which together with (4.4) and (3.1) gives

$$\|uC_{p,l}\gamma\|_{L_p} \le c\,\|\gamma\|_{MB_p^l}\|u\|_{B_p^l}.$$

Therefore, $\|C_{p,l}\gamma\|_{M(B_p^l\to L_p)} \le c\|\gamma\|_{MB_p^l}$ and, in view of (2.3), we obtain (4.3).

Suppose now that $l < m$. By (2.13)

$$\|\gamma C_{p,l}u\|_{L_p} \le \|\gamma\|_{M(B_p^{m-l}\to L_p)}\|J_{m-l}C_{p,l}\Lambda^{m-l}u\|_{B_p^{m-l}}. \tag{4.5}$$

Owing to Lemma 1, the last norm does not exceed

$$c\,\|C_{p,l}\Lambda^{m-l}u\|_{L_p} \le c\,\|\Lambda^{m-l}u\|_{B_p^l} \le c\,\|u\|_{B_p^m}$$

which in combination with (4.5) implies

$$\|\gamma C_{p,l}u\|_{L_p} \le c\,\|\gamma\|_{M(B_p^{m-l}\to L_p)}\|u\|_{B_p^m}. \tag{4.6}$$

Using (4.4), (4.6) and Lemma 8, we arrive at

$$\|uC_{p,l}\gamma\|_{L_p} \le c\,\|\gamma\|_{M(B_p^m\to B_p^l)}\|u\|_{B_p^m}.$$

Thus,

$$\|C_{p,l}\gamma\|_{M(B_p^m\to L_p)} \le c\,\|\gamma\|_{M(B_p^m\to B_p^l)}$$

which together with (2.3) gives (4.3).

(ii) Let $l = 1$. In view of the identity

$$\Delta_h^{(2)}(\gamma u) = \gamma\Delta_h^{(2)}u + u\Delta_h^{(2)}\gamma + \Delta_{2h}\gamma\Delta_{2h}u - 2\Delta_h\gamma\Delta_h u \tag{4.7}$$

one has

$$\|uC_{p,1}\gamma\|_{L_p} \leq \|\gamma u\|_{B_p^1} + \|\gamma C_{p,1}u\|_{L_p} + 4\Big(\int\int |\Delta_h\gamma(x)\Delta_h u(x)|^p|h|^{-n-p}dhdx\Big)^{1/p} \tag{4.8}$$

for any $u \in C_0^\infty$.

We proceed separately for $m = 1$ and $m > 1$. Let first $m = 1$. Using (2.15) with $k = 1$ and $\delta \in (0,1)$ together with (4.8) and (3.1), we find

$$\|uC_{p,1}\gamma\|_{L_p} \leq c\left(\|\gamma\|_{MB_p^1} + \sup_e \frac{\|C_{p,\delta}\gamma; e\|_{L_p}}{[\mathrm{cap}_{p,\delta}(e)]^{1/p}}\right)\|u\|_{B_p^1}. \tag{4.9}$$

In view of part (i) of this proof, the last supremum is majorized by $c\|\gamma\|_{MB_p^\delta}$. Hence (4.9) leads to the inequality

$$\sup_e \frac{\|C_{p,1}\gamma; e\|_{L_p}}{[\mathrm{cap}_{p,1}(e)]^{1/p}} \leq c\,(\|\gamma\|_{MB_p^1} + \|\gamma\|_{MB_p^\delta}). \tag{4.10}$$

Since by Corollary 2 there holds $\|\gamma\|_{MB_p^\delta} \leq c\|\gamma\|_{MB_p^1}$, we arrive at (4.3) for $m = l = 1$.

Next we estimate the right-hand side of (4.8) for $m > 1$. By (2.13), its second term is majorized by

$$\begin{aligned}\|\gamma J_{m-1}C_{p,1}\Lambda^{m-1}u\|_{L_p} &\leq c\,\|\gamma\|_{M(B_p^{m-1}\to L_p)}\|J_{m-1}C_{p,1}\Lambda^{m-1}u\|_{B_p^{m-1}} \\ &\leq c\,\|\gamma\|_{M(B_p^{m-1}\to L_p)}\|C_{p,1}\Lambda^{m-1}u\|_{L_p} \\ &\leq c\,\|\gamma\|_{M(B_p^{m-1}\to L_p)}\|\Lambda^{m-1}u\|_{B_p^1} \leq c\,\|\gamma\|_{M(B_p^m\to B_p^1)}\|u\|_{B_p^m}.\end{aligned} \tag{4.11}$$

The last inequality in this chain follows from (2.1) and (3.2). We estimate the third term in the right-hand side of (4.8) using (2.15) with $k = m > 1$ and (4.3) with $l = \delta < 1$. Then this term does not exceed

$$c\sup_e \frac{\|C_{p,\delta}\gamma; e\|_{L_p}}{[\mathrm{cap}_{p,m-1+\delta}(e)]^{1/p}}\|u\|_{B_p^m} \leq c\,\|\gamma\|_{M(B_p^{m-1+\delta}\to B_p^\delta)}\|u\|_{B_p^m}. \tag{4.12}$$

Furthermore, by Corollary 2

$$\|\gamma\|_{M(B_p^{m-1+\delta}\to B_p^\delta)} \leq c\,\|\gamma\|_{M(B_p^m\to B_p^1)}.$$

Hence, the third term on the right in (4.8) is dominated by $c\,\|\gamma\|_{M(B_p^m\to B_p^1)}\|u\|_{B_p^m}$. This along with (4.8) and (4.11) implies

$$\|uC_{p,1}\gamma\|_{L_p} \leq c\,\|\gamma\|_{M(B_p^m\to B_p^1)}\|u\|_{B_p^m}$$

and thus (4.3) holds for $l = 1$.

(iii) Suppose that l is a positive integer and that the lemma is proved for $\gamma \in M(B_p^m \to B_p^k)$, where k is any positive integer not exceeding $l-1$. Applying (4.7), we find

$$\|uC_{p,l}\gamma\|_{L_p} \leq \|\gamma u\|_{B_p^l} + c\sum_{j=0}^{l-1} \|\, |\nabla_j\gamma| C_{p,l-j}u\|_{L_p} + c\sum_{j=1}^{l-1} \|\, |\nabla_j u| C_{p,l-j}\gamma\|_{L_p}$$

$$+c\sum_{j=0}^{l-1}\Big(\int\int |\Delta_h \nabla_j\gamma(x)|^p |\Delta_h \nabla_{l-1-j}u|^p |h|^{-n-p} dh dx\Big)^{1/p}. \tag{4.13}$$

By (2.13) with $\alpha = l-j$, $\beta = m-l+j$ we have

$$(C_{p,l-j}u)(x) \leq (J_{m-l+j}C_{p,l-j}\Lambda^{m-l+j}u)(x).$$

Therefore, for $j=1,\dots,l-1$ and $m \geq l$,

$$\|\, |\nabla_j\gamma| C_{p,l-j}u\|_{L_p} \leq c\,\|\nabla_j\gamma\|_{M(B_p^{m-l+j}\to L_p)} \|J_{m-l+j}C_{p,l-j}\Lambda^{m-l+j}u\|_{B_p^{m-l+j}}$$

$$\leq c\,\|\nabla_j\gamma\|_{M(H_p^{m-l+j}\to L_p)} \|C_{p,l-j}\Lambda^{m-l+j}u\|_{L_p}. \tag{4.14}$$

According to (2.1),

$$\|C_{p,l-j}\Lambda^{m-l+j}u\|_{L_p} \leq \|\Lambda^{m-l+j}u\|_{B_p^{l-j}} \leq c\,\|u\|_{B_p^m}. \tag{4.15}$$

By Corollary 3,

$$\|\nabla_j\gamma\|_{M(H_p^{m-l+j}\to L_p)} \leq c\,\|\gamma\|_{M(B_p^m\to B_p^l)}, \quad j=1,\dots,l-1, \ \ m\geq l. \tag{4.16}$$

For $j=0$ by Lemma 8 we obtain

$$\|\gamma C_{p,l}u\|_{L_p} \leq \|\gamma\|_{M(B_p^m\to B_p^l)}\|u\|_{B_p^m}. \tag{4.17}$$

Unifying (4.14)–(4.17), we find that for all $j=0,\dots,l-1$ and $1\leq l\leq m$,

$$\|\, |\nabla_j\gamma| C_{p,l-j}u\|_{L_p} \leq c\,\|\gamma\|_{M(B_p^m\to B_p^l)}\|u\|_{B_p^m}. \tag{4.18}$$

For any $j=1,\dots,l-1$ we have

$$\|\, |\nabla_j u| C_{p,l-j}\gamma\|_{L_p} \leq c\sup_e \frac{\|C_{p,l-j}\gamma; e\|_{L_p}}{[\text{cap} - p, m-j(e)]^{1/p}}\|u\|_{B_p^m}. \tag{4.19}$$

From the induction assumption and Corollary 2 it follows that for $m \geq l$ one has

$$\sup_e \frac{\|C_{p,l-j}\gamma; e\|_{L_p}}{[\text{cap}_{p,m-j}(e)]^{1/p}} \leq c\,\|\gamma\|_{M(B_p^{m-j}\to B_p^{l-j})} \leq c\,\|\gamma\|_{M(B_p^m\to B_p^l)} \tag{4.20}$$

which together with (4.19) implies

$$\|\, |\nabla_j u| C_{p,l-j}\gamma\|_{L_p} \leq c\,\|\gamma\|_{M(B_p^m\to B_p^l)}\|u\|_{B_p^m}, \quad j=1,\dots,l-1. \tag{4.21}$$

Next we estimate the last sum in (4.13). Let $\delta \in (0,1)$ be such that $m+\delta$ is a noninteger. By (2.15) with γ replaced by $\nabla_j\gamma$, u replaced by $\nabla_{l-1-j}u$, and $k = m-l+j+1$ each term of the last sum in (4.13) does not exceed

$$c\sup_e \frac{\|C_{p,j+\delta}\gamma; e\|_{L_p}}{[\text{cap}_{p,m-l+j+\delta}(e)]^{1/p}}\|\nabla_{l-1-j}u\|_{B_p^{m-l+j+1}}. \tag{4.22}$$

By the induction assumption and Corollary 2 this implies

$$\Big(\int\!\!\int |\Delta_h\nabla_j\gamma(x)|^p|\Delta_h\nabla_{l-1-j}u|^p|h|^{-n-p}dhdx\Big)^{1/p}$$
$$\le c\|\gamma\|_{M(B_p^{m-l+j+\delta}\to B_p^{j+\delta})}\|u\|_{B_p^m}\le c\|\gamma\|_{M(B_p^m\to B_p^l)}\|u\|_{B_p^m}. \tag{4.23}$$

Combining this with (4.21) and (4.19), we obtain from (4.13)

$$\|uC_{p,l}\gamma\|_{L_p}\le c\,\|\gamma\|_{M(B_p^m\to B_p^l)}\|u\|_{B_p^m} \tag{4.24}$$

and thus (4.3) follows for all integer l.

(iv) Now let l be noninteger. Suppose that

$$\sup_e\frac{\|C_{p,l}\gamma;e\|_{L_p}}{[\mathrm{cap}_{p,m}(e)]^{1/p}}\le c\,\|\gamma\|_{M(B_p^m\to B_p^l)}$$

for all noninteger $l\in(0,N)$, where N is integer. Let $N<l<N+1$. In view of the equivalence $C_{p,l}\gamma\sim D_{p,l}\gamma$ we have

$$\|uD_{p,l}\gamma\|_{L_p}\le\|\gamma u\|_{B_p^l}+c\sum_{j=0}^{N}\|\,|\nabla_j\gamma|D_{p,l-j}u\|_{L_p}+c\sum_{j=1}^{N}\|\,|\nabla_ju|D_{p,l-j}\gamma\|_{L_p}. \tag{4.25}$$

Let $t\in(0,m-l+j)$ if $m>l$ or $m=l$, $j>0$ and let $t=0$ if $m=l$ and $j=0$. By (2.14) with $\alpha=l-j$ and $\beta=t$ one has

$$\big(D_{p,l-j}u\big)(x)\le\big(J_tD_{p,l-j}\Lambda^tu\big)(x).$$

Hence

$$\|\,|\nabla_j\gamma|D_{p,l-j}u\|_{L_p}\le\|\nabla_j\gamma\|_{M(W_p^{m-l+j}\to L_p)}\|J_tD_{p,l-j}\Lambda^tu\|_{W_p^{m-l+j}}$$
$$\le c\,\|\nabla_j\gamma\|_{M(B_p^{m-l+j}\to L_p)}\|D_{p,l-j}\Lambda^tu\|_{W_p^{m-l+j-t}}. \tag{4.26}$$

By definition of the operator $D_{p,l}$ and the space W_p^l,

$$\|D_{p,l-j}v\|_{W_p^{m-l+j-t}}=\|D_{p,m-l+j-t}D_{p,\{l\}}\nabla_{[l-j]}v\|_{L_p}+\|D_{p,l-j}v\|_{L_p}.$$

We use Lemma 7 with $\alpha=m-l+j-t$, $\beta=\{l\}$ assuming t to be so close to $m-l+j$ that $0<m-t-[l]+j<1$. Then

$$\|D_{p,m-l+j-t}D_{p,\{l\}}\nabla_{[l]-j}v\|_{L_p}\le c\|D_{p,m-t-[l]-j}\nabla_{[l]-j}v\|_{L_p}\le c\|v\|_{W_p^{m-t}}. \tag{4.27}$$

We may also choose t in such a way that $m-t$ is noninteger so that $W_p^{m-t}=B_p^{m-t}$. Then (4.26) and (4.27) with $v=\Lambda^tu$, together with Corollary 3 imply

$$\|\,|\nabla_j\gamma|D_{p,l-j}u\|_{L_p}\le c\,\|\nabla_j\gamma\|_{M(B_p^{m-l+j}\to L_p)}\|\Lambda^tu\|_{B_p^{m-t}}$$
$$\le c\,\|\gamma\|_{M(B_p^m\to B_p^l)}\|u\|_{B_p^m}. \tag{4.28}$$

By the induction hypothesis, we have for $j=1,\dots,N$

$$\|\,|\nabla_ju|D_{p,l-j}\gamma\|_{L_p}\le c\sup_e\frac{\|D_{p,l-j}\gamma;e\|_{L_p}}{[\mathrm{cap}_{p,m-j}(e)]^{1/p}}\|\nabla_ju\|_{B_p^{m-j}}$$
$$\le c\,\|\gamma\|_{M(B_p^{m-j}\to B_p^{l-j})}\|u\|_{B_p^m} \tag{4.29}$$

which together with Corollary 2 implies

$$\| \, |\nabla_j u| D_{p,l-j}\gamma\|_{L_p} \le c\, \|\gamma\|_{M(B_p^m \to B_p^l)} \|u\|_{B_p^m}.$$

Hence and by (4.28) it follows from (4.25) that

$$\|u D_{p,l}\gamma\|_{L_p} \le c\, \|\gamma\|_{M(B_p^m \to B_p^l)} \|u\|_{B_p^m}.$$

The proof is complete.

The following simple corollary contains the required lower estimate of the norm in $M(B_p^m \to B_p^l)$ in Theorem 1. It also finishes the proof of necessity in Theorem 1.

Corollary 4. *Let $\gamma \in M(B_p^m \to B_p^l)$, where $0 < l \le m$ and $p \in (1, \infty)$. Then*

$$c\Big(\sup_e \frac{\|C_{p,l}\gamma; e\|_{L_p}}{[\mathrm{cap}_{p,m}(e)]^{1/p}} + \sup_{x\in\mathbf{R}^n} \|\gamma; \mathcal{B}_1(x)\|_{L_p}\Big) \le \|\gamma\|_{M(B_p^m \to B_p^l)}. \tag{4.30}$$

For $m = l$ the second term on the left should be replaced by $\|\gamma\|_{L_\infty}$.

Proof. Since $\gamma \in M(B_p^m \to B_p^l)$ it follows that

$$\|\gamma\eta\|_{L_p} \le \|\gamma\|_{M(B_p^m \to B_p^l)} \|\eta\|_{B_p^m}$$

for any $\eta \in C_0^\infty(\mathcal{B}_2(x))$, $\eta = 1$ on $\mathcal{B}_1(x)$, where x is an arbitrary point of $\mathbf{R}^n$. Therefore,

$$\sup_{x\in\mathbf{R}^n} \|\gamma; \mathcal{B}_1(x)\|_{L_p} \le c\, \|\gamma\|_{M(B_p^m \to B_p^l)}.$$

The result follows by combining this with Lemma 10.

The next corollary contains one more lower estimate for the norm in the space $M(B_p^m \to B_p^l)$.

Corollary 5. *Let $\gamma \in M(B_p^m \to B_p^l)$, where $0 < l \le m$, $p \in (1, \infty)$. Then, for any $k = 0, \dots, [l]$ there holds the inclusion $C_{p,l-k}\gamma \in M(B_p^{m-k} \to L_p)$ and*

$$\|C_{p,l-k}\gamma\|_{M(B_p^{m-k} \to L_p)} \le c\, \|\gamma\|_{M(B_p^m \to B_p^l)}.$$

Proof. By Corollaries 4 and 2,

$$\sup_e \frac{\|C_{p,l-k}\gamma; e\|_{L_p}}{[\mathrm{cap}_{p,m-k}(e)]^{1/p}} \le c\, \|\gamma\|_{M(B_p^{m-k} \to B_p^{l-k})} \le c\, \|\gamma\|_{M(B_p^m \to B_p^l)}. \tag{4.31}$$

It remains to make use of (2.3).

5. Proof of sufficiency in Theorem 1

The aim of this section is to prove the upper estimate of $\|\gamma\|_{M(B_p^m \to B_p^l)}$ in (1.4).

Lemma 11. *Let $\gamma \in B_{p,loc}^l$, $p \in (1, \infty)$. Then for $m > l$*

$$c\, \|\gamma\|_{M(B_p^m \to B_p^l)} \le \sup_{e, \mathrm{diam}(e) \le 1} \Big(\frac{\|C_{p,l}\gamma; e\|_{L_p}}{[\mathrm{cap}_{p,m}(e)]^{1/p}} + \frac{\|\gamma, e\|_{L_p}}{[\mathrm{cap}_{p,m-l}(e)]^{1/p}} \Big). \tag{5.1}$$

For $m = l$ the second term should be replaced by $\|\gamma\|_{L_\infty}$.

Proof. It follows from the finiteness of the right-hand side of (5.1) that $\gamma \in L_{1,\mathrm{unif}}$. Let γ_ρ denote the mollifyer of γ with radius ρ. From $\gamma \in L_{1,\mathrm{unif}}$ it follows that all derivatives of γ_ρ are bounded. Hence $\gamma_\rho \in M(B_p^m \to B_p^l)$.

For integer l we find by (4.7) that there holds the estimate

$$\|\gamma_\rho u\|_{B_p^l} \le c\Big(\sum_{j=0}^{l-1} \| \, |\nabla_j \gamma_\rho| C_{p,l-j} u\|_{L_p} + \sum_{j=0}^{l-1} \| \, |\nabla_j u| C_{p,l-j} \gamma_\rho\|_{L_p}$$

$$+ \sum_{j=0}^{l-1} \Big(\int \int |\Delta_h \nabla_j \gamma_\rho(x)|^p |\Delta_h \nabla_{l-1-j} u|^p |h|^{-n-p} dh dx\Big)^{1/p}\Big). \tag{5.2}$$

By Corollary 3, for any $\alpha \in (0,1)$

$$\|\nabla_j \gamma_\rho\|_{M(B_p^{m-l+j} \to L_p)} \le c \, \|\gamma_\rho\|_{M(B_p^{m-l+j+\alpha} \to B_p^{j+\alpha})}. \tag{5.3}$$

In view of (2.10), for $m > l$ the right-hand side in (5.3) does not exceed

$$c \|\gamma_\rho\|^{(l-j-\alpha)/l}_{M(B_p^{m-l} \to L_p)} \|\gamma_\rho\|^{(j+\alpha)/l}_{M(B_p^m \to B_p^l)}.$$

Combining this with (4.14) and (4.15) we obtain

$$\| \, |\nabla_j \gamma_\rho| C_{p,l-j} u\|_{L_p} \le \big(\varepsilon \|\gamma_\rho\|_{M(B_p^m \to B_p^l)} + c(\varepsilon) \|\gamma_\rho\|_{M(B_p^{m-l} \to L_p)}\big) \|u\|_{B_p^m}, \tag{5.4}$$

where $j = 0, \ldots, l-1$, and ε is an arbitrary positive number.

In case $m = l$ inequalities (5.3) and (2.11) imply

$$\|\nabla_j \gamma_\rho\|_{M(B_p^j \to L_p)} \le c \|\gamma_\rho\|^{(l-j)/l}_{L_\infty} \|\gamma_\rho\|^{j/l}_{MB_p^l}.$$

unifying this with (4.14) and (4.15) for $m = l$ we obtain

$$\| \, |\nabla_j \gamma_\rho| C_{p,l-j} u\|_{L_p} \le \big(\varepsilon \|\gamma_\rho\|_{MB_p^l} + c(\varepsilon) \|\gamma_\rho\|_{L_\infty}\big) \|u\|_{B_p^l}. \tag{5.5}$$

It follows from (4.19), (4.20), and (2.10), (2.11) that for $j > 0$

$$\| \, |\nabla_j u| C_{p,l-j} \gamma_\rho\|_{L_p} \le \big(\varepsilon \|\gamma_\rho\|_{M(B_p^m \to B_p^l)} + c(\varepsilon) \|\gamma_\rho\|_{M(B_p^{m-l} \to L_p)}\big) \|u\|_{B_p^m}, \tag{5.6}$$

if $m > l$ and

$$\| \, |\nabla_j u| C_{p,l-j} \gamma_\rho\|_{L_p} \le \big(\varepsilon \|\gamma_\rho\|_{MB_p^l} + c(\varepsilon) \|\gamma_\rho\|_{L_\infty}\big) \|u\|_{B_p^l} \tag{5.7}$$

if $m = l$.

The third sum in the right-hand side of (5.2) is estimated by using (4.23) and (2.10), (2.11) and has the same majorant as the right-hand side of (5.6) for $m > l$ or (5.7) for $m = l$. Thus, for $m > l$ we find

$$\|\gamma_\rho u\|_{B_p^l} \le \Big(\varepsilon \|\gamma_\rho\|_{M(B_p^m \to B_p^l)} + c(\varepsilon) \|\gamma_\rho\|_{M(B_p^{m-l} \to L_p)}$$

$$+ c \sup_{e, \mathrm{diam}(e) \le 1} \frac{\|C_{p,l} \gamma_\rho; e\|_{L_p}}{[\mathrm{cap}_{p,m}(e)]^{1/p}}\Big) \|u\|_{B_p^m}. \tag{5.8}$$

Similarly, for $m = l$,

$$\|\gamma_\rho u\|_{B_p^l} \le \Big(\varepsilon \|\gamma_\rho\|_{MB_p^l} + c(\varepsilon) \|\gamma_\rho\|_{L_\infty} + c \sup_{e, \mathrm{diam}(e) \le 1} \frac{\|C_{p,l} \gamma_\rho; e\|_{L_p}}{[\mathrm{cap}_{p,l}(e)]^{1/p}}\Big) \|u\|_{B_p^l}. \tag{5.9}$$

For noninteger l the following estimate, simpler than (5.2), holds

$$\|\gamma_\rho u\|_{B_p^l} \le c\Big(\sum_{j=0}^{[l]-1} \| \, |\nabla_j \gamma_\rho| C_{p,l-j} u\|_{L_p} + \sum_{j=0}^{[l]-1} \| \, |\nabla_j u| C_{p,l-j}\gamma_\rho\|_{L_p}\Big).$$

Combining (4.28) with Corollary 3 and (2.10), (2.11), we arrive at (5.4) and (5.5) in the same way as for integer l. We also note that (4.29) and (2.10) for $m > l$ and (2.11) for $m = l$ imply (5.6) and (5.7) for noninteger l. Reference to (2.3) and Lemma 2 completes the proof.

The required upper estimate of $\|\gamma\|_{MB_p^l}$ in (1.4) is obtained in Lemma 11. In order to show that for $m > l$ the second term on the right in (5.1) can be replaced by $\|\gamma\|_{L_{1,\mathrm{unif}}}$, we need several auxiliary assertions. Let $\gamma(x, y)$ denote the Poisson integral of a function $\gamma \in L_{1,\mathrm{unif}}$.

Lemma 12. (see Lemma 5.1.2 [MS]) *Let l be noninteger and let $\gamma \in W^{[l]}_{1,\mathrm{loc}}$. Then*

$$\Big(\int_0^\infty \Big|\frac{\partial^{[l]+1}\gamma(x,y)}{\partial y^{[l]+1}}\Big|^p y^{p-1-p\{l\}}dy\Big)^{1/p} \le c\,(D_{p,l}\gamma)(x).$$

Lemma 13. (Verbitsky, see Sect. 2.6 [MS]) *For any $k = 0, 1, \ldots$ there holds the inequality*

$$|\gamma(x)| \le c\Big(\|\gamma\|_{L_{1,\mathrm{unif}}} + \int_0^1 \Big|\frac{\partial^{k+1}\gamma(x,y)}{\partial y^{k+1}}\Big| y^k dy\Big). \tag{5.10}$$

The following two lemmas are similar to those due to Verbitsky as presented in Sect. 2.6 [MS].

Lemma 14. *Let $\gamma \in W^{[l]}_{1,loc}$, $y \in (0,1]$. Then*

$$\Big|\frac{\partial^{[l]+1}\gamma(x,y)}{\partial y^{[l]+1}}\Big| \le c y^{\{l\}-m-1} \sup_{x\in\mathbf{R}^n, r\in(0,1)} r^{m-n/p}\|D_{p,l}\gamma; \mathcal{B}_r(x)\|_{L_p}.$$

Proof. We introduce the notation

$$K = \sup_{x\in\mathbf{R}^n, r\in(0,1)} r^{m-n/p}\|D_{p,l}\gamma; \mathcal{B}_r(x)\|_{L_p}. \tag{5.11}$$

Let $r \in (0, 1]$. By Lemma 12

$$\int_{\mathcal{B}_r(x)} \int_0^\infty \Big|\frac{\partial^{[l]+1}\gamma(x,y)}{\partial y^{[l]+1}}\Big|^p y^{p-1-p\{l\}}dy\, dt \le cK^p r^{n-mp}. \tag{5.12}$$

Applying the mean value theorem for harmonic functions we find for $\frac{r}{2} < y < \frac{2r}{3}$

$$\Big|\frac{\partial^{[l]+1}\gamma(x,y)}{\partial y^{[l]+1}}\Big| \le c\, r^{-n-1} \int_{\mathcal{B}_r(x)} \int_{r/4}^r \Big|\frac{\partial^{[l]+1}\gamma(t,\eta)}{\partial \eta^{[l]+1}}\Big|\, d\eta dt.$$

By Hölder's inequality the right-hand side is dominated by

$$c\, r^{\{l\}-1-n/p}\Big(\int_{\mathcal{B}_r(x)} \int_{r/4}^r \Big|\frac{\partial^{[l]+1}\gamma(t,\eta)}{\partial \eta^{[l]+1}}\Big|^p \eta^{p-1-p\{l\}}d\eta dt\Big)^{1/p}$$

which by (5.12) does not exceed $cr^{\{l\}-m-1}K$. The proof is complete.

Lemma 15. *Let $\gamma \in W^{[l]}_{1,loc}$. Then for all $x \in \mathbf{R}^n$ there holds inequality*

$$|\gamma(x)| \le c\Big(\Big(\sup_{x\in\mathbf{R}^n, r\in(0,1)} r^{m-n/p}\|D_{p,l}\gamma;\mathcal{B}_r(x)\|_{L_p}\Big)^{l/m}(D_{p,l}\gamma(x))^{(m-l)/m} + \|\gamma\|_{L_{1,\text{unif}}}\Big).$$

Proof. We put

$$v(y) = \begin{cases} \Big|\frac{\partial^{[l]+1}\gamma(x,y)}{\partial y^{[l]+1}}\Big|, & \text{for } 0 < y \le 1,\\ 0, & \text{for } y > 1.\end{cases}$$

Then, for any $R > 0$

$$\int_0^1 \Big|\frac{\partial^{[l]+1}\gamma(x,y)}{\partial y^{[l]+1}}\Big| y^{[l]}dy = \int_0^\infty v(y)y^{[l]}dy = \int_0^R v(y)y^{[l]}dy + \int_R^\infty v(y)y^{[l]}dy.$$

Applying Hölder's inequality, we find

$$\int_0^R v(y)y^{[l]}dy \le c\, R^l\Big(\int_0^R (v(y))^p y^{p-p\{l\}-1}dy\Big)^{1/p}.$$

By Lemma 14,

$$\Big|\frac{\partial^{[l]+1}\gamma(x,y)}{\partial y^{[l]+1}}\Big| \le cKy^{\{l\}-m-1},$$

where K is defined by (5.11). Hence

$$\int_0^\infty v(y)y^{[l]}dy \le c\Big(R^l\Big(\int_0^\infty (v(y))^p y^{p-p\{l\}-1}dy\Big)^{1/p} + R^{l-m}K\Big).$$

Putting here

$$R = K^{1/m}\Big(\int_0^\infty v(y)^p y^{p-p\{l\}-1}dy\Big)^{-1/pm},$$

we arrive at

$$\int_0^\infty v(y)y^{[l]}dy \le cK^{l/m}\Big(\int_0^\infty v(y)^p y^{p-p\{l\}-1}dy\Big)^{(m-l)/pm}.$$

Combining this with (5.10) for $k = [l]$ we arrive at

$$|\gamma(x)| \le \Big(K^{l/m}\Big(\int_0^\infty v(y)^p y^{p-p\{l\}-1}dy\Big)^{(m-l)/pm} + \|\gamma\|_{L_{1,\text{unif}}}\Big).$$

Reference to Lemma 12 completes the proof.

Now, we are in a position to prove the principle result of this section.

Lemma 16. *Let $0 < l < m$, $p \in (1,\infty)$. Then*

$$\|\gamma\|_{M(B^m_p\to B^l_p)} \le c\Big(\sup_{e,\text{diam}(e)\le 1} \frac{\|C_{p,l}\gamma;e\|_{L_p}}{[\text{cap}_{p,m}(e)]^{1/p}} + \|\gamma\|_{L_{1,\text{unif}}}\Big). \tag{5.13}$$

Proof. By (2.12) with $\varphi = |\gamma_\rho|^{\frac{1}{m-l}}$, $\lambda = m - l$, $\mu = m - \varepsilon$, where ε is a positive number less than l such that both $l - \varepsilon$ and $m - \varepsilon$ are nonintegers, we find

$$\sup_e \frac{\int_e |\gamma_\rho|^p(x)dx}{\mathrm{cap}_{p,m-l}(e)} \le c \sup_e \left(\frac{\int_e |\gamma_\rho|^{\frac{m-\varepsilon}{m-l}p}(x)dx}{\mathrm{cap}_{p,m-\varepsilon}(e)} \right)^{\frac{m-l}{m-\varepsilon}}. \tag{5.14}$$

Owing to Lemma 15 with l replaced by $l - \varepsilon$ and m replaced by $m - \varepsilon$

$$\int_e |\gamma_\rho|^{\frac{(m-\varepsilon)p}{m-l}} dx \le c\Big(\Big(\sup_{x\in\mathbf{R}^n, r\in(0,1)} r^{m-\varepsilon-\frac{n}{p}} \|D_{p,l-\varepsilon}\gamma_\rho; \mathcal{B}_r(x)\|_{L_p} \Big)^{\frac{(l-\varepsilon)p}{m-l}}$$

$$\times \int_e |D_{p,l-\varepsilon}\gamma_\rho|^p dx + \|\gamma_\rho\|_{L_{1,\mathrm{unif}}}^{\frac{(m-\varepsilon)p}{m-l}} \mathrm{mes}_n e \Big).$$

Hence

$$\left(\frac{\int_e |\gamma_\rho|^{\frac{(m-\varepsilon)p}{m-l}}(x)dx}{\mathrm{cap}_{p,m-\varepsilon}(e)} \right)^{\frac{m-l}{(m-\varepsilon)p}} \le c\Big\{ \Big(\sup_{x\in\mathbf{R}^n, r\in(0,1)} r^{m-\varepsilon-\frac{n}{p}} \|D_{p,l-\varepsilon}\gamma_\rho; \mathcal{B}_r(x)\|_{L_p} \Big)^{\frac{l-\varepsilon}{m-\varepsilon}}$$

$$\times \left(\sup_e \frac{\|D_{p,l-\varepsilon}\gamma_\rho; e\|_{L_p}}{[\mathrm{cap}_{p,m-\varepsilon}(e)]^{1/p}} \right)^{\frac{m-l}{m-\varepsilon}} + \|\gamma_\rho\|_{L_{1,\mathrm{unif}}} \Big\}. \tag{5.15}$$

By Corollary 2

$$\sup_e \frac{\|D_{p,l-\varepsilon}\gamma_\rho; e\|_{L_p}}{[\mathrm{cap}_{p,m-\varepsilon}(e)]^{1/p}} \le c \|\gamma_\rho\|_{M(W_p^{m-\varepsilon} \to W_p^{l-\varepsilon})}$$

$$= c \|\gamma_\rho\|_{M(B_p^{m-\varepsilon} \to B_p^{l-\varepsilon})} \le c \|\gamma_\rho\|_{M(B_p^m \to B_p^l)}.$$

Thus, the left-hand side of (5.15) has the majorant

$$c\Big(\Big(\sup_{x\in\mathbf{R}^n, r\in(0,1)} r^{m-\varepsilon-\frac{n}{p}} \|D_{p,l-\varepsilon}\gamma_\rho; \mathcal{B}_r(x)\|_{L_p} \Big)^{\frac{l-\varepsilon}{m-\varepsilon}} \|\gamma_\rho\|_{M(B_p^m \to B_p^l)}^{\frac{m-l}{m-\varepsilon}} + \|\gamma_\rho\|_{L_{1,\mathrm{unif}}} \Big)$$

which together with (5.14) implies the inequality

$$\sup_e \left(\frac{\int_e |\gamma_\rho|^p(x)dx}{\mathrm{cap}_{p,m-l}(e)} \right)^{1/p} \le c(\delta) \sup_{x\in\mathbf{R}^n, r\in(0,1)} r^{m-\varepsilon-\frac{n}{p}} \|D_{p,l-\varepsilon}\gamma_\rho; \mathcal{B}_r(x)\|_{L_p}$$

$$+\delta \|\gamma_\rho\|_{M(B_p^m \to B_p^l)} + c \|\gamma_\rho\|_{L_{1,\mathrm{unif}}}, \tag{5.16}$$

where δ is an arbitrary positive number.

Next we show that

$$\sup_{x\in\mathbf{R}^n, r\in(0,1)} r^{m-\varepsilon-\frac{n}{p}} \|D_{p,l-\varepsilon}\gamma_\rho; \mathcal{B}_r(x)\|_{L_p}$$

$$\le c(\sigma) \sup_{x\in\mathbf{R}^n, r\in(0,1)} r^{m-\frac{n}{p}} \|C_{p,l}\gamma_\rho; \mathcal{B}_r(x)\|_{L_p} + \sigma \|\gamma_\rho\|_{M(B_p^m \to B_p^l)} \tag{5.17}$$

where σ is an arbitrary positive number. We note that by (1.1) $D_{p,l-\varepsilon}\gamma_\rho$ can be replaced by $C_{p,l-\varepsilon}\gamma_\rho$. Let ω denote a positive number to be chosen later. Further, let $k = l - 1$ and $\lambda = 1$ for integer l and $k = [l]$ and $\lambda = \{l\}$ for noninteger l. We have

$$\int_{\mathcal{B}_r(x)} dy \int_{\mathcal{B}_{\omega r}} \frac{|\nabla_k \gamma_\rho(y+2h) - 2\nabla_k \gamma_\rho(y+h) + \nabla_k \gamma_\rho(y)|^p}{|h|^{n+p(\lambda-\varepsilon)}} dh$$

$$\leq (\omega r)^{p\varepsilon} \int_{\mathcal{B}_r(x)} dy \int_{\mathcal{B}_{\omega r}} \frac{|\nabla_k \gamma_\rho(y+2h) - 2\nabla_k \gamma_\rho(y+h) + \nabla_k \gamma_\rho(y)|^p}{|h|^{n+p\lambda}} dh$$

$$\leq (\omega r)^{p\varepsilon} \|C_{p,l}\gamma_\rho; \mathcal{B}_r(x)\|^p_{L_p}. \tag{5.18}$$

Besides,

$$\int_{\mathcal{B}_r(x)} dy \int_{\mathbf{R}^n \setminus \mathcal{B}_{\omega r}} \frac{|\nabla_k \gamma_\rho(y+2h) - 2\nabla_k \gamma_\rho(y+h) + \nabla_k \gamma_\rho(y)|^p}{|h|^{n+p(\lambda-\varepsilon)}} dh$$

$$\leq c\Big(\int_{\mathcal{B}_r(x)} dy \int_{\mathbf{R}^n \setminus \mathcal{B}_{\omega r}} \frac{|\nabla_k \gamma_\rho(y+2h)|^p}{|h|^{n+p(\lambda-\varepsilon)}} dh + \int_{\mathcal{B}_r(x)} dy \int_{\mathbf{R}^n \setminus \mathcal{B}_{\omega r}} \frac{|\nabla_k \gamma_\rho(y+h)|^p}{|h|^{n+p(\lambda-\varepsilon)}} dh$$

$$+ (\omega r)^{p(\varepsilon-\lambda)} \|\nabla_k \gamma_\rho; \mathcal{B}_r(x)\|^p_{L_p}\Big). \tag{5.19}$$

Further, we have

$$\int_{\mathcal{B}_r(x)} dy \int_{\mathbf{R}^n \setminus \mathcal{B}_{\omega r}} \frac{|\nabla_k \gamma_\rho(y+2h)|^p}{|h|^{n+p(\lambda-\varepsilon)}} dh$$

$$\leq \int_{\mathbf{R}^n \setminus \mathcal{B}_{\omega r}} \frac{dh}{|h|^{n+p(\lambda-\varepsilon)}} \int_{\mathcal{B}_r(x+2h)} |\nabla_k \gamma_\rho(z)|^p dz$$

$$\leq c\omega^{p(\varepsilon-\lambda)} r^{n-pm+p\varepsilon} \sup_{x\in\mathbf{R}^n, r\in(0,1)} r^{p(m-\lambda)-n} \|\nabla_k \gamma_\rho; \mathcal{B}_r(x)\|^p_{L_p}.$$

By (2.4)–(2.6) the last supremum is dominated by

$$c\, \|\nabla_k \gamma_\rho\|^p_{M(W^{m-\lambda}_p \to L_p)}$$

which by Corollary 3 does not exceed $c\,\|\gamma_\rho\|^p_{M(B^m_p \to B^l_p)}$.

Clearly, the second term in the right-hand side of (5.19) is estimated in the same way. Similarly, the third term does not exceed

$$c\, \omega^{p(\varepsilon-\lambda)} r^{n-pm+p\varepsilon} \|\gamma_\rho\|^p_{M(B^m_p \to B^l_p)}.$$

Hence

$$\int_{\mathcal{B}_r(x)} dy \int_{\mathbf{R}^n \setminus \mathcal{B}_{\omega r}} \frac{|\nabla_k \gamma_\rho(y+2h) - 2\nabla_k \gamma_\rho(y+h) + \nabla_k \gamma_\rho(y)|^p}{|h|^{n+p(\lambda-\varepsilon)}} dh$$

$$\leq c\, \omega^{p(\varepsilon-\lambda)} r^{n-pm+p\varepsilon} \|\gamma_\rho\|^p_{M(B^m_p \to B^l_p)}. \tag{5.20}$$

From (5.18) and (5.20) we obtain

$$r^{m-\varepsilon-n/p} \|D_{p,l-\varepsilon}\gamma_\rho\|_{L_p} \leq c\big(\omega^\varepsilon r^{m-n/p} \|C_{p,l}\gamma_\rho; \mathcal{B}_r(x)\|_{L_p} + \omega^{\varepsilon-\lambda} \|\gamma_\rho\|_{M(B^m_p \to B^l_p)}\big).$$

Setting $\sigma = c\omega^{\varepsilon-\lambda}$ we arrive at (5.17).

By (2.4)–(2.6) and (5.17),

$$\sup_{x\in\mathbf{R}^n, r\in(0,1)} r^{m-\varepsilon-\frac{n}{p}} \|D_{p,l-\varepsilon}\gamma_\rho; \mathcal{B}_r(x)\|_{L_p}$$

$$\leq c(\sigma) \sup_e \frac{\|C_{p,l}\gamma_\rho; e\|_{L_p}}{[\mathrm{cap}_{p,m}(e)]^{1/p}} + \sigma\|\gamma_\rho\|_{M(B_p^m\to B_p^l)},$$

which together with (5.16) and Lemma 11 results at

$$\|\gamma_\rho\|_{M(B_p^m\to B_p^l)} \leq c\Big(\sup_e \frac{\|C_{p,l}\gamma_\rho; e\|_{L_p}}{[\mathrm{cap}_{p,m}(e)]^{1/p}} + \|\gamma_\rho\|_{L_{1,\mathrm{unif}}}\Big). \tag{5.21}$$

Estimating the right-hand side of (5.21) by Lemma 2 and using the equivalence (see, Proposition 2.1. 5 [MS])

$$\mathrm{cap}_{p,m}(e) \sim \sum_{j\geq 1} \mathrm{cap}_{p,m}(e\cap \mathcal{B}^{(j)}),$$

where $\{\mathcal{B}^{(j)}\}_{j\geq 0}$ is a covering of $\mathbf{R}^n$ by balls of diameter one with multiplicity depending only on n, we complete the proof.

6. The case $mp > n$

For $mp > n$ Theorem 1 admits a simpler formulation.

Corollary 6. *Let $0 < l < m$, $mp > n$, and $p \in (1,\infty)$. Then*

$$\|\gamma\|_{M(B_p^m\to B_p^l)} \sim \sup_{x\in\mathbf{R}^n} \big(\|C_{p,l}\gamma; \mathcal{B}_1(x)\|_{L_p} + \|\gamma; \mathcal{B}_1(x)\|_{L_p}\big). \tag{6.1}$$

For $m = l$ the second term on the right should be replaced by $\|\gamma\|_{L_\infty}$.

Proof. The lower estimate of $\|\gamma\|_{M(B_p^m\to B_p^l)}$ follows from the relation

$$\mathrm{cap}_{p,m}(e) \sim 1 \tag{6.2}$$

valid for $mp > n$ and e with $\mathrm{diam}(e) \leq 1$, combined with Corollary 4. The upper estimate results from

$$\|\gamma\|_{M(B_p^m\to B_p^l)} \leq \|\gamma\|_{MB_p^l} \leq c\,\big(\sup_{e,\mathrm{diam}(e)\leq 1} \|C_{p,l}\gamma; e\|_{L_p} + \|\gamma\|_{L_\infty}\big)$$

$$\leq c \sup_{x\in\mathbf{R}^n} \big(\|C_{p,l}\gamma; \mathcal{B}_1(x)\| + \|\gamma; \mathcal{B}_1(x)\|_{L_p}\big).$$

The proof is complete.

Remark 1. One can easily verify that the right-hand side in (6.1) is equivalent to the norm of γ in $B^l_{p,\mathrm{unif}}$. Hence $M(B_p^m \to B_p^l)$ is isomorphic to $B^l_{p,\mathrm{unif}}$ for $0 < l < m$, $mp > n$, $p \in (1,\infty)$.

7. The space $M(W_p^m \to W_p^{-k})$

Let W_p^m denote the usual Sobolev space with $p \in (1, \infty)$ and integer m, and let W_p^{-k} stand for the dual space $(W_{p'}^k)^*$, $p+p' = pp'$. In [MS], the following sufficient condition for inclusion into the distribution space $M(W_p^m \to W_p^{-k})$ can be found. We supply it with the proof for completeness and reader's convenience.

Theorem 2. (see Sect. 1.5 [MS]) (i) *Let $p \in (1, \infty)$, $m \leq k$. If*

$$\gamma = \sum_{|\alpha| \leq k} D^\alpha \gamma_\alpha \tag{7.1}$$

with

$$\gamma_\alpha \in M(W_{p'}^k \to W_{p'}^{k-m}) \cap M(W_p^m \to L_p), \tag{7.2}$$

then $\gamma \in M(W_p^m \to W_p^{-k})$.

(ii) *Let $p \in (1, \infty)$, $m \geq k$. If*

$$\gamma = \sum_{|\alpha| \leq m} D^\alpha \gamma_\alpha$$

with

$$\gamma_\alpha \in M(W_p^m \to W_p^{m-k}) \cap M(W_{p'}^k \to L_{p'}),$$

then $\gamma \in M(W_p^m \to W_p^{-k})$.

Proof. It suffices to prove only (i) since (ii) follows from (i) by duality.

Let $u \in W_p^m$, $m \leq k$. Since

$$uD^\alpha \gamma_\alpha = \sum_{\lambda \leq \alpha} c_{\lambda\alpha} D^\lambda (\gamma_\alpha D^{\alpha-\lambda} u), \quad c_{\lambda\alpha} = \text{const},$$

we have

$$\|\gamma u\|_{W_p^{-k}} \leq c \sum_{|\lambda| \leq |\alpha| \leq k} \|\gamma_\alpha D^{\alpha-\lambda} u\|_{W_p^{|\lambda|-k}}$$

$$\leq c \sum_{|\lambda| \leq |\alpha| \leq k} \|\gamma_\alpha\|_{M(W_p^{m-k+|\lambda|} \to W_p^{|\lambda|-k})} \|u\|_{W_p^{m+|\alpha|+k}}. \tag{7.3}$$

Applying the interpolation inequality

$$\|\gamma_\alpha\|_{M(W_p^{m-k+|\lambda|} \to W_p^{|\lambda|-k})} \leq c \|\gamma_\alpha\|^{(k-|\lambda|)/k}_{M(W_p^{m-k} \to W_p^{-k})} \|\gamma_\alpha\|^{|\lambda|/k}_{M(W_p^m \to L_p)}$$

which follows from the interpolation property of Sobolev spaces (see [Tr], Sect. 2.4) we obtain from (7.3)

$$\|\gamma u\|_{W_p^{-k}} \leq c \left(\|\gamma_\alpha\|_{M(W_p^{m-k} \to W_p^{-k})} + \|\gamma_\alpha\|_{M(W_p^m \to L_p)} \right) \|u\|_{W_p^m}.$$

It remains to note that

$$\|\gamma_\alpha\|_{M(W_p^{m-k} \to W_p^{-k})} = \|\gamma_\alpha\|_{M(W_{p'}^k \to W_{p'}^{k-m})}.$$

The following assertion shows that this theorem provides a complete characterisation of $M(W_p^m \to W_p^{-k})$ which holds under some conditions involving k, m, p, and n.

Corollary 7. *Let k and m be positive integers and let either $k \geq m > 0$ and $k > n/p'$ or $m \geq k > 0$ and $m > n/p$. Then $\gamma \in M(W_p^m \to W_p^{-k})$ if and only if*

$$\gamma \in W_{p,\mathrm{unif}}^{-k} \cap W_{p',\mathrm{unif}}^{-m}. \tag{7.4}$$

In particular, if $\max\{k, m\} > n/2$ *then* $M(W_2^m \to W_2^{-k})$ *is isomorphic to* $W_2^{-\min\{m,k\}}$.

Proof. It suffices to consider the case $k \geq m > 0$, $k > n/p'$, because the case $m \geq k > 0$, $m > n/p$ results by duality.

Necessity. It follows from the inclusion $\gamma \in M(W_p^m \to W_p^{-k})$ that $\gamma \in W_{p,\mathrm{unif}}^{-k}$. Since $M(W_p^m \to W_p^{-k})$ is isomorphic to $M(W_{p'}^k \to W_{p'}^{-m})$, we have $\gamma \in W_{p',\mathrm{unif}}^{-m}$ as well.

Sufficiency. It is standard and easily proved (compare with Sect. 1.1.14 [M]) that $\gamma \in W_{p,\mathrm{unif}}^{-k} \cap W_{p',\mathrm{unif}}^{-m}$ if and only if (7.1) holds with $\gamma_\alpha \in L_{p,\mathrm{unif}} \cap W_{p',\mathrm{unif}}^{k-m}$. Since $M(W_{p'}^k \to W_{p'}^{k-m})$ is isomorphic to $W_{p',\mathrm{unif}}^{k-m}$ for $p'k > n$, it follows that $\gamma_\alpha \in M(W_{p'}^k \to W_{p'}^{k-m})$.

It remains to show that $\gamma_\alpha \in M(W_p^m \to L_p)$. We choose q and r to satisfy

$$1/q > \max\{0, 1/p - m/n\} > -\varepsilon + 1/q,$$

$$1/r > \max\{0, 1/p' - (k-m)/n\} > -\varepsilon + 1/r$$

with a sufficiently small ε. Since $1/p > 1 - k/n$, we have $1/p > 1/q + 1/r$. By Hölder's inequality

$$\|\gamma_\alpha u\|_{L_{p,\mathrm{unif}}} \leq c\, \|\gamma_\alpha\|_{L_{r,\mathrm{unif}}} \|u\|_{L_{q,\mathrm{unif}}}$$

and by Sobolev's embedding theorem

$$\|\gamma_\alpha u\|_{L_{p,\mathrm{unif}}} \leq c\, \|\gamma_\alpha\|_{W_{p',\mathrm{unif}}^{k-m}} \|u\|_{W_{p,\mathrm{unif}}^m}.$$

This means that $\gamma_\alpha \in M(W_p^m \to L_p)$. The proof is completed by reference to assertion (i) of Theorem 2.

Remark 2. Note that by Sobolev's embedding theorems $W_{p',\mathrm{unif}}^{-m} \subset W_{p,\mathrm{unif}}^{-k}$, $k \geq m$, if and only if either $n \leq (k-m)p$ or

$$n > (k-m)p, \quad \frac{k-m}{n} \geq \frac{2-p}{p}.$$

Under these conditions, $M(W_p^m \to W_p^{-k})$ is isomorphic to $W_{p',\mathrm{unif}}^{-m}$ if $kp' > n$. Analogously, if $m \geq k$, $mp > n$ and either $n \leq (m-k)p'$ or

$$n > (m-k)p', \quad \frac{m-k}{n} \geq \frac{p-2}{p},$$

then $M(W_p^m \to W_p^{-k})$ is isomorphic to $W_{p,\mathrm{unif}}^{-k}$.

We finish by stating a direct but important application of Corollary 8 to the theory of differential operators.

Corollary 9. *Let k and m be integers and let $\mathcal{L}_{m+k}(D)$ denote a differential operator of order $m+k$ with constant coefficients. If either $k \geq m$ and $kp' > n$, or $m \geq k > 0$ and $mp > n$ then the operator*

$$W_p^m \ni u \to \mathcal{L}(D)u + \gamma(x)u \in W_p^{-k}$$

is continuous if and only if

$$\gamma \in W_{p,\mathrm{unif}}^{-k} \cap W_{p',\mathrm{unif}}^{-m}.$$

References

[AH] Adams, D., Hedberg, L.I., *Function Spaces and Potential Theory*, Springer, 1996.

[KeS] Kerman, R., Saywer, E.T., *The trace inequality and eigenvalue estimates for Schrödinger operators*, Ann. Inst. Fourier (Grenoble), **36** (1986), 207–228.

[KS] Koch, H., Sickel, W., *Pointwise multipliers of Besov spaces of smoothness zero and spaces of continuous functions*, Rev. Mat. Iberoamericana, **18** (2002), 587–626.

[M] Maz'ya, V., *Sobolev Spaces*, Springer-Verlag, 1985.

[MS] Maz'ya, V., Shaposhnikova, T., *Theory of Multipliers in Spaces of Diffferentiable Functions*, Pitman, 1985.

[MV1] Maz'ya, V., Verbitsky, I., *Capacitary inequalities for fractional integrals, with applications to partial differential equations and Sobolev multipliers*, Ark. Mat. **33**:1 (1995), 81–115.

[MV2] Maz'ya, V., Verbitsky, I., *The Schrödinger operator on the energy space: boundedness and compactness criteria*, Acta Mathematica, **188** (2002), 263–302.

[MV3] Maz'ya, V., Verbitsky, I., *The form boundedness criterion for the relativistic Schrödinger operator*, to appear.

[MV4] Maz'ya, V., Verbitsky, I., *Boundedness and compactness criteria for the one-dimensional Schrödinger operator*, In: Function Spaces, Interpolation Theory and Related Topics. Proc. Jaak Peetre Conf., Lund, Sweden, August 17-22, 2000. Eds. M. Cwikel, A. Kufner, G. Sparr. De Gruyter, Berlin, 2002, 369–382.

[SS] Sickel, W., Smirnow, I., *Localization properties of Besov spaces and its associated multiplier spaces*, Jenaer Schriften Math/Inf 21/99, Jena, 1999.

[St] Stein, E.M., *Singular Integrals and Differentiability Properties of Functions*, Princeton Univ. Press, Princeton, 1970.

[Tr] Triebel, H., *Interpolation Theory. Function Spaces. Differential Operators*, VEB Deutscher Verlag der Wiss., Berlin, 1978.

Vladimir Maz'ya and Tatyana Shaposhnikova,
Department of Mathematics, University of Linköping
SE-581 83 Linköping, Sweden
e-mail: vlmaz@mai.liu.se
e-mail: tasha@mai.liu.se

Operator Theory
Advances and Applications, Vol. 147, 387–402

Boundary Variational Inequalities in the Theory of Interface Cracks

David Natroshvili and Wolfgang L. Wendland

Dedicated to the memory of Professor Dr. Erhard Meister

Abstract. We formulate three-dimensional interface crack problems in terms of Signorini conditions. With the help of potential theory we reduce these to variational inequalities which are employed on the two-dimensional boundary of a domain occupied by two homogeneous elastic anisotropic bodies. For the boundary variational inequalities, we prove uniqueness and existence. For a corresponding boundary element Galerkin approximation we also show convergence and asymptotic error estimates.

1. Introduction

Signorini contact problems as well as crack problems have been extensively studied for homogeneous elastic bodies (see [1], [14], [15], [19], [23], [26] and the references therein). There, the main tools to investigate these problems are the theory of spatial variational inequalities and convex analysis. In this framework, uniqueness and existence theorems have been proved and regularity properties of the solutions have been established in various function spaces.

The purpose of this paper is to formulate static three-dimensional *interface crack problems in terms of Signorini conditions* and to reduce these to boundary variational inequalities on the two-dimensional boundary surface of a domain occupied by piecewise homogeneous elastic *anisotropic* material. Then also standard crack problems are particular cases of these interface crack problems. The corresponding *spatial variational inequality* (SVI) (see also [12]) will be reduced to an equivalent *boundary variational inequality* (BVI). Uniqueness and existence for the solution of the BVI will be obtained and a boundary element Galerkin procedure will be presented along with an a priori error estimate.

2000 *Mathematics Subject Classification.* 35J85, 35J50, 74B05, 74M15.
Key words and phrases. Interface cracks, Signorini conditions, anisotropic elasticity.
This research was supported by the Collaborative Linkage Grant NATO PST.CLG. 976426/5437 and the Collaborative Research Center of the German Research Foundation DFG SFB 404 "Multifield Problems" at the University of Stuttgart.

It seems to us that our analysis could be extended to time harmonic loading in combination with the solution's explicit construction given by E. Meister and F.O. Speck in [29]; but this is yet to be done.

The formulation of the interface crack problem in terms of Signorini conditions avoids overlapping of the contacting materials in the contrary to oscillatory asymptotic behavior of solutions near to the crack edges in the linearized modelling (see [2] and the references therein). A similar approach for the Signorini problem in the *isotropic* case was considered in [18], where an unilateral boundary value problem is reduced to a BVI in combination with a singular boundary integral equation. Related problems such as contact between elastic and rigid bodies with friction are discussed in [9], [10] and [11] (see also the references therein).

2. The spatial variational inequality

2.1. Notations and preliminaries

Let two elastic homogeneous anisotropic bodies occupy two different compact regions $\overline{\Omega_i} = \overline{\Omega_i^+} \subset \mathbb{R}$, $i = 1,2$, with boundaries $S_i = \partial\Omega_i$. By $\Omega_i^- = \mathbb{R}^3 \setminus \overline{\Omega_i^+}$ we denote the corresponding exterior domains. For the corresponding elasticity constants $C^{(i)}_{kjpq}$ we assume symmetry [15]: $C^{(i)}_{kjpq} = C^{(i)}_{pqkj} = C^{(i)}_{jkpq}$. We consider linearized elasticity, where the stress $\{\sigma^{(i)}_{kj}\}$ and the strain tensor $\{\varepsilon^{(i)}_{kj}\}$ are related by Hooke's law,

$$\sigma^{(i)}_{kj}(u^{(i)}) = C^{(i)}_{kjpq}\varepsilon^{(i)}_{pq}(u^{(i)}), \quad \varepsilon^{(i)}_{kj}(u^{(i)}) = 2^{-1}\left(\partial_k u^{(i)}_j + \partial_j u^{(i)}_k\right),$$
$$k = 1,2,3, \quad i = 1,2.$$

Here $u^{(i)} = (u^{(i)}_1, u^{(i)}_2, u^{(i)}_3)^\top$ denotes the respective displacement vector and $\partial_k := \frac{\partial}{\partial x_k}$. We shall employ the Einstein summation convention, unless stated otherwise. The subscript $^\top$ denotes transposition. The symmetric form

$$E^{(i)}\left(u^{(i)}, v^{(i)}\right) := \sigma^{(i)}_{kj}\left(u^{(i)}\right)\varepsilon^{(i)}_{kj}\left(v^{(i)}\right) = C^{(i)}_{kjpq}\,\varepsilon^{(i)}_{pq}(u^{(i)})\,\varepsilon^{(i)}_{kj}(v^{(i)})$$

defines the density of virtual work. Then the density of potential energy corresponding to the displacement field $u^{(i)}$ is $\frac{1}{2}E^{(i)}\left(u^{(i)}, u^{(i)}\right)$. The energy density is supposed to be positive definite in the symmetric variables $\varepsilon^{(i)}_{kj} = \varepsilon^{(i)}_{jk}$ (cf., e.g., [15]), i.e.,

$$E^{(i)}\left(u^{(i)}, u^{(i)}\right) \geq \delta^{(i)}_1\,\varepsilon^{(i)}_{kj}\left(u^{(i)}\right)\,\varepsilon^{(i)}_{kj}\left(u^{(i)}\right), \; i = 1,2, \tag{2.1}$$

where $\delta^{(i)}_1$ are positive constants. By $T^{(i)}(\partial_x, n(x))$ we denote the stress operator acting on a surface element of S_i associated with the exterior unit normal vector $n = (n_1, n_2, n_3)$, which defines the stress vector

$$\left[T^{(i)}(\partial_x, n(x))u^{(i)}(x)\right]_k = \sigma^{(i)}_{kj}\left(u^{(i)}\right)n_j(x) = C^{(i)}_{kjpq}\,n_j(x)\,\partial_q u^{(i)}_p(x).$$

By $H^\alpha(\Omega_i^+)$, $H^\alpha_{loc}(\Omega_i^-)$ and $H^\alpha(\partial\Omega_i^\pm)$ we denote the real Sobolev-Slobodetski spaces with $\alpha \in \mathbb{R}$ (see, e.g., [25], [35]). For an open submanifold $S_i^* \subset S_i$, we denote by $\widetilde{H}^\alpha(S_i^*)$ the subspace $\widetilde{H}^\alpha(S_i^*) := \{\varphi \mid \varphi \in H^\alpha(S_i),\ \operatorname{supp}\varphi \subset \overline{S_i^*}\}$, while $H^\alpha(S_i^*)$ denotes the space of restrictions of functions from $H^\alpha(S_i)$ to S_i^* : $H^\alpha(S_i^*) := \{r_{S_i^*}\varphi \mid \varphi \in H^\alpha(S_i)\}$, where $r_{S_i^*}$ stands for the restriction operator to S_i^*.

2.2. Formulation of the contact problems

Let the domains Ω_1 and Ω_2 have a common boundary part $S_c \subset S_i$ for $i = 1, 2$, where the elastic bodies $\overline{\Omega_1}$ and $\overline{\Omega_2}$ may have contact. Moreover, let $S_i \setminus S_c$ be divided into the two disjoint parts S_i^D and S_i^N, where all the three subsurfaces S_c, S_i^D and S_i^N have positive measures, satisfying

$$S_i = \overline{S_c} \cup \overline{S_i^D} \cup \overline{S_i^N}, \quad \overline{S_c} \cap \overline{S_i^N} = \emptyset, \quad (S_1 \setminus \overline{S_c}) \cap (S_2 \setminus \overline{S_c}) = \emptyset.$$

The union $\overline{\Omega_1} \cup \overline{\Omega_2}$ can be considered as a piecewise homogeneous anisotropic body with the interface $\overline{S_c}$. Then the weak, variational form of the generalized interface crack problem reads as follows:

Problem (P) *Find a pair of vector functions* $\left(u^{(1)}, u^{(2)}\right) \in H^1(\Omega_1) \times H^1(\Omega_2)$ *satisfying*

(i) *the equilibrium equations in* Ω_i *for all* $v^{(i)} \in \overset{\circ}{H}{}^1(\Omega_i)$ *and* $i = 1, 2$,

$$\int_{\Omega_i} f^{(i)} \cdot v^{(i)}\, dx + \int_{\Omega_i} E^{(i)}\left(u^{(i)}, v^{(i)}\right) dx = 0\,; \tag{2.2}$$

(ii) *the Dirichlet condition*

$$\left[u^{(i)}(x)\right]^+ = 0 \quad \text{for } x \in S_i^D\,; \tag{2.3}$$

(iii) *the Neumann condition on* S_i^N,

$$\left\langle \left[T^{(i)}u^{(i)}\right]^+_{S_i^N}, \psi \right\rangle_{S_i^N} = \langle g^{(i)}, \psi\rangle_{S_i^N} \quad \text{for all } \psi \in \widetilde{H}^{\frac{1}{2}}(S_i^N)\,; \tag{2.4}$$

(iv) *the interface conditions on* S_c,

$$\left[u^{(1)}(x)\cdot n^{(1)}(x)\right]^+ + \left[u^{(2)}(x)\cdot n^{(2)}(x)\right]^+ \le 0 \quad \text{for all } x \in S_c\,; \tag{2.5}$$

$$\left\langle \left[T^{(i)}u^{(i)}\cdot n^{(i)}\right]^+_{S_c}, \nu \right\rangle_{S_c} \ge 0, \quad \text{for all } \nu \in \widetilde{H}^{\frac{1}{2}}(S_c) \quad \text{with } \nu \le 0; \tag{2.6}$$

$$\left\langle \left[T^{(1)}u^{(1)}\cdot n^{(1)}\right]^+_{S_c}, \left[u^{(1)}\cdot n^{(1)}\right]^+_{S_c} + \left[u^{(2)}\cdot n^{(2)}\right]^+_{S_c} \right\rangle_{S_c} = 0; \tag{2.7}$$

$$\left\langle \left[T^{(i)}u^{(i)} - n^{(i)}\left(T^{(i)}u^{(i)}\cdot n^{(i)}\right)\right]^+_{S_c}, \psi \right\rangle_{S_c} = 0 \tag{2.8}$$

$$\text{for all } \psi \in \widetilde{H}^{\frac{1}{2}}(S_c) \quad \text{with } \psi\cdot n^{(i)} = 0\,.$$

The symbols $[\cdot]^{\pm}$ denote the traces on S_i from $\Omega_i^{\pm}$. By $n^{(i)}(x)$ we denote the unit outward normal vector to S_i at the point $x \in S_i$, and $f^{(i)} = (f_1^{(i)}, f_2^{(i)}, f_3^{(i)})^\top \in L_2(\Omega_i)$ and $g^{(i)} = (g_1^{(i)}, g_2^{(i)}, g_3^{(i)})^\top \in H^{-\frac{1}{2}}(S_i^N)$ are given vector functions. By $\langle \cdot\,,\,\cdot\rangle_{S_c}$ and $\langle \cdot\,,\,\cdot\rangle_{S_i^N}$ we denote the duality brackets between the spaces $H^{-\frac{1}{2}}(S_c)$ and $\widetilde{H}^{\frac{1}{2}}(S_c)$, and $H^{-\frac{1}{2}}(S_i^N)$ and $\widetilde{H}^{\frac{1}{2}}(S_i^N)$, respectively, which extend the $L_2(S_i)$ scalar products.

Conditions (2.3) show that the bodies are fixed along the subsurface $S_1^D \cup S_2^D$, while (2.4) shows that the bodies are subjected to some assigned surface stresses on $S_1^N \cup S_2^N$.

The Signorini conditions on the interface part S_c describe a generalized contact interaction of the two elastic bodies Ω_1 and Ω_2 without friction [14] since the tangential stresses vanish on S_c. Condition (2.5) shows that there is no penetration along S_c due to (2.8), while (2.7) along with (2.5) and (2.6) implies that at each point of S_c either there is a gap between Ω_1 and Ω_2, where normal stresses vanish, or there is active contact between the two bodies on the *regions of coincidence* (cf. [15], [23]).

If not stated otherwise, we assume that the surfaces S_i are C^∞-smooth and ∂S_i^D, ∂S_i^N, and ∂S_c are C^∞-smooth curves.

Note that, for $u^{(i)} \in H^1(\Omega_i)$ and $(A^{(i)}(\partial)u^{(i)})_k := \partial_j C_{kjpq}^{(i)} \partial_q u_p^{(i)} \in \big(H^1(\Omega^{(i)})\big)'$, the trace $\big[u^{(i)}\big]_{S_i}^+$ belongs to the space $H^{\frac{1}{2}}(S_i)$, while, in general, $\big[T^{(i)}\left(\partial, n^{(i)}\right) u^{(i)}\big]_{S_i}^+$ is a functional belonging to $H^{-\frac{1}{2}}(S_i)$ determined by duality (see, e.g., [25], [28]) from

$$\begin{aligned} &\left\langle \left[T^{(i)}\left(\partial, n^{(i)}\right) u^{(i)}(x)\right]_{S_i}^+, \left[v^{(i)}\right]_{S_i}^+ \right\rangle_{S_i} \\ &\qquad = \int_{\Omega_i} f^{(i)} \cdot v^{(i)}\, dx + \int_{\Omega_i} E^{(i)}\left(u^{(i)}, v^{(i)}\right) dx\,. \end{aligned} \tag{2.9}$$

Note that if $u^{(i)} \in H^2(\Omega_i)$, then all the boundary and interface conditions of Problem (P) can be understood in the classical sense, i.e., pointwise.

2.3. Existence and uniqueness

Since the variational formulation of the contact problem turns out to be a classical minimization problem, let us set

$$\mathbf{H}(\Omega_1, \Omega_2) := \left\{ U = \left(u^{(1)}, u^{(2)}\right)^\top \mid u^{(i)} \in H^1(\Omega_i),\ \left[u^{(i)}\right]_{S_i^D}^+ = 0 \right\},$$

$$\mathbf{K}(\Omega_1, \Omega_2) := \left\{ U \in \mathbf{H}(\Omega_1, \Omega_2) \mid \left[u^{(1)} \cdot n^{(1)}\right]_{S_c}^+ + \left[u^{(2)} \cdot n^{(2)}\right]_{S_c}^+ \le 0 \right\}.$$

Further, for arbitrary U, $V \in \mathbf{H}(\Omega_1, \Omega_2)$ we put

$$(U, V)_{\mathbf{H}(\Omega_1,\Omega_2)} := \left(u^{(1)}, v^{(1)}\right)_{H^1(\Omega_1)} + \left(u^{(2)}, v^{(2)}\right)_{H^1(\Omega_2)}. \tag{2.10}$$

Evidently, $\mathbf{H}(\Omega_1, \Omega_2)$ is a Hilbert space with the scalar product (2.10), and $\mathbf{K}(\Omega_1, \Omega_2)$ is a closed, convex, nonempty cone in $\mathbf{H}(\Omega_1, \Omega_2)$.

We introduce the bilinear form

$$B(U,V) := B^{(1)}\left(u^{(1)}, v^{(1)}\right) + B^{(2)}\left(u^{(2)}, v^{(2)}\right) \quad \text{on } \left(\mathbf{H}(\Omega_1, \Omega_2)\right)^2 \tag{2.11}$$

with

$$B^{(i)}\left(u^{(i)}, v^{(i)}\right) := \int_{\Omega_i} E^{(i)}\left(u^{(i)}, v^{(i)}\right) dx. \tag{2.12}$$

With Korn's inequality for both the bilinear forms $B^{(i)}$ (see, e.g., [14], [32]) it follows that $B(\cdot,\cdot)$ is continuous and coercive on $\left(\mathbf{H}(\Omega_1, \Omega_2)\right)^2$ satisfying

$$\begin{aligned} |B(U,V)| &\le c_1 \|U\|_{\mathbf{H}(\Omega_1,\Omega_2)} \|V\|_{\mathbf{H}(\Omega_1,\Omega_2)} \quad \text{and} \\ B(U,U) &\ge c_2 \|U\|^2_{\mathbf{H}(\Omega_1,\Omega_2)} \end{aligned} \tag{2.13}$$

with $c_1, c_2 > 0$. From Green's formula we get

$$B(U,V) = -\sum_{i=1}^{2}\Big\{ \int_{\Omega_i} A^{(i)}(\partial)u^{(i)} \cdot v^{(i)}\, dx - \left\langle \left[T^{(i)}u^{(i)}\right]^+_{S_i}, \left[v^{(i)}\right]^+_{S_i} \right\rangle_{S_i} \Big\} \tag{2.14}$$

for arbitrary $U, V \in \mathbf{H}(\Omega_1, \Omega_2)$ and $A^{(i)}(\partial)u^{(i)} \in L_2(\Omega_i)$. Further, we introduce on $\mathbf{H}(\Omega_1, \Omega_2)$ the continuous, linear functional

$$\mathcal{F}(V) := \sum_{i=1}^{2}\Big\{ \left\langle g^{(i)}, \left[v^{(i)}\right]^+_{S_i^N} \right\rangle_{S_i^N} - \int_{\Omega_i} f^{(i)} \cdot v^{(i)}\, dx \Big\} \tag{2.15}$$

and the energy functional

$$J(U) := \tfrac{1}{2}B(U,U) - \mathcal{F}(U) \quad \text{for } U \in \mathbf{H}(\Omega_1, \Omega_2),$$

to formulate the following minimization problem:

Find the minimum $U_0 \in \mathbf{K}(\Omega_1, \Omega_2)$ of the energy functional

$$J(U_0) = \min_{U \in \mathbf{K}(\Omega_1,\Omega_2)} J(U). \tag{2.16}$$

This quadratic minimization problem (2.16) is equivalent to the spatial variational inequality (see, e.g., [14], [17]):

Find $U \in \mathbf{K}(\Omega_1, \Omega_2)$ such that

$$B(U, V-U) \ge \mathcal{F}(V-U) \quad \text{for all } V \in \mathbf{K}(\Omega_1, \Omega_2). \tag{2.17}$$

Clearly, the variational inequality (2.17) is the equivalent to Problem (P). The unique solvability of the SVI in (2.17) follows from the coerciveness property of the bilinear form $B(\cdot,\cdot)$ on $\left(\mathbf{H}(\Omega_1, \Omega_2)\right)^2$ and the general theory of variational inequalities in Hilbert spaces [36, Section 25.8.] (see also [8], [14], [17, Ch. 1, Theorems 2.1 and 2.2]). Thus, there holds the following assertion.

Theorem 2.1. *Let $f^{(i)} \in L_2(\Omega_i)$ and $g^{(i)} \in H^{-1/2}(S_i^N)$ be given. Then the SVI* (2.17) *has a unique solution which also defines the unique solution to the interface crack problem* (P).

3. Reduction to the boundary

3.1. Potential operators

From now on we assume that the volume forces are zero, i.e., $f^{(i)} = 0$. Then the spatial variational inequality (2.17) takes the form

$$B(U, V - U) \geq \sum_{i=1}^{2} \left\langle g^{(i)}, \left[v^{(i)} - u^{(i)} \right]_{S_i^N}^{+} \right\rangle_{S_i^N}. \tag{3.1}$$

Our goal is to reduce the SVI (3.1) to a boundary variational inequality which lives on the manifold $S_1 \cup S_2$ only. Since $A^{(i)}(\partial_x) u^{(i)}(x) = 0$ in Ω_i, the displacement field can be represented in terms of boundary potentials for $x \in \Omega_i$ (see [24]):

$$u^{(i)}(x) = SL^{(i)} \left(\left[T^{(i)} u^{(i)} \right]_{S_i}^{+} \right)(x) - DL^{(i)} \left(\left[u^{(i)} \right]_{S_i}^{+} \right)(x) \tag{3.2}$$

where $DL^{(i)}$ and $SL^{(i)}$ denote the double- and single-layer potentials, respectively,

$$\begin{aligned} DL^{(i)}(h)(x) &:= \int_{S_i} \left[T^{(i)} \left(\partial_y, n^{(i)}(y) \right) \Gamma^{(i)}(x-y) \right]^{\top} h(y)\, dS_y\,, \\ SL^{(i)}(h)(x) &:= \int_{S_i} \Gamma^{(i)}(x-y)\, h(y)\, dS_y \quad \text{for } x \in \mathbb{R}^3 \setminus S_i. \end{aligned} \tag{3.3}$$

Here h is some vector-valued density function, and $\Gamma^{(i)}$ is the fundamental matrix of the operator $A^{(i)}(\partial_x)$ (see e.g. [30]),

$$\Gamma^{(i)}(x) = \frac{1}{8\pi |x|} \int_0^{2\pi} \left[\widetilde{A}^{(i)}(a(x)\eta) \right]^{-1} d\varphi, \quad x \in \mathbb{R}^3 \setminus \{0\},$$

where $\widetilde{A}^{(i)}(\xi) = C_{kjpq}^{(i)} \xi_j \xi_q$ and $a(x) = [a_{kj}]_{3\times 3}$ is an orthogonal matrix associated with x, with the property $a^{\top} x = (0, 0, |x|)^{\top}$, and $\eta = (\cos\varphi, \sin\varphi, 0)^{\top}$.

The properties of the potentials in (3.3) and corresponding boundary integral and pseudodifferential operators in the Hölder ($C^{k+\alpha}$), Bessel potential (H_p^s) and Besov ($B_{p,q}^s$) spaces are studied in, e.g., [6], [7], [13], [21], [24], [27], [28], [31], (see also [3], [4], [5], [34], where the coerciveness properties of the boundary operators and also the case of Lipschitz domains are considered).

In what follows, we need some of these well-known results which we now shall recall.

Theorem 3.1. [24], [30] *Let $k \geq 0$ be an integer and $0 < \gamma < 1$. Then the operators*

$$SL^{(i)} : C^{k+\gamma}(S_i) \to C^{k+1+\gamma}(\overline{\Omega_i^{\pm}}),\ DL^{(i)} : C^{k+\gamma}(S_i) \to C^{k+\gamma}(\overline{\Omega_i^{\pm}}) \tag{3.4}$$

are bounded. For any $g \in C^{k+\gamma}(S_i)$ and any $x \in S_i$, the following "jump relations" are valid:

$$\left[SL^{(i)}(g)(x)\right]^{\pm} = \mathcal{V}^{(i)}\, g(x), \tag{3.5}$$

$$\left[T^{(i)}\left(\partial_x, n^{(i)}(x)\right) SL^{(i)}(g)(x)\right]^{\pm} = \left[\pm\tfrac{1}{2}I_3 + \mathcal{K}^{(i)*}\right] g(x) \tag{3.6}$$

$$\left[DL^{(i)}(g)(x)\right]^{\pm} = [\mp\tfrac{1}{2}I_3 + \mathcal{K}^{(i)}]\, g(x), \tag{3.7}$$

$$\left[T^{(i)}\left(\partial_x, n^{(i)}(x)\right) DL^{(i)}(g)(x)\right]^{\pm} = -\mathcal{D}^{(i)}\, g(x), \quad k \geq 1, \tag{3.8}$$

where for $x \in S_i$ and S_i smooth, $\mathcal{V}^{(i)} g(x)$ and $\mathcal{K}^{(i)} h(x)$are defined by Cauchy principal value integrals in (3.3), *and where*

$$\mathcal{D}^{(i)}\, g(x) := - \lim_{\Omega_i^{\pm} \ni z \to x \in S_i} T^{(i)}(\partial_z, n^{(i)}(x)) DL^{(i)}(g)(z)\,. \tag{3.9}$$

$\mathcal{K}^{(i)*}$ is the adjoint to $\mathcal{K}^{(i)}$.

Theorem 3.2. [3], [4], [7], [22], [34] *For $s \in \mathbb{R}$, the operators in* (3.4) *can be extended by continuity to bounded linear mappings:*

$$SL^{(i)} : H^{s-\frac{1}{2}}(S_i) \to H^{s+1}(\Omega_i^+) \ \text{ and } \ [H^{s-\frac{1}{2}}(S_i) \to H^{s+1}_{loc}(\Omega_i^-)]\,,$$
$$DL^{(i)} : H^{s+\frac{1}{2}}(S_i) \to H^{s+1}(\Omega_i^+) \ \text{ and } \ [H^{s+\frac{1}{2}}(S_i) \to H^{s+1}_{loc}(\Omega_i^-)]\,.$$

The jump relations (3.5)–(3.8) *on S_i remain valid for these extended operators.*

Theorem 3.3. [7], [22] *Let $k \geq 0$ be an integer, $0 < \gamma < 1$, and $s \in \mathbb{R}$. Then the operators*

$$\begin{aligned}
\mathcal{V}^{(i)} \quad &: C^{k+\gamma}(S_i) \to C^{k+1+\gamma}(S_i) \quad \text{and } H^{s-\frac{1}{2}}(S_i) \to H^{s+\frac{1}{2}}(S_i)\,,\\
\mathcal{K}^{(i)}, \mathcal{K}^{(i)*} \quad &: C^{k+\gamma}(S_i) \to C^{k+\gamma}(S_i) \quad \text{and } H^{s+\frac{1}{2}}(S_i) \to H^{s+\frac{1}{2}}(S_i)\,,\\
\mathcal{D}^{(i)} \quad &: C^{k+1+\gamma}(S_i) \to C^{k+\gamma}(S_i) \quad \text{and } H^{s+\frac{1}{2}}(S_i) \to H^{s-\frac{1}{2}}(S_i)
\end{aligned}$$

are bounded. In addition, the following solvability properties are valid:

i) *The operators $\pm\frac{1}{2}I_3 + \mathcal{K}^{(i)}$ and $\pm\frac{1}{2}I_3 + \mathcal{K}^{(i)*}$ are mutually adjoint singular integral operators of normal type with index zero and also are strongly elliptic pseudodifferential operators of order zero. The operators $\mathcal{V}^{(i)}$ and $\mathcal{D}^{(i)}$ are strongly elliptic self-adjoint pseudodifferential operators of order -1 and $+1$, respectively. The operators $\mathcal{V}^{(i)}$, $\frac{1}{2}I_3 - \mathcal{K}^{(i)}$ and $\frac{1}{2}I_3 - \mathcal{K}^{(i)*}$ are invertible.*

ii) *These operators satisfy the following Calderon projector identities:*

$$\begin{aligned}
\mathcal{K}^{(i)}\,\mathcal{V}^{(i)} &= \mathcal{V}^{(i)}\,\mathcal{K}^{(i)*}, & \mathcal{D}^{(i)}\,\mathcal{K}^{(i)} &= \mathcal{K}^{(i)*}\,\mathcal{D}^{(i)},\\
\mathcal{V}^{(i)}\,\mathcal{D}^{(i)} &= \tfrac{1}{4}\,I_3 - \left(\mathcal{K}^{(i)}\right)^2, & \mathcal{D}^{(i)}\,\mathcal{V}^{(i)} &= \tfrac{1}{4}\,I_3 - \left(\mathcal{K}^{(i)*}\right)^2.
\end{aligned} \tag{3.10}$$

iii) *The operators $\mathcal{V}^{(i)}$ and $\mathcal{D}^{(i)}$ are non-negative, i.e.,*

$$\left\langle \mathcal{V}^{(i)} h, h \right\rangle_{S_i} \geq 0 \quad \text{for all } h \in H^{-\frac{1}{2}}(S_i) \quad \text{with equality only for } h = 0\,,$$

$$\left\langle \mathcal{D}^{(i)} g, g \right\rangle_{S_i} \geq 0 \quad \text{for all } g \in H^{\frac{1}{2}}(S_i) \quad \text{with equality only for}$$

$$g \in \Re(S_i) = \left\{ \left[a \times x\right] + b \right]_{S_i}^{+} \mid a, b \in \mathbb{R}^3 \}\,, \tag{3.11}$$

i.e., the 6-dimensional space of the traces of rigid motions.

iv) *The kernels of $\frac{1}{2}I_3 + \mathcal{K}^{(i)}$ and $\mathcal{D}^{(i)}$ are given by $\Re(S_i)$.*

Theorem 3.4. [3], [4],[16], [22] *Let $f \in H^{\frac{1}{2}}(S_i)$ and $h \in H^{-\frac{1}{2}}(S_i)$ be related by the boundary integral equation*

$$\left[\tfrac{1}{2} I_3 + \mathcal{K}^{(i)} \right] f = \mathcal{V}^{(i)} h\,.$$

Then there exists exactly one displacement field $u^{(i)} \in H^1\left(\Omega_i^+\right)$ which satisfies the equilibrium equations $A^{(i)}(\partial)\, u^{(i)}(x) = 0$ in Ω_i^+, and has on S_i the Cauchy data $\left[u^{(i)}\right]_{S_i}^{+} = f$; $\left[T^{(i)}\left(\partial, n^{(i)}\right) u^{(i)}\right]_{S_i}^{+} = h$. Moreover, then $\mathcal{D}^{(i)} f = \left[\frac{1}{2} I_3 - \mathcal{K}^{(i)*}\right] h$.

Further, $H_*^r(S_i) := \left\{ \varphi \in H^r(S_i) \mid \langle \varphi, \chi \rangle_{S_i} = 0 \text{ for all } \chi \in \Re(S_i) \right\}$ is a closed subspace of any of the Hilbert spaces $H^r(S_i)$ with $r \in \mathbb{R}$.

Theorem 3.5. [3], [4], [21], [34]

i) *There exists a positive constant $\delta_2^{(i)}$ such that*

$$\left\langle \mathcal{V}^{(i)} \varphi, \varphi \right\rangle_{S_i} \geq \delta_2^{(i)} \|\varphi\|^2_{H^{-\frac{1}{2}}(S_i)} \quad \text{for all } \varphi \in H^{-\frac{1}{2}}(S_i).$$

ii) $\mathcal{D}^{(i)}\left(H^{\frac{1}{2}}(S_i)\right) = H_*^{-\frac{1}{2}}(S_i)$, *where $\mathcal{D}^{(i)}$ is given by (3.9). There exists a positive constant $\delta_3^{(i)}$ such that*

$$\left\langle \mathcal{D}^{(i)} \varphi, \varphi \right\rangle_{S_i} \geq \delta_3^{(i)} \|\varphi\|^2_{H^{\frac{1}{2}}(S_i)} \qquad \text{for all } \varphi \in H_*^{\frac{1}{2}}(S_i)$$

and $\mathcal{D}^{(i)} : H_*^{\frac{1}{2}}(S_i) \to H_*^{-\frac{1}{2}}(S_i)$ is an isomorphism.

We will also need some mapping and coerciveness properties of the Steklov-Poincaré operator defined by

$$\begin{aligned} \mathcal{M}^{(i)} &:= \left[\mathcal{V}^{(i)}\right]^{-1} \left[\tfrac{1}{2} I_3 + \mathcal{K}^{(i)}\right] = \left[\tfrac{1}{2} I_3 + \mathcal{K}^{(i)*}\right] \left[\mathcal{V}^{(i)}\right]^{-1} \\ &= \mathcal{D}^{(i)} + \left[\tfrac{1}{2} I_3 + \mathcal{K}^{(i)*}\right] \left[\mathcal{V}^{(i)}\right]^{-1} \left[\tfrac{1}{2} I_3 + \mathcal{K}^{(i)}\right] \end{aligned} \tag{3.12}$$

in view of (3.10).

Lemma 3.6. [16], [20], [34] *The operator $\mathcal{M}^{(i)} : H^{\frac{1}{2}}(S_i) \to H^{-\frac{1}{2}}(S_i)$ is a bounded, positive, self-adjoint and strongly elliptic pseudodifferential operator of order +1 with*

$$\ker \mathcal{M}^{(i)} = \ker \mathcal{D}^{(i)} = \ker \left[\tfrac{1}{2} I_3 + \mathcal{K}^{(i)} \right] = \Re(S_i) .$$

There exists a positive constant $\delta_4^{(i)}$ such that

$$\left\langle \mathcal{M}^{(i)} \varphi, \varphi \right\rangle_{S_i} \geq \delta_4^{(i)} \|\varphi\|^2_{H^{\frac{1}{2}}(S_i)} \quad \text{for all } \varphi \in H_*^{\frac{1}{2}}(S_i) .$$

Moreover,

$$\mathcal{M}^{(i)}\left(H^{\frac{1}{2}}(S_i)\right) = H_*^{-\frac{1}{2}}(S_i)$$

and the mapping $\mathcal{M}^{(i)} : H_*^{\frac{1}{2}}(S_i) \to H_*^{-\frac{1}{2}}(S_i)$ is an isomorphism.

3.2. The boundary variational inequality

For every solution $u^{(i)}$ of the homogeneous differential equation in Ω_i, we recall the Steklov-Poincaré relation between its Cauchy data,

$$\left[\tfrac{1}{2} I_3 + \mathcal{K}^{(i)} \right] \left[u^{(i)} \right]^+_{S_i} = \mathcal{V}^{(i)} \left[T^{(i)} u^{(i)} \right]^+_{S_i} \tag{3.13}$$

which implies $\left[T^{(i)} u^{(i)} \right]^+_{S_i} = \mathcal{M}^{(i)} \left[u^{(i)} \right]^+_{S_i}$ with the Steklov-Poincaré operator $\mathcal{M}^{(i)}$.

In order to formulate the boundary variational inequality, we introduce the trace space

$$\mathbf{H}(S_1, S_2) := \left\{ \Phi = \left(\varphi^{(1)}, \varphi^{(2)} \right)^\top \mid \varphi^{(i)} \in \widetilde{H}^{\frac{1}{2}}(S_i^N \cup S_c) \right\}$$

and the cone

$$\mathbf{K}(S_1, S_2) := \left\{ \Phi \in \mathbf{H}(S_1, S_2) \mid \varphi^{(1)} \cdot n^{(1)} + \varphi^{(2)} \cdot n^{(2)} \leq 0 \text{ a.e. on } S_c \right\} .$$

For the pairs $\Phi = (\varphi^{(1)}\ ,\ \varphi^{(2)})^\top$ and $\Psi = \left(\psi^{(1)}, \psi^{(2)} \right)^\top \in \mathbf{H}(S_1, S_2)$ we define the scalar product

$$(\Phi, \Psi)_{\mathbf{H}(S_1,S_2)} := \left(\varphi^{(1)}, \psi^{(1)} \right)_{H^{\frac{1}{2}}(S_1)} + \left(\varphi^{(2)}, \psi^{(2)} \right)_{H^{\frac{1}{2}}(S_2)}$$

and the induced norm $\|\Phi\|_{\mathbf{H}(S_1,S_2)}$, respectively.

By employing the Steklov-Poincaré operators, we define the bilinear form

$$\mathcal{B}(\Phi, \Psi) := \sum_{i=1}^{2} \langle \mathcal{M}^{(i)} \varphi^{(i,)} \psi^{(i)} \rangle_{S_i} \quad \text{for } \Phi, \Psi \in \mathbf{H}(S_1, S_2) . \tag{3.14}$$

Due to Lemma 3.6, the bilinear form $\mathcal{B}(\cdot, \cdot)$ is bounded and coercive, i.e.,

$$|\mathcal{B}(\Phi, \Psi)| \leq \kappa_1 \|\Phi\|_{\mathbf{H}(S_1,S_2)} \|\Psi\|_{\mathbf{H}(S_1,S_2)} \quad \text{and} \tag{3.15}$$

$$\mathcal{B}(\Phi, \Phi) \geq \kappa_2 \|\Phi\|^2_{\mathbf{H}(S_1,S_2)} \quad \text{for all } \Phi, \Psi \in \mathbf{H}(S_1, S_2) \tag{3.16}$$

with some positive constants κ_1, κ_2. Further, let (cf. (2.15))

$$\mathcal{F}(\Psi) := \left\langle g^{(1)}, \psi^{(1)} \right\rangle_{S_1^N} + \left\langle g^{(2)}, \psi^{(2)} \right\rangle_{S_2^N}, \tag{3.17}$$

where $g^{(i)} \in H^{-\frac{1}{2}}(S_i^N)$ and $\psi^{(i)} \in \widetilde{H}^{\frac{1}{2}}(S_i^N \cup S_c)$.

Now we formulate the following **boundary variational inequality:**

Find $\Phi \in \mathbf{K}(S_1, S_2)$ *such that*

$$\mathcal{B}(\Phi, \Psi - \Phi) \geq \mathcal{F}(\Psi - \Phi) \qquad \text{for all } \Psi \in \mathbf{K}(S_1, S_2). \tag{3.18}$$

Then the following theorem establishes the relation between the spatial and the boundary variational inequalities.

Theorem 3.7. *The spatial variational inequality* (3.1) *is equivalent to the boundary variational inequality* (3.18) *in the following sense:*

(a) *If* $U = \left(u^{(1)}, u^{(2)}\right)^\top$ *is a solution of the SVI* (3.1), *then* $\Phi := \left(\left[u^{(1)} \right]_{S_1}^+, \left[u^{(2)} \right]_{S_2}^+ \right)$ *solves the BVI* (3.18), *and*

(b) *if* $\Phi := \left(\varphi^{(1)}, \varphi^{(2)}\right)^\top$ *is a solution of the BVI* (3.18), *then* $U = \left(u^{(1)}, u^{(2)}\right)^\top$ *with* $u^{(i)}$ *for* $x \in \Omega_i$ *represented by*

$$u^{(i)}(x) = SL^{(i)}\left(\mathcal{M}^{(i)}\varphi^{(i)}\right)(x) - DL^{(i)}\left(\left[\varphi^{(i)}\right]_{S_i}\right)(x) \tag{3.19}$$

solves the SVI (3.1).

Proof. (a) Let $U = \left(u^{(1)}, u^{(2)}\right)^\top \in \mathbf{K}(\Omega_1, \Omega_2)$ be a solution to the SVI (3.1), and $\varphi^{(i)} := \left[u^{(i)} \right]_{S_i}^+$. By the representation formula (3.2) and the corresponding Steklov-Poincaré relation (3.13), $u^{(i)}$ can be represented in the form (3.19). For arbitrary $V = \left(v^{(1)}, v^{(2)}\right)^\top \in \mathbf{K}(\Omega_1, \Omega_2)$ we have $\Psi = \left([v^{(1)}]_{S_1}^+, [v_{S_2}^{(2)}]^+\right)^\top \in \mathbf{K}(S_1, S_2)$. Now, from (3.1) with the help of Theorem 3.4, and equalities (2.11), (2.12), (2.14) and (3.13) we get for U and every $V \in \mathbf{K}(\Omega_1, \Omega_2)$:

$$\begin{aligned} B(U, V - U) &= \sum_{i=1}^{2} \left\langle \left[T^{(i)}u^{(i)} \right]_{S_i}^+, \left[v^{(i)} - u^{(i)} \right]_{S_i}^+ \right\rangle_{S_i} \\ &= \sum_{i=1}^{2} \left\langle \mathcal{M}^{(i)}\varphi^{(i)}, \psi^{(i)} - \varphi^{(i)} \right\rangle_{S_i} \geq \sum_{i=1}^{2} \left\langle g^{(i)}, \psi^{(i)} - \varphi^{(i)} \right\rangle_{S_i^N}. \end{aligned} \tag{3.20}$$

From this inequality, the statement (a) follows since the trace operator $\tau : \mathbf{K}(\Omega_1, \Omega_2) \to \mathbf{K}(S_1, S_2)$ is surjective and, thus, (3.20) holds for arbitrary $\Psi = \left(\psi^{(1)}, \psi^{(2)}\right)^\top \in \mathbf{K}(S_1, S_2)$.

(b) Now, let $\Phi := \left(\varphi^{(1)}, \varphi^{(2)}\right)^\top$ solve the BVI (3.18) and $U = \left(u^{(1)}, u^{(2)}\right)^\top$ be given by formula (3.19). By Theorem 3.4 we then conclude that $\left[u^{(i)} \right]_{S_i}^+ = \varphi^{(i)}$ and $\left[T^{(i)}u^{(i)} \right]_{S_i}^+ = \mathcal{M}^{(i)}\varphi^{(i)}$.

Clearly, for arbitrary $\Psi = \left(\psi^{(1)}, \psi^{(2)}\right)^\top \in \mathbf{K}(S_1, S_2)$ there exists $V = \left(v^{(1)}, v^{(2)}\right)^\top$ such that $V \in \mathbf{K}(\Omega_1, \Omega_2)$ and $\left[v^{(i)}\right]_{S_i}^{+} = \psi^{(i)}$. Note that to any $\psi^{(i)}$ the corresponding vector $v^{(i)}$ is determined up to any element from the space $\overset{\circ}{H}{}^1(\Omega_i)$. By virtue of Theorem 3.4 and equation (2.14), we derive the relation

$$\mathcal{B}(\Phi, \Psi - \Phi) = \sum_{i=1}^{2} \left\langle \mathcal{M}^{(i)}\varphi^{(i)}, \psi^{(i)} - \varphi^{(i)} \right\rangle_{S_i} = B(U, V-U)$$
$$\geq \sum_{i=1}^{2} \left\langle g^{(i)}, \psi^{(i)} - \varphi^{(i)} \right\rangle_{S_i^N} = \sum_{i=1}^{2} \left\langle g^{(i)}, \left[v^{(i)} - u^{(i)}\right]_{S_i^N}^{+} \right\rangle_{S_i^N}.$$

As a consequence, we obtain the statement (b) since V involved is an arbitrary element of the cone $\mathbf{K}(\Omega_1, \Omega_2)$. □

Now we collect the previous results of Theorems 2.1 and 3.7 to formulate the main existence result for the BVI (3.18).

Theorem 3.8. *The BVI* (3.18) *is uniquely solvable. If* $\Phi := \left(\varphi^{(1)}, \varphi^{(2)}\right)^\top$ *is a solution to* (3.18), *then* $U = \left(u^{(1)}, u^{(2)}\right)^\top$ *given by* (3.19) *solves the interface problem with* $f_j^{(i)} = 0$ *and given* $g_j^{(i)} \in H^{-\frac{1}{2}}(S_i^N)$.

Remark 3.9. Assume that $g^{(i)} \in L_2(S_i^N)$ and let Φ be the corresponding solution of the BVI (3.18). Then one obtains the estimate

$$\left\|\Phi^{(i)}\right\|_{\mathbf{H}(S_1,S_2)} \leq \kappa_2^{-1} \left(\left\|g^{(1)}\right\|_{L_2(S_1^N)} + \left\|g^{(2)}\right\|_{L_2(S_2^N)} \right).$$

3.3. Some remarks concerning domains of less smoothness

Now we discuss whether our treatment of the interface crack problem (P) remains valid if the smoothness conditions on S_i are relaxed.

3.3.1. First, let us assume that

$$S_i \in C^{r+1,1}, \quad \partial S_i^D, \partial S_i^N, \partial S_c \in C^{r+1,1} \tag{3.21}$$

where $r \geq 0$ is some integer. In this case we can still apply the presented approach. Note that for such surfaces, the spaces $H^s(S_i)$, $H^s(S^*)$ and $\widetilde{H}^s(S^*)$ are well defined provided $|s| \leq r+2$ and $S^* \in \{S_i^D, S_i^N, S_c\}$ (see [32]). To formulate the interface problem (P) we have to employ the regularity $n^{(i)} \in C^{r,1}(S_i)$ of the normal vector fields. We also need the following technical property related to the product rule.

Lemma 3.10. *Let* $S^* \in \{S_i^D, S_i^N, S_c\}$ *and* $\varphi \in \widetilde{H}^{\frac{1}{2}}(S^*)$, $\psi \in H^{-\frac{1}{2}}(S^*)$, $\chi \in C^{0,1}(\overline{S^*})$. *Then* $\varphi\chi \in \widetilde{H}^{\frac{1}{2}}(S^*)$ *and* $\psi\chi \in H^{-\frac{1}{2}}(S^*)$.

In fact, all the arguments in Subsection 2.2 remain valid. Moreover, the mapping properties of the potential operators and corresponding boundary operators described in Theorems 3.2–3.5 and Lemma 3.6 remain valid in Bessel potential spaces in the range $-r-1 \leq s \leq r+1$. Therefore, if (3.21) is assumed, then Theorems 3.7 and 3.8 remain valid as before and assure the equivalence of the

BVI (3.18) and the interface problem (P) along with the corresponding existence and uniqueness results.

3.3.2. Now, let us require that S_i are just Lipschitz surfaces and S_i^D, S_i^N, and S_c are Lipschitz submanifolds with Lipschitz boundaries, i.e.,

$$S_i \in C^{0,1}, \quad \partial S_i^D, \partial S_i^N, \partial S_c \in C^{0,1}.$$

In this case, the domains Ω_i, are Lipschitz domains (see, e.g., [32]). Note that now the fields of normal vectors $n^{(i)}$ to the surfaces S_i satisfy the inclusions $n^{(i)} \in L_\infty(S_i)$. Then for $u^{(i)} \in H^1(\Omega_i)$ we have $\left[u^{(i)}\right]_{S_i}^+ \in H^{\frac{1}{2}}(S_i)$ and $\left[T^{(i)}u^{(i)}\right]_{S_i}^+ \in H^{-\frac{1}{2}}(S_i)$. However, in general, $\left[u^{(i)}\right]_{S_i}^+ \cdot n^{(i)} \notin H^{\frac{1}{2}}(S_i)$ and $\left[T^{(i)}u^{(i)}\right]_{S_i}^+ \cdot n^{(i)} \notin H^{-\frac{1}{2}}(S_i)$ and conditions (2.6)–(2.8) are not well defined. Therefore difficulties arise for Lipschitz domains, even in the weak interpretation of the interface conditions (2.6)–(2.8).

Nevertheless, the functional $\mathcal{F}(V)$ in (2.15) and the spatial variational inequality (2.17) still is well defined. The bilinear form $B(\cdot,\cdot)$ is bounded and coercive on $(\mathbf{K}(\Omega_1,\Omega_2))^2$, i.e., the inequalities (2.13) are satisfied. Consequently, the spatial variational inequalities (2.17) and (3.1) are still uniquely solvable.

On the other hand, the properties of potential operators and the corresponding boundary operators in Bessel potential spaces as in Theorems 3.2–3.5 and Lemma 3.6 remain valid within the range $-\frac{1}{2} < s < \frac{1}{2}$. This case was treated in [3], [28]. Therefore the boundary variational inequality (3.18) is well defined. As before, Theorem 3.7 holds and establishes the equivalence between the SVI (3.1) and the BVI (3.18), providing the existence and uniqueness results for the BVI (3.18).

Note that in the case of Lipschitz surfaces, of course, the operators $\mathcal{V}^{(i)}$, $\mathcal{K}^{(i)}$, $\mathcal{K}^{(i)*}$, and $\mathcal{D}^{(i)}$ are generalized, strongly singular integral operators, rather than pseudodifferential operators (see [3], [4], [5], [34]).

With the same arguments as in Section 2 we can show that if $u^{(1)}$ and $u^{(2)}$ are defined by the SVI (3.1) or by the BVI (3.18) (see Theorem 3.7), then conditions (2.2)–(2.5) of the interface problem (P) are fulfilled, but we cannot conclude anything concerning the conditions (2.6)–(2.8) (due to the above-mentioned nonsmoothness property of the normal vector $n^{(i)}$). However, if S_i is Lipschitz and piecewise smooth such that $S_c \in C^{1,1}$ and $n^{(i)} \in C^{0,1}(\overline{S_c})$, then with the help of Lemma 3.10 we can show that the conditions (2.6)–(2.8) are satisfied as well. This proves that under the above additional restriction, i.e., when S_i is a piecewise smooth Lipschitz surface with a smooth component S_c, the interface problem (P) (with $f^{(i)} = 0$), the SVI (3.1), and the BVI (3.18) are equivalent to each other according to Theorems 3.7–3.8.

The traditional Dirichlet and Neumann boundary value problems in *isotropic* linear elasticity for the limiting case $s = \pm\frac{1}{2}$ are considered in [5] on the basis of harmonic analysis techniques. There the Fredholm properties of the operators $\pm\frac{1}{2}I_3 + \mathcal{K}^{(i)}$ and $\pm\frac{1}{2}I_3 + \mathcal{K}^{(i)^*}$ and their invertibility in $L_2(S_i)$ are established. However, in the case of general *homogeneous anisotropy* and $s = \pm\frac{1}{2}$, even the Fredholm properties of these operators are still not justified.

4. Galerkin approximations

We now use $\mathbf{H}_h(S_1, S_2) \subset \mathbf{H}(S_1, S_2)$, a family of finite-dimensional subspaces with diminishing meshwidth h given by a sequence of boundary discretizations, and define by

$$\mathbf{K}_h(S_1, S_2) := \{\Psi_h = \left(\psi_h^{(1)}, \psi_h^{(2)}\right)^\top \in \mathbf{H}_h(S_1, S_2) \cap \mathbf{K}(S_1, S_2)\}$$

convex, closed, nonempty finite-dimensional subcones. Then an element $\Phi_h \in \mathbf{K}_h(S_1, S_2)$ defines an *approximate solution* of the BVI (3.18) if

$$\mathcal{B}(\Phi_h, \Psi_h - \Phi_h) \geq \mathcal{F}(\Psi_h - \Phi_h) \quad \text{for all } \Psi_h \in \mathbf{K}_h(S_1, S_2) \tag{4.1}$$

with $\mathcal{B}$ and $\mathcal{F}$ given by (3.14) and (3.17), respectively.

Existence and uniqueness for the approximate solutions to the BVI (4.1) are due to convex analysis and follow from Theorems 2.1 and 2.2 in the monograph [17]. Thus, for every $h > 0$, the BVI (4.1) has a unique solution Φ_h.

Theorem 4.1. *Let $\Phi \in \mathbf{K}(S_1, S_2)$ be a solution to the BVI* (3.18) *and $\Phi_h \in \mathbf{K}_h(S_1, S_2)$ be a solution to the BVI* (4.1).

Then, due to Ceea's lemma, we have the a priori error estimate

$$\begin{aligned} &\|\Phi - \Phi_h\|^2_{\mathbf{H}(S_1,S_2)} \\ &\leq c_0 \inf_{\Psi_h \in \mathbf{K}_h(S_1,S_2)} \left\{ \|\Phi - \Psi_h\|^2_{\mathbf{H}(S_1,S_2)} + \mathcal{B}(\Phi, \Psi_h - \Phi) - \mathcal{F}(\Psi_h - \Phi) \right\}, \end{aligned} \tag{4.2}$$

where c_0 is a constant independent of Φ, Φ_h, and $g^{(i)}$.

Proof. For the proof use coerciveness (3.16), insert $\Psi_h \in \mathbf{K}_h$, use continuity (3.15) and employ the variational inequalities (3.18) and (4.1); finally, exploit linearity of $\mathcal{F}$. □

The continuity of $\mathcal{B}$ and of $\mathcal{F}$ implies the following consequence.

Corollary 4.2. *Let Φ and Φ_h be as in Theorem* 4.1. *Then*

$$\begin{aligned} &\|\Phi - \Phi_h\|^2_{\mathbf{H}(S_1,S_2)} \\ &\leq c_1 \inf_{\Psi_h \in \mathbf{K}_h(S_1,S_2)} \Big\{ \|\Phi - \Psi_h\|^2_{\mathbf{H}(S_1,S_2)} + \|\Phi\|_{\mathbf{H}(S_1,S_2)} \|\Psi_h - \Phi\|_{\mathbf{H}(S_1,S_2)} \\ &\qquad + \sum_{i=1}^{2} \left\| g^{(i)} \right\|_{H^{-\frac{1}{2}}(S_i^N)} \left\| \psi_h - \varphi_h^{(i)} \right\|_{H^{\frac{1}{2}}(S_i^N)} \Big\}, \end{aligned}$$

where c_1 is a positive constant independent of Φ, Φ_h and $g^{(i)}$.

References

[1] Bach, M., Nazarov, S.A. and Wendland, W.L., *Stable propagation of a mode-1 planar crack in an isotropic elastic space. Comparison of the Irwin and the Griffith approaches.* In: Problemi Attuali dell'Analisi e della Fisica Matematica (P.E. Ricci ed.) MM, ARACNE EDITRICE, Roma 2000, 167–189.

[2] Comninou, M. and Dundurs, J., *On the behavior of interface crack*, Res Mechanica, **1** (1980), 249–264.

[3] Costabel, M., *Boundary integral operators on Lipschitz domains: elementary results*, SIAM J. Math. Anal., **19**(1988), 613–626.

[4] Costabel, M. and Wendland, W.L., *Strong ellipticity of boundary integral operators.* J. Reine Angew. Mathematik **372** (1986), 34–63.

[5] Dahlberg, B.E.J., Kenig, C.E. and Verchota, G.C., *Boundary value problems for the systems of elastostatics in Lipschitz domains,* Duke Math. J., **57** (1988), 795–818.

[6] Dautray, R. and Lions, J.L., *Mathematical Analysis and Numerical Methods for Science and Technology,* Vol. 4, Integral Equations and Numerical Methods, Springer-Verlag, Berlin–Heidelberg, 1990.

[7] Duduchava, R., Natroshvili, D. and Shargorodsky, E., *Basic boundary value problems of thermoelasticity for anisotropic bodies with cuts.* I, II, Georgian Mathematical Journal, **2** (1995), No. 2, 123–140, No. 3, 259–276.

[8] Duvaut, G. and Lions, J.L., *Les inéquations en méchanique et en physique*, Dunod, Paris, 1972.

[9] Eck, C., *Existenz und Regularität der Lösungen für Kontaktprobleme mit Reibung.* Doctoral Thesis, University of Stuttgart 1996.

[10] Eck, C., Steinbach, O. and Wendland, W.L., *A symmetric boundary element method for contact problems with friction,* Math. Comp. Simulation, **50** (1999), 1–4, 43–61.

[11] Eck, C. and Wendland, W.L., *A residual-based error estimator for BEM-discretizations of contact problems.* Numer. Math., to appear.

[12] Eck, C. and Wohlmuth, B., *Convergence of a contact-Neumann iteration for the solution of two-body contact problems.* Math. Models & Methods Appl. Sci. (2003), to appear.

[13] Eskin, G., *Boundary Value Problems for Elliptic Pseudodifferential Equations,* Transl. of Mathem. Monographs, Amer. Math. Soc., **52**, Providence, Rhode Island, 1981.

[14] Fichera, G., *Problemi elastostatici con vincoli unilaterali: il problema di Signorini con ambigue condicioni al contorno.* Accad. Naz. Lincei, **8** (1963–1964), 91–140.

[15] Fichera, G., *Existence Theorems in Elasticity.* Handb. der Physik, Bd. 6/2, Springer-Verlag, Heidelberg, 1973.

[16] Gachechiladze, A. and Natroshvili, D., *Boundary variational inequality approach in the anisotropic elasticity for the Signorini problem,* Georgian Math. J., **8** (2001), 462–692.

[17] Glowinski, R., Lions, J.L. and Tremolieres R., *Numerical Analysis of Variational Inequalities,* North-Holland, Amsterdam, 1981.

[18] Han, H., *A boundary element procedure for the Signorini problem in three-dimensional elasticity.* Numerical Mathematics, A Journal of Chinese Universities, **3** (1994), 104–117.

[19] Hlaváček, I., Haslinger, J., Nečas, J. and Loviček, J., *Solution of Variational Inequalities in Mechanics.* SNTL, Prague, 1983.

[20] Hsiao, G.C., Schnack, E. and Wendland, W.L., *Hybrid coupled finite-boundary element methods for elliptic systems of second order.* Comp. Methods Appl. Mech. Engrg., **190** (2000), 431–485.

[21] Hsiao, G.C. and Wendland, W.L., *A finite element method for some integral equations of the first kind.* J. Math. Anal. Appl., **58** (1977), 449–481.

[22] Hsiao, G.C. and Wendland, W.L., *Boundary Integral Equations: Variational Methods.* Springer-Verlag, Berlin, in preparation.

[23] Kinderlehrer, D., *Remarks about Signorini's problem in linear elasticity,* Ann. Sc. Norm. Sup. Pisa, Cl. Sci., S. IV, vol. VIII, **4** (1981), 605–645.

[24] Kupradze, V.D., Gegelia, T.G., Basheleishvili, M.O. and Burchuladze, T.V., *Three-dimensional Problems of the Mathematical Theory of Elasticity and Thermoelasticity,* (Russian), Nauka, Moscow, 1976 (English translation: North-Holland Publishing Company, Amsterdam-New York-Oxford, 1979.

[25] Lions, J.-L. and Magenes, E., *Problèmes aux limites non homogènes et applications,* Vol. 1, Dunod, Paris, 1968.

[26] Lions, J.L. and Stampacchia, G., *Variational inequalities,* Comm. Pure Appl. Math., **20** (1976), 493–519.

[27] Maz'ya, V.G., *Boundary integral equations,* in R.V. Gamkrelidze (ed.) Encyclopedia of Mathematical Sciences, **27**, Springer-Verlag, Heidelberg, 1991.

[28] McLean, W., *Strongly Elliptic Systems and Boundary Integral Equations,* Cambridge University Press, 2000.

[29] Meister, E. and Speck, F.-O., *The explicit solution of elastodynamic diffraction problems by symbol factorization.* Z. Anal. Anw., **8** (1989), 307–328.

[30] Natroshvili, D., *Boundary integral equation method in the steady state oscillation problems for anisotropic bodies,* Math. Methods Appl. Sci., **20** (1997), 95–119.

[31] Natroshvili, D., Chkadua, O. and Shargorodsky, E., *Mixed boundary value problems of the anisotropic elasticity,* Proc. I. Vekua Inst. Appl. Math. Tbilisi State University, **39** (1990), 133–181 (Russian).

[32] Nečas, J., *Méthodes Directes en Théorie des Équations Élliptique,* Masson, Paris, 1967.

[33] Steinbach, O., *Stability Estimates for Hybrid Coupled Domain Decomposition Methods.* Lecture Notes in Math., Springer-Verlag, Berlin, 2003.

[34] Steinbach, O. and Wendland, W.L., *On C. Neumann's method for second-order elliptic systems in domains with non-smooth boundaries.* J. Math. Anal. Appl., **262** (2001), 733–748.

[35] Triebel, H., *Theory of Function Spaces,* Birkhäuser Verlag, Basel-Boston-Stuttgart, 1983.

[36] Zeidler, E., *Nonlinear Functional Analysis and its Applications. Vol.II/B, Nonlinear Monotone Operators.* Springer-Verlag, New York, 1990.

David Natroshvili
Department of Mathematics
Georgian Technical University
Kostava str. 77
Tbilisi 380075
Georgia
e-mail: natrosh@hotmail.com

Wolfgang L. Wendland
Institut für Angewandte Analysis
und Numerische Simulation
Universität Stuttgart
Pfaffenwaldring 57
D-70550 Stuttgart
Germany
e-mail: wendland@mathematik.uni-stuttgart.de

Operator Theory
Advances and Applications, Vol. 147, 403–422

Limiting Absorption Principle for a Class of Difference Equations

Vladimir Rabinovich

To the memory of Professor Erhard Meister

Abstract. We consider the limiting absorption principle for a self-adjoint second order difference operator on the discrete half-space $\mathbb{Z}_+^n = \{x \in \mathbb{Z}^n : x_n \geq 0\}$. A characteristic feature of difference operators considered in the paper is the fact that their coefficients stabilize at infinity to functions depending on the single variable x_n. Such a statement is often needed in applications, for example, in problems involving the propagation of waves in perturbed stratified media and their difference approximations.

1. Introduction

The main aim of the paper is the limiting absorption principle for self-adjoint second order difference operators acting in functional spaces on the discrete half-space $\mathbb{Z}_+^n = \{x \in \mathbb{Z}^n : x_n \geq 0\}$. Note that the difference approximation is a very important and effective tool for the numerical analysis of problems of the acoustic waves propagation in perturbed stratified fluids (see for instance [22]). A characteristic feature of difference operators considered below is the fact that their coefficients stabilize at infinity to functions depending on the single variable x_n. Such a statement is often needed in applications, for example, in problems involving the propagation of waves in perturbed stratified media (see for instance [2], [22], [19], [20] and references given there). The limiting absorption principle and spectral properties of operators of boundary value problems for partial differential operators describing the acoustic wave propagation in perturbed stratified media were studied in many papers (see for instance [3], [4], [7], [17], [18], [19] and references cited there), but similar questions for the difference models of such problems have not been studied earlier. Note that there are essential technical distinctions between the continuous and discrete cases.

The other aim of the paper is the proof of the applicability of the finite sections method [9] for the numerical solution of problems under consideration.

The paper is organized as follows. Section 2 of the paper is devoted to the spectral decomposition and limiting absorption principle for a difference acoustic

---

2000 *Mathematics Subject Classification.* Primary 47B39, 39A470, Secondary 47A10.

operator which is characterized by the sound speed and density depending on the single variable x_n only. In Section 3 the limiting absorption principle and Fredholmness of perturbed operators are studied. Section 4 is devoted to the applicability of the finite section method [9] for the approximation of difference equations on $\mathbb{Z}_+^n$ simulating the process of wave propagation in a perturbed stratified medium by a finite system of linear equations.

2. Difference acoustic operator in a stratified half-space

2.1. Notations

We denote the discrete half-space by $\mathbb{Z}_+^n = \{x = (x_1, \dots, x_n) \in \mathbb{Z}^n : x_n \geq 0\}$. If a is a positive function on $\mathbb{Z}_+^n$ we denote by $l^2(\mathbb{Z}_+^n, a)$ a Hilbert space with the scalar product

$$(u, v)_{l^2(\mathbb{Z}_+^n, a)} = \sum_{x \in \mathbb{Z}_+^n} a(x) u(x) \bar{v}(x).$$

If $a(x) = \langle x \rangle^{\delta}$, where $\langle x \rangle = (1 + |x|^2)^{1/2}, \delta \in \mathbb{R}$ we will use the notation $l_\delta^2(\mathbb{Z}_+^n)$ and $l^2(\mathbb{Z}_+^n)$ for $\delta = 0$. We set

$$\nabla_j = \tau_j - I, \nabla_j^* = \tau_j^* - I, j = 1, \dots, n$$

where τ_j is the shift operator

$$\tau_j f(x) = f(x - e_j), e_j = (0, \dots, 1_j, \dots, 0), \tau_j^* f(x) = f(x + e_j).$$

We will use the difference analogs of the gradient and the divergence:

$$\nabla = (\nabla_1, \dots, \nabla_n), \nabla^* \cdot g = (\nabla^*, g) = \sum_{j=1}^n \nabla_j^* g_j.$$

In what follows, if X and Y are Banach spaces, then $\mathcal{B}(X, Y)$ denotes the space of bounded linear operators acting from X into Y with the uniform operator topology.

2.2. Formulation of the main results of the section

We consider a difference acoustic operator

$$A_0 u_+(x) = c_0(x_n) \rho_0(x_n) \nabla^* \cdot (\rho_0^{-1}(x_n) \nabla u_+(x)), x \in \mathbb{Z}_+^n, \tag{2.1}$$

where u_+ is the extension by zero on $\mathbb{Z}^n$ of a given function u on $\mathbb{Z}_+^n$. We add the condition on the boundary $\mathbb{Z}^{n-1}$ to the operator A_0

$$u(x', 0) = 0, x' = (x_1, \dots, x_{n-1}) \in \mathbb{Z}^{n-1}. \tag{2.2}$$

We suppose that c_0, ρ_0 are positive-valued functions on $\mathbb{Z}_+^n$ depending on the variable x_n only, and such that

$$c_m = \min_{x_n \geq 0} c_0(x_n) > 0, \rho_m = \min_{x_n \geq 0} \rho_0(x_n) > 0.$$

Moreover $c_0(x_n) = c_+, \rho_0(x_n) = \rho_+$ for $x_n \geq h > 0$.

We denote by $\mathbf{H}_{0,n}$ the subspace in $l^2(\mathbb{Z}_+^n, c_0^{-2}\rho_0^{-1})$ of functions satisfying the condition

$$u(x',0) = 0, x' \in \mathbb{Z}^{n-1}.$$

It is easy to check that the operator $\mathcal{A}_0$ generated by difference expression (2.1) and boundary condition (2.2) is a self-adjoint and nonnegative operator on $\mathbf{H}_{0,n}$. We will denote the resolvent of $\mathcal{A}_0$ by $R_0(\zeta) = (\mathcal{A}_0 - \zeta I)^{-1}$.

The following theorems are the main results of this section.

Theorem 1.

(i) *The spectrum of $\mathcal{A}_0$ coincides with the segment $[0, 4c_+^2 n]$.*

(ii) *For each point $\lambda \in [0, 4c_+^2 n] \setminus \Lambda = \mathcal{E}$, where Λ is a finite set, there exists a uniform limit on each segment in $\mathcal{E}$,*

$$R_0^{\pm}(\lambda) = \lim_{\varepsilon \to +0} R_0(\lambda \pm i\varepsilon),$$

in the topology of $\mathcal{B}(l^2_\delta(\mathbb{Z}_+^n), l^2_{-\delta}(\mathbb{Z}_+^n)), \delta > 1/2$. Moreover $R_0(\lambda)$ is a Hölder function on $\mathcal{E}$ with values in $\mathcal{B}(l^2_\delta(\mathbb{Z}_+^n), l^2_{-\delta}(\mathbb{Z}_+^n))$, $\delta > 1/2$.

Remark 2. *The set Λ will be indicated below.*

Theorem 3. *Let $\lambda \in \mathcal{E}$, $\delta > 3/2$, $f \in l^2_\delta(\mathbb{Z}_+^n)$ and let one of the following conditions hold:*

$$\Im \left(R_0^{\pm}(\lambda) f, f\right)_{\mathbf{H}_{0,n}} = 0. \tag{2.3}$$

Then the other condition holds also, and

$$R_0^+(\lambda) f = R_0^-(\lambda) f \in l^2(\mathbb{Z}_+^n).$$

Moreover for each segment $[a, b] \subset \mathcal{E}$ there exists a constant $C > 0$ such that

$$\sup_{\lambda \in [a,b]} \left\| R_0^{\pm}(\lambda) f \right\|_{l^2(\mathbb{Z}_+^n)} \le C \left\| f \right\|_{l^2_\delta(\mathbb{Z}_+^n)}$$

for all functions f satisfying (2.3).

2.3. Difference operator on the half-axis and its spectral properties

We associate with the operator $\mathcal{A}_0$ a difference operator on $\mathbb{Z}_+ = \{y \in \mathbb{Z} : y \ge 0\}$ depending on a parameter $\mu \in \mathbb{C}$

$$L_0(\mu) v(y) = c_0^2(y) \rho_0(y) \nabla^* \cdot (\rho_0^{-1}(y) \nabla v_+(y)) + \mu c_0^2(y) v_+(y), y \in \mathbb{Z}_+ \tag{2.4}$$

with the boundary condition

$$v(0) = 0. \tag{2.5}$$

As above v_+ is the extension by zero of the function v given on $\mathbb{Z}_+$.

Below we will give some spectral properties of the operator $L(\mu)$ following the paper [15], see also the books [1], [10]. These properties are similar to the property of the *Sturm-Liouville* problem on the half-axis (see, for instance, [13], [5]). Note that the detailed analysis of the *Sturm-Liouville* problem on $\mathbb{R}$ connected with wave propagation problems in stratified fluids was presented in the book [20].

It is easy to obtain the following estimate

$$c_m^2\mu\,\|v\|^2_{\mathbf{H}_{0,1}} \leq (L_0(\mu)v,v)_{\mathbf{H}_{0,1}} \leq (4+\mu)\,c_+^2\,\|v\|^2_{\mathbf{H}_{0,1}} \tag{2.6}$$

which implies that the spectrum of $L_0(\mu)$ belongs to the segment $\left[c_m^2\mu,(4+\mu)\,c_+^2\right]$.

We will describe the structure of *eigenfunctions* and *generalized eigenfunctions* of the self-adjoint operator $L_0(\mu)$ in a convenient form in what follows. The definition of *generalized eigenfunctions* of a self-adjoint operator can be found in [5], but we need a concrete form of generalized eigenfunctions, which will be presented below.

Let $\psi = \psi(y,\alpha,\mu)$ be a solution of the equation

$$L_0(\mu)\psi(y,\alpha,\mu) = \lambda(\alpha,\mu)\psi(y,\alpha,\mu), y \in [0,h] \tag{2.7}$$

satisfying the initial conditions:

$$\psi(0,\alpha,\mu) = 0,\ \psi(1,\alpha,\mu) = 1$$

where

$$\lambda(\alpha,\mu) = \left(4\sin^2\alpha/2+\mu\right)c_+^2, \alpha \in [0,\pi]. \tag{2.8}$$

Such a solution exists, is unique and depends analytically on the parameters α,μ. Let $\alpha \in (0,\pi)$, $\mu \in [0,\infty)$. Then the general solution of the equation $L_0(\mu)\psi = \lambda(\alpha,\mu)\psi$ for $y > h$ has the form

$$\psi(y,\alpha,\mu) = C_+(\alpha,\mu)e^{i\alpha(y-h)} + C_-(\alpha,\mu)e^{-i\alpha(y-h)} \tag{2.9}$$

where $C_\pm(\alpha,\mu)$ can be found from the conditions:

$$\begin{aligned}\psi(h,\alpha,\mu) &= C_+(\alpha,\mu)+C_-(\alpha,\mu),\\ \psi(h+1,\alpha,\mu) &= C_+(\alpha,\mu)e^{i\alpha}+C_-(\alpha,\mu)e^{-i\alpha}.\end{aligned}$$

It implies that

$$C_\pm(\alpha,\mu) = \pm\frac{\psi(h+1,\alpha,\mu)-\psi(h,\alpha,\mu)e^{\mp i\alpha}}{2i\sin\alpha}. \tag{2.10}$$

The function $\psi(y,\alpha,\mu)$ is real-valued and, for $y > h$, can be written in the form

$$\begin{aligned}\psi(y,\alpha,\mu) &= \psi(h,\alpha,\mu)\cos\alpha(y-h)\\ &\quad+\frac{\psi(h+1,\alpha,\mu)-\psi(h,\alpha,\mu)\cos\alpha}{\sin\alpha}\sin\alpha(y-h).\end{aligned}$$

We set

$$\varphi_0(y,\alpha,\mu) = \frac{1}{M(\alpha,\mu)}\psi(y,\alpha,\mu), \tag{2.11}$$

where

$$M(\alpha,\mu) = \sqrt{\frac{2\rho_+c_+}{\pi}}\left((\psi^2(h,\alpha,\mu)) + \left(\frac{\psi(h+1,\alpha,\mu)-\psi(h,\alpha,\mu)\cos\alpha}{\sin\alpha}\right)^2\right)^{1/2}. \tag{2.12}$$

By means of simple calculations one can check that the family $\varphi_0(y,\alpha,\mu)$ has the following properties: (i) it satisfies

$$\tilde{u}(\alpha,\mu)=(u(\cdot),\varphi_0(\cdot,\mu,\alpha))_{\mathbf{H}_{0,1}}\in L^2((0,\pi))$$

for every $u\in\mathbf{H}_{0,1}$, (ii) the system $\varphi_0(\cdot,\alpha,\mu),\alpha\in(0,\pi)$ is orthonormal in $\mathbf{H}_{0,1}$, that is, for every $f\in C_0^\infty((0,\pi))$,

$$\left(\int_0^\pi f(\alpha)\varphi_0(\cdot,\alpha,\mu)d\alpha,\varphi_0(\cdot,\beta,\mu)\right)_{\mathbf{H}_{0,1}}=f(\beta),\beta\in(0,\pi)$$

Thus the family $\varphi_0(\cdot,\alpha,\mu),\alpha\in(0,\pi)$ is a system of generalized eigenfunctions of $L_0(\mu)$ according to the definition in the book [5], Section 1.2.

Let $\alpha=i\beta,\beta>0$. Then the function $\varphi(\cdot,i\beta,\mu)$ defined on $\mathbb{Z}_+$ as

$$\varphi(y,i\beta,\mu)=\begin{cases}\psi(y,i\beta,\mu),0\le y\le h,\\ C_+(i\beta,\mu)e^{-\beta(y-h)},y>h\end{cases}$$

is an eigenfunction of $L_0(\mu)$ if and only if β is a solution of the dispersion equation

$$C_-(i\beta,\mu)=F(\beta,\mu)=f(h+1,i\beta,\mu)-f(h,i\beta,\mu)e^{-\beta}=0. \tag{2.13}$$

This equation has a finite number of positive solutions $\beta_j(\mu),j=1,\dots,k(\mu)$ such that the eigenvalues of $L_0(\mu)$ are given by the formulas

$$\lambda_j(\mu)=\lambda(i\beta_j,\mu)=\left(-4\sinh^2\beta_j/2+\mu\right)c_+^2$$

and the normed eigenfunctions of $L_0(\mu)$ are:

$$\varphi_j(y,\mu)=\frac{1}{M_j(\mu)}\begin{cases}\psi(y,i\beta_j,\mu),y\in[0,h]\cap\mathbb{Z},\\ C_+(i\beta,\mu)e^{-\beta_j(y-h)},y\in(h,\infty)\cap\mathbb{Z},\end{cases}$$

where

$$M_j(\mu)=\left(\sum_{m=0}^{a}c_0^{-2}(m)\rho_0(m)\left|\psi(m,i\beta_j,\mu)\right|^2+c_+^{-2}\rho_+\left(\frac{C_+(i\beta_j,\mu)}{1-e^{-2\beta_j(\mu)}}\right)\right)^{1/2}.$$

The next proposition (see [15]) describes the spectral properties of $L_0(\mu)$.

Proposition 4.

(i) *The operator $L_0(\mu),\mu\in[0,\infty)$ has an absolutely continuous spectrum filling the segment $\left[c_+^2\mu,(4+\mu)c_+^2\right]$. Moreover, if $\alpha\in[0,\pi]$, then $\lambda(\alpha,\mu)=\left(4\sin^2\alpha/2+\mu\right)c_+^2\in\left[c_+^2\mu,(4+\mu)c_+^2\right]$ is a point of the absolutely continuous spectrum.*

(ii) *The operator $L_0(\mu),\mu\in[0,\infty)$ has a finite set of points of the single discrete spectrum $\lambda_j(\mu),j=1,\dots,k(\mu)$ such that*

$$c_m^2\mu\le\lambda_1(\mu)\le\cdots\le\lambda_k(\mu)<c_+^2\mu.$$

(iii) *Let $\{\varphi_1(\cdot,\mu),\dots,\varphi_k(\cdot,\mu)\}$ be the orthonormal system of eigenfunctions of $L(\mu)$, and $\{\varphi_0(\cdot,\mu,\alpha),\alpha\in(0,\pi)\}$ be the orthonormal system of generalized eigenfunctions of $L_0(\mu)$. Then the system*

$$\{\varphi_1(\cdot,\mu),\dots,\varphi_k(\cdot,\mu)\},\{\varphi_0(\cdot,\mu,\alpha),\alpha\in(0,\pi)\}$$

is complete in $\mathbf{H}_{0,1}$. *It means that, if* $u \in \mathbf{H}_{0,1}$ *and*

$$\begin{aligned} u_0(\mu,\alpha) &= (u,\varphi_0(\mu,\alpha))_{\mathbf{H}_{0,1}}, \\ u_j(\mu) &= (u,\varphi_j(\mu))_{\mathbf{H}_{0,1}},\ j=1,\dots,k(\mu), \end{aligned} \tag{2.14}$$

then

$$u(y) = \int_0^{\pi} u_0(\alpha,\mu)\varphi_0(y,\mu,\alpha)d\alpha + \sum_{j=1}^{k} u_j(\mu)\varphi_j(y,\mu)$$

and

$$\|u\|^2_{\mathbf{H}_{0,1}} = \int_0^{\pi} |u_0(\alpha,\mu)|^2 \, d\alpha + \sum_{j=1}^{k} |u_j(\mu)|^2$$

(the Parseval equality).

It follows from Proposition 4 that the mapping

$$\Psi_\mu : \mathbf{H}_{0,1} \to L^2((0,\pi)) \oplus \mathbb{C}^k$$

defined by the formula (2.14) is a unitary operator.

We say that μ_j is a *critical value of the parameter* μ if

$$\lim_{\mu\to\mu_j+0} \beta_j(\mu) = 0.$$

Hence if $\mu \to \mu_j + 0$ then $\lambda_j(\mu) \to c_+^2\mu_j$ and $M_j(\mu) \to \infty$. It means that, if $\mu \to \mu_j+0$, the eigenvalue $\lambda_j(\mu)$ is absorbed by the end of the continuous spectrum.

It follows from (2.13) that the critical values μ_j are nonnegative solutions of the equation

$$\psi(h+1,0,\mu) - \psi(h,0,\mu) = 0.$$

Note that $\lambda_j(\mu) \in C^\infty((\mu_j,\infty))$ since it is a simple eigenvalue of the operator $L(\mu)$ with C^∞-dependence on the parameter (see, [11]).

It is easy to check that $F'_\beta(0,\mu_j) \neq 0$. Hence, applying the theorem on implicit functions, one can prove that $\beta_j(\mu)$ belongs C^∞ in a neighborhood of the point μ_j. It implies that the eigenvalue $\lambda_j(\mu)$ can be continued to a function of the class $C^\infty([\mu_j,\infty))$.

Then, applying the well-known *mini-max variational principle* for increasing enumeration $\alpha_1 \leq \cdots \leq \alpha_N \leq \cdots$ of all eigenvalues of a self-adjoint operator A in a Hilbert space H (see, for instance [12], Theorem 4.22), which states that

$$\alpha_N = \max_{S_N} \min_{x\in S_N} \frac{(Ax,x)_H}{(x,x)_H},$$

where S_N is an N-dimensional subspace of H, we obtain that $\lambda_j(\mu)$ is a function which increases monotonically on (μ_j,∞), that is $\lambda'_j(\mu) > 0$ on (μ_j,∞).

But we will suppose that the following stronger condition is fulfilled:

$$\lambda'_j(\mu) > 0, \mu \in [\mu_j,\infty).$$

Note that

$$M_j^{-1}(\mu) \sim c_+ \left(\frac{2\beta_j(\mu)}{\rho_+\psi(h,0,\mu_j)}\right)^{1/2}, \mu \to \mu_j + 0.$$

Thus

$$\varphi_j(y,\mu) = (\beta_j(\mu))^{1/2}\,\psi_j(y,\mu),$$

where $\psi_j(y,\mu) \in C^\infty([\mu_j,\infty))$ for each $y \in \mathbb{Z}_+^n$. Moreover for each $d > 0$ there exist positive constants C_j such that

$$\sup_{\mu\in[d,\infty)} \left|\frac{\partial\psi_j(y,\mu)}{\partial\mu}\right| \le C_j y.$$

Proposition 5. *Let $u \in l_\delta^2(\mathbb{Z}_+), \delta > 1/2$. Then for each $b > 0$ there exists a number $C > 0$ such that:*

(i)

$$\sup_{\mu\in[0,b],\alpha\in(0,\pi)} |u_0(\alpha,\mu)| \le C\,\|u\|_{l_\delta^2(\mathbb{Z}_+^n)};$$

(ii)

$$\sup_{\mu\in[\mu_j,b]} |u_j(\mu)| \le C\,\|u\|_{l_\delta^2(\mathbb{Z}_+^n)};$$

(iii)

$$\begin{aligned} &\sup_{\mu\in[0,b]} |u_0(\alpha',\mu) - u_0(\alpha'',\mu)| \\ \le\ & C(\sin^{-\gamma}\alpha' + \sin^{-\gamma}\alpha'')\,|\alpha' - \alpha''|^\gamma\,\|u_+\|_{l_\delta^2(\mathbb{Z}_+^n)}, \\ \alpha',\alpha'' \in\ & (0,\pi), 0 < \gamma \le \min\{1, \delta - 1/2\}; \end{aligned}$$

(iv) *Let $v_j(\mu) = (u_+(\cdot), \psi_j(\cdot,\mu))_{\mathbf{H}_{0,1}}$. Then*

$$\begin{aligned} |v_j(\mu') - v_j(\mu'')| &\le C\,|\mu' - \mu''|^\gamma\,\|u_+\|_{l_\delta^2(\mathbb{Z}_+^n)}, \mu',\mu'' \in [\mu_j, b], \\ 0 &< \gamma \le \min\{1, \delta - 1/2\}; \end{aligned}$$

(v) *For every $b > 0$ and $\varepsilon > 0$ there exists a number $C_\varepsilon > 0$ such that*

$$\sup_{\alpha\in[\varepsilon,\pi-\varepsilon]} |\hat{u}_0(\alpha,\mu') - \hat{u}_0(\alpha,\mu'')| \le C_\varepsilon\,|\mu' - \mu''|\,\|u\|_{l_\delta^2(\mathbb{Z}_+^n)}, \mu',\mu'' \in [0,b].$$

Proof. Statements (i), (ii) follow easily from the estimates:

$$\sup_{\mathbb{Z}_+^n\times[0,\pi]\times[0,b]} |\varphi_0(y,\alpha,\mu)| < \infty, \quad \sup_{\mathbb{Z}_+^n\times[\mu_j,b]} |\varphi_j(y,\mu)| < \infty, j = 1,\ldots,k. \tag{2.15}$$

Let us prove (iii). By differentiating (2.11) with respect to α we obtain the estimate

$$\sup_{\mu\in[0,b]} \left|\frac{\partial\varphi_0(y,\alpha,\mu)}{\partial\alpha}\right| \le \frac{Cy}{\sin\alpha}, y \in \mathbb{Z}_+^n, \alpha \in (0,\pi). \tag{2.16}$$

Applying the relations (2.15), (2.16), the Lagrange formula and interpolation we obtain the estimate (iii);

Statements (iv) and (v) also are proved by applying estimates (2.15) together with

$$\sup_{\alpha\in[\varepsilon,\pi-\varepsilon],\mu\in[0,b]}\left|\frac{\partial\varphi_0(y,\alpha,\mu)}{\partial\alpha}\right| \le C_\varepsilon y,$$
$$\sup_{y\in\mathbb{Z}_+^n,\mu\in[\mu_j,b]}\left|\frac{\partial\psi_j(y,\mu)}{\partial\mu}\right| \le C,$$

the Lagrange formula and interpolation. □

2.4. Diagonalization of the operator $\mathcal{A}_0$

We use the notations:

$$\tilde{\mu}(\xi') = 4\sum_{k=1}^{n-1}\sin^2\frac{\xi_k}{2}, \mu(\xi) = 4\sum_{k=1}^{n}\sin^2\frac{\xi_k}{2}.$$

The real number t is called a critical value of the trigonometric polynomial $\mu(\xi)$ if $t = \mu(a)$ and $(\nabla\mu)(a) = 0$. Note that the polynomial $\mu(\xi)$ has a finite set $\mathcal{K} = \{0, 4, 8, \ldots, 4n\}$ of critical values.

Let

$$\begin{aligned}\mathcal{K}_j &= \{0,4,8,\ldots,4(n-1)\}\cap\{\mu\in\mathbb{R}:\mu>\mu_j\},\\ \Lambda_0 &= \{0,4c_+^2,8c_+^2,\ldots,4nc_+^2\},\Lambda_j=\lambda_j(\mathcal{K}_j),j=1,\ldots,N\end{aligned}$$

where $N = \max\{j\in\mathbb{Z}_+ : \mu_j > 4(n-1)\}$, and let $\Lambda = \cup_{j=0}^N\Lambda_j$.

Note that Λ *is the set in the statement of Theorem* 1.

We will use the following notations:

$$\begin{aligned}\Omega_0 &= (-\pi,\pi)^{n-1}\times(0,\pi),\\ \Omega_j &= \left\{\xi'=(\xi_1,\ldots,\xi_{n-1})\in(-\pi,\pi)^{n-1}:\tilde{\mu}(\xi')\ge\mu_j, j=1,\ldots,N\right\},\\ \varkappa_0(\xi) &= c_+^2\mu(\xi),\ \varkappa_j(\xi')=\lambda_j(\tilde{\mu}(\xi')),\ j=1,\ldots,N.\end{aligned}$$

We define a unitary operator

$$\Phi_0 : \mathbf{H}_{0,n} \to \hat{H}_0 = \oplus_{j=0}^N L^2(\Omega_j)$$

acting as

$$\Phi_0(f) = (\check{f}_0, \check{f}_1, \ldots, \check{f}_N),$$

where

$$\begin{aligned}\check{f}_0(\xi) &= (2\pi)^{-\frac{n-1}{2}}\left(\hat{f}(\xi',\cdot),\varphi_0(\cdot,\xi_n,\tilde{\mu}(\xi'))\right)_{\mathbf{H}_{0,1}},\\ \check{f}_j(\xi') &= (2\pi)^{-\frac{n-1}{2}}\left(\hat{f}(\xi',\cdot),\varphi_j(\cdot,\tilde{\mu}(\xi'))\right)_{\mathbf{H}_{0,1}}, j=1,..,N\end{aligned}$$

and $\hat{f}(\xi',\cdot)$ is the partial discrete Fourier transform of f

$$\begin{aligned}\hat{f}(\xi',\cdot) &= \left(F_{x'\to\xi'}f\right)(\xi',\cdot) \\ &= \left(\frac{1}{2\pi}\right)^{\frac{n-1}{2}} \sum_{x'\in\mathbb{Z}^{n-1}} f(x',\cdot)e^{i(x',\xi')}, \xi'\in\mathbb{Z}^{n-1}.\end{aligned}$$

It is clear that $\Phi_0 = \Psi_{\tilde{\mu}(\xi')}F_{x'\to\xi'}$ is a unitary operator and

$$\hat{\mathcal{A}}_0 = \left(\Phi_0\mathcal{A}_0\Phi_0^* f\right)(\xi) = \left(\varkappa_0(\xi)\check{f}_0, \varkappa_1(\xi')\check{f}_1(\xi'),\ldots,\varkappa_N(\xi')\check{f}_N(\xi')\right). \tag{2.17}$$

Thus the operator $\mathcal{A}_0 : \mathbf{H}_{0,n} \to \mathbf{H}_{0,n}$ is unitarily equivalent to the operator $\hat{\mathcal{A}}_0 = \oplus_{j=0}^N \mathcal{A}_{0j} : \hat{H}_0 \to \hat{H}_0$, where $\mathcal{A}_{0j}$ are the operators of multiplication by the real-valued functions $\varkappa_j, j=0,1,\ldots,N$, that is, formula (2.17) gives a diagonalization of $\mathcal{A}_0$.

2.5. Proofs of Theorem 1 and Theorem 3

Now we are in a position to prove Theorem 1 and Theorem 3.

2.5.1. Proof of Theorem 1

Proof. (i) Note that the closure of the set of values of $\varkappa_0$ is the segment $\left[0,4c_+^2 n\right]$.

Let $\lambda_j^{cr} = c_+^2\mu_j$. Then

$$0 < \lambda_j^{cr} \le \lambda_j(\mu) \le c_+^2\mu.$$

Setting $\mu = \tilde{\mu}(\xi')$ in these inequalities we obtain that

$$0 < \lambda_j^{cr} \le \varkappa_j(\xi') \le c_+^2\tilde{\mu}(\xi') \le 4c_+^2 n.$$

Hence all values of $\varkappa_j(\xi')$ are situated on $\left[0,4c_+^2 n\right]$.

(ii) Let $\mathcal{E} = \left[0,4c_+^2 n\right]\setminus\Lambda$, $\chi\in C_0^\infty(\mathbb{R}), 0\le\chi(t)\le 1$ and χ equals 1 in a small neighborhood of the segment $[a,b]\subset\mathcal{E}$, and equals zero outside another small neighborhood of $[a,b]$, $\phi = 1-\chi$. Then

$$R_0(\lambda\pm i\varepsilon) = \chi(\mathcal{A}_0)R_0(\lambda\pm i\varepsilon) + \phi(\mathcal{A}_0)R_0(\lambda\pm i\varepsilon).$$

It is evident that for $\lambda\in[a,b]$

$$\lim_{\varepsilon\to 0}\phi(\mathcal{A}_0)R_0(\lambda\pm i\varepsilon) = \phi(\mathcal{A}_0)R_0(\lambda)$$

in the topology of $\mathcal{B}(l^2(\mathbb{Z}_+^n), l^2(\mathbb{Z}_+^n))$. Moreover $\phi(\mathcal{A}_0)R_0(\lambda)$ is a C^∞-function on $[a,b]$ with values in $\mathcal{B}(l_\delta^2(\mathbb{Z}_+^n), l_{-\delta}^2(\mathbb{Z}_+^n)), \delta\ge 0$.

Let $f,g\in l_\delta^2(\mathbb{Z}_+^n), \delta > 1/2$. Then

$$\begin{aligned}\left(\chi(\mathcal{A}_0)R_0(\lambda\pm i\varepsilon)f,g\right)_{\mathbf{H}_{0,n}} &= \int_0^{d_0}\frac{\chi(t)\left(\mathcal{A}_{00}(t)f,g\right)_{\mathbf{H}_{0,n}}}{t-(\lambda\pm i\varepsilon)}dt \\ &+\sum_{j=1}^N\int_{\lambda_j^{cr}}^{d_j}\frac{\chi(t)\left(\mathcal{A}_{0j}(t)f,g\right)_{\mathbf{H}_{0,n}}}{t-(\lambda\pm i\varepsilon)}dt,\end{aligned} \tag{2.18}$$

where $d_0 = 4c_+^2 n, d_j = \lambda_j(4(n-1)),\ j = 1, \dots, N,$

$$\begin{aligned}(\mathcal{A}_{00}(t)f, g)_{\mathbf{H}_0,n} &= \int_{\Gamma_0(t)} \frac{\check{f}_0(\xi)\overline{\check{g}_0(\xi)}d\sigma_0(t)}{|(\nabla \varkappa_0)(\xi)|}, \\ \Gamma_0(t) &= \{\xi \in \Omega_0 : \varkappa_0(\xi) = t\},\end{aligned} \tag{2.19}$$

$d\sigma_0(t)$ is the Euclidean measure of the $(n-1)$-dimensional surface $\Gamma_0(t)$,

$$\begin{aligned}(\mathcal{A}_{0j}(t)f, g)_{\mathbf{H}_0,n} &= \int_{\Gamma_j(t)} \frac{\check{f}_j(\xi')\overline{\check{g}_j(\xi')}d\sigma_j(t)}{|(\nabla \varkappa_j)(\xi')|}, \\ \Gamma_j(t) &= \{\xi' \in \Omega_j : \varkappa_j(\xi') = t\},\end{aligned} \tag{2.20}$$

and $d\sigma_j(t)$ is the Euclidean measure of the $(n-2)$-dimensional surface $\Gamma_j(t)$.

The integrals in (2.18) are Cauchy type integrals and, if their densities are Hölder functions, then there exist the limits in (2.18) for $\varepsilon \to 0$ and these limits are Hölder functions, see for instance [14]. Hence the statement of Theorem 1 follows from the next proposition. □

Proposition 6. *Let $f, g \in l_\delta^2(\mathbb{Z}_+^n), \delta > 1/2$. Then*

(i)

$$\sup_t \left|(\chi(t)\mathcal{A}_{0j}(t)f, g)_{\mathbf{H}_0,n}\right| \le C\|f\|_{l_\delta^2(\mathbb{Z}_+^n)} \|g\|_{l_\delta^2(\mathbb{Z}_+^n)}, \quad j = 0, 1, \dots, N \tag{2.21}$$

(ii) *there exists a constant $C > 0$ such that for arbitrary $t', t'' \in [a, b]$*

$$\begin{aligned} &\left|((\chi(t')\mathcal{A}_{0j}(t') - \chi(t'')\mathcal{A}_{0j}(t''))\, f, g)_{\mathbf{H}_0,n}\right| \\ \le\ & C|t' - t''|^\gamma \|f\|_{l_\delta^2(\mathbb{Z}_+^n)} \|g\|_{l_\delta^2(\mathbb{Z}_+^n)}, \\ j\ =\ & 0, 1, \dots, N, 0 < \gamma < \min\{1/2, \delta - 1/2\}. \end{aligned} \tag{2.22}$$

Proof. 1) First we consider the case $j = 0$. It follows from the condition: $f \in l_\delta^2(\mathbb{Z}_+^n), \delta > 1/2$ and Proposition 5 (i) that the function $\check{f}_0$ has a restriction $\gamma_{k\tau}^0(\check{f}_0)$ on the plane $\{\xi \in \Omega_0, \xi_k = \tau\}$ and the following estimates hold:

$$\sup_{\tau \in [-\pi,\pi]} \left\|\gamma_{k\tau}^0(\check{f}_0)\right\|_{L^2([-\pi,\pi]^{n-2}\times[0,\pi])} \le C\|f\|_{l_\delta^2(\mathbb{Z}_+^n)}, \quad k = 1, \dots, n-1, \tag{2.23}$$

$$\sup_{\tau \in [0,\pi]} \left\|\gamma_{n\tau}^0(\check{f}_0)\right\|_{L^2([-\pi,\pi]^{n-1})} \le C\|f\|_{l_\delta^2(\mathbb{Z}_+^n)}. \tag{2.24}$$

The first one (2.21) follows from (2.23), (2.24) by means of a local rectification of the surface $\Gamma_0(t)$.

Let us prove (2.22). Denote by ψ a function in $C_0^\infty(\mathbb{R})$ with *supp* $\psi = [\varepsilon, \pi - \varepsilon]$, where $\varepsilon > 0$ is small. Applying Proposition 5 (i), (v) we obtain the estimate

$$\left\|\psi(\xi_n)\left(\gamma_{k\tau'}^0(\check{f}_0) - \gamma_{k\tau''}^0(\check{f}_0)\right)\right\|_{L^2([-\pi,\pi]^{n-2}\times[0,\pi])} \le C\|f\|_{l_\delta^2(\mathbb{Z}_+^n)} |\tau' - \tau''|^\gamma, \tag{2.25}$$

$$0 < \gamma \le \min\{1, \delta - 1/2\}, \tau', \tau'' \in (-\pi, \pi), \quad j = 1, \dots, n-1.$$

The same relation holds with respect to ξ_n

$$\left\|\psi(\tau')\gamma^0_{n\tau'}\left(\check{f}_0\right)-\psi(\tau'')\gamma^0_{n\tau''}\left(\check{f}_0\right)\right\|_{L^2([-\pi,\pi]^{n-1})}\leq C\left\|f\right\|_{l^2_\delta(\mathbb{Z}^n_+)}\left|\tau'-\tau''\right|^\gamma; \quad (2.26)$$

$$\tau',\tau''\in(0,\pi), 0<\gamma\leq\min\left\{1,\delta-1/2\right\}.$$

Let

$$\left(\mathcal{A}^\psi_{00}(t)f,g\right)_{\mathbf{H}_{0,n}}=\int_{\Gamma_0(t)}\frac{\psi(\xi_n)\check{f}_0(\xi)\overline{\check{g}_0(\xi)}d\sigma_0(t)}{\left|(\nabla\varkappa_0)(\xi)\right|}.$$

Then estimate (2.22) for $\left(\mathcal{A}^\psi_{00}(t)f,g\right)_{\mathbf{H}_{0,n}}$, $\operatorname{supp}\psi\subset[\varepsilon,\pi-\varepsilon]$ follows from (2.23)–(2.26).

Let us establish the Hölder conditions of $\left(\mathcal{A}^\psi_{00}(t)f,g\right)_{\mathbf{H}_{0,n}}$ if the support of ψ is situated in a small neighborhood of 0 or π.

We set

$$\begin{aligned}\varkappa_0(\xi) &= c_+^2\mu(\xi)=t,\tilde{\mu}(\xi')=\rho,\xi_n=2\arcsin\frac{1}{2}\left(c_+^{-2}t-\rho\right)^{1/2}, \quad (2.27)\\ F(\xi',t,\rho) &= (2\pi)^{-\frac{n-1}{2}}\left(\hat{f}(\xi',\cdot),\varphi_0(\cdot,2\arcsin\frac{1}{2}\left(c_+^{-2}t-\rho\right)^{1/2},\rho)\right)_{\mathbf{H}_{0,1}},\\ t &\geq c_+\rho,\rho\geq 0,\\ G(\xi',t,\rho) &= (2\pi)^{-\frac{n-1}{2}}\left(\check{g}(\xi',\cdot),\varphi_0(\cdot,2\arcsin\frac{1}{2}\left(c_+^{-2}t-\rho\right)^{1/2},\rho)\right)_{\mathbf{H}_{0,1}},\\ t &\geq c_+\rho,\rho\geq 0.\end{aligned}$$

Note that if t is not a critical point of $\varkappa_0(\xi)$, then ρ is not a critical point of $\tilde{\mu}(\xi')$ for $\rho\in[c_+^{-2}t-\varepsilon,c_+^{-2}t]$,where $\varepsilon>0$ is small enough. Then the surface $S_\rho=\{\xi':\tilde{\mu}(\xi')=\rho\}$ for such ρ is smooth and $\nabla\tilde{\mu}(\xi')\neq 0,\xi'\in S_\rho$. A straight application of Proposition 5 gives the following estimates

$$\left\|F(\cdot,t,\rho)\right\|_{L^2(S_\rho)}\leq C\left\|f\right\|_{l^2_\delta(\mathbb{Z}^n_+)},\delta>1/2. \quad (2.28)$$

$$\begin{aligned}&\left\|F(\cdot,t',\rho)-F(\cdot,t'',\rho)\right\|_{L^2(S_\rho)} \quad (2.29)\\ \leq\ & C\left\|f\right\|_{l^2_\delta(\mathbb{Z}^n_+)}\left|t'-t''\right|^\gamma\left(\left|c_+^{-2}t'-\rho\right|^{-\gamma}-\left|c_+^{-2}t''-\rho\right|^{-\gamma}\right)\end{aligned}$$

where $t',t''\in\left\{t:0<c_+^{-2}t-\rho<\varepsilon\right\},0<\gamma<\min\left\{1,\delta-1/2\right\}$ with a constant C depending on $\varepsilon>0$.

Fix now a function $\psi\in C_0^\infty(\mathbb{R})$ with support in a small neighborhood of 0. Then

$$\begin{aligned}\left(\mathcal{A}^\psi_{00}(t)f,g\right)_{\mathbf{H}_{0,n}} &= \int_{\Gamma_0(t)}\frac{\psi(\xi_n)\check{f}_0(\xi)\overline{\check{g}_0(\xi)}d\sigma_0(t)}{\left|(\nabla\varkappa_0)(\xi)\right|}\\ &= \int_{c_+^{-2}t-\varepsilon}^{c_+^{-2}t}d\rho\int_{S_\rho}\frac{\psi(\xi_n)F(\xi',t,\rho)\bar{G}(\xi',t,\rho)d\sigma_\rho}{\left|\nabla\varkappa_0(\xi)\right|\left|\nabla\tilde{\mu}(\xi')\right|}, \quad (2.30)\end{aligned}$$

where $d\sigma_\rho$ is the Euclidean measure on S_ρ. We can choose $\varepsilon>0$ small enough such that $\nabla\tilde{\mu}(\xi')\neq 0$ if $\left\{\xi'\in S_\rho:\rho\in\left[c_+^{-2}t-\varepsilon,c_+^{-2}t\right]\right\}$.

Applying formula (2.29) we obtain estimate (2.22) for $\left(\mathcal{A}_{00}^{\psi}(t)f,g\right)_{\mathbf{H}_0,n}$ with $\operatorname{supp}\psi\subset[0,\varepsilon]$ from (2.30).

In the same way we obtain estimate (2.22) for $\left(\mathcal{A}_{00}^{\psi}(t)f,g\right)_{\mathbf{H}_0,n}$ with $\operatorname{supp}\psi\subset[\pi-\varepsilon,\pi)$.

2) Let $j\geq 1$. We denote by $\gamma_{k\tau}^{j}$ the operator of restriction on the plane $\{\xi\in\Omega_j,\xi_k=\tau\}$. The following inequality

$$\left\|\gamma_{k\tau}^{j}\left(\check{f}_j\right)\right\|_{L^2([-\pi,\pi]^{n-2})}\leq C\left\|f\right\|_{l_\delta^2(\mathbb{Z}_+^n)},\delta>1/2,\quad j=1,\dots,n-1 \tag{2.31}$$

is a consequence of Proposition 5 (ii), (iv), and it implies estimate (2.21).

Let

$$F_j(\xi')=\left(\hat{f}(\xi',\cdot)\right),\psi_j(\cdot,\tilde{\mu}(\xi')))_{\mathbf{H}_{0,1}},G_j(\xi')=(\hat{g}(\xi',\cdot),\psi_j(\cdot,\tilde{\mu}(\xi')))_{\mathbf{H}_{0,1}}.$$

Then the property (iv) of Proposition 5 implies that

$$\left\|\gamma_{k\tau'}^{j}\left(F_j\right)-\gamma_{k\tau''}^{j}\left(F_j\right)\right\|_{L^2([-\pi,\pi]^{n-2})}\leq C\left|\tau'-\tau''\right|^{\gamma}\left\|f\right\|_{l_\delta^2(\mathbb{Z}_+^n)},$$

$$0<\gamma\leq\min\left\{1,\delta-1/2\right\},\tau',\tau''\in(-\pi,\pi),\quad k=1,\dots,n-1$$

and the same estimates hold for G_j.

We have $\check{f}_j(\xi')=\sqrt{\beta_j(\tilde{\mu}(\xi'))}F_j(\xi')$, where $\beta_j\in C^\infty([\mu_j,b])$ and $\beta_j(\mu_j)=0,\beta_j'(\mu_j)\neq 0$. It yields that β_j satisfies the Hölder condition with the exponent $1/2$, and

$$\left\|\gamma_{k\tau'}^{j}(\check{f}_j)-\gamma_{k\tau''}^{j}(\check{f}_j)\right\|_{L^2([-\pi,\pi]^{n-2})}\leq C\left|\tau'-\tau''\right|^{\gamma}\left\|f\right\|_{l_\delta^2(\mathbb{Z}_+^n)},$$

$$0<\gamma\leq\min\left\{1/2,\delta-1/2\right\}.\tau',\tau''\in\left[-\pi,\pi\right],k=1,\dots,n-1.$$

Similar relations hold for $\check{g}$.

The final result (2.22) follows from these estimates by means of a passage to the local coordinates on the surface $\Gamma_j(t)$. □

2.5.2. Proof of Theorem 3

Proof. Applying the Plemeli-Sockhotski formulas (see [14]) for Cauchy type integrals in (2.18) we obtain that

$$\Im(R_0^{\pm}(\lambda)f,f)_{\mathbf{H}_{0,n}}=\pm\pi\sum_{j=0}^{N}(\mathcal{A}_{0j}(\lambda)f,f)_{\mathbf{H}_{0,n}}. \tag{2.32}$$

It follows from (2.32) that $\Im(R_0^{+}(\lambda)f,f)_{\mathbf{H}_{0,n}}=0$ if and only if $\Im(R_0^{-}(\lambda)f,f)_{\mathbf{H}_{0,n}}=0$. Let $f\in l_\delta^2(\mathbb{Z}_+^n)$ where $\delta>3/2$. Then we can prove as above that $(\mathcal{A}_{0j}(t)f,f)_{\mathbf{H}_{0,n}}$

is a differentiable function, and

$$\left(\mathcal{A}_{0j}'(t)f,f\right)_{\mathbf{H}_0,n}\leq C\left\|f\right\|_{l_\delta^2(\mathbb{Z}_+^n)}^2,\quad j=0,1,\ldots,N, \tag{2.33}$$

$$\begin{aligned}&\left(\left(\mathcal{A}_{0j}'(t')-\mathcal{A}_{0j}'(t'')\right)f,f\right)_{\mathbf{H}_0,n} \\ \leq\;& C\left|t'-t''\right|^{\gamma}\left\|f\right\|_{l_\delta^2(\mathbb{Z}_+^n)}^2,\quad 0<\gamma\leq\min\left\{1,\delta-3/2\right\}.\end{aligned} \tag{2.34}$$

Let $(\mathcal{A}_{0j}(\lambda)f,f)_{\mathbf{H}_{0,n}}=0$. Then, because $(\mathcal{A}_{0j}(t)f,f)_{\mathbf{H}_{0,n}}\geq 0$ for all t, we obtain that $\left(\mathcal{A}_{0j}'(\lambda)f,f\right)_{\mathbf{H}_{0,n}}=0$. It follows from (2.33), (2.34) that

$$(\mathcal{A}_{0j}(t)-\mathcal{A}_{0j}(\lambda)f,f)_{\mathbf{H}_{0,n}}\leq C\left|t-\lambda\right|^{1+\gamma}. \tag{2.35}$$

Further, by the Parseval equality

$$\left\|\chi(\mathcal{A}_0)R_0^{\pm}(\lambda)f\right\|_{\mathbf{H}_{0,n}}^2=\sum_{j=0}^{N}\int_{\lambda_j^{cr}}^{d_j}\frac{\chi^2(t)\left(\mathcal{A}_{0j}(t)f,f\right)_{\mathbf{H}_{0,n}}dt}{\left|t-\lambda\right|^2},\quad\lambda\in[a,b]\,,\lambda_0^{cr}=0.$$

Applying (2.35) we obtain that

$$\left\|\chi(\mathcal{A}_0)R_0^{\pm}(\lambda)f\right\|_{\mathbf{H}_{0,n}}\leq C\left\|f\right\|_{l_\delta^2(\mathbb{Z}_+^n)},\quad\delta>3/2,$$

and hence

$$\left\|R_0^{\pm}(\lambda)f\right\|_{\mathbf{H}_{0,n}}\leq C\left\|f\right\|_{l_\delta^2(\mathbb{Z}_+^n)},\quad\delta>3/2.$$

□

In conclusion of this chapter we will give some results without proof, which are standard corollaries of Theorems 1 and 3, see for instance [21], p. 43.

Proposition 7. *Let $E_0(\lambda)$ be a canonical spectral measure of $\mathcal{A}_0$. Then $E_0(\lambda)$ is weakly differentiable on the set E. If $f\in l_\delta^2(\mathbb{Z}_+^n),\delta>1/2$ then*

$$\frac{d(E_0(\lambda)f,f)_{\mathbf{H}_{0,n}}}{d\lambda}=\frac{1}{\pi}\Im(R_0^{+}(\lambda)f,f)_{\mathbf{H}_{0,n}}=\sum_{j=0}^{N}\int_{\Gamma_j(\lambda)}\frac{\left|\check{f}_j\right|^2d\sigma_j(\lambda)}{\left|\varkappa_j\right|}.$$

We define the operator $F_0(\lambda)$ for functions $f\in l_\delta^2(\mathbb{Z}_+^n)$, $\delta>1/2$, by the formula

$$F_0(\lambda)f=\left(\check{f}_j|_{\Gamma_j(\lambda)}\right)_{j=0}^{N}.$$

Proposition 8. *The operator*

$$F_0(\lambda):l_\delta^2(\mathbb{Z}_+^n)\to\hat{H}_0(\lambda)=\oplus_{j=0}^{M(\lambda)}L^2(\Gamma_j(\lambda),\left|\varkappa_j\right|^{-1}d\sigma_j(\lambda))$$

is bounded if $\delta>1/2$.

Proposition 9. *The space $\hat{H}_0=\oplus_{j=0}^{N}L^2(\Omega_j)$ is a direct integral*

$$\hat{H}_0=\int_{[0,4c_+^2n]}^{\oplus}\hat{H}_0(\lambda).$$

The operator F_0 can be represented in the form

$$F_0 = \int_{[0,4c_+^2 n]}^{\oplus} F_0(\lambda) d\lambda,$$

and

$$\|F_0 f\|_{\hat{H}_0}^2 = \int_{[0,4c_+^2 n]}^{\oplus} \|F_0(\lambda) f\|_{\hat{H}_0(\lambda)}^2 \, d\lambda, \; f \in l_\delta^2(\mathbb{Z}_+^n), \delta > 1/2.$$

3. Limiting absorption principle for a perturbed difference acoustic operator

We consider a difference acoustic operator of the form

$$Au(x) = c^2(x)\rho(x)\nabla^* \cdot \rho^{-1}(x)\nabla u_+(x), x \in \mathbb{Z}_+^n, \tag{3.1}$$

with the boundary conditions

$$u(x', 0) = 0, x' \in \mathbb{Z}^{n-1}. \tag{3.2}$$

We suppose that c and ρ are bounded real-valued functions on $\mathbb{Z}_+^n$ such that

$$\inf_{\mathbb{Z}_+^n} \rho(x) > 0, \inf_{\mathbb{Z}_+^n} c(x) > 0$$

and

$$\begin{aligned} |\rho(x) - \rho_0(x_n)| &\leq C(1 + |x|)^{-\delta_0}, \\ |c(x) - c_0(x_n)| &\leq C(1 + |x|)^{-\delta_0}, \delta_0 > 0. \end{aligned}$$

We denote by $\mathbf{H}$ the subspace in $l^2(\mathbb{Z}_+^n, c^{-2}\rho^{-1})$ of functions satisfying the boundary condition $u(x', 0) = 0, x' \in \mathbb{Z}^{n-1}$. Further we denote by $\mathcal{A}$ the operator generated by A and boundary condition (3.2). It is easy to check that $\mathcal{A}$ is a self-adjoint operator on $\mathbf{H}$.

Note that the operators $\mathcal{A} - \varsigma I$ and $\mathcal{A}_0 - \varsigma I$, $\varsigma \in \mathbb{C}$ are connected by the following relation

$$\mathcal{A} - \varsigma I = \Phi \left(\mathcal{A}_0 - \varsigma I\right) + V(\varsigma) \tag{3.3}$$

where $\Phi = \frac{c^2 \rho}{c_0^2 \rho_0}$ and

$$V(\varsigma)u_+(x) = c^2 \rho \nabla^* \cdot \left(\rho^{-1} - \rho_0^{-1}\right) \nabla u_+(x) + \varsigma(\Phi - 1)u_+(x), \quad x \in \mathbb{Z}_+^n.$$

It is easy to see that $V(\varsigma)$ is a bounded operator from $l_\delta^2(\mathbb{Z}_+^n)$ into $l_{\delta+\delta_0}^2(\mathbb{Z}_+^n)$ and, since $l_{\delta+\delta_0}^2(\mathbb{Z}_+^n)$ is embedded compactly in $l_\delta^2(\mathbb{Z}_+^n)$, the operator $V(\varsigma) : l_\delta^2(\mathbb{Z}_+^n) \to l_\delta^2(\mathbb{Z}_+^n)$ is a compact operator. This implies the following result.

Theorem 10. *The essential spectra of $\mathcal{A}$ and $\mathcal{A}_0$ coincide.*

Formula (3.3) yields the following relation between the resolvent $R_0(\varsigma)$ and $R(\varsigma)$ of the operators $\mathcal{A}_0$ and $\mathcal{A}$

$$R_0(\varsigma) = R(\varsigma)Q(\varsigma), R(\varsigma) = R_0(\varsigma)Q_0(\varsigma), \tag{3.4}$$

where

$$Q_0(\varsigma) = \Phi + V(\varsigma)R_0(\varsigma), Q(\varsigma) = \Phi^{-1}(I - V(\varsigma)R(\varsigma)). \tag{3.5}$$

It easy to check that

$$Q_0(\varsigma)Q(\varsigma) = Q(\varsigma)Q_0(\varsigma) = I. \tag{3.6}$$

In formulas (3.4)–(3.6) the spaces $\mathbf{H}_0$ and $\mathbf{H}$ are identified as spaces with the same elements.

Definition 11. *Let $\lambda \in \mathcal{E}$. We denote by $D_{+,\delta}(\lambda)$, $\delta > 1/2$ the space of functions defined on $\mathbb{Z}_+^n$ represented in the form*

$$u_+ = R_0^+(\lambda)v, \quad v \in l_\delta^2(\mathbb{Z}_+^n)$$

with norm

$$\|u_+\|_{D_{+,\delta}(\lambda)} = \left\|R_0^+(\lambda)v\right\|_{l_\delta^2(\mathbb{Z}_+^n)}.$$

Theorem 12. *Let $\lambda \in \mathcal{E}$, $\delta \in (1/2, \delta_0/2), \delta_0 > 1$. Then:*

(i) *the operator $\mathcal{A} - \lambda I : l_\delta^2(\mathbb{Z}_+^n) \to D_{+,\delta}(\lambda)$ is a Fredholm operator of index zero,*

(ii) *the operator $\mathcal{A}$ has at least a countable set $\sigma_p(\mathcal{A}) \subset \mathcal{E}$ of eigenvalues. The limit points of the set $\sigma_p(\mathcal{A})$ are contained in the set Λ only,*

(iii) *the limiting absorption principle holds on the set $\mathcal{E}\backslash\sigma_p(\mathcal{A})$, that is in the space $\mathcal{B}(l_\delta^2(\mathbb{Z}_+^n), l_{-\delta}^2(\mathbb{Z}_+^n))$ there exists the limit*

$$R^\pm(\lambda) = \lim_{\varepsilon\to+0} R(\lambda \pm i\varepsilon), \quad \lambda \in \mathcal{E}\backslash\sigma_p(\mathbf{A}),$$

and $R^\pm(\lambda)$ is a Hölder function on $\mathcal{E}\backslash\sigma_p(\mathcal{A})$,

(iv) *the singular spectrum of the operator $\mathcal{A}$ has no points in $\mathcal{E}\backslash\sigma_p(\mathcal{A})$.*

Proof. (i) Relation (3.3) implies that

$$\mathcal{A} - \lambda I = \Phi\left(\mathcal{A}_0 - \lambda I\right) + V(\lambda), \quad \lambda \in \mathcal{E}.$$

Note that, if $\lambda \in \mathcal{E}$, the operator $\mathcal{A}_0 - \lambda I : D_{+,\delta}(\lambda) \to l_\delta^2(\mathbb{Z}_+^n)$ is invertible with inverse $R^+(\lambda)$ and the operator $V(\lambda)$ is compact in the spaces $l_\delta^2(\mathbb{Z}_+^n)$ and $D_{+,\delta}(\lambda), \delta \in (1/2, \delta_0/2)$.

Thus $\mathcal{A} - \lambda I$ is a sum of an invertible operator and a compact operator that implies the statement (i).

(ii) We will show that a solution $u_+ \in D_{+,\delta}(\lambda)$ of the equation

$$(\mathcal{A} - \lambda I)\, u_+ = 0, \quad \lambda \in \mathcal{E}, \tag{3.7}$$

in fact, belongs to $l^2(\mathbb{Z}_+^n)$.

Indeed, equation (3.7) is equivalent to the equation

$$(I + K_+(\lambda))\, v = 0, \quad v \in l_\delta^2(\mathbb{Z}_+^n) \tag{3.8}$$

where $K_+(\lambda) = \Phi^{-1}V(\lambda)R_0^+(\lambda)$, $u_+ = R_0^+(\lambda)v$. It follows from (3.8) that

$$\left(R_0^+(\lambda)v, v\right)_{\mathbf{H}_{0,n}} = -\left(R_0^+(\lambda)v, \Phi^{-1}V(\lambda)R_0^+(\lambda)v\right)_{\mathbf{H}_{0,n}}.$$

The operator $\Phi^{-1}V(\lambda)$ is self-adjoint on $\mathbf{H}_{0,n}$ hence

$$\Im\left(R_0^+(\lambda)v, v\right)_{\mathbf{H}_{0,n}} = 0. \tag{3.9}$$

We denote by $\mathring{l}^2_\delta(\mathbb{Z}^n_+), \delta > 3/2$ the subspace in $l^2_\delta(\mathbb{Z}^n_+)$ consisting of the functions for which (3.9) holds. Let

$$K_+^\varepsilon(\lambda) = (1+|x|)^\varepsilon K_+(\lambda), \quad \varepsilon = \delta_0 - \delta.$$

By Theorem 1, $K_+^\varepsilon(\lambda) \in \mathcal{B}(\mathring{l}^2_\delta(\mathbb{Z}^n_+), l^2_\delta(\mathbb{Z}^n_+))$, $\delta > 1/2$, and by Theorem 3, $R_0^+(\lambda)v \in l^2\left(\mathbb{Z}^n_+\right)$ if $v \in l^2_\delta(\mathbb{Z}^n_+)$, $\delta > 1/2$ and

$$\sup_{\lambda\in[a,b]\subset\mathcal{E}} \left\|R_0^\pm(\lambda)f\right\|_{l^2(\mathbb{Z}^n_+)} \leq C\left\|f\right\|_{l^2_{\delta+1}(\mathbb{Z}^n_+)}.$$

It implies that $K_+^\varepsilon(\lambda) \in \mathcal{B}(\mathring{l}^2_{\delta+1}(\mathbb{Z}^n_+), l^2_{\delta+1}(\mathbb{Z}^n_+))$, $\delta > 1/2$. The Riesz-Thorin interpolation theorem [6] yields

$$K_+^\varepsilon(\lambda) \in \mathcal{B}(\mathring{l}^2_{\delta+\theta}(\mathbb{Z}^n_+), l^2_{\delta+\theta}(\mathbb{Z}^n_+)),\ \delta > 1/2,\ 0 \leq \theta \leq 1.$$

Hence $K_+(\lambda)$ acts from $\mathring{l}^2_\gamma(\mathbb{Z}^n_+)$ in $l^2_{\gamma+\varepsilon}(\mathbb{Z}^n_+)$, $\gamma > 1/2$. So we have proved that each solution of equation (3.8) belongs to $\mathring{l}^2_\delta(\mathbb{Z}^n_+)$, $\delta > 1/2$. But it follows from equation (3.8) that $v \in \mathring{l}^2_{\delta+\varepsilon}(\mathbb{Z}^n_+)$. By iteration of these arguments we obtain that $v \in \mathring{l}^2_{\delta+1}(\mathbb{Z}^n_+)$, $\delta > 1/2$. Moreover the estimate

$$\|v\|_{l^2_{\delta+1}(\mathbb{Z}^n_+)} \leq C\|v\|_{l^2_\delta(\mathbb{Z}^n_+)}, \quad \delta > 1/2 \tag{3.10}$$

is uniform on each segment $[a,b] \subset \mathcal{E}$.

Hence by Theorem 3 the solution $u_+ = R_0^+(\lambda)v$ of equation (3.7) is such that $v \in l^2_{\delta+1}(\mathbb{Z}^n_+)$, $\delta > 1/2$, and

$$\left\|R_0^+(\lambda)v\right\|_{l^2(\mathbb{Z}^n_+)} \leq C\|v\|_{l^2_{\delta+1}(\mathbb{Z}^n_+)}, \tag{3.11}$$

and estimate (3.11) is uniform on $[a,b] \subset \mathcal{E}$.

Let $\lambda_j \in \mathcal{E}$ be a sequence of eigenvalues of $\mathcal{A}$ converging to a point $\lambda \in \mathcal{E}$ and $\lambda_j, \lambda \in [a,b] \subset \mathcal{E}$, and let u_{+j} be the corresponding sequence of eigenfunctions. Equation (3.7) is equivalent to the equation

$$u_+ + R_0^+(\lambda)\Phi^{-1}V(\lambda)u_+ = 0. \tag{3.12}$$

Let $u_{+j} = R_0^+(\lambda)v_j$ with norms $\|u_{+j}\|_{l^2_{-\delta}(\mathbb{Z}^n_+)} = 1$. Then $v_j = -\Phi^{-1}V(\lambda)u_{+j}$. The function $v_j \in l^2_{\delta+1}(\mathbb{Z}^n_+), \delta > 1/2$ and

$$\Im(R_0^+(\lambda)v_j, v_j)_{\mathbf{H}_{0,n}} = 0. \tag{3.13}$$

It follows from (3.10) and (3.13) that

$$\|v_j\|_{l^2_{\delta+1}(\mathbb{Z}^n_+)} \leq C, \quad \delta > 1/2.$$

Applying (3.11) we obtain that $\|u_{+j}\|_{l^2(\mathbb{Z}_+^n)} \leq C$. The compactness of the embedding $l^2(\mathbb{Z}_+^n)$ in $l^2_{-\delta}(\mathbb{Z}_+^n)$ implies that the sequence u_{+j} has a precompact subsequence in $l^2_{-\delta}(\mathbb{Z}_+^n)$. Let the sequence u_{+j} be such a subsequence. Then the sequence v_j is precompact in $l^2_{\delta+1}(\mathbb{Z}_+^n)$. Hence by estimate (3.11) the sequence u_{+j} is precompact in $\mathbf{H}$. Since $\mathcal{A}$ is a self-adjoint operator in $\mathbf{H}$ the sequence u_{+j} is orthogonal, hence $\lim_{j\to\infty} \|u_{+j}\|_{\mathbf{H}} = 0$. But

$$1 = | u_{+j}\|_{l^2_{-\delta}(\mathbb{Z}_+^n)} \leq C \|u_{+j}\|_{\mathbf{H}} \to 0.$$

This contradiction proves statement (ii).

(iii) The proof of (iii) is based on the next proposition, which can be shown following the proof of Theorem 14.5.2 in [8]. □

Proposition 13. *Let $\lambda \in \mathcal{E}$, and $v_1, \dots, v_r$ be a base of the space of solutions of the equation*

$$(I + K_+(\lambda))\, v = 0, v \in l^2_\delta(\mathbb{Z}_+^n).$$

Then the system of functions $u_{+j} = R_0^+(\lambda)v_j$ is a base of the space of solutions in $D_{+,\delta}(\lambda)$ of the equation

$$(\mathcal{A} - \lambda I)\, u_+ = 0, u_+ \in D_{+,\delta}(\lambda).$$

The equation

$$(\mathcal{A} - \lambda I)\, u_+ = f,\ f \in l^2_\delta(\mathbb{Z}_+^n)$$

is solvable in $D_{+,\delta}(\lambda)$ if and only if

$$(f, u_{+j})_{\mathbf{H}} = 0, j = 1, \dots, r.$$

Proof. We will continue the proof of (iii). Proposition 13 implies that if $\lambda \in \mathcal{E}\backslash\sigma_p(\mathcal{A})$, then $I + K_+(\lambda)$ is invertible on $l^2_\delta(\mathbb{Z}_+^n)$. Since $K_+(\lambda) = \lim_{\varepsilon\to 0} K(\lambda + i\varepsilon)$ in the space $\mathcal{B}(l^2_\delta(\mathbb{Z}_+^n), l^2_\delta(\mathbb{Z}_+^n))$ then

$$(I + K_+(\lambda))^{-1} = \lim_{\varepsilon\to 0} (I + K(\lambda + i\varepsilon))^{-1}. \tag{3.14}$$

It follows from (3.4), (3.5), (3.14) that for sufficiently small $\varepsilon > 0$

$$R(\lambda + i\varepsilon) = R_0(\lambda + i\varepsilon)(I + K(\lambda + i\varepsilon))^{-1}\Phi^{-1}.$$

Applying Theorem 1 we obtain that there exists the limit

$$\lim_{\varepsilon\to 0} R(\lambda + i\varepsilon) = R(\lambda + i0) = R_0^+(\lambda)(I + K(\lambda))^{-1}\Phi^{-1}$$

in the topology of $\mathcal{B}\left(l^2_\delta(\mathbb{Z}_+^n), l^2_{-\delta}(\mathbb{Z}_+^n)\right), 1/2 < \delta < \delta_0/2$. It follows from Theorem 1 that $R(\lambda + i0)$ is a Hölder operator-function on $\mathcal{E}\backslash\sigma_p(\mathcal{A})$ with values in $\mathcal{B}\left(l^2_\delta(\mathbb{Z}_+^n), l^2_{-\delta}(\mathbb{Z}_+^n)\right)$.

(iv) By Theorem XIII.20 in [16] the operator $\mathcal{A}$ does not have points of the singular spectrum in $\mathcal{E}\backslash\sigma_p(\mathcal{A})$. □

The following theorem gives some sufficient conditions for the absence of eigenvalues of $\mathcal{A}$ in $\mathcal{E}$.

Theorem 14. *Let the following conditions be satisfied:*

(i) *there exists a number* $h \in \mathbb{N}$ *such that* $c(x) = c_+, \rho(x) = \rho_+$ *for* $x_n > h$;

(ii)

$$|c(x) - c_0(x_n)| \leq e^{-\beta|x'|}, 0 \leq x_n \leq h, \beta > 0;$$

(iii) *there exists a function* $a \in l^\infty(\mathbb{Z}^n_+)$ *such that*

$$\rho(x) = a(x)\rho_0(x_n), |a(x) - 1| \leq Ce^{-\beta|x'|},\ 0 \leq x_n \leq h, \beta > 0.$$

Then $\sigma_p(\mathcal{A}) \cap \mathcal{E} = \emptyset$.

Proof. The idea of the proof is the following. Let $\lambda \in \mathcal{E}$ and $u_+ \in \mathbf{H}$ be a solution of the equation $(\mathcal{A} - \lambda I)u_+ = 0$. Then estimates (i), (ii) provide that the partial Fourier transform $\hat{u}(\xi', x_n)$ is a real analytical function with respect to $\xi' \in \mathbb{R}^{n-1}$ for each $x_n \in \mathbb{Z}_+$. Let $x_n > h$. Then

$$\hat{u}(\xi', x_n) = C_1(\xi')e^{ix_n\alpha(\xi')} + C_2(\xi')e^{-ix_n\alpha(\xi')},$$

where $\alpha(\xi') = 2\arcsin\left(\frac{1}{2}\sqrt{\lambda - \tilde{\mu}(\xi')c_+}\right), \xi' \in (-\pi, \pi)^{n-1}$. Hence $\hat{u}(\xi', x_n) = 0$ if $x_n > h$ and $\lambda \geq \tilde{\mu}(\xi')c_+$. Since $\hat{u}(\xi', x_n)$ is an analytic function with respect to ξ' for each x_n, we obtain that $\hat{u}(\xi', x_n) = 0$ for all $\xi' \in \mathbb{R}^{n-1}$ if $x_n > h$. Hence $u_+(x', x_n) = 0$ for all $x_n > h$ and $x' \in \mathbb{Z}^{n-1}$.

It is easy to see that the initial conditions

$$u_+(x', h+1) = \hat{u}(x', h+2) = 0$$

for solutions of the homogeneous difference equation

$$(A - \lambda I)\, u_+(x) = c^2(x)\rho(x)\nabla^* \cdot \rho^{-1}(x)\nabla u_+(x) - \lambda u_+(x) = 0, x \in \mathbb{Z}^n_+,$$

imply that $u_+(x) = 0$ for all $x' \in \mathbb{Z}^{n-1}, x_n \leq h$. Thus $u_+(x', x_n) = 0$ for all points $x = (x', x_n) \in \mathbb{Z}^n_+$. □

4. Finite section method

We consider here the question of the applicability of the finite section method (the reduction method) for the solution of the infinite system on $\mathbb{Z}^n_+$

$$\begin{aligned} (A - \lambda I)\, u_+(x) &= c^2(x)\rho(x)\nabla^* \cdot \rho^{-1}(x)\nabla u_+(x) - \lambda u_+(x) = f(x), \quad (4.1) \\ x &\in \mathbb{Z}^n_+;\ u(x', 0) = 0,\ x' \in \mathbb{Z}^{n-1}. \end{aligned}$$

We suppose that all the assumptions of Section 3 for c and ρ are satisfied. Let

$$\begin{aligned} Q_m &= \left\{x \in \mathbb{Z}^n_+ : |x_j| \leq m,\ j = 1, \dots, n\right\}, \\ u_m(x) &= \begin{cases} u_+(x),\ x \in Q_m, \\ 0,\ x \notin Q_m. \end{cases} \end{aligned}$$

Theorem 15. *Let* $1/2 < \delta < \delta_0, \lambda \in \mathcal{E}\backslash\sigma_p(\mathbf{A}),\ f \in l^2_\delta(\mathbb{Z}^n_+)$. *Then, for any* $\varepsilon > 0$ *and* $m \in \mathbb{N}$, *the equation*

$$(A - (\lambda + i\varepsilon) I)\, u_m(x) = f_m(x), x \in Q_m, u_m|_{\partial Q_m} = 0 \tag{4.2}$$

has a unique solution u^ε_m.

If $u_+ = R^+(\lambda)f$ *is a solution of problem* (4.1), *then*

$$u_+ = \lim_{\varepsilon\to 0}\lim_{m\to\infty} u^\varepsilon_m \tag{4.3}$$

in the sense of the convergence in the space $l^2_{-\delta}(\mathbb{Z}^n_+)$.

Proof. Let u^ε_+ be a solution of the problem

$$\begin{aligned} (A - (\lambda + i\varepsilon) I)\, u^\varepsilon_+(x) &= f(x), x \in \mathbb{Z}^n_+, \\ u^\varepsilon_+(x', 0) &= 0, x' \in \mathbb{Z}^{n-1}. \end{aligned} \tag{4.4}$$

where $f \in l^2_\delta(\mathbb{Z}^n_+), 1/2 < \delta < \delta_0, \lambda \in \mathcal{E}\backslash\sigma_p(\mathbf{A})$. By Theorem 12 (iii)

$$\lim_{\varepsilon\to 0} \left\| u_+ - u^\varepsilon_+ \right\|_{l^2_{-\delta}(\mathbb{Z}^n_+)} = 0. \tag{4.5}$$

Since $A - \lambda I$ is a self-adjoint operator on $\mathbf{H}$ we obtain that

$$\Im((A - (\lambda + i\varepsilon))u_+, u_+)_{\mathbf{H}} = \varepsilon \left\| u_+ \right\|^2_{\mathbf{H}}, \varepsilon > 0.$$

Thus $A - (\lambda + i\varepsilon) I$ is an accretive operator on $\mathbf{H}$. It implies that the equation (4.4) has an unique solution in $\mathbf{H}$ for any f. Moreover (see, [9], p. 32), for any fixed $\varepsilon > 0$ and for any $m \in \mathbb{N}$, equation (4.2) has a unique solution u^ε_m and

$$\lim_{m\to\infty} \left\| u^\varepsilon_+ - u^\varepsilon_m \right\|_{l^2(\mathbb{Z}^n_+)} = 0. \tag{4.6}$$

Since $l^2(\mathbb{Z}^n_+) \subset l^2_{-\delta}(\mathbb{Z}^n_+), \delta > 0$, formulas (4.5), (4.6) yield (4.3). □

References

[1] F.V. Atkinson, *Discrete and Continuous Boundary Problems*, Academic Press, Inc., New York, 1964.

[2] L.M. Brekhovskih and O.A. Godin, *The Acoustic of Layered Media*, Nauka, Moskow, 1989; English transl. Springer-Verlag, Berlin, 1990.

[3] M. Ben-Artizi, Y. Dermenjian, J.-C. Gulliot, *Acoustic waves in perturbed stratified fluids: a spectral theory*, Comm. Partial Dif. Eq., **14**, (1989), 479–517.

[4] A. Boutet de Monvel, D. Manda, *Spectral and scattering theory for wave propagation in perturbed stratified media*, J. Math. Appl., **191**, (1995), 137–167.

[5] F.A. Berezin, M.A. Shubin, *The Schrödinger Equation*, Izd. Moscow Univ., Moscow, 1983.

[6] Y. Bergh, J. Löfström, *Interpolation Spaces, An Introduction*, Springer-Verlag, New York, 1976.

[7] Y. Dermenjian, J.-C. Gulliot, *Théorie spectrale de la propagation des ondes acoustiques dans un milieu stratifié perturbé*, J. Diff. Equat., **62**, No.3, (1986), 57–72.

[8] L. Hörmander, *The Analysis of Linear Partial Differential Operators*, Vol. 3, Springer-Verlag, New York, 1985.

[9] R. Hagen, S. Roch, B. Silbermann, *C^*-Algebras and Numerical Analysis*, Marcel Dekker, Inc., New York, 2001.

[10] A. Jirari, *Second Order Sturm-Liouville Difference Equations and Orthogonal Polinomials*, Memories of AMS, No. 542, 1995.

[11] T. Kato, *Perturbation Theory for Linear Operators*, Springer-Verlag, New York, 1976.

[12] M. Dimassi, J. Sjöstrand, *Spectral Asymptotics in the Semi-Classical Limit*, Cambridge Univ. Press, 1999.

[13] V.A. Marchenko, *Spectral Theory of Sturm-Liouville Operators*, Naukova Dumka, Kiev, 1972 (in Russian).

[14] N.I. Muskhelishvili, *Singular Integral Equations.* P. Nordholf, Groningen, 1953.

[15] E.M. Nikishin, *Discrete Sturm-Liouville operator*, Trudi Sem. im. Petrovskogo, **10**, (1984), 3–76.

[16] M. Reed, B. Simon, *Methods of Modern Mathematical Physics*, Vol. IV, Academic Press, New York, 1978.

[17] R. Weder, *Spectral and scattering theory in perturbed stratified fluids*, J. Math. Pures Appl, **64**, (1985), 149–173.

[18] R. Weder, *The limiting absorption principle at thresholds,* J. Math. Pures Appl., **67**, (1988), 313–338.

[19] R. Weder, *Spectral and Scattering Theory for Wave Propagation in Perturbed Stratified Media,* Appl. Math. Sci, Vol. 87, Springer-Verlag, 1991.

[20] C.H. Wilcox, *Sound Propagation in Stratified Fluids*, Appl. Math. Sci., Vol. 50, Springer-Verlag, 1984.

[21] D.R. Yafaev, *Mathematical Theory of Scattering*, St. Petersburg, 1994.

[22] V.Yu. Zavadskii, *The Finite Difference Method in Wave Problems of Acoustics*, Nauka, Moskow, 1982 (in Russian).

Vladimir Rabinovich
Instituto Politecnico Nacional
ESIME-Zacatenco
Depto. de Telecomunicaciones, Edif. 1, 2do piso
Av. IPN, Mexico, D.F., 07738
Mexico
e-mail: rabinov@maya.esimez.ipn.mx

Operator Theory
Advances and Applications, Vol. 147, 423–455

Fredholmness of Convolution Type Operators

Vladimir S. Rabinovich and Steffen Roch

Dedicated to the memory of Erhard Meister

Abstract. We study the Fredholmness on $L^p(D)$ of operators of convolution type. Here D is an unbounded measurable domain in $\mathbb{R}^N$, and an operator A on $L^p(D)$ is of convolution type if it is constituted by operators of the form $aC(k)bI$ where $C(k)$ is the operator of convolution by the $L^1(\mathbb{R}^N)$-function k and where a and b are bounded and uniformly continuous functions. The domains under consideration include, for example, curved layers, curved cylinders, and cones with angular or cuspidal edges. The criterion for the Fredholmness of the operator A is formulated in terms of limit operators of A.

The topic of this paper is the Fredholmness of compressions of operators of convolution type. An operator of convolution type is an operator on $L^p(\mathbb{R}^N)$ which belongs to the smallest norm-closed subalgebra of $L(L^p(\mathbb{R}^N))$ which contains the identity operator, all compact operators, and all operators of the form $aC(k)bI$ where $C(k)$ is the operator of convolution by the $L^1(\mathbb{R}^N)$-function k and where a and b are bounded and uniformly continuous functions on $\mathbb{R}^N$. For an unbounded measurable subset D of $\mathbb{R}^N$, the compression of the operator A onto D is the operator

$$B := \chi_D A \chi_D I|_{L^p(D)} : L^p(D) \to L^p(D) \tag{0.1}$$

where χ_D is the characteristic function of the set D.

There are many papers which are devoted to the Fredholmness of compressions of operators of convolution type for concrete classes of coefficients a, b and concrete domains D; see, e.g., [2, 3, 5, 6, 7, 14]. For example, the multidimensional Wiener-Hopf operators

$$\chi_D(\gamma I + C(k))\chi_D I|_{L^p(D)} : L^p(D) \to L^p(D)$$

where $\gamma \in \mathbb{C}$ and $k \in L^1(\mathbb{R}^N)$ are considered in [3] for D being a half-space and in [14] in case D a cone in $\mathbb{R}^N$ with smooth cross section, whereas the quarter plane case is the topic of [2, 5, 7]. Operators on 3D wedge shaped domains are studied in [6].

2000 *Mathematics Subject Classification.* Primary 45E10; Secondary 45A05.
Key words and phrases. Convolution operators, Fredholmness, limit operators.
Supported by the German Research Foundation (DFG) under Grant Nr. 436 RUS 17/24/01 and by the CONACYT project 32424-E.

We will consider the Fredholm problem for these operators, for example, in case when D is a curved layer, a curved cylinder, a cone with angular or cuspidal edges, or the epigraph of a certain function. In each of these cases, our solution to the Fredholm problem will be as follows. We associate with the operator B in (0.1) its so-called operator spectrum. This is a family of operators on $L^p(\mathbb{R}^N)$ which describes the behavior of the operator B at infinity. Then the main result says that the operator B is Fredholm if and only if each operator in its operator spectrum is invertible, and if the norms of these inverses are uniformly bounded. Moreover, it turns out that in many cases (for example, if the coefficients a, b are slowly oscillating functions) the operators in the operator spectrum are much simpler objects than the operator B itself. This fact allows us to study their invertibility effectively. It will also turn out that the condition of *uniform* invertibility is redundant in many cases.

We will prove these Fredholm criteria by having recourse to the results of [8, 9]. In these papers we considered band-dominated operators on discrete l^p-spaces of sequences with values in a Banach space X. We showed that a band-dominated operator is invertible at infinity (in case X has finite dimension this simply means that the operator is Fredholm, i.e., that its kernel and cokernel have finite dimension) if the operators in its operator spectrum are uniformly invertible. Thus, the proof of the desired Fredholm criteria for compressions of operators of convolution type rests on two basic steps: we prove that a suitable discretization of that operator leads to a band-dominated operator on an discrete l^p-space, and we compute its operator spectrum.

The paper is organized as follows. In its first two sections we recall some basic definitions and facts on operators of convolution type as well as on band-dominated operators. Then we study discretizations of operators of convolution type. In particular, we show that the discretization of every convolution type operator B is band-dominated and that this discretization is invertible at infinity if and only if B is Fredholm. Finally, we specify the general Fredholm criterion to the concrete cases mentioned above where the operator spectrum can be explicitly computed.

Throughout this paper, we let $1 < p < \infty$, $q := p/(p-1)$, and N a positive integer.

1. Operators of convolution

In this section, we collect some basic facts on convolution operators on L^p-spaces. Theorem 1.2 goes back to [15], and the compactness of commutators of operators of multiplication by slowly oscillating function with convolution operators has been verified in [1]. Our presentation follows [11] where the results mentioned in this section are proved in the more general context of operators on locally compact groups.

Let $k \in L^1(\mathbb{R}^N)$ and $u \in L^p(\mathbb{R}^N)$. Then Young's inequality implies that the convolution

$$(k * u)(x) := \int_{\mathbb{R}^N} k(x-y)u(y)dy, \quad x \in \mathbb{R}^N, \tag{1.1}$$

belongs to $L^p(\mathbb{R}^N)$, and that $\|k*u\|_p \le \|k\|_1 \|u\|_p$ ([10], IX.4). Hence, the operator $C(k)u := k * u$ of convolution by $k \in L^1(\mathbb{R}^N)$ acts boundedly on $L^p(\mathbb{R}^N)$, and

$$\|C(k)\|_{L^p(\mathbb{R}^N)} \le \|k\|_1. \tag{1.2}$$

Let $\mathcal{C}_p$ denote the closure in $L(L^p(\mathbb{R}^N))$ of the set of all convolution operators $C(k)$ with kernels $k \in L^1(\mathbb{R}^N)$. Then $\mathcal{C}_p$ is a closed and commutative subalgebra of $L(L^p(\mathbb{R}^N))$ without identity. Its maximal ideal space can be identified with $\mathbb{R}^N$ (with its usual topology) in such a way that the Gelfand transform $\hat{C}$ of $C \in \mathcal{C}_p$ coincides with the Fourier transform of k if $C = C(k)$ (see [14]). Consequently, an operator $\gamma I + C$ in the unitization $\mathbb{C}I + \mathcal{C}_p$ of $\mathcal{C}_p$ is invertible if and only if

$$\inf_{\xi \in \mathbb{R}^N} |\gamma + \hat{C}(\xi)| > 0. \tag{1.3}$$

Note that $\gamma + \hat{C}$ is just the Gelfand transform of $\gamma I + C$ in $\mathbb{C}I + \mathcal{C}_p$.

A *semi-commutator* is an operator of the form $aC(k)$ or $C(k)aI$ where k is in $L^1(\mathbb{R}^N)$ and $a \in L^\infty(\mathbb{R}^N)$. The functions a for which the semi-commutators $aC(k)$ and $C(k)aI$ are compact for every function $k \in L^1(\mathbb{R}^N)$ can be characterized as follows.

Definition 1.1. Let $Q_{SC}(\mathbb{R}^N)$ refer to the set of all functions $a \in L^\infty(\mathbb{R}^N)$ such that

$$\limsup_{t \to \infty} \int_M |a(t+s)|\, ds = 0$$

for every compact subset M of $\mathbb{R}^N$.

For example, the class $Q_{SC}(\mathbb{R}^N)$ contains all functions $a \in L^\infty(\mathbb{R}^N)$ with

$$\lim_{R \to \infty} \text{ess sup}_{|x| \ge R} |a(x)| = 0$$

and, in particular, all compactly supported functions. The characteristic function of the set $\cup_{n \ge 2}[n - \frac{1}{n}, n + \frac{1}{n}] \subset \mathbb{R}$ is an example of a function in $Q_{SC}(\mathbb{R}^N)$ which does not vanish at infinity.

Theorem 1.2. *The following conditions are equivalent for a bounded measurable function a:*

(a) *the operators BaI and aB are compact on $L^p(\mathbb{R}^N)$ for every $B \in \mathcal{C}_p$ and every $1 < p < \infty$,*

(b) $a \in Q_{SC}(\mathbb{R}^N)$,

(c) *There is a bounded open set $D \subset \mathbb{R}^N$ such that $\lim_{t \to \infty} \int_D |a(t+s)|\, ds = 0$.*

Consequently, $Q_{SC}(\mathbb{R}^N)$ is a closed ideal in $L^\infty(\mathbb{R}^N)$.

The next goal is to characterize those functions $a \in L^\infty(\mathbb{R}^N)$ for which the *commutators* $aC(k) - C(k)aI$ are compact for every function $k \in L^1(\mathbb{R}^N)$. We start with defining two related subclasses of functions in $L^\infty(\mathbb{R}^N)$.

Definition 1.3. Let $SO(\mathbb{R}^N)$ denote the set of all bounded continuous functions a on $\mathbb{R}^N$ such that, for every compact subset M of $\mathbb{R}^N$,

$$\lim_{t\to\infty} \sup_{h\in M} |a(t) - a(t+h)| = 0.$$

The class $SO(\mathbb{R}^N)$ is a unital commutative C^*-subalgebra of $BUC(\mathbb{R}^N)$, the algebra of the bounded and uniformly continuous functions on $\mathbb{R}^N$. Functions in $SO(\mathbb{R}^N)$ are called *slowly oscillating* on $\mathbb{R}^N$. Examples of slowly oscillating functions are provided by the continuous functions which possess a finite limit at infinity and by the differentiable functions the derivative of which tends to zero at infinity.

Definition 1.4. A function $a \in L^\infty(\mathbb{R}^N)$ belongs to the class $Q_C(\mathbb{R}^N)$ if, for every open and bounded subset M of $\mathbb{R}^N$, the function

$$t \mapsto \int_M (a(t) - a(t+s))\, ds$$

lies in $Q_{SC}(\mathbb{R}^N)$.

The following result does not only solve the commutator problem; it moreover verifies the relation between the classes $Q_{SC}(\mathbb{R}^N)$, $SO(\mathbb{R}^N)$ and $Q_C(\mathbb{R}^N)$.

Theorem 1.5. *The following assertions are equivalent for $a \in L^\infty(\mathbb{R}^N)$:*

(a) *the operators $BaI - aB$ are compact on $L^p(\mathbb{R}^N)$ for every $B \in \mathcal{C}_p$,*

(b) *the function a belongs to $Q_C(\mathbb{R}^N)$,*

(c) *the function a belongs to $Q_{SC}(\mathbb{R}^N) + SO(\mathbb{R}^N)$.*

As a consequence one gets that $Q_C(\mathbb{R}^N) = Q_{SC}(\mathbb{R}^N) + SO(\mathbb{R}^N)$ is a unital commutative C^*-subalgebra of $L^\infty(\mathbb{R}^N)$ and that $Q_{SC}(\mathbb{R}^N)$ is a closed ideal of that algebra. Moreover, one can show that the intersection $Q_{SC}(\mathbb{R}^N) \cap SO(\mathbb{R}^N)$ consists of all continuous functions which tend to zero at infinity.

2. Band-dominated operators on l^p-spaces

Given a complex Banach space X, consider the Banach spaces $l^p(\mathbb{Z}^N, X)$ and $l^\infty(\mathbb{Z}^N, X)$ of all functions $f : \mathbb{Z}^N \to X$ such that

$$\|f\|_p^p := \sum_{x\in\mathbb{Z}^N} \|f(x)\|_X^p < \infty \quad \text{and} \quad \|f\|_\infty := \sup_{x\in\mathbb{Z}^N} \|f(x)\|_X < \infty,$$

respectively. Let E stand for one of the spaces $l^p(\mathbb{Z}^N, X)$ with $p \in (1, \infty)$. Every function $a \in l^\infty(\mathbb{Z}^N, L(X))$ gives rise to a multiplication operator on E on defining

$$(af)(x) := a(x)f(x), \quad x \in \mathbb{Z}^N.$$

We denote this operator by aI. Evidently, $aI \in L(E)$ and $\|aI\|_{L(E)} = \|a\|_\infty$. Finally, for $\alpha \in \mathbb{Z}^N$, let V_α refer to the shift operator

$$(V_\alpha f)(x) := f(x-\alpha), \quad x \in \mathbb{Z}^N,$$

which also belongs to $L(E)$ and has norm 1.

Definition 2.1. A *band operator* on E is a finite sum of the form $\sum_\alpha a_\alpha V_\alpha$ where $\alpha \in \mathbb{Z}^N$ and $a_\alpha \in l^\infty(\mathbb{Z}^N, L(X))$. A *band-dominated operator* on E is the norm limit of a sequence of band operators.

The band-dominated operators on E form a closed subalgebra of $L(E)$ which we denote by $\mathcal{A}_E$. One can show that an operator $A \in L(E)$ is band-dominated if and only if, for every function $\varphi \in BUC(\mathbb{R}^N)$,

$$\lim_{t\to 0} \|A\hat{\varphi}_{t,r}I - \hat{\varphi}_{t,r}A\|_{L(E))} = 0 \quad \text{uniformly with respect to } r \in \mathbb{R}^N$$

where, for r, t, $x \in \mathbb{R}^N$,

$$\varphi_{t,r}(x) := \varphi_t(x-r) \quad \text{and} \quad \varphi_t(x) := \varphi(tx) := \varphi(t_1x_1, \ldots, t_Nx_N)$$

and where $\hat{a}$ refers to the restriction of the function $a : \mathbb{R}^N \to \mathbb{C}$ onto $\mathbb{Z}^N$. For this and the following facts we refer to [8, 9].

For $n \geq 0$, define $\hat{P}_n : E \to E$ by

$$(\hat{P}_n f)(x) = \begin{cases} f(x) & \text{if } |x|_\infty \leq n \\ 0 & \text{if } |x|_\infty > n, \end{cases}$$

set $\hat{Q}_n := I - \hat{P}_n$, and let $\hat{\mathcal{P}}$ refer to the family $(\hat{P}_n)$.

Definition 2.2. An operator $K \in L(E)$ is *$\hat{\mathcal{P}}$-compact* if

$$\|K\hat{Q}_n\| \to 0 \quad \text{and} \quad \|\hat{Q}_nK\| \to 0 \quad \text{as } n \to \infty.$$

By $K(E, \hat{\mathcal{P}})$ we denote the set of all $\hat{\mathcal{P}}$-compact operators on E, and by $L(E, \hat{\mathcal{P}})$ the set of all operators $A \in L(E)$ for which both AK and KA are $\hat{\mathcal{P}}$-compact whenever K is $\hat{\mathcal{P}}$-compact.

It turns out that $L(E, \hat{\mathcal{P}})$ is a closed subalgebra of $L(E)$, $K(E, \hat{\mathcal{P}})$ a closed two-sided ideal of $L(E, \hat{\mathcal{P}})$, and $K(E, \hat{\mathcal{P}}) \subset \mathcal{A}_E \subset L(E, \hat{\mathcal{P}})$. Operators $A \in L(E, \hat{\mathcal{P}})$ for which the coset $A + K(E, \hat{\mathcal{P}})$ is invertible in the quotient algebra $L(E, \hat{\mathcal{P}})/K(E, \hat{\mathcal{P}})$ are called *$\hat{\mathcal{P}}$-Fredholm*. If X is a finite-dimensional space, then $L(E, \hat{\mathcal{P}}) = L(E)$, $K(E, \hat{\mathcal{P}})$ is the ideal of the compact operators on E, and the $\hat{\mathcal{P}}$-Fredholm operators are just the Fredholm operators in the common sense.

Let $\mathcal{H}$ stand for the set of all sequences $h : \mathbb{N} \to \mathbb{Z}^N$ which tend to infinity.

Definition 2.3. Let $A \in L(E, \hat{\mathcal{P}})$ and $h \in \mathcal{H}$. The operator $A_h \in L(E)$ is called the *limit operator of A with respect to h* if

$$\lim_{n\to\infty} \|(V_{-h(n)}AV_{h(n)} - A_h)\hat{P}_m\| = \lim_{n\to\infty} \|\hat{P}_m(V_{-h(n)}AV_{h(n)} - A_h)\| = 0 \quad (2.1)$$

for every $\hat{P}_m \in \hat{\mathcal{P}}$. The set $\sigma_{op}(A)$ of all limit operators of A is called the *operator spectrum* of A.

Let finally refer $\mathcal{L}_E^{rich}$ to the set of all operators $A \in L(E, \mathcal{P})$ enjoying the following property: every sequence $h \in \mathcal{H}$ possesses a subsequence g for which the limit operator A_g exists. Set $\mathcal{A}_E^{rich} := \mathcal{A}_E \cap \mathcal{L}_E^{rich}$. Then the main result of [9] can be stated as follows.

Theorem 2.4. *An operator $A \in \mathcal{A}_E^{rich}$ is $\hat{\mathcal{P}}$-Fredholm if and only if all of its limit operators are invertible and if*

$$\sup\{\|(A_h)^{-1}\| : A_h \in \sigma_{op}(A)\} < \infty. \tag{2.2}$$

The condition of *uniform* invertibility can be weakened by employing local techniques. To describe some typical ideas and results we have to introduce some more notations. Let S^{N-1} denote the unit sphere $\{\eta \in \mathbb{R}^N : |\eta| = 1\}$ where $|\eta|$ stands for the Euclidean norm of η. Given a 'radius' $R > 0$, a 'direction' $\eta \in S^{N-1}$, and a neighborhood $U \subseteq S^{N-1}$ of η, we set

$$W_{R,U} := \{z \in \mathbb{Z}^N : |z| > R \text{ and } z/|z| \in U,\} \tag{2.3}$$

and we call $W_{R,U}$ a *neighborhood at infinity of* η. A sequence $h \in \mathcal{H}$ is said to *tend into the direction of* $\eta \in S^{N-1}$ if, for every neighborhood at infinity W of η, there is an m_0 such that $h(m) \in W$ for all $m \geq m_0$.

Definition 2.5. Let $\eta \in S^{N-1}$ and $A \in L(E, \hat{\mathcal{P}})$.

(a) The *local operator spectrum* $\sigma_\eta(A)$ *of* A *at* η is the set of all limit operators A_h of A with respect to sequences h tending into the direction of η.

(b) The operator A is *locally invertible at* η if there are operators $B, C \in L(E, \hat{\mathcal{P}})$ and a neighborhood at infinity W of η such that

$$BA\hat{\chi}_W I = \hat{\chi}_W AC = \hat{\chi}_W I$$

where $\hat{\chi}_W$ refers to the characteristic function of W.

Theorem 2.6. *Let $A \in \mathcal{A}_E^{rich}$ and $\eta \in S^{N-1}$. Then the operator A is locally invertible at η if and only if all limit operators in $\sigma_\eta(A)$ are invertible and if*

$$\sup\{\|(A_h)^{-1}\| : A_h \in \sigma_\eta(A)\} < \infty.$$

Corollary 2.7. *An operator $A \in \mathcal{A}_E^{rich}$ is $\hat{\mathcal{P}}$-Fredholm if and only if all of its limit operators are invertible, and if*

$$\sup\{\|(A_h)^{-1}\| : A_h \in \sigma_\eta(A)\} < \infty \quad \textit{for all } \eta \in S^{N-1}.$$

This result is indeed a generalization of Theorem 2.4: It does not require that the suprema are uniformly bounded with respect to η.

3. Band-dominated operators on $L^p(\mathbb{R}^N)$ and their discretizations

$\mathcal{P}$-Fredholmness. We start with adapting the notion of $\hat{\mathcal{P}}$-Fredholmness introduced in the previous section to the context of L^p-spaces.

Let P_n stand for the operator of multiplication by the characteristic function of the cube $[-n, n]^N$ acting on $L^p(\mathbb{R}^N)$, and set $\mathcal{P} := (P_n)_{n=1}^\infty$ and $Q_n := I - P_n$.

Further we introduce the set $K(L^p(\mathbb{R}^N), \mathcal{P})$ of the $\mathcal{P}$-compact operators, i.e., of the operators $K \in L(L^p(\mathbb{R}^N))$ such that

$$\lim_{n\to\infty} \|KQ_n\| = \lim_{n\to\infty} \|Q_nK\| = 0,$$

and the set $L(L^p(\mathbb{R}^N), \mathcal{P})$ of all operators $A \in L(L^p(\mathbb{R}^N))$ such that AK and KA are $\mathcal{P}$-compact whenever K is $\mathcal{P}$-compact. Then $L(L^p(\mathbb{R}^N), \mathcal{P})$ is a closed unital subalgebra of $L(L^p(\mathbb{R}^N))$ which contains $K(L^p(\mathbb{R}^N), \mathcal{P})$ as its closed ideal. Further, since both the operators P_n and their adjoints converge strongly to the identity operators on $L^p(\mathbb{R}^N)$ and $L^q(\mathbb{R}^N)$, respectively, one gets that $K(L^p(\mathbb{R}^N), \mathcal{P})$ contains the ideal $K(L^p(\mathbb{R}^N))$ of the compact operators on $L^p(\mathbb{R}^N)$ (but the ideal $K(L^p(\mathbb{R}^N), \mathcal{P})$ is strictly larger than $K(L^p(\mathbb{R}^N))$ since the operators P_n are not compact).

Our earlier definitions of generalized Fredholmness, invertibility at infinity and local invertibility at infinity specify as follows to the present context.

Definition 3.1. The operator $A \in L(L^p(\mathbb{R}^N), \mathcal{P})$ is $\mathcal{P}$-*Fredholm* if the coset $A + K(L^p(\mathbb{R}^N), \mathcal{P})$ is invertible in the quotient algebra $L(L^p(\mathbb{R}^N), \mathcal{P})/K(L^p(\mathbb{R}^N), \mathcal{P})$, that is if there exist operators $B, C \in L(L^p(\mathbb{R}^N), \mathcal{P})$ such that

$$BA - I \in K(L^p(\mathbb{R}^N), \mathcal{P}) \quad \text{and} \quad AC - I \in K(L^p(\mathbb{R}^N), \mathcal{P}). \tag{3.1}$$

Equivalently, an operator $A \in L(L^p(\mathbb{R}^N), \mathcal{P})$ is $\mathcal{P}$-Fredholm if and only if it is *invertible at infinity* in the sense that there exist an $m \in \mathbb{N}$ and operators $B, C \in L(L^p(\mathbb{R}^N), \mathcal{P})$ such that

$$BAQ_m = Q_m \quad \text{and} \quad Q_mAC = Q_m.$$

Local invertibility. There is also an adequate notion of local invertibility at an infinitely distant point $\eta \in S^{N-1}$. Given $R > 0$ and a neighborhood $U \subseteq S^{N-1}$ of η, we set

$$V_{R,U} := \{x \in \mathbb{R}^N : |x| > R \text{ and } x/|x| \in U\}$$

and call $V_{R,U}$ again a *neighborhood at infinity of* η. Then an operator A is called *locally invertible at* η if there exist a neighborhood V at infinity of η and operators $B, C \in L(L^p(\mathbb{R}^N), \mathcal{P})$ such that

$$BA\chi_V I = \chi_V I \quad \text{and} \quad \chi_V AC = \chi_V I.$$

Shifts and limit operators. For $\alpha \in \mathbb{Z}^N$, we consider the operator

$$U_\alpha : L^p(\mathbb{R}^N) \to L^p(\mathbb{R}^N), \quad (U_\alpha f)(t) := f(t - \alpha)$$

of shift by α. In accordance with the definitions from Section 2, we call the operator A_h a *limit operator of* $A \in L(L^p(\mathbb{R}^N), \mathcal{P})$ with respect to the sequence $h \in \mathcal{H}$ if

$$\lim_{m\to\infty} \|(U_{-h(m)}AU_{h(m)} - A_h)P_m\| = \lim_{m\to\infty} \|P_m(U_{-h(m)}AU_{h(m)} - A_h)\| = 0$$

for every $P_m \in \mathcal{P}$. The set $\sigma_{op}(A)$ of all limit operators of A is the *operator spectrum* of A. Further we denote by $\mathcal{L}_p^{rich}$ the subalgebra of $L(L^p(\mathbb{R}^N), \mathcal{P})$ which consists of all operators with rich operator spectrum. The latter means for an operator

$A \in L(L^p(\mathbb{R}^N), \mathcal{P})$, that every sequence $h \in \mathcal{H}$ has a subsequence g such that the limit operator A_g with respect to g exists.

Discretization. Let χ_0 denote the characteristic function of the cube $I_0 := [0, 1)^N$, and set $X := L^p(I_0)$ and $E := l^p(\mathbb{Z}^N, X)$. Then the mapping G which maps the function $f \in L^p(\mathbb{R}^N)$ to the sequence

$$Gf = ((Gf)_\alpha)_{\alpha \in \mathbb{Z}^N} \quad \text{with} \quad (Gf)_\alpha := \chi_0 U_{-\alpha} f \tag{3.2}$$

is an isometry from $L^p(\mathbb{R}^N)$ onto $l^p(\mathbb{Z}^N, X)$, the inverse of which is given by

$$G^{-1} : u = (u_\alpha)_{\alpha \in \mathbb{Z}^N} \mapsto \sum_{\alpha \in \mathbb{Z}^N} U_\alpha u_\alpha \chi_0 \tag{3.3}$$

where the series converges in the norm in $L^p(\mathbb{R}^N)$. Thus, the mapping

$$\Gamma : L(L^p(\mathbb{R}^N)) \to L(l^p(\mathbb{Z}^N, X)), \quad A \mapsto GAG^{-1}$$

is an isometric algebra isomorphism. Obviously, $\Gamma(P_m)$ is the projection $\hat{P}_m$, and $\Gamma(U_\alpha)$ is the shift V_α, both introduced in Section 2.

Proposition 3.2. *The isometry Γ maps the ideal $K(L^p(\mathbb{R}^N), \mathcal{P})$ onto $K(E, \hat{\mathcal{P}})$ and the algebra $L(L^p(\mathbb{R}^N), \mathcal{P})$ onto $L(E, \hat{\mathcal{P}})$.*

Proof. Since

$$\|K - KP_n\| = \|\Gamma(K - KP_n)\| = \|\Gamma(K) - \Gamma(K)\hat{P}_n\|$$

and $\|K - P_n K\| = \|\Gamma(K) - \hat{P}_n \Gamma(K)\|$, we get $\Gamma(K(L^p(\mathbb{R}^N), \mathcal{P})) = K(E, \hat{\mathcal{P}})$. Similarly, the second assertion follows if one takes into account that an operator $A \in L(L^p(\mathbb{R}^N))$ belongs to $L(L^p(\mathbb{R}^N), \mathcal{P})$ if and only if

$$\|P_k A Q_n\| \to 0 \quad \text{and} \quad \|Q_n A P_k\| \to 0 \quad \text{as } n \to \infty$$

and that an analogous result holds for operators on E. □

A consequence is that an operator $A \in L(L^p(\mathbb{R}^N), \mathcal{P})$ is $\mathcal{P}$-Fredholm if and only if $\Gamma(A)$ is $\hat{\mathcal{P}}$-Fredholm. A similar result holds for the local invertibility at $\eta \in S^{N-1}$. However, here the situation is a little bit more involved since, if $V \subseteq \mathbb{R}^N$ is a neighborhood at infinity of η, then $\Gamma(\chi_V I) \neq \hat{\chi}_{V \cap \mathbb{Z}^N} I$ in general. Nevertheless, the local invertibility of A at η is equivalent to that of $\Gamma(A)$, which can be seen as follows. Given a neighborhood $V \subseteq \mathbb{R}^N$ at infinity of η, there is evidently a neighborhood $W \subseteq \mathbb{Z}^N$ at infinity of η such that $\Gamma(\chi_V I)\hat{\chi}_W I = \hat{\chi}_W I$. Thus, if $BA\chi_V I = \chi_V I$, then

$$\Gamma(B)\Gamma(A)\Gamma(\chi_V I) = \Gamma(\chi_V I),$$

and after multiplication by $\hat{\chi}_W I$ from the right-hand side we get

$$\Gamma(B)\Gamma(A)\hat{\chi}_W I = \hat{\chi}_W I,$$

whence the local invertibility at η of $\Gamma(A)$. The reverse implication follows similarly.

The next result shows that also the limit operators behave nicely under discretization.

Proposition 3.3. *Let $A \in L(L^p(\mathbb{R}^N), \mathcal{P})$ and $h \in \mathcal{H}$. Then the limit operator A_h of A exists (with respect to $\mathcal{P}$) if and only if the limit operator $(\Gamma(A))_h$ of $\Gamma(A)$ exists (with respect to $\Gamma(\mathcal{P}) = \hat{\mathcal{P}}$), and*

$$\Gamma(A_h) = (\Gamma(A))_h. \tag{3.4}$$

In particular, A belongs to $\mathcal{L}_p^{rich}$ if and only if $\Gamma(A)$ belongs to $\mathcal{L}_E^{rich}$.

Proof. Let the limit operator A_h of A exist, i.e., let

$$\lim_{n\to\infty} \|(U_{-h(n)}AU_{h(m)} - A_h)P_m\| = \lim_{n\to\infty} \|P_m(U_{-h(n)}AU_{h(m)} - A_h)\| = 0$$

for all m. Since $\Gamma(U_\alpha) = V_\alpha$ and $\Gamma(P_m) = \hat{P}_m$, and since Γ is an isometrical algebra isomorphism, we conclude that

$$\lim_{n\to\infty} \|\Gamma((U_{-h(n)}AU_{h(m)} - A_h)P_m)\| = \lim_{n\to\infty} \|(V_{-h(n)}\Gamma(A)V_{h(m)} - \Gamma(A_h))\hat{P}_m\| = 0$$

and, analogously, $\hat{P}_m(\|(V_{-h(n)}\Gamma(A)V_{h(m)} - \Gamma(A_h))\| \to 0$ for every m. Thus, the limit operator of $\Gamma(A)$ with respect to h exists and (3.4) holds. The reverse implication follows analogously. □

In particular, an operator B belongs to the operator spectrum of A if and only if the operator $\Gamma(B)$ belongs to the operator spectrum of $\Gamma(A)$. An analogous relation holds for the local operator spectra.

Band-dominated operators on $L^p(\mathbb{R}^N)$. The following definition is motivated by the characterization of band-dominated operators on $l^p(\mathbb{Z}^N, X)$ mentioned in Section 2.

Definition 3.4. An operator $A \in L(L^p(\mathbb{R}^N))$ is *band-dominated* if, for every function $\varphi \in BUC(\mathbb{R}^N)$,

$$\lim_{t\to 0} \|A\varphi_{t,r}I - \varphi_{t,r}A\|_{L(L^p(\mathbb{R}^N))} = 0 \quad \text{uniformly with respect to } r \in \mathbb{R}^N. \tag{3.5}$$

The set of all band-dominated operators in $L(L^p(\mathbb{R}^N))$ will be denoted by $\mathcal{B}_p$, and we write $\mathcal{B}_p^{rich}$ instead of $\mathcal{B}_p \cap \mathcal{L}_p^{rich}$.

Clearly, $\mathcal{B}_p$ and $\mathcal{B}_p^{rich}$ are closed unital subalgebras of $L(L^p(\mathbb{R}^N))$, and the set $K(L^p(\mathbb{R}^N), \mathcal{P})$ is a closed two-sided ideal of both algebras. The latter can be checked, for example, by means of the following proposition.

Proposition 3.5. *$\Gamma(\mathcal{B}_p)$ coincides with the algebra $\mathcal{A}_E$ of the band-dominated operators on $E = l^p(\mathbb{Z}^N, L^p(I_0))$, and $\Gamma(\mathcal{B}_p^{rich}) = \mathcal{A}_E^{rich}$.*

Proof. If $A \in \mathcal{B}_p$ then, for every function $\varphi \in BUC(\mathbb{R}^N)$,

$$\lim_{t\to 0} \|[A, \varphi_{t,r}I]\|_{L(L^p(\mathbb{R}^N))} = 0$$

(with [., .] referring to the commutator) and, consequently,

$$\lim_{t\to 0} \|[\Gamma(A), \Gamma(\varphi_{t,r}I)]\|_{L(E)} = 0 \tag{3.6}$$

uniformly with respect to $r \in \mathbb{R}^N$. We claim that

$$\lim_{t\to 0} \|\hat{\varphi}_{t,r} I - \Gamma(\varphi_{t,r} I)\|_{L(E)} = 0 \tag{3.7}$$

uniformly with respect to $r \in \mathbb{R}^N$. Indeed,

$$\begin{aligned} \sup_{r\in\mathbb{R}^N} \|(\hat{\varphi}_{t,r} I - \Gamma(\varphi_{t,r} I)\|_{L(E)} &= \sup_{r\in\mathbb{R}^N} \sup_{\alpha\in\mathbb{Z}^N} \sup_{x\in I_0} |\hat{\varphi}_{t,r}(\alpha) - (\Gamma(\varphi_{t,r} I)_\alpha)(x)| \\ &= \sup_{r\in\mathbb{R}^N} \sup_{\alpha\in\mathbb{Z}^N} \sup_{x\in I_0} |\varphi(t(\alpha - r)) - \varphi(t(x + \alpha - r))| \\ &\le \sup_{\beta\in\mathbb{R}^N} \sup_{x\in I_0} |\varphi(t\beta) - \varphi(t(x+\beta))| \to 0 \end{aligned}$$

as $t \to 0$ due to the uniform continuity of φ. By (3.7) and (3.6),

$$\lim_{t\to 0} \|[\Gamma(A),\, \hat{\varphi}_{t,r} I]\|_{L(E)} = 0$$

uniformly with respect to $r \in \mathbb{R}^N$. Thus, $\Gamma(\mathcal{B}_p) \subseteq \mathcal{A}_E$. The reverse inclusion follows analogously. The second assertion is a consequence of the first one, together with Proposition 3.3. □

As an immediate consequence of Theorems 2.4 and 2.6 and of Propositions 3.3 and 3.5 we finally get the following result.

Theorem 3.6. *Let $A \in \mathcal{B}_p^{rich}$. Then the operator A is*

(a) *locally invertible at point $\eta \in S^{N-1}$ if and only if all limit operators $A_h \in \sigma_\eta(A)$ are uniformly invertible.*

(b) *invertible at infinity if and only if, for every $\eta \in S^{N-1}$, all limit operators $A_h \in \sigma_\eta(A)$ are uniformly invertible.*

4. Fredholmness of convolution type operators

Now we will apply the results of the preceding sections to examine the Fredholm properties of operators on $L^p(\mathbb{R}^N)$ which are constituted by convolution operators with kernels in $L^1(\mathbb{R}^N)$ and by operators of multiplication by functions in suitable subclasses of $L^\infty(\mathbb{R}^N)$.

4.1. Operators of convolution type

Given a subalgebra $\mathcal{E}$ of $L^\infty(\mathbb{R}^N)$, we let $\mathcal{A}(\mathcal{E},\, \mathcal{C}_p)$ denote the smallest closed subalgebra of $L(L^p(\mathbb{R}^N))$ which contains the identity operator, all compact operators, and all operators of the form

$$aKbI \quad \text{where} \quad a,\, b \in \mathcal{E} \text{ and } K \in \mathcal{C}_p, \tag{4.1}$$

and we call the elements of $\mathcal{A}(L^\infty(\mathbb{R}^N),\, \mathcal{C}_p)$ *convolution type operators*. Thus, every convolution type operator can be approximated as closely as desired by operators of the form

$$A := \gamma I + \sum \prod a_{ij} K_{ij} b_{ij} I + T \tag{4.2}$$

where a_{ij}, $b_{ij} \in L^\infty(\mathbb{R}^N)$, $K_{ij} \in \mathcal{C}_p$, $\gamma \in \mathbb{C}$ and T is compact, and where the sum and all products are finite.

Proposition 4.1. $\mathcal{A}(L^\infty(\mathbb{R}^N), \mathcal{C}_p) \subseteq \mathcal{B}_p$.

The proof is based on the following norm estimate which is known as Schur's lemma ([16], Appendix A, Proposition 5.1).

Proposition 4.2. *Let l be a measurable function on $\mathbb{R}^N \times \mathbb{R}^N$ with*

$$M_1 := \sup_{x\in\mathbb{R}^N} \int_{\mathbb{R}^N} |l(x,y)|dy < \infty \quad \text{and} \quad M_2 := \sup_{y\in\mathbb{R}^N} \int_{\mathbb{R}^N} |l(x,y)|dx < \infty.$$

Then the operator

$$(Lu)(x) := \int_{\mathbb{R}^N} l(x,y)u(y)dy, \quad x \in \mathbb{R}^N$$

acts boundedly on $L^p(\mathbb{R}^N)$, and $\|L\|_{L(L^p(\mathbb{R}^N))} \leq M_1^{1/q} M_2^{1/p}$.

Proof of Proposition 4.1. The algebra $\mathcal{B}_p$ contains the ideal $K(L^p(\mathbb{R}^N), \mathcal{P})$ and, hence, the ideal of the compact operators. Clearly, this algebra also contains all operators of multiplication by a bounded measurable function. Thus, and since $\mathcal{B}_p$ is a closed algebra, the result will follow once we have shown that $\mathcal{B}_p$ also contains a dense subset of $\mathcal{C}_p$. Actually, we will check that

$$\lim_{t\to 0} \sup_{h\in\mathbb{R}^N} \|[\varphi_{t,h}I, C(k)]\| = 0 \tag{4.3}$$

for every function $k \in L^1(\mathbb{R}^N)$ with compact support and every $\varphi \in BUC(\mathbb{R}^N)$. For definiteness, let the support of k be contained in a ball with center 0 and radius R. Since

$$([\varphi_{t,h}I, C(k)]u)(x) = \int_{\mathbb{R}^N} (\varphi_{t,h}(x) - \varphi_{t,h}(y))\, k(x-y)u(y)\, dy,$$

Proposition 4.2 implies

$$\begin{aligned} \|[\varphi_{t,h}I, C(k)]\|_{L(L^p)} &\leq \|k\|_1 \sup_{x,\, y\in\mathbb{R}^N: |x-y|\leq R} |\varphi_{t,h}(x) - \varphi_{t,h}(y)| \\ &= \|k\|_1 \sup_{x,\, y\in\mathbb{R}^N: |x-y|\leq R} |\varphi(t(x-h)) - \varphi(t(y-h))|. \end{aligned}$$

For $|x-y| \leq R$, we have

$$|t(x-h) - t(y-h)| \leq |t|R \to 0 \quad \text{as} \quad t \to 0.$$

Since $\varphi \in BUC(\mathbb{R}^N)$, we obtain (4.3). □

A striking property of operators of convolution type is that their $\mathcal{P}$-Fredholmness coincides with common Fredholmness.

Proposition 4.3. *An operator in $\mathcal{A}(L^\infty(\mathbb{R}^N), \mathcal{C}_p)$ is Fredholm if and only if it is $\mathcal{P}$-Fredholm.*

Proof. Let $\mathcal{J}$ refer to the closed ideal of $\mathcal{A}(L^\infty(\mathbb{R}^N), \mathcal{C}_p)$ which contains all operators in $\mathcal{C}_p$ and all compact operators. It is easy to check that, whenever $J \in \mathcal{J}$, the operator JP_k is compact for every k. Indeed, every operator $J \in \mathcal{J}$ can be approximated as closely as desired by a sum of a compact operator T and of products of operators of the form $aKbI$ where a and b are bounded measurable functions and $K \in \mathcal{C}_p$. Then TP_k is compact, and the compactness of $aKbP_k = aKP_kbI$ follows from Theorem 1.2.

Since P_k fails to be compact, we have $I \notin \mathcal{J}$, and the algebra $\mathcal{A}(L^\infty(\mathbb{R}^N), \mathcal{C}_p)$ decomposes into the direct sum $\mathbb{C}I + \mathcal{J}$. In particular, every operator A in this algebra can be uniquely written as $\gamma_A I + K_A$ where $\gamma_A \in \mathbb{C}$ and $K_A \in \mathcal{J}$, and it turns out that the mapping $A \mapsto \gamma_A$ is a continuous algebra homomorphism.

In the next step we will show that

$$\mathcal{J} \cap K(L^p(\mathbb{R}^N), \mathcal{P}) = K(L^p(\mathbb{R}^N)).$$

The inclusion $\supseteq$ follows from the definitions. If, conversely, $J \in \mathcal{J} \cap K(L^p(\mathbb{R}^N), \mathcal{P})$, then JP_k is compact for every k as we have just seen. On the other hand, since $J \in K(L^p(\mathbb{R}^N), \mathcal{P})$, one has $\|J - JP_k\| \to 0$ as $k \to \infty$. Thus, being the norm limit of compact operators, the operator J is compact.

Since $K(L^p(\mathbb{R}^N)) \subseteq K(L^p(\mathbb{R}^N), \mathcal{P})$, it is clear that every Fredholm operator is also $\mathcal{P}$-Fredholm. Let, conversely, $A \in \mathcal{A}(L^\infty(\mathbb{R}^N), \mathcal{C}_p)$ be a $\mathcal{P}$-Fredholm operator. Then there are an operator $L' \in \mathcal{B}_p$ and an operator $T \in K(L^p(\mathbb{R}^N), \mathcal{P})$ such that $L'A = I + T$. We claim that $\gamma_A \neq 0$. Contrary to what we want, assume that $\gamma_A = 0$. Then $A \in \mathcal{J}$. Choose $m > 0$ and $n \in \mathbb{Z}^N$ such that $\|P_m U_{-n} T U_n P_m\| < 1/2$ (which can be done since T can be approximated by an operator of the form $P_k T$ as closely as desired). Then, by Neumann series, the right-hand side of

$$P_m U_{-n} L' A U_n P_m = P_m + P_m U_{-n} T U_n P_m$$

is an invertible operator on the range of P_m, whence

$$P_m = (P_m + P_m U_{-n} T U_n P_m)^{-1} P_m U_{-n} L' A U_n P_m. \tag{4.4}$$

Since $U_n P_m U_{-n}$ is the operator of multiplication by a compactly supported function, the operator $AU_n P_m = A(U_n P_m U_{-n}) U_n$ and, hence, the operator on the right-hand side of (4.4) are compact. But P_m is not compact, and this contradiction proves the claim.

Now write A as $\gamma_A I + K_A$ and set $L := K_A L' + I$. Then

$$LA - \gamma_A I = \gamma_A L'A - AL'A + A - \gamma_A I = (\gamma_A I - A)(L'A - I).$$

Since $L'A - I \in K(L^p(\mathbb{R}^N), \mathcal{P})$ and $\gamma_A I - A = K_A \in \mathcal{J}$, the operator $LA - \gamma_A I$ is compact. Similarly, one shows that $AR - \gamma_A I$ is compact for a certain operator $R \in \mathcal{B}_p$. Hence, and because of $\gamma_A \neq 0$, the operator A is Fredholm. □

Corollary 4.4. $\mathcal{A}(L^\infty(\mathbb{R}^N), \mathcal{C}_p) \cap K(L^p(\mathbb{R}^N), \mathcal{P}) = K(L^p(\mathbb{R}^N))$.

There are operators in $\mathcal{A}(L^\infty(\mathbb{R}^N), \mathcal{C}_p)$ which do not possess a rich operator spectrum. The next result identifies a subalgebra of $\mathcal{A}(L^\infty(\mathbb{R}^N), \mathcal{C}_p) \cap \mathcal{B}_p^{rich}$ which contains sufficiently many interesting operators.

Proposition 4.5. $\mathcal{A}(BUC(\mathbb{R}^N), \mathcal{C}_p) \subseteq \mathcal{B}_p^{rich}$.

Proof. It is easy to see that every compact operator T belongs to $\mathcal{B}_p^{rich}$ and that the limit operator T_h exists with respect to every sequence $h \in \mathcal{H}$ and is equal to zero.

Next, let $a \in BUC(\mathbb{R}^N)$, and let h be a sequence which tends to infinity. The family of all functions $x \mapsto a(x + h(m))$ is bounded and equicontinuous on every compact subset M of $\mathbb{R}^N$. Hence, by the Arzelà-Ascoli theorem, there are a subsequence g of h and a continuous bounded function a_h on $\mathbb{R}^N$ such that, for every compact $M \subset \mathbb{R}^N$,

$$\lim_{m\to\infty} \sup_{x\in M} |a(x + g(m)) - a_h(x)| = 0.$$

Thus, the operators $U_{-g(m)} a U_{g(m)}$ of multiplication by the function $x \mapsto a(x + g(m))$ converge $*$-strongly to the operator of multiplication by the function a_h.

Let A be an operator of the form (4.2), but with a_{ij}, $b_{ij} \in BUC$. As we have just seen, given a sequence h tending to infinity, we can choose a subsequence g of h such that the operators $U_{-g(m)} a_{ij} U_{g(m)}$ and $U_{-g(m)} b_{ij} U_{g(m)}$ converge $*$-strongly to certain multiplication operators $(a_{ij})_h I$ and $(b_{ij})_h I$, respectively. Then

$$\begin{aligned} &U_{-g(m)} A U_{g(m)} P_k \\ &\quad = (\gamma I + \sum \prod (U_{-g(m)} a_{ij} U_{g(m)}) K_{ij} (U_{-g(m)} b_{ij} U_{g(m)}) + U_{-g(m)} T U_{g(m)}) P_k \\ &\quad = \gamma P_k + \sum \prod (U_{-g(m)} a_{ij} U_{g(m)}) K_{ij} P_k (U_{-g(m)} b_{ij} U_{g(m)}) + U_{-g(m)} T U_{g(m)} P_k \end{aligned}$$

converges in the norm to

$$\gamma P_k + \sum \prod (a_{ij})_h K_{ij} P_k (b_{ij})_h) I = (\gamma I + \sum \prod (a_{ij})_h K_{ij} (b_{ij})_h I) P_k$$

for every P_k and that the operators $K_{ij} P_k$ are compact due to Theorem 1.2).

Hence, all operators of the form (4.2) with a_{ij}, $b_{ij} \in BUC$ possess a rich operator spectrum. Since the operators of this form lie densely in $\mathcal{A}(BUC(\mathbb{R}^N), \mathcal{C}_p)$, and since $\mathcal{B}_p^{rich}$ is a closed algebra, this yields the assertion. □

4.2. Fredholmness

Due to Proposition 4.5, the operators in $\mathcal{A}(BUC(\mathbb{R}^N), \mathcal{C}_p)$ are subject to Theorem 3.6. In combination with Proposition 4.3, we obtain the following result.

Theorem 4.6. *Let $A \in \mathcal{A}(BUC(\mathbb{R}^N), \mathcal{C}_p)$. Then A is*

(a) *Fredholm (i.e., locally invertible at infinity) if and only if all limit operators of A are uniformly invertible.*

(b) *locally invertible at the infinitely distant point $\eta \in S^{N-1}$ if and only if all operators in the local operator spectrum $\sigma_\eta(A)$ are uniformly invertible.*

Corollary 4.7. *An operator $A \in \mathcal{A}(BUC(\mathbb{R}^N), \mathcal{C}_p)$ is a Fredholm operator if and only if, for each point $\eta \in S^{N-1}$, all operators in $\sigma_\eta(A)$ are uniformly invertible.*

We are going to specialize these results to operators with coefficients in certain subalgebras of $L^\infty(\mathbb{R}^N)$.

Slowly oscillating coefficients. Since slowly oscillating functions are uniformly continuous, one has $\mathcal{A}(SO(\mathbb{R}^N), \mathcal{C}_p) \subseteq \mathcal{A}(BUC(\mathbb{R}^N), \mathcal{C}_p)$, and Theorem 4.6 and its corollary apply to operators in the algebra $\mathcal{A}(SO(\mathbb{R}^N), \mathcal{C}_p)$. Limit operators of operators in $\mathcal{A}(SO(\mathbb{R}^N), \mathcal{C}_p)$ are of a particularly simple form which allows us to check their invertibility effectively via (1.3).

Proposition 4.8. *Every limit operator of an operator in $\mathcal{A}(SO(\mathbb{R}^N), \mathcal{C}_p)$ lies in $\mathbb{C}I + \mathcal{C}_p$.*

Proof. Every operator in $\mathcal{A}(SO(\mathbb{R}^N), \mathcal{C}_p)$ can be uniformly approximated by operators of the form (4.2) where a_{ij}, $b_{ij} \in SO(\mathbb{R}^N)$. If $K \in \mathcal{C}_p$ then, clearly, the limit operator K_h exists with respect to every sequence $h \in \mathcal{H}$, and $K_h = K$. Further, if T is compact, then the limit operator T_h also exists with respect to every sequence $h \in \mathcal{H}$, and $T_h = 0$. Thus, in view of the proof of Proposition 4.5, it remains to check the following: If $a \in SO(\mathbb{R}^N)$, and if $h \in \mathcal{H}$ is a sequence such that the operators of multiplication $U_{-h(n)}aU_{h(n)}$ converge *-strongly to $a_h I$ as $n \to \infty$, then a_h is a constant function. This can be done as follows. Let $a \in SO(\mathbb{R}^N)$. Then

$$\lim_{k\to\infty} (a(x' + h(k)) - a(x'' + h(k))) = 0$$

for all sequences h tending to infinity and for all x', $x'' \in \mathbb{R}^N$. Hence, if h is a sequence such that the limit operator $(aI)_h$ exists, then $\lim_{k\to\infty} a(x + h_k)$ is independent of $x \in \mathbb{R}^N$. □

Corollary 4.9. *Let A be an operator of the form (4.2) with a_{ij}, $b_{ij} \in SO(\mathbb{R}^N)$. Then A is Fredholm if and only if all limit operators of A are invertible.*

Thus, the uniformity of the invertibility is not required.

Proof. We conclude from the previous proposition that every limit operator of A is a linear combination of the operators $\prod_{j=1}^{n_i} K_{ij}$ with $i = 1, \ldots, n$. Thus, $\sigma_{op}(A)$ lies in a finite-dimensional subspace of $L(L^p(\mathbb{R}^N))$. Then a simple compactness argument yields the assertion. □

Remark. The algebra $\mathcal{A}(Q_C(\mathbb{R}^N), \mathcal{C}_p)$ which is apparently larger than the algebra $\mathcal{A}(SO(\mathbb{R}^N), \mathcal{C}_p)$ actually coincides with the latter algebra. Indeed, by Theorem 1.5, every operator aK with $a \in Q_C(\mathbb{R}^N)$ and $K \in \mathcal{C}_p$ is the sum of an operator a_1K with $a_1 \in SO(\mathbb{R}^N)$ and an operator a_2K with $a_2 \in Q_{SC}(\mathbb{R}^N)$. Since slowly oscillating functions are uniformly continuous and since a_2K is compact (Theorem 1.2), one has $aK \in \mathcal{A}(SO(\mathbb{R}^N), \mathcal{C}_p)$.

Coefficients stabilizing at infinity. Theorem 4.6 and its corollary attain their most simple form for operators with coefficients which stabilize at infinity in the following sense. This class has been introduced in [4] in case $N = 1$.

Definition 4.10. We say that the function $a \in L^\infty(\mathbb{R}^N)$ *stabilizes at infinity* if, for every infinitely distant point $\eta \in S^{N-1}$, there is a constant $y \in \mathbb{C}$ such that, for every $\varepsilon > 0$, there exists a neighborhood $U = U_{\eta,\varepsilon}$ at infinity of η such that

$$\operatorname{mes}\{x \in U_{\eta,\varepsilon} : |a(x) - y| > \varepsilon\} < \varepsilon. \tag{4.5}$$

The class of all functions which stabilize at infinity will be denoted by $L^\infty_{stab}(\mathbb{R}^N)$.

If a stabilizes at infinity and η is an infinitely distant point, then the constant y which satisfies (4.5) is uniquely determined. We denote it by $\hat{a}(\eta)$.

Lemma 4.11. *Let $a \in L^\infty_{stab}(\mathbb{R}^N)$ and $\eta \in S^{N-1}$ be an infinitely distant point. Then $|\hat{a}(\eta)| \le \|a\|_\infty$.*

Proof. Let $\varepsilon > 0$ and choose a neighborhood U of infinity such that

$$\operatorname{mes}\{x \in U : |a(x) - \hat{a}(\eta)| > \varepsilon\} < \varepsilon.$$

Then

$$\operatorname{mes}\{x \in U : |\,|a(x)| - |\hat{a}(\eta)|\,| > \varepsilon\} < \varepsilon.$$

Since the measure of U is infinite, there is a subset $M \subset U$ of measure 1 such that

$$|a(x)| - \varepsilon < |\hat{a}(\eta)| < |a(x)| + \varepsilon \quad \text{for all } x \in M.$$

This yields the assertion. □

Theorem 4.12. *$L^\infty_{stab}(\mathbb{R}^N)$ is a C^*-subalgebra of $Q_C(\mathbb{R}^N)$.*

Proof. First we will show that $L^\infty_{stab}(\mathbb{R}^N)$ is closed in $L^\infty(\mathbb{R}^N)$. Let $a_n \in L^\infty_{stab}(\mathbb{R}^N)$ and $a \in L^\infty(\mathbb{R}^N)$ such that $\lim \|a_n - a\|_\infty = 0$. Fix $\varepsilon > 0$, and choose $n_0 \in \mathbb{N}$ such that

$$\|a_n - a_m\|_\infty < \varepsilon \quad \text{for all } n,\, m \ge n_0.$$

Further, let $U_{\eta,\varepsilon,n}$ be a neighborhood at infinity of η such that

$$\operatorname{mes}\{x \in U_{\eta,\varepsilon,n} : |a_n(x) - \widehat{a_n}(\eta)| > \varepsilon\} < \varepsilon,$$

and set

$$U'_{\eta,\varepsilon,n} := \{x \in U_{\eta,\varepsilon,n} : |a_n(x) - \widehat{a_n}(\eta)| \le \varepsilon\}.$$

Then, for $x \in U'_{\eta,\varepsilon,n} \cap U'_{\eta,\varepsilon,m}$ and $m,\, n > n_0$,

$$|\widehat{a_n}(\eta) - \widehat{a_m}(\eta)| \le |\widehat{a_n}(\eta) - a_n(x)| + |a_n(x) - a_m(x)| + |a_m(x) - \widehat{a_m}(\eta)| \le 3\,\varepsilon.$$

Thus, $(\widehat{a_n}(\eta))_{n\in\mathbb{N}}$ is a Cauchy sequence, and we let $\hat{a}(\eta)$ denote its limit.

Now we fix $n > n_0$ such that

$$\|a_n - a\|_\infty < \varepsilon/3 \quad \text{and} \quad |\widehat{a_n}(\eta) - \hat{a}(\eta)| < \varepsilon/3.$$

The estimate

$$|a_n(x) - \hat{a_n}(\eta)| \ge |a(x) - \hat{a}(\eta)| - |a(x) - a_n(x)| - |a_n(x) - \widehat{a_n}(\eta)|$$

implies that $|a_n(x) - \widehat{a_n}(\eta)| > \varepsilon/3$ whenever $|a(x) - \hat{a}(\eta)| > \varepsilon$. Since a_n stabilizes at infinity, there is a neighborhood $U_{\eta,\,\varepsilon/3,\,n}$ such that

$$\text{mes}\,\{x \in U_{\eta,\,\varepsilon/3,\,n} : |a_n(x) - \widehat{a_n}(\eta)| > \varepsilon/3\} < \varepsilon/3.$$

Thus,

$$\text{mes}\,\{x \in U_{\eta,\,\varepsilon/3,\,n} : |a(x) - \hat{a}(\eta)| > \varepsilon\} < \varepsilon/3 < \varepsilon,$$

whence $a \in L^\infty_{stab}(\mathbb{R}^N)$.

In the next step we show that $L^\infty_{stab}(\mathbb{R}^N)$ is a *-algebra. The symmetry is obvious. Let $a, b \in L^\infty_{stab}(\mathbb{R}^N)$, and let η be an infinitely distant point. We choose neighborhoods at infinity $U_{\eta,\,\varepsilon/2,\,a}$ and $U_{\eta,\,\varepsilon/2,\,b}$ of η such that

$$\text{mes}\,\{x \in U_{\eta,\,\varepsilon/2,\,a} : |a(x) - \hat{a}(\eta)| > \varepsilon/2\} < \varepsilon/2 \tag{4.6}$$

and

$$\text{mes}\,\{x \in U_{\eta,\,\varepsilon/2,\,b} : |b(x) - \hat{b}(\eta)| > \varepsilon/2\} < \varepsilon/2. \tag{4.7}$$

Set $W_\eta := U_{\eta,\,\varepsilon/2,\,a} \cap U_{\eta,\,\varepsilon/2,\,b}$. Then W_η is a neighborhood at infinity of η, and it follows from

$$\begin{aligned}&\{x \in W_\eta : |a(x) + b(x) - \hat{a}(\eta) - \hat{b}(\eta)| > \varepsilon\}\\ &\qquad \subseteq \{x \in W_\eta : |a(x) - \hat{a}(\eta)| > \varepsilon/2\} \cup \{x \in W_\eta : |b(x) - \hat{b}(\eta)| > \varepsilon/2\}\end{aligned}$$

and from (4.6), (4.7) that

$$\text{mes}\,\{x \in W_\eta : |a(x) + b(x) - \hat{a}(\eta) - \hat{b}(\eta)| > \varepsilon\} < \varepsilon.$$

Thus, $a + b \in L^\infty_{stab}(\mathbb{R}^N)$ and

$$\widehat{(a+b)}(\eta) = \hat{a}(\eta) + \hat{b}(\eta) \quad \text{for all } \eta \in S^{N-1}.$$

In order to show that $ab \in L^\infty_{stab}(\mathbb{R}^N)$, too, we can assume that $a, b \neq 0$ (otherwise the assertion is obvious). Choose $m \in \mathbb{N}$ such that $m\|a\|_\infty > 1$ and $m\|b\|_\infty > 1$. Given an infinitely distant point η and $\varepsilon > 0$ a, choose neighborhoods at infinity of η such that

$$\text{mes}\,\{x \in U_{\eta,\,a} : |a(x) - \hat{a}(\eta)| > \varepsilon/(2m\|b\|_\infty)\} < \varepsilon/(2m\|b\|_\infty)$$

and

$$\text{mes}\,\{x \in U_{\eta,\,b} : |b(x) - \hat{b}(\eta)| > \varepsilon/(2m\|a\|_\infty)\} < \varepsilon/(2m\|a\|_\infty).$$

Set $W_\eta := U_{\eta,\,a} \cap U_{\eta,\,b}$. Then W_η is a neighborhood at infinity of η, and

$$\begin{aligned}&\text{mes}\,\{x \in W_\eta : |(ab)(x) - \hat{a}(\eta)\hat{b}(\eta)| > \varepsilon\}\\ &\qquad = \text{mes}\,\{x \in W_\eta : |(a(x) - \hat{a}(\eta))b(x) + \hat{a}(\eta))(b(x) - \hat{b}(\eta))| > \varepsilon\}\\ &\qquad \leq \text{mes}\,\{x \in W_\eta : |a(x) - \hat{a}(\eta)|\,\|b\|_\infty + \|a\|_\infty\,|b(x) - \hat{b}(\eta)| > \varepsilon\}\end{aligned}$$

$$\begin{aligned}
&\leq \operatorname{mes}\{x \in W_\eta : |a(x) - \hat{a}(\eta)|\, \|b\|_\infty > \varepsilon/2\} \\
&\qquad + \operatorname{mes}\{x \in W_\eta : |b(x) - \hat{b}(\eta)|\, \|a\|_\infty > \varepsilon/2\} \\
&\leq \operatorname{mes}\{x \in U_{\eta,a} : |a(x) - \hat{a}(\eta)|\, \|b\|_\infty > \varepsilon/(2m)\} \\
&\qquad + \operatorname{mes}\{x \in U_{\eta,b} : |b(x) - \hat{b}(\eta)|\, \|a\|_\infty > \varepsilon/(2m)\} \\
&\leq \operatorname{mes}\{x \in U_{\eta,a} : |a(x) - \hat{a}(\eta)| > \varepsilon/(2m\|b\|_\infty)\} \\
&\qquad + \operatorname{mes}\{x \in U_{\eta,b} : |b(x) - \hat{b}(\eta)| > \varepsilon/(2m\|a\|_\infty)\} \\
&< \varepsilon/(2m\|b\|_\infty) + \varepsilon/(2m\|a\|_\infty) < \varepsilon/2 + \varepsilon/2 = \varepsilon.
\end{aligned}$$

Consequently, $ab \in L^\infty_{stab}(\mathbb{R}^N)$ and

$$\widehat{(ab)}(\eta) = \hat{a}(\eta)\,\hat{b}(\eta) \quad \text{for all } \eta \in S^{N-1}.$$

It remains to show the inclusion $L^\infty_{stab}(\mathbb{R}^N) \subseteq Q_C(\mathbb{R}^N)$. Thus, if $a \in L^\infty_{stab}(\mathbb{R}^N)$, we have to show that, for every open bounded set $M \subset \mathbb{R}^N$, there is an open bounded set $D \subset \mathbb{R}^N$ such that

$$\lim_{t\to\infty} \int_D \left| \int_M (a(t+h) - a(t+h+s))\, ds \right| dh = 0 \tag{4.8}$$

(Definition 1.4). Let $M \subset \mathbb{R}^N$ be open and bounded, choose D as the open unit disk in $\mathbb{R}^N$, and let $d > 0$ be the radius of a disk with center 0 which contains $M + D$. Let further $\varepsilon > 0$. Then, for every infinitely distant point η, there is a neighborhood at infinity of η such that

$$\operatorname{mes}\{x \in U_{\eta,\varepsilon} : |a(x) - \hat{a}(\eta)| > \varepsilon\} < \varepsilon.$$

Each neighborhood $U_{\eta,\varepsilon}$ is of the form

$$U_{\eta,\varepsilon} = \{y \in \mathbb{R}^N : |y| > R_{\eta,\varepsilon} \text{ and } y/|y| \in W_{\eta,\varepsilon}\}$$

where $R_{\eta,\varepsilon} \geq 0$ and $W_{\eta,\varepsilon} \subseteq S^{N-1}$ is an open neighborhood of η. In particular, $\{W_{\eta,\varepsilon}\}_{\eta \in S^{N-1}}$ is an open cover of the unit sphere, from which we can choose a finite subcover $\{W_{\eta_i,\varepsilon}\}_{i=1}^k$. Set

$$R_0 := \max\{R_{\eta_i,\varepsilon} : i = 1, \ldots, k\} + d.$$

Further, since the function $f : S^{N-1} \to \mathbb{R}^N$,

$$f(x) := \max\{\operatorname{dist}(x, S^{N-1} \setminus W_{\eta_i,\varepsilon}) : i = 1, \ldots, k\},$$

is positive for every x (every x belongs to one of the sets $W_{\eta_i,\varepsilon}$) and continuous on the compact set S^{N-1}, there is a $\delta > 0$ such that $f(x) \geq \delta$ for all $x \in S^{N-1}$. Thus, for every $x \in S^{N-1}$, there is an $i \in \{1, \ldots, k\}$ such that

$$x \in W_{\eta_i,\varepsilon} \quad \text{and} \quad \operatorname{dist}(x, \partial W_{\eta_i,\varepsilon}) \geq \delta.$$

Consequently, there is an $R_1 \geq R_0$ such that, for every $y \in \mathbb{R}^N$ with $|y| \geq R_1$, there is an $i \in \{1, \ldots, k\}$ such that

$$y \in U_{\eta_i,\varepsilon} \quad \text{and} \quad \operatorname{dist}(y, \partial U_{\eta_i,\varepsilon}) \geq d.$$

Let now $t \in \mathbb{R}^N$ with $|t| \geq R_1$. By what we have just seen, there is an $i \in \{1, \ldots, k\}$ such that $t + D$ and $t + M + D$ are contained in $U_{\eta_i,\varepsilon}$. Thus,

$$\text{mes}\,\{x \in t + D : |a(x) - \hat{a}(\eta)| > \varepsilon\} < \varepsilon$$

and

$$\text{mes}\,\{x \in t + D + M : |a(x) - \hat{a}(\eta)| > \varepsilon\} < \varepsilon.$$

This implies

$$\begin{aligned}
&\int_D \left| \int_M (a(t+h) - a(t+h+s))\, ds \right| dh \\
&\leq \int_D \int_M |a(t+h) - \hat{a}(\eta)|\, ds\, dh + \int_D \int_M |a(t+h+s) - \hat{a}(\eta)|\, ds\, dh \\
&\leq \text{mes}\, D \int_{t+M} |a(h) - \hat{a}(\eta)|\, dh + \text{mes}\, D \int_{t+D+M} |a(h) - \hat{a}(\eta)|\, dh \\
&\leq \text{mes}\, D\, (\text{mes}\, M \cdot \varepsilon + 2\varepsilon \|a\|_\infty) + \text{mes}\, D\, (\text{mes}\,(D+M) \cdot \varepsilon + 2\varepsilon\|a\|_\infty) \\
&\leq \varepsilon\, \text{mes}\, D\, (\text{mes}\, M + \text{mes}\,(D+M) + 4\|a\|_\infty),
\end{aligned}$$

whence the assertion (4.8). □

Proposition 4.13. *Let $a \in L^\infty_{stab}(\mathbb{R}^N)$, and let h be a sequence which tends to infinity into the direction of $\eta \in S^{N-1}$. Then*

$$V_{-h(n)} a V_{h(n)} \to \hat{a}(\eta) I \quad \textit{strongly on } L^p(\mathbb{R}^N) \textit{ as } n \to \infty.$$

Proof. Given $\varepsilon > 0$, we find a neighborhood $U_{\eta,\varepsilon}$ at infinity of η such that

$$\text{mes}\,\{x \in U_{\eta,\varepsilon} : |a(x) - \hat{a}(\eta)| > \varepsilon\} < \varepsilon.$$

Let f be a continuous function with compact support. Then

$$\|(V_{-h(n)} a V_{h(n)} - \hat{a}(\eta)) f\|_p = \|(a - \hat{a}(\eta)) V_{h(n)} f\|_p.$$

Clearly, there exists an n_0 such that $\text{supp}\,(V_{h(n)} f) \subset U_{\eta,\varepsilon}$ for all $n \geq n_0$. Thus, if $n \geq n_0$, then

$$\|(V_{-h(n)} a V_{h(n)} - \hat{a}(\eta)) f\|_p \leq \varepsilon \|f\|_p + 2\|a\|_\infty \|u\|_\infty \varepsilon^{1/p}.$$

This proves the strong convergence on a dense subset of $L^p(\mathbb{R}^N)$. Since the operators $V_{-h(n)} a V_{h(n)}$ are uniformly bounded, we get the assertion. □

An obvious consequence of this proposition is that the local operator spectrum $\sigma_\eta(A)$ for operators $A \in \mathcal{A}(L^\infty_{stab}(\mathbb{R}^N), \mathcal{C}_p)$ is a singleton for every infinitely distant point $\eta \in S^{N-1}$, say $\sigma_\eta(A) = \{A_\eta\}$. Moreover, every limit operator A_η belongs to $\mathbb{C}I + \mathcal{C}_p$ since $\mathcal{A}(L^\infty_{stab}(\mathbb{R}^N), \mathcal{C}_p)$ is a subalgebra of $\mathcal{A}(SO(\mathbb{R}^N), \mathcal{C}_p)$, and by Proposition 4.8. Thus, the invertibility of A_η can be effectively checked via (1.3).

Corollary 4.14. *An operator $A \in \mathcal{A}(L^\infty_{stab}(\mathbb{R}^N), \mathcal{C}_p)$ is Fredholm if and only if every limit operator A_η (with $\eta \in S^{N-1}$) of A is invertible.*

5. Compressions of operators of convolution type

In this section we are going to study the Fredholm properties of compressions of operators of convolution type. If A is a linear bounded operator on $L^p(\mathbb{R}^N)$ and D is a measurable subset of $\mathbb{R}^N$, then the *compression of A onto D* is the operator

$$\chi_D A \chi_D I|_{L^p(\mathbb{R}^N)} : L^p(D) \to L^p(D).$$

The archetypical example is the *Wiener-Hopf operator* $W(k)$ on $L^p(\mathbb{R}^+)$ which is the compression of the convolution operator $\gamma I + C(k)$ with $k \in L^1(\mathbb{R})$ onto $\mathbb{R}^+$. Thus,

$$W(k) = \chi_+(\gamma I + C(k))\chi_+ I|_{L^p(\mathbb{R}^+)},$$

where χ_+ refers to the characteristic function of $\mathbb{R}^+$. Clearly, this operator is Fredholm on $L^p(\mathbb{R}^+)$ is and only if the operator $\gamma I + \chi_+ C(k)\chi_+ I$ is Fredholm on $L^p(\mathbb{R})$. Let f be the function with $f(x) = 0$ if $x < 0$, $f(x) = x$ on $[0, 1]$ and $f(x) = 1$ for $x > 1$. Then the function $\chi_+ - f$ has a compact support. Thus, the operator $\chi_+ C(k)\chi_+ I - fC(k)fI$ is compact on $L^p(\mathbb{R})$, and the operator $W(k)$ is Fredholm on $L^p(\mathbb{R}^+)$ if and only if the operator $\gamma I + fC(k)fI$ is Fredholm on $L^p(\mathbb{R})$. The latter operator is subject to Corollary 4.9 which says that this operator (hence, the Wiener-Hopf operator $W(k)$) is Fredholm if and only if the convolution operator $\gamma I + C(k)$ is invertible. This simple reduction is no longer possible for compressions of operators onto more involved sets.

5.1. Compressions of operators in $\mathcal{A}(BUC(\mathbb{R}^N), \mathcal{C}_p)$

Let $1 < p < \infty$, and let D be a measurable subset of $\mathbb{R}^N$ whose associated multiplication operator belongs to $\mathcal{B}_p^{rich}$. Such subsets will be called *rich* .

We will consider compressions of operators in $\mathcal{A}(BUC(\mathbb{R}^N), \mathcal{C}_p)$ onto the rich set D. Clearly, the compression $\chi_D A \chi_D I$ is invertible (Fredholm) on $L^p(D)$ if and only if its extension $(1 - \chi_D)I + \chi_D A \chi_D I$ is an invertible (Fredholm) operator on $L^p(\mathbb{R}^N)$. Each such extension can be considered as an element of the algebra $\mathcal{A}(BUC(\mathbb{R}^N), \chi_D, \mathcal{C}_p)$ which is the smallest closed subalgebra of the algebra $L(L^p(\mathbb{R}^N))$ which contains the algebra $\mathcal{A}(BUC(\mathbb{R}^N), \mathcal{C}_p)$ as well as the multiplication operator $\chi_D I$. As a consequence of Proposition 4.5 we get

$$\mathcal{A}(BUC(\mathbb{R}^N), \chi_D, \mathcal{C}_p) \subseteq \mathcal{B}_p^{rich}, \tag{5.1}$$

and from Proposition 4.3 we conclude that an operator $A \in \mathcal{A}(BUC(\mathbb{R}^N), \chi_D, \mathcal{C}_p)$ is Fredholm if and only if it is $\mathcal{P}$-Fredholm. Thus, Theorem 3.6 (b) implies the following result.

Theorem 5.1. *Let $A \in \mathcal{A}(BUC(\mathbb{R}^N), \mathcal{C}_p)$, and let D be a rich subset of $\mathbb{R}^N$. Then the compression of A onto D is Fredholm on $L^p(D)$ if and only if, for each point $\eta \in S^{N-1}$, all limit operators in $\sigma_\eta((1 - \chi_D)I + \chi_D A \chi_D I)$ are uniformly invertible on $L^p(\mathbb{R}^N)$.*

In the following subsections we will give some examples of unbounded rich domains D for which the limit operators of $\chi_D I$ can be explicitly calculated and for which, thus, explicit criteria for the Fredholmness of the compressions of operators from $\mathcal{A}(BUC(\mathbb{R}^N), \mathcal{C}_p)$ onto D can be derived.

5.2. Compressions to a half-space

Given a non-zero vector $a \in \mathbb{R}^N$, consider the half-space

$$\mathbf{H}(a) := \{x \in \mathbb{R}^N : \langle x, a \rangle > 0\}. \tag{5.2}$$

Let further $h \in \mathcal{H}$ be a sequence which tends to infinity into the direction of $\eta \in S^{N-1}$. We distinguish several cases.

- If $\langle \eta, a \rangle > 0$, then $\langle h(n), a \rangle \to +\infty$, and the limit operator of $\chi_{\mathbf{H}(a)} I$ exists and is equal to the identity operator.
- If $\langle \eta, a \rangle < 0$, then $\langle h(n), a \rangle \to -\infty$, and the limit operator of $\chi_{\mathbf{H}(a)} I$ exists and is equal to the zero operator.
- If $\langle \eta, a \rangle = 0$, then h has a subsequence $g \in \mathcal{H}$ such that either the numbers $\langle g(n), a \rangle$ tend to $+\infty$, or to $-\infty$, or to a finite limit $b_g \in \mathbb{R}$. In each of these cases, the limit operator of $\chi_{\mathbf{H}(a)} I$ with respect to g exists, and it is equal to the identity operator in the first case, to the zero operator in the second case and to the operator of multiplication by the characteristic function of the shifted half-space

$$\mathbf{H}(a, b_g) := \{x \in \mathbb{R}^N : \langle x, a \rangle > -b_g\}$$

  in the third case.

Let $\mathcal{H}_\eta(A)$ stand for the set of all sequences $h \in \mathcal{H}$ which tend to infinity into the direction of $\eta \in S^{N-1}$ and for which the limit operator A_h exists. Further, we denote by $\mathcal{H}_{\eta,\infty}(A)$ and $\mathcal{H}_{\eta,b}(A)$ the set of all sequences $h \in \mathcal{H}_\eta(A)$ such that $\langle h(n), a \rangle \to \infty$ and $\langle h(n), a \rangle \to b \in \mathbb{R}^N$, respectively. Then Theorem 5.1 gives the following result.

Theorem 5.2. *Let $A \in \mathcal{A}(BUC(\mathbb{R}^N), \mathcal{C}_p)$ and $D = \mathbf{H}(a)$ with $a \in \mathbb{R}^N \setminus \{0\}$. Then the compression of A onto D is Fredholm on $L^p(D)$ if and only if the following conditions are satisfied:*

(a) *for each point $\eta \in S^{N-1}$ with $\langle \eta, a \rangle > 0$, the set $\{A_h : h \in \mathcal{H}_\eta(A)\}$ of limit operators of A is uniformly invertible.*

(b) *for each point $\eta \in S^{N-1}$ with $\langle \eta, a \rangle = 0$, the set $\{A_h : h \in \mathcal{H}_{\eta,\infty}(A)\}$ of limit operators of A is uniformly invertible.*

(c) *for each point $\eta \in S^{N-1}$ with $\langle \eta, a \rangle = 0$ and each $b \in \mathbb{R}^N$, the set*

$$\{(1 - \chi_{\mathbf{H}(a,b)})I + \chi_{\mathbf{H}(a,b)} A_h \chi_{\mathbf{H}(a,b)} I : h \in \mathcal{H}_{\eta,b}(A)\}$$

of extended compressions of limit operators of A is uniformly invertible.

5.3. Compressions to curved half-spaces

Let $N > 1$ and $f \in BUC(\mathbb{R}^{N-1})$. We consider the *curved half-space*

$$\mathbf{P}(f) := \{x = (x', x_N) \in \mathbb{R}^{N-1} \times \mathbb{R} : x_N > f(x')\} \subseteq \mathbb{R}^N. \tag{5.3}$$

Let further $h \in \mathcal{H}$ be a sequence which tends to infinity into the direction of $\eta = (\eta', \eta_N) \in S^{N-1} \subset \mathbb{R}^{N-1} \times \mathbb{R}$. Again, we distinguish several cases.

- If $\eta_N > 0$, then the limit operator of $\chi_{\mathbf{P}(f)} I$ exists and is equal to the identity operator.
- If $\eta_N < 0$, then the limit operator of $\chi_{\mathbf{P}(f)} I$ exists and is equal to the zero operator.
- Now let $\eta_N = 0$. Then h has a subsequence $g \in \mathcal{H}$ such that either the numbers $g(n)_N$ tend to $+\infty$, or to $-\infty$, or that the sequence $(g(n)_N)_{n \geq 1}$ is bounded. In the first two cases, the limit operator of $\chi_{\mathbf{P}(f)} I$ with respect to g exists, and it is equal to the identity operator in the first case and to the zero operator in the second case. In the third case, there exists a subsequence k of g, a real number b_k and a function $f_k : \mathbb{R}^{N-1} \to \mathbb{R}$ such that

$$\lim_{n \to \infty} k(n)_N = b_k \quad \text{and} \quad \lim_{n \to \infty} f(x' + k(n)') = f_k(x')$$

in the sense of the uniform convergence on compact subsets of $\mathbb{R}^{N-1}$. In this case, the limit operator of $\chi_{\mathbf{P}(f)} I$ exists, too, and it is equal to the operator of multiplication by the characteristic function of

$$\mathbf{P}(f_k - b_k) = \{x \in \mathbb{R}^N : x_N > f_k(x') - b_k\}$$

in the third case.

Let $\mathcal{H}_\eta(A)$ stand for the set of all sequences $h \in \mathcal{H}$ which tend to infinity into the direction of $\eta \in S^{N-1}$ and for which the limit operator A_h exists. Further, given a real number b and a function $g : \mathbb{R}^{N-1} \to \mathbb{R}$, we denote by $\mathcal{H}_{\eta,\infty}(A)$ and $\mathcal{H}_{\eta,g,b}(A)$ the set of all sequences $h \in \mathcal{H}_\eta(A)$ such that $h(n)_N \to \infty$ and

$$h(n)_N \to b \quad \text{and} \quad f(x' + h(n)') \to g(x')$$

uniformly on compact subsets of $\mathbb{R}^{N-1}$, respectively. If the set $\mathcal{H}_{\eta,g,b}(A)$ is not empty, then we call g a *limit function* with respect to η. Then Theorem 5.1 implies the following result.

Theorem 5.3. *Let $A \in \mathcal{A}(BUC(\mathbb{R}^N), \mathcal{C}_p)$ and $D = \mathbf{P}(f)$ with $f \in BUC(\mathbb{R}^{N-1})$. Then the compression of A onto D is Fredholm on $L^p(D)$ if and only if the following conditions are satisfied:*

(a) *for each point $\eta \in S^{N-1}$ with $\eta_N > 0$, the set $\{A_h : h \in \mathcal{H}_\eta(A)\}$ of limit operators of A is uniformly invertible.*

(b) *for each point $\eta \in S^{N-1}$ with $\eta_N = 0$, the set $\{A_h : h \in \mathcal{H}_{\eta,\infty}(A)\}$ of limit operators of A is uniformly invertible.*

(c) *for each point $\eta \in S^{N-1}$ with $\eta_N = 0$, each $b \in \mathbb{R}$, and each limit function $g : \mathbb{R}^{N-1} \to \mathbb{R}$, the set $\{(1-\chi_{\mathbf{P}(g-b)})I + \chi_{\mathbf{P}(g-b)} A_h \chi_{\mathbf{P}(g-b)} I : h \in \mathcal{H}_{\eta,g,b}(A)\}$ of extended compressions of limit operators of A is uniformly invertible.*

This result gets a particular simple form if $f \in SO(\mathbb{R}^N)$. In the setting of assertion (c) of the theorem, this hypothesis implies that all functions g are constant (their possible values are just the partial limits of $f(x')$ as $x' \to \infty$). Thus, all possible limit domains $\mathcal{P}(g-b)$ are (uncurved) half-spaces.

5.4. Compressions to curved layers

Let again $N > 1$, and let $f_1, f_2 \in BUC(\mathbb{R}^{N-1})$ be such that $f_1(x') < f_2(x')$ for all $x' \in \mathbb{R}^{N-1}$. Then we call the set

$$\mathbf{L}(f_1, f_2) := \{x = (x', x_N) \in \mathbb{R}^{N-1} \times \mathbb{R} : f_1(x') < x_N < f_2(x')\} \tag{5.4}$$

a *curved layer*. Let $h \in \mathcal{H}$ be a sequence which tends to infinity into the direction of $\eta \in S^{N-1}$. If $\eta_N \neq 0$, then the limit operator of $\chi_{\mathbf{L}(f_1, f_2)}I$ with respect to h exists and it is equal to 0. The same happens if $\eta_N = 0$ and the sequence $(h(n)_N)_{n\geq 1}$ tends to $\pm\infty$. Thus, the only non-trivial case is when $\eta_N = 0$ and the sequence $(h(n)_N)_{n\geq 1}$ is bounded. Then, as in the previous subsection, there is a subsequence k of h, a real number b_k as well as functions $f_{1k}, f_{2k} : \mathbb{R}^{N-1} \to \mathbb{R}$ such that the limit operator of $\chi_{\mathbf{L}(f_1, f_2)}I$ with respect to k exists and is equal to $\chi_{\mathbf{L}(f_{1k}-b_k, f_{2k}-b_k)}I$.

Let again $\mathcal{H}_\eta(A)$ stand for the set of all sequences $h \in \mathcal{H}$ which tend to infinity into the direction of $\eta \in S^{N-1}$ and for which the limit operator A_h exists, and denote by $\mathcal{H}_{\eta, g_1, g_2, b}(A)$ the set of all sequences $h \in \mathcal{H}_\eta(A)$ such that

$$h(n)_N \to b \quad \text{and} \quad f_i(x' + h(n)') \to g_i(x') \quad (i = 1, 2)$$

uniformly on compact subsets of $\mathbb{R}^{N-1}$.

Theorem 5.4. *Let $A \in \mathcal{A}(BUC(\mathbb{R}^N), \mathcal{C}_p)$ and $D = \mathbf{L}(f_1, f_2)$ with functions f_1, f_2 in $BUC(\mathbb{R}^{N-1})$ and $f_1 < f_2$. Then the compression of A onto D is Fredholm on $L^p(D)$ if and only if, for each point $\eta \in S^{N-1}$ with $\eta_N = 0$, each $b \in \mathbb{R}$, and all limit functions $g_1, g_2 : \mathbb{R}^{N-1} \to \mathbb{R}$, the set*

$$\{(1 - \chi_{\mathbf{L}(g_1-b, g_2-b)})I + \chi_{\mathbf{L}(g_1-b, g_2-b)}A_h\chi_{\mathbf{L}(g_1-b, g_2-b)}I : h \in \mathcal{H}_{\eta, g_1, g_2, b}(A)\}$$

of extended compressions of limit operators of A is uniformly invertible.

If $f_1, f_2 \in SO(\mathbb{R}^{N-1})$, then the functions f_{1k}, f_{2k} are constant, and $\mathbf{L}(f_{1k} - b_k, f_{2k} - b_k)$ is a usual layer bounded by two parallel planes.

Corollary 5.5. *In addition to the hypothesis from Theorem 5.4, let*

$$\lim_{x'\to\infty} (f_1(x') - f_2(x')) = 0.$$

Then all limit operators of $\chi_{\mathbf{L}(f_1, f_2)}I$ are zero, and the compression of A onto $\mathbf{L}(f_1, f_2)$ is Fredholm on $L^p(\mathbf{L}(f_1, f_2))$.

5.5. Compressions to curved cylinders

Let $N > 1$, $\Omega \subset \mathbb{R}^{N-1}$ be a bounded domain, and $f \in BUC(\mathbb{R})$ a positive function, and consider the *curved cylinder*

$$\mathbf{Z}_\Omega(f) := \{x = (x', x_N) \in \mathbb{R}^{N-1} \times \mathbb{R} : x' \in f(x_N)\Omega\}. \tag{5.5}$$

Let $h \in \mathcal{H}$. If $h(n)' \to \infty$, then the limit operator of $\chi_{\mathbf{Z}_\Omega(f)} I$ with respect to h exists and is equal to the zero operator. Thus, nontrivial limit operators of $\chi_{\mathbf{Z}_\Omega(f)} I$ with respect to h exist only if the sequence $(h(n)')_{n \geq 1}$ is bounded and $h(n)_N \to \pm\infty$. In this case, there is a subsequence k of h, a point $b_k \in \mathbb{Z}^{N-1}$, and a function f_k on $\mathbb{R}$ such that

$$g(n)' \to b_k \quad \text{and} \quad f(x_N + k(n)_N) \to f_k(x_N) \quad \text{as } n \to \infty$$

uniformly on compact subsets of $\mathbb{R}$. Then the limit operator of $\chi_{\mathbf{Z}_\Omega(f)} I$ with respect to the sequence k exists, and it is equal to the operator of multiplication by the characteristic function of the shifted curved cylinder

$$\mathbf{Z}_\Omega(f_k, b_k) := \{x = (x', x_N) \in \mathbb{R}^{N-1} \times \mathbb{R} : x' \in f_k(x_N)\Omega - b_k\}.$$

Let $\mathcal{H}_\eta(A)$ denote the set of all sequences $h \in \mathcal{H}$ which tend to infinity into the direction of $\eta \in S^{N-1}$ and for which the limit operator A_h exists, and write $\mathcal{H}_{\eta, g, b}(A)$ for the set of all sequences $h \in \mathcal{H}_\eta(A)$ such that

$$h(n)_N \to b \in \mathbb{Z}^{N-1} \quad \text{and} \quad f(x_N + h(n)_N) \to g(x_N)$$

uniformly on compact subsets of $\mathbb{R}$.

Theorem 5.6. *Let $A \in \mathcal{A}(BUC(\mathbb{R}^N), \mathcal{C}_p)$ and $D = \mathbf{Z}_\Omega(f)$ with $f \in BUC(\mathbb{R})$. Then the compression of A onto D is Fredholm on $L^p(D)$ if and only if, for each point $\eta \in S^{N-1}$ with $\eta' = 0$, each $b \in \mathbb{Z}^{N-1}$, and all limit functions $g : \mathbb{R} \to \mathbb{R}$, the set*

$$\{(1 - \chi_{\mathbf{Z}_\Omega(g, b)})I + \chi_{\mathbf{Z}_\Omega(g, b)} A_h \chi_{\mathbf{Z}_\Omega(g, b)} I : h \in \mathcal{H}_{\eta, g, b}(A)\}$$

of extended compressions of limit operators of A is uniformly invertible.

If $f \in SO(\mathbb{R})$, then the function f_g is constant and, thus, $\mathbf{Z}_\Omega(f_g, b)$ is a usual straight cylinder.

Corollary 5.7. *In addition to the hypothesis from Theorem 5.6, let the ends of the cylinder be cuspidal, i.e., let*

$$\lim_{x_N \to \pm\infty} f(x_N) = 0.$$

Then all limit operators of $\chi_{\mathbf{Z}_\Omega(f)} I$ are zero, and the compression of A onto $\mathbf{Z}_\Omega(f)$ is Fredholm on $L^p(\mathbf{Z}_\Omega(f))$.

5.6. Compressions to cones with smooth cross section

Let $\Omega \subseteq \mathbb{R}^N$ be an open domain with C^1-boundary $\partial\Omega$ in case $N \geq 2$ or an open interval in $\mathbb{R}^1$. By $\mathbf{C}_\Omega$, we denote the cone in $\mathbb{R}^{N+1}$ generated by Ω,

$$\mathbf{C}_\Omega := \{(y, y_{N+1}) \in \mathbb{R}^N \times [0, \infty) : y \in y_{N+1}\Omega\}. \tag{5.6}$$

Given $x \in \mathbb{R}^N$, let $\eta_x \in S^N$ be the point which lies on the ray in $\mathbb{R}^{N+1}$ starting at the origin and passing through the point $(x, 1)$, i.e.,

$$\eta_x = \frac{(x, 1)}{\sqrt{\|x\|^2 + 1}}.$$

Let $h \in \mathcal{H}$ be a sequence which tends to infinity into the direction of $\eta \in S^N$. Again there are two trivial cases: If η is not of the form η_x with some $x \in \overline{\Omega}$, then the limit operator of $\chi_{\mathbf{C}_\Omega} I$ exists and is equal to the zero operator. If $\eta = \eta_x$ with $x \in \Omega$, then the limit operator of $\chi_{\mathbf{C}_\Omega} I$ exists, too, and is equal to the identity operator.

Let now $x \in \partial\Omega$ and $\eta = \eta_x$. We denote by $T_x\Omega$ the tangential space and by ν_x the interior normal unit vector to $\partial\Omega$ at x. Further, we write $\mathbb{H}_x$ for the closed half-space in $\mathbb{R}^N$ which is bounded by $T_x\Omega$ and for which ν_x is an interior normal unit vector to $\partial\mathbb{H}_x$ at x. Finally, we let $\mathbf{H}_x$ refer to the half-space in $\mathbb{R}^{N+1}$ which is generated by $\mathbb{H}_x$,

$$\mathbf{H}_x := \{(y, y_{N+1}) \in \mathbb{R}^N \times \mathbb{R} : y \in \mathbb{H}_x + (y_{N+1} - 1)x\}.$$

Further, we write the sequence h as

$$h(n) := \alpha_n(\nu_x, 0) + (r_n, 0) + \beta_n(x, 1) \tag{5.7}$$

where $r_n \in T_x\Omega$ and $\alpha_n, \beta_n \in \mathbb{R}$. The following lemma claims the conditions under which the sequence (5.7) tends to infinity in the direction of η_x.

Lemma 5.8. *The sequence h defined by* (5.7) *tends to infinity in the direction of η_x if and only if $\beta_n \to +\infty$ and*

$$\alpha_n/\beta_n \to 0 \quad \text{and} \quad r_n/\beta_n \to 0 \quad \text{as} \quad n \to \infty. \tag{5.8}$$

Proof. The sequence h tends to infinity if and only if

$$|\alpha_n|^2 + \|r_n\|^2 + |\beta_n|^2 \to \infty, \tag{5.9}$$

and then it converges in the direction of η_x if and only if

$$\frac{(\alpha_n\nu_x + r_n + \beta_n x, \beta_n)}{\sqrt{\|\alpha_n\nu_x + r_n + \beta_n x\|^2 + |\beta_n|^2}} \to \frac{(x, 1)}{\sqrt{\|x\|^2 + 1}}. \tag{5.10}$$

The convergence of the last component of (5.10) tells us that $\beta_n > 0$ for all sufficiently large n. Thus, (5.9) implies

$$\frac{(\frac{\alpha_n}{\beta_n}\nu_x + \frac{r_n}{\beta_n} + x, 1)}{\sqrt{\|\frac{\alpha_n}{\beta_n}\nu_x + \frac{r_n}{\beta_n} + x\|^2 + 1}} \to \frac{(x, 1)}{\sqrt{\|x\|^2 + 1}}. \tag{5.11}$$

From the convergence of the last component of (5.11) we conclude that

$$\left\| \frac{\alpha_n}{\beta_n}\nu_x + \frac{r_n}{\beta_n} + x \right\| \to \|x\|.$$

This implies for the first component of (5.11) that

$$\frac{\alpha_n}{\beta_n}\nu_x + \frac{r_n}{\beta_n} + x \to x$$

whence (5.8) since $\nu_x \perp r_n$. Writing (5.9) as

$$\beta_n^2 \left(\left| \frac{\alpha_n}{\beta_n} \right|^2 + \left\| \frac{r_n}{\beta_n} \right\|^2 + 1 \right) \to \infty$$

and taking into account (5.8), we finally get $\beta_n \to +\infty$. The reverse implications can be checked similarly. □

In order to compute the limit operators into the direction of η_x for $x \in \partial\Omega$, we assume for simplicity that $x = 0$ (which can be reached by shifting Ω) and that $T_x\Omega = T_0\Omega = \mathbb{R}^{N-1} \times \{0\}$ (which can be reached by rotating the shifted Ω). Then, since Ω has a C^1-boundary, there is an open neighborhood $U \subseteq \mathbb{R}^{N-1}$ of 0, an open interval $I \subseteq \mathbb{R}$ which contains 0, and a continuously differentiable function $g : U \to I$ such that

$$\partial\Omega \cap (U \times I) = \{(x, g(x)) \in \mathbb{R}^{N-1} \times \mathbb{R} : x \in U\}$$

and

$$\Omega \cap (U \times I) = \{(x, x_N) \in U \times I : x_N > g(x)\}.$$

Thus, if $\beta > 0$, then the part of the boundary of $\beta\Omega$ which lies in $\beta U \times \beta I$ is just the graph of the function

$$\beta U \to \beta I, \quad x \mapsto \beta g(x/\beta).$$

Let h be as in (5.7), and assume that the limit

$$\delta^* := \lim(\beta_n g(r_n/\beta_n) - \alpha_n) \in \mathbb{R} \cup \{\pm\infty\}$$

exists (otherwise we pass to a suitable subsequence of h). Let further $d > 0$ and $K_d^N := [-d, d]^N$, and set $\mathbf{C}_{n,\Omega} := V_{-h(n)}\mathbf{C}_\Omega$. We consider the intersection of the shifted cone $\mathbf{C}_{n,\Omega}$ with $\mathbb{R}^N \times \{0\}$ and identify this intersection with a subset of $\mathbb{R}^N$. Since $(y + r_n)/\beta_n \in U$ for all $y \in K_d^{N-1}$ and for all sufficiently large n, the boundary of $\mathbf{C}_{n,\Omega} \cap (\mathbb{R}^N \times \{0\})$ can be locally described as the graph of the function

$$G_n : K_d^{N-1} \to \mathbb{R}, \quad y \mapsto \beta_n g((y + r_n)/\beta_n) - \alpha_n.$$

Then, for every $y \in K_d^{N-1}$, we have

$$\lim(G_n(y) - \delta^*) = \lim(G_n(y) - G_n(0))$$

with

$$\begin{aligned} |G_n(y) - G_n(0)| &\le \max_{\xi\in[0,\,y]} \|G'(\xi)\| \, \|y-0\| \\ &= \max_{\xi\in[0,\,y]} \|g'((\xi+r_n)/\beta_n)\| \, \|y\|. \end{aligned}$$

Since g is continuously differentiable with $g'(0) = 0$, and since

$$\|(\xi + r_n)/\beta_n\| \le (d + \|r_n\|)/\beta_n \to 0$$

by Lemma 5.8, we conclude that $G_n(y) \to \delta^*$ for every $y \in K_d^{N-1}$. Thus, if $(y,\, y_N) \in K_d^N$, then

$$\chi_{\mathbf{C}_{n,\Omega}\cap(\mathbb{R}^N\times\{0\})}(y) \to \begin{cases} 1 & \text{if} \quad y_N > \delta^* \\ 0 & \text{if} \quad y_N < \delta^* \end{cases}$$

An analogous result holds of the sequence (β_n) is replaced by $(\beta_n + \beta')$ with $\beta' \in [-d,\, d]$. This shows that

$$\chi_{\mathbf{C}_{n,\Omega}}(y) \to \begin{cases} \chi_{\mathbb{R}^{N+1}}(y) = y & \text{if} \quad \delta^* = -\infty \\ \chi_{\mathbf{H}_x+\delta^*(\nu_x,\,0)}(y) & \text{if} \quad \delta^* \in \mathbb{R} \\ \chi_{\emptyset}(y) = 0 & \text{if} \quad \delta^* = +\infty \end{cases}$$

almost everywhere on K_d^{N+1}. By the dominated convergence theorem, this implies that

$$\chi_{\mathbf{C}_{n,\Omega}}\,\chi_{K_d^{N+1}} \to \begin{cases} \chi_{K_d^{N+1}} & \text{if} \quad \delta^* = -\infty \\ \chi_{\mathbf{H}_x+\delta^*(\nu_x,\,0)}\,\chi_{K_d^{N+1}} & \text{if} \quad \delta^* \in \mathbb{R} \\ 0 & \text{if} \quad \delta^* = +\infty \end{cases}$$

with respect to the L^1-norm and, hence, also with respect to every L^p-norm with $1 \le p < \infty$ (the occurring functions take values in $\{-1,\, 0,\, 1\}$ almost everywhere). Since d is arbitrary, this finally yields that

$$\chi_{\mathbf{C}_{n,\Omega}} I \to \begin{cases} I & \text{if} \quad \delta^* = -\infty \\ \chi_{\mathbf{H}_x+\delta^*(\nu_x,\,0)} I & \text{if} \quad \delta^* \in \mathbb{R} \\ 0 & \text{if} \quad \delta^* = +\infty \end{cases}$$

strongly on $L^p(\mathbb{R}^{N+1})$.

Given $A \in \mathcal{A}(BUC(\mathbb{R}^{N+1}),\, \mathcal{C}_p(\mathbb{R}^{N+1}))$, let $\mathcal{H}_\eta(A)$ denote the set of all sequences $h \in \mathcal{H}$ which tend to infinity into the direction of $\eta \in S^N$ and for which the limit operator A_h exists. Further, write $\mathcal{H}_{\eta,-\infty}(A)$ and $\mathcal{H}_{\eta,\delta^*}(A)$ with $\delta^* \in \mathbb{R}$ for the set of all sequences $h \in \mathcal{H}_\eta(A)$ with

$$\lim(\beta_n g(r_n/\beta_n) - \alpha_n) = -\infty \quad \text{and} \quad \lim(\beta_n g(r_n/\beta_n) - \alpha_n) = \delta^*,$$

respectively. Finally, we abbreviate the shifted half-space $\mathbf{H}_x + \delta^*(\nu_x,\, 0)$ to $\mathbf{H}_{x,\delta^*}$. Then Theorem 5.1 has the following consequence.

Theorem 5.9. *Let $A \in \mathcal{A}(BUC(\mathbb{R}^{N+1}),\, \mathcal{C}_p(\mathbb{R}^{N+1}))$ and $D = \mathbf{C}_\Omega$ with $\Omega \in \mathbb{R}^N$ an open domain with C^1 boundary. Then the compression of A onto D is Fredholm on $L^p(D)$ if and only if the following conditions are satisfied:*

(a) *for each point* $x \in \Omega$, *the set* $\{A_h : h \in \mathcal{H}_{\eta_x}(A)\}$ *of limit operators of* A *is uniformly invertible.*

(b) *for each point* $x \in \partial\Omega$, *the set* $\{A_h : h \in \mathcal{H}_{\eta_x, -\infty}(A)\}$ *of limit operators of* A *is uniformly invertible.*

(c) *for each point* $x \in \partial\Omega$, *the set*

$$\{(1 - \chi_{\mathbf{H}_{x,\delta^*}})I + \chi_{\mathbf{H}_{x,\delta^*}} A_h \chi_{\mathbf{H}_{x,\delta^*}} I : h \in \mathcal{H}_{\eta,\delta^*}(A),\ \delta^* \in \mathbb{R}\}$$

of extended compressions of limit operators of A *is uniformly invertible.*

We still mention some special situations in which the conditions of Theorem 5.9 take a very simple form.

Let $A \in \mathcal{A}(SO(\mathbb{R}^{N+1}), \mathcal{C}_p(\mathbb{R}^{N+1}))$. Then all limit operators of A belong to $\mathbb{C}I + \mathcal{C}_p(\mathbb{R}^{N+1})$. In this case, the invertibility of the compressions in condition (c) can be effectively checked. Let, for example, A_h be the operator $\gamma I + C(k)$ with $\gamma \in \mathbb{C}$ and $k \in L^1(\mathbb{R}^{N+1})$. Since $C(k)$ is shift invariant, the corresponding compression (c) is invertible if and only if the operator

$$(1 - \chi_{\mathbf{H}_x})I + \chi_{\mathbf{H}_x}(\gamma I + C(k))\chi_{\mathbf{H}_x} I \tag{5.12}$$

is invertible. Further, given an orthogonal mapping S on $\mathbb{R}^{N+1}$, we write R_S for the rotation operator $(R_S f)(x) = f(Sx)$, and for $k \in L^1(\mathbb{R}^{N+1})$, we let k_S be the function $k_S(x) = k(S^T x)$ with S^T referring to the transposed of S. Then convolution operators are rotation invariant in the sense that

$$C(k)R_S = R_S C(k_S).$$

Thus, if we choose S such that it rotates $\mathbf{H}_x$ to the half-space $\mathbf{H} := \{(x_1, x) \in \mathbb{R} \times \mathbb{R}^N : x_1 \geq 0\}$, then the compression (5.12) is invertible if and only if the operator

$$(1 - \chi_{\mathbf{H}})I + \chi_{\mathbf{H}}(\gamma I + C(k_S))\chi_{\mathbf{H}} I \tag{5.13}$$

is invertible. Finally, the compression (5.13) is invertible if and only if the operator

$$(1 - \chi_{\mathbf{H}})I + (\gamma I + C(k_S))\chi_{\mathbf{H}} I$$

is invertible. This follows easily from the identities $AP+Q = (PAP+Q)(I+QAP)$ and $(I+QAP)^{-1} = I-QAP$ which hold for arbitrary operators A and idempotents P, Q with $P + Q = I$.

The same is true if $C(k)$ is replaced by an arbitrary operator in $\mathcal{C}_p(\mathbb{R}^{N+1})$. For the invertibility of the resulting compressions, one has the following remarkable fact from [14] (Theorem 1.4).

Theorem 5.10. *Let* $B \in \mathbb{C}I + \mathcal{C}_p(\mathbb{R}^{N+1})$. *Then the operator* $(1 - \chi_{\mathbf{H}})I + B\chi_{\mathbf{H}} I$ *is invertible on* $L^p(\mathbb{R}^{N+1})$ *if and only if the operator* B *is invertible on* $L^p(\mathbb{R}^{N+1})$.

Note once more that the invertibility of $B \in \mathbb{C}I + \mathcal{C}_p(\mathbb{R}^{N+1})$ can be effectively checked via (1.3).

With these remarks, we get the following corollaries to Theorem 5.9.

Corollary 5.11. *Let $A \in \mathcal{A}(SO(\mathbb{R}^{N+1}), \mathcal{C}_p(\mathbb{R}^{N+1}))$ and $D = \mathbf{C}_\Omega$ with $\Omega \in \mathbb{R}^N$ an open domain with C^1 boundary. Then the compression of A onto D is Fredholm on $L^p(D)$ if and only if, for each point $x \in \overline{\Omega}$, the set $\{A_h : h \in \mathcal{H}_{\eta_x}(A)\}$ of limit operators of A (= the local operator spectrum at η_x) is uniformly invertible.*

Corollary 5.12. *Let $A \in \mathcal{A}(L^\infty_{stab}(\mathbb{R}^{N+1}), \mathcal{C}_p(\mathbb{R}^{N+1}))$ and $D = \mathbf{C}_\Omega$ with $\Omega \in \mathbb{R}^N$ an open domain with C^1 boundary. Then the compression of A onto D is Fredholm on $L^p(D)$ if and only if, for each point $x \in \overline{\Omega}$, the limit operator A_{η_x} of A is invertible.*

Remark. Let $f : [0, \infty) \to \mathbb{R}$ be a slowly oscillating function, and let $\Omega \in \mathbb{R}^N$ be an open domain with C^1 boundary. We consider the slowly oscillating cone,

$$\mathbf{C}_{\Omega, f} := \{(y, y_{N+1}) \in \mathbb{R}^N \times [0, \infty) : y \in (y_{N+1} + f(y_{N+1}))\Omega\}. \tag{5.14}$$

In a similar way as above, one can show that the limit operators of the multiplication operator $\chi_{\mathbf{C}_{\Omega, f}} I$ are the same as in case of the unperturbed cone $\mathbf{C}_\Omega$, and that the analogue of Theorem 5.9 holds.

5.7. Compressions to cones with edges

Here we are going to consider compressions of convolution operators to cones which are allowed to have a finite number of edges. For simplicity, we restrict ourselves to the case $N = 2$.

More precisely, we let Ω be an open domain in $\mathbb{R}^2$ the boundary $\partial\Omega$ of which is C^1 up to a finite set M of singular points (i.e., $\partial\Omega$ is not C^1 in any neighborhood of $x \in M$). For each point $x \in M$ we suppose that there are an open neighborhood $U_x \subseteq \mathbb{R}^2$ of x as well as two open domains $\Omega_{x,l}$ and $\Omega_{x,r}$ with C^1-boundary such that either

$$U_x \cap \Omega = U_x \cap (\Omega_{x,l} \cap \Omega_{x,r}) \tag{5.15}$$

or

$$U_x \cap \Omega = U_x \setminus (\Omega_{x,l} \cap \Omega_{x,r}). \tag{5.16}$$

If the tangent spaces $T_x\Omega_{x,l}$ and $T_x\Omega_{x,r}$ do not coincide, then we call x an outward angular point in case of (5.15) and an inward angular point in case of (5.16). If these tangent spaces coincide, then x is called an outward resp. inward cuspidal point.

As in the previous section, we consider the cone generated by Ω,

$$\mathbf{C}_\Omega := \{(y, y_3) \in \mathbb{R}^2 \times [0, \infty) : y \in y_3\Omega\} \tag{5.17}$$

and, for $x \in \mathbb{R}^2$ and $\delta \in \mathbb{R}$, the half-spaces $\mathbf{H}_x$ and $\mathbf{H}_{x,\delta}$. Further we set $\mathbf{H}_{x,-\infty} := \mathbb{R}^3$ and $\mathbf{H}_{x,+\infty} := \emptyset$ and, for $\delta, \epsilon \in \mathbb{R} \cup \{\pm\infty\}$ and $x \in M$,

$$\mathbf{K}_{x,\delta,\epsilon} := \mathbf{H}_{x,\delta,l} \cap \mathbf{H}_{x,\epsilon,r}$$

where $\mathbf{H}_{x,\delta,l}$ and $\mathbf{H}_{x,\epsilon,r}$ are half-spaces belonging to $\Omega_{x,l}$ and $\Omega_{x,r}$, respectively.

Proposition 5.13. *Let $x \in \mathbb{R}^2$. Then the local operator spectrum $\sigma_{\eta_x}(\chi_{\mathbf{C}_\Omega} I)$ is equal to*

(a) $\{0\}$ *if* $x \notin \overline{\Omega}$.

(b) $\{I\}$ *if* $x \in \Omega$.

(c) $\{\chi_{\mathbb{H}_{x,\delta}} I : \delta \in \mathbb{R} \cup \{\pm\infty\}\}$ *if* $x \in \partial\Omega \setminus M$.

(d) $\{\chi_{\mathbb{K}_{x,\delta,\epsilon}} I : \delta, \epsilon \in \mathbb{R} \cup \{\pm\infty\}\}$ *if* $x \in \partial\Omega \cap M$ *is an angular point.*

(e) $\{0\}$ *if* $x \in \partial\Omega \cap M$ *is an outward cuspidal point.*

(f) $\{I\}$ *if* $x \in \partial\Omega \cap M$ *is an inward cuspidal point.*

Proof. The proof for (a)–(c) is the same as in the previous section. The results of the previous section also show that the local operator spectrum $\sigma_{\eta_x}(\chi_{\mathbf{C}_\Omega})$ is contained in the set (*d*) if $x \in \partial\Omega \cap M$ is an angular point. That it is actually equal to this set can be seen as follows. Since x is an angular point, we can independently shift the half-space $\mathbf{H}_{x,0,l}$ by a sequence which tends into the direction of η_x and comes closer and closer to $\partial\mathbf{H}_{x,0,r}$ and the half-space $\mathbf{H}_{x,0,r}$ by a sequence which also tends into the direction of η_x and comes closer and closer to $\partial\mathbf{H}_{x,0,l}$. Since each of these sequences influences the associated limit operators of only one of the half-spaces, we get any desired combination of shifts of the half-spaces $\mathbf{H}_{x,0,l}$ and $\mathbf{H}_{x,0,r}$ in this way. This shows (*d*), and (*e*) and (*f*) can be proved as in the previous section. (See the discussion before Theorem 5.9. The obvious point is that, in case of a cuspidal point, the half-spaces $\mathbf{H}_{x,0,l}$ and $\mathbf{H}_{x,0,r}$ cannot be shifted independently of each other.) □

Given $A \in \mathcal{A}(BUC(\mathbb{R}^3), \mathcal{C}_p(\mathbb{R}^3))$, let $\mathcal{H}_\eta(A)$ denote the set of all sequences $h \in \mathcal{H}$ which tend to infinity into the direction of $\eta \in S^2$ and for which the limit operator A_h exists. Further, if $x \in \partial\Omega \setminus M$ and $\delta \in \mathbb{R} \cup \{\pm\infty\}$, write $\mathcal{H}_{x,\delta}(A)$ for the set of all sequences $h \in \mathcal{H}_{\eta_x}(A)$ such that the limit operator of $\chi_{\mathbf{C}_\Omega} I$ exists and is equal to $\chi_{\mathbb{H}_{x,\delta}} I$. Finally, if $x \in \partial\Omega \cap M$ is an angular point and $\delta, \epsilon \in \mathbb{R} \cup \{\pm\infty\}$, then let $\mathcal{H}_{x,\delta,\epsilon}(A)$ stand for the set of all sequences $h \in \mathcal{H}_{\eta_x}(A)$ such that the limit operator of $\chi_{\mathbf{C}_\Omega} I$ exists and is equal to $\chi_{\mathbb{K}_{x,\delta,\epsilon}} I$. With these notations, we have the following consequence of Theorem 5.1.

Theorem 5.14. *Let $A \in \mathcal{A}(BUC(\mathbb{R}^3), \mathcal{C}_p(\mathbb{R}^3))$ and $D = \mathbf{C}_\Omega$ with $\Omega \in \mathbb{R}^2$ an open domain with piecewise C^1 boundary as above. Then the compression of A onto D is Fredholm on $L^p(D)$ if and only if the following conditions are satisfied:*

(a) *for each point $x \in \Omega$, the set $\{A_h : h \in \mathcal{H}_{\eta_x}(A)\}$ of limit operators of A is uniformly invertible.*

(b) *for each point $x \in \partial\Omega \setminus M$, the set*

$$\{(1 - \chi_{\mathbf{H}_{x,\delta}})I + \chi_{\mathbf{H}_{x,\delta}} A_h \chi_{\mathbf{H}_{x,\delta}} I : h \in \mathcal{H}_{x,\delta}(A),\ \delta \in \mathbb{R} \cup \{\pm\infty\}\}$$

of extended compressions of limit operators of A is uniformly invertible.

(c) *for each angular point* $x \in \partial\Omega \cap M$, *the set*

$$\{(1-\chi_{\mathbf{K}_{x,\delta,\epsilon}})I + \chi_{\mathbf{K}_{x,\delta,\epsilon}} A_h \chi_{\mathbf{K}_{x,\delta,\epsilon}} I : h \in \mathcal{H}_{x,\delta,\epsilon}(A),\ \delta, \epsilon \in \mathbb{R} \cup \{\pm\infty\}\}$$

of extended compressions of limit operators of A *is uniformly invertible.*

(d) *for each inward cuspidal point* $x \in \partial\Omega \cap M$, *the set* $\{A_h : h \in \mathcal{H}_{\eta_x}(A)\}$ *of limit operators of* A *is uniformly invertible.*

Note that the conditions in (*b*) and (*c*) get a simpler form if one of the shift parameters δ and ϵ is $\pm\infty$. Let us also emphasize that, if x is an outward cuspidal point, then the local invertibility at η_x of the compression of A onto D is trivially satisfied since all limit operators of $(1-\chi_D)I + \chi_D A \chi_D I$ with respect to sequences which tend to infinity into the direction of η_x are equal to I.

Again we mention some special situations in which the conditions of Theorem 5.14 can be readily verified.

Corollary 5.15. *Let* $A \in \mathcal{A}(SO(\mathbb{R}^3), \mathcal{C}_p(\mathbb{R}^3))$ *and* D *be as in Theorem* 5.14. *Then the compression of* A *onto* D *is Fredholm on* $L^p(D)$ *if and only if the following conditions are satisfied:*

(a) *for each point* $x \in \overline{\Omega}$ *which is neither angular nor outward cuspidal, the set* $\{A_h : h \in \mathcal{H}_{\eta_x}(A)\}$ *of limit operators of* A *is uniformly invertible.*

(b) *for each angular point* $x \in M$, *the set*

$$\{(1-\chi_{\mathbf{K}_{x,0,0}})I + \chi_{\mathbf{K}_{x,0,0}} A_h \chi_{\mathbf{K}_{x,0,0}} I : h \in \mathcal{H}_{x,\delta,\epsilon}(A),\ \delta, \epsilon \in \mathbb{R}\}$$

of extended compressions of limit operators of A *is uniformly invertible.*

(c) *for each angular point* $x \in M$, *the set*

$$\{A_h : h \in \mathcal{H}_{x,-\infty,\epsilon}(A) \cup \mathcal{H}_{x,\delta,-\infty}(A) \cup \mathcal{H}_{x,-\infty,-\infty}(A) : \delta, \epsilon \in \mathbb{R}\}$$

of limit operators of A *is uniformly invertible.*

Here we have used the shift invariance of the limit operators of A as well as Simonenko's Theorem 5.10 again.

Corollary 5.16. *Let* $A \in \mathcal{A}(L^\infty_{stab}(\mathbb{R}^3), \mathcal{C}_p(\mathbb{R}^3))$ *and* D *be as in Theorem* 5.14. *Then the compression of* A *onto* D *is Fredholm on* $L^p(D)$ *if and only if the following conditions are satisfied:*

(a) *for each point* $x \in \overline{\Omega}$ *which is neither angular nor outward cuspidal, the limit operator* A_{η_x} *of* A *is invertible.*

(b) *for each angular point* $x \in M$, *the extended compression*

$$(1-\chi_{\mathbf{K}_{x,0,0}})I + \chi_{\mathbf{K}_{x,0,0}} A_{\eta_x} \chi_{\mathbf{K}_{x,0,0}}$$

of the limit operator A_h *of* A *is invertible.*

Indeed, this result follows from the fact that each local operator spectrum is a singleton under the hypothesis of the corollary. Furthermore, one shows by choosing suitable sequences tending to infinity that every operator in condition (*c*) of Corollary 5.15 is a limit operator of the operator in (*b*) (compare the proof of

Theorem 2.33 in [9]). Thus, the invertibility of that operator already implies the invertibility of all operators in Corollary 5.15 (*c*).

5.8. Compressions to epigraphs of functions

We let $f : \mathbb{R} \to \mathbb{R}$ be a continuously differentiable function with

$$\lim_{t\to\pm\infty} f(t) = +\infty \quad \text{and} \quad \lim_{t\to\pm\infty} f'(t) = 0 \tag{5.18}$$

and consider its epigraph

$$\mathbf{E}_f := \{(x_1, x_2) \in \mathbb{R}^2 : x_2 > f(x_1)\}. \tag{5.19}$$

Let $h = (h_1, h_2) \in \mathcal{H}$ be a sequence which tends to infinity into the direction of $\eta = (\eta_1, \eta_2) \in S^1$. It is evident that the limit operator of $\chi_{\mathbf{E}_f} I$ exists and is equal to the identity operator if $\eta_2 > 0$, whereas the limit operator of $\chi_{\mathbf{E}_f} I$ exists and is equal to zero if $\eta_2 < 0$. Now let $\eta = (1, 0)$, and let h be a sequence for which the limit operator of $\chi_{\mathbf{E}_f}$ exists. We write

$$h_2(n) =: f(h_1(n)) + d_n$$

and choose a subsequence of h (which we denote by h again) such that the sequence (d_n) becomes convergent with limit $\delta \in \mathbb{R} \cup \{\pm\infty\}$. Further, for $\delta \in \mathbb{R} \cup \{\pm\infty\}$, we let

$$\mathbf{H}_\delta := \{(x_1, x_2) \in \mathbb{R}^2 : x_2 > \delta\}.$$

Then it is easy to check that the limit operator of $\chi_{\mathbf{E}_f} I$ coincides with $\chi_{\mathbf{H}_\delta} I$ and that, conversely, every operator of this form appears as a limit operator of $\chi_{\mathbf{E}_f} I$. The same holds for $\eta = (-1, 0)$.

Given $A \in \mathcal{A}(BUC(\mathbb{R}^2), \mathcal{C}_p(\mathbb{R}^2))$, let $\mathcal{H}_\eta(A)$ denote the set of all sequences $h \in \mathcal{H}$ which tend to infinity into the direction of $\eta \in S^1$ and for which the limit operator A_h exists. Further, for $\delta \in \mathbb{R} \cup \{\pm\infty\}$, write $\mathcal{H}_{\pm 1, \delta}(A)$ for the set of all sequences $h \in \mathcal{H}_{(\pm 1, 0)}(A)$ such that the limit operator of $\chi_{\mathbf{E}_f} I$ exists and is equal to $\chi_{\mathbb{H}_\delta} I$. Then Theorem 5.1 yields, for example, the following.

Theorem 5.17. *Let $A \in \mathcal{A}(L^\infty_{stab}(\mathbb{R}^2), \mathcal{C}_p(\mathbb{R}^2))$, and let $D = \mathbf{E}_f$ be the epigraph of the function f satisfying* (5.18). *Then the compression of A onto D is Fredholm on $L^p(D)$ if and only if, for each point $\eta = (\eta_1, \eta_2) \in S^1$ with $\eta_2 \geq 0$, the limit operator A_η of A is invertible.*

The proof is the same as for Corollaries 5.12 and 5.16. □

Finally, let $f_\pm : \mathbb{R} \to \mathbb{R}$ be continuously differentiable functions with

$$\lim_{t\to+\infty} f_\pm(t) = \lim_{t\to-\infty} f_\pm(t) = \pm\infty \quad \text{and} \quad \lim_{t\to+\infty} f'_\pm(t) = \lim_{t\to-\infty} f'_\pm(t) = 0, \tag{5.20}$$

and let

$$\mathbf{E}_{f_+, f_-} := \{(x_1, x_2) \in \mathbb{R}^2 : f_-(x_1) < x_2 < f_+(x_1)\}. \tag{5.21}$$

As before one can check that every limit operator of $\chi_{\mathbf{E}_{f_+, f_-}} I$ is of the form $\chi_{\mathbf{H}_\delta} I$ with $\delta \in \mathbb{R} \cup \{\pm\infty\}$ and that, conversely, every operator of this form is a limit operator of $\chi_{\mathbf{H}_\delta} I$ if h tends into the direction of $(\pm 1, 0) \in S^1$. If h tends to infinity into the direction of $\eta \in S^1$ with $\eta_2 \neq 0$ then, necessarily, the limit operator of $\chi_{\mathbf{H}_\delta} I$ with respect to h exists and is equal to the zero operator.

Theorem 5.18. *Let $A \in \mathcal{A}(L^\infty_{stab}(\mathbb{R}^2), \mathcal{C}_p(\mathbb{R}^2))$, and let $D = \mathbf{E}_{f_+, f_-}$ with functions $f_\pm$ satisfying* (5.20). *Then the compression of A onto D is Fredholm on $L^p(D)$ if and only if the limit operators A_η of A with $\eta = (\pm 1, 0) \in S^1$ are invertible.*

References

[1] H.O. Cordes, On compactness of commutators of multiplications and convolutions and boundedness of pseudodifferential operators. – J. Fctl. Anal. **18**(1975), 2, 115–131.

[2] R.G. Douglas, R. Howe, On the C^*-algebra of Toeplitz operators on the quarter plane. – Trans. Am. Math. Soc. **158**(1971), 203–217.

[3] L.S. Goldenstein, I.C. Gohberg, On a multidimensional integral equation on a half-space whose kernel is a function of the difference of the arguments, and on a discrete analogue of this equation. – Dokl. AN SSSR **131**(1960), 1, 4–12 (Russian, Engl. transl. Sov. Math. **1**(1960), 1, 173–176).

[4] N. Karapetiants, S. Samko, Equations with Involutive Operators. – Birkhäuser, Boston, Basel, Berlin, 2001.

[5] E. Meister, F.-O. Speck, Some multidimensional Wiener-Hopf equations with applications. – *In*: Trends in Applications of Pure Mathematics to Mechanics, Vol. II, Proc. Symp. Kozubnik, Poland, September 1977 (Ed. H. Zorski), Pitman, London et al. 1979, 217–262.

[6] E. Meister, F.-O. Speck, Wiener-Hopf operators on three-dimensional wedge-shaped regions. – Appl. Anal. **10**(1980), 31–45.

[7] E. Meister, F.-O. Speck, A contribution to the quarter-plane problem in diffraction theory. – J. Math. Anal. Appl. **130**(1988), 223–236.

[8] V.S. Rabinovich, S. Roch, B. Silbermann, Fredholm theory and finite section method for band-dominated operators. – Integral Equations Oper. Theory **30**(1998), 4, 452–495.

[9] V.S. Rabinovich, S. Roch, B. Silbermann, Band-dominated operators with operator-valued coefficients, their Fredholm properties and finite sections. – Integral Equations Oper. Theory **40**(2001), 3, 342–381.

[10] M. Reed, B. Simon, Methods of Modern Mathematical Physics, II. – Academic Press, New York, San Francisco, London 1975.

[11] B.Ya. Shteinberg, Operators of convolution type on locally compact groups. – Rostov-na-Donu 1979, Dep. at VINITI 715–80, 65 p. (Russian).

[12] B.Ya. Shteinberg, On convolution operators on locally compact groups. – Funkts. Anal. prilozh. **15**(1981), 95–96 (Russian).

[13] B.Ya. Shteinberg, Boundedness and compactness of convolution operators on locally compact groups. – Mat. Zametki **38**(1985), 2, 278–292 (Russian).

[14] I.B. Simonenko, Operators of convolution type in cones. – Mat. Sbornik **74** **(116)**(1967), 2, 567–586 (Russian, Engl. transl. Math. USSR-Sbornik **3**(1967), 2, 279–293).

[15] F.-O. SPECK, Eine Erweiterung des Satzes von Rakovšik und ihre Anwendung in der Simonenko-Theorie. – Math. Ann. **228**(1977), 2, 93–100.

[16] M.E. TAYLOR, Partial Differential Equations I. – Springer-Verlag, New York 1996.

Vladimir S. Rabinovich
Instituto Politechnico National
ESIME-Zacatenco, Ed.1, 2-do piso
Av. IPN, Mexico, D.F., 07738
Mexico
e-mail: rabinov@maya.esimez.ipn.mx

Steffen Roch
Technische Universität Darmstadt
Fachbereich Mathematik
Schlossgartenstrasse 7
D-64289 Darmstadt
Germany
e-mail: roch@mathematik.tu-darmstadt.de

Operator Theory
Advances and Applications, Vol. 147, 457–473

The Distribution of Eigenvalues of Some Fractal Elliptic Operators and Weyl Measures

Hans Triebel

Dedicated to the memory of Erhard Meister

Abstract. Let μ be a compactly supported Radon measure in $\mathbb{R}^n$ and let id^μ be the related operator in the framework of $S'(\mathbb{R}^n)$. Then μ is called a *Weyl measure* if $B = (-\Delta + id)^{-\frac{n}{2}} \circ id^\mu$ generates a nonnegative, compact, self-adjoint operator in $H^{\frac{n}{2}}(\mathbb{R}^n)$ and if its positive eigenvalues ϱ_k, where $k \in \mathbb{N}$, behave like k^{-1}. The paper deals with criteria under which μ is a Weyl measure.

1. Introduction

Let μ be a finite compactly supported Radon measure in $\mathbb{R}^n$ and let

$$id^\mu : \quad \varphi \mapsto \varphi\mu, \quad \varphi \in S(\mathbb{R}^n),$$

be the related operator considered in the framework of $S'(\mathbb{R}^n)$. The paper deals with the distinguished operator

$$B = (-\Delta + id)^{-\frac{n}{2}} \circ id^\mu. \tag{1.1}$$

It comes out that for a large class of measures μ the operator B is self-adjoint, nonnegative and compact in the Sobolev space $H^{\frac{n}{2}}(\mathbb{R}^n)$. Let ϱ_k be the positive eigenvalues counted with respect of multiplicity and ordered by decreasing order of magnitude. Then μ is called a *Weyl measure* if there are two positive numbers c_1 and c_2 such that

$$c_1\, k^{-1} \le \varrho_k \le c_2\, k^{-1}, \quad k \in \mathbb{N}. \tag{1.2}$$

This is the typical Weyl behavior of eigenvalues of the inverse of an elliptic operator of order n (say, in a smooth bounded domain in $\mathbb{R}^n$, complemented by suitable boundary conditions). We studied operators of type (1.1) and their spectra in a more general context in [11], Ch. V, and [12], Ch. III. The interest in the peculiar operator B comes from the physical interpretation in case of $n = 2$, where (1.1), (1.2) is related to the eigenfrequencies of a vibrating fractal drum. We refer for details and results obtained so far to [11], Section 30; and [12], Section 19. It is the main aim of this paper to continue these considerations describing new classes

---

2000 *Mathematics Subject Classification.* 46E35; 42B35, 28A80.

of Weyl measures. For example, as a consequence of our main assertion it comes out that any compactly supported Radon measure μ with

$$\mu(\, B(\gamma, r)\,) \sim r^d \Psi(r), \quad \gamma \in \operatorname{supp} \mu, \quad 0 < r \le 1,$$

is a Weyl measure if $0 < d \le n$ and if $\Psi(r)$ is a positive differentiable function on the interval $(0, 1]$ such that

$$r \frac{\Psi'(r)}{\Psi(r)} \to 0 \quad \text{for} \quad r \to 0,$$

(Corollary 1). Here $B(\gamma, r)$ is a ball centred at γ and of radius r. A typical example is given by $\Psi(r) = (1 + |\log r|)^b$ where $b \in \mathbb{R}$ (with $b \ge 0$ if $d = n$).

The plan of the paper is the following. We collect in Section 2 the necessary definitions, notation, and some prerequisites. We follow closely [11], [12]. Section 3 contains the results. Proofs are shifted to Section 4.

We use $\sim$ for two positive functions $a(x)$ and $b(x)$ or for two sequences of positive numbers a_k and b_k (say, $k \in \mathbb{N}$) if there are two positive numbers c and C such that

$$c\, a(x) \le b(x) \le C\, a(x) \quad \text{or} \quad c\, a_k \le b_k \le C\, a_k \tag{1.3}$$

for all admitted values of the variables x or k.

2. Some prerequisites

2.1. Basic notation

Let $\mathbb{N}$ be the collection of all natural numbers and $\mathbb{N}_0 = \mathbb{N} \cup \{0\}$. Let $\mathbb{R}^n$ be euclidean n-space, where $n \in \mathbb{N}$; put $\mathbb{R} = \mathbb{R}^1$, whereas $\mathbb{C}$ is the complex plane. Let $S(\mathbb{R}^n)$ be the Schwartz space of all complex-valued, rapidly decreasing, infinitely differentiable functions on $\mathbb{R}^n$. By $S'(\mathbb{R}^n)$ we denote its topological dual, the space of tempered distributions on $\mathbb{R}^n$.

Furthermore, $L_2(\mathbb{R}^n)$ is the standard complex Hilbert space with respect to the Lebesgue measure μ_L, normed by

$$\| f \,| L_2(\mathbb{R}^n) \| = \left(\int_{\mathbb{R}^n} |f(x)|^2 \, \mu_L(dx) \right)^{\frac{1}{2}},$$

where we prefer $\mu_L(dx)$ in place of dx. As usual, the identification of $f \in L_2(\mathbb{R}^n)$ with a tempered distribution is given by

$$id_{\mu_L} \; : \quad L_2(\mathbb{R}^n) \hookrightarrow S'(\mathbb{R}^n)$$

with

$$id_{\mu_L} f \; : \quad \varphi \mapsto \int_{\mathbb{R}^n} f(x)\, \varphi(x)\, \mu_L(dx), \quad \varphi \in S(\mathbb{R}^n).$$

We call id_{μ_L} the respective identification operator.

As usual, $\mathbb{Z}$ is the collection of all integers; and $\mathbb{Z}^n$, where $n \in \mathbb{N}$, denotes the lattice of all points $m = (m_1, \ldots, m_n) \in \mathbb{R}^n$ with $m_j \in \mathbb{Z}$. Let $\mathbb{N}_0^n$, where $n \in \mathbb{N}$, be the set of all multi-indices

$$\alpha = (\alpha_1, \ldots, \alpha_n) \quad \text{with} \quad \alpha_j \in \mathbb{N}_0 \quad \text{and} \quad |\alpha| = \sum_{j=1}^{n} \alpha_j.$$

If $x = (x_1, \ldots, x_n) \in \mathbb{R}^n$ and $\beta = (\beta_1, \ldots, \beta_n) \in \mathbb{N}_0^n$ then we put

$$x^\beta = x_1^{\beta_1} \cdots x_n^{\beta_n} \quad \text{(monomial)}. \tag{2.1}$$

2.2. Measures

We always assume that μ stands for a Radon measure in $\mathbb{R}^n$ with

$$\Gamma = \operatorname{supp} \mu \quad \text{compact and} \quad \mu(\mathbb{R}^n) < \infty. \tag{2.2}$$

Then $L_2(\Gamma, \mu)$ is the usual complex Hilbert space, normed by

$$\|f \,|L_2(\Gamma, \mu)\| = \left(\int_\Gamma |f(\gamma)|^2 \, \mu(d\gamma) \right)^{\frac{1}{2}}$$

and interpreted as a subspace of $S'(\mathbb{R}^n)$ via the identification operator id_μ,

$$id_\mu : \quad L_2(\Gamma, \mu) \hookrightarrow S'(\mathbb{R}^n) \tag{2.3}$$

given by

$$id_\mu f : \quad \varphi \mapsto \int_\Gamma f(\gamma)\, \varphi(\gamma)\, \mu(d\gamma), \quad \varphi \in S(\mathbb{R}^n). \tag{2.4}$$

Since μ is Radon, $S(\mathbb{R}^n)$ (or, what is the same, its restriction to Γ) is dense in $L_2(\Gamma, \mu)$. Then the identification (2.3), (2.4) can be done (in analogy to the Lebesgue measure) without ambiguity. We refer to [12], 9.2, pp. 122–125, for some details.

2.3. Sobolev spaces and traces

As usual, $-\Delta = -\sum_{j=1}^n \frac{\partial^2}{\partial x_j^2}$ stands for the Laplacian in $\mathbb{R}^n$. Let $s \in \mathbb{R}$. Then

$$H^s(\mathbb{R}^n) = (-\Delta + id)^{-\frac{s}{2}} \, L_2(\mathbb{R}^n),$$

where *id* is the identity, are the well-known Sobolev spaces. Of interest for us is the critical value $s = \frac{n}{2}$ (where *critical* refers to the peculiar behavior of $H^{\frac{n}{2}}(\mathbb{R}^n)$ in connection with limiting embeddings and traces). Let μ be a Radon measure in $\mathbb{R}^n$ according to 2.2. If there is a positive constant c such that

$$\|\varphi \,|L_2(\Gamma, \mu)\| \le c \, \|\varphi \,|H^{\frac{n}{2}}(\mathbb{R}^n)\|, \quad \varphi \in S(\mathbb{R}^n), \tag{2.5}$$

then one can define a trace operator tr_μ by completion,

$$tr_\mu : \quad H^{\frac{n}{2}}(\mathbb{R}^n) \hookrightarrow L_2(\Gamma, \mu). \tag{2.6}$$

In particular, (2.5) can be extended to

$$\|tr_\mu f \,|L_2(\Gamma, \mu)\| \le c \, \left\| f \,|H^{\frac{n}{2}}(\mathbb{R}^n) \right\|, \quad f \in H^{\frac{n}{2}}(\mathbb{R}^n). \tag{2.7}$$

We refer to [12], 9.2–9.8, pp. 122–131, where we dealt in detail with traces of function spaces on sets Γ in $\mathbb{R}^n$ in a more general context. Restricted to our special case one has the following assertion: Let Q_{jm} be a cube in $\mathbb{R}^n$ centred at $2^{-j}m$ where $j \in \mathbb{N}_0$ and $m \in \mathbb{Z}^n$ and with side-length 2^{-j+1}. Let

$$f_{jm} = \int\limits_{Q_{jm}} f(\gamma)\,\mu(d\gamma); \quad j \in \mathbb{N}_0, \quad m \in \mathbb{Z}^n, \quad f \in L_1(\Gamma,\mu).$$

By [12], Theorem 9.3, p. 125, the trace operator tr_μ with (2.6) exists if, and only if,

$$\sup \sum_{j=0}^{\infty} \sum_{m \in \mathbb{Z}^n} f_{jm}^2 < \infty,$$

where the supremum is taken over all

$$f \in L_2(\Gamma,\mu) \quad \text{with} \quad f \geq 0 \quad \text{and} \quad \|f\,|L_2(\Gamma,\mu)\| \leq 1.$$

By [12], Corollary 9.8, pp. 129–130, one gets as an easy consequence of this criterion that tr_μ according to (2.6) exists if

$$\sum_{j=0}^{\infty} \sup_{m \in \mathbb{Z}^n} \mu(Q_{jm}) < \infty. \tag{2.8}$$

2.4. The operator B: The set-up

We assume that the trace operator tr_μ according to (2.6), (2.7) exists. Let id_μ be the identification operator introduced in (2.3), (2.4). By the arguments in [12], 9.2, especially on p. 124, these two operators are dual to each other,

$$tr'_\mu = id_\mu \quad \text{and} \quad id'_\mu = tr_\mu,$$

where id_μ can also be considered as the extension by continuity from $tr_\mu S(\mathbb{R}^n)$ (which is dense in $L_2(\Gamma,\mu)$) to $L_2(\Gamma,\mu)$. By the usual identification of $L_2(\Gamma,\mu)$ with its dual and the duality assertions for function spaces within the dual pairing $(S(\mathbb{R}^n), S'(\mathbb{R}^n))$ it follows that

$$id_\mu : \quad L_2(\Gamma,\mu) \hookrightarrow H^{-\frac{n}{2}}(\mathbb{R}^n) = H^{\frac{n}{2}}(\mathbb{R}^n)'.$$

In particular, $H^{-\frac{n}{2}}(\mathbb{R}^n)$ is the target space of id_μ. Since all spaces involved are reflexive Banach spaces (even Hilbert spaces), $id'_\mu = tr_\mu$ follows from $tr'_\mu = id_\mu$. We have

$$id^\mu = id_\mu \circ tr_\mu : \quad H^{\frac{n}{2}}(\mathbb{R}^n) \hookrightarrow H^{-\frac{n}{2}}(\mathbb{R}^n); \tag{2.9}$$

and, hence, B,

$$B = (-\Delta + id)^{-\frac{n}{2}} \circ id^\mu : \quad H^{\frac{n}{2}}(\mathbb{R}^n) \hookrightarrow H^{\frac{n}{2}}(\mathbb{R}^n), \tag{2.10}$$

is a linear and bounded operator in the Hilbert spaces $H^{\frac{n}{2}}(\mathbb{R}^n)$. Next we equip $H^{\frac{n}{2}}(\mathbb{R}^n)$ with the scalar product

$$(f,g)_{H^{\frac{n}{2}}(\mathbb{R}^n)} = \int\limits_{\mathbb{R}^n} (-\Delta + id)^{\frac{n}{4}} f(x) \cdot (-\Delta + id)^{\frac{n}{4}} \overline{g(x)}\, \mu_L(dx), \tag{2.11}$$

where both f and g are elements of $H^{\frac{n}{2}}(\mathbb{R}^n)$. By the same reasoning as in [12], p. 257, and in [11], 28.6, 30.2, pp. 226, 234, it follows that

$$(Bf, g)_{H^{\frac{n}{2}}(\mathbb{R}^n)} = \int_\Gamma (tr_\mu f)(\gamma) \cdot \overline{(tr_\mu g)(\gamma)}\, \mu(d\gamma). \tag{2.12}$$

Hence, B is the linear and bounded operator in $H^{\frac{n}{2}}(\mathbb{R}^n)$ generated by the scalar product of $L_2(\Gamma, \mu)$, considered as a bounded, symmetric, nonnegative bi-linear form in $H^{\frac{n}{2}}(\mathbb{R}^n)$. Then it comes out that B is a bounded, nonnegative, self-adjoint operator in $H^{\frac{n}{2}}(\mathbb{R}^n)$ and, as a consequence,

$$\left\| \sqrt{B} f \,|H^{\frac{n}{2}}(\mathbb{R}^n) \right\| = \| tr_\mu f \,|L_2(\Gamma, \mu)\|, \quad f \in H^{\frac{n}{2}}(\mathbb{R}^n). \tag{2.13}$$

2.5. Weyl measures

Let μ be a Radon measure with (2.2). We say that the corresponding operator B is compact, if the trace operator according to (2.6), (2.7) exists and if, in addition, the resulting bounded, nonnegative, self-adjoint operator B in $H^{\frac{n}{2}}(\mathbb{R}^n)$ is compact. In this case we denote the positive eigenvalues of B by ϱ_k, repeated according to multiplicity and ordered so that

$$\varrho_1 \geq \varrho_2 \geq \cdots > 0, \quad \varrho_k \to 0 \quad \text{if} \quad k \to \infty. \tag{2.14}$$

Definition 1. *Let μ be a Radon measure in $\mathbb{R}^n$ with (2.2). Then μ is called a Weyl measure if the corresponding operator B is compact and*

$$\varrho_k \sim k^{-1} \quad \textit{where} \quad k \in \mathbb{N}. \tag{2.15}$$

Remark 1. Recall that we use $\sim$ according to (1.3). In rough terms, the regular component of the operator B, given by (2.10), is the inverse of the elliptic operator $(-\Delta + id)^{\frac{n}{2}}$ of order n. Then the exponent -1 in (2.15) reflects the classical Weyl behavior of an elliptic (pseudo-)differential operator of order n in (bounded smooth) domains in $\mathbb{R}^n$ as it may be found in many books. We refer, for example, to [10], 5.6.2, pp. 395–397, where we also collected the relevant historical literature. Corresponding considerations with respect to fractal elliptic (pseudo-)differential operators have been given in [11], Sect. 28, especially 28.9–28.11, p. 230, where we discussed Weylian, sub-Weylian, and super-Weylian spectral behavior. In other words, we ask for fractal elliptic operators of type (2.10), preserving the Weylian behavior in a limiting case. The notation *Weyl measure* has been introduced in [12], 19.13, pp. 274–275, in connection with fractal drums in the plane $\mathbb{R}^2$ (hence $n = 2$). There is an immaterial technical difference between [12], 19.13, and the above Definition 1. In [12], 19.13, we dealt with the Dirichlet Laplacian $-\Delta$ with respect to a bounded smooth domain Ω in the plane $\mathbb{R}^2$, a finite Radon measure μ and a corresponding operator B such that

$$\operatorname{supp} \mu \subset \Omega \quad \text{and} \quad B = (-\Delta)^{-1} \circ id^\mu. \tag{2.16}$$

Then μ was called a Weyl measure if B makes sense, is compact (in $\overset{\circ}{H}{}^1(\Omega)$) and if we have (2.15). This setting was caused by the physical background interpreting

(2.16) as the adequate description of a fractal drum in the plane. But as far as the notation of a Weyl measure is concerned, [12], 19.13, and the above Definition 1 (with $n = 2$) describe the same class of Radon measures. The results obtained so far in case of fractal drums may be found in [11], Sect. 30, and in [12], Sect. 19, with Theorem 19.17 on p. 280, as the most general assertion. The discussion given in [12], 19.18, pp. 288–291, has a direct counterpart, mutatis mutandis, to the situation considered in this paper. But we do not stress this point.

2.6. Isotropic measures

The problem to characterize all Weyl measures according to Definition 1 seems to be beyond the techniques available so far (at least we could not see how to do it). We refer again to [11], Sect. 30, and in particular to [12], 19.18, pp. 288–291, where we discussed this problem in a slightly different context. We restrict ourselves mostly, but not exclusively, to isotropic measures which we are going to define now. Let μ be a Radon measure in $\mathbb{R}^n$ with (2.2). Then μ is called an *isotropic Radon measure* if there is a positive continuous non-decreasing function h on the interval $(0,1]$ such that

$$\mu\big(B(\gamma,r)\big) \sim h(r); \quad 0 < r \le 1 \quad \text{and} \quad \gamma \in \Gamma, \tag{2.17}$$

where $B(\gamma, r)$ is a ball in $\mathbb{R}^n$ centred at γ and of radius r. By (1.3) the equivalence $\sim$ in (2.17) is independent of γ and r. The best-known examples are the so-called d-sets in $\mathbb{R}^n$ with $0 \le d \le n$ and

$$\mu\big(B(\gamma,r)\big) \sim r^d, \qquad 0 < r \le 1. \tag{2.18}$$

Then one may choose $\mu = \mathcal{H}^d|\Gamma$, the restriction of the Hausdorff measure $\mathcal{H}^d$ in $\mathbb{R}^n$ to Γ. This is just the case treated in detail in [11], Sections 28–30. The first step beyond (2.18) by asking for isotropic measures μ and related operators B was done in [5], [6], [7], [8], replacing (2.18) by

$$\mu\big(B(\gamma,r)\big) \sim r^d\,\Psi(r), \qquad 0 < r \le 1, \tag{2.19}$$

where again $0 \le d \le n$ and where $\Psi(r)$ might be considered as a perturbation. A short description may also be found in [12], Sections 22 and 23. Finally the problem for which functions h there is a Radon measure μ with (2.2), (2.17), was solved in [1], [2], [3]. We do not go into details and refer to the quoted papers where one finds also non-standard examples. It is the main aim of this paper to ask for conditions ensuring that a given isotropic Radon measure is a Weyl measure.

3. Results

3.1. The main assertion

Let μ be an isotropic Radon measure in $\mathbb{R}^n$ with (2.2), (2.17). Let

$$\sum_{j=0}^{\infty} h\left(2^{-j}\right) < \infty. \tag{3.1}$$

Then, by (2.8), the trace operator tr_μ according to (2.6), (2.7), and hence B, given by (2.9), (2.10), exist, and B is a bounded, nonnegative, self-adjoint operator in $H^{\frac{n}{2}}(\mathbb{R}^n)$.

Theorem 1. *Let μ be an isotropic Radon measure in $\mathbb{R}^n$ with* (2.2), (2.17),

$$\sup_{L\in\mathbb{N}} \sum_{l=0}^{L} \frac{h(2^{-L})}{h(2^{-l})} \cdot \log \frac{h(2^{-l})}{h(2^{-L})} < \infty, \tag{3.2}$$

and

$$\sup_{L\in\mathbb{N}} \sum_{l=L}^{\infty} \left(\frac{h(2^{-l})}{h(2^{-L})} \right)^{\frac{1}{2}} < \infty. \tag{3.3}$$

Then μ is a Weyl measure.

Remark 2. Obviously, (3.1) is an immediate consequence of (3.3) with, say, $L = 1$. Hence B exists. If $\varkappa$ is a number with $0 < \varkappa < 1$ such that

$$\sup_{L\in\mathbb{N}} \sum_{l=0}^{L} \left(\frac{h(2^{-L})}{h(2^{-l})} \right)^{\varkappa} < \infty, \tag{3.4}$$

then (3.2) is satisfied. Hence, also (3.3), (3.4), ensure that μ is a Weyl measure. In particular, if $0 < d \le n$, then μ with

$$h(r) \sim r^d; \qquad 0 < r \le 1,$$

is a Weyl measure in $\mathbb{R}^n$. But this is essentially not new. If $n = 2$ then this assertion is covered by [12], Theorems 19.7 and 19.17, pp. 264–265, 280. As for the general case $n \in \mathbb{N}$ one can modify the proof of [11], Theorem 28.6, p. 226, appropriately. The above assertion can be reformulated as follows: If Γ is a compact d-set in $\mathbb{R}^n$ with $0 < d \le n$, then $\mu = \mathcal{H}^d|\Gamma$ is a Weyl measure.

3.2. Two corollaries

First we generalize the above assertion in Remark 2.

Corollary 1. *Let μ be an isotropic Radon measure in $\mathbb{R}^n$ with* (2.2), (2.17), *and let*

$$h(r) = r^d\, \Psi(r), \qquad 0 < r \le 1, \tag{3.5}$$

where $\Psi(r)$ is a positive differentiable function on $(0,1]$,

$$0 < d \le n \quad \textit{and} \quad \lim_{r\to 0} r\, \frac{\Psi'(r)}{\Psi(r)} = 0. \tag{3.6}$$

Then μ is a Weyl measure.

Remark 3. One might think of $\Psi(r)$ as a perturbation of r^d. If $b \in \mathbb{R}$ then $\Psi(r) = (1 + |\log r|)^b$ are typical examples (with $b \ge 0$ if $d = n$). This type of perturbation is covered by the above references in connection with (2.19). As for more general examples we refer to [1], [2], [3].

Corollary 2. *Let μ_j with $j = 1, \ldots, N$, be isotropic Weyl measures covered by Theorem 1 with*

$$\mu_j\big(B(\gamma, r)\big) \sim h_j(r); \quad 0 < r \le 1 \quad \text{and} \quad \gamma \in \Gamma_j = \operatorname{supp} \mu_j.$$

Let

$$0 < h_1(r) \le h_2(r) \le \cdots \le h_N(r) \quad \text{if} \quad 0 < r \le 1,$$

and

$$\frac{h_l(r)}{h_{l+1}(r)} \to 0 \quad \text{if} \quad r \to 0, \qquad l = 1, \ldots, N-1. \tag{3.7}$$

Then

$$\mu = \sum_{j=1}^{N} \mu_j \tag{3.8}$$

is a Weyl measure.

Remark 4. The measure μ in (3.8) need not be isotropic according to (2.17). A first assertion of this type, restricted to $n = 2$, d_j-sets, and disjoint supports Γ_j may be found in [12], 19.12, p. 274. However it comes out that the supports Γ_j need not be disjoint.

3.3. Problems and generalizations

As said this paper should be considered as a complement to [11], Chapter V, and, in particular, to [12], Chapter III, especially Sect. 19. There one may find also the necessary references to the literature. The only additional reference we wish to add here is [9]. Reformulated in our notation this paper deals for even n with operators B of type (2.10) where the fractal measure μ is replaced by the weighted Lebesgue measure $V(x)\,\mu_L$. Sufficient conditions are given ensuring the Weyl behavior (2.15). There are also further references with respect to preceding papers of this type.

Problem. Theorem 1, the Corollaries 1 and 2, and [9] might be considered as contributions to the challenging problem:

Calculate all Weyl measures in $\mathbb{R}^n$,

abandoning (2.2) as a convenient prerequisite for the underlying Radon measure in $\mathbb{R}^n$. We refer to [12], 19.18, 19.19, pp. 288–294, where we discussed this problem in some detail in case of the plane $\mathbb{R}^2$.

Generalizations. Our arguments are qualitative. The estimates from above of the eigenvalues ϱ_k of B,

$$0 < \varrho_k \le c\, k^{-1}, \qquad k \in \mathbb{N},$$

for some $c > 0$, are based on estimates for entropy numbers of compact embeddings between function spaces and on the mapping property

$$(-\Delta + id)^{-\frac{n}{2}} : \quad H^{-\frac{n}{2}}(\mathbb{R}^n) \hookrightarrow H^{\frac{n}{2}}(\mathbb{R}^n). \tag{3.9}$$

However (3.9) holds for large classes of fractional powers of elliptic operators and for pseudodifferential operators of order $-n$,

$$b(x,D) \in \Psi^{-\frac{n}{2}}_{1,\delta}(\mathbb{R}^n): \quad H^{-\frac{n}{2}}(\mathbb{R}^n) \hookrightarrow H^{\frac{n}{2}}(\mathbb{R}^n), \tag{3.10}$$

where $0 \le \delta \le 1$, including even the exotic case $\delta = 1$. We dealt in some detail in [11], Sections 28 and 29, with references to the Sections 26 and 27, with (fractal) pseudodifferential operators and fractional powers of elliptic differential operators. There one finds also the necessary explanations, covering in particular (3.10), which will not be repeated here. Let

$$B^b = b(\cdot,D) \circ id^\mu \quad \text{with} \quad b(\cdot,D) \in \Psi^{-\frac{n}{2}}_{1,\delta}(\mathbb{R}^n) \tag{3.11}$$

where $0 \le \delta \le 1$, and where id^μ is given by (2.9). Then B^b is a bounded operator in $H^{\frac{n}{2}}(\mathbb{R}^n)$. If B^b is compact then we denote the non-vanishing (complex) eigenvalues by ϱ_k, repeated according to algebraic multiplicity and ordered so that

$$|\varrho_1| \ge |\varrho_2| \ge \cdots > 0.$$

Then the arguments given below cover also the following assertion:

Let μ be a Weyl measure according to Theorem 1 *or to the Corollaries* 1 *or* 2. *Then B^b given by* (3.11) *is compact and there is a number $c > 0$ such that*

$$|\varrho_k| \le c\,k^{-1}, \qquad k \in \mathbb{N}.$$

(If there are only $K \in \mathbb{N}$ non-vanishing eigenvalues ϱ_k then we put $\varrho_k = 0$ if $k > K$.) The converse assertion,

$$\varrho_k \ge c\,k^{-1}, \qquad k \in \mathbb{N},$$

for some $c > 0$, in Theorem 1 and in the Corollaries 1 and 2 with respect to the operator B given by (2.10), relies on estimates of approximation numbers of compact embeddings between function spaces based on (2.11), (2.12), and in particular on (2.13). This makes clear that again $(-\Delta + id)^{-\frac{n}{2}}$ in (2.10) can be replaced by fractional powers of suitable (regular) elliptic differential operators or by special pseudodifferential operators of type (3.10) with the following properties: The operator B^b is bounded, nonnegative and self-adjoint with respect to the equivalent norm $\|\cdot\,|H^{\frac{n}{2}}(\mathbb{R}^n)\|_b$, generated by the scalar product,

$$(f,g)^b_{H^{\frac{n}{2}}(\mathbb{R}^n)} = \int_{\mathbb{R}^n} b(x,D)^{-\frac{1}{2}} f(x) \cdot \overline{b(x,D)^{-\frac{1}{2}} g(x)}\, \mu_L(dx) \tag{3.12}$$

such that

$$\left(B^b f, g\right)^b_{H^{\frac{n}{2}}(\mathbb{R}^n)} = \int_\Gamma (tr_\mu f)(\gamma)\, \overline{(tr_\mu g)(\gamma)}\, \mu(d\gamma) \tag{3.13}$$

and

$$\left\| \sqrt{B^b} f\,|H^{\frac{n}{2}}(\mathbb{R}^n) \right\|_b = \|tr_\mu f\,|L_2(\Gamma,\mu)\|, \quad f \in H^{\frac{n}{2}}(\mathbb{R}^n). \tag{3.14}$$

These are the counterparts of (2.11), (2.12), (2.13). They are sufficient to get the following assertion as a by-product of our arguments below:

Let B^b be given by (3.11) *with* (3.12), (3.13), (3.14). *If μ is a Weyl measure according to Theorem 1, or to the Corollaries* 1 *or* 2, *then B^b is a nonnegative, compact, self-adjoint operator in $H^{\frac{n}{2}}(\mathbb{R}^n)$ with*

$$\varrho_k \sim k^{-1}; \qquad k \in \mathbb{N},$$

where ϱ_k are now the respective eigenvalues of B^b ordered by (2.14).

Weight functions. Following again our considerations in [11], Section 28, and, especially, Section 29, one may generalize (3.11) by

$$B^{b,V} = b(\cdot, D) \circ id_\mu \circ V \circ tr_\mu, \qquad b(\cdot, D) \in \Psi^{-n}_{1,\delta}(\mathbb{R}^n), \tag{3.15}$$

where V is a nonnegative weight function on $\Gamma = \operatorname{supp} \mu$ with

$$V \in L_r(\Gamma, \mu) \quad \text{or} \quad V \in L_r(\log L)_a(\Gamma, \mu).$$

We do not go into detail. But the point of interest in connection with the above problem to calculate all Weyl measures, is the possibility to step from isotropic measures μ to measures $V\mu$, which need not be isotropic.

Comment. By the above considerations and the outlined further possibilities it is quite clear that B given by (2.10) is the *primus inter pares* within a larger class of operators of type B^b or even $B^{b,V}$ according to (3.11) and (3.15). But we restrict ourselves here exclusively to the distinguished operator B.

4. Proofs

4.1. Proof of Theorem 1

Step 1. By Remark 2 we have to prove that the bounded nonnegative, self-adjoint operator B, given by (2.10), is compact and that we have (2.15) where ϱ_k are the positive eigenvalues of B. In this step we show that there is a positive number c such that

$$\varrho_k \geq c\, k^{-1}; \qquad k \in \mathbb{N}. \tag{4.1}$$

We rely on the same arguments as in Step 4 of the proof of Theorem 28.6 in [11], pp. 228–229. There are

$$M_j \sim h(2^{-j})^{-1}; \qquad j \in \mathbb{N}_0, \tag{4.2}$$

disjoint balls $B\left(\gamma^{j,l}, 2^{-j}\right)$ centred at $\gamma^{j,l} \in \Gamma = \operatorname{supp} \mu$ and of radius 2^{-j}. Let φ be a non-trivial C^∞ function in $\mathbb{R}^n$ with support in the unit ball in $\mathbb{R}^n$ and let

$$\varphi_{jl}(x) = \varphi\big(2^j(x - \gamma^{j,l})\big); \qquad j \in \mathbb{N}_0; \quad l = 1, \ldots, M_j.$$

By the localization property in [4], 2.3.2, p. 35–36, we have

$$\left\| \sum_{l=1}^{M_j} c_{jl}\, \varphi_{jl} \,|H^{\frac{n}{2}}(\mathbb{R}^n) \right\| \sim \left(\sum_{l=1}^{M_j} |c_{jl}|^2 \right)^{\frac{1}{2}}, \quad j \in \mathbb{N}_0, \tag{4.3}$$

where $c_{jl} \in \mathbb{C}$. The equivalence constants may be chosen independently of $j \in \mathbb{N}_0$. Furthermore, we have

$$\left(\int_\Gamma \left| \sum_{l=1}^{M_j} c_{jl}\, \varphi_{jl}(\gamma) \right|^2 \mu(d\gamma) \right)^{\frac{1}{2}} \sim h^{\frac{1}{2}} \left(2^{-j}\right) \left(\sum_{l=1}^{M_j} |c_{jl}|^2 \right)^{\frac{1}{2}} . \tag{4.4}$$

Then we obtain by (2.13) and (4.3), (4.4) that

$$\left\| \sqrt{B} \left(\sum_{l=1}^{M_j} c_{jl}\, \varphi_{jl} \right) |H^{\frac{n}{2}}(\mathbb{R}^n) \right\| \sim h^{\frac{1}{2}}(2^{-j}) \left\| \sum_{l=1}^{M_j} c_{jl}\, \varphi_{jl}\, |H^{\frac{n}{2}}(\mathbb{R}^n) \right\| . \tag{4.5}$$

Let $a_k(\sqrt{B})$ be the approximation numbers of the operator $\sqrt{B}$. It follows by (4.5) that for some $c > 0$,

$$a_{M_j}(\sqrt{B}) \geq c\, h^{\frac{1}{2}}(2^{-j}) \sim M_j^{-\frac{1}{2}}; \quad j \in \mathbb{N}_0. \tag{4.6}$$

We refer to [11], p. 220, for this type of argument. A ball of radius 2^{-j} can be covered by N balls of radius 2^{-j-1} with a natural number $N \in \mathbb{N}$ which is independent of $j \in \mathbb{N}_0$. As a consequence it follows by (2.17) that an isotropic measure satisfies the doubling condition, and that, in particular,

$$h\left(2^{-j-1}\right) \sim h\left(2^{-j}\right) \quad \text{and} \quad M_{j+1} \sim M_j, \quad \text{where} \quad j \in \mathbb{N}_0. \tag{4.7}$$

Hence, by (4.6),

$$\varrho_k^{\frac{1}{2}} = a_k\left(\sqrt{B}\right) \geq c\, k^{-\frac{1}{2}}, \qquad k \in \mathbb{N}.$$

The first equality is a well-known property of approximation numbers in Hilbert spaces which may be found, for example, in [11], pp. 191–192, including the necessary references. This proves (4.1).

Step 2. The proof of the converse assertion,

$$\varrho_k \leq c\, k^{-1}; \qquad k \in \mathbb{N}, \tag{4.8}$$

is more complicated and we need some preparations. Recall that $\Gamma = \operatorname{supp} \mu$. By (2.10), (2.9), and (2.13) the null space (or kernel) $N(B)$ is given by

$$N(B) = \left\{ f \in H^{\frac{n}{2}}(\mathbb{R}^n),\ tr_\mu f = 0 \right\}.$$

Furthermore, in modification of our notation introduced in [12], Definition 9.29, pp. 148–149, and 9.33, pp. 151–152, we have

$$H^t(\Gamma, \mu) = tr_\mu\, H^{\frac{n}{2}}(\mathbb{R}^n) \hookrightarrow L_2(\Gamma, \mu) \tag{4.9}$$

with

$$t = \sup \left\{ \tau : \sup_{k \in \mathbb{N}_0} 2^{\tau k}\, h\left(2^{-k}\right)^{\frac{1}{2}} < \infty \right\}. \tag{4.10}$$

The latter is covered by [12], (9.143), p. 151, and 9.25, p. 145. However we will not need the number t in (4.9), (4.10) in the sequel. By the usual Weyl decomposition and (4.9) we have

$$H^{\frac{n}{2}}(\mathbb{R}^n) = N(B) \oplus H^t(\Gamma,\mu),$$

where we identified the orthogonal complement of $N(B)$ in $H^{\frac{n}{2}}(\mathbb{R}^n)$ with the trace space in (4.9). Let B^μ be the restriction of B to $H^t(\Gamma,\mu)$. Then B^μ is a positive self-adjoint bounded operator on $H^t(\Gamma,\mu)$ with the same eigenvalues ϱ_k according to (2.14) as B. Furthermore by (2.13) we have

$$\left\| \sqrt{B^\mu}\, f \,|H^t(\Gamma,\mu)\right\| = \|f\,|L_2(\Gamma,\mu)\|\,. \tag{4.11}$$

In Steps 3 and 4 below we prove that the embedding in (4.9) is compact. Hence, B and B^μ are compact, too. Then by the same reasoning as in [12], p. 286, B^μ can be extended to a positive self-adjoint bounded (even compact) operator B_μ in $L_2(\Gamma,\mu)$ with the same eigenvalues ϱ_k. We have (4.11) with $\sqrt{B_\mu}$ in place of $\sqrt{B^\mu}$, now for all $f \in L_2(\Gamma,\mu)$. Hence,

$$\sqrt{B_\mu}\,(L_2 \hookrightarrow H^t) : \quad L_2(\Gamma,\mu) \hookrightarrow H^t(\Gamma,\mu),$$

is an isomorphic map. Let

$$id_t : \quad H^t(\Gamma,\mu) \hookrightarrow L_2(\Gamma,\mu)$$

be the corresponding embedding operator. Then $\sqrt{B_\mu}$, considered as a map in $L_2(\Gamma,\mu)$, can be factorized as

$$\sqrt{B_\mu} = id_t \circ \sqrt{B_\mu}\,(L_2 \hookrightarrow H^t). \tag{4.12}$$

Recall Carl's inequality,

$$\varrho_k^{\frac{1}{2}} \le \sqrt{2}\, e_k\left(\sqrt{B_\mu}\right), \quad k \in \mathbb{N},$$

where $e_k(\sqrt{B_\mu})$ are the corresponding entropy numbers. We refer for details and explanations to [4], 1.3.4, pp. 18–22. Then it follows by (4.12) that

$$\varrho_k^{\frac{1}{2}} \le c\, e_k(id_t), \quad k \in \mathbb{N}.$$

Using in addition (4.9), we obtain that

$$\varrho_k^{\frac{1}{2}} \le c\, e_k\left(tr_\mu : \ H^{\frac{n}{2}}(\mathbb{R}^n) \hookrightarrow L_2(\Gamma,\mu)\right), \quad k \in \mathbb{N}. \tag{4.13}$$

Step 3. The estimate of the entropy numbers e_k in (4.13) relies on quarkonial decompositions of $H^{\frac{n}{2}}(\mathbb{R}^n)$ adapted to $\Gamma = \operatorname{supp}\mu$ and on the technique developed in connection with the proof of Theorem 19.17 in [12]. Let $\gamma^{j,l} \in \Gamma$ with $j \in \mathbb{N}_0$ and $l = 1,\ldots,M_j$ be such that

$$\left|\gamma^{j,l_1} - \gamma^{j,l_2}\right| \ge c_1\, 2^{-j}, \qquad j \in \mathbb{N}_0, \quad l_1 \ne l_2, \tag{4.14}$$

and that for fixed j the union of all balls $B\left(\gamma^{j,l}, c_2\, 2^{-j}\right)$ with $l = 1,\ldots,M_j$ covers a 2^{-j}-neighborhood of Γ, where c_1 and c_2 are two appropriate positive numbers which are independent of j. We may assume that M_j are the same numbers as in

(4.2). We extend $\{\gamma^{j,l}\}$ for fixed j to a corresponding (irregular) lattice $\{x^{j,l}\}$ on $\mathbb{R}^n$ with the counterpart of (4.14) and

$$\bigcup_{l=1}^{\infty} B\left(x^{j,l}, c_2\, 2^{-j}\right) = \mathbb{R}^n.$$

We may assume that only those balls centred at $\gamma^{j,l} = x^{j,l}$ with $l = 1, \ldots, M_j$, have a non-empty intersection with Γ. Let $\{\psi^{j,l}\} \subset S(\mathbb{R}^n)$ be a subordinated resolution of unity,

$$\sum_{l=1}^{\infty} \psi^{j,l}(x) = 1, \qquad x \in \mathbb{R}^n, \quad j \in \mathbb{N}_0,$$

with

$$\operatorname{supp} \psi^{j,l} \subset B\left(x^{j,l}, c_2\, 2^{-j}\right), \quad \left|D^\alpha \psi^{j,l}(x)\right| \le c_\alpha\, 2^{j|\alpha|},$$

for some $c_\alpha > 0$. Then

$$(\beta\text{-qu})_{jl}(x) = 2^{j|\beta|}\,(x - x^{j,l})^\beta\, \psi^{j,l}(x), \qquad j \in \mathbb{N}_0, \quad l \in \mathbb{N},$$

are (generalized) β-quarks adapted to our situation. Here $(x - x^{j,l})^\beta$ are monomials according to (2.1). By [12], 2.14, 2.16, any $f \in H^{\frac{n}{2}}(\mathbb{R}^n)$ can be represented as

$$\begin{aligned} f &= \sum_{\beta\in\mathbb{N}_0^n} \sum_{j=0}^{\infty} \sum_{l=0}^{\infty} \lambda^\beta_{j,l}(f)\, (\beta\text{-qu})_{jl}(x) \\ &= \sum_{\beta\in\mathbb{N}_0^n} \sum_{j=0}^{\infty} \sum_{l=1}^{M_j} \lambda^\beta_{j,l}(f)\, (\beta\text{-qu})_{jl}(x) + \cdots, \end{aligned} \tag{4.15}$$

absolute convergence in $L_2(\mathbb{R}^n)$, where $\lambda^\beta_{j,l}(f) \in \mathbb{C}$ are optimal coefficients, depending linearly on f, with

$$\left\| f \,|H^{\frac{n}{2}}(\mathbb{R}^n)\right\| \sim \sup_{\beta\in\mathbb{N}_0^n} 2^{\varkappa|\beta|} \left(\sum_{j=0}^{\infty} \sum_{l=1}^{\infty} \left|\lambda^\beta_{j,l}(f)\right|^2 \right)^{\frac{1}{2}}, \tag{4.16}$$

where $\varkappa > \varkappa_0$ for some $\varkappa_0 > 0$ is at our disposal (the equivalence constants in (4.16) depend on $\varkappa$, but not on f). Only the terms with $l = 1, \ldots, M_j$ are of interest for us, what we indicated in the second line in (4.15). Let

$$\lambda = \left\{ \lambda^\beta_{j,l} \in \mathbb{C} : \quad \beta \in \mathbb{N}_0^n,\ l = 1, \ldots, M_j \right\}.$$

We put, adapting our notation used in [12], p. 284,

$$\left\| \lambda \,|\ell_{\infty,\varkappa} \left[\ell_2(\ell_2^{M_j})\right] \right\| = \sup_{\beta\in\mathbb{N}_0^n} 2^{\varkappa|\beta|} \left(\sum_{j=0}^{\infty} \left(\sum_{l=1}^{M_j} |\lambda^\beta_{j,l}|^2 \right) \right)^{\frac{1}{2}}$$

and

$$\left\|\lambda\,|\ell_{\infty,\varkappa}\left[\ell_1\left(M_j^{-\frac{1}{2}}\,\ell_2^{M_j}\right)\right]\right\| = \sup_{\beta\in\mathbb{N}_0^n} 2^{\varkappa|\beta|}\sum_{j=0}^{\infty} M_j^{-\frac{1}{2}}\left(\sum_{l=1}^{M_j}|\lambda_{j,l}^{\beta}|^2\right)^{\frac{1}{2}}.$$

We denote by $\ell_{\infty,\varkappa}\left[\ell_2\left(\ell_2^{M_j}\right)\right]$ and $\ell_{\infty,\varkappa}\left[\ell_1\left(M_j^{-\frac{1}{2}}\,\ell_2^{M_j}\right)\right]$ the spaces of all sequences where the respective norms are finite. We decompose the trace operator tr_μ on the right-hand side of (4.13) by

$$tr_\mu = T\circ Id\circ S$$

with

$$\begin{aligned} S &: H^{\frac{n}{2}}(\mathbb{R}^n)\mapsto \ell_{\infty,\varkappa_1}\left[\ell_2\left(\ell_2^{M_j}\right)\right],\\ Id &: \ell_{\infty,\varkappa_1}\left[\ell_2\left(\ell_2^{M_j}\right)\right]\hookrightarrow \ell_{\infty,\varkappa_2}\left[\ell_1\left(M_j^{-\frac{1}{2}}\,\ell_2^{M_j}\right)\right],\\ T &: \ell_{\infty,\varkappa_2}\left[\ell_1\left(M_j^{-\frac{1}{2}}\,\ell_2^{M_j}\right)\right]\mapsto L_2(\Gamma,\mu). \end{aligned}$$

Here S, given by

$$S:\quad f\mapsto \lambda(f)=\left\{\lambda_{j,l}^{\beta}(f)\;:\;\beta\in\mathbb{N}_0^n,\;j\in\mathbb{N}_0,\;l=1,\ldots,M_j\right\}$$

according to (4.15), (4.16), is a linear and bounded map. Furthermore, by (3.3) and (4.2) the embedding Id is linear and (as will be shown) compact. Finally, T is given by

$$T\lambda=\sum_{\beta\in\mathbb{N}_0^n}\sum_{j=0}^{\infty}\sum_{l=1}^{M_j}\lambda_{j,l}^{\beta}(\beta\text{-qu})_{jl}(\gamma),\qquad \gamma\in\Gamma.$$

Taking the traces of the β-quarks on Γ pointwise it follows that $T\circ Id\circ S$ is the trace for finite linear combinations of these β-quarks. The rest will be a matter of completion. Following [12], p. 285, we prove that T is bounded. First we remark that there are numbers $c>0$ and $r\in\mathbb{R}$ such that

$$|(\beta\text{-qu})_{jl}(x)|\le c\,2^{r|\beta|};\qquad \beta\in\mathbb{N}_0^n,\quad j\in\mathbb{N}_0,\quad l\in\mathbb{N}.$$

We always assume that

$$\varkappa_1>\varkappa_2>r>0.$$

For fixed $\beta\in\mathbb{N}_0^n$ we have

$$\begin{aligned}\left\|\sum_{j=0}^{\infty}\sum_{l=1}^{M_j}\lambda_{j,l}^{\beta}\,(\beta\text{-qu})_{jl}\,|L_2(\Gamma,\mu)\right\| &\le \sum_{j=0}^{\infty}\left(\int_\Gamma\left|\sum_{l=1}^{M_j}\lambda_{j,l}^{\beta}\,(\beta\text{-qu})_{jl}(\gamma)\right|^2\mu(d\gamma)\right)^{\frac{1}{2}}\\ &\le c\,2^{r|\beta|}\sum_{j=0}^{\infty}\left(\sum_{l=1}^{M_j}\left|\lambda_{j,l}^{\beta}\right|^2 h(2^{-j})\right)^{\frac{1}{2}}.\end{aligned}$$

We used the support properties of $(\beta\text{-qu})_{jl}$ and (4.7). By (4.2) and $\varkappa_2 > r$ it follows that T is bounded. By well-known properties of entropy numbers we reduce (4.13) to

$$\varrho_k^{\frac{1}{2}} \le c\, e_k(Id); \qquad k \in \mathbb{N}. \tag{4.17}$$

We need a further reduction clarifying the influence of $\beta \in \mathbb{N}_0^n$. Let

$$Id^\circ : \quad \ell_2\left(\ell_2^{M_j}\right) \hookrightarrow \ell_1\left(M_j^{-\frac{1}{2}}\, \ell_2^{M_j}\right); \quad j \in \mathbb{N}_0.$$

We prove in the next step that

$$e_k(Id^\circ) \le c\, k^{-\frac{1}{2}}; \qquad k \in \mathbb{N}. \tag{4.18}$$

Temporarily we take this for granted. Let $\varkappa_1 - \varkappa_2 = \varkappa_3$ and $0 < \delta < \varkappa_3$. For given $k \in \mathbb{N}$ let N be the smallest number $N \in \mathbb{N}$ such that $2^{-\delta N} \le k^{-\frac{1}{2}}$. Let $k_\beta = k\, 2^{-2\delta|\beta|}$ if $|\beta| \le N$. Then

$$k \le K = \sum_{|\beta|<N} [k_\beta] \sim k; \quad k \in \mathbb{N}. \tag{4.19}$$

By the assumption (4.18) and well-known properties of entropy numbers we obtain that

$$\begin{aligned} e_K(Id) &\le \sum_{|\beta|<N} e_{[k_\beta]}(Id^\circ)\, 2^{-\varkappa_3|\beta|} + c \sum_{|\beta|\ge N} 2^{-\varkappa_3|\beta|} \\ &\le c'\, k^{-\frac{1}{2}} \sum_{|\beta|<N} 2^{-(\varkappa_3-\delta)|\beta|} + c'\, 2^{-\delta N} \qquad (4.20) \\ &\le c'\, k^{-\frac{1}{2}}, \end{aligned}$$

where c and c' are independent of k. Hence, by (4.17), (4.19), (4.20), we have

$$\varrho_k^{\frac{1}{2}} \le c\, k^{-\frac{1}{2}}; \qquad k \in \mathbb{N}. \tag{4.21}$$

This proves (4.8) and, together with (4.1),

$$\varrho_k \sim k^{-1}; \qquad k \in \mathbb{N}. \tag{4.22}$$

Hence, by Definition 1, μ is a Weyl measure.

Step 4. It remains to prove (4.18). Let

$$Id_l : \quad \ell_2^{M_l} \hookrightarrow M_l^{-\frac{1}{2}}\, \ell_2^{M_l}; \qquad l \in \mathbb{N}_0. \tag{4.23}$$

Then we have in slight abuse of notation,

$$Id^\circ = \sum_{l=0}^{\infty} Id_l. \tag{4.24}$$

We fix $L \in \mathbb{N}$ and choose

$$k_l = 3M_l\left(1 + \log\frac{M_L}{M_l}\right), \qquad l = 0, \ldots, L.$$

Then we have

$$M_L \le \sum_{l=0}^{L} k_l \le c\, M_L, \tag{4.25}$$

where we used (3.2), (4.2). It follows by (4.23), (4.24), (4.25),

$$\begin{aligned} e_{cM_L}(Id^\circ) &\le \sum_{l=0}^{L} e_{k_l}(Id_l) + \sum_{l=L+1}^{\infty} M_l^{-\frac{1}{2}} \\ &\le \sum_{l=0}^{L} e_{k_l}(Id_l) + c'\, M_L^{-\frac{1}{2}}, \end{aligned} \tag{4.26}$$

where we used (3.3). (We assume tacitly $cM_L \in \mathbb{N}$ and $k_l \in \mathbb{N}$, otherwise one may choose nearby natural numbers.) We have by [11], Theorem 7.3, p. 37,

$$e_{k_l}(Id_l) \le c\, M_l^{-\frac{1}{2}}\, 2^{-\frac{3}{2}\log\frac{M_L}{M_l}} \le c'\, M_L^{-\frac{1}{2}}\, \frac{M_l}{M_L}.$$

Inserting this estimate in (4.26) and using again (3.2) we obtain that

$$e_{cM_L}(Id^\circ) \le c'\, M_L^{-\frac{1}{2}}; \qquad L \in \mathbb{N}. \tag{4.27}$$

Now, (4.18) is a consequence of (4.27) and (4.7).

4.2. Proof of Corollary 1

We check that h given by (3.5), (3.6) satisfies (3.2), (3.3). Without restriction of generality we may assume that

$$\left|\frac{\Psi'(r)}{\Psi(r)}\right| \le \frac{\varepsilon}{r}; \qquad 0 < r \le 1,$$

for some ε with $0 < \varepsilon < d$. Let $0 < t_2 < t_1 \le 1$. It follows that

$$\left| \int_{t_2}^{t_1} (\log \Psi(r))'\, dr \right| \le \int_{t_2}^{t_1} (\log r^\varepsilon)'\, dr,$$

and, hence,

$$\frac{\Psi(t_1)}{\Psi(t_2)}, \frac{\Psi(t_2)}{\Psi(t_1)} \le \left(\frac{t_1}{t_2}\right)^\varepsilon; \quad 0 < t_2 < t_1 \le 1. \tag{4.28}$$

Now both (3.2) and (3.3) are easy consequences of (3.5), (4.28), and $0 < \varepsilon < d$.

4.3. Proof of Corollary 2

Let

$$M_j^l \sim h_l(2^{-j})^{-1}; \qquad j \in \mathbb{N}_0; \quad l = 1, \ldots, N;$$

according to (4.2). Then we have by (3.7) that

$$M_j = M_j^1 - \sum_{l=2}^{N} M_j^l \sim M_j^1 \quad \text{if} \quad j \ge J.$$

Now we can apply the arguments from Step 1 of the proof of Theorem 1 in 4.1 restricted to balls having an empty intersection with the supports of $\mu_2, \ldots, \mu_N$. This proves (4.1). As for the reverse inequality one has N estimates of type (4.20) for each measure μ_j separately. They can be added according to (3.8). This proves (4.21) and finally (4.22).

References

[1] M. Bricchi. Existence and properties of h-sets. Georgian Math. Journ. **9** (2002), 13–32.

[2] M. Bricchi. Tailored Besov spaces and h-sets. Math. Nachr., (to appear).

[3] M. Bricchi. Complements and results on h-sets. Proceedings Conf. *Function spaces, differential operators and nonlinear analysis*, Teistungen 2001, (to appear).

[4] D.E. Edmunds, H. Triebel. Function spaces, entropy numbers, differential operators. Cambridge Univ.Press, 1996.

[5] D.E. Edmunds, H. Triebel. Spectral theory for isotropic fractal drums. C. R. Acad. Sci. Paris **326**, Sér. I, (1998), 1269–1274.

[6] D.E. Edmunds, H. Triebel. Eigenfrequencies of isotropic fractal drums. Operator Theory: Advances Appl. **110** (1999), 81–102.

[7] S. Moura. Function spaces of generalised smoothness, entropy numbers, applications. PhD Thesis, Coimbra, 2001.

[8] S. Moura. Function spaces of generalised smoothness. Dissertationes Math. **398** (2001).

[9] M. Solomyak. Spectral problems related to the critical exponent in the Sobolev embedding theorem. Proc. London Math. Soc. (3), **71** (1995), 53–75.

[10] H. Triebel. Interpolation theory, function spaces, differential operators. North-Holland, Amsterdam, 1978. [Sec. ed., Barth Verlag, Heidelberg, 1995]

[11] H. Triebel. Fractals and spectra. Birkhäuser, Basel, 1997.

[12] H. Triebel. The structure of functions. Birkhäuser, Basel, 2001.

Hans Triebel
Mathematisches Institut
Fakultät für Mathematik und Informatik
Friedrich-Schiller-Universität Jena
D-07740 Jena
Germany
e-mail: `triebel@minet.uni-jena.de`

MIX
Papier aus verantwortungsvollen Quellen
Paper from responsible sources
FSC® C105338
www.fsc.org

Printed by Libri Plureos GmbH
in Hamburg, Germany